HANDBOOK OF

Chemical and Biological Warfare Agents

SECOND EDITION

HANDBOOK OF

Chemical and Biological Warfare Agents

SECOND EDITION

D. HANK ELLISON

CRC Press is an imprint of the
Taylor & Francis Group, an informa business

CRC Press
Taylor & Francis Group
6000 Broken Sound Parkway NW, Suite 300
Boca Raton, FL 33487-2742

Printed in the United States of America on acid-free paper
10 9 8 7 6 5 4 3 2 1

International Standard Book Number-10: 0-8493-1434-8 (Hardcover)
International Standard Book Number-13: 978-0-8493-1434-6 (Hardcover)

Library of Congress Cataloging-in-Publication Data

Ellison, D. Hank.
Handbook of chemical and biological warfare agents / D. Hank Ellison. -- 2nd ed.
p. cm.
Includes bibliographical references and index.
ISBN-13: 978-0-8493-1434-6 (alk. paper)
ISBN-10: 0-8493-1434-8 (alk. paper)
1. Chemical warfare--Handbooks, manuals, etc. 2. Chemical agents (Munitions)--Handbooks, manuals, etc. 3. Biological warfare--Handbooks, manuals, etc. I. Title.

UG447.E44 2007
358'.34--dc22 2007005392

Visit the Taylor & Francis Web site at
http://www.taylorandfrancis.com

and the CRC Press Web site at
http://www.crcpress.com

Contents

Preface

The first edition of this handbook was written primarily as an aid to first responders and meant to help bridge the gap between what was known about responding to industrial hazardous materials and responding to military chemical and biological agents. At the time it was written, it was one of only a few readily available resources that addressed the topic. This is no longer the case. There are numerous books that have been written on virtually every aspect of a potential response; some are specialized and directed toward a specific audience while others are written for the average citizen. However, the world of military agents goes well beyond the "dirty thirty" that are usually discussed and there are still only a limited number of references that provide rapid access to technical data on a wider range of agents.

On the basis of this assessment of the current literature and on comments received on the first edition of this book, this volume has been written to focus on these details. With this new information, additional classes of agents have been added. Where it provided clarity, multiple classes have been consolidated into a single class. The information in existing classes has been updated and expanded. There is a significant increase in the number of agents described, as well as in the number of components, precursors, and decomposition products. There is more information on health effects and on the chemical, physical, and biological properties of these materials.

As in the first edition, all materials listed in this handbook have been used on the battlefield, stockpiled as weapons, received significant interest by research programs, used or threatened to be used by terrorists, or have been assessed by qualified law enforcement and response organizations as agents of significant concern. To assure accuracy, all data have been cross-checked over the widest variety of military, scientific, and medical sources available.

Finally, in presenting this broad spectrum, I do not offer an evaluation of the efficacy or viability of the agent classes or any of the individual agents. I have included agents classified by the military as obsolete along with those that are still considered a major threat. It is important to remember that while an agent may have been a failure on the battlefield, it could still be a very successful weapon in the hands of a terrorist.

Acknowledgments

I would like to once again thank Pam Ellison, DVM, for her assistance on the biological sections of this book. She was a wealth of information on both the technical and practical aspects of response to biological agents. She spent a great deal of time challenging my ideas, providing guidance, and correcting my drafts. Without her, the quality and content of this handbook would have suffered greatly.

There are numerous others out there who have provided comments, insights, and suggestions, both on the first edition and on the manuscript for this one. I appreciate them all. I have tried to address each of them, and incorporated changes that I believe have improved this edition. Any failures or omissions are mine and not due to a lack of vigilance or effort on the part of others.

Finally, I would like to thank my wife and children for their patience as I worked on this project. They have endured not only my mental absences during family events, but also my attempts to sneak off and get back on the computer. This second edition took far longer than I had anticipated and they have suffered the brunt of it.

Author

D. Hank Ellison served in the United States Army as a chemical officer and has worked for the U.S. Environmental Protection Agency as both a remedial project manager and federal on-scene coordinator under the Superfund Program. He currently is president of Cerberus & Associates, Inc., a consulting firm that specializes in response to technological disasters.

As a private consultant, Ellison has responded to hazardous material incidents involving highly poisonous materials, chemical fires, water reactive substances, and shock-sensitive materials throughout the state of Michigan. He has provided chemical and biological counterterrorism training to members of EMS units, hazmat teams, police SWAT teams, and bomb squads. During the anthrax events of 2001, he helped state and local governments as well as Fortune 500 companies to develop and implement response plans for biological threats. He currently advises clients on issues of hazardous materials, and related safety and security concerns. In addition, he is a member of the Department of Health and Human Services DMORT-WMD emergency response team, which has the primary mission for recovery and decontamination of fatalities contaminated with radiological, biological, or chemical materials.

Ellison earned a master of science in chemistry from the University of California, Irvine. His graduate research involved methods to synthesize poisons extracted from Colombian poison dart frogs. He has a bachelor of science in chemistry from the Georgia Institute of Technology. He is a member of the American Chemical Society and Federation of American Scientists. In addition to his works on weapons of mass destruction, he is the author of a chapter on the hazardous properties of materials in the sixth edition of the *Handbook on Hazardous Materials Management*, a textbook published in 2002 by the Institute of Hazardous Materials Management.

Explanatory Notes

In this handbook, information about the agents is divided into classes based on the common military groupings of chemical (i.e., nerve, vesicant, blood, pulmonary, incapacitating, and riot control), biological (i.e., bacterial, viral, rickettsial, and fungal), and toxin agents. In instances where the divisions are too broad to allow appropriate identification of the chemical or physiological properties of the individual agents, additional classes (e.g., organophosphorus nerve agents and carbamate nerve agents) are provided. There are also classes for nontraditional agents that do not fit neatly into one of the common military groupings (e.g., convulsants), and for industrial materials that could be used as improvised agents. Classes are identified by a number that corresponds to the first 20 chapters in this handbook (i.e., C01–C20). Classes contain general information about that specific group of agents. Although this book covers most of the major classes of chemical, biological, and toxin agents, it does not deal with antiplant chemicals, antimaterial agents, bioregulators or modulators, or incendiary and smoke agents.

At the end of each class is detailed technical information about individual agents, components, or decomposition products within that class. Each of these individual materials is assigned a handbook number to allow for rapid identification and cross-referencing throughout the book. The first three characters identify the agent class (e.g., **C01**). The letter following the hyphen (e.g. C01-**A**) indicates that the material is primarily an agent (A), component or precursor of that class of agents (C), or is a significant decomposition product or impurity of that class of agents (D). The three digits that follow the letter indicate the specific agent in the order that it appears in the class (e.g., C01-A**001**).

Chapter 21 contains four indices to allow easy access to specific agents in this handbook. These indices are the Alphabetical Index of names, the Chemical Abstract Service (CAS) numbers index, the International Classification of Diseases (ICD-10) numbers index, and the Organization for the Prohibition of Chemical Weapons (OPCW) agent numbers index. These indices contain synonyms and identifying numbers for the agents in this handbook that are cross-referenced to the individual agents via the handbook number.

Information in classes for chemical agents and toxins is in the following general format:

- General information
- Toxicology (effects, pathways and routes of exposure, general exposure hazards, latency period)
- Characteristics (physical appearance/odor, stability, persistency, environmental fate)
- Additional hazards (exposure, livestock/pets, fire, reactivity, hazardous decomposition products)
- Protection (evacuation recommendations, personal protective requirements, decontamination)
- Medical [Centers for Disease Control and Prevention (CDC), case definition, differential diagnosis, signs and symptoms, mass-casualty triage recommendations, casualty management, fatality management]

Information in classes for biological (i.e., pathogen) agents is in the following general format:

General information

Response (personal protective requirements, decontamination, fatality management)

Information on the individual chemical agents is in the following general format:

Handbook number

Name and reference numbers (CAS, RTECS, UN, ERG)

Formula

Description of the agent

Additional information including mixtures with other agents, industrial uses, threat, or treaty listing

Exposure hazards

Properties

AEGLs status and exposure values

Information on the individual toxins is in the following general format:

Handbook number

Name and reference numbers (CAS, RTECS)

Formula and molecular weight (if known)

Description of the toxin and source

Routes of exposure and signs and symptoms

Additional information including medicinal uses, threat, or treaty listing

Exposure hazards

Information on the individual pathogens is in the following general format:

Handbook number

Name, disease, and ICD-10

Description of the disease including natural transmission, natural reservoir, and a biosafety level if established

Additional information including threat or treaty listing

The disease as it appears in people including the CDC case definition, communicability, normal routes of exposure, infectious dose, secondary hazards, incubation period, signs and symptoms, suggested alternatives for differential diagnosis, and the untreated mortality rate

The disease as it appears in animals including agricultural target species, communicability, normal routes of exposure, secondary hazards, incubation period, signs and symptoms, suggested alternatives for differential diagnosis, and the untreated mortality rate

The disease as it appears in plants including agricultural target species, normal routes of exposure, secondary hazards, and signs

Abbreviations used in identifying individual agent are listed below.

CAS: Chemical Abstracts Service registry number. It is unique for each chemical without inherent meaning that and is assigned by the Chemical Abstracts Service, a division of the American Chemical Society. It allows for efficient searching of computerized databases.

ICD-10: Tenth revision of the International Statistical Classification of Diseases and Related Health Problems. It is the international standard diagnostic classification for all general epidemiological and many health management purposes.

RTECS: Registry of Toxic Effects of Chemical Substances number is a unique and unchanging number used to cross-reference the RTECS database, which is a compendium of data extracted from the open scientific literature. Six types of toxicity data are included in each file: (1) primary irritation, (2) mutagenic effects, (3) reproductive effects, (4) tumorigenic effects, (5) acute toxicity, and (6) other multiple dose toxicity.

UN: United Nations identification number used in transportation of hazardous materials.

ERG: 2004 Emergency Response Guidebook number. As in the Guidebook, the letter "P" following the guide number indicates that the material has a significant risk of violent polymerization if not properly stabilized.

Unless otherwise indicated, exposure hazards are for a "standard" man (i.e., a male weighing 70 kg/154 lbs) with a respiratory tidal volume of 15 L/min (i.e., involved in light activity). If a different breathing rate is used, then it is indicated in parentheses. If temperature is a factor, then the critical values are indicated. The military typically classifies moderate temperatures as 65–85°F. Temperatures above 85°F are classified as hot. For any given parameters, a dash (i.e., —) means that the value is unavailable because it has not been determined or has not been published.

Conversion Factor: Ratio of parts per million to milligrams per cubic meter at 77°F.

LCt_{50}: Is an expression of the dose of vapor or aerosolized agent necessary to kill half of the exposed population. These values are expressed as milligram-minute per cubic meter (mg-min/m^3). The lethal concentration (LC_{50}) is determined by dividing the LCt_{50} by the duration of exposure in minutes. Values are for inhalation (Inh) and percutaneous (Per) exposures. These dose–response values are not universally valid over all exposure periods. For inhalation of agent, time parameters are generally 2–8 min. For percutaneous absorption of agent, time parameters are generally 30 min to 6 h. Typically, a lethal concentration in parts per million (ppm) for a set exposure time is included in parentheses following the mg-min/m^3 value.

LC_{50}: Concentration of vapor or aerosolized agent necessary to kill half of the exposed population. Used when a specific set of exposure conditions (i.e., concentration and duration of exposure) are known but a generalized dose–response (LCt_{50}) is not available.

LD_{50}: Amount of liquid or solid material required to kill half of the exposed population. Values are for ingestion (Ing), percutaneous (Per) exposures, and subcutaneous injection (Sub). These values are expressed as total grams per individual.

Miosis: Concentration in parts per million (ppm) required to induce significant constriction of the pupil of the eye following a 2-min exposure to the agent.

ICt_{50}: Is an expression of the dose of vapor or aerosolized agent necessary to incapacitate half of the exposed population. These values are expressed as milligram-minute per cubic meter (mg-min/m^3). The incapacitating concentration (IC_{50}) is determined by dividing the ICt_{50} by the duration of exposure in minutes. Values are for inhalation (Inh) and percutaneous (Per) exposures; and in the case of vesicants, damage to the skin (Skin) and eyes (Eyes). These dose–response values are not universally valid over all exposure periods. For inhalation of agent, time parameters are generally 2–8 min. For percutaneous absorption of agent, as well as damage to the skin and eyes, time parameters are generally 2 min to 6 h. Typically, an incapacitating concentration in parts per million (ppm) for a set exposure time is included in parentheses following the mg-min/m^3 value.

Irritation values for eyes, skin, and respiratory system. These values are expressed as a concentration (ppm for gases, mg/m^3 for aerosols) for a 2-min exposure. "Intolerable" concentrations cited in the literature are also noted.

Vomiting: Inhaled concentration of vapor or aerosolized agent necessary to induce significant nausea and vomiting in half of the exposed population. These values are expressed as a concentration (ppm for gases, mg/m^3 for aerosols) for a 2-min exposure.

MEG: Military exposure guidelines for deployed personnel. Levels reported in this handbook are for 1-h exposures and consider three health endpoints. *Minimal (Min)*: Continuous exposure to concentrations above these levels could produce mild, transient, reversible effects but should not impair military operational performance. *Significant (Sig)*: Continuous exposure to concentrations above these levels could produce irreversible, permanent, or serious health effects, and could degrade military operational performance and even incapacitate some individuals. *Severe (Sev)*: Continuous exposure to concentrations above these levels could produce life-threatening or lethal effects in some individuals.

WPL AEL: Worker Population Airborne Exposure Limits developed for the military by the CDC. They are based on a time-weighted average exposure over an 8-h period and 40-h work week.

OSHA PEL: Federal Permissible Exposure Limits based on a time-weighted average exposure over an 8-h period and 40-h work week. A [Skin] notation indicates that percutaneous absorption of the material is a potential hazard and may contribute to the overall exposure.

ACGIH TLV: American Conference of Governmental Industrial Hygienists recommended Threshold Limit Values based on a time-weighted average exposure over an 8-h period and 40-h work week. A [Skin] notation indicates that percutaneous absorption of the material is a potential hazard and may contribute to the overall exposure.

AIHA WEEL: American Industrial Hygiene Association recommended Workplace Environmental Exposure Levels based on a time-weighted average exposure over an 8-h period and 40-h work week.

STEL: Short-Term Exposure Limits based on a time-weighted average exposure of 15 min (unless otherwise noted). A [Skin] notation indicates that percutaneous absorption of the material is a potential hazard and may contribute to the overall exposure.

Ceiling: Exposure limit that specifies the concentration of vapor, dust, or aerosol that should not be exceeded at any time during the workday. In some instances, a time limit for exposure to the ceiling value is established and is indicated in parentheses. A [Skin] notation indicates that percutaneous absorption of the material is a potential hazard and may contribute to the overall exposure.

IDLH: Immediately Dangerous to Life or Health levels indicate that exposure to the listed concentrations of airborne contaminants is likely to cause death, immediate or delayed permanent adverse health effects, or prevent escape from the contaminated environment in a short period of time, typically 30 min or less. These values constitute a hazardous materials emergency in the workplace and require the use of a supplier air respirator (e.g., SCBA).

Unless otherwise indicated, chemical and physical properties are for the pure or production quality material. Properties of mixed, binary, thickened, or dusty agents, even those in solutions, will have physical and chemical properties that vary from the listed values. These variations will depend on the proportion of agent to other materials (e.g., solvents, thickener, etc.) and the properties of these other materials. If available, data on mixtures or modified agents (e.g., salts) are included. For any given parameters, a dash (i.e., —) means that the value is unavailable because it has not been determined or has not been published.

MW: Molecular or formula weight of the material.

D: Density of the solid or liquid material at 68°F. If the density is reported for another temperature, it is indicated in parentheses following the density. Liquefied gases are also indicated.

MP: Melting point of the material in degrees Fahrenheit. A designation of "decomposes" indicates that the agent will thermally decompose before it reaches its melting point.

BP: Boiling point of the material in degrees Fahrenheit at standard pressure (760 mmHg). If a reduced pressure is reported, it is indicated in parentheses following the boiling point. The designation "decomposes" indicates that the agent will thermally decompose before it reaches its boiling point. The designation "sublimes" indicates that the material goes directly from a solid to a gas without melting.

Vsc: Dynamic viscosity of the material in centistokes (cS). Obtained when the absolute viscosity, expressed as centipoise, is divided by the specific gravity of the material.

VP: Vapor pressure of the material in millimeters of mercury (mmHg) at 68°F. If another temperature is used, it is indicated in parentheses following the vapor pressure.

VD: Relative vapor density of the gaseous agent as compared to air. Unless otherwise indicated, these values are calculated based on the standard reference weight of air (i.e., 29).

Vlt: Volatility is the mass of agent in a unit of air that is saturated with the agent vapor. The volatility of an agent varies with temperature and it is often used to estimate the tendency of a chemical to vaporize or give off fumes. It can be calculated using the following formula:

$$\mathrm{Vlt} = 16{,}040\left(\frac{Mp}{K}\right),$$

where Vlt is the volatility of the agent in mg/m^3, M is the gram molecular weight of the agent, p is the vapor pressure of the agent in mmHg at the ambient temperature, and K is the ambient temperature in degrees Kelvin.

Volatility is also sometimes used to estimate the persistency of an agent. However, it does not account for the migration (diffusion) of that vapor out of the area to allow more agent to evaporate. A better estimate of persistency is relative persistency (RP).

H_2O: Solubility of the agent in water. Solubilities are generally given in percentages, indicating the weight of agent that will dissolve in the complementary amount of water. When quantitative solubility data are not available, qualitative terms (e.g., negligible, slight) are used to provide an intuitive evaluation of agent solubility. The designation "miscible" indicates that the agent is soluble in water in all proportions. The designation "insoluble" indicates that no appreciable amount of the agent will dissolve in water. The designation "decomposes" indicates that the agent reacts with water and will decompose into other materials which may or may not be hazardous. If known, the rate of decomposition is indicated.

Sol: List of common organic solvents in which the material has appreciable solubility.

FlP: Flashpoint of the material. The flashpoint is the temperature in degrees Fahrenheit at which the liquid phase gives off enough vapor to flash when exposed to an ignition source.

LEL: Lower explosive limit in air, expressed as a percentage by volume.

UEL: Upper explosive limit in air, expressed as a percentage by volume.

RP: Relative persistency is a mathematical comparison of the evaporation and diffusion rates of water at 68°F to the evaporation and diffusion rates of the agent. The value represents an estimate of the ratio of the time required for a liquid or solid agent to dissipate as compared to the amount of time required for an equal amount of liquid water to dissipate. A value of "1" indicates that puddle of the agent and a similar puddle of water will evaporate at about the same rate. The greater the value, the greater the proportional amount of time required for the agent to evaporate. For example, a value of "2" would mean that it would take about twice as long for the puddle of agent to evaporate as compared to a similar puddle of water. Relative persistency is calculated by the following formula:

$$\mathrm{RP} = \left(\frac{4.34}{p}\right)\left(\frac{K}{M}\right)^{1/2},$$

where RP is the relative persistency of the agent, p is the vapor pressure of the agent in mmHg at the temperature K, M is the gram molecular weight of the agent, and K is the ambient temperature in degrees Kelvin.

The relative persistency value does not account for additional factors that could impact the stability and persistence of a given agent, such as decomposition due to reaction with other chemicals in the environment (e.g., water).

IP: Ionization potential is the amount of energy needed to remove an electron from a molecule of chemical vapor. The resultant ion is a charged particle that is detectable by various instrumentation such as photo ionization or flame ionization detectors.

AEGL: Acute Exposure Guideline Levels describe the risk from single, nonrepetitive exposures to airborne chemicals in a once-in-a-lifetime event. They represent threshold exposure limits for the general public and are applicable to emergency exposure periods ranging from 10 min to 8 h. In this handbook, only values for 1, 4, and 8 h are denoted. Values that have not been developed or not recommended are annotated as "Not Developed."

AEGLs appear in three categories. The *AEGL-1* is the airborne concentration above which the general population could experience notable discomfort, irritation, or certain asymptomatic nonsensory effects. Effects are not disabling; they are transient and reversible. The *AEGL-2* is the airborne concentration above which the general population could experience irreversible or other serious, long-lasting adverse health effects or an impaired ability to escape. The *AEGL-3* is the airborne concentration above which the general population could experience life-threatening health effects or death.

The status indicates where the chemical is in the review process and includes Draft, Proposed, Interim, and Final. For more information, see the United State Environmental Protection Agency website at http://www.epa.gov/oppt/aegl/index.htm.

Section I

Nerve Agents

1

Organophosphorus Nerve Agents

1.1 General Information

There are four major series of organophosphorus nerve agents.

1.1.1 G-Series Nerve Agents

These agents are alkyl phosphonofluoridates, alkyl phosphoramidocyanidates, and alkyl phosphonofluoridothiates. They are second generation chemical warfare agents. The original agents in this series—tabun (C01-A001), sarin (C01-A002), and soman (C01-A003)—were developed by German scientists during the 1930s. Both tabun and sarin were stockpiled by Nazi Germany during World War II but were never used. Since the end of World War II, modern weapons researchers have evaluated numerous other variations of the basic phosphonofluoridate and phosphoramidocyanidate structures. Although G-series agents have been stockpiled by most countries that have pursued a chemical weapons program, they have been used only a limited number of times on the battlefield. They have also been used by terrorists as a mass casualty agent.

The majority of G-series agents are listed in Schedule 1 of the Chemical Weapons Convention (CWC) as long as they are within the following limitations:

O-Alkyl (less than or equal to C10, including cycloalkyl) alkyl (methyl, ethyl, propyl, or isopropyl) phosphonofluoridates, or

O-Alkyl (less than or equal to C10, including cycloalkyl) *N,N*-dialkyl (methyl, ethyl, propyl, or isopropyl) phosphoramidocyanidates.

1.1.2 V-Series Nerve Agents

These agents are alkyl *S*-2-dialkylaminoethyl alkylphosphonothiolates, dialkyl *S*-2-dialkylaminoethyl phosphorothiolates, and alkyl *S*-2-dialkylaminoethyl alkylphosphonoselenoates. They are third generation chemical warfare agents. The first of these agents—Amiton (C01-A013)—was developed and patented in Britain in the early 1950s by Imperial Chemicals Industries Limited as a pesticide. Since that time, modern weapons researchers have evaluated numerous other variations of the basic phosphorothiolate structure. Researchers have also developed quaternary amine salts of agents to enhance their ability to penetrate into neuromuscular junctions. Although stockpiled by a number of countries,

V-series agents have never been used on the battlefield. They have been used in a very limited way by terrorists to assassinate individuals.

A majority of V-series agents are listed in Schedule 1 of the CWC as long as they are within the following limitations:

> *O*-Alkyl (H or less than or equal to C10, including cycloalkyl) *S*-2-dialkyl (Me, Et, n-Pr, or i-Pr)-aminoethyl alkyl (Me, Et, n-Pr, or i-Pr) phosphonothiolate and corresponding alkylated or protonated salts.
>
> Amiton (C01-A013) is listed under Schedule 2 of the CWC.

1.1.3 GV-Series Nerve Agents

These agents are alkyl phosphoramidofluoridates. They are fourth generation chemical warfare agents that were developed by the United States in the 1970s. GV-series agents were designed to possess the key advantages of the higher volatility of G-series agents and the high percutaneous toxicity of V-series agents. Researchers have also developed quaternary amine salts of agents to enhance their ability to penetrate into neuromuscular junctions. Based on available information, GV-series agents have only been used for research purposes.

GV-series agents are not specifically listed in the CWC, nor are they covered by the language of the general definitions. However, because of their toxicity and lack of commercial application, they would be prohibited based on the Guidelines for Schedules of Chemicals.

1.1.4 Novichok-Series Nerve Agents

These agents are carbonimidic phosphorohalides. They are fourth generation chemical warfare agents that were developed by the former Soviet Union. Research on novichok agents began in the 1960s and continued through the early 1990s under the Foliant program. Minimal information about these agents or this research program has been published in the unclassified literature. Although production of several thousand tons for testing purposes has been reported, there is no information to indicate that these agents were ever stockpiled in the Soviet arsenal. They have never been used on the battlefield.

Novichok agents are not specifically listed in the CWC, or are they covered by the language of the general definitions in the Schedules. However, because of their toxicity and lack of commercial application, they would be prohibited based on the Guidelines for Schedules of Chemicals.

1.1.5 Comments

With the exception of the novichok series, organophosphorus nerve agents are relatively easy to synthesize and disperse. Novichok agents are moderately difficult to synthesize. For information on some of the chemicals used to manufacture nerve agents, see the Component Section (C01-C) following information on the individual agents in this chapter. Overall, nerve agents are easy to deliver. Although no specific information is available on dispersing novichok agents, they should perform similar to other organophosphorus nerve agents.

In addition to the agents detailed in this handbook, the Organization for the Prohibition of Chemical Weapons (OPCW) identifies in its *Declaration Handbook 2002 for the Convention on the Prohibition of the Development, Production, Stockpiling, and Use of Chemical Weapons and on their Destruction* numerous other G-series and V-series organophosphorus nerve agents. However, no information is available in the unclassified literature concerning the physical, chemical, or toxicological properties of these additional agents.

1.2 Toxicology

1.2.1 Effects

Nerve agents are the most toxic of formerly stockpiled man-made chemical warfare agents. These compounds are similar to, but much more deadly than, agricultural organophosphate pesticides. Nerve agents disrupt the function of the nervous system by interfering with the enzyme acetylcholinesterase. Serious effects are on skeletal muscles and the central nervous system. Nerve agents also affect glands that discharge secretions to the outside of the body causing discharge of mucous, saliva, sweat, and gastrointestinal fluids. Exposure to solids, liquids, or vapors from these agents is hazardous and can result in death within minutes of exposure.

1.2.2 Pathways and Routes of Exposure

Nerve agents are hazardous through any route of exposure including inhalation, exposure of the skin and eye to either liquid or vapor, ingestion, and broken, abraded, or lacerated skin (e.g., penetration of skin by debris). Thickened agents primarily pose a hazard through skin absorption. Dusty agents are primarily an inhalation hazard although percutaneous absorption is possible especially if there is contact with bulk agent.

1.2.3 General Exposure Hazards

Organophosphorus nerve agents do not have good warning properties. They have little or no odor, and, other than causing miosis, the vapors do not irritate the eyes. Contact with liquid agent neither irritates the skin nor causes cutaneous injuries. Exposure to GV-series agents produces fewer, milder, and more transient physiological symptoms than seen with either the G-series or V-series agents.

The rate of detoxification of organophosphorus nerve agents by the body is very low. Exposures are essentially cumulative.

1.2.3.1 G-Series

Lethal concentrations (LC_{50}s) for inhalation of G-series agents are as low as 2 ppm for a 2-minutes exposure.

Lethal percutaneous exposures (LD_{50}s) to liquid G-series agents are as low as 0.17 grams per individual.

Miosis from exposure of the eyes to G-series agent vapors occurs after exposures to as little as 0.01 ppm for 2 minutes.

1.2.3.2 V-Series

Lethal concentrations (LC_{50}s) for inhalation of V-series agents are as low as 0.69 ppm for a 2-minutes exposure.

Lethal percutaneous exposures (LD_{50}s) to liquid V-series agents are as low as 0.005 grams per individual.

Miosis from exposure of the eyes to V-series agent vapors occurs after exposures to as little as 0.005 ppm for 2 minutes.

1.2.3.3 GV-Series

Human toxicity data for these nerve agents have not been published or have not been established.

1.2.3.4 Novichok Series

Human toxicity data for the Novichok series nerve agents have not been published or have not been established. However, available information indicates that under optimum conditions novichok agents are 5–10 times more toxic than nerve agent VX (C01-A016).

1.2.4 Latency Period

1.2.4.1 Vapor/Aerosols (Mists or Dusts)

Depending on the concentration of agent vapor, the effects begin to appear from 30 seconds to 2 minutes after initial exposure.

1.2.4.2 Liquids

Typically, there is a latent period with no visible effects between the time of exposure and the sudden onset of symptoms. This latency can range from 1 minutes to 18 hours. Some factors affecting the length of time before the onset of symptoms are the amount of agent involved, the amount of skin surface in contact with the agent, previous exposure to materials that chap or dry the skin (e.g., organic solvents such as gasoline or alcohols), and addition of additives designed to enhance the rate of percutaneous penetration by the agents.

Another key factor affecting the rate of percutaneous penetration by the agent is the part of the body that is exposed. It takes the agent longer to penetrate thicker and tougher skin. The regions of the body that allow the fastest percutaneous penetration are the groin, head, and neck. The least susceptible body regions are the hands, feet, front of the knee, and outside of the elbow.

1.2.4.3 Solids (Nonaerosol)

Typically, there is a latent period with no visible effects between the time of exposure and the sudden onset of symptoms. This latency can range from 1 minutes to 18 hours and is affected by such factors as the amount of agent involved, the amount of skin surface in contact with the agent, and the area of the body exposed (see Liquids). Moist, sweaty areas of the body are more susceptible to percutaneous penetration by solid nerve agents.

1.3 Characteristics

1.3.1 Physical Appearance/Odor

1.3.1.1 Laboratory Grade

Laboratory grade agents are typically colorless with a consistency ranging from water to motor oil. Salts are colorless to white crystalline solids. Neither solids nor liquids have any significant odor when pure.

1.3.1.2 Munition Grade

Munition grade agents are typically yellow to brown liquids or solids. As the agent ages and decomposes it continues to discolor until it may appear black. Production impurities and decomposition products in these agents may give them an odor. Odors for G-series agents have been described as fruity, camphor-like, or similar to bitter almonds. Odors for V-series agents have been described as resembling rotten fish or smelling of sulfur. Odors for all agents may become more pronounced during storage.

1.3.1.3 Binary Agents

Binary versions of all series of organophosphorus nerve agents have been developed. A binary agent consists of two components, either solids or liquids, that form a standard nerve agent when they are mixed. Since the agent is formed just prior to or as a result of deployment, the product formed by mixing the components is crude and will consist of the agent, the individual components, and any by-products formed during the reaction. The color, odor, and consistency of the resulting agent will vary depending on the quality of the components and the degree of mixing.

The components, by-products of the reaction or solvents used to facilitate mixing the components may have their own toxic properties and could present additional hazards. They may also change the rate that the binary nerve agent volatilizes or penetrates the skin. Residual components may react with common materials, such as alcohols, to produce other nerve agents. For data on binary components, see the Component Section (C01-C) following information on the individual agents.

1.3.1.4 Modified Agents

Solvents have been added to nerve agents to facilitate handling, to stabilize the agents, or to increase the ease of percutaneous penetration by the agents. Percutaneous enhancement solvents include dimethyl sulfoxide, *N,N*-dimethylformamide, *N,N*-dimethylpalmitamide, *N,N*-dimethyldecanamide, and saponin. Color and other properties of these solutions may vary from the pure agent. Odors will vary depending on the characteristics of the solvent(s) used and concentration of nerve agent in the solution.

Conversely, nerve agents have also been thickened with various substances to enhance deployment, increase their persistency, and increase the risk of percutaneous exposure. Thickeners include polyalkyl methacrylates (methyl, ethyl, butyl, isobutyl), polyvinyl acetate, polystyrene, plexiglas, alloprene, polychlorinated isoprene, nitrocellulose, as well as bleached montan and lignite waxes. Military thickener K125 is a mixture of methyl, ethyl, and butyl polymethacrylates. When thickened, agents become sticky with a consistency similar to honey. Typically, not enough thickener is added to significantly affect either the color or odor of the agent.

Nerve agents have also been converted to a "dusty" form by adsorbing the liquid agent onto a solid carrier. Dusty carriers include aerogel, talc, alumina, silica gel, diatomite, kaolinite, fuller's earth, and pumice. Dusty agents appear as finely ground, free-flowing powders with individual particles in the range of 10 μm or less and are dispersed as a particulate cloud. Particles in this range can penetrate clothing and breathable protective gear, such as U.S. military mission-oriented protective posture (MOPP) garments. Dusty agents pose both an inhalation and contact hazard. Color and other physical properties of dusty agents will depend on the characteristics of the carrier. Odors may vary from the unmodified agent.

1.3.2 Stability

Crude G-series agents are relatively stable at low to moderate temperatures. Stability increases with purity and distilled materials can be stored even under tropical conditions. The presence of moisture facilitates decomposition of agents during storage. Stabilizers may be added to remove moisture and to react with acidic decomposition products. Stabilizers are typically diisopropylcarbodiimide or dicyclohexylcarbodiimide, but can also be tributylamine or dibutylchloramine. For additional information on these stabilizers, see the Component Section (C01-C) following information on the individual agents in this chapter. Agents can be stored in glass or steel containers. Aluminum is acceptable if the agent has been stabilized.

Crude V-series agents are relatively stable at low to moderate temperatures. Stability increases with purity, and distilled materials can be stored even under tropical conditions. Agents undergo gradual deterioration during storage that is autocatalytic and accelerated by moisture or impurities normal to agent production. Stabilizers may be added to remove moisture and to react with acidic decomposition products. Stabilizers are carbodiimides such as diisopropylcarbodiimide and dicyclohexylcarbodiimide. For additional information on these stabilizers, see the Component Section (C01-C) following information on the individual agents in this chapter. Agents can be stored in aluminum, glass, or steel containers. Agents can deteriorate in the presence of silver found in some solders or brazes, and in the presence of rust.

Liquid GV-series agents are not as stable as either G-series or V-series agents and tend to decompose during storage. Purified salts are stable over extended periods. Agents can be stored in glass containers.

Information on the stability of novichok agents has not been published.

1.3.3 Persistency

Depending on the properties of the specific agent, unmodified G-series agents are classified as either nonpersistent or moderately persistent by the military. Evaporation rates range from near that of water down to that of light machine oil.

Unmodified V-series agents are classified as persistent by the military. Evaporation rates range from near that of light machine oil down to that of motor oil.

Depending on the properties of the specific agent, unmodified GV-series agents are classified as either moderately persistent or persistent by the military. Evaporation rates range from near that of light machine oil down to that of motor oil.

Information on the persistency of novichok agents has not been published.

Addition of solvents, thickeners, or conversion to a dusty form may alter the persistency of these agents. Thickened agents will persist significantly longer after dispersal than unmodified agents. Dusty agents can be very persistent depending on the carrier employed and can be reaerosolized by ground traffic or strong winds.

Salts of agents have negligible vapor pressure and will not evaporate. Depending on the size of the individual particles and on any encapsulation or coatings applied to the particles, they can be reaerosolized by ground traffic or strong winds.

1.3.4 Environmental Fate

Organophosphorus nerve agent vapors have a density greater than air and tend to collect in low places. Porous material, including painted surfaces, will absorb both liquid and gaseous agent. After the initial surface contamination has been removed, agent that has been absorbed into porous material can migrate back to the surface posing both a contact and vapor hazard. Clothing may emit trapped agent vapor for up to 30 minutes after contact with a vapor cloud.

With the exception of sarin (C01-A002), which is miscible with water, most of these agents are only slightly soluble or insoluble in water. V-series agents are unusual in that they tend to be more soluble in cold water than in hot water. However, the solubility of any agent may be modified (either increased or decreased) by solvents, components, or impurities. The specific gravities of unmodified liquid agents are slightly greater than that of water. Nerve agents are soluble in most organic solvents including gasoline, alcohols, and oils. V-series and GV-series agents are also soluble in mild aqueous acidic solutions. Salts of organophosphorus nerve agents are water soluble.

1.4 Additional Hazards

1.4.1 Exposure

Individuals who have had previous exposure to materials that chap or dry the skin, such as alcohols, gasoline, or paint thinners, may be more susceptible to percutaneous penetration of liquid agents. In these situations, the rate of percutaneous penetration of the agent is greatly increased resulting in a decrease in the survival time that would otherwise be expected.

All foodstuffs in the area of a release should be considered contaminated. Unopened items packaged in glass, metal, or heavy duty plastic and exposed only to agent vapors may be used after decontamination of the container. Unopened items exposed to liquid or solid agents should be decontaminated within a few hours postexposure or destroyed. Opened or unpackaged items, or those packaged only in paper or cardboard, should be destroyed.

Meat from animals that have suffered only mild to moderate effects from exposure to nerve agents should be safe to consume. Milk should be discarded for the first 7 days postexposure and then should be safe to consume. Meat, milk, and animal products, including hides, from animals severely affected or killed by nerve agents should be destroyed.

Plants, fruits, vegetables, and grains should be quarantined until tested and determined to be safe to consume. In addition to a direct contact hazard, some G-series agents can translocate from roots to other areas (e.g., leaves, fruits, etc.) of some plants and may pose an ingestion hazard.

1.4.2 Livestock/Pets

Animals can be decontaminated with shampoo/soap and water, or a 0.5% household bleach solution (see Section 1.5). If the animals' eyes have been exposed to agent, they should be irrigated with water or saline solution for a minimum of 30 minutes.

The topmost layer of unprotected feedstock (e.g., hay or grain) should be destroyed. The remaining material should be quarantined until tested. Leaves of forage vegetation could still retain sufficient nerve agent to produce mild to moderate effects for several weeks post release, depending on the level of contamination and the weather conditions. G-series nerve agents that have translocated to the leaves or fruit of plants may pose an extended ingestion hazard.

1.4.3 Fire

Heat from a fire will increase the amount of agent vapor in the area. A significant amount of the agent could be volatilized and escape into the surrounding environment before it is consumed by the fire. Actions taken to extinguish the fire can also spread the agent. With the exception of sarin, most G-series agents are only slightly soluble or insoluble in water. However, because of their extreme toxicity, runoff from firefighting efforts will still pose a significant threat. Some of the decomposition products resulting from hydrolysis or combustion of nerve agents are water soluble and highly toxic (see Section 1.4.5). Other potential decomposition products include toxic and/or corrosive gases.

1.4.4 Reactivity

Most organophosphorus nerve agents decompose slowly in water. Raising the pH of an aqueous solution of these agents significantly increases the rate of decomposition. Reaction with dry bleach may produce toxic gases.

1.4.5 Hazardous Decomposition Products

For information on individual impurities and decomposition products, see the Decomposition Products and Impurities section (C01-D) at the end of this chapter.

1.4.5.1 Hydrolysis

G-series nerve agents: Agents produce hydrogen fluoride (HF) or hydrogen cyanide (HCN) when hydrolyzed. Some may produce hydrogen chloride (HCl).

V-series nerve agents: Extremely toxic decomposition products, including *S*-[2-dialkylaminoethyl] alkylphosphonothioic acids and alkyl pyrophosphonates, may be produced by hydrolysis.

GV-series nerve agents: Agents produce HF when hydrolyzed. Additional products, depending on the pH, include amines and complex organophosphates that should be considered to be extremely toxic.

Novichok agents: Agents produce HF, HCl, or HCN when hydrolyzed. They may also produce highly toxic oximes.

1.4.5.2 Combustion

G-series nerve agents: Volatile decomposition products may include HF, HCl, HCN, sulfur oxides (SO_x), phosphorous oxides (PO_x), as well as potentially toxic organophosphates. In addition, toxic phosphate residue may remain.

V-series nerve agents: Volatile decomposition products may include SO_x, NO_x, PO_x, as well as potentially toxic organophosphates. In addition, toxic phosphate residue may remain.

GV-series nerve agents: Volatile decomposition products may include HF, NO_x, PO_x, as well as potentially toxic organophosphates. In addition, toxic phosphate residue may remain.

Novichok agents: Volatile decomposition products may include HF, HCl, HCN, PO_x as well as potentially toxic organophosphates. In addition, toxic phosphate residue may remain.

1.5 Protection

1.5.1 Evacuation Recommendations

Isolation and protective action distances listed below are taken from Argonne National Laboratory Report No. ANL/DIS-00-1, *Development of the Table of Initial Isolation and*

Protective Action Distances for the 2000 Emergency Response Guidebook, which is still the basis for the "when used as a weapon" scenarios in the *2004 Emergency Response Guidebook* (ERG). For organophosphorus nerve agents, these recommendations are based on a release scenario involving either a spray or explosively generated mist of nerve agent that quickly settles to the ground and soaks in to a depth of no more than 0.25 millimeters. A secondary cloud will then be generated by evaporation of this deposited material. Under these conditions, the difference between a small and a large release of nerve agent is not based on the standard 200 liters spill used for commercial hazardous materials listed in the ERG. A small release involves 2 kilograms (approximately 2.0 liters) of liquid agent and a large release involves 100 kilograms (approximately 26 gallons) of liquid agent.

	Initial isolation (feet)	Downwind day (miles)	Downwind night (miles)
GA (Tabun) *C01-A001*			
Small device (2 kilograms)	100	0.2	0.4
Large device (100 kilograms)	500	1.0	1.9
GB (Sarin) *C01-A002*			
Small device (2 kilograms)	500	1.0	2.1
Large device (100 kilograms)	3000	7+	7+
GD (Soman) *C01-A003*			
Small device (2 kilograms)	300	0.5	1.1
Large device (100 kilograms)	2500	4.2	6.5
GF (Cyclosarin) *C01-A004*			
Small device (2 kilograms)	100	0.2	0.4
Large device (100 kilograms)	800	1.4	3.2
VX *C01-A016*			
Small device (2 kilograms)	100	0.1	0.1
Large device (100 kilograms)	200	0.4	0.6

1.5.2 Personal Protective Requirements

1.5.2.1 Structural Firefighters' Gear

Structural firefighters' protective clothing is recommended for fire situations only; it is not effective in spill situations or release events. If chemical protective clothing is not available and it is necessary to rescue casualties from a contaminated area, then structural firefighters' gear will provide very limited skin protection against nerve agent vapors. Contact with liquids, solids, and solutions should be avoided.

In the event entry is made into an environment contaminated with nerve agent vapor, responders are at a significantly greater risk of exposure and resultant adverse effects. Using the 3/30 Rule, operations in structural firefighters' gear should only be undertaken if there are known living victims and vapor concentrations have peaked. Entries must be limited to 30 minutes. If there are no known living victims, then entries must be limited to 3 minutes. Even operating within these constraints, up to 50% of the firefighters who enter a contaminated area wearing only structural firefighters' gear may experience mild to moderate effects. In addition, since exposures to nerve agents are essentially cumulative, all responders who enter the hot zone without appropriate chemical protective clothing are at increased risk during the remainder of the emergency.

1.5.2.2 Respiratory Protection

Self-contained breathing apparatuses (SCBAs) or air purifying respirators (APRs) should have a National Institute for Occupational Safety and Health (NIOSH) Chemical/Biological/Radiological/Nuclear (CBRN) certification since nerve agents can degrade the materials used to make some respirators. However, during emergency operations, other NIOSH approved SCBAs or APRs that have been specifically tested by the manufacturer against chemical warfare agents may be used if deemed necessary by the Incident Commander. APRs should be equipped with a NIOSH approved CBRN filter or a combination organic vapor/acid gas/particulate cartridge.

Immediately dangerous to life or health (IDLH) levels are the ceiling limit for respirators other than SCBAs. Any exposures approaching the IDLH level should be regarded with extreme caution and the use of SCBAs for respiratory protection should be considered.

1.5.2.3 Chemical Protective Clothing

Use only chemical protective clothing that has undergone material and construction performance testing against the series of nerve agent that has been released. However, there is currently no information on performance testing of chemical protective clothing against either GV-series or novichok agents. Reported permeation rates may be affected by solvents, components, or impurities in munition grade or binary agents.

In addition to the risk of percutaneous migration of agent following dermal exposure to liquid agents, nerve agent vapors can also penetrate the skin and produce a toxic effect. However, these concentrations are significantly greater than concentrations that will produce similar effects if inhaled. If the concentration of vapor exceeds the level necessary to produce effects through dermal exposure, then responders should wear a Level A protective ensemble.

1.6 Decontamination

1.6.1 General

1.6.1.1 G-Series Nerve Agents

These agents are readily destroyed by high pH (i.e., basic solutions). Use an aqueous caustic solution (minimum of 10% by weight sodium hydroxide or sodium carbonate) or use undiluted household bleach. However, hydrolysis of G-series agents produces acidic by-products; so a large excess of base will be needed to ensure complete destruction of the agents. Due to the extreme volume required, household bleach is not an efficient means of decontaminating large quantities of these agents.

Avoid using highly concentrated caustic solutions because agents can become insoluble. Any agent that does not dissolve in the aqueous caustic solution will not be hydrolyzed. An agent will also be insoluble if it is dissolved in an immiscible organic solvent or if it is thickened such that it forms a protective layer at the agent/water interface. Addition of solvents or mechanical mixing may be required to overcome insolubility problems.

It is also important to ensure that the pH of the final waste solution remains basic. If the pH is below 7, fluoride ions can react with methylphosphonate diesters, a common impurity in these agents, to regenerate the nerve agent.

Solid hypochlorites [e.g., high test hypochlorite (HTH), super tropical bleach (STB), and Dutch powder] are also effective in destroying G-series agents. Reaction with hypochlorites, including household bleach, may produce toxic gases such as chlorine.

Reactive oximes and their salts, such as potassium 2,3-butanedione monoximate found in commercially available Reactive Skin Decontaminant Lotion (RSDL), are extremely effective at rapidly detoxifying nerve agents. Basic peroxides (e.g., a solution of baking soda, 30–50% hydrogen peroxide and an alcohol) also rapidly detoxify G-series agents. Some chloroisocyanurates, similar to those found in the Canadian Aqueous System for Chemical-Biological Agent Decontamination (CASCAD), are effective at detoxifying G-series agents.

1.6.1.2 V-Series Nerve Agents

It is difficult to detoxify V-series agents with aqueous caustic solutions alone due to their limited solubility at high pH. Solubility of an agent can be further reduced if it is dissolved in an immiscible organic solvent or if it has been thickened such that it forms a protective layer at the agent/water interface. In addition, V-series agents can react with aqueous caustic to produce stable and highly toxic *S*-[2-dialkylaminoethyl] alkylphosphonothiolates as by-products. These alkylphosphonothiolates have toxicities near those of the original agents. They are persistent and resist further hydrolysis. For additional information on these hazardous decomposition products, see the Decomposition Products and Impurities Section (C01-D) at the end of this chapter. To prevent the formation of these alkylphosphonothiolates, the reaction must be heated in excess of 180°F. Anhydrous solutions of caustic in alcohol are effective for destroying V-series agents without the production of alkylphosphonothiolates.

Household bleach is not an efficient means of decontaminating large quantities of V-series agents. In addition to limited solubility in commercial bleach due to the high pH, a minimum 10-fold excess of active chlorine to nerve agent is required to ensure destruction of the agent.

V-series agents react vigorously with caustics producing heat and off-gassing. Under these conditions, the amount of vapor given off by unreacted agent can increase significantly.

V-series agents can be destroyed by aqueous solutions of strong oxidizing agents. If a weak oxidizer or if an insufficient amount of a strong oxidizer is used, then toxic by-products will remain. However, they cannot be rendered nontoxic by oxidation without the presence of water. Basic peroxides (e.g., a solution of baking soda, 30–50% hydrogen peroxide, and an alcohol) also rapidly detoxify V-series agents.

An effective decontamination solution for V-series agents is prepared by mixing aqueous HTH slurry (10% by weight HTH) with alcohol in the ratio of 9:1. The solution should be prepared just before use. Use a large excess of neutralizing solution to ensure agent destruction. The reaction will produce heat and a significant amount of gas. Allow the neutralizing agent to remain in contact with the agent for a minimum of 1 hours. Adjust the pH of the expended neutralizing solution to 12.5 or more using a 10% aqueous sodium hydroxide solution.

Reactive oximes and their salts, such as potassium 2,3-butanedione monoximate found in commercially available RSDL, are extremely effective at rapidly detoxifying nerve agents. Some chloroisocyanurates, similar to those found in the CASCAD, are effective at detoxifying V-series agents and so is oxone, a peroxymonosulfate triple salt.

1.6.1.3 GV-Series Nerve Agents

These agents are readily destroyed by high pH (i.e., basic solutions). Use an aqueous caustic solution (minimum of 10% by weight sodium hydroxide or sodium carbonate) containing 20% alcohol or use undiluted household bleach. However, hydrolysis of GV-series agents produces acidic by-products; therefore, a large excess of base will be needed to ensure complete destruction of the agents. Due to the extreme volume required, household bleach is not an efficient means of decontaminating large quantities of these agents.

Avoid using highly concentrated caustic solutions because agents can become insoluble. Any agent that does not dissolve in the aqueous caustic solution will not be hydrolyzed. An agent will also be insoluble if it is dissolved in an immiscible organic solvent. Addition of solvents or mechanical mixing may be required to overcome insolubility problems.

Solid hypochlorites (e.g., HTH, STB, and Dutch powder) are also effective in destroying GV-series agents. Reaction with hypochlorites, including household bleach, may produce toxic gases such as chlorine.

Although specific data have not been published in the unclassified literature, preliminary studies indicate that reactive oximes and their salts, such as potassium 2,3-butanedione monoximate found in commercially available RSDL, are extremely effective at rapidly detoxifying GV-series nerve agents. Basic peroxides (e.g., a solution of baking soda, 30–50% hydrogen peroxide, and an alcohol) also rapidly detoxify GV-series agents.

1.6.1.4 Novichok Agents

Information on decontaminating novichok agents has not been published. However, on the basis of similarities to other organophosphates, it is likely that these agents will be destroyed by high pH (i.e., basic solutions). Use an aqueous caustic solution (minimum of 10% by weight sodium hydroxide or sodium carbonate) or use undiluted household bleach. Hydrolysis of novichok agents will produce acidic by-products; therefore, a large excess of base will be needed to ensure complete destruction of the agents. Due to the extreme volume required, household bleach is not an efficient means of decontaminating large quantities of these agents.

Solid hypochlorites (e.g., HTH, STB, and Dutch powder) should also be effective in destroying novichok series nerve agents. Reaction with hypochlorites, including household bleach, may produce toxic gases such as chlorine.

Although specific data have not been published in the unclassified literature, preliminary studies indicate that reactive oximes and their salts, such as potassium 2,3-butanedione monoximate found in commercially available RSDL, are extremely effective at rapidly detoxifying novichok series nerve agents. Also, based on similarities to other organophosphates, basic peroxides (e.g., a solution of baking soda, 30–50% hydrogen peroxide, and an alcohol) should rapidly detoxify novichok agents.

1.6.1.5 Vapors

Casualties/personnel: Remove all clothing as it may continue to emit "trapped" agent vapor after contact with the vapor cloud has ceased. Shower using copious amounts of soap and water. Ensure that the hair has been washed and rinsed to remove potentially trapped vapor. If there is a potential that the eyes have been exposed to nerve agents, irrigate with water or 0.9% saline solution for a minimum of 15 minutes.

Small areas: Ventilate to remove the vapors. If condensation is present, decontaminate with copious amounts of a decontamination solution as described in Section 1.6.1. Collect and place into containers lined with high-density polyethylene. Wash the area with copious amounts of soap and water. Collect and containerize the rinseate. Removal of porous material, including painted surfaces, may be required because the nerve agent that has been absorbed into these materials can migrate back to the surface posing both a contact and vapor hazard.

1.6.1.6 Liquids, Solutions, or Liquid Aerosols

Casualties/personnel: Remove all clothing immediately. Even clothing that has not come into direct contact with the agent may contain "trapped" vapor. To avoid further exposure of the head, neck, and face to the agent, cut off potentially contaminated clothing that must be pulled over the head. Remove as much of the nerve agent from the skin as fast as possible. If water is not immediately available, the agent can be absorbed with any convenient material such as paper towels, toilet paper, flour, talc, and so on. To minimize both spreading the agent and abrading the skin, do not rub the agent with the absorbent. Blot the contaminated skin with the absorbent.

Use a sponge or cloth with liquid soap and copious amounts of water to wash the skin surface and hair at least three times. Do not delay decontamination to find warm or hot water if it is not readily available. Avoid rough scrubbing as this could abrade the skin and increase percutaneous absorption of residual agent. Rinse with copious amounts of water. If there is a potential that the eyes have been exposed to nerve agents, irrigate with water or 0.9% saline solution for a minimum of 15 minutes.

Alternatively, a household bleach solution can be used instead of soap and water. The bleach solution should be no more than one part household bleach in nine parts water (i.e., 0.5% sodium hypochlorite) to avoid damaging the skin. Avoid any contact with sensitive areas such as the eyes. Rinse with copious amounts of water.

Small areas: Puddles of liquid can be absorbed by covering with absorbent material such as vermiculite, diatomaceous earth, clay, sponges, or towels. Place the absorbed material into containers lined with high-density polyethylene. Before sealing the container, cover the contents with a decontamination solution as described in Section 1.6.1. Decontaminate the area and the exterior of the container with copious amounts of the neutralizing agent. Allow it to stand for a minimum of 5 minutes before rinsing with water. Collect and containerize the rinseate. Ventilate the area to remove vapors. Removal of porous material, including painted surfaces, may be required because the nerve agent that has been absorbed into these materials can migrate back to the surface posing both a contact and vapor hazard.

1.6.1.7 Solids, Dusty Agents, or Particulate Aerosols

Casualties/personnel: Do not attempt to brush the agent off the individual or their clothing as this can aerosolize the agent. Remove all clothing immediately. To avoid further exposure of the head, neck, and face to the agent, cut off potentially contaminated clothing that must be pulled over the head. Wash the skin surface and hair at least three times with copious amounts of soap and water. Do not delay decontamination to find warm or hot water if it is not readily available. Rinse with copious amounts of water. If there is a potential that the eyes have been exposed to nerve agents, irrigate with water or 0.9% saline solution for a minimum of 15 minutes.

Small areas: If indoors, close windows and doors in the area and turn off anything that could create air currents (e.g., fans, air conditioner, etc.). Avoid actions that could aerosolize the agent such as sweeping or brushing. Collect the agent using a vacuum cleaner equipped with a high-efficiency particulate air (HEPA) filter. Do not use a standard home or industrial vacuum. Do not allow the vacuum exhaust to stir the air in the contaminated area. Vacuum all surfaces with extreme care in a very slow and controlled manner to minimize aerosolizing the agent. Place the collected material into containers lined with high-density polyethylene. Before sealing the container, cover the contents with a decontamination solution described in Section 1.6.1. Decontaminate the area with copious amounts of the neutralizing agent. Allow it to stand for a minimum of 5 minutes before rinsing with water. Collect and containerize the rinseate.

1.7 Medical

1.7.1 CDC Case Definition

1) A case in which nerve agents are detected in the urine. Decreased plasma or red blood cell cholinesterase levels based on a specific commercial laboratory reference range might indicate a nerve agent or organophosphate exposure; however, the normal range levels for cholinesterase are wide, which makes interpretation of levels difficult without a baseline measurement or repeat measurements over time. 2) Detection of organophosphates in environmental samples. The case can be confirmed if laboratory testing is not performed because either a predominant amount of clinical and nonspecific laboratory evidence is present or an absolute certainty of the etiology of the agent is known.

1.7.2 Differential Diagnosis

The following factors have been suggested as alternatives to consider when presented with a potential case of exposure to nerve agents: carbamate and organophosphate pesticides; alkaloids such as nicotine or coniine; ingestion of mushrooms containing muscarine; and medicinals such as carbamates, cholinomimetic compounds, and neuromuscular blocking drugs.

1.7.3 Signs and Symptoms

1.7.3.1 Vapors/Aerosols

Pinpointing of pupils (miosis) and extreme nasal discharge (rhinorrhea) may be the first indications of exposure. Miosis usually indicates exposure to vapors or aerosols unless the individual has had liquid agent in or around their eyes. The casualty may also experience difficulty in breathing with a feeling of shortness of breath or tightness of the chest. In cases of exposure to high-vapor concentrations, the gastrointestinal tract may be affected producing vomiting, urination, or defecation. Inhalation of lethal amounts of nerve agent can cause loss of consciousness and convulsions in as little as 30 seconds, followed by cessation of breathing and flaccid paralysis after several more minutes.

In contrast to either the G-series or V-series agents, the observable signs and symptoms of exposure to the GV-series agents are more insidious and tend to be very mild and transient. Even convulsions occurring just prior to death are usually milder than with G-series or V-series agents.

1.7.3.2 Liquids/Solids

General signs and symptoms of small to moderate exposure include localized sweating, nausea, vomiting, involuntary urination/defecation, and a feeling of weakness. The casualty may also experience difficulty in breathing with a feeling of shortness of breath or tightness of the chest. Vomiting and uncontrolled urination/defecation generally indicates an exposure to liquid or solid agent and not just agent vapors. Miosis usually does not occur unless the individual has had agent in or around their eyes. Exposure to a large amount of agent causes copious secretions, loss of consciousness, convulsions progressing into flaccid paralysis, and cessation of breathing.

In contrast to either the G-series or V-series agents, the observable signs and symptoms of exposure to the GV-series agents are more insidious and tend to be very mild and transient. Even convulsions occurring just prior to death are usually milder. In addition, the progression of typical signs and symptoms of exposure to cholinesterase inhibiting substances may be atypical.

1.7.4 Mass-Casualty Triage Recommendations

1.7.4.1 Priority 1

A casualty with symptoms in two or more organ systems (not including miosis or rhinorrhea), who has a heartbeat and a palpable blood pressure. The casualty may or may not be conscious and/or breathing.

1.7.4.2 Priority 2

A casualty with known exposure to liquid agent but no apparent signs or symptoms, or a casualty who is recovering from a severe exposure after receiving treatment.

1.7.4.3 Priority 3

A casualty who is walking and talking, although miosis and/or rhinorrhea may be present.

1.7.4.4 Priority 4

A casualty who is not breathing and does not have a heartbeat or palpable blood pressure.

1.7.5 Casualty Management

Decontaminate the casualty ensuring that all nerve agents have been removed. If nerve agents have gotten into the eyes, irrigate the eyes with water or 0.9% saline solution for at least 15 minutes. Irrigate open wounds with water or 0.9% saline solution for at least 10 minutes. However, do not delay treatment if thorough decontamination cannot be undertaken immediately.

Once the casualty has been decontaminated, including the removal of foreign matter from wounds, medical personnel do not need to wear a chemical-protective mask.

Ventilate the patient. There may be an increase in airway resistance due to constriction of the airway and the presence of secretions. If breathing is difficult, administer oxygen. As soon as possible administer of atropine alone or in combination with pralidoxime chloride (2-PAMCl) or other appropriate oxime. Diazepam may be required to prevent or control severe convulsions. If diazepam is not administered within 40-minutes postexposure, then its effectiveness at controlling seizures is minimal.

Over time, the nerve agent enzyme complex undergoes an irreversible biochemical decomposition known as aging. After aging occurs, the nerve agent molecule can no longer be removed from the enzyme by treatment with an oxime. The rate of aging is dependent on the chemical structure of the specific nerve agent and ranges from several minutes to several days. Soman (C01-A003) ages in about 2 minutes, making treatment for exposure to this particular agent difficult. The former Soviet Union is reported to have developed other agents that age rapidly, but the specific agents have not been identified in unclassified literature.

1.8 Fatality Management

Remove all clothing and personal effects segregating them as either durable or nondurable items. Although it may be possible to decontaminate durable items, it may be safer and more efficient to destroy nondurable items rather than attempt to decontaminate them. Items that will be retained for further processing should be double sealed in impermeable containers, ensuring that the inner container is decontaminated before placing it in the outer one.

Nerve agents that have entered the body are metabolized, hydrolyzed, or bound to tissue and pose little threat of off-gassing. To remove agents on the outside of the body, wash the remains with a 2% sodium hypochlorite bleach solution (i.e., 2 gallons of water for every gallon of household bleach) ensuring the solution is introduced into the ears, nostrils, mouth, and any wounds. This concentration of bleach will not affect remains but will neutralize organophosphorus nerve agents. Higher concentrations of bleach can harm remains. Pay particular attention to areas where agent may get trapped, such as hair, scalp, pubic areas, fingernails, folds of skin, and wounds. The bleach solution should remain on the cadaver for a minimum of 5 minutes. Wash with soap and water. Ensure that all the bleach solution is removed before embalming as it will react with embalming fluid. All wash and rinse waste must be contained for proper disposal. Screen the remains for agent vapors and residual liquid at the conclusion of the decontamination process. If the remains must be stored before embalming, then place them inside body bags designed to contain contaminated bodies or in double body bags. If double body bags are used, seal the inner bag with duct tape, rinse, then place in the second bag. After embalming is complete, place the remains in body bags designed to contain contaminated bodies or in double body bags. Body fluids removed during the embalming process do not pose any additional risks, and should be contained and handled according to established procedures.

Standard burials are acceptable when contamination levels are low enough to allow bodies to be handled without wearing additional protective equipment. Cremation may be required if remains cannot be completely decontaminated. Although organophosphorus nerve agents are destroyed after 15 minutes at the operating temperature of a commercial crematorium (i.e., above 1000°F), the initial heating phase may volatilize some of the agents and allow vapors to escape.

C01-A

G-SERIES AGENTS

C01-A001

Tabun (Agent GA)

CAS: 77-81-6; 93957-09-6 (Isomer); 93957-08-5 (Isomer)

RTECS: TB4550000

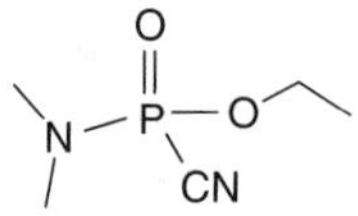

O
N — P — O
CN

$C_5H_11N_2O_2P$

Colorless to brown liquid that is odorless when pure; impurities may give a faintly fruity or bitter almonds odor. Undergoes considerable decomposition when explosively disseminated.

Exposure Hazards

Conversion Factor: 1 ppm = 6.63 mg/m^3 at 77°F

$LCt_{50(Inh)}$: 70 mg-min/m^3 (5 ppm for a 2-min exposure)

$LCt_{50(Per)}$: 15,000 mg-min/m^3 (2 ppm for a 30-min exposure)

LD_50: 1 g

Miosis: 0.03 ppm for a 2-min exposure
$MEG_{(1\ h)}$*Min*: 0.00042 ppm; *Sig*: 0.0053 ppm; *Sev*: 0.039 ppm
WPL AEL: 0.000005 ppm
STEL: 0.000015 ppm
IDLH: 0.015 ppm

Properties:

MW: 162.1	*VP*: 0.037 mmHg	*FlP*: 172°F
D: 1.073 g/mL (77°F)	*VD*: 5.6 (calculated)	*LEL*: –
MP: −58°F	*Vlt*: 49 ppm	*UEL*: –
BP: 473°F	H_2O: 7.2%	*RP*: 210
Vsc: 2.18 cS (77°F)	*Sol:* Most organic solvents	*IP*: < 10.6 eV

Final AEGLs

AEGL-1: 1 h, 0.0004 ppm	4 h, 0.0002 ppm	8 h, 0.0002 ppm
AEGL-2: 1 h, 0.005 ppm	4 h, 0.003 ppm	8 h, 0.002 ppm
AEGL-3: 1 h, 0.04 ppm	4 h, 0.02 ppm	8 h, 0.02 ppm

C01-A002

Sarin (Agent GB)
CAS: 107-44-8; 6171-94-4 (Isomer)
RTECS: TA8400000

$C_4H_{10}FO_2P$

Colorless liquid that is odorless when pure.

Also reported stockpiled as a mixture with Cyclosarin (C01-A004).

Exposure Hazards

Conversion Factor: 1 ppm = 5.73 mg/m^3 at 77°F
$LCt_{50(Inh)}$: 35 mg-min/m^3 (3 ppm for a 2-min exposure)
$LCt_{50(Per)}$: 12,000 mg-min/m^3 (1 ppm for a 30-min exposure)
LD_{50}: 1.7 g
Miosis: 0.03 ppm for a 2-min exposure
$MEG_{(1\ h)}$*Min*: 0.00048 ppm; *Sig*: 0.0060 ppm; *Sev*: 0.022 ppm
WPL AEL: 0.000005 ppm
STEL: 0.00002 ppm
IDLH: 0.02 ppm

Properties:

MW: 140.1	*VP*: 2.9 mmHg (77°F)	*FlP*: None
D: 1.102 g/mL	*VD*: 4.8 (calculated)	*LEL*: None
MP: −69°F	*Vlt*: 2800 ppm	*UEL*: None
BP: 316°F	H_2O: Miscible	*RP*: 3
Vsc: 1.28 cS (77°F)	*Sol*: Most organic solvents	*IP*: ~10.6 eV

Final AEGLs

AEGL-1: 1 h, 0.0005 ppm	4 h, 0.0002 ppm	8 h, 0.0002 ppm
AEGL-2: 1 h, 0.006 ppm	4 h, 0.003 ppm	8 h, 0.002 ppm
AEGL-3: 1 h, 0.02 ppm	4 h, 0.01 ppm	8 h, 0.009 ppm

C01-A003

Soman (Agent GD)

CAS: 96-64-0; 22956-47-4 (Isomer); 255842-00-3 (Isomer); 255841-99-7 (Isomer); 89254-46-6 (Isomer); 89254-45-5 (Isomer); 66429-60-5 (Isomer); 66429-59-2 (Isomer); 24753-16-0 (Isomer); 24753-15-9 (Isomer); 22956-48-5 (Isomer)

RTECS: TA8750000

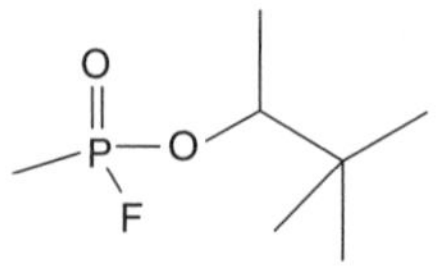

$C_7H_{16}FO_2P$

Colorless to brown liquid that is relatively odorless when pure; impurities may give a fruity or camphor odor.

Exposure Hazards

Conversion Factor: 1 ppm = 7.45 mg/m^3 at 77°F

This agent "ages" rapidly.

LCt$_{50(Inh)}$: 35 mg-min/m^3 (2.3 ppm for a 2-min exposure)
LCt$_{50(Per)}$: 3000 mg-min/m^3 (0.2 ppm for a 30-min exposure)
LD$_{50}$: 0.35 g
Miosis: 0.01 ppm for a 2-min exposure
MEG$_{(1\ h)}$*Min*: 0.00018 ppm; *Sig*: 0.0022 ppm; *Sev*: 0.017 ppm
WPL AEL: 0.000004 ppm
STEL: 0.000007 ppm
IDLH: 0.007 ppm

Properties:

MW: 182.2	*VP*: 0.40 mmHg (77°F)	*FlP*: 250°F
D: 1.022 g/mL (77°F)	*VD*: 6.2 (calculated)	*LEL*: –
MP: −44°F	*Vlt*: 520 ppm (calculated)	*UEL*: –
BP: 388°F	H_2O: 2.1%	*RP*: 19 (77°F)
Vsc: 3.10 cS (77°F)	*Sol*: Hydrocarbons; Alcohols	*IP*: <10.6 eV

Final AEGLs

AEGL-1: 1 h, 0.0002 ppm	4 h, 0.00009 ppm	8 h, 0.00007 ppm
AEGL-2: 1 h, 0.002 ppm	4 h, 0.001 ppm	8 h, 0.0009 ppm
AEGL-3: 1 h, 0.02 ppm	4 h, 0.009 ppm	8 h, 0.007 ppm

C01-A004

Cyclosarin (Agent GF)

CAS: 329-99-7; 111422-21-0 (Isomer); 111422-20-9 (Isomer)

RTECS: —

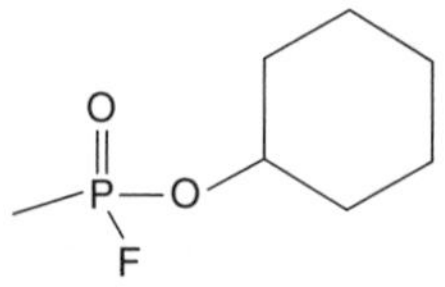

$C_7H_{14}FO_2P$

Clear, colorless liquid that is odorless.
Also reported stockpiled as a mixture with Sarin (C01-A002).

Exposure Hazards

Conversion Factor: 1 ppm = 7.37 mg/m^3 at 77°F
LCt$_{50(Inh)}$: 35 mg-min/m^3 (2 ppm for a 2-min exposure)
LCt$_{50(Per)}$: 3000 mg-min/m^3 (0.2 ppm for a 30-min exposure)
LD$_{50}$: 0.35 g
Miosis: 0.03 ppm for a 2-min exposure
MEG$_{(1\ h)}$Min: 0.00020 ppm; *Sig*: 0.0024 ppm; *Sev*: 0.018 ppm
WPL AEL: 0.000004 ppm
STEL: 0.000007 ppm
IDLH: 0.007 ppm

Properties:

MW: 180.2	*VP*: 0.044 mmHg	*FlP*: 201°F
D: 1.133 g/mL	*VD*: 6.2 (calculated)	*LEL*: —
MP: 10°F	*Vlt*: 59 ppm	*UEL*: —
BP: 462°F	*H_2O*: 3.7%	*RP*: 110
Vsc: 4.27 cS (77°F)	*Sol*: Most organic solvents	*IP*: <10.6 eV

Final AEGLs

AEGL-1: 1 h, 0.0002 ppm	4 h, 0.0001 ppm	8 h, 0.00007 ppm
AEGL-2: 1 h, 0.002 ppm	4 h, 0.001 ppm	8 h, 0.0009 ppm
AEGL-3: 1 h, 0.02 ppm	4 h, 0.01 ppm	8 h, 0.007 ppm

C01-A005

Ethyl sarin (Agent GE)
CAS: 1189-87-3
RTECS:—

$C_5H_{12}FO_2P$

Specific information on physical appearance is not available for this agent.

Exposure Hazards

Conversion Factor: 1 ppm = 6.30 mg/m^3 at 77°F

Human toxicity values have not been established or have not been published. However, based on available information, this agent appears to be approximately 90% as toxic as Sarin (C01-A002).

Properties:

MW: 154.1	*VP*: —	*FlP*: —
D: —	*VD*: 5.3 (calculated)	*LEL*: —
MP: —	*Vlt*: —	*UEL*: —
BP: 165°F (16 mmHg)	*H_2O*: —	*RP*: —
Vsc: —	*Sol*: —	*IP*: —

C01-A006

***O*-Cyclohexyl methylphosphonofluoridothiate** (Agent EA 2223)
CAS: 4241-34-3
RTECS: —

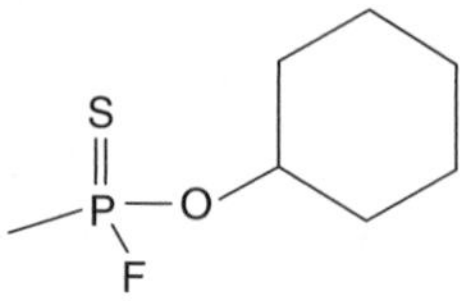

$C_7H_{14}FOPS$

Water-white to yellow liquid that is odorless. This agent will hydrolyze to produce Cyclosarin (C01-A004).

Exposure Hazards

Conversion Factor: 1 ppm = 8.02 mg/m^3 at 77°F

This agent is refractory to treatment.

$LCt_{50(Inh)}$: 100 mg-min/m^3 (6 ppm for a 2-min exposure)
LD_{50}: 0.18 g

Properties:

MW: 196.2	*FlP*: 234°F	*FlP*: 234°F
D: 1.12 g/mL (77°F)	*LEL*: —	*LEL*: —
MP: –16°F	*UEL*: —	*UEL*: —
BP: 441°F	*RP*: 60	*RP*: 60
Vsc: 3.53 cS (77°F)	*IP*: —	*IP*: —

C01-A007

2-Methylcyclohexyl methylphosphonofluoridate (Agent EA 1356 or Agent EA 3534)
CAS: 85473-32-1; 193090-56-1 (Isomer); 193090-30-1 (Isomer)
RTECS: —

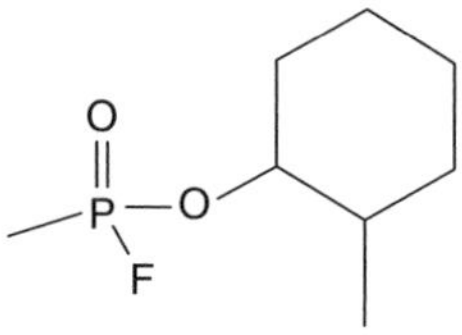

$C_8H_{16}FO_2P$

Water-white to light straw colored liquid that is odorless. There are two configurational isomers of this agent that have been studied.

Exposure Hazards

Conversion Factor: 1 ppm = 7.94 mg/m^3 at 77°F

This agent is refractory to treatment.

$LCt_{50(Inh)}$: 70 mg-min/m^3 (4 ppm for a 2-min exposure)
LD_{50}: 0.17 g

Properties:

MW: 194.2	*VP*: 0.055 mmHg (77°F)	*FlP*: —
D: 1.1 g/mL (77°F)	*VD*: 6.7 (calculated)	*LEL*: —
MP: 15°F	*Vlt*: 64 ppm	*UEL*: —
BP: 455°F	*H_2O*: 1%	*RP*: 130
Vsc: 4.78 cS (77°F)	*Sol*: Most organic solvents	*IP*: —

C01-A008

Isopropyl dimethylamidocyanidophosphate (Agent EA 4352)
CAS: 63815-55-4
RTECS: —

$C_6H_{13}N_2O_2P$

Odorless liquid.

Exposure Hazards
Conversion Factor: 1 ppm = 7.21 mg/m^3 at 77°F

Human toxicity values have not been established or have not been published. However, based on available information, this agent appears to be as toxic as Sarin (C01-A002).

Properties:

MW: 176.2	*VP*: 0.055 mmHg (77°F)	*FlP*: —
D: 1.0425 g/mL (77°F)	*VD*: 6.1 (calculated)	*LEL*: —
MP: —	*Vlt*: 76 ppm	*UEL*: —
BP: 453°F	H_2O: —	*RP*: 130
Vsc: —	*Sol*: —	*IP*: —

C01-A009

Methyl ethylphosphonofluoridate
CAS: 665-03-2
RTECS: —

$C_3H_8FO_2P$

Specific information on physical appearance is not available for this agent.

Exposure Hazards
Conversion Factor: 1 ppm = 5.16 mg/m^3 at 77°F

Human toxicity values have not been established or have not been published. However, this agent is a powerful cholinesterase inhibitor.

Properties:

MW: 126.1	*VP*: —	*FlP*: —
D: —	*VD*: 4.3 (calculated)	*LEL*: —
MP: —	*Vlt*: —	*UEL*: —
BP: 117°F (12 mmHg)	H_2O: —	*RP*: —
Vsc: —	*Sol*: —	*IP*: —

C01-A010

Ethyl methylphosphonofluoridate
CAS: 673-97-2
RTECS: —

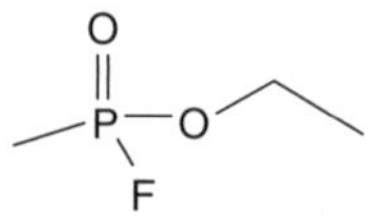

$C_3H_8FO_2P$

Specific information on physical appearance is not available for this agent.

Exposure Hazards
Conversion Factor: 1 ppm = 5.16 mg/m^3 at 77°F

Human toxicity values have not been established or have not been published. However, this agent is a powerful cholinesterase inhibitor.

Properties:

MW: 126.1	*VP*: —	*FlP*: —
D: —	*VD*: 4.3 (calculated)	*LEL*: —
MP: —	*Vlt*: —	*UEL*: —
BP: 127°F (12 mmHg)	H_2O: —	*RP*: —
Vsc: —	*Sol*: —	*IP*: —

C01-A011

Ethyl ethylphosphonofluoridate
CAS: 650-20-4
RTECS: —

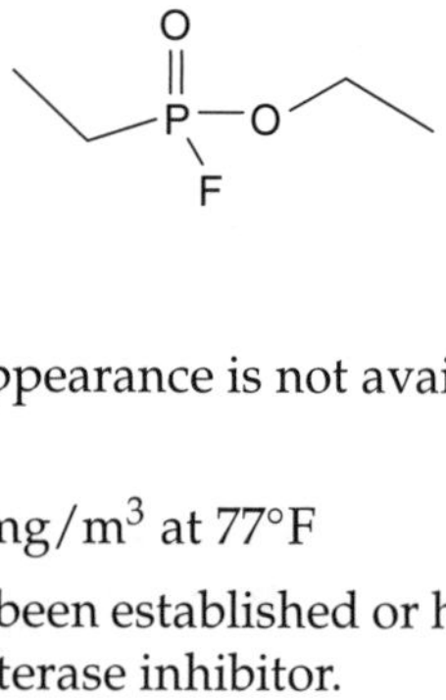

$C_4H_{10}FO_2P$

Specific information on physical appearance is not available for this agent.

Exposure Hazards
Conversion Factor: 1 ppm = 5.73 mg/m^3 at 77°F

Human toxicity values have not been established or have not been published. However, this agent is a powerful cholinesterase inhibitor.

Properties:

MW: 140.1	*VP*: —	*FlP*: —
D: —	*VD*: 4.8 (calculated)	*LEL*: —
MP: —	*Vlt*: —	*UEL*: —
BP: 115°F (9 mmHg)	H_2O: —	*RP*: —
Vsc: —	*Sol*: —	*IP*: —

C01-A012

Fluorotabun
CAS: 358-29-2
RTECS: —

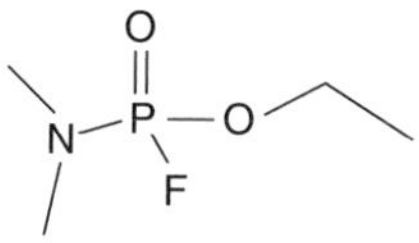

$C_4H_{11}FNO_2P$

Specific information is not available for this agent.

Exposure Hazards

Conversion Factor: 1 ppm = 6.34 mg/m^3 at 77°F

Human toxicity values have not been established or have not been published. However, based on available information, this agent appears to be less than half as toxic as Tabun (C01-A001).

Properties:

MW: 155.1	*VP*: —	*FlP*: —
D: —	*VD*: 5.3 (calculated)	*LEL*: —
MP: —	*Vlt*: —	*UEL*: —
BP: 167°F (18 mmHg)	H_2O: —	*RP*: —
Vsc: —	*Sol*: —	*IP*: —

V-SERIES AGENTS

C01-A013

Amiton (Agent VG)

CAS: 78-53-5; 3734-96-1 (*p*-Toluenesulfonate salt); 3734-97-2 (Oxalate salt)
RTECS: —

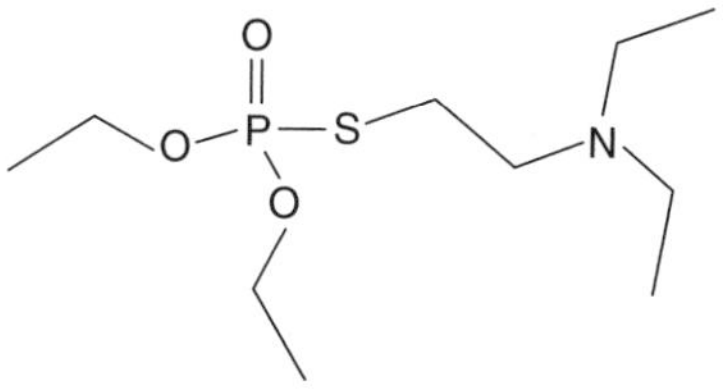

$C_{10}H_{24}NO_3PS$

Oily liquid. Various salts (solids) have been reported.

Exposure Hazards

Conversion Factor: 1 ppm = 11.01 mg/m^3 at 77°F

Human toxicity values have not been established or have not been published. However, this agent is a powerful cholinesterase inhibitor.

Properties:

MW: 269.3	*VP*: 0.01 mmHg (176°F)	*FlP*: —
D: —	*VD*: 9.3 (calculated)	*LEL*: —
MP: —	*Vlt*: 11 ppm	*UEL*: —
BP: 230°F (0.2 mmHg)	H_2O: "Highly soluble"	*RP*: 730 (176°F)
Vsc: —	*Sol*: Most organic solvents	*IP*: —

Oxalate salt
MW: 359.4
MP: 190°F

***p*-Toluenesulfonate salt**
MW: 441.5
MP: 205°F

C01-A014

***O*-Ethyl *S*-(2-diethylaminoethyl) methyl-phosphonothiolate** (Agent VM)
CAS: 21770-86-5; 107059-49-4 (*p*-Toluenesulfonate salt)
RTECS: —

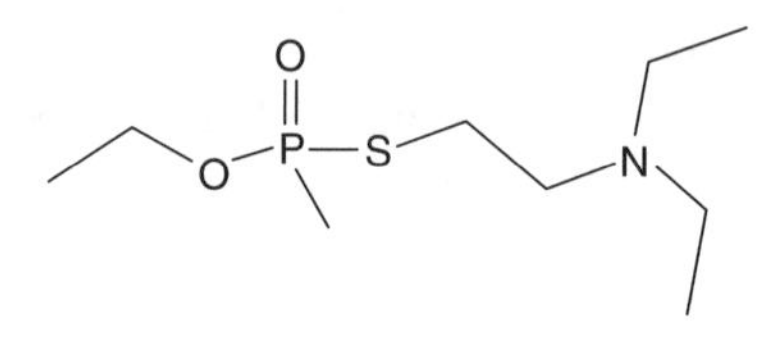

$C_9H_{22}NO_2PS$

Water-white to dark-yellow oily liquid that is odorless. Various salts (solids) have been reported.

Exposure Hazards
Conversion Factor: 1 ppm = 9.79 mg/m^3 at 77°F
$LCt_{50(Inh)}$: 50 mg-min/m^3 (3 ppm for a 2-min exposure). This value is for resting individuals.

Properties:
MW: 239.3
D: 1.0312 g/mL (77°F)
MP: −58°F
BP: "Near 560°F"
Vsc: 5.67 cS (77°F)

VP: 0.0021 mmHg (77°F)
VD: 8.3 (calculated)
Vlt: 2.9 ppm
H_2O: Miscible
Sol: Most organic solvents; Dilute mineral acids

FlP: 457°F
LEL: —
UEL: —
RP: 3100
IP: <10.6 eV

***p*-Toluenesulfonate salt**
MW: 411.5
MP: 147°F

C01-A015

***O*-Isobutyl *S*-2-diethylaminoethyl methylphosphonothiolate** (Agent VR)
CAS: 159939-87-4
RTECS: —

$C_{11}H_{26}NO_2PS$

Specific information on physical appearance is not available for this agent. Various salts (solids) have been reported.

Exposure Hazards
Conversion Factor: 1 ppm = 10.94 mg/m^3 at 77°F

Human toxicity values have not been established or have not been published. However, this agent is a powerful cholinesterase inhibitor.

Properties:

MW: 267.4
D: 1.003 g/mL
MP: —
BP: 187° F (0.001 mmHg)
Vsc: —

VP: 0.00062 mmHg (77°F)
VD: 9.0 (calculated)
Vlt: 0.81 ppm
H_2O: —
Sol: —

FlP: —
LEL: —
UEL: —
RP: 9900
IP: —

C01-A016

***O*-Ethyl *S*-(2-diisopropylaminoethyl) methylphosphonothiolate** (Agent VX)
CAS: 50782-69-9; 65167-63-7 (Isomer); 65167-64-8 (Isomer)
RTECS: TB1090000

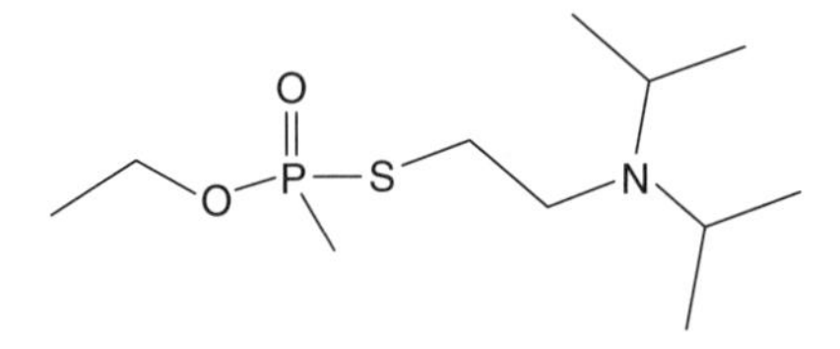

$C_{11}H_{26}NO_2PS$

Colorless oily liquid that is odorless. Similar in appearance to motor oil. Various salts (solids) have been reported.

Exposure Hazards

Conversion Factor: 1 ppm = 10.94 mg/m^3 at 77°F
$LCt_{50(Inh)}$: 15 mg-min/m^3 (0.7 ppm for a 2-min exposure)
$LCt_{50(Per)}$: 150 mg-min/m^3 (0.5 ppm for a 30-min exposure)
LD_{50}: 0.005 g
Miosis: 0.005 ppm for a 2-min exposure
$MEG_{(1\,h)}$*Min*: 0.000016 ppm; *Sig*: 0.00027 ppm; *Sev*: 0.00091 ppm
WPL AEL: 0.00000008 ppm
STEL: 0.0000009 ppm
IDLH: 0.0003 ppm

Properties:

MW: 267.4
D: 1.0083 g/mL (77°F)
MP: –38.2°F
MP: <–60° F with impurities
BP: Decomposes
Vsc: 9.96 cS (77°F)

VP: 0.0007 mmHg
VP: 0.00004 mmHg (32°F)
VD: 9.2 (calculated)
Vlt: 1.2 ppm (77°F)
H_2O: 3% (77°F)
H_2O: Miscible (<49°F)
Sol: Most organic solvents; Dilute mineral acids

FlP: 318°F
LEL: —
UEL: —
RP: 8800
IP: <10.6 eV

Final AEGLs

AEGL-1: 1 h, 0.000016 ppm; 4 h, 0.0000091 ppm; 8 h, 0.0000065 ppm
AEGL-2: 1 h, 0.00027 ppm; 4 h, 0.00014 ppm; 8 h, 0.000095 ppm
AEGL-3: 1 h, 0.00091 ppm; 4 h, 0.00048 ppm; 8 h, 0.00035 ppm

C01-A017

***O*-Ethyl *S*-2-dimethylaminoethyl methylphosphonothiolate** (Agent Vx)
CAS: 20820-80-8
RTECS: —

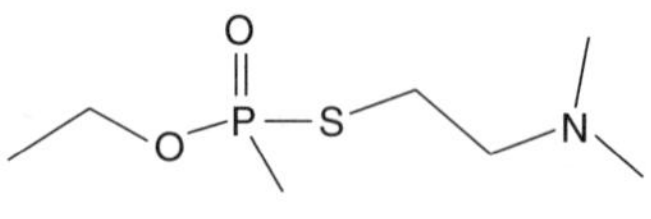

$C_7H_{18}NO_2PS$

Amber colored oily liquid that is odorless. Various salts (solids) have been reported.

Exposure Hazards

Conversion Factor: 1 ppm = 8.64 mg/m^3 at 77°F

Human toxicity values have not been established or have not been published. However, this agent is a powerful cholinesterase inhibitor.

Properties:

MW: 211.3	*VP*: 0.0042 mmHg	*FlP*: —
D: 1.062 g/mL (77°F)	*VD*: 7.3 (calculated)	*LEL*: —
MP: —	*Vlt*: 5.6 ppm	*UEL*: —
BP: 490°F (approx.)	*H_2O*: "Slight"	*RP*: 1600
Vsc: —	*H_2O*: Miscible ("Cooler temperatures")	*IP*: <10.6 eV
	Sol: Most organic solvents	

C01-A018

***O*-Cyclopentyl *S*-(2-diethylaminoethly) methylphosphonothiolate** (Agent EA 3148)

CAS: 93240-66-5

RTECS: —

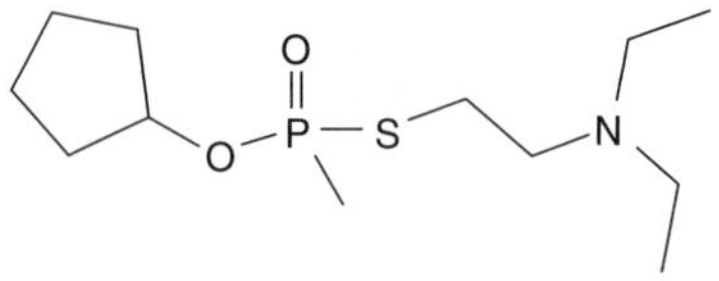

$C_{12}H_{26}NO_2PS$

Colorless to pale yellow liquid that is odorless.

Exposure Hazards

Conversion Factor: 1 ppm = 11.43 mg/m^3 at 77°F

Human toxicity values have not been established or have not been published. However, this agent is a powerful cholinesterase inhibitor.

Properties:

MW: 279.4	*VP*: 0.0004 mmHg (77°F)	*FlP*: >500°F
D: 1.05 g/mL (77°F)	*VD*: 9.6 (calculated)	*LEL*: —
MP: —	*Vlt*: 0.5 ppm (77°F)	*UEL*: —
BP: 232°F (0.05 mmHg)	*H_2O*: "Low"	*RP*: 15,000
Vsc: 1.96 cS (77°F)	*Sol*: Most organic solvents	*IP*: —

C01-A019

***O*-Cyclopentyl *S*-(2-diisopropylaminoethyl) methylphosphonothiolate** (Agent EA 3317)

CAS: 85473-33-2; 102490-57-3 (Isomer); 102490-59-5 (Isomer)

RTECS: —

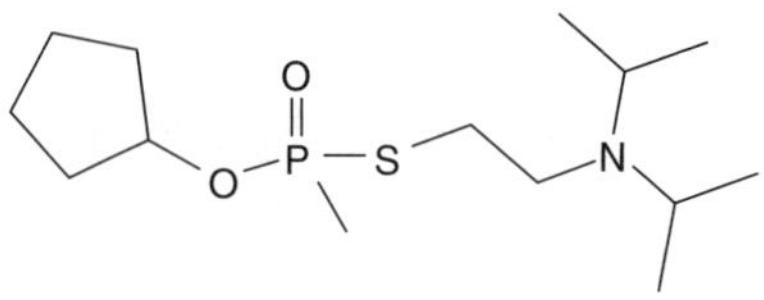

$C_{14}H_{30}NO_2PS$

Odorless liquid.

Exposure Hazards

Conversion Factor: 1 ppm = 12.57 mg/m^3 at 77°F

Human toxicity values have not been established or have not been published. However, this agent is a powerful cholinesterase inhibitor.

Properties:

MW: 307.4	*VP*: 0.00014 mmHg (77°F)	*FlP*: —
D: 1.02 g/mL (77°F)	*VD*: 11 (calculated)	*LEL*: —
MP: —	*Vlt*: 0.2 ppm (77°F)	*UEL*: —
BP: 235°F (0.08 mmHg)	*H_2O*: —	*RP*: 41,000
Vsc: 35.1 cS	*Sol*: —	*IP*: —

C01-A020

***O*-Cyclohexyl *S*-[2-(diethylamino)ethyl] methylphosphonothiolate**

CAS: 71293-89-5

RTECS: —

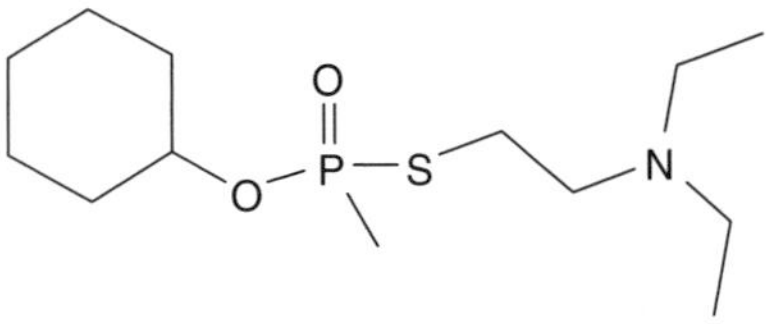

$C_{13}H_{28}NO_2PS$

Specific information on physical appearance is not available for this agent.

Exposure Hazards

Conversion Factor: 1 ppm = 12.00 mg/m^3 at 77°F

Human toxicity values have not been established or have not been published. However, this agent is a powerful cholinesterase inhibitor.

Properties:

MW: 293.4	*VP*: —	*FlP*: —
D: 0.9451 g/mL	*VD*: 10 (calculated)	*LEL*: —
MP: —	*Vlt*: —	*UEL*: —
BP: 194°F (0.3 mmHg)	*H_2O*: —	*RP*: —
Vsc: —	*Sol*: —	*IP*: —

C01-A021

***O*-Ethyl *S*-[2-(dimethylamino)ethyl] ethylphosphonothiolate**

CAS: 98543-25-0; 110422-92-9 (Oxalate salt)

RTECS: —

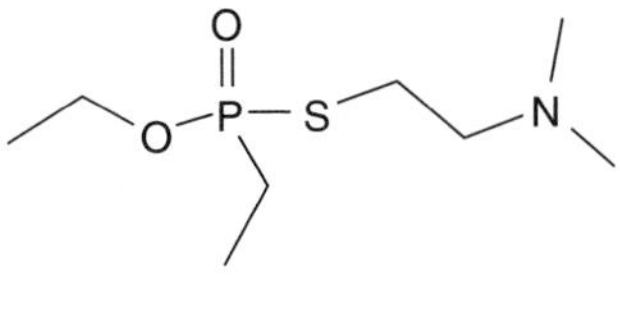

$C_8H_{20}NO_2PS$

Oily liquid. Various salts (solids) have been reported.

Exposure Hazards

Conversion Factor: 1 ppm = 9.21 mg/m^3 at 77°F

Human toxicity values have not been established or have not been published. However, this agent is a powerful cholinesterase inhibitor.

Properties:

MW: 225.3	*VP*: —	*FlP*: —
D: —	*VD*: 7.8 (calculated)	*LEL*: —
MP: —	*Vlt*: —	*UEL*: —
BP: 167°F (0.005 mmHg)	*H_2O*: —	*RP*: —
Vsc: —	*Sol*: —	*IP*: —

Oxalate Salt

MW: 315.3

MP: 226°F

C01-A022

***O*-Isopropyl *S*-[2-(diethylamino)ethyl] methylphosphonothiolate**

CAS: 91134-95-1

RTECS: —

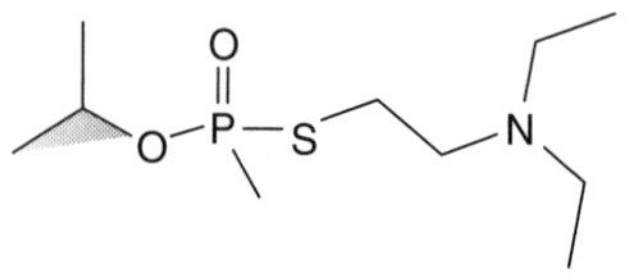

$C_{10}H_{24}NO_2PS$

Specific information on physical appearance is not available for this agent.

Exposure Hazards

Conversion Factor: 1 ppm = 10.36 mg/m^3 at 77°F

Human toxicity values have not been established or have not been published. However, this agent is a powerful cholinesterase inhibitor.

Properties:

MW: 253.3	*VP*: —	*FlP*: —
D: —	*VD*: 8.7 (calculated)	*LEL*: —
MP: —	*Vlt*: —	*UEL*: —
BP: 250°F (2 mmHg)	*H_2O*: —	*RP*: —
Vsc: —	*Sol*: —	*IP*: —

C01-A023

***O*-Ethyl *S*-[2-(piperidylamino)ethyl] ethylphosphonothiolate**

CAS: 108753-95-3; 109100-20-1 (Oxalate salt)

RTECS: —

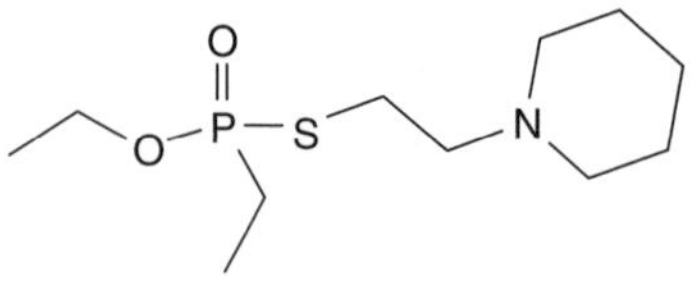

$C_{11}H_{24}NO_2PS$

Oily liquid. Various salts (solids) have been reported.

Exposure Hazards

Conversion Factor: 1 ppm = 10.85 mg/m^3 at 77°F

Human toxicity values have not been established or have not been published. However, this agent is a powerful cholinesterase inhibitor.

Properties:

MW: 265.4	*VP*: —	*FlP*: —
D: —	*VD*: 9.2 (calculated)	*LEL*: —
MP: —	*Vlt*: —	*UEL*: —
BP: 230°F (0.006 mmHg)	H_2O: —	*RP*: —
Vsc: —	*Sol*: —	*IP*: —

Oxalate salt

MW: 355.4
MP: 264°F

C01-A024

***O*-Ethyl *S*-[2-(diethylamino)ethyl] isopropylphosphonothiolate**

CAS: 99991-06-7; 101884-85-9 (*p*-Toluenesulfonate salt)
RTECS: —

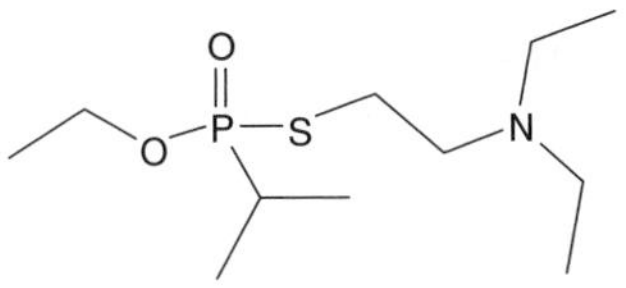

$C_{11}H_{26}NO_2PS$

Oily liquid. Various salts (solids) have been reported.

Exposure Hazards

Conversion Factor: 1 ppm = 10.94 mg/m^3 at 77°F

Human toxicity values have not been established or have not been published. However, this agent is a powerful cholinesterase inhibitor.

Properties:

MW: 267.4	*VP*: —	*FlP*: —
D: —	*VD*: 9.2 (calculated)	*LEL*: —
MP: —	*Vlt*: —	*UEL*: —
BP: 180°F (0.0007 mmHg)	H_2O: —	*RP*: —
Vsc: —	*Sol*: —	*IP*: —

***p*-Toluenesulfonate salt**

MW: 439.6
MP: 241°F

C01-A025

***O*-Ethyl *S*-[2-(diethylamino)ethyl] propylphosphonothiolate**

CAS: 99991-07-8; 101884-86-0 (*p*-Toluenesulfonate salt)
RTECS: —

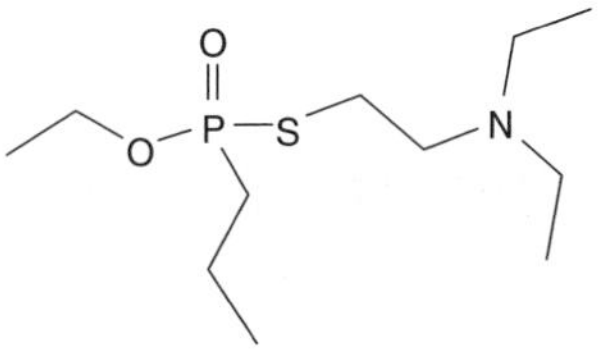

$C_{11}H_{26}NO_2PS$

Oily liquid. Various salts (solids) have been reported.

Exposure Hazards

Conversion Factor: 1 ppm = 10.94 mg/m^3 at 77°F

Human toxicity values have not been established or have not been published. However, this agent is a powerful cholinesterase inhibitor.

Properties:

MW: 267.4	*VP*: —	*FlP*: —
D: —	*VD*: 9.2 (calculated)	*LEL*: —
MP: —	*Vlt*: —	*UEL*: —
BP: 212°F (0.015 mmHg)	*H_2O*: —	*RP*: —
Vsc: —	*Sol*: —	*IP*: —

***p*-Toluenesulfonate salt**

MW: 439.6

MP: 232°F

C01-A026

***O*-Ethyl *S*-[2-(diethylamino)ethyl] butylphosphonothiolate**

CAS: 100454-47-5; 102180-38-1 (*p*-Toluenesulfonate salt)

RTECS: —

$C_{12}H_{28}NO_2PS$

Oily liquid. Various salts (solids) have been reported.

Exposure Hazards

Conversion Factor: 1 ppm = 11.51 mg/m^3 at 77°F

Human toxicity values have not been established or have not been published. However, this agent is a powerful cholinesterase inhibitor.

Properties:

MW: 281.4	*VP*: —	*FlP*: —
D: —	*VD*: 9.7 (calculated)	*LEL*: —
MP: —	*Vlt*: —	*UEL*: —
BP: 194°F (0.0025 mmHg)	*H_2O*: —	*RP*: —
Vsc: —	*Sol*: —	*IP*: —

***p*-Toluenesulfonate salt**

MW: 453.6

MP: 226°F

C01-A027

***O*-Ethyl *S*-[2-(diethylamino)ethyl] hexylphosphonothiolate**

CAS: 102444-87-1; 109644-82-8 (Oxalate Salt); 102444-88-2 (*p*-Toluenesulfonate salt)

RTECS: —

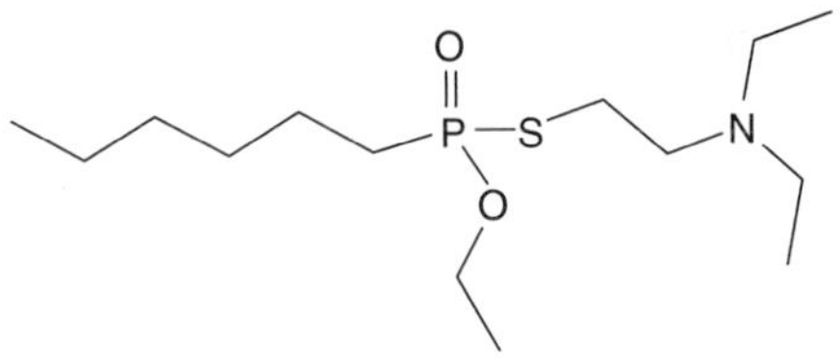

$C_{14}H_{32}NO_2PS$

Oily liquid. Various salts (solids) have been reported.

Exposure Hazards

Conversion Factor: 1 ppm = 12.66 mg/m^3 at 77°F

Human toxicity values have not been established or have not been published. However, this agent is a powerful cholinesterase inhibitor.

Properties:

MW: 309.5	*VP*: —	*FlP*: —
D: —	*VD*: 11 (calculated)	*LEL*: —
MP: —	*Vlt*: —	*UEL*: —
BP: 244°F (0.003 mmHg)	*H_2O*: —	*RP*: —
Vsc: —	*Sol*: —	*IP*: —

Oxalate salt	***p*-Toluenesulfonate salt**
MW: 399.5	*MW*: 481.7
MP: 185°F	*MP*: 149°F

GV-SERIES AGENTS

C01-A028

2-Dimethylaminoethyl *N, N*-dimethylphosphoramidofluoridate (Agent GP)

CAS: 141102-74-1

RTECS: —

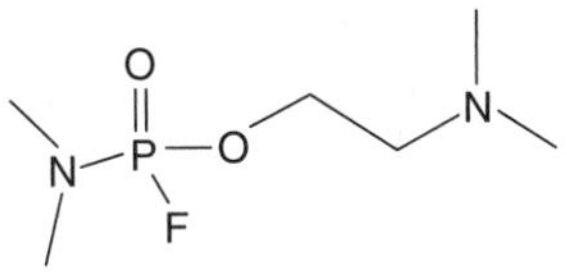

$C_6H_{16}FN_2O_2P$

Colorless liquid to white semisolid depending on purity. Salts are odorless white solids.

Exposure Hazards

Conversion Factor: 1 ppm = 8.15 mg/m^3 at 77°F

Human toxicity values have not been established or have not been published. However, based on available information, this agent appears to be slightly less toxic than VX (C01-A016).

Properties:

MW: 198.2	*VP*: 0.049 mmHg (77°F)	*FlP*: —
D: 1.15 g/mL (77°F)	*VD*: 6.8 (calculated)	*LEL*: —
MP: –166°F	*Vlt*: 64 ppm (77°F)	*UEL*: —
BP: 102°F (0.015 mmHg)	*H_2O*: "Very soluble"	*RP*: 150 (77°F)
Vsc: —	*Sol*: —	*IP*: —

Methyl Iodide Quaternary Amine salt
MW: 340.1
MP: 223°F (decomposes)

C01-A029

3-Quinuclidyl *N, N*-dimethylphosphoramidofluoridate (Agent EA 5488)
CAS: —
RTECS: —

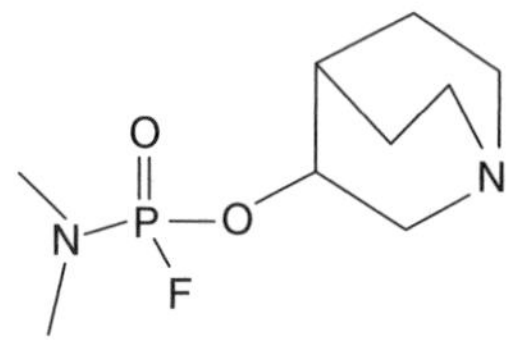

$C_9H_{18}FN_2O_2P$

Liquid. Other descriptive information has not been published.

Exposure Hazards
Conversion Factor: 1 ppm = 9.66 mg/m^3 at 77°F

Human toxicity values have not been established or have not been published. However, based on available information, this agent appears to be less than one half as toxic as VX (C01-A016).

Properties:

MW: 236.3	*VP*: 0.0044 mmHg (77°F)	*FlP*: —
D: —	*VD*: 8.2 (calculated)	*LEL*: —
MP: —	*Vlt*: 5.8 ppm	*UEL*: —
BP: 581°F	*H_2O*: —	*RP*: 1500
Vsc: —	*Sol*: —	*IP*: —

C01-A030

2-Dimethylaminoethyl *N, N*-diethylphosphoramidofluoridate (Agent GV1)
CAS: 141102-75-2
RTECS: —

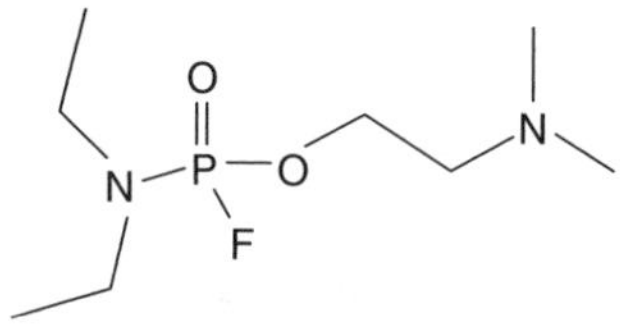

$C_8H_{20}FN_2O_2P$

Specific information on physical appearance is not available for this agent.

Exposure Hazards
Conversion Factor: 1 ppm = 9.25 mg/m^3 at 77°F

Human toxicity values have not been established or have not been published. However, based on available information, this agent appears to be less than one half as toxic as VX (C01-A016).

Properties:

MW: 226.2	*VP*: —	*FlP*: —
D: 1.072 g/mL	*VD*: 7.8 (calculated)	*LEL*: —
MP: –140°F	*Vlt*: —	*UEL*: —
BP: 134°F (0.049 mmHg)	*H_2O*: —	*RP*: —
Vsc: —	*Sol*: —	*IP*: —

C01-A031

2-Diethylaminoethyl *N, N*-dimethylphosphoramidofluoridate (Agent GV2)
CAS: 141102-77-4
RTECS: —

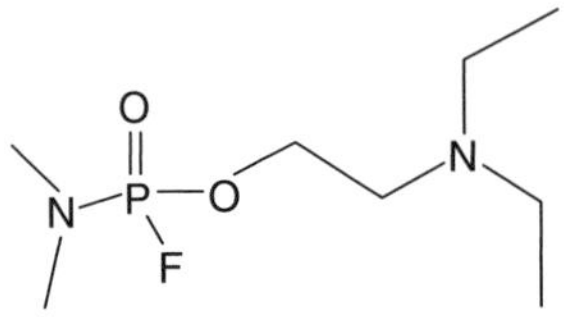

$C_8H_{20}FN_2O_2P$

Specific information on physical appearance is not available for this agent.

Exposure Hazards
Conversion Factor: 1 ppm = 9.25 mg/m^3 at 77°F

Human toxicity values have not been established or have not been published. However, based on available information, this agent appears to be approximately one-tenth as toxic as VX (C01-A016).

Properties:

MW: 226.2	*VP*: —	*FlP*: —
D: 1.046 g/mL	*VD*: 7.8 (calculated)	*LEL*: —
MP: –119°F	*Vlt*: —	*UEL*: —
BP: 127°F (0.002 mmHg)	*H_2O*: —	*RP*: —
Vsc: —	*Sol*: —	*IP*: —

C01-A032

2-Diethylaminoethyl *N, N*-diethylphosphoramidofluoridate (Agent GV3)
CAS: 141102-78-5
RTECS: —

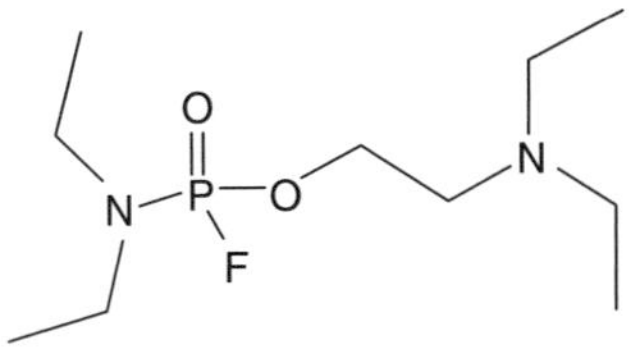

$C_{10}H_{24}FN_2O_2P$

Specific information on physical appearance is not available for this agent.

Exposure Hazards
Conversion Factor: 1 ppm = 10.40 mg/m^3 at 77°F

Human toxicity values have not been established or have not been published. However, based on available information, this agent appears to be approximately one-twentieth as toxic as VX (C01-A016).

Properties:

MW: 254.3	*VP*: —	*FlP*: —
D: 1.018 g/mL	*VD*: 8.8 (calculated)	*LEL*: —
MP: –132°F	*Vlt*: —	*UEL*: —
BP: 133°F (0.0008 mmHg)	*H_2O*: —	*RP*: —
Vsc: —	*Sol*: —	*IP*: —

C01-A033

3-Dimethylamionpropyl *N, N*-dimethylphosphoramidofluoridate (Agent EA 5414)
CAS: 158847-17-7
RTECS: —

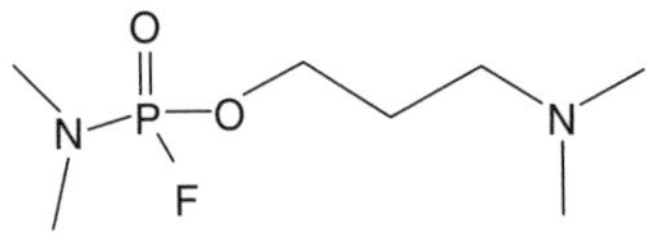

$C_7H_{18}FN_2O_2P$

Liquid. Other descriptive information has not been published.

Exposure Hazards

Conversion Factor: 1 ppm = 6.68 mg/m^3 at 77°F

Human toxicity values have not been established or have not been published. However, based on available information, this agent appears to be approximately one-fifth as toxic as VX (C01-A016).

Properties:

MW: 212.2	*VP*: 0.014 mmHg (77°F)	*FlP*: —
D: 1.037 g/mL	*VD*: 7.3 (calculated)	*LEL*: —
MP: –116°F	*Vlt*: 24 ppm (77°F)	*UEL*: —
BP: 435°F	*H_2O*: —	*RP*: 100
Vsc: —	*Sol*: —	*IP*: —

C01-A034

2-Dimethylaminopropyl *N, N*-diethylphosphoramidofluoridate (Agent GV5)
CAS: 158847-18-8
RTECS: —

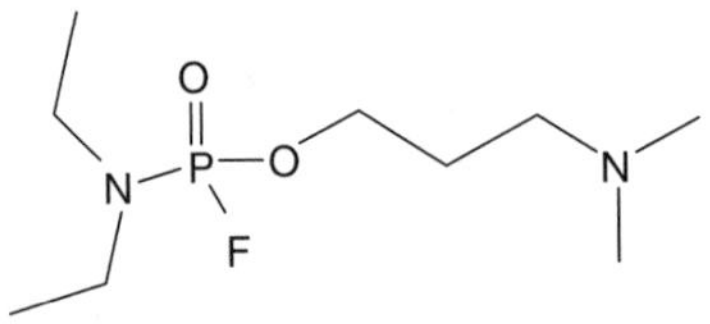

$C_9H_{22}FN_2O_2P$

Specific information on physical appearance is not available for this agent.

Exposure Hazards

Conversion Factor: 1 ppm = 9.83 mg/m^3 at 77°F

Human toxicity values have not been established or have not been published. However, based on available information, this agent appears to be less than one-twentieth as toxic as VX (C01-A016).

Properties:

MW: 240.3	*VP*: —	*FlP*: —
D: 1.019 g/mL	*VD*: 8.3 (calculated)	*LEL*: —
MP: –122°F	*Vlt*: —	*UEL*: —
BP: 136°F (0.034 mmHg)	H_2O: —	*RP*: —
Vsc: —	*Sol*: —	*IP*: —

NOVICHOK AGENTS

C01-A035

[(Fluoromethoxyphosphinyl)oxy]carbonimidic dichloride

CAS: 17642-31-8

RTECS: —

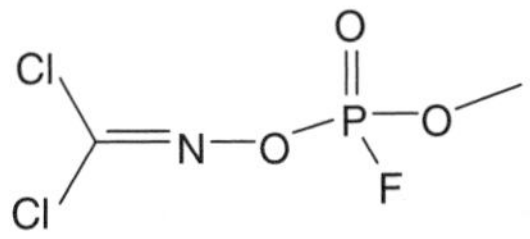

$C_2H_3Cl_2FNO_3P$

Specific information on physical appearance is not available for this agent.

Exposure Hazards

Conversion Factor: 1 ppm = 8.59 mg/m^3 at 77°F

Human toxicity values have not been established or have not been published.

Properties:

MW: 209.9	*VP*: —	*FlP*: —
D: 1.488 g/mL	*VD*: 7.2 (calculated)	*LEL*: —
MP: —	*Vlt*: 600 ppm (calculated)	*UEL*: —
BP: 135°F (2 mmHg)	H_2O: —	*RP*: —
Vsc: —	*Sol*: —	*IP*: —

C01-A036

[(Fluoromethoxyphosphinyl)oxy]carbonimidic chloride fluoride

CAS: 17642-26-1

RTECS: —

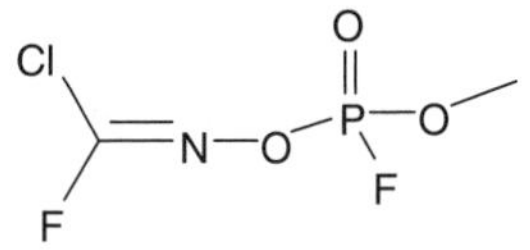

$C_2H_3ClF_2NO_3P$

Specific information on physical appearance is not available for this agent.

Exposure Hazards

Conversion Factor: 1 ppm = 7.91 mg/m^3 at 77°F

Human toxicity values have not been established or have not been published.

Properties:

MW: 193.5	*VP*: —	*FlP*: —
D: 1.51 g/mL	*VD*: 6.7 (calculated)	*LEL*: —
MP: —	*Vlt*: 2000 ppm (calculated)	*UEL*: —
BP: 158°F (2 mmHg)	H_2O: —	*RP*: —
Vsc: —	*Sol*: —	*IP*: —

C01-A037

[(Fluoromethoxyphosphinyl)oxy]carbonimidic difluoride

CAS: 18016-10-9

RTECS: —

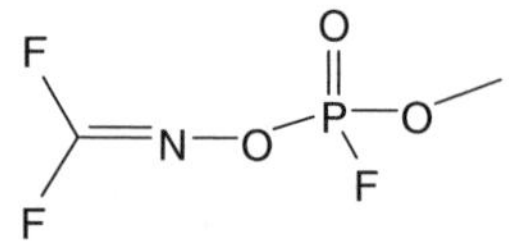

$C_2H_3F_3NO_3P$

Specific information on physical appearance is not available for this agent.

Exposure Hazards

Conversion Factor: 1 ppm = 7.24 mg/m^3 at 77°F

Human toxicity values have not been established or have not been published.

Properties:

MW: 177.0	*VP*: —	*FlP*: —
D: 1.424 g/mL	*VD*: 6.1 (calculated)	*LEL*: —
MP: —	*Vlt*: 10,000 ppm (calculated)	*UEL*: —
BP: 136°F (20 mmHg)	H_2O: —	*RP*: —
Vsc: —	*Sol*: —	*IP*: —

C01-A038

2,2-Difluoro-*N*-[(fluoromethoxyphosphinyl)oxy]-2-nitroethanimidoyl fluoride

CAS: 17642-29-4

RTECS: —

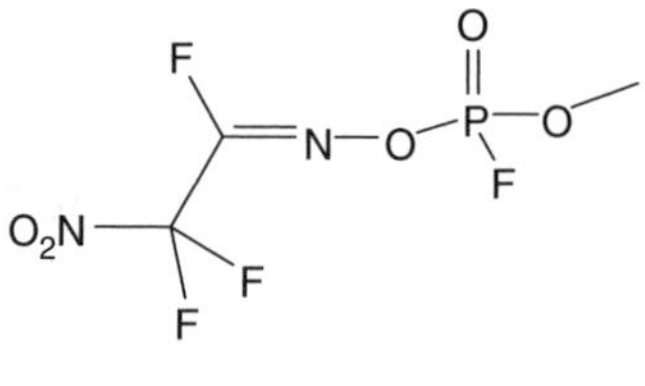

$C_3H_3F_4N_2O_5P$

Specific information on physical appearance is not available for this agent.

Exposure Hazards

Conversion Factor: 1 ppm = 10.39 mg/m^3 at 77°F

Human toxicity values have not been established or have not been published.

Properties:

MW: 254.0	*VP*: —	*FlP*: —
D: 1.639 g/mL	*VD*: 8.8 (calculated)	*LEL*: —
MP: —	*Vlt*: 900 ppm (calculated)	*UEL*: —

BP: 169°F (3 mmHg)	*H_2O*: —	*RP*: —
Vsc: —	*Sol*: —	*IP*: —

C01-A039

[(Ethoxyfluorophosphinyl)oxy]carbonimidic dichloride

CAS: 17642-32-9
RTECS: —

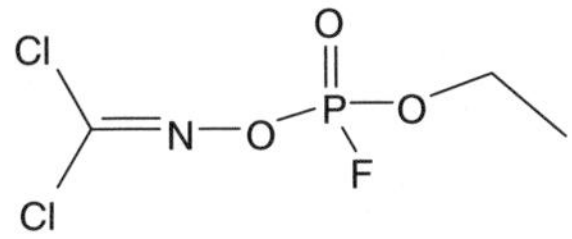

$C_3H_5Cl_2FNO_3P$

Specific information on physical appearance is not available for this agent.

Exposure Hazards

Conversion Factor: 1 ppm = 9.16 mg/m^3 at 77°F

Human toxicity values have not been established or have not been published.

Properties:

MW: 224.0	*VP*: —	*FlP*: —
D: 1.366 g/mL	*VD*: 7.7 (calculated)	*LEL*: —
MP: —	*Vlt*: 200 ppm (calculated)	*UEL*: —
BP: 189°F (4 mmHg)	*H_2O*: —	*RP*: —
Vsc: —	*Sol*: —	*IP*: —

C01-A040

[(Ethoxyfluorophosphinyl)oxy]carbonimidic chloride fluoride

CAS: 17642-27-2
RTECS: —

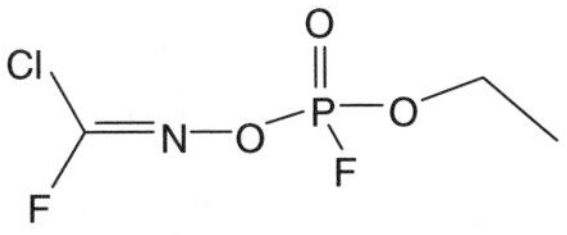

$C_3H_5ClF_2NO_3P$

Specific information on physical appearance is not available for this agent.

Exposure Hazards

Conversion Factor: 1 ppm = 8.49 mg/m^3 at 77°F

Human toxicity values have not been established or have not been published.

Properties:

MW: 207.5	*VP*: —	*FlP*: —
D: 1.45 g/mL	*VD*: 7.2 (calculated)	*LEL*: —
MP: —	*Vlt*: 1000 ppm (calculated)	*UEL*: —
BP: 167°F (4 mmHg)	*H_2O*: —	*RP*: —
Vsc: —	*Sol*: —	*IP*: —

C01-A041

[(Ethoxyfluorophosphinyl)oxy]carbonimidic difluoride

CAS: 17642-28-3
RTECS: —

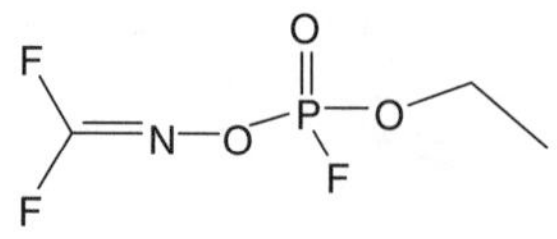

$C_3H_5F_3NO_3P$

Specific information on physical appearance is not available for this agent.

Exposure Hazards

Conversion Factor: 1 ppm = 7.81 mg/m^3 at 77°F

Human toxicity values have not been established or have not been published.

Properties:

MW: 191.0	*VP*: —	*FlP*: —
D: —	*VD*: 6.6 (calculated)	*LEL*: —
MP: —	*Vlt*: 4000 ppm (calculated)	*UEL*: —
BP: 129°F (5 mmHg)	*H_2O*: —	*RP*: —
Vsc: —	*Sol*: —	*IP*: —

C01-A042

***N*-[(Ethoxyfluorophosphinyl)oxy]-2,2-difluoro-2-nitroethanimidoyl fluoride**

CAS: 17642-30-7

RTECS: —

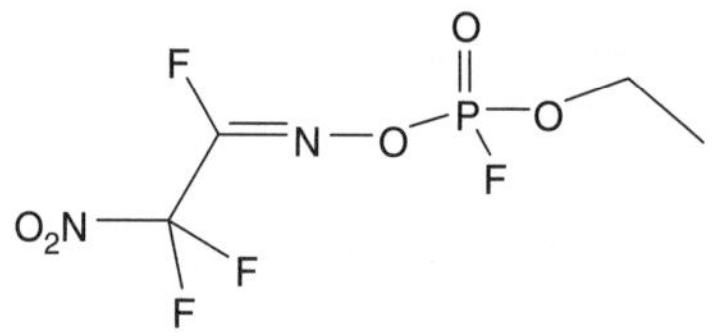

$C_4H_5F_4N_2O_5P$

Specific information on physical appearance is not available for this agent.

Exposure Hazards

Conversion Factor: 1 ppm = 10.96 mg/m^3 at 77°F

Human toxicity values have not been established or have not been published.

Properties:

MW: 268.1	*VP*: —	*FlP*: —
D: 1.506 g/mL	*VD*: 9.2 (calculated)	*LEL*: —
MP: —	*Vlt*: 200 ppm (calculated)	*UEL*: —
BP: 178°F (5 mmHg)	*H_2O*: —	*RP*: —
Vsc: —	*Sol*: —	*IP*: —

C01-A043

[[(2-Chloroethoxy)fluorohydroxyphosphinyl]oxy]carbonimidic chloride fluoride

CAS: 26102-97-6

RTECS: —

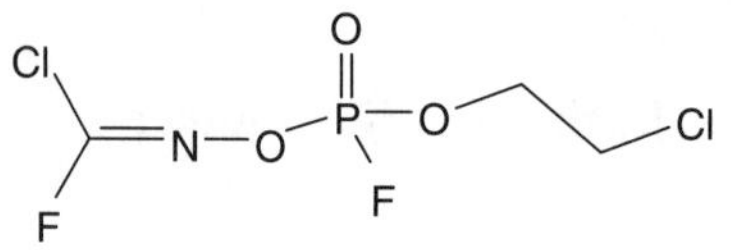

$C_3H_4Cl_2F_2NO_3P$

Specific information on physical appearance is not available for this agent.

Exposure Hazards

Conversion Factor: 1 ppm = 9.90 mg/m^3 at 77°F

Human toxicity values have not been established or have not been published.

Properties:

MW: 241.9	*VP*: —	*FlP*: —
D: —	*VD*: 8.3 (calculated)	*LEL*: —
MP: —	*Vlt*: 60 ppm (calculated)	*UEL*: —
BP: —	*H_2O*: —	*RP*: —
Vsc: —	*Sol*: —	*IP*: —

C01-A044

[[(2-Chloro-1-methylethoxy)fluorophosphinyl]oxy]carbonimidic chloride fluoride

CAS: 26102-98-7

RTECS: —

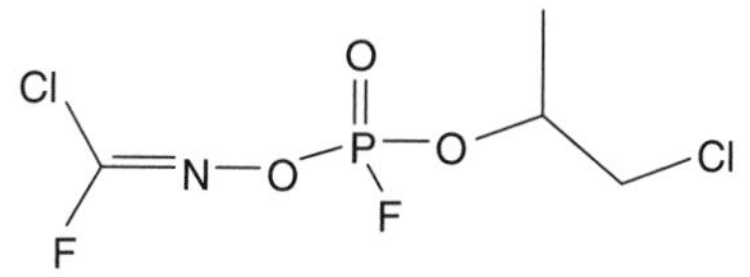

$C_4H_6Cl_2F_2NO_3P$

Specific information on physical appearance is not available for this agent.

Exposure Hazards

Conversion Factor: 1 ppm = 10.47 mg/m^3 at 77°F

Human toxicity values have not been established or have not been published.

Properties:

MW: 256.0	*VP*: —	*FlP*: —
D: —	*VD*: 8.8 (calculated)	*LEL*: —
MP: —	*Vlt*: 40 ppm (calculated)	*UEL*: —
BP: —	*H_2O*: —	*RP*: —
Vsc: —	*Sol*: —	*IP*: —

C01-A045

[[(2-Chloro-1-methylpropoxy)fluorophosphinyl]oxy]carbonimidic chloride fluoride

CAS: 26102-99-8

RTECS: —

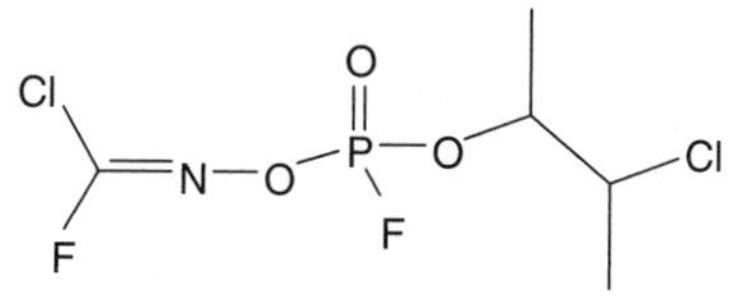

$C_5H_8Cl_2F_2NO_3P$

Specific information on physical appearance is not available for this agent.

Exposure Hazards

Conversion Factor: 1 ppm = 11.04 mg/m^3 at 77°F

Human toxicity values have not been established or have not been published.

Properties:

MW: 270.0	*VP*: —	*FlP*: —
D: —	*VD*: 9.3 (calculated)	*LEL*: —
MP: —	*Vlt*: 20 ppm (calculated)	*UEL*: —
BP: —	H_2O: —	*RP*: —
Vsc: —	*Sol*: —	*IP*: —

C01-C
COMPONENTS AND PRECURSORS

C01-A046

Methylphosphonic dichloride (DC)

CAS: 676-97-1
RTECS: —
UN: 9206
ERG: 137

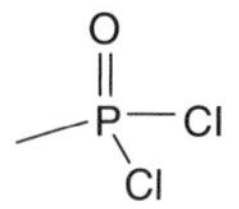

CH_3Cl_2OP

Clear solid or liquid with a pungent, stinging, disagreeable odor. This material is hazardous through inhalation and ingestion, and produces local skin/eye impacts.

May cause severe and painful irritation of the eyes, nose, throat, and lungs. Severe exposure can cause accumulation of fluid in the lungs (pulmonary edema). Inhalation toxicity similar to hydrogen chloride and hydrogen fluoride. May cause second or third degree burns upon short contact with skin surfaces. Oral ingestion may result in tissue destruction of the gastrointestinal tract. Decreased blood cholinesterase levels have been reported in animals.

This material is on the Australia Group Export Control list and Schedule 2 of the CWC.

This material is a general precursor for nerve agents. It may appear as a mixture with Methylphosphonic difluoride (C01-C047), known as Didi, that is used as a constituent for binary G-series nerve agents.

Exposure Hazards

Conversion Factor: 1 ppm = 5.44 mg/m^3 at 77°F
WPL AEL: 0.006 ppm

Properties:

MW: 132.9	*VP*: 10 mmHg (122°F)	*FlP*: 300°F
D: 1.64 g/mL (79°F)	*VD*: 4.6 (calculated)	*LEL*: —
MP: 88°F	*Vlt*: 12,000 ppm (122°F)	*UEL*: —
BP: 331°F	H_2O: Decomposes	*RP*: 0.95 (122°F)
BP: 127°F (15 mmHg)	*Sol*: —	*IP*: —
Vsc: —		

C01-A047

Methylphosphonic difluoride (DF)
CAS: 676-99-3
RTECS:T01840700

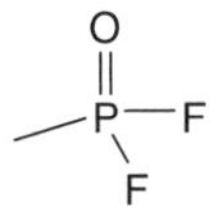

CH_3F_2OP

Liquid with a pungent, acid-like odor. This material is hazardous through inhalation and ingestion, and produces local skin/eye impacts.

May cause severe and painful irritation of the eyes, nose, throat, and lungs. Severe exposure can cause accumulation of fluid in the lungs (pulmonary edema). Inhalation toxicity similar to hydrogen chloride and hydrogen fluoride. May cause second or third degree burns upon short contact with skin surfaces. Oral ingestion may result in tissue destruction of the gastrointestinal tract. High overexposure may inhibit cholinesterase.

This material is on the Australia Group Export Control list and Schedule 1 of the CWC.

This material is a constituent in GB2, which is the binary version of Sarin (C01-A002). It is also commonly found as a decomposition product/impurity in unitary sarin. It may appear as a mixture with Methylphosphonic dichloride (C01-C046), known as Didi, that is also used as a constituent for binary G-series nerve agents.

Exposure Hazards
Conversion Factor: 1 ppm = 4.09 mg/m^3 at 77°F

Human toxicity values have not been established or have not been published.

Properties:

MW: 100.1	*VP*: 36 mmHg (77°F)	*FlP*: None
D: 1.359 g/mL (77°F)	*VD*: 3.5 (calculated)	*LEL*: None
MP: –35°F	*Vlt*: 36,000 ppm	*UEL*: None
BP: 212°F	H_2O: Decomposes	*RP*: 0.28
Vsc: —	*Sol*: —	*IP*: —

C01-A048

Diisopropyl methylphosphonate (DIMP)
CAS: 1445-75-6
RTECS: SZ9090000

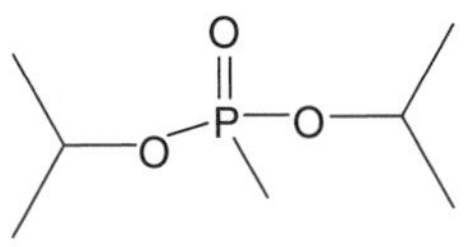

$C_7H_{17}O_3P$

Colorless liquid. This material is hazardous through inhalation, skin absorption, penetration through broken skin, and ingestion.

This material is on Schedule 2 of the CWC.

This material is a precursor for Sarin (C01-A002) and is also commonly found as a decomposition product/impurity (up to 20%) in sarin. If fluoride ion is present and the pH falls below 7, sarin will be formed. This material has been used as a simulant for nerve agents in government tests.

Exposure Hazards

Conversion Factor: 1 ppm = 7.37 mg/m^3 at 77°F

Human toxicity values have not been established or have not been published.

Properties:

MW: 180.2
D: 0.976 g/mL (77°F)
MP: —
BP: 345°F
BP: 250°F (10 mmHg)
Vsc: —
VP: 0.17 mmHg (77°F)
VD: 6.2 (calculated)
Vlt: 220 ppm (77°F)
H_2O: 1.6% (77°F)
Sol: —
FlP: —
LEL: —
UEL: —
RP: 44
IP: —

C01-A049

Dicyclohexylcarbodiimide (DCCDI)

CAS: 538-75-0

RTECS: FF2160000

$C_{13}H_{22}N_2$

Colorless to white crystalline solid. This material is hazardous through inhalation, skin absorption, penetration through broken skin, and ingestion, and produces local skin/eye impacts.

Causes severe eye irritation, skin irritation, nausea, headache, and vomiting. Inhalation is irritating to the mucous membranes and upper respiratory tract. May cause sensitization by skin contact.

This material is on the FBI threat list.

This material is a stabilizer for G-series and V-series nerve agents, usually within the range of 2–10% by weight.

Exposure Hazards

Conversion Factor: 1 ppm = 8.44 mg/m^3 at 77°F

Human toxicity values have not been established or have not been published.

Properties:

MW: 206.3
D: —
MP: 93°F
BP: 316°F (12 mmHg)
Vsc: —
VP: —
VD: 7.1 (calculated)
Vlt: —
H_2O: Decomposes
Sol: Methylene chloride
FlP: 235°F
LEL: —
UEL: —
RP: —
IP: —

C01-A050

Diisopropylcarbodiimide (DIPC)

CAS: 693-13-0

RTECS: FF2175000

$C_7H_{14}N_2$

Clear, colorless to yellow to faint brown liquid with a foul odor. This material is hazardous through inhalation, skin absorption, penetration through broken skin, and ingestion, and produces local skin/eye impacts.

Causes severe eye irritation that can progress to severe corneal edema. Temporary blindness has been reported. Causes skin irritation, nausea, headache, and vomiting. Inhalation is irritating to the mucous membrane and upper respiratory tract. May cause sensitization by skin contact.

Used industrially as an activating reagent in solid-phase synthesis.

This material is on the FBI threat list.

This material is a stabilizer for G-series and V-series nerve agents, usually within the range of 2–10% by weight.

Exposure Hazards

Conversion Factor: 1 ppm = 5.16 mg/m^3 at 77°F

Human toxicity values have not been established or have not been published.

Properties:

MW: 126.2	*VP*: —	*FlP*: 91°F
D: 0.815 g/mL	*VD*: 4.4 (calculated)	*LEL*: —
MP: —	*Vlt*: —	*UEL*: —
BP: 296°F	H_2O: Decomposes	*RP*: —
Vsc: —	*Sol*: —	*IP*: eV

C01-A051

Methanephosphonic acid (MPA)

CAS: 993-13-5

RTECS: —

CH_5O_3P

White solid. This material is hazardous through inhalation and ingestion, and produces local skin/eye impacts.

Used industrially for organic synthesis and in the manufacture of lubricant additives and for treating textiles.

This material is on Schedule 2 of the CWC.

This material is a general precursor for nerve agents and is also commonly found as a decomposition product/impurity resulting from hydrolysis of G-series nerve agents. This material has been used as a simulant for nerve agents in government tests.

Exposure Hazards

Conversion Factor: 1 ppm = 3.93 mg/m^3 at 77°F

Human toxicity values have not been established or have not been published.

Properties:

MW: 96.0	*VP*: 0.000002 mmHg	*FlP*: —
D: —	*VD*: —	*LEL*: —
MP: 221°F	*Vlt*: —	*UEL*: —
BP: Decomposes	H_2O: >100%	*RP*: —
Vsc: —	*Sol*: Ethanol; Ether	*IP*: —

C01-A052

Dimethyl methylphosphonate (DMMP)
CAS: 756-79-6
RTECS: SZ9120000

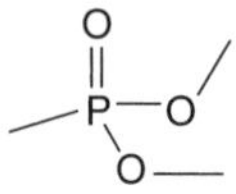

$C_3H_9O_3PS$

Description of the agent including appearance including odor and colorless liquid. This material is hazardous through inhalation, skin absorption, penetration through broken skin, and ingestion.

Used industrially as a fire retardant, gasoline additive, antifoam agent, plasticizer, plastic stabilizer, textile conditioner, antistatic agent, and hydraulic fluid additive.

This material is on the Australia Group Export Control list and Schedule 2 of the CWC.

This material is a precursor for G-series nerve agents and has been used as a simulant for nerve agents in government tests.

Exposure Hazards
Conversion Factor: 1 ppm = 5.08 mg/m^3 at 77°F
Human toxicity values have not been established or have not been published.

Properties:

MW: 124.1	*VP*: 0.61 mmHg (77°F)	*FlP*: —
D: 1.1589 g/mL	*VD*: 4.3 (calculated)	*LEL*: —
MP: —	*Vlt*: 800 ppm (77°F)	*UEL*: —
BP: 360°F	H_2O: Soluble	*RP*: —
BP: 145°F (10 mmHg)	*Sol*: Alcohol; Ether	*IP*: —
Vsc: —		

C01-A053

Sulfur (NE)
CAS: 7704-34-9
RTECS: WS4250000
UN: 1350
ERG: 133

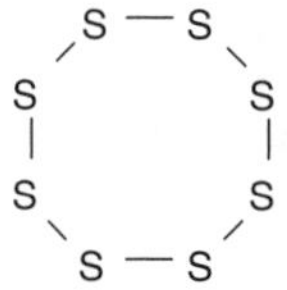

S_8

Fine, pale yellow, amorphous, or microcrystalline powder. May come as sublimed, washed, or precipitated. Pure sulfur exists in two stable crystalline forms and at least two amorphous (liquid) forms. The rhombic form is the principle material used in binary V-series agents. Pure sulfur is odorless but traces of hydrocarbon impurities may impart an oily and/or rotten egg odor to commercial material. This material is hazardous through inhalation and produces local skin/eye impacts.

Used industrially in manufacturing drugs and pharmaceuticals, medicated cosmetics and shampoos, sulfuric acid, carbon disulfide, sulfur dioxide, phosphorus pentasulfide, sulfites, insecticides, plastics, enamels, metal–glass cements, dyes, detergents, gunpowder, pyrotechnics, explosives, matches, and phosphatic fertilizers. Used in petroleum refining, vulcanizing rubber, photographic film, cement sealant, binder and asphalt extender in road paving, for bleaching dried fruits, wood pulp, straw, wool, silk, felt, and linen. Used as a medication, fungicide, acaricide, and as an electrical insulator.

This material is a constituent in VX2, which is the binary version of VX (C01-A016).

Exposure Hazards
Conversion Factor: 1 ppm = 10.49 mg/m^3 at 77°F
Human toxicity values have not been established or have not been published.

Rhombic Sulfur Properties:

MW: 256.5	*VP*: 0.00000395 mmHg (87°F)	*FlP*: 405°F
D: 2.07 g/cm^3	*VD*: —	*LEL*: 35 g/m^3
MP: 235°F	*Vlt*: 0.01 ppm (87° F)	*UEL*: 1,400 g/m^3
BP: 832°F	H_2O: Insoluble	*RP*: 680,000
Vsc: —	*Sol*: Carbon disulfide; Aromatic hydrocarbons	*IP*: —

C01-A054

Dimethylpolysulfide (NM)
CAS: 73062-48-3
RTECS: —

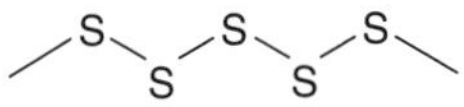

$C_2H_6S_5$

Liquid with a very noxious odor. This material poses a "considerable risk" from inhalation of high concentrations.

This material is a constituent in VX2, which is the binary version of VX (C01-A016).

Exposure Hazards
Conversion Factor: 1 ppm = 7.79 mg/m^3 at 77°F (based on average molecular weight)
Human toxicity values have not been established or have not been published.

Properties:

MW: 190.4 (average)	*VP*: —	*FlP*: 221°F
D: 1.3895 g/mL (77°F)	*VD*: 6.6 (calculated)	*LEL*: —
MP: –40°F	*Vlt*: —	*UEL*: —
BP: 243°F	H_2O: Insoluble	*RP*: "Very persistent"
Vsc: —	*Sol*: —	*IP*: —

C01-A055

Isopropyl alcohol and isopropyl amine mixture (OPA)
CAS: —
RTECS: —

Mixture

Clear liquid with an odor that is a mixture of alcohol and ammonia. This material is hazardous through inhalation and ingestion, and produces local skin/eye impacts.

Inhalation of the agent may cause irritation of the lower respiratory tract, coughing, difficulty in breathing and, in high concentration, loss of consciousness. It causes severe irritation in contact with the skin and eyes. If ingested it causes nausea, salivation, and severe irritation of the mouth and stomach.

A similar mixture is used industrially in processing cutting oils.

This material is a constituent in GB2, which is the binary version of Sarin (C01-A002).

Exposure Hazards

Human toxicity values have not been established or have not been published.

Properties:

MW: Mixture	*VP*: 197 mmHg (77°F)	*FlP*: 15°F
D: 0.7443 g/mL (77°F)	*VD*: 2.1 (calculated)	*LEL*: —
MP: <–126°F	*Vlt*: —	*UEL*: —
BP: 140°F	*H_2O*: Soluble	*RP*: —
Vsc: —	*Sol*: —	*IP*: —

C01-A056

***O*-Ethyl 2-diisopropylaminoethyl methylphosphonite** (QL)

CAS: 57856-11-8

RTECS: —

$C_{11}H_{26}NO_2P$

Viscous liquid with a strong fishy odor. This material is hazardous through inhalation and ingestion, and produces local skin/eye impacts. Hydrolysis product TR is formed when there is only a limited amount of water present. It is highly reactive and toxic.

This material is on the Australia Group Export Control list and Schedule 1 of the CWC.

This material is a constituent in VX2, which is the binary version of VX (C01-A016).

Exposure Hazards

Conversion Factor: 1 ppm = 9.62 mg/m^3 at 77°F

Human toxicity values have not been established or have not been published.

Properties:

MW: 235.3	*VP*: 0.01 mmHg (77°F)	*FlP*: 192°F
D: 0.908 g/mL (77°F)	*VD*: 8.1 (calculated)	*LEL*: —
MP: —	*Vlt*: 13 ppm (77°F)	*UEL*: —
BP: 450°F	*H_2O*: Decomposes	*RP*: 660
Vsc: 2.237 cS (77°F)	*Sol*: —	*IP*: —

Hydrolysis product TR:

MW: —	*VP*: 11 mmHg	*FlP*: 82°F
MP: —	*VD*: —	Autoignition temperature: 104°F
BP: 248°F	*Vlt*: —	

C01-A057

Methylphosphinyl dichloride (SW)
CAS: 676-83-5
RTECS: —

CH_3Cl_2P

Clear colorless liquid. This material is hazardous through inhalation, skin absorption, penetration through broken skin, and ingestion, and produces local skin/eye impacts. Spontaneously flammable in air at or slightly above normal temperature.

Used industrially for organic synthesis.

This material is on the Australia Group Export Control list and Schedule 2 of the CWC.

This material is used to synthesize precursors for all series of nerve agents.

Exposure Hazards
Conversion Factor: 1 ppm = 4.78 mg/m^3 at 77°F
Human toxicity values have not been established or have not been published.

Properties:

MW: 116.9	*VP*: —	*FlP*: 118°F
D: 1.2941 g/mL	*VD*: 4.0 (calculated)	*LEL*: —
MP: —	*Vlt*: —	*UEL*: —
BP: 176°F	H_2O: Decomposes	*RP*: —
Vsc: —	*Sol*: —	*IP*: 9.85 eV

C01-A058

Methylphosphonothioic dichloride (SWS)
CAS: 676-98-2
RTECS: TB2100000

CH_3Cl_2PS

Specific information is not available for this agent.

This material is on Schedule 2 of the CWC.

This material is used to synthesize precursors to V-series nerve agents.

Exposure Hazards
Conversion Factor: 1 ppm = 6.09 mg/m^3 at 77°F
Human toxicity values have not been established or have not been published.

Properties:

MW: 149.0	*VP*: —	*FlP*: —
D: 1.422 g/mL	*VD*: 5.1 (calculated)	*LEL*: —
MP: —	*Vlt*: —	*UEL*: —
BP: 111°F (9 mmHg)	H_2O: Decomposes	*RP*: —
Vsc: —	*Sol*: —	*IP*: —

C01-A059

Tributylamine (TBA)
CAS: 102-82-9
RTECS: YA0350000

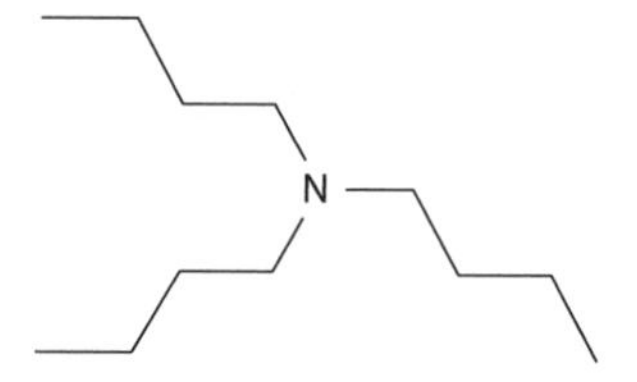

$C_{12}H_{27}N$

Pale yellow hygroscopic liquid with an ammonia odor. This material is hazardous through inhalation, skin absorption, penetration through broken skin, and ingestion, and produces local skin/eye impacts.

Causes irritation to skin, eyes, and respiratory system, CNS stimulation, skin irritation, sensitization. Causes severe eye and skin burns. May cause severe tearing, conjunctivitis, and corneal edema. Inhalation may cause difficulties ranging from coughing and nausea to accumulation of fluid in the lungs (pulmonary edema).

Used industrially as a solvent, inhibitor in hydraulic fluids, polymerization catalyst, insecticide, emulsifying agent, and as a chemical intermediate.

This material is a stabilizer for nerve agents.

Exposure Hazards
Conversion Factor: 1 ppm = 7.58 mg/m^3 at 77°F
Human toxicity values have not been established or have not been published.

Properties:

MW: 185.4	*VP*: 0.0934 mmHg (77°F)	*FlP*: 187°F
D: 0.7782 g/mL	*VD*: 6.4 (calculated)	*LEL*: —
MP: –94°F	*Vlt*: 120 ppm (77° F)	*UEL*: —
BP: 422°F	*H_2O*: 0.014% (77° F)	*RP*: 79
BP: 244°F (25 mmHg)	*H_2O*: 0.004% (64° F)	*IP*: —
Vsc: 1.7 cS (77°F)	*Sol*: Most organic solvents	
Vsc: 0.91 cS (60°F)		

C01-A060

Isopropylphosphonic difluoride
CAS: 677-42-9
RTECS: —

$C_3H_7F_2OP$

Specific information on physical appearance is not available for this material. This material is hazardous through inhalation and ingestion, and produces local skin/eye impacts.

May cause severe and painful irritation of the eyes, nose, throat, and lungs. Severe exposure can cause accumulation of fluid in the lungs (pulmonary edema). Inhalation toxicity similar to HCl and HF. May cause second or third degree burns upon short contact with skin surfaces. Oral ingestion may result in tissue destruction of the gastrointestinal tract. High overexposure may inhibit cholinesterase.

This material is on Schedule 1 of the CWC.

This material is a binary constituent in G-series nerve agents.

Exposure Hazards

Conversion Factor: 1 ppm = 5.24 mg/m^3 at 77°F

Human toxicity values have not been established or have not been published.

Properties:

MW: 128.1	*VP*: —	*FlP*: —
D: 1.1748 g/mL	*VD*: 4.4 (calculated)	*LEL*: —
MP: —	*Vlt*: —	*UEL*: —
BP: 234°F	H_2O: —	*RP*: —
Vsc: —	*Sol*: —	*IP*: —

C01-A061

Ethyl methylphosphonothioic acid

CAS: 18005-40-8; 22307-81-9 (Sodium salt); 73790-51-9 (Dicyclohexylamine salt)

RTECS: —

$C_3H_9O_2PS$

Specific information on physical appearance is not available for this material. Various salts (solids) have been reported.

Used industrially for organic synthesis and the manufacturing of pesticides.

This material is on Schedule 2 of the CWC.

This material is a binary constituent in V-series nerve agents. It is also commonly found as a decomposition product/impurity from hydrolysis of V-series agents.

Exposure Hazards

Conversion Factor: 1 ppm = 5.73 mg/m^3 at 77°F

Human toxicity values have not been established or have not been published.

Properties:

MW: 140.1	*VP*: 0.043 mmHg	*FlP*: —
D: —	*VD*: 4.8 (calculated)	*LEL*: —
MP: —	*Vlt*: 58 ppm	*UEL*: —
BP: —	H_2O: 0.11%	*RP*: 200
Vsc: —	*Sol*: —	*IP*: —

C01-A062

Diethylphosphoramidic difluoride

CAS: 359-94-4

RTECS: —

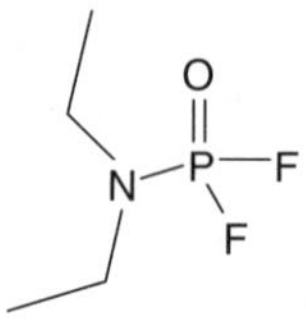

$C_4H_{10}F_2NOP$

Specific information on physical appearance is not available for agent material. This material is hazardous through inhalation and ingestion, and produces local skin/eye impacts.

May cause severe and painful irritation of the eyes, nose, throat, and lungs. Severe exposure can cause accumulation of fluid in the lungs (pulmonary edema). Inhalation toxicity similar to HCl and HF. May cause second or third degree burns upon short contact with skin surfaces. Oral ingestion may result in tissue destruction of the gastrointestinal tract. High overexposure may inhibit cholinesterase.

This material is a binary constituent in G-series and GV-series nerve agents.

Exposure Hazards

Conversion Factor: 1 ppm = 6.43 mg/m^3 at 77°F

Human toxicity values have not been established or have not been published.

Properties:

MW: 157.1	*VP*: —	*FIP*: —
D: —	*VD*: 5.4 (calculated)	*LEL*: —
MP: —	*Vlt*: —	*UEL*: —
BP: 113°F (15 mmHg)	H_2O: Decomposes	*RP*: —
Vsc: —	*Sol*: —	*IP*: —

C01-A063

Potassium fluoride

CAS: 7789-23-3
RTECS: TT0700000
UN: 1812
ERG: 154

KF

Colorless to white hygroscopic crystals that are odorless. This material is hazardous through inhalation, skin absorption, penetration through broken skin, and ingestion, and produces local skin/eye impacts.

Used industrially as a cleaning agent, disinfecting agent, insecticide, to etch glass, and as a welding flux.

This material is on the Australia Group Export Control list.

This material is a precursor for G-series and GV-series nerve agents.

Exposure Hazards

Conversion Factor: 1 ppm = mg/m^3 at 77°F

Human toxicity values have not been established or have not been published.

Properties:

MW: 58.1	*VP*: —	*FIP*: None
D: 2.48 g/cm^3	*VD*: —	*LEL*: None

MP: 1576°F	*Vlt*: —	*UEL*: None
BP: 2741°F	H_2O: 92.3% (64°F)	*RP*: —
Vsc: —	*Sol*: —	*IP*: —

C01-A064

Sodium fluoride
CAS: 7681-49-4
RTECS: WB0350000
UN: 1690
ERG: 154

NaF

Colorless to white hygroscopic crystals that are odorless. Pesticide grade is often dyed blue. This material is hazardous through inhalation, skin absorption, penetration through broken skin, and ingestion, and produces local skin/eye impacts.

Used industrially as a fungicide, rodenticide, fluoridation agent in drinking water, a toothpaste ingredient, a glass frosting agent, an agent in ore flotation, a stainless steel pickling agent, a component of bitreous enamels, in removal of HF from exhaust gases, in the manufacture of coated paper, for disinfecting fermentation apparatus in breweries and distilleries, in electroplating, as a component of glues and adhesives, a component of wood preservatives, as a "sour" in laundering cloth, and as a flux in the manufacture of rimmed steel, aluminum, and magnesium. Used medically as an anticoagulant for blood.

This material is on the Australia Group Export Control list.

This material is a precursor for G-series and GV-series nerve agents.

Exposure Hazards
OSHA PEL: 2.5 mg/m^3 as fluoride
IDLH: 250 mg/m^3 as fluoride

Properties:

MW: 42.0	*VP*: —	*FlP*: None
D: 2.78 g/cm^3	*VD*: —	*LEL*: None
MP: 1819°F	*Vlt*: —	*UEL*: None
BP: 3099°F	H_2O: 4%	*RP*: —
Vsc: —	*Sol*: —	*IP*: —

C01-A065

Potassium bifluoride
CAS: 7789-29-9
RTECS: TS6650000
UN: 1811
ERG: 154

KHF_2

White to light gray hygroscopic crystals that are odorless. This material is hazardous through inhalation, skin absorption, penetration through broken skin, and ingestion, and produces local skin/eye impacts.

Used industrially as welding flux, wood preservative, and in the ceramic industry.

This material is on the Australia Group Export Control list.

This material is a precursor for G-series and GV-series nerve agents.

Exposure Hazards

Human toxicity values have not been established or have not been published.

Properties:

MW: 78.1	*VP*: —	*FlP*: —
D: 2.37 g/cm^3	*VD*: —	*LEL*: —
MP: 460°F	*Vlt*: —	*UEL*: —
BP: —	H_2O: 39%	*RP*: —
Vsc: —	*Sol*: —	*IP*: —

C01-A066

Sodium bifluoride

CAS: 1333-83-1
RTECS: WB0350010
UN: 2439
ERG: 154

$NaHF_2$

Colorless to white crystalline powder. This material is hazardous through inhalation, skin absorption, penetration through broken skin, and ingestion, and produces local skin/eye impacts.

Used industrially for etching glass, as an antiseptic and disinfectant, as a leather bleach, in the production of tin plate, for rust removal, as a welding flux, as a neutralizer in laundry rinsing operations, and as a cleaner for stone and brick building faces. Used to preserve zoological and anatomical specimens.

This material is on the Australia Group Export Control list.

This material is a precursor for G-series and GV-series nerve agents.

Exposure Hazards

Human toxicity values have not been established or have not been published.

Properties:

MW: 62.0	*VP*: —	*FlP*: —
D: 2.08 g/cm^3	*VD*: —	*LEL*: —
MP: —	*Vlt*: —	*UEL*: —
BP: —	H_2O: "Soluble"	*RP*: —
Vsc: —	*Sol*: —	*IP*: —

C01-A067

Ammonium bifluoride

CAS: 1341-49-7
RTECS: BQ9200000
UN: 1727
ERG: 154

$(NH_4)HF_2$

White deliquescent crystalline solid that is odorless. This material is hazardous through inhalation and ingestion, and produces local skin/eye impacts.

Used industrially as a fungicide, in the manufacture magnesium fluoride, ceramics and in the production of fluorine, used industrially in electroplating and treating metals, in oil well acidizing, as a "sour" in laundering cloth, for brightening aluminum, etching glass, processing of beryllium, for cleaning and sterilizing beer, dairy, and other food equipment.

This material is on the Australia Group Export Control list.

This material is a precursor for G-series and GV-series nerve agents.

Exposure Hazards

Human toxicity values have not been established or have not been published.

Properties:

MW: 57.0	*VP*: —	*FlP*: —
D: 1.50 g/cm^3	*VD*: —	*LEL*: —
MP: 258°F	*Vlt*: —	*UEL*: —
BP: 463°F	H_2O: 41.5% (77° F)	*RP*: —
Vsc: —	*Sol*: Slightly in Ethanol	*IP*: —

C01-A068

Phosphorous trichloride

CAS: 7719-12-
RTECS: TH3675000
UN: 1809
ERG: 137

PCl_3

Colorless to yellow, fuming liquid with an irritating, pungent, acrid odor like hydrochloric acid. This material is hazardous through inhalation and ingestion, and produces local skin/eye impacts.

Used industrially for the manufacture of phosphorus oxychloride, phosphorus pentachloride, phosphites, organophosphorus pesticides, surfactants, gasoline additives, plasticizers, dyestuffs; used as a chlorinating agent and catalyst. Used to prepare rubber surfaces for electrodeposition of metal. Used as an ingredient of textile finishing agents.

This material is on the ITF-25 high threat list, Australia Group Export Control list, and Schedule 3 of the CWC.

This material is a general precursor for nerve agents and a chlorinating agent for sulfur and nitrogen vesicants.

Exposure Hazards

Conversion Factor: 1 ppm = mg/m^3 at 77°F
$LD_{50(Ing)}$: 1 g (estimate)
$MEG_{(1h)}$*Min*: —; *Sig*: —; *Sev*: 0.87 ppm
OSHA PEL: 0.5 ppm
ACGIH TLV: 0.2 ppm
ACGIH STEL: 0.5 ppm
NIOSH STEL: 0.5 ppm
IDLH: 25 ppm

Properties:

MW: 137.4	*VP*: 100 mmHg (70°F)	*FlP*: None
D: 1.574 g/mL (70°F)	*VD*: 4.8 (calculated)	*LEL*: None
MP: –170°F	*Vlt*: 130,000 ppm (70°F)	*UEL*: None
BP: 169°F	H_2O: Decomposes	*RP*: 0.085
Vsc: —	*Sol*: Benzene; Chloroform; Ether	*IP*: 9.91 eV

Interim AEGLs

AEGL-1: 1 h, 0.62 ppm	4 h, 0.39 ppm	8 h, 0.26 ppm
AEGL-2: 1 h, 2.0 ppm	4 h, 1.3 ppm	8 h, 0.83 ppm
AEGL-3: 1 h, 5.6 ppm	4 h, 3.5 ppm	8 h, 1.8 ppm

C01-A069

Phosphorous pentachloride
CAS: 10026-13-8
RTECS: TB6125000
UN: 1806
ERG: 137

PCl_5

White to pale yellow, crystalline solid with a pungent, unpleasant odor. This material is hazardous through inhalation and ingestion, and produces local skin/eye impacts.

Used industrially as a chlorinating agent, dehydrating agent, catalyst, in the manufacture of pharmaceuticals, and in aluminum metallurgy.

This material is on the Australia Group Export Control list and Schedule 3 of the CWC.

This material is a general precursor for nerve agents and a chlorinating agent for sulfur and nitrogen vesicants.

Exposure Hazards
Conversion Factor: 1 ppm = 8.52 mg/m^3 at 77°F
OSHA PEL: 0.12 ppm
ACGIH TLV: 0.10 ppm
IDLH: 8.2 ppm

Lethal human toxicity values have not been established or have not been published. However, based on available information, this material appears to have approximately the same toxicity as Phosgene (C10-A003).

Properties:

MW: 208.2	*VP*: 0.012 mmHg	*FlP*: None
D: 3.60 g/cm^3	*VP*: 1 mmHg (132°F)	*LEL*: None
MP: 324°F	*VD*: 7.2 (calculated)	*UEL*: None
BP: Sublimes	*Vlt*: 16 ppm	*RP*: 574
Vsc: —	*H_2O*: Decomposes	*IP*: —
	Sol: Carbon disulfide; Carbon tetrachloride	

C01-A070

Phosphorous oxychloride
CAS: 10025-87-3
RTECS: TH4897000
UN: 1810
ERG: 137

$POCl_3$

Clear, colorless to yellow, oily fuming liquid with a pungent, musty odor that is disagreeable and lingering. This material is hazardous through inhalation and ingestion, and produces local skin/eye impacts. May ignite combustible materials.

Used industrially for the manufacture of organophosphorus compounds (Insecticides, dyes, pharmaceuticals, defoliants) as well as esters for plasticizers, gasoline additives, and hydraulic fluids; used in industry as a chlorinating agent, catalyst, dopant for semiconductor grade silicon, fire retarding agent, and solvent in cryoscopy.

This material is on the ITF-25 medium threat list, Australia Group Export Control list, and Schedule 3 of the CWC.

This material is a general precursor for nerve agents and a chlorinating agent for sulfur and nitrogen vesicants.

Exposure Hazards

Conversion Factor: 1 ppm = 6.27 mg/m^3 at 77°F
$MEG_{(1h)}$*Min*: —; *Sig*: —; *Sev*: 0.85 ppm
ACGIH TLV: 0.1 ppm
NIOSH STEL: 0.5 ppm

Properties:

MW: 153.3	*VP*: 40 mmHg (81°F)	*FlP*: None
D: 1.645 g/mL (77°F)	*VD*: 5.3 (calculated)	*LEL*: None
MP: 34°F	*Vlt*: 52,000 ppm	*UEL*: None
BP: 222°F	H_2O: Decomposes	*RP*: 0.21
Vsc: —	*Sol*: —	*IP*: —

Interim AEGLs

AEGL-1: Not Developed
AEGL-2: Not Developed
AEGL-3: 1 h, 0.85 ppm 4 h, 0.54 ppm 8 h, 0.27 ppm

C01-A071

Phosphorus pentasulfide

CAS: 1314-80-3
RTECS: TH4375000
UN: 1340
ERG: 139

P_4S_{10}

Greenish-gray to yellow, deliquescent crystalline solid, with an odor of rotten eggs due to the formation of hydrogen sulfide. Olfactory fatigue may occur at high concentrations. This material is hazardous through inhalation and ingestion, and produces local skin/eye impacts.

Phosphorus pentasulfide autoignites at 287°F. It may spontaneously ignite in the presence of moisture.

Used industrially for manufacture of pyrotechnics, safety matches, lubricating oil additive, pesticides, and in organic synthesis.

This material is on the Australia Group Export Control list.

This material is a general precursor for nerve agents.

Exposure Hazards

Conversion Factor: 1 ppm = 18.18 mg/m^3 at 77°F
OSHA PEL: 1 mg/m^3
ACGIH TLV: 1 mg/m^3

ACGIH STEL: 3 mg/m^3
NIOSH STEL: 3 mg/m^3
IDLH: 250 mg/m^3

Properties:

MW: 444.6	*VP*: 1 mmHg (572°F)	*FlP*: —
D: 2.090 g/cm^3	*VD*: 15 (calculated)	*LEL*: —
MP: 550°F	*Vlt*: Negligible	*UEL*: —
BP: 957°F	H_2O: Decomposes	*RP*: —
Vsc: —	*Sol*: Carbon disulfide; aqueous bases	*IP*: —

C01-A072

Dimethyl phosphite
CAS: 868-85-9
RTECS: SZ7710000

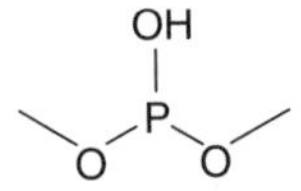

OH
|
P
O O

$C_2H_7O_3P$

Mobile, colorless liquid with a "mild" odor. This material is hazardous through inhalation, skin absorption, penetration through broken skin, and ingestion, and produces local skin/eye impacts.

Used industrially as a lubricant additive, flame retardant in textiles, to manufacture adhesives, and as a chemical intermediate.

This material is on the Australia Group Export Control list and Schedule 3 of the CWC.

This material is a general precursor for nerve agents. This material has been used as a simulant for nerve agents in government tests.

Exposure Hazards
Conversion Factor: 1 ppm = 4.50 mg/m^3 at 77°F

Human toxicity values have not been established or have not been published.

Properties:

MW: 110.1	*VP*: —	*FlP*: 85°F
D: 1.1941 g/mL	*VD*: 3.8 (calculated)	*LEL*: —
MP: —	*Vlt*: —	*UEL*: —
BP: 338°F	H_2O: "Soluble"	*RP*: —
BP: 162°F (25 mmHg)	*Sol*: Most Organic solvents	*IP*: —
Vsc: —		

C01-C073

Trimethyl phosphite
CAS: 121-45-9
RTECS: TH1400000
UN: 2329
ERG: 130

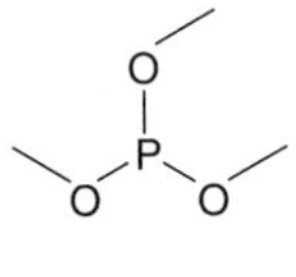

O
|
P
O O

$C_3H_9O_3P$

Colorless liquid with a distinctive, pungent, irritating, oily odor that smells like pyridine at higher concentrations. The odor is detectable at 0.1 ppb. This material is hazardous through inhalation and ingestion, and produces local skin/eye impacts.

Used industrially as a chemical intermediate in the manufacture of pesticides and phosphosilicate glass. Used as a gasoline additive, catalyst, and as a fireproofing agent in the production of textiles and flame-retardant polymers for polyurethane foams.

This material is on the Australia Group Export Control list and Schedule 3 of the CWC.

This material is a general precursor for nerve agents.

Exposure Hazards

Conversion Factor: 1 ppm = 5.08 mg/m^3 at 77°F

ACGIH TLV: 2 ppm

Properties:

MW: 124.1	*VP*: 24 mmHg (77°F)	*FlP*: 130°F
D: 1.0520 g/mL	*VD*: 4.3 (calculated)	*LEL*: —
MP: −108°F	*Vlt*: 32,000 ppm (77°F)	*UEL*: —
BP: 232°F	*H_2O*: Decomposes	*RP*: 0.38
Vsc: —	*Sol*: Hydrocarbons; Ethanol; Ether; Acetone	*IP*: —

C01-C074

Diethyl phosphite

CAS: 762-04-9

RTECS: TG7875000

OH P O O

$C_4H_{11}O_3P$

Specific information on physical appearance is not available for this material. This material is hazardous through inhalation and ingestion, and produces local skin/eye impacts.

Used industrially as a textile finishing agent, antioxidant, paint solvent, additive for adhesives, additive for extreme pressure lubricant; chemical intermediate for organic phosphorus compounds.

This material is on the Australia Group Export Control list and Schedule 3 of the CWC.

This material is a general precursor for nerve agents. This material has been used as a simulant for nerve agents in government tests.

Exposure Hazards

Conversion Factor: 1 ppm = 5.65 mg/m^3 at 77°F

Human toxicity values have not been established or have not been published.

Properties:

MW: 138.1	*VP*: —	*FlP*: 180°F
D: 1.073 g/mL	*VD*: 4.8 (calculated)	*LEL*: —
MP: —	*Vlt*: —	*UEL*: —

BP: 160°F (10 mmHg)	H_2O: —	*RP*: —
BP: 122°F (2 mmHg)	*Sol*: —	*IP*: —
Vsc: —		

C01-C075

Triethyl phosphite
CAS: 122-52-1
RTECS: TH1130000
UN: 2323
ERG: 130

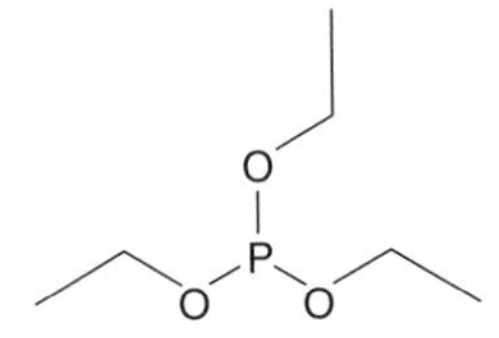

$C_6H_{15}O_3P$

Colorless liquid. This material is hazardous through inhalation, skin absorption, penetration through broken skin, and ingestion, and produces local skin/eye impacts.

Used industrially as a plasticizer, vinyl stabilizer, grease additive, color inhibitor for resins, and as a chemical intermediate for insecticides. Used in agriculture to ripen sugarcane.

This material is on the Australia Group Export Control list and Schedule 3 of the CWC.

This material is a general precursor for nerve agents. This material has been used as a simulant for nerve agents in government tests.

Exposure Hazards
Conversion Factor: 1 ppm = 6.80 mg/m^3 at 77°F

Human toxicity values have not been established or have not been published.

Properties:

MW: 166.2	*VP*: —	*FlP*: 130°F
D: 0.9629 g/mL	*VD*: 5.7 (calculated)	*LEL*: —
MP: —	*Vlt*: —	*UEL*: —
BP: 313°F	H_2O: "Insoluble"	*RP*: —
BP: 144°F (25 mmHg)	*Sol*: Ethanol; Ether	*IP*: —
Vsc: —		

C01-C076

Diethyl methylphosphonate
CAS: 683-08-9
RTECS: SZ9085000

$C_5H_{13}O_3P$

Clear, colorless liquid. This material is hazardous through inhalation and produces local skin/eye impacts.

Used in organic synthesis.

This material is on the Australia Group Export Control list and Schedule 2 of the CWC.

This material is a general precursor for nerve agents. It is also commonly found as a decomposition product/impurity and degradation product from hydrolysis of nerve agents.

Exposure Hazards

Conversion Factor: 1 ppm =mg/m^3 at 77°F

Human toxicity values have not been established or have not been published.

Properties:

MW: 152.1	*VP*: —	*FlP*: 167°F
D: 1.041 g/mL	*VD*: 5.2 (calculated)	*LEL*: —
MP: —	*Vlt*: —	*UEL*: —
BP: 374°F	H_2O: —	*RP*: —
BP: 165°F (10 mmHg)	*Sol*: —	*IP*: —
BP: 122°F (1 mmHg)		
Vsc: —		

C01-C077

Diethyl ethylphosphonate

CAS: 78-38-6

RTECS: SZ7925000

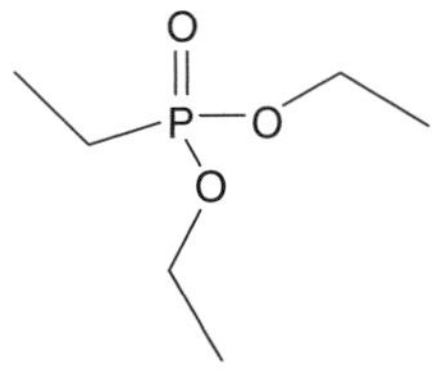

$C_6H_{15}O_3P$

Colorless liquid with a sweet odor. This material is hazardous through inhalation, skin absorption, penetration through broken skin, and ingestion, and produces local skin/eye impacts.

Used industrially as a solvent to extract heavy metals, a gasoline additive, antifoam agent, plasticizer, chelating agent, and as a textile conditioner and antistatic agent.

This material is on the Australia Group Export Control list and Schedule 2 of the CWC.

This material is a general precursor for nerve agents. It is also commonly found as a decomposition product/impurity and degradation product from hydrolysis of some nerve agents.

Exposure Hazards

Conversion Factor: 1 ppm = 6.80 mg/m^3 at 77°F

Human toxicity values have not been established or have not been published.

Properties:

MW: 166.2	*VP*: —	*FlP*: 221°F
D: 1.0259 g/mL	*VD*: 5.7 (calculated)	*LEL*: —
MP: —	*Vlt*: —	*UEL*: —
BP: 388°F	*H_2O*: "Slightly"	*RP*: —
BP: 162°F (10 mmHg)	*Sol*: Most organic solvents	*IP*: —
Vsc: —		

C01-C078

Diethyl isopropylphosphonate
CAS: 1538-69-8
RTECS: —

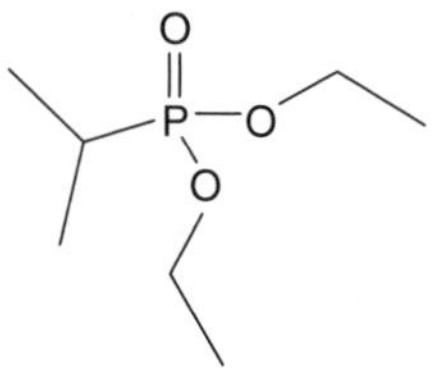

$C_7H_{17}O_3P$

Specific information on physical appearance is not available for this agent material. This material is hazardous through inhalation, skin absorption, penetration through broken skin, and ingestion, and produces local skin/eye impacts.

This material is on Schedule 2 of the CWC.

This material is a general precursor for nerve agents. It is also commonly found as a decomposition product/impurity and degradation product from hydrolysis of some nerve agents.

Exposure Hazards
Conversion Factor: 1 ppm = 7.37 mg/m^3 at 77°F

Human toxicity values have not been established or have not been published.

Properties:

MW: 180.2	*VP*: —	*FlP*: —
D: —	*VD*: 6.2 (calculated)	*LEL*: —
MP: —	*Vlt*: —	*UEL*: —
BP: 138°F (25 mmHg)	*H_2O*: —	*RP*: —
Vsc: —	*Sol*: —	*IP*: —

C01-C079

Dibutyl methylphosphonate
CAS: 2404-73-1
RTECS: —

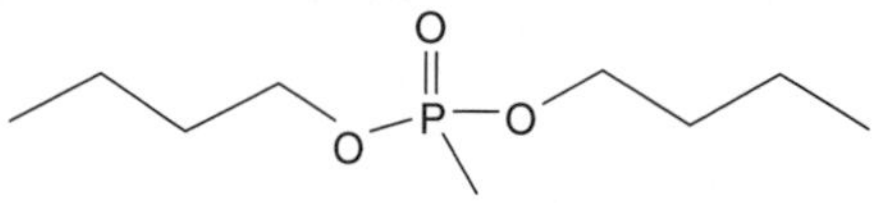

$C_9H_{21}O_3P$

Specific information on physical appearance is not available for this material. This material is hazardous through inhalation, skin absorption, penetration through broken skin, and ingestion, and produces local skin/eye impacts.

This material is on Schedule 2 of the CWC.
This material is a general precursor for nerve agents.

Exposure Hazards

Conversion Factor: 1 ppm = 8.52 mg/m^3 at 77°F
Human toxicity values have not been established or have not been published.

Properties:

MW: 208.2	*VP*: 0.0106 mmHg (77°F)	*FlP*: —
D: —	*VD*: 7.2 (calculated)	*LEL*: —
MP: —	*Vlt*: 14 ppm	*UEL*: —
BP: —	H_2O: 0.8% (77°F)	*RP*: 660
Vsc: —	*Sol*: —	*IP*: —

C01-C080

Ethyldichlorophospine

CAS: 1498-40-4
RTECS: TB2465000

$C_2H_5Cl_2P$

Colorless to yellow liquid with a strong disagreeable odor (stench). This material is hazardous through inhalation and ingestion, and produces local skin/eye impacts.

Used industrially for organic synthesis.

This material is on the Australia Group Export Control list and Schedule 2 of the CWC.

This material is a general precursor for nerve agents.

Exposure Hazards

Conversion Factor: 1 ppm = 5.36 mg/m^3 at 77°F
Human toxicity values have not been established or have not been published.

Properties:

MW: 130.9	*VP*: —	*FlP*: 90°F
D: 1.26 g/mL	*VD*: 4.5 (calculated)	*LEL*: —
MP: —	*Vlt*: —	*UEL*: —
BP: 239°F	H_2O: Decomposes	*RP*: —
Vsc: —	*Sol*: —	*IP*: —

C01-C081

Ethylphosphonyl dichloride

CAS: 1066-50-8
RTECS: TA1780000

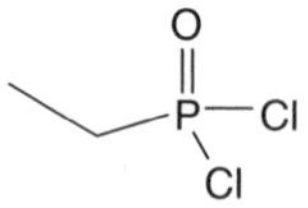

$C_2H_5Cl_2OP$

Clear colorless to light yellow liquid. This material is hazardous through inhalation, skin absorption, penetration through broken skin, and ingestion, and produces local skin/eye impacts.

Used industrially for organic synthesis.

This material is on the ITF-25 low threat list, Australia Group Export Control list and Schedule 2 of the CWC.

This material is a precursor for general nerve agents.

Exposure Hazards

Conversion Factor: 1 ppm = 6.01 mg/m^3 at 77°F

Human toxicity values have not been established or have not been published.

Properties:

MW: 146.9	*VP*: —	*FlP*: >230°F
D: 1.3760 g/mL	*VD*: 5.1 (calculated)	*LEL*: —
MP: —	*Vlt*: —	*UEL*: —
BP: 216°F (68 mmHg)	H_2O: Decomposes	*RP*: —
BP: 160°F (12 mmHg)	*Sol*: —	*IP*: —
Vsc: —		

C01-C082

Propylphosphonic dichloride

CAS: 4708-04-7

RTECS: —

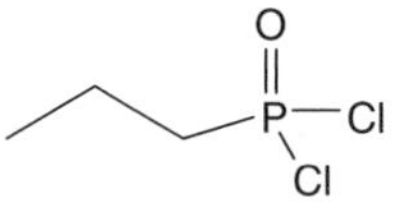

$C_3H_7Cl_2OP$

Clear yellow liquid. This material is hazardous through inhalation, skin absorption, penetration through broken skin, and ingestion, and produces local skin/eye impacts.

This material is on Schedule 2 of the CWC.

This material is a general precursor for nerve agents.

Exposure Hazards

Conversion Factor: 1 ppm = 6.58 mg/m^3 at 77°F

Human toxicity values have not been established or have not been published.

Properties:

MW: 161.0	*VP*: —	*FlP*: >230°F
D: 1.29 g/mL	*VD*: 5.6 (calculated)	*LEL*: —
MP: —	*Vlt*: —	*UEL*: —
BP: 190°F (50 mmHg)	H_2O: Decomposes	*RP*: —
Vsc: —	*Sol*: —	*IP*: —

C01-C083

2-Propanephosphonyl chloride

CAS: 1498-46-0
RTECS: —

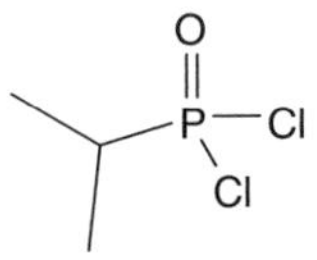

$C_3H_7Cl_2OP$

Specific information on physical appearance is not available for this agent material. This material is hazardous through inhalation and ingestion, and produces local skin/eye impacts.

This material is on Schedule 2 of the CWC.

This material is a general precursor for nerve agents.

Exposure Hazards

Conversion Factor: 1 ppm = 6.58 mg/m^3 at 77°F

Human toxicity values have not been established or have not been published.

Properties:

MW: 161.0	*VP*: —	*FlP*: —
D: 1.3 g/mL	*VD*: 5.6 (calculated)	*LEL*: —
MP: —	*Vlt*: —	*UEL*: —
BP: 176°F (20 mmHg)	H_2O: Decomposes	*RP*: —
Vsc: —	*Sol*: —	*IP*: —

C01-C084

Dimethylamine

CAS: 124-40-3; 506-59-2 (Hydrochloride Salt)
RTECS: IP8750000
UN: 1032
ERG: 118

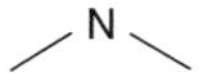

C_2H_7N

Colorless gas with an ammonia or fish like odor detectable at 0.53 ppm. Shipped as a liquefied compressed gas. Various salts have been reported. This material is hazardous through inhalation and produces local skin/eye impacts.

Used industrially for manufacture of detergents, pesticides, and pharmaceuticals. Used as a gasoline additive and solvent.

This material is on the Australia Group Export Control list.

This material is a precursor for some G-series, GV-series, and V-series nerve agents and is also commonly found as a decomposition product/impurity resulting from hydrolysis of Tabun (C01-A001).

Exposure Hazards

Conversion Factor: 1 ppm = 1.84 mg/m^3 at 77°F
OSHA PEL: 10 ppm

ACGIH TLV: 5 ppm
ACGIH STEL: 15 ppm
IDLH: 500 ppm

Properties:

MW: 45.1	*VP*: 1300 mmHg	*FlP*: 20°F
D: 0.67 g/mL (liq. gas, 44°F)	*VP*: 1520 mmHg (77°F)	*LEL*: 2.8%
MP: –134°F	*VD*: 1.6 (calculated)	*UEL*: 14.4%
BP: 44°F	*Vlt*: —	*RP*: 0.001
Vsc: —	H_2O: 24% (140°F)	*IP*: 8.24 eV
	Sol: Alcohols; Ether	

Proposed AEGLs

AEGL-1: 1 h, 10 ppm	4 h, 10 ppm	8 h, 10 ppm
AEGL-2: 1 h, 120 ppm	4 h, 74 ppm	8 h, 59 ppm
AEGL-3: 1 h, 460 ppm	4 h, 280 ppm	8 h, 220 ppm

C01-C085

Dimethylaminoethyl chloride

CAS: 107-99-3; 4584-46-7 (Hydrochloride salt); 69153-76-0 (Sulfate salt)
RTECS: KQ9020000

$C_4H_{10}ClN$

Oil is unstable and decomposes on storage. Hydrochloride is colorless to light beige crystalline powder with a "characteristic" odor. Various salts (solids) have been reported. This material is hazardous through inhalation, ingestion, and produces local skin/eye impacts.

This material is on Schedule 2 of the CWC.

This material is a precursor for some GV-series and V-series nerve agents.

Exposure Hazards

Conversion Factor: 1 ppm = 4.40 mg/m^3 at 77°F

Human toxicity values have not been established or have not been published.

Properties:

MW: 107.6	*VP*: —	*FlP*: —
D: —	*VD*: —	*LEL*: —
MP: —	*Vlt*: —	*UEL*: —
BP: —	H_2O: Decomposes	*RP*: —
Vsc: —	*Sol*: —	*IP*: —

Hydrochloride salt

MW: 144.1
MP: 397°F
H_2O: 200%

C01-C086

Methyl t-butyl ketone

CAS: 75-97-8
RTECS: EL7700000

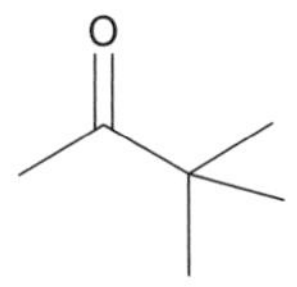

$C_6H_{12}O$

Colorless liquid with an odor like peppermint or camphor. This material is hazardous through inhalation, skin absorption, and ingestion.

Used industrially for manufacture of fungicides and herbicides.

This material is on the Australia Group Export Control list.

This material can be reduced to pinacolyl alcohol (C01-C087), which is an intermediate for production of Soman (C01-A003).

Exposure Hazards

Conversion Factor: 1 ppm = 4.10 mg/m^3 at 77°F

Human toxicity values have not been established or have not been published.

Properties:

MW: 100.2	*VP*: 31.5 mmHg (77°F)	*FlP*: 63°F
D: 0.7229 g/mL (77°F)	*VD*: 3.5 (calculated)	*LEL*: —
MP: –63°F	*Vlt*: 41,000 ppm (77°F)	*UEL*: —
BP: 223°F	*H_2O*: 2.44% (59° F)	*RP*: 0.32 (77°F)
Vsc: 3.42 cS (–63°F)	*Sol*: Alcohols; Ether; Acetone	*IP*: 9.14 eV

C01-C0087

Pinacolyl alcohol

CAS: 464-07-3

RTECS: EL2276000

$C_6H_{14}O$

Clear, colorless liquid. Hazardous routes of exposure for this material have not been established.

Used industrially for organic synthesis.

This material is on the Australia Group Export Control list and Schedule 2 of the CWC.

This material is an intermediate for production of Soman (C01-A003).

Exposure Hazards

Conversion Factor: 1 ppm = 4.81 mg/m^3 at 77°F

Human toxicity values have not been established or have not been published.

Properties:

MW: 102.2	*VP*: 8.81 mmHg (77°F)	*FlP*: 82°F
D: 0.812 g/mL	*VD*: 3.5 (calculated)	*LEL*: —
MP: 41°F	*Vlt*: 12,000 ppm (77°F)	*UEL*: —
BP: 246°F	*H_2O*: 2.43% (77° F)	*RP*: 1.1 (77° F)
Vsc: —	*Sol*: —	*IP*: —

C01-C088

Diethylaminoethanol
CAS: 100-37-8
RTECS: KK5075000
UN: 2686
ERG: 132

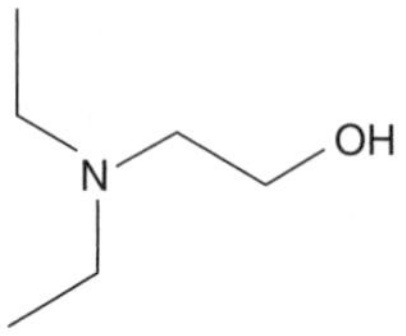

$C_6H_{15}NO$

Colorless hygroscopic liquid with a nauseating odor similar to ammonia and detectable at 0.011–0.04 ppm. This material is hazardous through inhalation, skin absorption, penetration through broken skin, and ingestion, and produces local skin/eye impacts.

Used industrially for the manufacture of emulsifying agents, flocculants, soaps, textiles softeners, cosmetics, in pharmaceuticals and crop protection agents, in the preparation of chemicals for the paper and leather industries, and in the production of plastics.

This material is on the Australia Group Export Control list.

This material is a precursor for production of GV and V series nerve agents.

Exposure Hazards
Conversion Factor: 1 ppm = 4.79 mg/m^3 at 77°F
OSHA PEL: 10 ppm [Skin]
ACGIH TLV: 2 ppm [Skin]
IDLH: 100 ppm

Properties:

MW: 117.2	*VP*: 1.4 mmHg	*FlP*: 140°F
D: 0.8921 g/mL	*VD*: 4.0 (calculated)	*LEL*: 6.7%
MP: –94°F	*Vlt*: 1900 ppm	*UEL*: 11.7%
BP: 325°F	H_2O: Miscible	*RP*: 6.6
BP: 212°F (80 mmHg)	*Sol*: Alcohols; Acetone; Ether; Benzene	*IP*: 8.58 eV
BP: 131°F (10 mmHg)		
Vsc: 5.6 cS		

C01-C089

2-(Diisopropylamino)ethanol
CAS: 96-80-0
RTECS: KK5950000

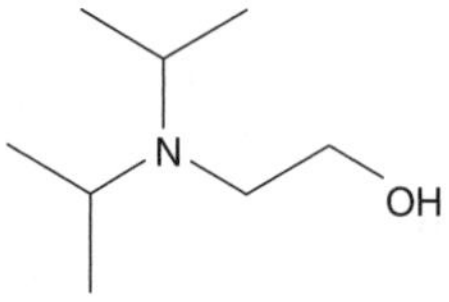

$C_8H_{19}NO$

Clear colorless liquid with a nauseating odor similar to ammonia. Various salts (solids) have been reported. This material is hazardous through inhalation, skin absorption, penetration through broken skin, and ingestion, and produces local skin/eye impacts.

Used industrially as a catalyst; emulsifying agent; and for manufacture of pharmaceuticals.

This material is on the Australia Group Export Control list and Schedule 2 of the CWC.

This material is an intermediate for production of VX (C01-A016) and is also commonly found as a decomposition product resulting from hydrolysis of VX.

Exposure Hazards

Conversion Factor: 1 ppm = 5.94 mg/m^3 at 77°F
Human toxicity values have not been established or have not been published.

Properties:

MW: 145.2	*VP*: 1.8 mmHg	*FlP*: 316°F
D: 0.826 g/mL	*VD*: 5.0 (calculated)	*LEL*: —
MP: —	*Vlt*: 2400 ppm	*UEL*: —
BP: 374°F	H_2O: —	*RP*: 4.6
Vsc: —	*Sol*: —	*IP*: —

C01-C090

[(Dibromophosphinyl)oxy]carbonimidic chloride fluoride
CAS: 18262-26-5
RTECS: —

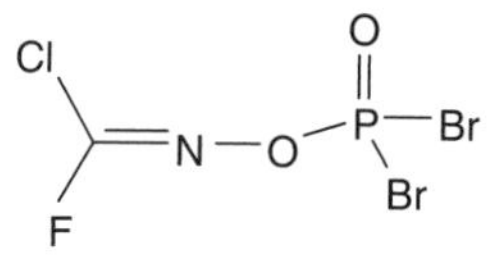

CBr_2ClFNO_2P

Specific information on physical appearance is not available for this material. This material is hazardous through inhalation, skin absorption, penetration through broken skin, and ingestion, and produces local skin/eye impacts.

This material is a precursor for Novichok series agents.

Exposure Hazards

Conversion Factor: 1 ppm = 12.40 mg/m^3 at 77°F
Human toxicity values have not been established or have not been published.

Properties:

MW: 303.3	*VP*: 0.03 mmHg (77°F) (estimate)	*FlP*: —
D: 2.2030 g/mL	*VD*: 10 (calculated)	*LEL*: —
MP: —	*Vlt*: —	*UEL*: —
BP: 140°F (0.1 mmHg)	H_2O: "Sparingly soluble"	*RP*: —
Vsc: —	*Sol*: —	*IP*: —

C01-C091

[(Dichlorophosphinyl)oxy]carbonimidic difluoride
CAS: 18262-24-3
RTECS: —

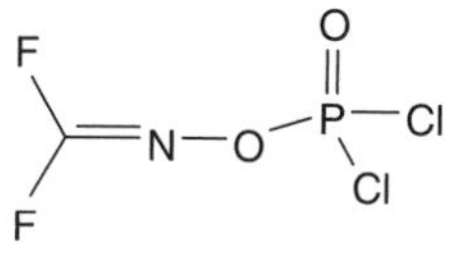

$CCl_2F_2NO_2P$

Specific information on physical appearance is not available for this material. This material is hazardous through inhalation, skin absorption, penetration through broken skin, and ingestion, and produces local skin/eye impacts.

This material is a precursor for Novichok series agents.

Exposure Hazards

Conversion Factor: 1 ppm = 8.09 mg/m^3 at 77°F

Human toxicity values have not been established or have not been published.

Properties:

MW: 197.9	*VP*: 0.9 mmHg (77°F) (estimate)	*FlP*: —
D: 1.6360 g/mL	*VD*: 6.8 (calculated)	*LEL*: —
MP: —	*Vlt*: —	*UEL*: —
BP: 149°F (20 mmHg)	H_2O: "Slightly soluble"	*RP*: —
Vsc: —	*Sol*: —	*IP*: —

C01-C092

[(Dichlorophosphinyl)oxy]carbonimidic chloride fluoride

CAS: 18425-23-5

RTECS: —

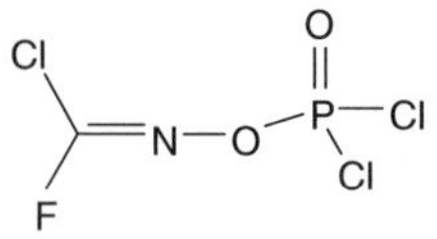

CCl_3FNO_2P

Specific information on physical appearance is not available for this material. This material is hazardous through inhalation, skin absorption, penetration through broken skin, and ingestion, and produces local skin/eye impacts.

This material is a precursor for Novichok series agents.

Exposure Hazards

Conversion Factor: 1 ppm = 8.77 mg/m^3 at 77°F

Human toxicity values have not been established or have not been published.

Properties:

MW: 214.4	*VP*: 0.2 mmHg (77°F) (estimate)	*FlP*: —
D: 1.7080 g/mL	*VD*: 7.4 (calculated)	*LEL*: —
MP: —	*Vlt*: —	*UEL*: —
BP: 142°F (6 mmHg)	H_2O: "Sparingly soluble"	*RP*: —
Vsc: —	*Sol*: —	*IP*: —

C01-C093

[(Dichlorophosphinyl)oxy]carbonimidic dichloride

CAS: 17642-35-2

RTECS: —

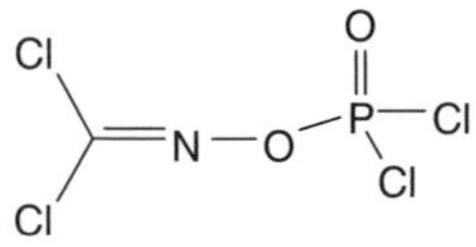

CCl_4NO_2P

Specific information on physical appearance is not available for this material. This material is hazardous through inhalation, skin absorption, penetration through broken skin, and ingestion, and produces local skin/eye impacts.

This material is a precursor for Novichok series agents.

Exposure Hazards

Conversion Factor: 1 ppm = 9.44 mg/m^3 at 77°F

Human toxicity values have not been established or have not been published.

Properties:

MW: 230.8	*VP*: 0.04 mmHg (77°F) (estimate)	*FlP*: —
D: 1.7360 g/mL	*VD*: 8.0 (calculated)	*LEL*: —
MP: —	*Vlt*: —	*UEL*: —
BP: 167°F (5 mmHg)	H_2O: "Sparingly soluble"	*RP*: —
Vsc: —	*Sol*: —	*IP*: —

C01-C094

***N*-[(Dichlorophosphinyl)oxy]-2,2-difluoro-2-nitroethanimidoyl fluoride**

CAS: 18262-25-4

RTECS: —

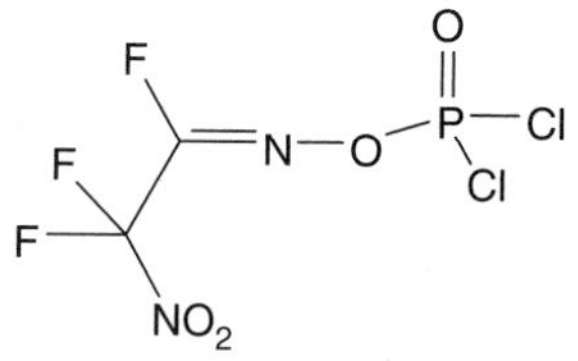

$C_2Cl_2F_3N_2O_4P$

Specific information on physical appearance is not available for this material. This material is hazardous through inhalation, skin absorption, penetration through broken skin, and ingestion, and produces local skin/eye impacts.

This material is a precursor for Novichok series agents.

Exposure Hazards

Conversion Factor: 1 ppm = 11.24 mg/m^3 at 77°F

Human toxicity values have not been established or have not been published.

Properties:

MW: 274.9	*VP*: 0.08 mmHg (77°F) (estimate)	*FlP*: —
D: 1.7470 g/mL	*VD*: 9.5 (calculated)	*LEL*: —
MP: —	*Vlt*: —	*UEL*: —
BP: 149°F (3 mmHg)	H_2O: "Sparingly soluble"	*RP*: —
Vsc: —	*Sol*: —	*IP*: —

C01-C095

2-Chloro-*N*-[(dichlorophosphinyl)oxy]ethanimidoyl chloride

CAS: 111203-62-4

RTECS: —

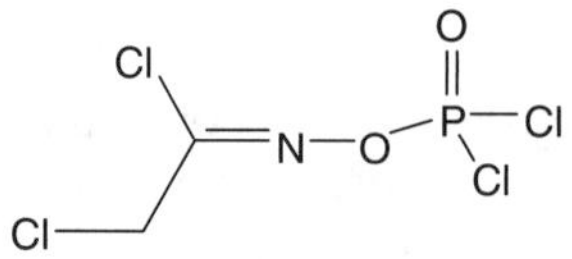

$C_2H_2Cl_4NO_2P$

Specific information on physical appearance is not available for this material. This material is hazardous through inhalation, skin absorption, penetration through broken skin, and ingestion, and produces local skin/eye impacts.

This material is a precursor for Novichok series agents.

Exposure Hazards

Conversion Factor: 1 ppm = 10.01 mg/m^3 at 77°F

Human toxicity values have not been established or have not been published.

Properties:

MW: 244.8	*VP*: 0.01 mmHg (77°F) (estimate)	*FlP*: —
D: —	*VD*: 8.4 (calculated)	*LEL*: —
MP: —	*Vlt*: —	*UEL*: —
BP: 210°F (2 mmHg)	H_2O: "Sparingly soluble"	*RP*: —
Vsc: —	*Sol*: —	*IP*: —

C01-C096

[(Chloromethoxyphosphinyl)oxy]carbonimidic chloride fluoride

CAS: 17650-48-5

RTECS: —

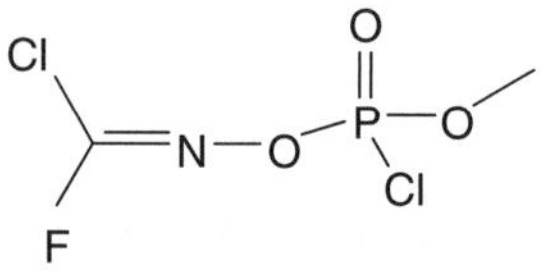

$C_2H_3Cl_2FNO_3P$

Specific information on physical appearance is not available for this material. This material is hazardous through inhalation, skin absorption, penetration through broken skin, and ingestion, and produces local skin/eye impacts.

This material is a precursor for Novichok series agents.

Exposure Hazards

Conversion Factor: 1 ppm = 8.58 mg/m^3 at 77°F

Human toxicity values have not been established or have not been published.

Properties:

MW: 209.9	*VP*: 0.5 mmHg (77°F) (estimate)	*FlP*: —
D: 1.5790 g/mL	*VD*: 7.2 (calculated)	*LEL*: —
MP: —	*Vlt*: —	*UEL*: —
BP: 203°F (6 mmHg)	H_2O: "Sparingly soluble"	*RP*: —
Vsc: —	*Sol*: —	*IP*: —

C01-C097

***N*-[(Dichlorophosphinyl)oxy]ethanimidoyl chloride**

CAS: 120932-13-0

RTECS: —

$C_2H_3Cl_3NO_2P$

Specific information on physical appearance is not available for this material. This material is hazardous through inhalation, skin absorption, penetration through broken skin, and ingestion, and produces local skin/eye impacts.

This material is a precursor for Novichok series agents.

Exposure Hazards

Conversion Factor: 1 ppm = 8.61 mg/m^3 at 77°F

Human toxicity values have not been established or have not been published.

Properties:

MW: 210.4	*VP*: 0.1 mmHg (77°F) (estimate)	*FlP*: —
D: —	*VD*: 7.3 (calculated)	*LEL*: —
MP: —	*Vlt*: —	*UEL*: —
BP: 160°F (2 mmHg)	H_2O: "Sparingly soluble"	*RP*: —
Vsc: —	*Sol*: —	*IP*: —

C01-C098

[(Chloromethoxyphosphinyl)oxy]carbonimidic dichloride

CAS: 17642-33-0

RTECS: —

$C_2H_3Cl_3NO_3P$

Specific information on physical appearance is not available for this material. This material is hazardous through inhalation, skin absorption, penetration through broken skin, and ingestion, and produces local skin/eye impacts.

This material is a precursor for Novichok series agents.

Exposure Hazards

Conversion Factor: 1 ppm = 9.26 mg/m^3 at 77°F

Human toxicity values have not been established or have not been published.

Properties:

MW: 226.4	*VP*: 0.1 mmHg (77°F) (estimate)	*FlP*: —
D: 1.5830 g/mL	*VD*: 7.8 (calculated)	*LEL*: —
MP: —	*Vlt*: —	*UEL*: —
BP: 180°F (1 mmHg)	H_2O: "Sparingly soluble"	*RP*: —
Vsc: —	*Sol*: —	*IP*: —

C01-C099

[(Chloromethoxyphosphinyl)oxy]carbonimidic difluoride

CAS: 18262-30-1

RTECS: —

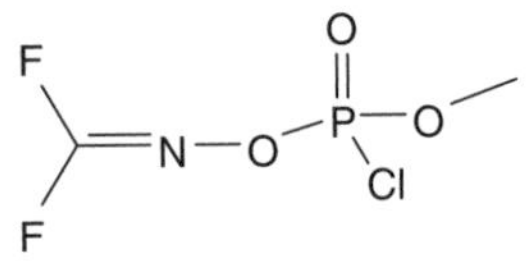

$C_2H_3ClF_2NO_3P$

Specific information on physical appearance is not available for this material. This material is hazardous through inhalation, skin absorption, penetration through broken skin, and ingestion, and produces local skin/eye impacts.

This material is a precursor for Novichok series agents.

Exposure Hazards

Conversion Factor: 1 ppm = 7.91 mg/m^3 at 77°F

Human toxicity values have not been established or have not been published

Properties:

MW: 193.5	*VP*: 2 mmHg (77°F) (estimate)	*FlP*: —
D: 1.5490 g/mL	*VD*: 6.7 (calculated)	*LEL*: —
MP: —	*Vlt*: —	*UEL*: —
BP: 138°F (2 mmHg)	H_2O: "Sparingly soluble"	*RP*: —
Vsc: —	*Sol*: —	*IP*: —

C01-C100

2,2-Dichloro-*N*-[(dichlorophosphinyl)oxy]ethanimidoyl chloride

CAS: 114700-94-6

RTECS: —

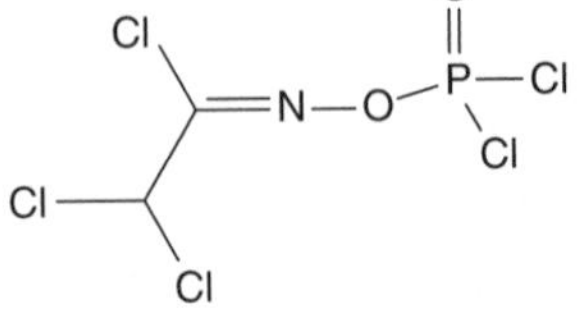

$C_2HCl_5NO_2P$

Specific information on physical appearance is not available for this material. This material is hazardous through inhalation, skin absorption, penetration through broken skin, and ingestion, and produces local skin/eye impacts.

This material is a precursor for Novichok series agents.

Exposure Hazards

Conversion Factor: 1 ppm = 11.42 mg/m^3 at 77°F

Human toxicity values have not been established or have not been published.

Properties:

MW: 279.3	*VP*: 0.008 mmHg (77°F) (estimate)	*FlP*: —
D: —	*VD*: 9.6 (calculated)	*LEL*: —
MP: —	*Vlt*: —	*UEL*: —
BP: 201°F (1.5 mmHg)	H_2O: "Sparingly soluble"	*RP*: —
Vsc: —	*Sol*: —	*IP*: —

C01-C101

***N*-[(Chloromethoxyphosphinyl)oxy]-2,2-difluoro-2-nitroethanimidoyl fluoride**

CAS: 18262-33-4

RTECS: —

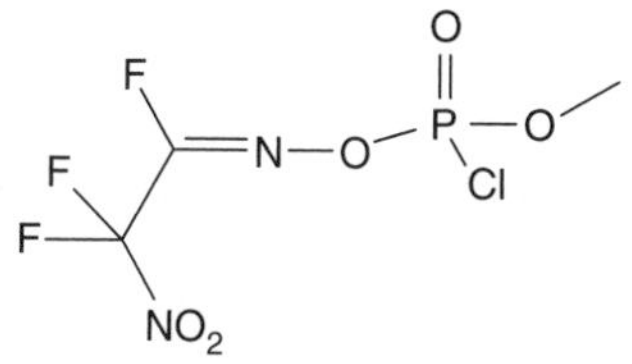

$C_3H_3ClF_3N_2O_5P$

Specific information on physical appearance is not available for this material. This material is hazardous through inhalation, skin absorption, penetration through broken skin, and ingestion, and produces local skin/eye impacts.

This material is a precursor for Novichok series agents.

Exposure Hazards

Conversion Factor: 1 ppm = 11.06 mg/m^3 at 77°F

Human toxicity values have not been established or have not been published.

Properties:

MW: 270.5	*VP*: 0.2 mmHg (77°F) (estimate)	*FlP*: —
D: 1.6270 g/mL	*VD*: 9.3 (calculated)	*LEL*: —
MP: —	*Vlt*: —	*UEL*: —
BP: 189°F (3 mmHg)	H_2O: "Sparingly soluble"	*RP*: —
Vsc: —	*Sol*: —	*IP*: —

C01-C102

[(Chloro-2-chloroethoxyphosphinyl)oxy]carbonimidic chloride fluoride

CAS: 23233-25-2

RTECS: —

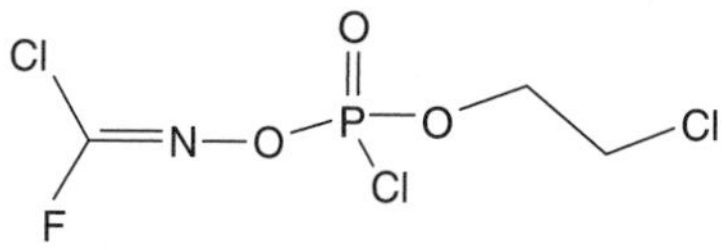

$C_3H_4Cl_3FNO_3P$

Specific information on physical appearance is not available for this material. This material is hazardous through inhalation, skin absorption, penetration through broken skin, and ingestion, and produces local skin/eye impacts.

This material is a precursor for Novichok series agents.

Exposure Hazards

Conversion Factor: 1 ppm = 10.57 mg/m^3 at 77°F

Human toxicity values have not been established or have not been published.

Properties:

MW: 258.4	*VP*: 0.01 mmHg (77°F) (estimate)	*FlP*: —
D: 1.5980 g/mL	*VD*: 8.9 (calculated)	*LEL*: —
MP: —	*Vlt*: —	*UEL*: —
BP: 163°F (0.01 mmHg)	H_2O: "Sparingly soluble"	*RP*: —
Vsc: —	*Sol*: —	*IP*: —

C01-C103

[(Bromoethoxyphosphinyl)oxy]carbonimidic chloride fluoride

CAS: 17642-25-0
RTECS: —

$C_3H_5BrClFNO_3P$

Specific information on physical appearance is not available for this material. This material is hazardous through inhalation, skin absorption, penetration through broken skin, and ingestion, and produces local skin/eye impacts.

This material is a precursor for Novichok series agents.

Exposure Hazards

Conversion Factor: 1 ppm = 10.98 mg/m^3 at 77°F

Human toxicity values have not been established or have not been published.

Properties:

MW: 268.4	*VP*: 0.07 mmHg (77°F) (estimate)	*FlP*: —
D: 1.6970 g/mL	*VD*: 9.3 (calculated)	*LEL*: —
MP: —	*Vlt*: —	*UEL*: —
BP: 214°F (1 mmHg)	H_2O: "Sparingly soluble"	*RP*: —
Vsc: —	*Sol*: —	*IP*: —

C01-C104

[(Chloroethoxyphosphinyl)oxy]carbonimidic chloride fluoride

CAS: 18262-27-6
RTECS: —

$C_3H_5Cl_2FNO_3P$

Specific information on physical appearance is not available for this material. This material is hazardous through inhalation, skin absorption, penetration through broken skin, and ingestion, and produces local skin/eye impacts.

This material is a precursor for Novichok series agents.

Exposure Hazards

Conversion Factor: 1 ppm = 9.16 mg/m^3 at 77°F

Human toxicity values have not been established or have not been published.

Properties:

MW: 224.0	*VP*: 0.2 mmHg (77°F) (estimate)	*FlP*: —
D: 1.4490 g/mL	*VD*: 7.2 (calculated)	*LEL*: —
MP: —	*Vlt*: —	*UEL*: —
BP: 170°F (3 mmHg)	*H_2O*: "Sparingly soluble"	*RP*: —
Vsc: —	*Sol*: —	*IP*: —

C01-C105

N-[(Dichlorophosphinyl)oxy]propanimidoyl chloride

CAS: 121951-54-0

RTECS: —

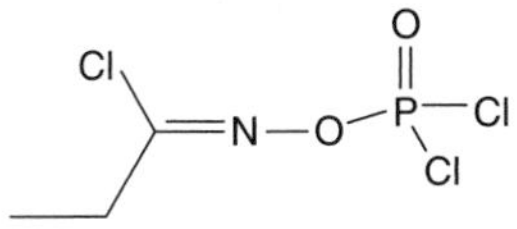

$C_3H_5Cl_3NO_2P$

Specific information on physical appearance is not available for this material. This material is hazardous through inhalation, skin absorption, penetration through broken skin, and ingestion, and produces local skin/eye impacts.

This material is a precursor for Novichok series agents.

Exposure Hazards

Conversion Factor: 1 ppm = 9.18 mg/m^3 at 77°F

Human toxicity values have not been established or have not been published.

Properties:

MW: 224.4	*VP*: 0.04 mmHg (77°F) (estimate)	*FlP*: —
D: —	*VD*: 7.7 (calculated)	*LEL*: —
MP: —	*Vlt*: —	*UEL*: —
BP: 153°F (2 mmHg)	*H_2O*: "Sparingly soluble"	*RP*: —
Vsc: —	*Sol*: —	*IP*: —

C01-C106

2-Chloro-*N*-[(chloromethoxyphosphinyl)oxy]ethanimidoyl chloride

CAS: 114700-91-3

RTECS: —

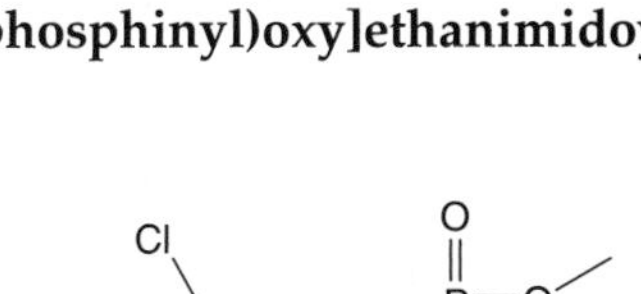

$C_3H_5Cl_3NO_3P$

Specific information on physical appearance is not available for this material. This material is hazardous through inhalation, skin absorption, penetration through broken skin, and ingestion, and produces local skin/eye impacts.

This material is a precursor for Novichok series agents.

Exposure Hazards

Conversion Factor: 1 ppm = 9.83 mg/m^3 at 77°F

Human toxicity values have not been established or have not been published.

Properties:

MW: 240.4	*VP*: 0.03 mmHg (77°F) (estimate)	*FlP*: —
D: —	*VD*: 8.3 (calculated)	*LEL*: —
MP: —	*Vlt*: —	*UEL*: —
BP: 244°F (2 mmHg)	*H_2O*: "Sparingly soluble"	*RP*: —
Vsc: —	*Sol*: —	*IP*: —

C01-C107

[(Chloroethoxyphosphinyl)oxy]carbonimidic difluoride

CAS: 18262-31-2

RTECS: —

$C_3H_5ClF_2NO_3P$

Specific information on physical appearance is not available for this material. This material is hazardous through inhalation, skin absorption, penetration through broken skin, and ingestion, and produces local skin/eye impacts.

This material is a precursor for Novichok series agents.

Exposure Hazards

Conversion Factor: 1 ppm = 8.49 mg/m^3 at 77°F

Human toxicity values have not been established or have not been published.

Properties:

MW: 207.5	*VP*: 0.8 mmHg (77°F) (estimate)	*FlP*: —
D: 1.4050 g/mL	*VD*: 7.2 (calculated)	*LEL*: —
MP: —	*Vlt*: —	*UEL*: —
BP: 149°F (2 mmHg)	*H_2O*: "Sparingly soluble"	*RP*: —
Vsc: —	*Sol*: —	*IP*: —

C01-C108

***N*-[(Chloroethoxyphosphinyl)oxy]-2,2-difluoro-2-nitroethanimidoyl fluoride**

CAS: 18262-34-5

RTECS: —

$C_4H_5ClF_3N_2O_5P$

Specific information on physical appearance is not available for this material. This material is hazardous through inhalation, skin absorption, penetration through broken skin, and ingestion, and produces local skin/eye impacts.

This material is a precursor for Novichok series agents.

Exposure Hazards

Conversion Factor: 1 ppm = 11.64 mg/m^3 at 77°F

Human toxicity values have not been established or have not been published.

Properties:

MW: 284.5	*VP*: 0.05 mmHg (77°F) (estimate)	*FlP*: —
D: 1.5140 g/mL	*VD*: 9.8 (calculated)	*LEL*: —
MP: —	*Vlt*: —	*UEL*: —
BP: 163°F (0.1 mmHg)	*H_2O*: "Sparingly soluble"	*RP*: —
Vsc: —	*Sol*: —	*IP*: —

C01-C109

2,2-Dichloro-*N*-[(chloroethoxyphosphinyl)oxy]ethanimidoyl chloride

CAS: 114700-93-5

RTECS: —

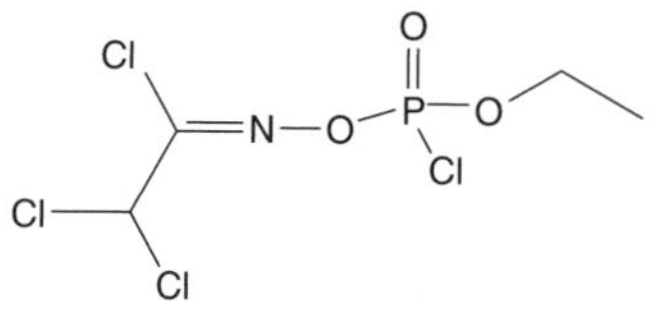

$C_4H_6Cl_4NO_3P$

Specific information on physical appearance is not available for this material. This material is hazardous through inhalation, skin absorption, penetration through broken skin, and ingestion, and produces local skin/eye impacts.

This material is a precursor for Novichok series agents.

Exposure Hazards

Conversion Factor: 1 ppm = 11.82 mg/m^3 at 77°F

Human toxicity values have not been established or have not been published.

Properties:

MW: 288.9	*VP*: 0.006 mmHg (77°F) (estimate)	*FlP*: —
D: —	*VD*: 10 (calculated)	*LEL*: —
MP: —	*Vlt*: —	*UEL*: —
BP: 234°F (1.5 mmHg)	*H_2O*: "Sparingly soluble"	*RP*: —
Vsc: —	*Sol*: —	*IP*: —

C01-C110

[[Chloroisopropylphosphinyl]oxy]carbonimidic chloride fluoride

CAS: 18262-28-7

RTECS: —

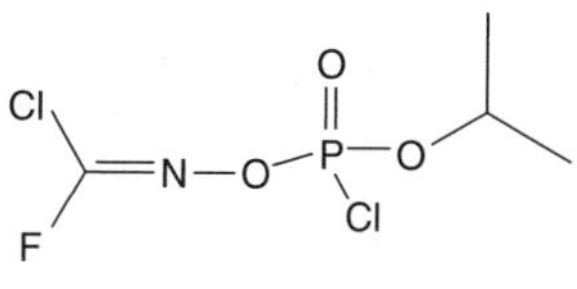

$C_4H_7Cl_2FNO_3P$

Specific information on physical appearance is not available for this material. This material is hazardous through inhalation, skin absorption, penetration through broken skin, and ingestion, and produces local skin/eye impacts.

This material is a precursor for Novichok series agents.

Exposure Hazards

Conversion Factor: 1 ppm = 9.73 mg/m^3 at 77°F

Human toxicity values have not been established or have not been published.

Properties:

MW: 238.0	*VP*: 0.09 mmHg (77°F) (estimate)	*FlP*: —
D: 1.3690 g/mL	*VD*: 8.2 (calculated)	*LEL*: —
MP: —	*Vlt*: —	*UEL*: —
BP: 147°F (2 mmHg)	H_2O: "Sparingly soluble"	*RP*: —
Vsc: —	*Sol*: —	*IP*: —

C01-C111

[[Chloropropylphosphinyl]oxy]carbonimidic chloride fluoride

CAS: 18262-29-8

RTECS: —

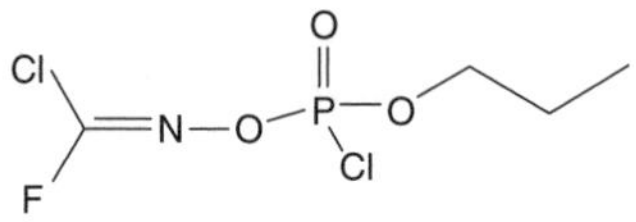

$C_4H_7Cl_2FNO_3P$

Specific information on physical appearance is not available for this material. This material is hazardous through inhalation, skin absorption, penetration through broken skin, and ingestion, and produces local skin/eye impacts.

This material is a precursor for Novichok series agents.

Exposure Hazards

Conversion Factor: 1 ppm = 9.73 mg/m^3 at 77°F

Human toxicity values have not been established or have not been published.

Properties:

MW: 238.0	*VP*: 0.06 mmHg (77°F) (estimate)	*FlP*: —
D: 1.3480 g/mL	*VD*: 8.2 (calculated)	*LEL*: —
MP: —	*Vlt*: —	*UEL*: —
BP: 212°F (10 mmHg)	H_2O: "Sparingly soluble"	*RP*: —
Vsc: —	*Sol*: —	*IP*: —

C01-C112

***N*-[(Dichlorophosphinyl)oxy]-2-methylpropanimidoyl chloride**

CAS: 121951-56-2

RTECS: —

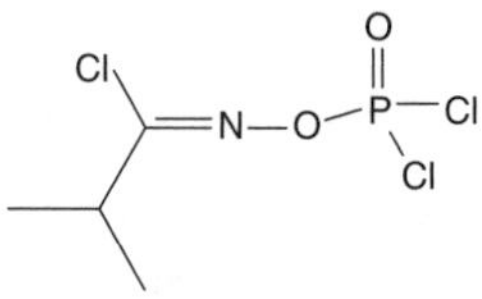

$C_4H_7Cl_3NO_2P$

Specific information on physical appearance is not available for this material. This material is hazardous through inhalation, skin absorption, penetration through broken skin, and ingestion, and produces local skin/eye impacts.

This material is a precursor for Novichok series agents.

Exposure Hazards

Conversion Factor: 1 ppm = 9.75 mg/m^3 at 77°F

Human toxicity values have not been established or have not been published.

Properties:

MW: 238.4	*VP*: 0.02 mmHg (77°F) (estimate)	*FlP*: —
D: —	*VD*: 8.2 (calculated)	*LEL*: —
MP: —	*Vlt*: —	*UEL*: —
BP: 154°F (2 mmHg)	H_2O: "Sparingly soluble"	*RP*: —
Vsc: —	*Sol*: —	*IP*: —

C01-C113

***N*-[(Dichlorophosphinyl)oxy]butanimidoyl chloride**

CAS: 121951-55-1

RTECS: —

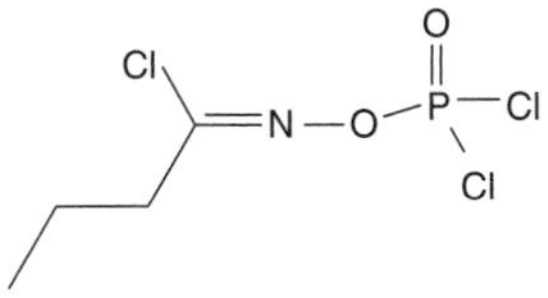

$C_4H_7Cl_3NO_2P$

Specific information on physical appearance is not available for this material. This material is hazardous through inhalation, skin absorption, penetration through broken skin, and ingestion, and produces local skin/eye impacts.

This material is a precursor for Novichok series agents.

Exposure Hazards

Conversion Factor: 1 ppm = 9.75 mg/m^3 at 77°F

Human toxicity values have not been established or have not been published.

Properties:

MW: 238.4	*VP*: 0.02 mmHg (77°F) (estimate)	*FlP*: —
D: —	*VD*: 8.2 (calculated)	*LEL*: —
MP: —	*Vlt*: —	*UEL*: —
BP: 180°F (3 mmHg)	H_2O: "Sparingly soluble"	*RP*: —
Vsc: —	*Sol*: —	*IP*: —

C01-C114

[[Chloroisopropylphosphinyl]oxy]carbonimidic dichloride

CAS: 17642-34-1

RTECS: —

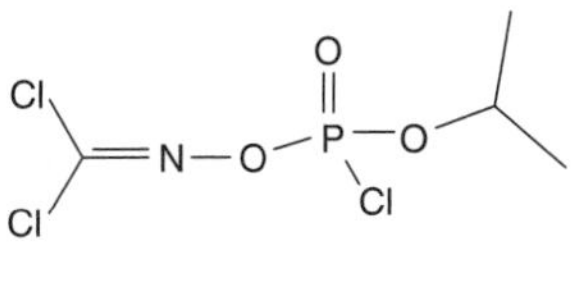

$C_4H_7Cl_3NO_3P$

Specific information on physical appearance is not available for this material. This material is hazardous through inhalation, skin absorption, penetration through broken skin, and ingestion, and produces local skin/eye impacts.

This material is a precursor for Novichok series agents.

Exposure Hazards

Conversion Factor: 1 ppm = 10.40 mg/m^3 at 77°F

Human toxicity values have not been established or have not been published.

Properties:

MW: 254.4	*VP*: 0.02 mmHg (77°F) (estimate)	*FlP*: —
D: 1.40 g/mL	*VD*: 8.8 (calculated)	*LEL*: —
MP: —	*Vlt*: —	*UEL*: —
BP: 194°F (2 mmHg)	H_2O: "Sparingly soluble"	*RP*: —
Vsc: —	*Sol*: —	*IP*: —

C01-C115

2-Chloro-*N*-[(chloroethoxyphosphinyl)oxy]ethanimidoyl chloride

CAS: 114700-92-4
RTECS: —

$C_4H_7Cl_3NO_3P$

Specific information on physical appearance is not available for this material. This material is hazardous through inhalation, skin absorption, penetration through broken skin, and ingestion, and produces local skin/eye impacts.

This material is a precursor for Novichok series agents.

Exposure Hazards

Conversion Factor: 1 ppm = 10.40 mg/m^3 at 77°F

Human toxicity values have not been established or have not been published.

Properties:

MW: 254.4	*VP*: 0.01 mmHg (77°F) (estimate)	*FlP*: —
D: —	*VD*: 8.8 (calculated)	*LEL*: —
MP: —	*Vlt*: —	*UEL*: —
BP: 253°F (1.5 mmHg)	H_2O: "Sparingly soluble"	*RP*: —
Vsc: —	*Sol*: —	*IP*: —

C01-C116

[(Chloropropoxyphosphinyl)oxy]carbonimidic dichloride

CAS: 37990-97-9
RTECS: —

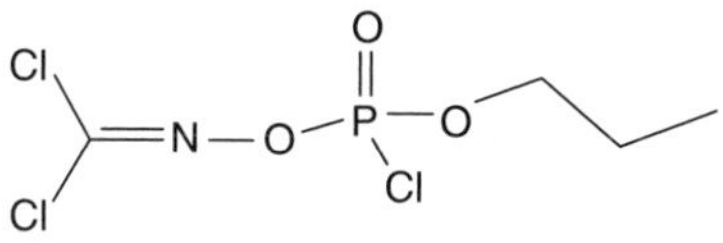

$C_4H_7Cl_3NO_3P$

Specific information on physical appearance is not available for this material. This material is hazardous through inhalation, skin absorption, penetration through broken skin, and ingestion, and produces local skin/eye impacts.

This material is a precursor for Novichok series agents.

Exposure Hazards

Conversion Factor: 1 ppm = 10.40 mg/m^3 at 77°F

Human toxicity values have not been established or have not been published.

Properties:

MW: 254.4	*VP*: 0.01 mmHg (77°F) (estimate)	*FlP*: —
D: —	*VD*: 8.8 (calculated)	*LEL*: —
MP: —	*Vlt*: —	*UEL*: —
BP: —	H_2O: "Sparingly soluble"	*RP*: —
Vsc: —	*Sol*: —	*IP*: —

C01-C117

[(Chloropropoxyphosphinyl)oxy]carbonimidic difluoride

CAS: 37990-84-4

RTECS: —

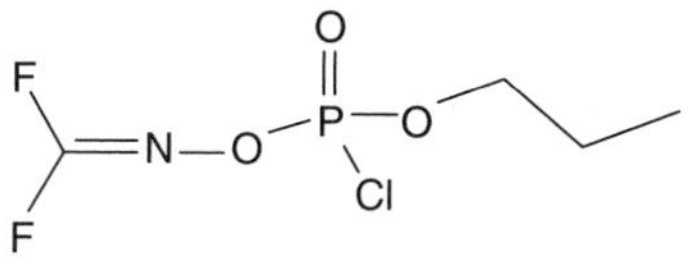

$C_4H_7ClF_2NO_3P$

Specific information on physical appearance is not available for this material. This material is hazardous through inhalation, skin absorption, penetration through broken skin, and ingestion, and produces local skin/eye impacts.

This material is a precursor for Novichok series agents.

Exposure Hazards

Conversion Factor: 1 ppm = 9.06 mg/m^3 at 77°F

Human toxicity values have not been established or have not been published.

Properties:

MW: 221.5	*VP*: 0.3 mmHg (77°F) (estimate)	*FlP*: —
D: —	*VD*: 7.6 (calculated)	*LEL*: —
MP: —	*Vlt*: —	*UEL*: —
BP: —	H_2O: "Sparingly soluble"	*RP*: —
Vsc: —	*Sol*: —	*IP*: —

C01-C118

[(Chloroisopropoxyphosphinyl)oxy]carbonimidic difluoride
CAS: 18262-32-3
RTECS: —

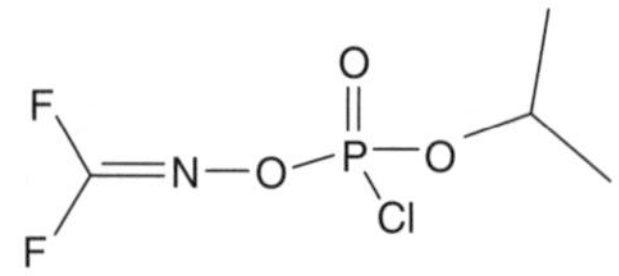

$C_4H_7ClF_2NO_3P$

Specific information on physical appearance is not available for this material. This material is hazardous through inhalation, skin absorption, penetration through broken skin, and ingestion, and produces local skin/eye impacts.

This material is a precursor for Novichok series agents.

Exposure Hazards
Conversion Factor: 1 ppm = 9.06 mg/m^3 at 77°F

Human toxicity values have not been established or have not been published.

Properties:

MW: 221.5	*VP*: 0.4 mmHg (77°F) (estimate)	*FlP*: —
D: 1.3720 g/mL	*VD*: 7.6 (calculated)	*LEL*: —
MP: —	*Vlt*: —	*UEL*: —
BP: 142°F (2 mmHg)	H_2O: "Sparingly soluble"	*RP*: —
Vsc: —	*Sol*: —	*IP*: —

C01-C119

***N*-[(Chloroethoxyphosphinyl)oxy]ethanimidoyl chloride**
CAS: 114700-90-2
RTECS: —

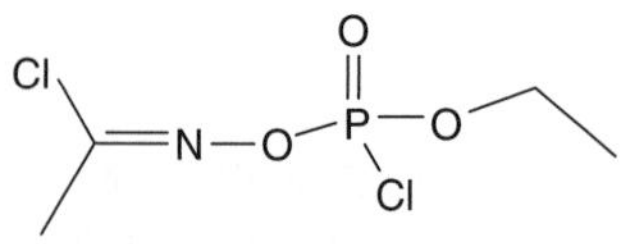

$C_4H_8Cl_2NO_3P$

Specific information on physical appearance is not available for this material. This material is hazardous through inhalation, skin absorption, penetration through broken skin, and ingestion, and produces local skin/eye impacts.

This material is a precursor for Novichok series agents.

Exposure Hazards
Conversion Factor: 1 ppm = 9.00 mg/m^3 at 77°F

Human toxicity values have not been established or have not been published.

Properties:

MW: 220.0	*VP*: 0.08 mmHg (77°F) (estimate)	*FlP*: —
D: —	*VD*: 7.6 (calculated)	*LEL*: —
MP: —	*Vlt*: —	*UEL*: —

BP: 205°F (3 mmHg)	H_2O: "Sparingly soluble"	*RP*: —
Vsc: —	*Sol*: —	*IP*: —

C01-C120

***N*-[(Chloropropoxyphosphinyl)oxy]ethanimidoyl chloride**

CAS: 129003-90-3

RTECS: —

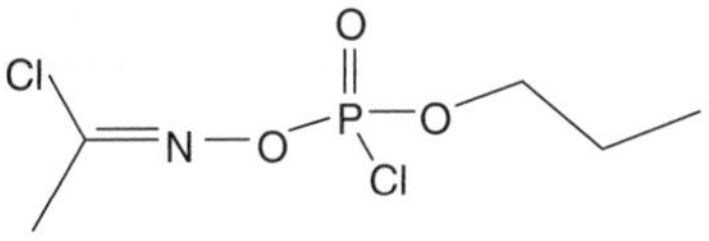

$C_5H_{10}Cl_2NO_3P$

Specific information on physical appearance is not available for this material. This material is hazardous through inhalation, skin absorption, penetration through broken skin, and ingestion, and produces local skin/eye impacts.

This material is a precursor for Novichok series agents.

Exposure Hazards

Conversion Factor: 1 ppm = 9.57 mg/m^3 at 77°F

Human toxicity values have not been established or have not been published.

Properties:

MW: 234.0	*VP*: 0.03 mmHg (77°F) (estimate)	*FlP*: —
D: —	*VD*: 8.1 (calculated)	*LEL*: —
MP: —	*Vlt*: —	*UEL*: —
BP: 221°F (2 mmHg)	H_2O: "Sparingly soluble"	*RP*: —
Vsc: —	*Sol*: —	*IP*: —

C01-C121

***N*-[[Chloro(1-methylethoxy)phosphinyl]oxy]ethanimidoyl chloride**

CAS: 128981-16-8

RTECS: —

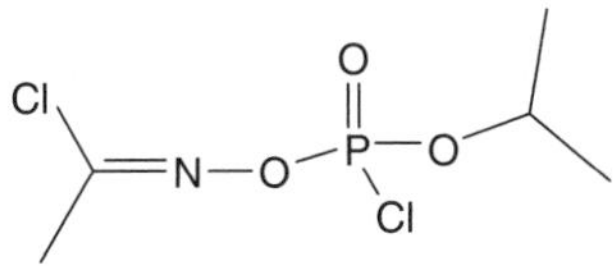

$C_5H_{10}Cl_2NO_3P$

Specific information on physical appearance is not available for this material. This material is hazardous through inhalation, skin absorption, penetration through broken skin, and ingestion, and produces local skin/eye impacts.

This material is a precursor for Novichok series agents.

Exposure Hazards

Conversion Factor: 1 ppm = 9.57 mg/m^3 at 77°F

Human toxicity values have not been established or have not been published.

Properties:

MW: 234.0	*VP*: 0.04 mmHg (77°F) (estimate)	*FlP*: —
D: —	*VD*: 8.1 (calculated)	*LEL*: —
MP: —	*Vlt*: —	*UEL*: —
BP: 210°F (2 mmHg)	H_2O: "Sparingly soluble"	*RP*: —
Vsc: —	*Sol*: —	*IP*: —

C01-C122

***N*-[(Chloroethoxyphosphinyl)oxy]propanimidoyl chloride**

CAS: 128981-18-0

RTECS: —

$C_5H_{10}Cl_2NO_3P$

Specific information on physical appearance is not available for this material. This material is hazardous through inhalation, skin absorption, penetration through broken skin, and ingestion, and produces local skin/eye impacts.

This material is a precursor for Novichok series agents.

Exposure Hazards

Conversion Factor: 1 ppm = 9.57 mg/m^3 at 77°F

Human toxicity values have not been established or have not been published.

Properties:

MW: 234.0	*VP*: 0.03 mmHg (77°F) (estimate)	*FlP*: —
D: —	*VD*: 8.1 (calculated)	*LEL*: —
MP: —	*Vlt*: —	*UEL*: —
BP: 219°F (2.5 mmHg)	H_2O: "Sparingly soluble"	*RP*: —
Vsc: —	*Sol*: —	*IP*: —

C01-C123

2-Chloro-*N*-[[chloro(2,2,3,3-tetrafluoropropoxy)phosphinyl]oxy]ethanimidoyl chloride

CAS: 114967-39-4

RTECS: —

$C_5H_5Cl_3F_4NO_3P$

Specific information on physical appearance is not available for this material. This material is hazardous through inhalation, skin absorption, penetration through broken skin, and ingestion, and produces local skin/eye impacts.

This material is a precursor for Novichok series agents.

Exposure Hazards

Conversion Factor: 1 ppm = 13.92 mg/m^3 at 77°F

Human toxicity values have not been established or have not been published.

Properties:

MW: 340.4	*VP*: 0.003 mmHg (77°F) (estimate)	*FlP*: —
D: —	*VD*: 12 (calculated)	*LEL*: —
MP: —	*Vlt*: —	*UEL*: —
BP: 244°F (3 mmHg)	H_2O: "Sparingly soluble"	*RP*: —
Vsc: —	*Sol*: —	*IP*: —

C01-C124

[[2-Chloro-1-methylpropylphosphinyl]oxy]carbonimidic chloride fluoride

CAS: 24946-19-8

RTECS: —

$C_5H_8Cl_3FNO_3P$

Specific information on physical appearance is not available for this material. This material is hazardous through inhalation, skin absorption, penetration through broken skin, and ingestion, and produces local skin/eye impacts.

This material is a precursor for Novichok series agents.

Exposure Hazards

Conversion Factor: 1 ppm = 11.72 mg/m^3 at 77°F

Human toxicity values have not been established or have not been published.

Properties:

MW: 286.5	*VP*: 0.004 mmHg (77°F) (estimate)	*FlP*: —
D: 1.4811 g/mL	*VD*: 9.9 (calculated)	*LEL*: —
MP: —	*Vlt*: —	*UEL*: —
BP: 183°F (0.3 mmHg)	H_2O: "Sparingly soluble"	*RP*: —
Vsc: —	*Sol*: —	*IP*: —

C01-C125

***N*-[[Chloro(1-methylethoxy)phosphinyl]oxy]propanimidoyl chloride**

CAS: 128981-20-4

RTECS: —

$C_6H_{12}Cl_2NO_3P$

Specific information on physical appearance is not available for this material. This material is hazardous through inhalation, skin absorption, penetration through broken skin, and ingestion, and produces local skin/eye impacts.

This material is a precursor for Novichok series agents.

Exposure Hazards

Conversion Factor: 1 ppm = 10.14 mg/m^3 at 77°F

Human toxicity values have not been established or have not been published.

Properties:

MW: 248.0	*VP*: 0.02 mmHg (77°F) (estimate)	*FlP*: —
D: —	*VD*: 8.6 (calculated)	*LEL*: —
MP: —	*Vlt*: —	*UEL*: —
BP: 207°F (2 mmHg)	H_2O: "Sparingly soluble"	*RP*: —
Vsc: —	*Sol*: —	*IP*: —

C01-C126

N-[[Chloro(2-methylpropoxy)phosphinyl]oxy]ethanimidoyl chloride

CAS: 128981-17-9

RTECS: —

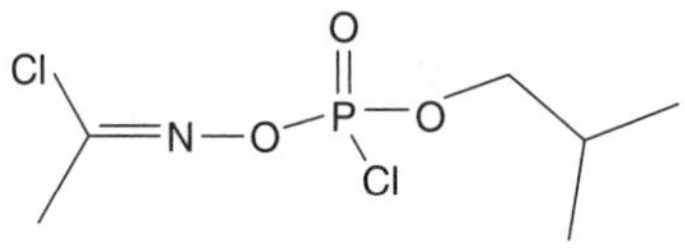

$C_6H_{12}Cl_2NO_3P$

Specific information on physical appearance is not available for this material. This material is hazardous through inhalation, skin absorption, penetration through broken skin, and ingestion, and produces local skin/eye impacts.

This material is a precursor for Novichok series agents.

Exposure Hazards

Conversion Factor: 1 ppm = 10.14 mg/m^3 at 77°F

Human toxicity values have not been established or have not been published.

Properties:

MW: 248.0	*VP*: 0.02 mmHg (77°F) (estimate)	*FlP*: —
D: —	*VD*: 8.6 (calculated)	*LEL*: —
MP: —	*Vlt*: —	*UEL*: —
BP: 226°F (2 mmHg)	H_2O: "Sparingly soluble"	*RP*: —
Vsc: —	*Sol*: —	*IP*: —

C01-C127

N-[(Chloropropoxyphosphinyl)oxy]propanimidoyl chloride

CAS: 128981-19-1

RTECS: —

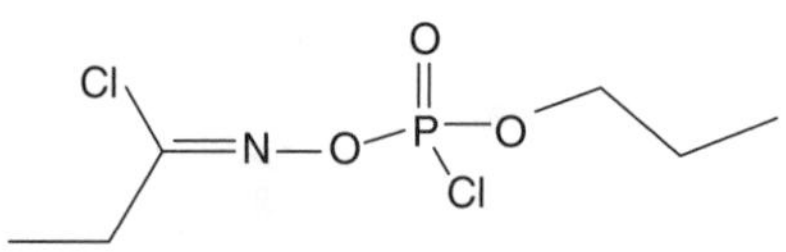

$C_6H_{12}Cl_2NO_3P$

Specific information on physical appearance is not available for this material. This material is hazardous through inhalation, skin absorption, penetration through broken skin, and ingestion, and produces local skin/eye impacts.

This material is a precursor for Novichok series agents.

Exposure Hazards

Conversion Factor: 1 ppm = 10.14 mg/m^3 at 77°F

Human toxicity values have not been established or have not been published.

Properties:

MW: 248.0	*VP*: 0.01 mmHg (77°F) (estimate)	*FlP*: —
D: —	*VD*: 8.6 (calculated)	*LEL*: —
MP: —	*Vlt*: —	*UEL*: —
BP: 216°F (2 mmHg)	H_2O: "Sparingly soluble"	*RP*: —
Vsc: —	*Sol*: —	*IP*: —

C01-C128

***N*-[(Chloroethoxyphosphinyl)oxy]-2-methylpropanimidoyl chloride**

CAS: 128981-25-9

RTECS: —

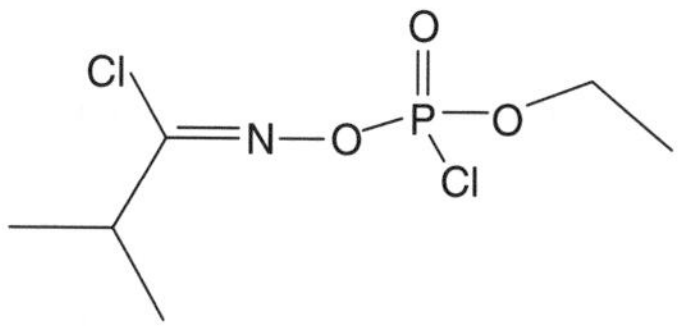

$C_6H_{12}Cl_2NO_3P$

Specific information on physical appearance is not available for this material. This material is hazardous through inhalation, skin absorption, penetration through broken skin, and ingestion, and produces local skin/eye impacts.

This material is a precursor for Novichok series agents.

Exposure Hazards

Conversion Factor: 1 ppm = 10.14 mg/m^3 at 77°F

Human toxicity values have not been established or have not been published.

Properties:

MW: 248.0	*VP*: 0.02 mmHg (77°F) (estimate)	*FlP*: —
D: —	*VD*: 8.6 (calculated)	*LEL*: —
MP: —	*Vlt*: —	*UEL*: —
BP: 207°F (2 mmHg)	H_2O: "Sparingly soluble"	*RP*: —
Vsc: —	*Sol*: —	*IP*: —

C01-C129

***N*-[(Chloropropoxyphosphinyl)oxy]butanimidoyl chloride**

CAS: 128981-22-6

RTECS: —

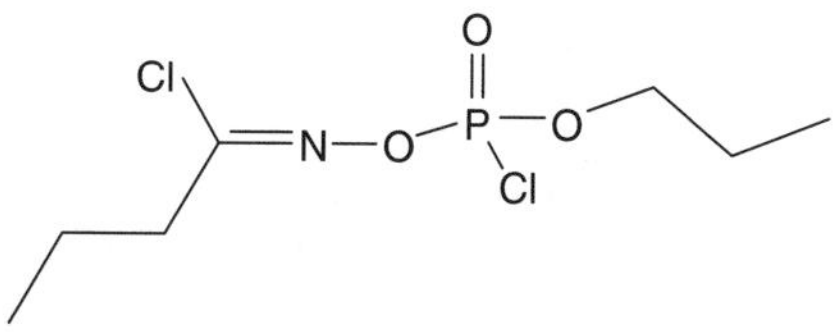

$C_7H_{14}Cl_2NO_3P$

Specific information on physical appearance is not available for this material. This material is hazardous through inhalation, skin absorption, penetration through broken skin, and ingestion, and produces local skin/eye impacts.

This material is a precursor for Novichok series agents.

Exposure Hazards

Conversion Factor: 1 ppm = 10.72 mg/m^3 at 77°F

Human toxicity values have not been established or have not been published.

Properties:

MW: 262.1	*VP*: 0.004 mmHg (77°F) (estimate)	*FlP*: —
D: —	*VD*: 9.0 (calculated)	*LEL*: —
MP: —	*Vlt*: —	*UEL*: —
BP: 226°F (2 mmHg)	H_2O: "Sparingly soluble"	*RP*: —
Vsc: —	*Sol*: —	*IP*: —

C01-C130

***N*-[[Chloro(2-methylpropoxy)phosphinyl]oxy]propanimidoyl chloride**

CAS: 128981-21-5

RTECS: —

$C_7H_{14}Cl_2NO_3P$

Specific information on physical appearance is not available for this material. This material is hazardous through inhalation, skin absorption, penetration through broken skin, and ingestion, and produces local skin/eye impacts.

This material is a precursor for Novichok series agents.

Exposure Hazards

Conversion Factor: 1 ppm = 10.72 mg/m^3 at 77°F

Human toxicity values have not been established or have not been published.

Properties:

MW: 262.1	*VP*: 0.006 mmHg (77°F) (estimate)	*FlP*: —
D: —	*VD*: 9.0 (calculated)	*LEL*: —
MP: —	*Vlt*: —	*UEL*: —
BP: 228°F (2 mmHg)	H_2O: "Sparingly soluble"	*RP*: —
Vsc: —	*Sol*: —	*IP*: —

C01-C131

***N*-[[Chloro(1-methylethoxy)phosphinyl]oxy]butanimidoyl chloride**

CAS: 128981-23-7

RTECS: —

$C_7H_{14}Cl_2NO_3P$

Specific information on physical appearance is not available for this material. This material is hazardous through inhalation, skin absorption, penetration through broken skin, and ingestion, and produces local skin/eye impacts.

This material is a precursor for Novichok series agents.

Exposure Hazards

Conversion Factor: 1 ppm = 10.72 mg/m^3 at 77°F

Human toxicity values have not been established or have not been published.

Properties:

MW: 262.1	*VP*: 0.006 mmHg (77°F) (estimate)	*FlP*: —
D: —	*VD*: 9.0 (calculated)	*LEL*: —
MP: —	*Vlt*: —	*UEL*: —
BP: 221°F (2 mmHg)	H_2O: "Sparingly soluble"	*RP*: —
Vsc: —	*Sol*: —	*IP*: —

C01-C132

***N*-[[Chloro(1-methylethoxy)phosphinyl]oxy]-2-methylpropanimidoyl chloride**

CAS: 128981-27-1

RTECS: —

$C_7H_{14}Cl_2NO_3P$

Specific information on physical appearance is not available for this material. This material is hazardous through inhalation, skin absorption, penetration through broken skin, and ingestion, and produces local skin/eye impacts.

This material is a precursor for Novichok series agents.

Exposure Hazards

Conversion Factor: 1 ppm = 10.72 mg/m^3 at 77°F

Human toxicity values have not been established or have not been published.

Properties:

MW: 262.1	*VP*: 0.008 mmHg (77°F) (estimate)	*FlP*: —
D: —	*VD*: 9.0 (calculated)	*LEL*: —
MP: —	*Vlt*: —	*UEL*: —
BP: 214°F (2 mmHg)	H_2O: "Sparingly soluble"	*RP*: —
Vsc: —	*Sol*: —	*IP*: —

C01-C133

***N*-[(Chloropropoxyphosphinyl)oxy]-2-methylpropanimidoyl chloride**

CAS: 128981-26-0

RTECS: —

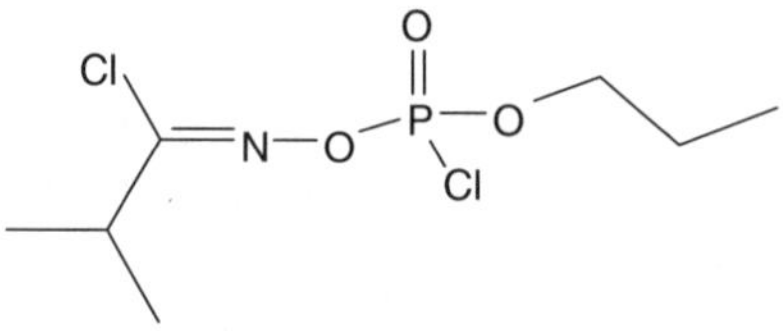

$C_7H_{14}Cl_2NO_3P$

Specific information on physical appearance is not available for this material. This material is hazardous through inhalation, skin absorption, penetration through broken skin, and ingestion, and produces local skin/eye impacts.

This material is a precursor for Novichok series agents.

Exposure Hazards

Conversion Factor: 1 ppm = 10.72 mg/m^3 at 77°F

Human toxicity values have not been established or have not been published.

Properties:

MW: 262.1	*VP*: 0.006 mmHg (77°F) (estimate)	*FlP*: —
D: —	*VD*: 9.0 (calculated)	*LEL*: —
MP: —	*Vlt*: —	*UEL*: —
BP: 225°F (2 mmHg)	H_2O: "Sparingly soluble"	*RP*: —
Vsc: —	*Sol*: —	*IP*: —

C01-C134

***N*-[[Chloro(2-methylpropoxy)phosphinyl]oxy]butanimidoyl chloride**

CAS: 128981-24-8

RTECS: —

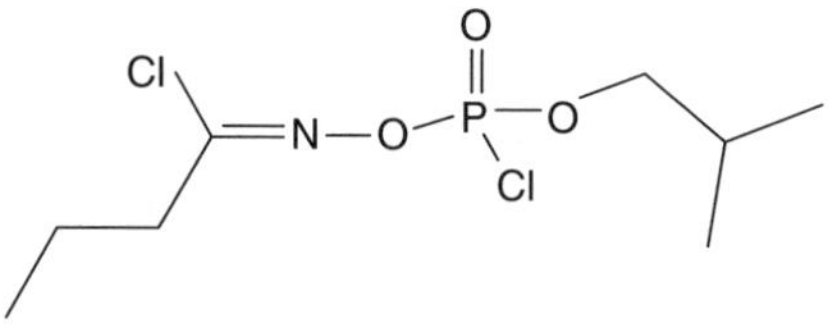

$C_8H_{16}Cl_2NO_3P$

Specific information on physical appearance is not available for this material. This material is hazardous through inhalation, skin absorption, penetration through broken skin, and ingestion, and produces local skin/eye impacts.

This material is a precursor for Novichok series agents.

Exposure Hazards

Conversion Factor: 1 ppm = 11.29 mg/m^3 at 77°F

Human toxicity values have not been established or have not been published.

Properties:

MW: 276.1	*VP*: 0.002 mmHg (77°F) (estimate)	*FlP*: —
D: —	*VD*: 9.5 (calculated)	*LEL*: —
MP: —	*Vlt*: —	*UEL*: —
BP: 226°F (1.5 mmHg)	H_2O: "Sparingly soluble"	*RP*: —
Vsc: —	*Sol*: —	*IP*: —

C01-C135

***N*-[[Chloro(2-methylpropoxy)phosphinyl]oxy]-2-methylpropanimidoyl chloride**
CAS: 128981-28-2
RTECS: —

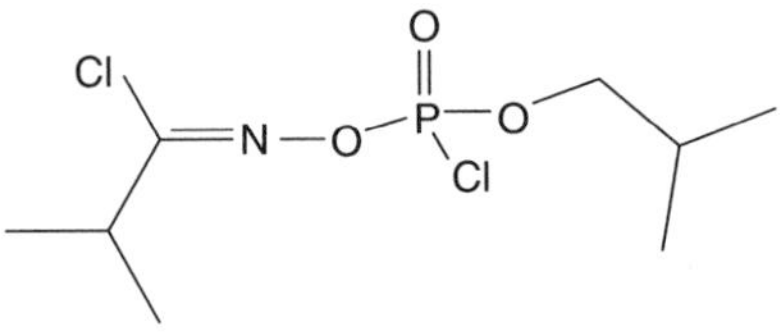

$C_8H_{16}Cl_2NO_3P$

Specific information on physical appearance is not available for this material. This material is hazardous through inhalation, skin absorption, penetration through broken skin, and ingestion, and produces local skin/eye impacts.

This material is a precursor for Novichok series agents.

Exposure Hazards
Conversion Factor: 1 ppm = 11.29 mg/m^3 at 77°F

Human toxicity values have not been established or have not been published.

Properties:

MW: 276.1	*VP*: 0.003 mmHg (77°F) (estimate)	*FlP*: —
D: —	*VD*: 9.5 (calculated)	*LEL*: —
MP: —	*Vlt*: —	*UEL*: —
BP: 228°F (2 mmHg)	H_2O: "Sparingly soluble"	*RP*: —
Vsc: —	*Sol*: —	*IP*: —

C01-D
DECOMPOSITION PRODUCTS AND IMPURITIES

C01-D136

Bis(2-diisopropylaminoethyl)sulfide ([DE2]S)
CAS: 110501-56-9
RTECS: —

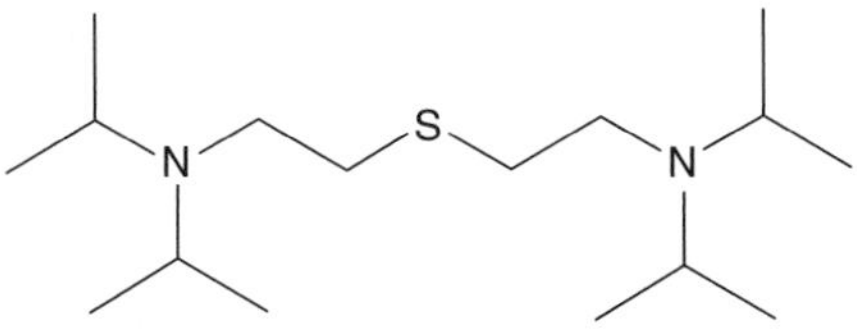

$C_{16}H_{36}N_2S$

Specific information on physical appearance is not available for this material.

This material is a degradation product from hydrolysis of V-series nerve agents.

Exposure Hazards
Human toxicity values have not been established or have not been published.

Properties:

MW: 288.5	*VP*: 0.00000027 mmHg	*FlP*: —
D: —	*VD*: 10 (calculated)	*LEL*: —
MP: —	*Vlt*: —	*UEL*: —
BP: —	*H_2O*: 0.00012%	*RP*: —
Vsc: —	*Sol*: —	*IP*: —

C01-D137

Bis[2-(diisopropylamino)ethyl] disulfide (Agent EA 4196)
CAS: 65332-44-7
RTECS: —

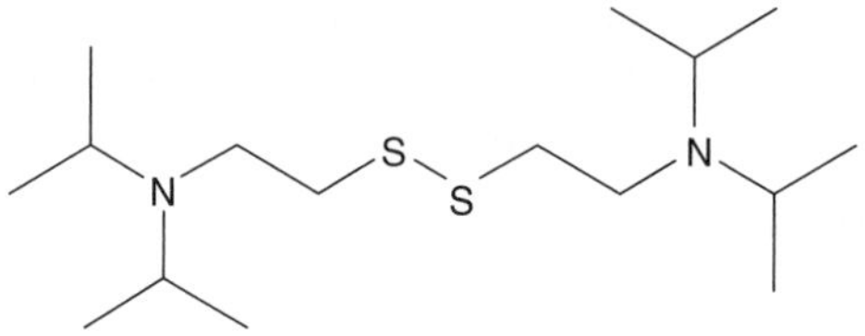

$C_{16}H_{36}N_2S_2$

Specific information on physical appearance is not available for this material.

This material is a degradation product from hydrolysis of V-series nerve agents.

Exposure Hazards
Human toxicity values have not been established or have not been published.

Properties:

MW: 320.6	*VP*: 0.0000000059 mmHg	*FlP*: —
D: —	*VD*: 11 (calculated)	*LEL*: —
MP: —	*Vlt*: —	*UEL*: —
BP: —	*H_2O*: 0.00095%	*RP*: —
Vsc: —	*Sol*: —	*IP*: —

C01-D138

Ethyl methylphosphonic acid (EMPA)
CAS: 1832-53-7
RTECS: —

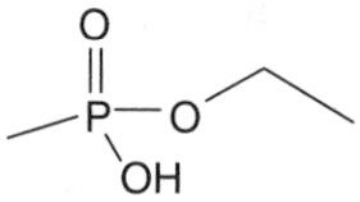

$C_3H_9O_3P$

Specific information on physical appearance is not available for this material. Various salts (solids) have been reported.

This material is a degradation product from hydrolysis of VX (C01-A016) and Ethyl Sarin (C01-A005).

Exposure Hazards
Human toxicity values have not been established or have not been published.

Properties:

MW: 124.1	*VP*: 0.00036 mmHg	*FlP*: —
D: —	*VD*: 4.3 (calculated)	*LEL*: —
MP: —	*Vlt*: —	*UEL*: —
BP: 226°F (0.4 mmHg)	H_2O: 18%	*RP*: —
Vsc: —	*Sol*: —	*IP*: —

C01-D139

***O,S*-Diethyl methylphosphonothioate** (Agent EA 5533)

CAS: 2511-10-6

RTECS: —

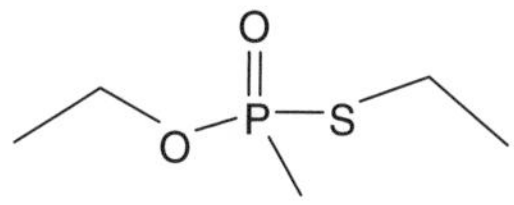

$C_5H_{13}O_2PS$

Specific information on physical appearance is not available for this material.

This material is commonly found as an impurity and degradation product from hydrolysis of VX (C01-A016). This material has been used as a simulant in government tests.

Exposure Hazards

Conversion Factor: 1 ppm = 6.88 mg/m^3 at 77°F

Human toxicity values have not been established or have not been published.

Properties:

MW: 168.2	*VP*: 0.081 mmHg (77°F)	*FlP*: —
D: —	*VD*: 5.8 (calculated)	*LEL*: —
MP: 17°F	*Vlt*: 100 ppm (77°F)	*UEL*: —
BP: 445°F	H_2O: 1.55%	*RP*: —
Vsc: —	*Sol*: —	*IP*: —

C01-D140

***N,N′*-Dicyclohexylurea**

CAS: 2387-23-7

RTECS: —

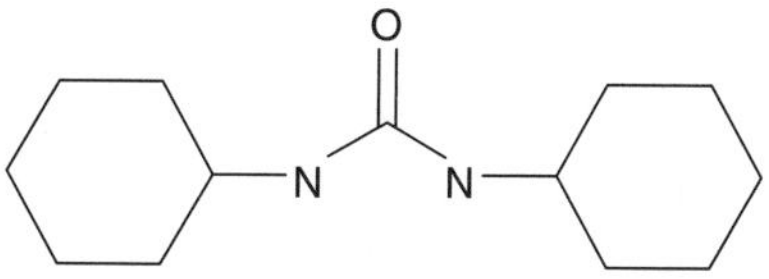

$C_{13}H_{24}N_2O$

White crystalline powder.

This material is commonly found as a degradation product from hydrolysis of VX stabilizer Dicyclohexyl carbodiimide (C01-C049).

Exposure Hazards

Human toxicity values have not been established or have not been published.

Properties:

MW: 224.3	*VP*: —	*FlP*: —
D: 1.34 g/cm^3	*VD*: —	*LEL*: —
MP: 448°F	*Vlt*: —	*UEL*: —
BP: —	H_2O: —	*RP*: —
Vsc: —	*Sol*: —	*IP*: —

C01-D141

***N*-Chloroisopropylamine**

CAS: 26245-56-7

RTECS: —

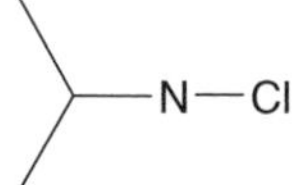

C_3H_8ClN

Specific information on physical appearance is not available for this material.

This material is a degradation product from the reaction of VX (C01-A016) with hypochlorites.

Exposure Hazards

Human toxicity values have not been established or have not been published.

Properties:

MW: 93.6	*VP*: —	*FlP*: —
D: 1.019 g/mL (77°F)	*VD*: —	*LEL*: —
MP: —	*Vlt*: —	*UEL*: —
BP: 99°F (100 mmHg)	H_2O: —	*RP*: —
Vsc: —	*Sol*: —	*IP*: —

C01-D142

Diethyl hydrogen phosphate

CAS: 598-02-7

RTECS: TC0665000

$C_4H_{11}O_4P$

Clear, colorless liquid.

This material is commonly found as a degradation product from hydrolysis of Tabun (C01-A001).

Exposure Hazards

Conversion Factor: 1 ppm = 6.30 mg/m^3 at 77°F

Human toxicity values have not been established or have not been published.

Properties:

MW: 154.1	*VP*: —	*FlP*: 196° F
D: 1.29 g/mL	*VD*: —	*LEL*: —
MP: 43°F	*Vlt*: —	*UEL*: —
BP: 397°F	H_2O: "Slightly soluble"	*RP*: —
BP: 212°F (0.1 mmHg)	*Sol*: —	*IP*: —
Vsc: —		

C01-D143

Diisopropylamine

CAS: 108-18-9; 819-79-4 (Hydrochloride Salt); 30321-74-5 (Hydrobromide Salt); 6143-52-8 (Nitrate Salt); 65087-26-5 (Sulfate Salt)

RTECS: IM4025000

UN: 1158

ERG: 132

$C_6H_{15}N$

Colorless liquid with an odor like ammonia or fish. Various salts (solids) have been reported. This material is hazardous through inhalation, skin absorption, penetration through broken skin, and ingestion, and produces local skin/eye impacts. Concentrations between 25 and 50 ppm can cause vision disturbances (e.g., colored haloes to be seen around lights).

Used industrially for organic synthesis; as an antifoam agent, as a stabilizer for mesityl oxide; and as a chemical intermediate for detergents; dyes; pesticides; and pharmaceuticals.

This material is on the Australia Group Export Control list.

This is commonly found as an impurity in VX (C01-A016) and from decomposition of VX with hypochlorite.

Exposure Hazards

Conversion Factor: 1 ppm = 4.14 mg/m^3 at 77°F

OSHA PEL: 5 ppm [Skin]

ACGIH TLV: 5 ppm [Skin]

IDLH: 200 ppm

Properties:

MW: 101.2	*VP*: 70 mmHg	*FlP*: 30°F
D: 0.7169 g/mL	*VP*: 79.4 mmHg (77°F)	*LEL*: 1.1%
MP: –78°F	*VD*: 3.5 (calculated)	*UEL*: 7.1%
BP: 183°F	*Vlt*: 100,000 ppm (77°F)	*RP*: 0.13
Vsc: 0.558 cS (77°F)	H_2O: Miscible	*IP*: 7.73 eV
Sol: Acetone; Benzene; Ether; Ethanol		

C01-D144

Triethyl phosphate

CAS: 78-40-0

RTECS: —

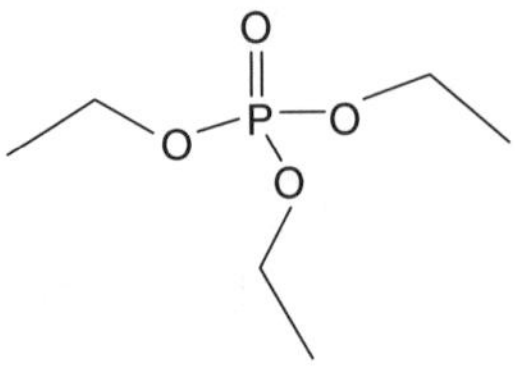

$C_6H_{15}O_4P$

Clear, colorless liquid with a mild odor. This material is hazardous through inhalation and ingestion, and produces local skin/eye impacts.

Used industrially as a chemical intermediate for organophosphorus insecticides, as a catalyst in the production of acetic anhydride by the ketene process; and used in industry as a plasticizer, fire-retarding agent, antifoaming agent, and desensitizing agent for peroxides.

This material is commonly found as degradation product of Tabun (C01-A001). This material has been used as a simulant in government tests.

Exposure Hazards

Conversion Factor: 1 ppm = 7.45 mg/m^3 at 77°F

Human toxicity values have not been established or have not been published.

Properties:

MW: 182.2	*VP*: 0.39 mmHg (77°F)	*FlP*: 240°F
D: 1.0689 g/mL	*VD*: 6.3 (calculated)	*LEL*: —
D: 1.0725 g/mL (66°F)	*Vlt*: 510 ppm (77°F)	*UEL*: —
MP: −70°F	H_2O: 50%	*RP*: 19
BP: 419°F	*Sol*: Most organic solvents	*IP*: ~10 eV
Vsc: —		

C01-D145

***N,N*-Diisopropylethylamine**

CAS: 7087-68-5

RTECS: —

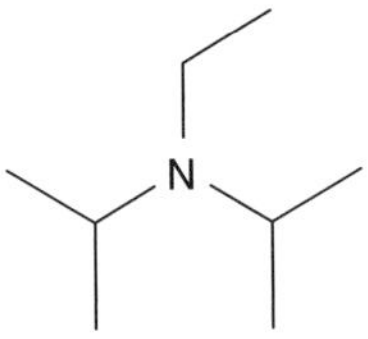

$C_8H_{19}N$

Clear colorless to yellow liquid with an amine odor. Various salts (solids) have been reported. This material is hazardous through inhalation and ingestion, and produces local skin/eye impacts.

Used as a reagent in organic chemistry.

This material is commonly found as an impurity and degradation product in VX (C01-A016).

Exposure Hazards

Conversion Factor: 1 ppm = 5.28 mg/m^3 at 77°F

Human toxicity values have not been established or have not been published.

Properties:

MW: 129.2	*VP*: 31 mmHg (100°F)	*FlP*: 43° F
D: 0.782 g/mL	*VD*: 4.5 (calculated)	*LEL*: 3%
MP: –51°F	*Vlt*: 39,000 ppm (100°F)	*UEL*: 17%
BP: 261°F	*H_2O*: Miscible	*RP*: 0.30
Vsc: —	*Sol*: —	*IP*: —

C01-D146

Hydrogen *S*-2-diisopropylaminoethyl methylphosphonothiolate (Agent EA 2192)
CAS: 73207-98-4
RTECS: —

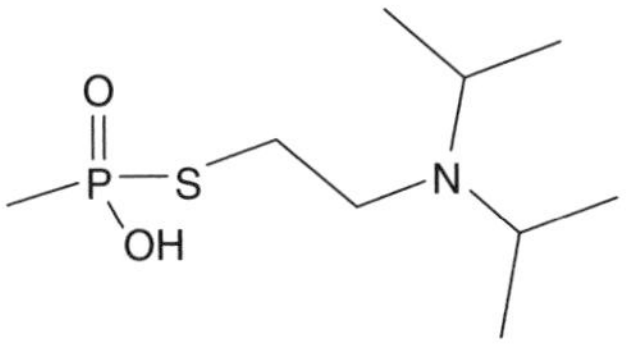

$C_9H_{22}NO_2PS$

Specific information on physical appearance is not available for this material. This material is hazardous through inhalation, skin absorption, penetration through broken skin, and ingestion.

This material is on Schedule 1 of the CWC.

This material is a degradation product from hydrolysis of VX (C01-A016), a V-series nerve agent.

Exposure Hazards

Conversion Factor: 1 ppm = 9.79 mg/m^3 at 77°F

Human toxicity values have not been established or have not been published. However, based on available information, this material appears to have nearly the same toxicity as the parent compound.

Properties:

MW: 239.3	*VP*: —	*FlP*: —
D: —	*VD*: 8.3 (calculated)	*LEL*: —
MP: —	*Vlt*: —	*UEL*: —
BP: —	*H_2O*: Miscible	*RP*: —
Vsc: —	*Sol*: —	*IP*: —

C01-D147

Hydrogen *S*-2-dimethylaminoethyl methylphosphonothiolate
CAS: 34256-71-8
RTECS: —

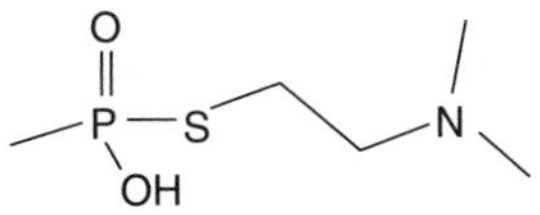

$C_5H_{14}NO_2PS$

This material is hazardous through inhalation, skin absorption, penetration through broken skin, and ingestion.

Specific information on physical appearance is not available for this material.

This material is on Schedule 1 of the CWC.

This material is a degradation product from hydrolysis of VX (C01-A017), a V-series nerve agent.

Exposure Hazards

Conversion Factor: 1 ppm = 7.49 mg/m^3 at 77°F

Human toxicity values have not been established or have not been published. However, based on available information, this material appears to have nearly the same toxicity as the parent compound.

Properties:

MW: 183.2	*VP*: —	*FlP*: —
D: —	*VD*: 6.3 (calculated)	*LEL*: —
MP: —	*Vlt*: —	*UEL*: —
BP: —	H_2O: Soluble	*RP*: —
Vsc: —	*Sol*: —	*IP*: —

C01-D148

Hydrogen *S*-2-dimethylaminoethyl ethylphosphonothiolate

CAS: —

RTECS: —

$C_6H_{16}NO_2PS$

Specific information on physical appearance is not available for this material. This material is hazardous through inhalation, skin absorption, penetration through broken skin, and ingestion.

This material is on Schedule 1 of the CWC.

This material is a degradation product from hydrolysis of ethyl *S*-[2-(dimethylamino)ethyl] ethylphosphonothiolate (C01-A021), a V-series nerve agent.

Exposure Hazards

Conversion Factor: 1 ppm = 8.07 mg/m^3 at 77°F

Human toxicity values have not been established or have not been published. However, based on available information, this material appears to have nearly the same toxicity as the parent compound.

Properties:

MW: 197.2	*VP*: —	*FlP*: —
D: —	*VD*: 6.8 (calculated)	*LEL*: —
MP: —	*Vlt*: —	*UEL*: —
BP: —	H_2O: Soluble	*RP*: —
Vsc: —	*Sol*: —	*IP*: —

References

Agency for Toxic Substances and Disease Registry. "Diisopropyl Methylphosphonate ToxFAQs." August 1999.

———. "Fluorides, Hydrogen Fluoride, and Fluorine ToxFAQs." September 2003.

———. *Managing Hazardous Materials Incidents Volume III—Medical Management Guidelines for Acute Chemical Exposures.* Rev Edn. Washington, DC: Government Printing Office, 2000.

———. "Nerve Agents (GA, GB, GD, VX) ToxFAQs." April 2002.

———. *Toxicological Profile for Diisopropyl Methylphosphonate.* Washington, DC: Government Printing Office, August 1998.

———. *Toxicological Profile for Fluorides, Hydrogen Fluoride, and Fluorine.* Washington, DC: Government Printing Office, 2003.

Bajgar, J., J. Fusek, V. Hrdina, J. Patocka, and J. Vachek (Purkyne Military Medical Academy, Prague, Czech Republic). "Acute toxicities of 2-dialkylaminoalkyl-(dialyklamido)-fluorophosphates." *Physiological Research* 41 (1992): 399–402.

Brown, D.F., A.J. Policastro, W.E. Dunn, R.A. Carhart, M.A. Lazaro, W.A. Freeman, and M. Krumpolc. *Development of the Table of Initial Isolation and Protective Action Distances for the 2000 Emergency Response Guidebook,* Argonne National Laboratory Report No. ANL/DIS-00-1. October 2000.

Buckles, L.C. and S.M. Lewis. "S-(2-Diisopropylamino-ethyl)O-ethyl methylphosphonothioate Stabilized with Soluble Carbodiimides," United States Patent 4012464, September 24, 1965.

Centers for Disease Control and Prevention. "Case Definition: Nerve Agents or Organophosphates." March 15, 2005.

———. "Counter Terrorism Card for GF." 2000.

———. "Counter Terrorism Card for Sarin." 2000.

———. "Counter Terrorism Card for Soman." 2000.

———. "Counter Terrorism Card for Tabun." 2000.

———. "Counter Terrorism Card for VX." 2000.

———. "Facts About Sarin." March 7, 2003.

———. "Facts About Soman." March 7, 2003.

———. "Facts About Tabun." March 7, 2003.

———. "Facts About VX." March 7, 2003.

Cohen, L., P.B. Coulter, and B.M. Zeffert. "Thickened Phosphorus Esters," United States Patent 3868446, February 25, 1975.

Compton, J.A.F. *Military Chemical and Biological Agents: Chemical and Toxicological Properties.* Caldwell, NJ: The Telford Press, 1987.

Cooper, G.H., I.W. Lawston, R.L. Rickard, and T.D. Inch. "Structure-Activity Relations in 2,6,7-Trioxa-1-phosphabicyclo[2.2.2]octanes and Related Compounds." *European Journal of Medicinal Chemistry* 13 (1978): 207–212.

Eckhaus, S.R., J.C. Davis, Jr., B.M. Zeffert, and T.R. Moore. "Preparation of Alkylphosphonothiolates," United States Patent 3911059, October 7, 1975.

Edgewood Research Development, and Engineering Center, Department of the Army. *Material Safety Data Sheet (MSDS) for Lethal Nerve Agent (GA).* Aberdeen Proving Ground, MD: Chemical Biological Defense Command, Revised February 28, 1996.

———. *Material Safety Data Sheet (MSDS) for Lethal Nerve Agent (GB).* Aberdeen Proving Ground, MD: Chemical Biological Defense Command, Revised February 21, 2004.

———. *Material Safety Data Sheet (MSDS) for Lethal Nerve Agent (GD).* Aberdeen Proving Ground, MD: Chemical Biological Defense Command, Revised February 28, 1996.

———. *Material Safety Data Sheet (MSDS) for Lethal Nerve Agent (GF).* Aberdeen Proving Ground, MD: Chemical Biological Defense Command, Revised June 30, 1994.

———. *Material Safety Data Sheet (MSDS) for Lethal Nerve Agent (VX).* Aberdeen Proving Ground, MD: Chemical Biological Defense Command, Revised December 21, 2004.

Epstein, J., K.E. Levy, and H.O. Michel. "Production of Toxic Organophosphorus Compounds," United States Patent 3903210, September 2, 1975.

"Final Recommendations for Protecting Human Health From Potential Adverse Effects of Exposure to Agents GA (Tabun), GB (Sarin), and VX." *Federal Register* 68, No. 196 (October 9, 2003): 58348–58351.

Fusek, J. and J. Bajgar. "Antidotal Treatment of Intoxication with New Nerve Agent GV." In *Proceedings of the CB Medical Treatment Symposium: an Exploration of Present Capabilities and Future Requirements for Chemical and Biological Medical Treatment*. Edgewood, MD: Battelle Memorial Institute, 1994, 6.6–6.8.

Ghosh, R. "New Basic Esters of Thiophosphonic Acids and Salts Thereof," Great Britain Patent 797603, July 12, 1958.

Gordon, J.J., R.H. Inns, M.K. Johnson, L. Leadbeater, M.P. Maidment, D.G. Upshall, G.H. Cooper, and R.L. Rickard. "The Delayed Neuropathic Effects of Nerve Agents and Some Other Organophosphorus Compounds." *Archives of Toxicology* 52 (1983): 71–82.

International Association of Fire Fighters (IAFF) Division of Occupational Health, Safety and Medicine. *Position on the U.S. Army Soldier & Biological Chemical Command's (SBCCOM), Withdrawn "3/30 Rule" and the New Guide: "The Risk Assessment of Using Firefighter Protective Ensemble with Self-Contained Breathing Apparatus for Rescue Operations During a Terrorist Chemical Agent Incident."* February 13, 2004.

Koblin, A. "Persistent Incapacitating Chemical Warfare Composition and Its Use," United States Patent 4708869, November 24, 1987.

Kruglyak, Yu.L., S.I. Malekin, and I.V. Martynov. "Phosphorylated Oximes. XII. Reactions of 2-Halophospholanes with Dichlorofluoronitrosomethane." *Zhurnal Obshchei Khimii* 42 (1972): 811–814 (In Russian).

Lyashenko, Yu.E. and V.B. Sokolov. "Reactions of O-(Alkylchloroformoimino)trichloromethylphosphoranes With S-Nucleophiles." *Phosphorus, Sulfur and Silicon and the Related Elements* 69 (1992): 153–161.

Malekin, S.I., V.I. Yakutin, M.A. Sokal'skii, Yu.L. Kruglyak, and I.V. Martynov. "Mechanism of the Reaction of α-Chloronitrosoalkanes with Phosphorus (III) Compounds." *Zhurnal Obshchei Khimii* 42 (1972): 807–811 (In Russian).

Marrs, T.C., R.L. Maynard, and F.R. Sidell. *Chemical Warfare Agents: Toxicology and Treatment*. Chichester, England: John Wiley & Sons, 1997.

Martynov, I.V., A.N. Ivanov, T.A. Epishina, and V.B. Sokolov. "Reaction of 1,1-Dichloro-1-nitrosoalkanes with Phosphorus (III) Chlorides." *Seriya Khimicheskaya* 9 (1988): 2128–2132 (In Russian).

———. "Reaction of Phosphorus Trichloride with 1,1,2-Trichloro-1-nitrosoethane in Sulfur Dioxide." *Seriya Khimicheskaya* 9 (1986): 2158 (In Russian).

———. "Reaction of Polychloronitrosoethanes with Phosphorous Acid Derivatives." *Seriya Khimicheskaya* 5 (1987): 1086–1089 (In Russian).

Martynov, I.V., Yu.L. Kruglyak, and S.I. Malekin. "Carbonyl Halide Oxime β-Chloroalkyl Fluorophosphates," Soviet Patent 241433, April 18, 1969 (In Russian).

Martynov, I.V., Yu.L. Kruglyak, and N.F. Privezentseva. "Phosphorylated Oximes. I. Reaction of Halodialkyl Phosphites with α-Chlorofluoronitroso Alkanes." *Zhurnal Obshchei Khimii* 37 (1967): 1125–1130 (In Russian).

———. "Phosphorylated Oximes. II. Reaction of Halodialkyl Phosphites with α-Chloronitro Alkanes." *Zhurnal Obshchei Khimii* 37 (1967): 1130–1132 (In Russian).

———. "Reaction of Alkyl Phosphites with α-Halonitroso and α-Halonitro Compounds." Edited by Kabachnik, M.I. Khim. Primen. Fosfororg. Soedin., Tr. Vses. Konf., 3rd (1972), Meeting Date 1965, 273-9. Publisher: "Nauka", Moscow, USSR (In Russian).

Martynov, I.V., Yu.L. Kruglyak, G.A. Leibovskaya, Z.I. Khromova, and O.G. Strukov. "Phosphorylated Oximes. IV. Reactions of 1,3,2-Dioxaphospholanes and 1,3,2-Oxathiaphospholanes with α-Halonitrosoalkanes." *Zhurnal Obshchei Khimii* 39 (1969): 996–999 (In Russian).

Matousek, J. "On New Potential Supertoxic Lethal Organophosphorous Chemical Warfare Agents with Intermediary Volatility." In *Proceedings of the CB Medical Treatment Symposium: an Exploration of Present Capabilities and Future Requirements for Chemical and Biological Medical Treatment*. Edgewood, MD: Battelle Memorial Institute, 1994, pp. 5.3–5.5.

———. "On Some New Trends in the Development of Potential Chemical Warfare Agents." In *Proceedings of the CB Medical Treatment Symposium: an Exploration of Present Capabilities and Future Requirements for Chemical and Biological Medical Treatment*. Edgewood, Maryland: Battelle Memorial Institute, 1994, p. 5.6.

Mioduszewski, R.J., S.A. Reutter, L.L. Miller, E.J. Olajos, and S.A. Thomson. *Evaluation of Airborne Exposure Limits for G-Agents: Occupational and General Population Exposure Criteria, Edgewood Arsenal Report No. ERDEC-TR-489*. April 1998; with Erratum Sheet dated April 17, 2000.

Munro, N.B., S.S. Talmage, G.D. Griffin, L.C. Waters, A.P. Watson, J.F. King, and V. Hauschild. "The Sources, Fate and Toxicity of Chemical Warfare Agent Degradation Products." *Environmental Health Perspectives* 107 (1999): 933–974.

National Institute of Health. *Hazardous Substance Data Bank (HSDB)*. (http://toxnet.nlm.nih.gov/cgi-bin/sis/htmlgen?HSDB/). 2005.

National Institute for Occupational Safety and Health. "Emergency Response Card for Sarin." Interim Document March 20, 2003.

———. *NIOSH Pocket Guide to Chemical Hazards*. Washington, DC: Government Printing Office, September 2005.

National Research Council. *Review of Acute Human-Toxicity Estimates for Selected Chemical-Warfare Agents*. Washington, DC: National Academy Press, 1997.

Occupational Safety and Health Administration, National Institute for Occupational Safety and Health. Interim Guidance on Personal Protective Equipment Selection for Emergency Responders: Nerve Agents, August 30, 2004 (http://www.osha.gov/SLTC/emergencypreparedness/cbrnmatrix/nerve.html). 2005.

Privezentseva, N.F., F.N. Chelobov, Yu.L. Kruglyak, and I.V. Martynov. "Phosphorylated Oximes. XI. Oxime Tertrachlorophosphoranes." *Zhurnal Obshchei Khimii* 42 (1972): 305–307 (In Russian).

Rautio, M. "An International Interlaboratory Round-Robin Test for Chemical Warfare Agents." *Kemia-Kemi* 18 (1991): 19–21 (In Finnish).

Rohrbaugh, D.K., D.I. Rossman, W.P. Ashman, J.J. DeFrank, H.D. Durst, V.K. Rastogi, and S. Munavalli. "GC-MS Characterization and Synthesis of VX Simulants. Part 2. Ethyl S-(3-Methylbutyl) Methylphosphonothiolate and its Analogs" in *Proceedings of the ERDEC Scientific Conference on Chemical and Biological Defense Research*. Aberdeen Proving Ground, MD, United States, November 17–20, 1998. Edited by Dorothy A. Berg. Springfield, VA: National Technical Information Service, 1999: 439–447.

Sartori, M.F. "New Developments in the Chemistry of War Gases." *Chemical Reviews* 48 (1951): 225–257.

Schegk, E., H. Schloer, and G. Schrade. "Phosphonic Acid Esters," United States Patent 3014943, December 26, 1961.

Sidell, F.R. *Medical Management of Chemical Warfare Agent Casualties: A Handbook for Emergency Medical Services*. Bel Air, MD: HB Publishing, 1995.

Sidell, F.R., E.T. Takafuji, and D.R. Franz, ed. *Medical Aspects of Chemical and Biological Warfare, Textbook of Military Medicine Series, Part 1, Warfare, Weaponry, and the Casualty*. Washington, DC: Office of the Surgeon General, Department of the Army, 1997.

Sifton, D.W. ed. *PDR Guide to Biological and Chemical Warfare Response*. Montvale, NJ: Thompson/Physicians Desk Reference, 2002.

Sokolov, V.B., T.A. Epishina, and I.V. Martynov. "Reaction of 1,1-dichloro-1-nitrosoethane with phosphorus oxychloride in the presence of zinc." *Seriya Khimicheskaya* 7 (1988): 1691 (In Russian).

Sokolov, V.B., A.N. Ivanov, T.A. Epishina, and I.V. Martynov. "Interaction of 2,2,3,3-Tetrafluoropropyl Dichlorophosphite with 1,1,2-Trichloro-1-nitrosoethane." *Seriya Khimicheskaya* 6 (1987): 1422–1423 (In Russian).

———. "Reaction of Dichloromethylphosphine with 1,1-Dichloro-1-nitrosoalkanes." *Zhurnal Obshchei Khimii* 57 (1987): 1659–1660 (In Russian).

Sokolov, V.B., A.N. Ivanov, T.V. Goreva, T.A. Epishina, and I.V. Martynov. "O-(Alkylchloroformimino)-O-alkylphosphoric Acid Chlorides." *Seriya Khimicheskaya* 5 (1990): 1122–1125 (In Russian).

Somani, S.M., ed. *Chemical Warfare Agents*. New York: Academic Press, 1992.

Somani, S.M. and J.A. Romano, Jr., ed. *Chemical Warfare Agents: Toxicity at Low Levels*. Boca Raton, FL: CRC Press, 2001.

Smith, A., P. Heckelman, and M.J. Oneil, ed. *The Merck Index: An Encyclopedia of Chemicals, Drugs, & Biologicals.* 13th Edn. Rahway, NJ: Merck & Co., Inc., 2001.
Trapp, R. *The Detoxification and Natural Degradation of Chemical Warfare Agents.* Volume 3 of *SIPRI Chemical & Biological Warfare Studies.* London: Taylor & Francis, 1985.
United States Air Force. *Development of Candidate Chemical Simulant List: The Evaluation of Candidate Chemical Simulants Which May Be Used in Chemically Hazardous Operations, Technical Report AFAMRL-TR-82-87,* Washington, DC: Government Printing Office, 1982.
United States Army Headquarters. *Chemical Agent Data Sheets Volume I, Edgewood Arsenal Special Report No. EO-SR-74001.* Washington, DC: Government Printing Office, December 1974.
———. *Chemical Agent Data Sheets Volume II, Edgewood Arsenal Special Report No. EO-SR-74002.* Washington, DC: Government Printing Office, December 1974.
———. *Potential Military Chemical/Biological Agents and Compounds, Field Manual No. 3-11.9.* Washington, DC: Government Printing Office, January 10, 2005.
United States Army Medical Research Institute of Chemical Defense. *Medical Management of Chemical Casualties Handbook.* 3rd Edn. Aberdeen Proving Ground, MD: United States Army Medical Research Institute of Chemical Defense, July 2000.
United States Army Soldiers and Biological Chemical Command. *Risk Assessment of Using Firefighter Protective Ensemble and Self-Contained Breathing Apparatus for Rescue Operations During a Terrorist Chemical Agent Incident.* August 2003.
United States Coast Guard. *Chemical Hazards Response Information System (CHRIS) Manual, 1999 Edition.* (http://www.chrismanual.com/Default.htm). March 2004.
Wagner, G.W. and Y.-C. Yang. "Universal Decontaminating Solution for Chemical Warfare Agents," United States Patent 6245957, June 12, 2001.
———. "Rapid Nucleophilic/Oxidative Decontamination of Chemical Warfare Agents." *Industrial & Engineering Chemistry Research* 41 (2002): 1925–1928.
Wardrop, A.W.H. and C. Stratford. "Improvements in the Manufacture of Organic Phosphorus Compounds Containing Sulfur," Great Britain Patent 1346410, February 13, 1974.
Watson, A., D. Opresko, and V. Hauschild. *Evaluation of Chemical Warfare Agent Percutaneous Vapor Toxicity: Derivation of Toxicity Guidelines for Assessing Chemical Protective Ensembles, ORNL/TM-2003/180.* July 2003.
Watson, A.P. and N.B. Munro. *Reentry Planning: The Technical Basis for Offsite Recovery Following Warfare Agent Contamination, ORNL-6628.* April 1990.
Wiener, S.W. and R.S. Hoffman. "Nerve Agents: a Comprehensive Review." *Journal of Intensive Care Medicine* 19 (2004): 22–37.
Williams, K.E. *Detailed Facts About Nerve Agent GA.* Aberdeen Proving Ground, MD: United States Army Center for Health Promotion and Preventive Medicine, 1996.
———. *Detailed Facts About Nerve Agent GB.* Aberdeen Proving Ground, MD: United States Army Center for Health Promotion and Preventive Medicine, 1996.
———. *Detailed Facts About Nerve Agent GD.* Aberdeen Proving Ground, MD: United States Army Center for Health Promotion and Preventive Medicine, 1996.
———. *Detailed Facts About Nerve Agent VX.* Aberdeen Proving Ground, MD: United States Army Center for Health Promotion and Preventive Medicine, 1996.
Wise, D. *Cassidy's Run: The Secret Spy War Over Nerve Gas.* New York: Random House, 2000.
World Health Organization. *Health Aspects of Chemical and Biological Weapons.* Geneva: World Health Organization, 1970.
———. *International Chemical Safety Cards (ICSCs)* (http://www.cdc.gov/niosh/ipcs/icstart.html). 2004.
———. *Public Health Response to Biological And Chemical Weapons: WHO Guidance.* Geneva: World Health Organization, 2004.
Yang, Y.-C. "Chemical Reactions for Neutralizing Chemical Warfare Agents." *Chemistry & Industry* (May 1, 1995): 334–337.
Yang, Y.-C., J.A. Baker, and J.R. Ward. "Decontamination of Chemical Warfare Agents." *Chemical Reviews* 92 (1992): 1729–1743.
Yaws, C.L. *Matheson Gas Data Book.* 7th Edn. Parsippany, NJ: Matheson Tri-Gas, 2001.

2

Carbamate Nerve Agents

2.1 General Information

This class of agents is not covered by the Chemical Weapons Convention. Because of the toxicity of the agents and lack of commercial application, carbamate nerve agents would be prohibited based on the Guidelines for Schedules of Chemicals.

These materials are fourth generation chemical warfare agents. They contain one or more quaternary amine centers that increase the ability of the agent to penetrate into neuromuscular junctions. They are relatively simple to synthesize although the starting materials are not commonly available. Because they produce negligible vapor, they are somewhat difficult to deliver in a manner that will produce immediate casualties.

Although the United States began investigating carbamates as warfare agents in the late 1950s, there is no information to indicate that these agents have ever been used other than for research purposes.

2.2 Toxicology

2.2.1 Effects

Nerve agents are the most toxic of formerly stockpiled man-made chemical warfare agents. These compounds are similar to, but much more deadly than, agricultural carbamate pesticides. Nerve agents disrupt the function of the nervous system by interfering with the enzyme acetylcholinesterase. Serious effects are on skeletal muscles and the central nervous system. Nerve agents also affect glands that discharge secretions to the outside of the body causing discharge of mucous, saliva, sweat, and gastrointestinal fluids. Exposure to solids, solutions, or aerosols from these agents is hazardous and can result in death within minutes of exposure.

2.2.2 Pathways and Routes of Exposure

Nerve agents are hazardous through any route of exposure including inhalation, exposure of the skin and eye, ingestion, and broken, abraded, or lacerated skin (e.g., penetration of skin by debris).

2.2.3 General Exposure Hazards

Carbamate nerve agents do not have good warning properties. They have no odor, and, other than causing miosis, aerosols do not irritate the eyes. Contact neither irritates the skin nor causes cutaneous injuries.

Human toxicity data for the carbamate nerve agents has not been published or has not been established.

These agents are rapidly detoxified or eliminated from the body and there is little or no cumulative toxicity. Some agents are refractory to treatment.

2.2.4 Latency Period

2.2.4.1 Aerosols (Mists or Dusts)

Depending on the concentration of agent aerosols, the effects begin to appear 30 seconds to 2 minutes after initial exposure.

2.2.4.2 Solids/Solutions (Nonaerosol)

Typically, there is a latent period with no visible effects between the time of exposure and the sudden onset of symptoms. This latency can range from 1 minutes to 18 hours and is affected by factors such as the amount of agent involved, the amount of skin surface in contact with the agent, and previous exposure to materials that chap or dry the skin (e.g., organic solvents such as gasoline or alcohols). Moist, sweaty areas of the body are more susceptible to percutaneous penetration by solid nerve agents.

Another key factor is the part of the body that is exposed to the agent. It takes more time for the agent to penetrate areas of the body that are covered by thicker and tougher skin. The regions of the body that allow the fastest percutaneous penetration are the groin, head, and neck. The least susceptible body regions are the hands, feet, front of the knee, and outside of the elbow.

2.3 Characteristics

2.3.1 Physical Appearance/Odor

2.3.1.1 Laboratory Grade

Laboratory grade agents are typically white, to pale yellow crystalline solids. Some are hygroscopic or deliquescent. They have little or no odor.

2.3.1.2 Modified Agents

Carbamate nerve agents have been dissolved in both water and organic solvents to facilitate handling, enhance dispersal, or increase the ease of percutaneous penetration by the agents. Percutaneous enhancement solvents include dimethyl sulfoxide, *N*,*N*-dimethylformamide, *N*,*N*-dimethylpalmitamide, *N*,*N*-dimethyldecanamide, and saponin. Color and other properties of these solutions may vary from the pure agent. Odors will be dependent on the characteristics of the solvent(s) used.

2.3.2 Stability

Dry carbamate nerve agents are stable over a wide range of temperatures. Stabilizers are not required. Agents can be stored in glass, steel, stainless steel, or aluminum containers. Agents in solution are much more susceptible to hydrolysis and decomposition.

2.3.3 Persistency

For military purposes, unmodified carbamate nerve agents are classified as extremely persistent. Agents have negligible vapor pressure and they will not evaporate. Depending on the size of the individual particles and on any encapsulation or coatings applied to the particles, they can be reaerosolized by ground traffic or strong winds.

2.3.4 Environmental Fate

Carbamate nerve agents are nonvolatile and do not pose a vapor hazard. Although these agents may be dissolved in volatile solvents, evaporation of the solvent does not increase the evaporation of the agent itself. Porous material, including painted surfaces, may absorb solutions of agents. After the initial surface contamination has been removed, the agent that has been absorbed into porous material can migrate back to the surface posing a contact hazard.

Although these nerve agents are water soluble, agent solubility may be modified (either increased or decreased) by solvents. These agents are also soluble in some polar organic solvents such as alcohols and acetonitrile.

2.4 Additional Hazards

2.4.1 Exposure

Individuals who have had previous exposure to materials that chap or dry the skin, such as alcohols, gasoline, or paint thinners, may be more susceptible to percutaneous exposure from dermal contact with these agents. In these situations, the rate of percutaneous penetration of the agent is greatly increased resulting in a decrease in the survival time that would otherwise be expected.

All foodstuffs in the area of release should be considered contaminated. Unopened items packaged in glass, metal, or heavy duty plastic and exposed only to agent aerosols solid agents may be used after decontamination of the container. Unopened items exposed to solid agents or solutions of agents should be decontaminated within a few hours postexposure or destroyed. Opened or unpackaged items, or those packaged only in paper or cardboard, should be destroyed.

Meat from animals that have suffered only mild to moderate effects from exposure to nerve agents should be safe to consume. Milk should be discarded for the first 7 days postexposure and then should be safe to consume. Meat, milk, and animal products, including hides, from animals severely affected or killed by nerve agents should be destroyed.

Plants, fruits, vegetables, and grains exposed to carbamate nerve agents should be quarantined until tested.

2.4.2 Livestock/Pets

Animals can be decontaminated with shampoo/soap and water (see Section 2.5.3). If the animals' eyes have been exposed to the agent, they should be irrigated with water or saline solution for a minimum of 30 minutes.

The topmost layer of unprotected feedstock (e.g., hay or grain) should be destroyed. The remaining material should be quarantined until tested. Carbamate nerve agents are very persistent and forage vegetation could still retain sufficient agent to produce severe effects for several weeks postrelease, depending on the level of contamination and the weather conditions.

2.4.2.1 *Fire*

Heat from a fire will destroy carbamate nerve agents before generating any significant concentration of agent vapor. However, actions taken to extinguish the fire can spread the agent. Carbamates are water-soluble and runoff from firefighting efforts will pose a significant threat. Some of the potential decomposition products include toxic or corrosive gases or both.

2.4.2.2 *Reactivity*

Carbamate nerve agents are stable in water. However, at high-pH they are rapidly destroyed.

2.4.3 Hazardous Decomposition Products

2.4.3.1 *Hydrolysis*

Agents produce dimethylamine [$(CH_3)_2NH$], carbon dioxide (CO_2), and complex organic compounds.

2.4.3.2 *Combustion*

Volatile decomposition products may include hydrogen chloride (HCl), hydrogen bromide (HBr), hydrogen iodide (HI), CO_2, aromatic hydrocarbons, and nitrogen oxides (NO_x).

2.5 Protection

2.5.1 Evacuation Recommendations

There are no published recommendations for isolation or protective action distances for carbamate nerve agents released in mass casualty situations.

2.5.2 Personal Protective Requirements

2.5.2.1 *Structural Firefighters' Gear*

Structural firefighters' protective clothing is recommended for fire situations only; it is not effective in spill situations or release events. However, since carbamate nerve agents have negligible vapor pressure, they do not pose a vapor hazard. The primary risk of exposure is through contact with aerosolized agents, solids, or solutions of agents. If chemical protective clothing is not available and it is necessary to rescue casualties from a contaminated area, then structural firefighters' gear will provide some skin protection against nerve agent aerosols. Contact with solids and solutions should be avoided.

Even though carbamate nerve agents are rapidly detoxified or eliminated from the body, exposures may have a cumulative effect in the short-term, placing all responders who entered the hot zone without appropriate chemical protective clothing at increased risk during the remainder of the emergency.

2.5.2.2 Respiratory Protection

Self-contained breathing apparatuses (SCBAs) or air purifying respirators (APRs) should have a National Institute for Occupational Safety and Health (NIOSH) and Chemical/Biological/Radiological/Nuclear (CBRN) certification. However, during emergency operations, other NIOSH approved SCBAs or APRs that have been specifically tested by the manufacturer against chemical warfare agents may be used if deemed necessary by the Incident Commander. APRs should be equipped with a NIOSH approved CBRN filter or a combination organic vapor/acid gas/particulate cartridge.

Immediately dangerous to life or health (IDLH) levels are the ceiling limit for respirators other than SCBAs. However, IDLH levels have not been established for carbamate nerve agents. Therefore, any potential exposure to aerosols of these agents should be regarded with extreme caution and the use of SCBAs for respiratory protection should be considered.

2.5.2.3 Chemical Protective Clothing

Currently, there is no information on performance testing of chemical protective clothing against carbamate nerve agents. Evaluation of fabrics used to prevent exposure to carbamate pesticides may provide guidance on selection of appropriate protective clothing.

Because these agents do not produce any significant concentration of vapor, the primary risk of exposure is through inhalation of aerosols or by the percutaneous migration of agents following dermal exposure to solid agents or solutions containing these agents. However, aerosolized agents can also penetrate the skin and produce a toxic effect. If there is any possibility of contact with aerosolized agent, then responders should consider wearing a Level A protective ensemble.

2.5.3 Decontamination

2.5.3.1 General

Carbamate nerve agents are readily destroyed by high pH (i.e., basic solutions). Use an aqueous caustic solution (minimum of 10% by weight sodium hydroxide or sodium carbonate) or use undiluted household bleach. Basic peroxides also rapidly detoxify carbamate nerve agents.

2.5.3.2 Solutions or Liquid Aerosols

Casualties/personnel: Remove all clothing immediately. To avoid further exposure of the head, neck, and face to the agent, cut off potentially contaminated clothing that must be pulled over the head. Remove as much of the nerve agent from the skin as fast as possible. If water is not immediately available, the agent can be absorbed with any convenient material such as paper towels, toilet paper, flour, or talc. To minimize both spreading the agent and abrading the skin, do not rub the agent with the absorbent. Blot the contaminated skin with the absorbent.

Use a sponge or cloth with liquid soap and copious amounts of water to wash the skin surface and hair at least three times. Do not delay decontamination to find warm or hot water if it is not readily available. Avoid rough scrubbing as this could abrade the skin and increase percutaneous absorption of residual agent. Rinse with copious amounts of water. If there is a potential that the eyes have been exposed to nerve agents, irrigate with water or 0.9% saline solution for a minimum of 15 minutes.

Small areas: Puddles of liquid can be contained by covering with absorbent material such as vermiculite, diatomaceous earth, clay, sponges, or towels. Place the absorbed material into containers lined with high-density polyethylene. Before sealing the container, cover the contents with an aqueous caustic solution or undiluted household bleach (see Section 2.5.3.1). Wash the area with copious amounts of soap and water. Collect and containerize the rinseate. Removal of porous material, including painted surfaces, may be required because the nerve agent that has been absorbed into these materials can migrate back to the surface and pose a contact hazard.

2.5.3.3 Solids or Particulate Aerosols

Casualties/personnel: Do not attempt to brush the agent off the individual or their clothing as this can aerosolize the agent. Remove all clothing immediately. To avoid further exposure of the head, neck, and face to the agent, cut off potentially contaminated clothing that must be pulled over the head. Wash the skin surface and hair at least three times with copious amounts of soap and water. Do not delay decontamination to find warm or hot water if it is not readily available. Rinse with copious amounts of water. If there is a potential that the eyes have been exposed to nerve agents, then irrigate with water or 0.9% saline solution for a minimum of 15 minutes.

Small areas: If indoors, close windows and doors in the area and turn off anything that could create air currents (e.g., fans, air conditioner, etc.). Avoid actions that could aerosolize the agent such as sweeping or brushing. Collect the agent using a vacuum cleaner equipped with a high-efficiency particulate air (HEPA) filter. Do not use a standard home or industrial vacuum. Do not allow the vacuum exhaust to stir the air in the affected area. Vacuum all surfaces with extreme care in a very slow and controlled manner to minimize aerosolizing the agent. Place the collected material into containers lined with high-density polyethylene. Before sealing the container, cover the contents with an aqueous caustic solution or undiluted household bleach (see Section 2.5.3.1). Wash the area with copious amounts of the soap and water. Collect and containerize the rinseate.

2.6 Medical

2.6.1 CDC Case Definition

The Centers for Disease Control and Prevention (CDC) has not published a specific case definition for intoxication by carbamates. However, the case definition for nerve agents and organophosphates states

> 1) A case in which nerve agents are detected in the urine. Decreased plasma or red blood cell cholinesterase levels based on a specific commercial laboratory reference range might indicate a nerve agent or organophosphate exposure; however, the normal range levels for cholinesterase are wide, which makes interpretation of levels difficult without a baseline measurement or repeat measurements over time. 2) Detection of organophosphates in environmental samples. The case can be confirmed if laboratory testing is not performed because either a predominant amount of clinical and nonspecific laboratory evidence is present or an absolute certainty of the etiology of the agent is known.

2.6.2 Differential Diagnosis

The following factors have been suggested as alternatives to consider when presented with a potential case of exposure to nerve agents: carbamate and organophosphate pesticides; alkaloids such as nicotine or coniine; ingestion of mushrooms containing muscarine; and medicinals such as carbamates, cholinomimetic compounds, and neuromuscular blocking drugs.

2.6.3 Signs and Symptoms

2.6.3.1 Aerosols

Pinpointing of pupils (miosis) and extreme nasal discharge (rhinorrhea) may be the first indications of exposure. The casualty may also experience difficulty in breathing with a feeling of shortness of breath or tightness of the chest. In cases of exposure to high aerosol concentrations, the gastrointestinal tract may be affected, resulting in vomiting, urination, or defecation. Inhalation of lethal amounts of nerve agent can cause loss of consciousness and convulsions in as little as 30 seconds, followed by cessation of breathing and flaccid paralysis after several more minutes.

2.6.3.2 Solutions/Solids

Localized sweating, nausea, vomiting, involuntary urination/defecation, and a feeling of weakness are signs of small to moderate nerve agent exposure. Miosis usually does not occur unless the individuals have had agent in or around their eyes. Exposure to a large amount of agent causes copious secretions, loss of consciousness, convulsions progressing into flaccid paralysis, and cessation of breathing.

2.6.4 Mass-Casualty Triage Recommendations

2.6.4.1 Priority 1

A casualty with symptoms in two or more organ systems (not including miosis or rhinorrhea), who has a heartbeat and a palpable blood pressure. The casualty may or may not be conscious and/or breathing.

2.6.4.2 Priority 2

A casualty with a known expossure to a solid agent or solution but no apparent signs or symptoms, or a casualty who is recovering from a severe exposure after receiving treatment.

2.6.4.3 Priority 3

A casualty who is walking and talking, although miosis and/or rhinorrhea may be present.

2.6.4.4 Priority 4

A casualty who is not breathing and does not have a heartbeat or palpable blood pressure.

2.6.5 Casualty Management

Decontaminate the casualty ensuring that all nerve agents have been removed. If nerve agents have gotten into the eyes, irrigate the eyes with water or 0.9% saline solution for at least 15 minutes. Irrigate open wounds with water or 0.9% saline solution for at least 10 minutes. However, do not delay treatment if thorough decontamination cannot be undertaken immediately.

Although these agents do not produce any significant vapor, aerosolization of residual dusts on casualties could cause impacts to medical responders. Once the casualty has been decontaminated, including the removal of foreign matter from wounds, medical personnel do not need to wear a chemical-protective mask.

Ventilate the patient. There may be an increase in airway resistance due to constriction of the airway and the presence of secretions. If breathing is difficult, administer oxygen. As soon as possible administer atropine. Severely poisoned individuals may exhibit tolerance to atropine and require large doses. Oximes such as pralidoxime chloride (2-PAMCl) do not significantly increase the effectiveness of atropine and in some cases may be contraindicated. Diazepam may be required to prevent or control severe convulsions or both. If diazepam is not administered within 40 minutes postexposure, then its effectiveness at controlling seizures is minimal. These agents are rapidly detoxified or eliminated from the body and aging of these agents is not an issue.

2.7 Fatality Management

Remove all clothing and personal effects segregating them as either durable or nondurable items. While it may be possible to decontaminate durable items, it may be safer and more efficient to destroy nondurable items rather than attempt to decontaminate them. Items that will be retained for further processing should be double sealed in impermeable containers, ensuring that the inner container is decontaminated before placing it in the outer one.

Carbamate nerve agents that have entered the body are metabolized, hydrolyzed, or bound to tissue and pose little threat of off-gassing. To remove agents on the outside of the body, wash the remains with copious amounts of soap and water. Pay particular attention to areas where agent may get trapped, such as hair, scalp, pubic areas, fingernails, folds of skin, and wounds. All wash and rinse waste must be contained for proper disposal. Body fluids removed during the embalming process do not pose any additional risks and should be contained and handled according to established procedures.

Use standard burial procedures.

C02-A

AGENTS

C02-A001

1,10-Bis[methyl-2-(3-dimethyl-carbamoxypyridyl)methylamino]decane dimethobromide (Agent EA 3887)

CAS: 110913-97-8

RTECS: —

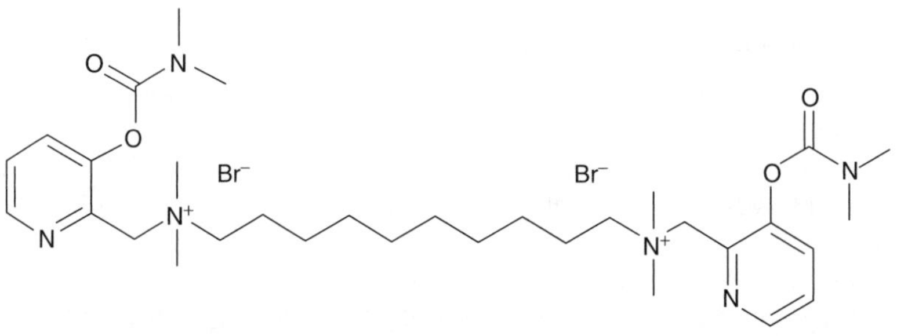

$C_{32}H_{54}N_6O_4 \cdot Br_2$
White odorless solid. Various other salts have been reported.

Exposure Hazards
This agent is refractory to treatment.
Human toxicity values have not been established or have not been published. However, based on available information, this agent appears to be approximately twice as toxic as VX (C01-A016). The iodine salt is slightly more toxic than VX.

Properties:

MW: 746.7	*VP*: Negligible	*FlP*: —
D: 1.4 g/cm^3	*VD*: —	*LEL*: —
MP: Decomposes	*Vlt*: —	*UEL*: —
BP: —	*H_2O*: "Soluble"	*RP*: 'Extremely persistent"
Vsc: —	*Sol*: Alcohols; Acetonitrile	*IP*: —

Iodide salt
MW: 840.6
MP: 320°F

C02-A002

1,6-Bis[methyl-2-(3-dimethylcarbamoxypyridyl)methylamino]hexane dimethobromide
(Agent EA 3948)
CAS: 110913-93-4
RTECS: —

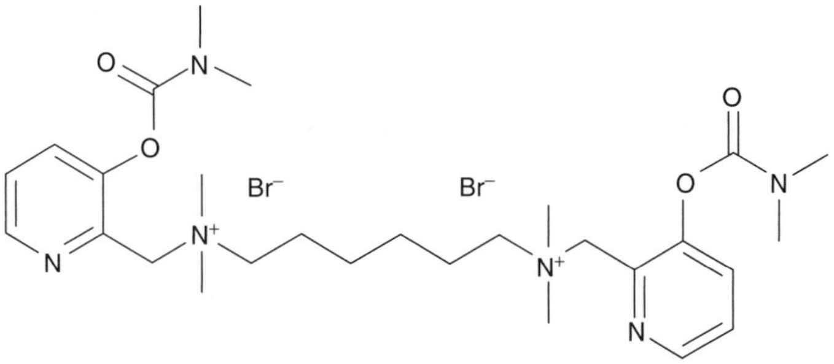

$C_{28}H_{46}N_6O_4 \cdot Br_2$
Crystalline solid. Various other salts have been reported.

Exposure Hazards
Human toxicity values have not been established or have not been published. However, based on available information, this agent appears to be approximately half as toxic as VX (C01-A016).

Properties:

MW: 690.5	*VP*: Negligible	*FlP*: —
D: —	*VD*: —	*LEL*: —
MP: 293–297°F	*Vlt*: —	*UEL*: —
BP: —	*H_2O*: Soluble	*RP*: Persistent
Vsc: —	*Sol*: Alcohols	*IP*: —

C02-A003

1-(*N*,*N*-Dimethylamino)-10-[*N*-(3-dimethylcarbamoxy-2-pyridylmethyl)-*N*-methylamino]decane dimethobromide (Agent EA 3966)

CAS: 110913-86-5

RTECS: —

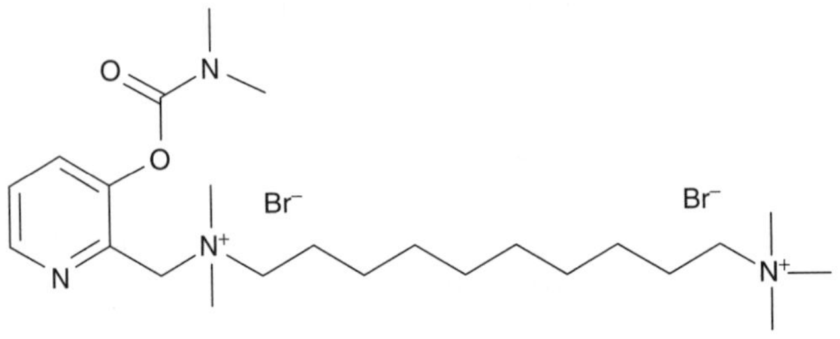

$C_{24}H_{46}N_4O_2 \cdot Br_2$

Crystalline solid. Various other salts have been reported.

Exposure Hazards

Human toxicity values have not been established or have not been published.

Properties:

MW: 582.5	*VP*: Negligible	*FlP*: —
D:— *MP*: Decomposes	*VD*: —	*LEL*: —
BP: —	*Vlt*: —	*UEL*: —
Vsc: —	H_2O: Soluble	*RP*: Persistent
	Sol: Acetonitrile	*IP*: —

C02-A004

1,8-Bis[methyl-2(3-dimethylcarbamoxypyridyl)methylamino]octane dimethobromide (Agent EA 3990)

CAS: 110913-95-6

RTECS: —

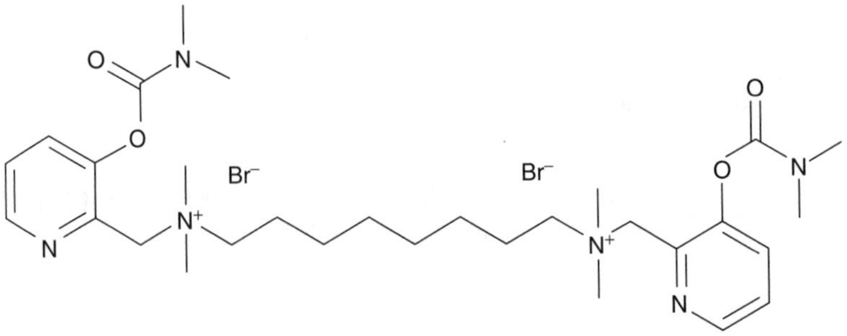

$C_{30}H_{50}N_6O_4 \cdot Br_2$

White, odorless crystalline solid. Various other salts have been reported.

Exposure Hazards

Human toxicity values have not been established or have not been published. However, based on available information, this agent appears to be approximately three times more toxic than VX (C01-A016).

Properties:

MW: 718.7	*VP*: Negligible	*FlP*: —
D: 1.33 g/cm^3	*VD*: —	*LEL*: —
MP: Decomposes	*Vlt*: —	*UEL*: —
BP: —	H_2O: 82% (77°F)	*RP*: "Extremely persistent"
Vsc: —	*Sol*: Alcohols; Acetic acid; Chloroform	*IP*: —

C02-A005

1,5-Bis[methyl-2(3-dimethylcarbamoxypyridyl)methylamino]pentane dimethobromide (Agent EA 4026)

CAS: 110913-92-3

RTECS: —

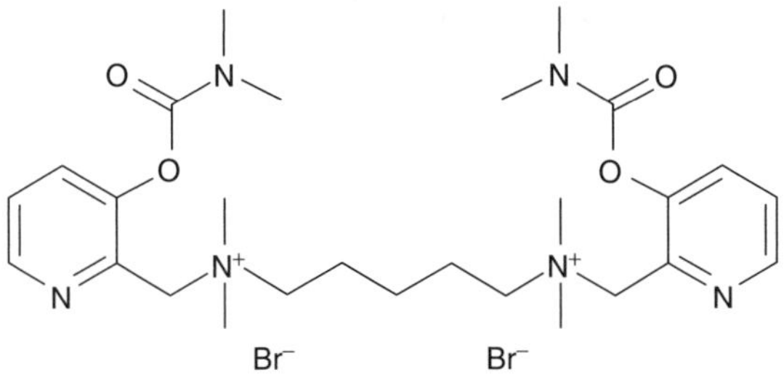

$C_{27}H_{44}N_6O_4 \cdot Br_2$

Crystalline solid. Various other salts have been reported.

Exposure Hazards

Human toxicity values have not been established or have not been published. However, based on available information, this agent appears to be approximately one-fifth as toxic as VX (C01-A016).

Properties:

MW: 676.5	*VP*: Negligible	*FlP*: —
D: —	*VD*: —	*LEL*: —
MP: 370–381°F	*Vlt*: —	*UEL*: —
BP: —	H_2O: Soluble	*RP*: Persistent
Vsc: —	*Sol*: Alcohols	*IP*: —

C02-A006

1,4-Bis[methyl-2(3-dimethylcarbamoxypyridyl)methylamino]butane dimethobromide (Agent EA 4038)

CAS: 110913-91-2

RTECS: —

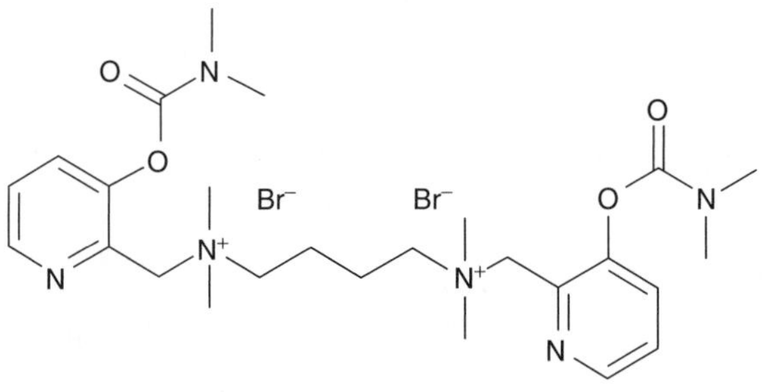

$C_{26}H_{42}N_6O_4 \cdot Br_2$

Crystalline solid. Various other salts have been reported.

Exposure Hazards

Human toxicity values have not been established or have not been published. However, based on available information, this agent appears to be approximately 200 times less toxic than VX (C01-A016).

Properties:

MW: 662.5	*VP*: Negligible	*FlP*: —
D:—	*VD*: —	*LEL*: —
MP: 338–347°F	*Vlt*: —	*UEL*: —
BP: —	H_2O: Soluble	*RP*: Persistent
Vsc: —	*Sol*: Alcohols	*IP*: —

C02-A007

1,3-Bis[methyl-2(3-dimethylcarbamoxypyridyl)methylamino]propane dimethobromide (Agent EA 4048)

CAS: 110913-90-1

RTECS: —

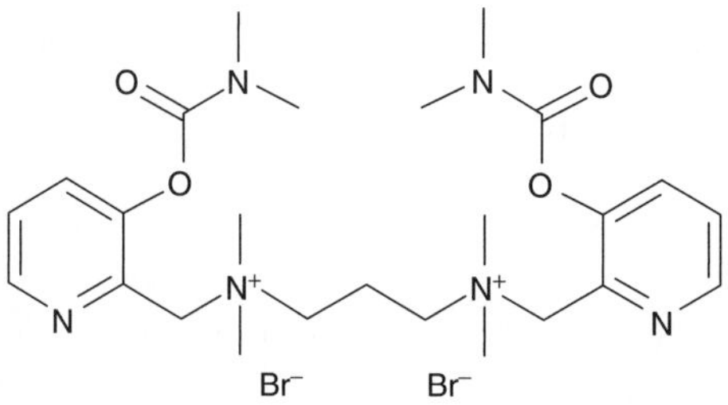

$C_{25}H_{40}N_6O_4 \cdot Br_2$

Crystalline solid. Various other salts have been reported.

Exposure Hazards

Human toxicity values have not been established or have not been published. However, based on available information, this agent appears to have minimal toxicity.

Properties:

MW: 648.4	*VP*: Negligible	*FlP*: —
D:—	*VD*: —	*LEL*: —
MP: 388°F	*Vlt*: —	*UEL*: —
BP: —	H_2O: Soluble	*RP*: Persistent
Vsc: —	*Sol*:—	*IP*: —

C02-A008

1,9-Bis[methyl-2(3-dimethylcarbamoxypyridyl)methylamino]nonane dimethobromide (Agent EA 4056)

CAS: 110913-96-7

RTECS: —

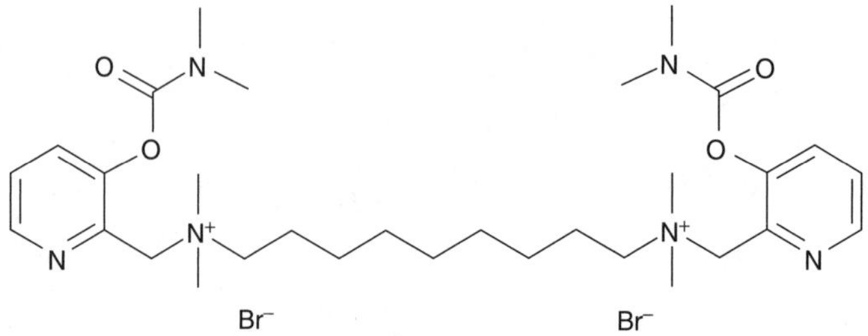

$C_{31}H_{52}N_6O_4 \cdot Br_2$
Crystalline solid. Various other salts have been reported.

Exposure Hazards
Human toxicity values have not been established or have not been published. However, based on available information, this agent appears to be approximately three times more toxic than VX (C01-A016).

Properties:
MW: 732.6
D: —
MP: 212–221°F
BP: —
Vsc: —
VP: Negligible
VD: —
Vlt: —
H_2O: Soluble
Sol: Alcohols
FlP: —
LEL: —
UEL: —
RP: Persistent
IP: —

C02-A009

1,11-Bis[methyl-2(3-dimethylcarbamoxypyridyl)methylamino]undecane dimethobromide (Agent EA 4057)
CAS: 110913-98-9
RTECS: —

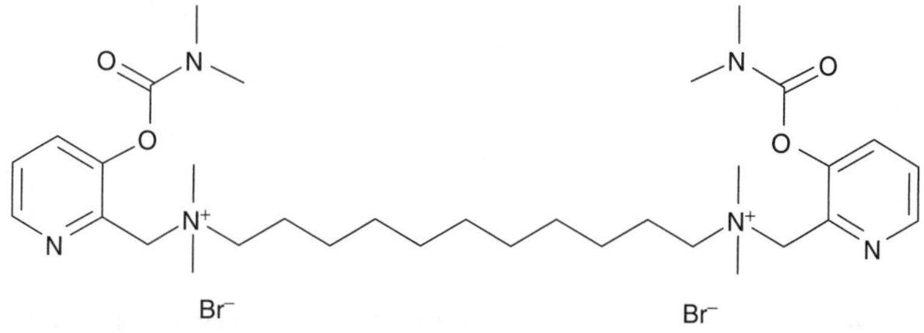

$C_{33}H_{56}N_6O_4 \cdot Br_2$
Crystalline solid. Various other salts have been reported.

Exposure Hazards
Human toxicity values have not been established or have not been published. However, based on available information, this agent appears to be slightly more toxic than VX (C01-A016).

Properties:
MW: 760.7
D: —
MP: 266–270°F
BP: —
Vsc: —
VP: Negligible
VD: —
Vlt: —
H_2O: Soluble
Sol: Alcohols
FlP: —
LEL: —
UEL: —
RP: Persistent
IP: —

C02-A010

1,7-Bis[methyl-2(3-dimethylcarbamoxypyridyl)methylamino]heptane dimethobromide (Agent EA 4181)
CAS: 110913-94-5
RTECS: —

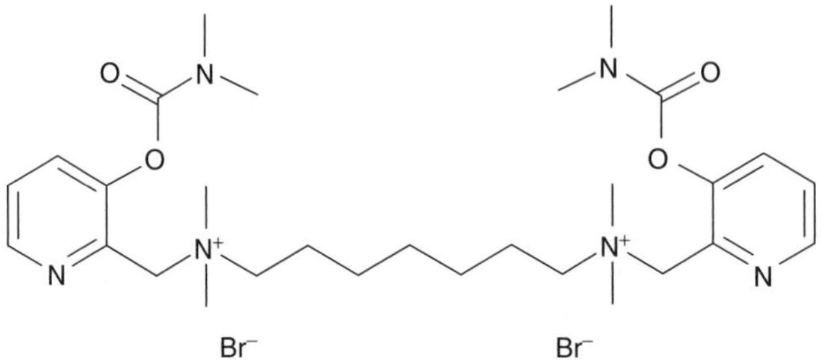

$C_{29}H_{48}N_6O_4 \cdot Br_2$
Crystalline solid. Various other salts have been reported.

Exposure Hazards
Human toxicity values have not been established or have not been published. However, based on available information, this agent appears to be slightly more toxic than VX (C01-A016).

Properties:

MW: 704.5	*VP*: Negligible	*FlP*: —
D: —	*VD*: —	*LEL*: —
MP: 320–325°F	*Vlt*: —	*UEL*: —
BP: —	H_2O: Soluble	*RP*: Persistent
Vsc: —	*Sol*: Alcohols	*IP*: —

C02-A011

1-(4-Dimethylaminophenoxy)-2-(3-dimethylamino-5-dimethylcarbamoxyphenoxy)-ethane dimethiodide
CAS: 57169-76-3
RTECS: —

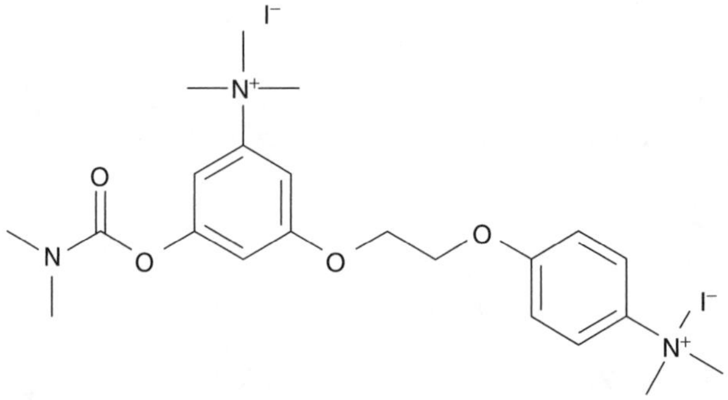

$C_{23}H_{35}N_3O_4 \cdot I_2$
Pale yellow crystalline solid. Various other salts have been reported.

Exposure Hazards
Human toxicity values have not been established or have not been published. However, based on available information, this agent appears to be just as toxic as VX (C01-A016).

Properties:

MW: 671.4	*VP*: Negligible	*FlP*: —
D: —	*VD*: —	*LEL*: —
MP: 266°F	*Vlt*: —	*UEL*: —
BP: —	H_2O: —	*RP*: Persistent
Vsc: —	*Sol*: —	*IP*: —

C02-A012

1-(3-Dimethylaminophenoxy)-3-(3-dimethylamino-5-dimethylcarbamoxyphenoxy)-propane dimethiodide

CAS: 57168-28-2

RTECS: —

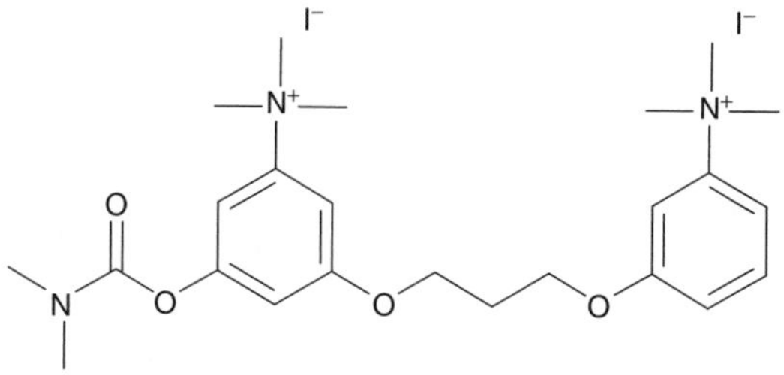

$C_{24}H_{37}N_3O_4 \cdot I_2$

Crystalline solid. Various other salts have been reported.

Exposure Hazards

Human toxicity values have not been established or have not been published. However, based on available information, this agent appears to be just as toxic as VX (C01-A016).

Properties:

MW: 685.4	*VP*: Negligible	*FlP*: —
D: —	*VD*: —	*LEL*: —
MP: 360°F	*Vlt*: —	*UEL*: —
BP: —	H_2O: Soluble	*RP*: Persistent
Vsc: —	*Sol*: —	*IP*: —

C02-A013

Decamethylene-(3-hydroxyquinuclidinium bromide) [(2-dimethylcarbamoxy-ethyl)-dimethylammonium bromide]

CAS: 58619-61-7

RTECS: —

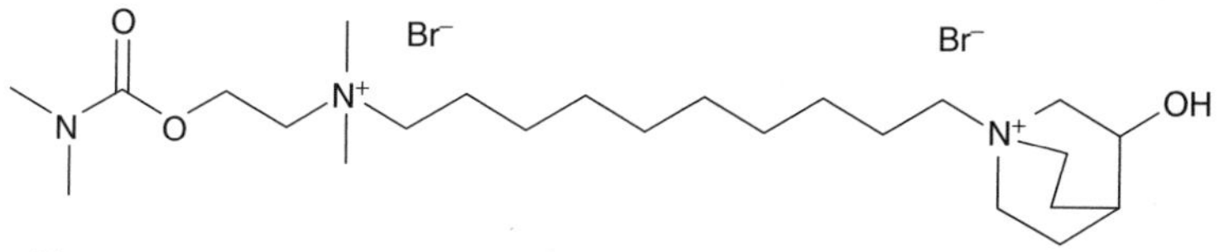

$C_{24}H_{49}N_3O_3 \cdot Br_2$

Specific information on physical appearance is not available for this agent. Various other salts have been reported.

Exposure Hazards

Human toxicity values have not been established or have not been published. However, based on available information, this agent appears to be approximately 50 times less toxic than VX (C01-A016).

Properties:

MW: 587.5	*VP*: Negligible	*FIP*: —
D: —	*VD*: —	*LEL*: —
MP: 286°F	*Vlt*: —	*UEL*: —
BP: —	*H_2O*: Soluble	*RP*: Persistent
Vsc: —	*Sol*: Alcohols	*IP*: —

C02-A014

1,8-Bis[*N*-(3-dimethylcarbamoxy-α-picolyl)-*N*,*N*-dimethylammonio]octane-2,7-dione dibromide

CAS: 77104-01-9
RTECS: —

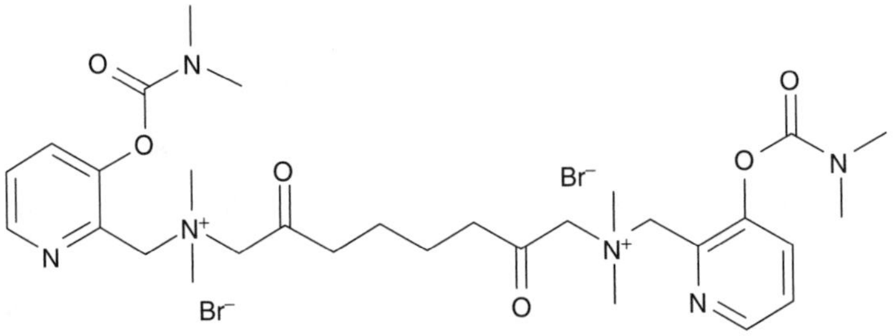

$C_{30}H_{46}N_6O_6 \cdot Br_2$
Crystalline solid. Various other salts have been reported.

Exposure Hazards
Human toxicity values have not been established or have not been published. However, based on available information, this agent appears to be approximately three times more toxic than VX (C01-A016).

Properties:

MW: 746.5	*VP*: Negligible	*FIP*: —
D:—	*VD*: —	*LEL*: —
MP: 410°F	*Vlt*: —	*UEL*: —
BP: —	*H_2O*: Soluble	*RP*: Persistent
Vsc: —	*Sol*: Alcohols	*IP*: —

C02-A015

Octamethylene-bis(5-dimethylcarbamoxyisoquinolinium bromide)

CAS: 110203-40-2
RTECS: —

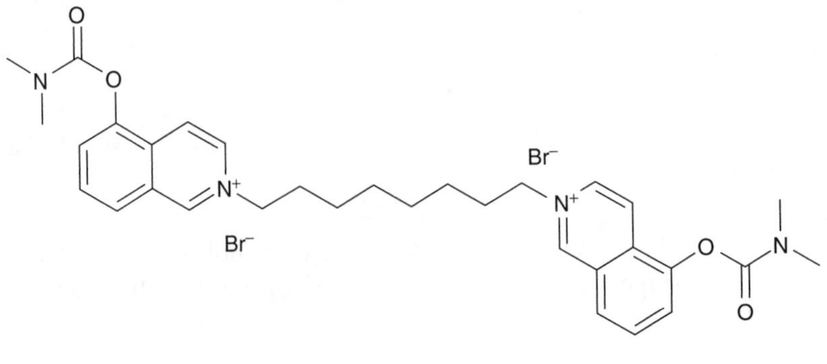

$C_{32}H_{40}N_4O_4 \cdot Br_2$
Solid. Various other salts have been reported.

Exposure Hazards
Human toxicity values have not been established or have not been published. However, based on available information, this agent appears to be just as toxic as VX (C01-A016).

Properties:

MW: 704.5	*VP*: Negligible	*FlP*: —
D: —	*VD*: —	*LEL*: —
MP: 250–257°F	*Vlt*: —	*UEL*: —
BP: —	H_2O: Soluble	*RP*: Persistent
Vsc: —	*Sol*: —	*IP*: —

C02-A016

1,8-Bis[*N*-(2-dimethylcarbamoxybenzyl)-*N*,*N*-dimethylammonio]octane-2,7-dione dibromide

CAS: 110801-39-3
RTECS: —

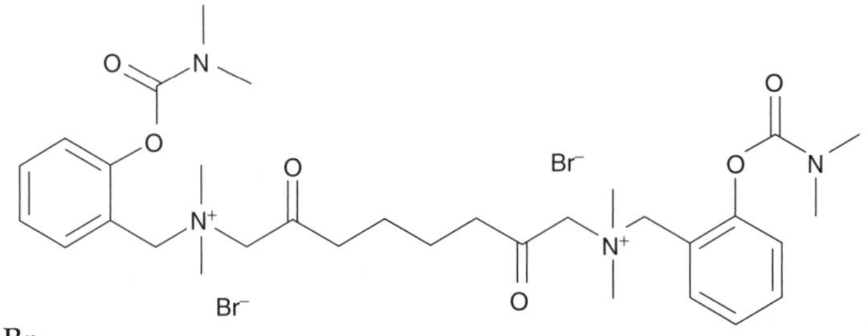

$C_{32}H_{48}N_4O_6 \cdot Br_2$
White crystalline solid. Various other salts have been reported.

Exposure Hazards
Human toxicity values have not been established or have not been published. However, based on available information, this agent appears to be just as toxic as VX (C01-A016).

Properties:

MW: 744.6	*VP*: Negligible	*FlP*: —
D: —	*VD*: —	*LEL*: —
MP: 329–336°F	*Vlt*: —	*UEL*: —
BP: —	H_2O: Soluble	*RP*: Persistent
Vsc: —	*Sol*: Acetonitrile	*IP*: —

C02-A017

1,10-Bis[*N*-(3-dimethylcarbamoxy-α-picolyl)-*N*,*N*-dimethylammonio]decane-2,9-dione dibromide

CAS: 77103-99-2
RTECS: —

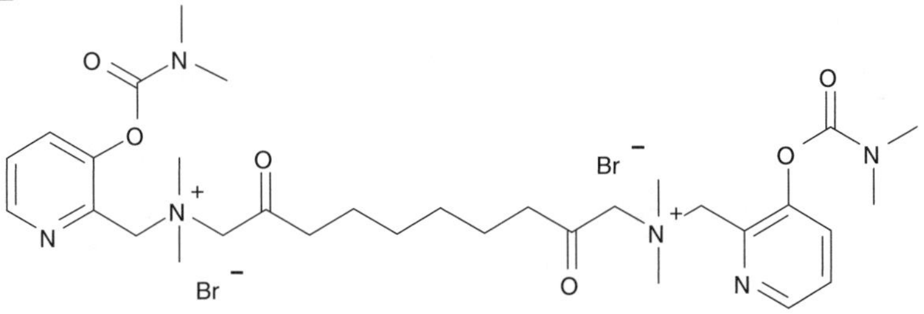

$C_{32}H_{50}N_6O_6 \cdot Br_2$
Crystalline solid. Various other salts have been reported.

Exposure Hazards
Human toxicity values have not been established or have not been published. However, based on available information, this agent appears to be approximately three times more toxic than VX (C01-A016).

Properties:

MW: 774.6	*VP*: Negligible	*FlP*: —
D: —	*VD*: —	*LEL*: —
MP: 336°F	*Vlt*: —	*UEL*: —
BP: —	H_2O: Soluble	*RP*: Persistent
Vsc: —	*Sol*: —	*IP*: —

C02-A018

1,8-Bis[(3-dimethylcarbamoxy-α-picolinyl)ethylamino]octane dimethobromide
CAS: 113402-83-8
RTECS: —

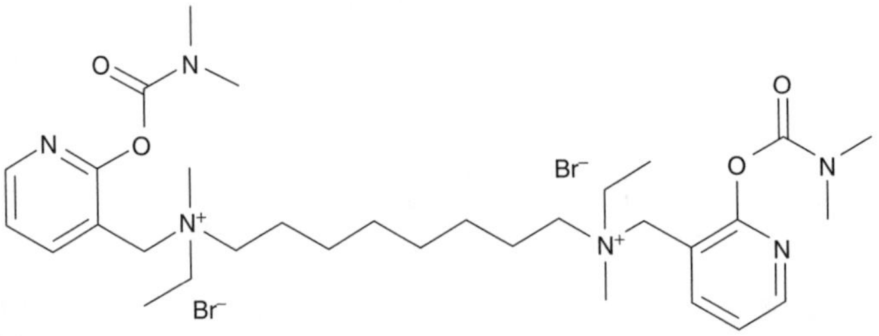

$C_{32}H_{54}N_6O_4 \cdot Br_2$
White crystalline solid. Various other salts have been reported.

Exposure Hazards
Human toxicity values have not been established or have not been published. However, based on available information, this agent appears to be approximately twice as toxic as VX (C01-A016).

Properties:

MW: 746.3	*VP*: Negligible	*FlP*: —
D: —	*VD*: —	*LEL*: —
MP: 282°F	*Vlt*: —	*UEL*: —
BP: —	H_2O: Soluble	*RP*: Persistent
Vsc: —	*Sol*: Acetonitrile	*IP*: —

C02-A019

1,10-Bis[*N*-(2-dimethylcarbamoxybenzyl)-*N*,*N*-dimethylammonio]decane-2,9-dione dibromide
CAS: 110801-36-0
RTECS: —

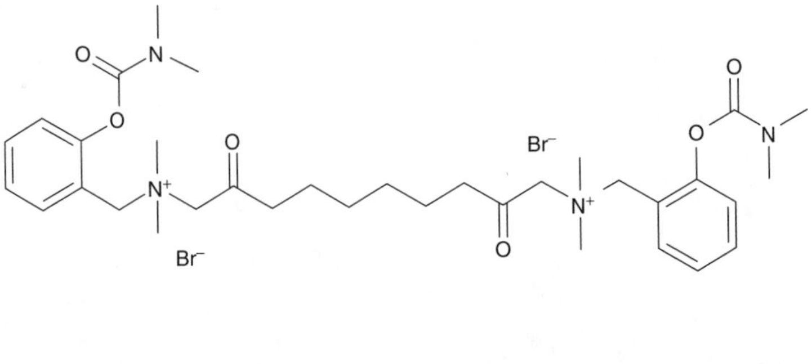

$C_{34}H_{52}N_4O_6 \cdot Br_2$
White crystalline solid. Various other salts have been reported.

Exposure Hazards
Human toxicity values have not been established or have not been published. However, based on available information, this agent appears to be slightly more toxic than VX (C01-A016).

Properties:

MW: 772.6	*VP*: Negligible	*FlP*: —
D: —	*VD*: —	*LEL*: —
MP: 372–376°F	*Vlt*: —	*UEL*: —
BP: —	H_2O: Soluble	*RP*: Persistent
Vsc: —	*Sol*: Acetonitrile	*IP*: —

C02-A020

1,10-Bis[*N*-(3-dimethylcarbamoxy-α-picolyl)-*N*-ethyl-*N*-methylammonio]decane-2,9-dione dibromide
CAS: 77104-00-8
RTECS: —

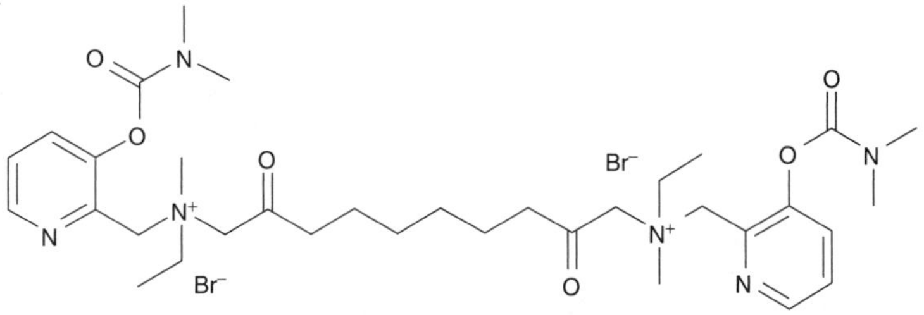

$C_{34}H_{54}N_6O_6 \cdot Br_2$
Crystalline solid. Various other salts have been reported.

Exposure Hazards
Human toxicity values have not been established or have not been published. However, based on available information, this agent appears to be approximately twice as toxic as VX (C01-A016).

Properties:

MW: 802.7	*VP*: Negligible	*FlP*: —
D: —	*VD*: —	*LEL*: —
MP: Decomposes	*Vlt*: —	*UEL*: —
BP: —	H_2O: Soluble	*RP*: Persistent
Vsc: —	*Sol*: Acetonitrile	*IP*: —

C02-A021

1,8-Bis[(2-dimethylcarbamoxybenzyl)ethylamino]octane dimethobromide
CAS: 117585-55-4
RTECS: —

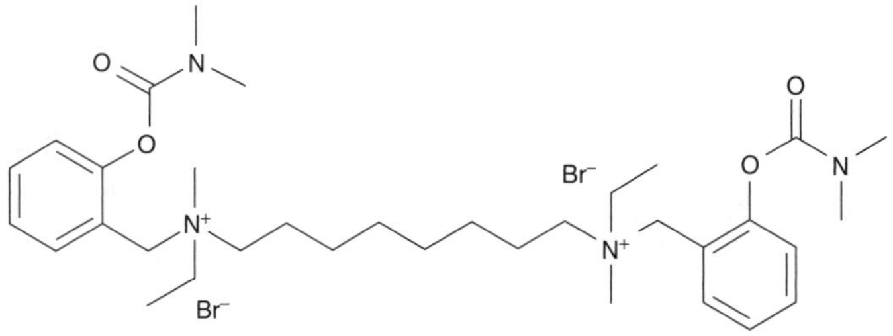

$C_{34}H_{56}N_4O_4 \cdot Br_2$
White crystalline solid. Various other salts have been reported.

Exposure Hazards
Human toxicity values have not been established or have not been published. However, based on available information, this agent appears to be slightly more toxic than VX (C01-A016).

Properties:

MW: 744.3	*VP*: Negligible	*FlP*: —
D: —	*VD*: —	*LEL*: —
MP: 349°F	*Vlt*: —	*UEL*: —
BP: —	H_2O: Soluble	*RP*: Persistent
Vsc: —	*Sol*:Acetonitrile	*IP*: —

C02-A022

1,10-Bis[(3-dimethylcarbamoxy-α-picolinyl)ethylamino]decane dimethobromide
CAS: 113402-82-7
RTECS: —

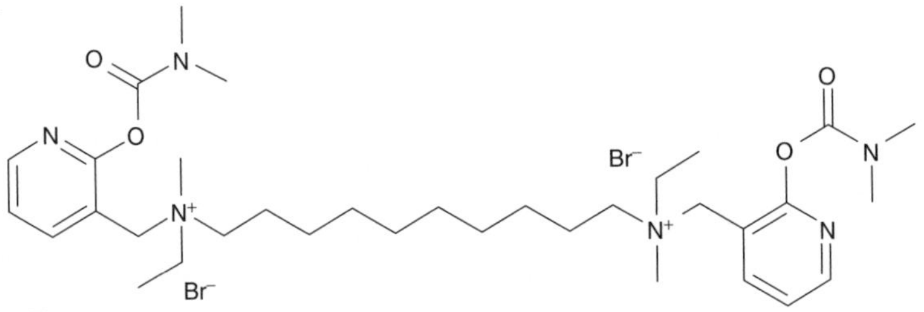

$C_{34}H_{58}N_6O_4 \cdot Br_2$
White crystalline solid. Various other salts have been reported.

Exposure Hazards
Human toxicity values have not been established or have not been published. However, based on available information, this agent appears to be approximately twice as toxic as VX (C01-A016).

Properties:

MW: 774.3	*VP*: Negligible	*FlP*: —
D:—	*VD*: —	*LEL*: —
MP: 343°F	*Vlt*: —	*UEL*: —
BP: —	H_2O: Soluble	*RP*: Persistent
Vsc: —	*Sol*: Acetonitrile	*IP*: —

C02-A023

1,10-Bis{*N*-[1-(2-dimethylcarbamoxyphenyl)ethyl]-*N*,*N*-dimethylammonio}decane-2,9-dione tetraphenylboronate
CAS: 110801-38-2
RTECS: —

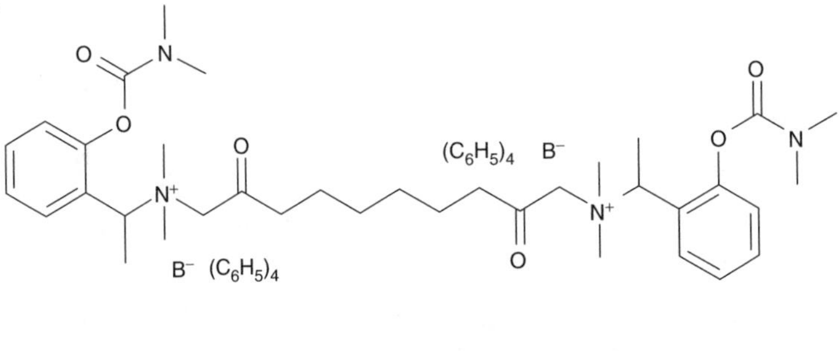

$C_{36}H_{56}N_4O_6 \cdot B_2C_{48}H_{40}$
Crystalline solid. Bromide salt is a hygroscopic white solid. Various other salts have been reported.

Exposure Hazards
Human toxicity values have not been established or have not been published. However, based on available information, this agent appears to be slightly more toxic than VX (C01-A016).

Properties:

MW: 1279.3	*VP*: Negligible	*FlP*: —
D: —	*VD*: —	*LEL*: —
MP: 180–183°F	*Vlt*: —	*UEL*: —
BP: —	H_2O: Insoluble	*RP*: Persistent
Vsc: —	*Sol*: —	*IP*: —

Bromide Salt
MW: 800.7
H_2O: Soluble
Sol: Acetonitrile

C02-A024

Bis{α-[(3-dimethylcarbamoxyphenyl)methylamino]}-4,4′-biacetophenone dimethobromide
CAS: 113402-23-6
RTECS: —

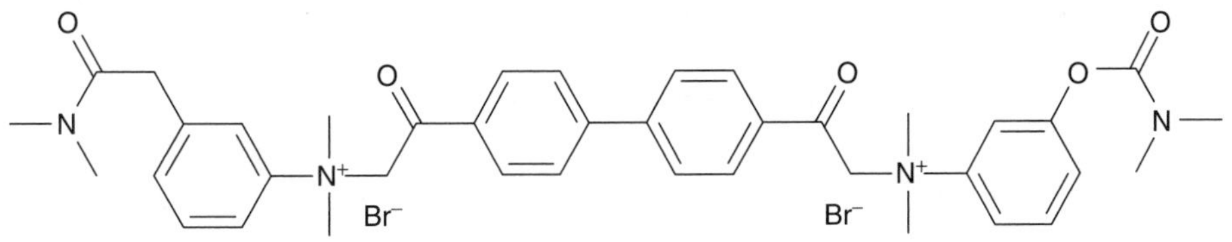

$C_{38}H_{44}N_4O_6 \cdot Br_2$
Light yellow crystalline solid. Various other salts have been reported.

Exposure Hazards
Human toxicity values have not been established or have not been published. However, based on available information, this agent appears to be approximately half as toxic as VX (C01-A016).

Properties:

MW: 812.2	*VP*: Negligible	*FlP*: —
D: —	*VD*: —	*LEL*: —
MP: 302°F	*Vlt*: —	*UEL*: —
BP: —	H_2O: Soluble	*RP*: Persistent
Vsc: —	*Sol*: Alcohols	*IP*: —

C02-A025

1,10-Bis[(2-dimethylcarbamoxybenzyl)propylamino]decane dimethobromide

CAS: 117569-53-6
RTECS: —

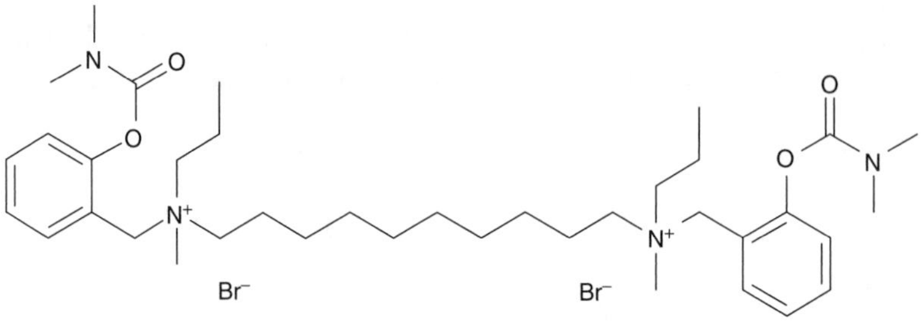

$C_{38}H_{64}N_4O_4 \cdot Br_2$
White crystalline solid. Various other salts have been reported.

Exposure Hazards
Human toxicity values have not been established or have not been published. However, based on available information, this agent appears to be just as toxic as VX (C01-A016).

Properties:

MW: 800.4	*VP*: Negligible	*FlP*: —
D: —	*VD*: —	*LEL*: —
MP: 180°F	*Vlt*: —	*UEL*: —
BP: —	H_2O: Soluble	*RP*: Persistent
Vsc: —	*Sol*: Acetonitrile	*IP*: —

C02-A026

Bis{α-[(3-dimethylcarbamoxy-α-picolinyl)pyrrolidinio]}4,4′-biacetophenone dibromide

CAS: 113402-25-8
RTECS: —

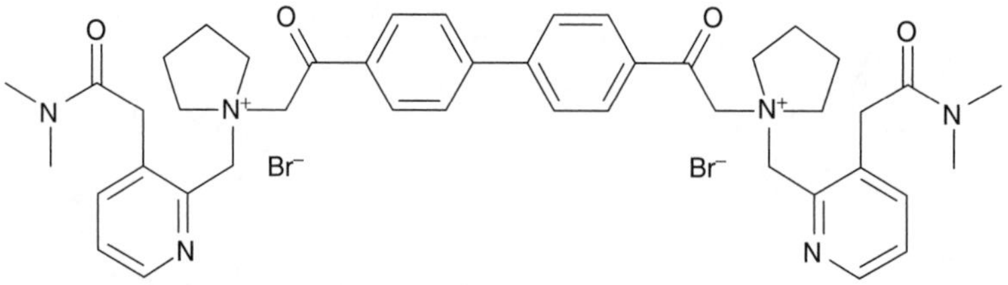

$C_{42}H_{50}N_6O_6 \cdot Br_2$
Crystalline solid. Various other salts have been reported.

Exposure Hazards
Human toxicity values have not been established or have not been published. However, based on available information, this agent appears to be approximately half as toxic as VX (C01-A016).

Properties:

MW: 893.3	*VP*: Negligible	*FlP*: —
D: —	*VD*: —	*LEL*: —
MP: Decomposes	*Vlt*: —	*UEL*: —
BP: —	H_2O: Soluble	*RP*: Persistent
Vsc: —	*Sol*: Alcohols	*IP*: —

C02-A027

1-(4-Dimethylcarbamoxy-2-dimethylaminophenoxy)-3-(4-dimethylaminophenoxy)-propane dimethiodide

CAS: 58149-55-6
RTECS: —

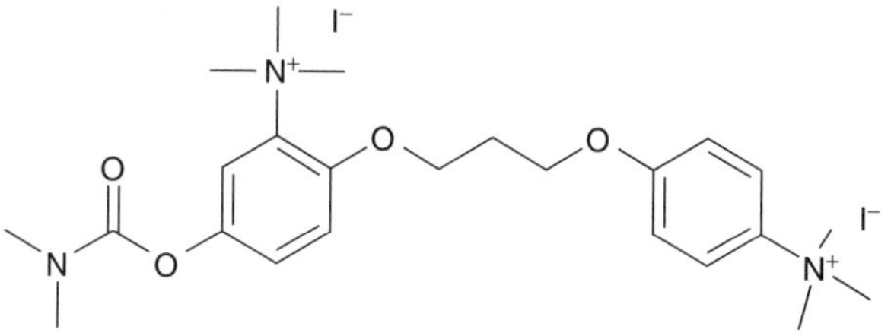

$C_{24}H_{37}N_3O_4 \cdot I_2$
Solid. Various other salts have been reported.

Exposure Hazards
Human toxicity values have not been established or have not been published. However, based on available information, this agent appears to be approximately one-fifth as toxic as VX (C01-A016).

Properties:

MW: 685.4
D: —
MP: 349–360°F
BP: —
Vsc: —
VP: Negligible
VD: —
Vlt: —
H_2O: Soluble
Sol: —
FlP: —
LEL: —
UEL: —
RP: Persistent
IP: —

Oxalate salt
MW: 591.4
MP: 329°F

C02-A028

1-(*N*,*N*,*N*-Trimethylammonio)-8-[*N*-(2-dimethylcarbamoxybenzyl)-*N*,*N*-dimethylammonio]octane dibromide

CAS: 77104-70-2
RTECS: —

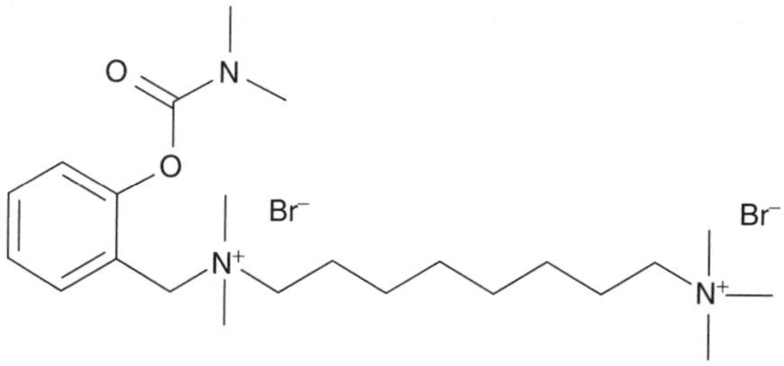

$C_{23}H_{43}N_3O_2 \cdot Br_2$
Hygroscopic white crystalline solid. Various other salts have been reported.

Exposure Hazards
Human toxicity values have not been established or have not been published. However, based on available information, this agent appears to be just as toxic as VX (C01-A016).

Properties:

MW: 553.4	*VP*: Negligible	*FlP*: —
D: —	*VD*: —	*LEL*: —
MP: —	*Vlt*: —	*UEL*: —
BP: —	*H_2O*: Soluble	*RP*: Persistent
Vsc: —	*Sol*: Acetonitrile	*IP*: —

C02-A029

1-(*N*,*N*,*N*-Trimethylammonio)-10-[*N*-(5-dimethylcarbamoxy)isoquinolinio]decane dibromide

CAS: 77223-00-8
RTECS: —

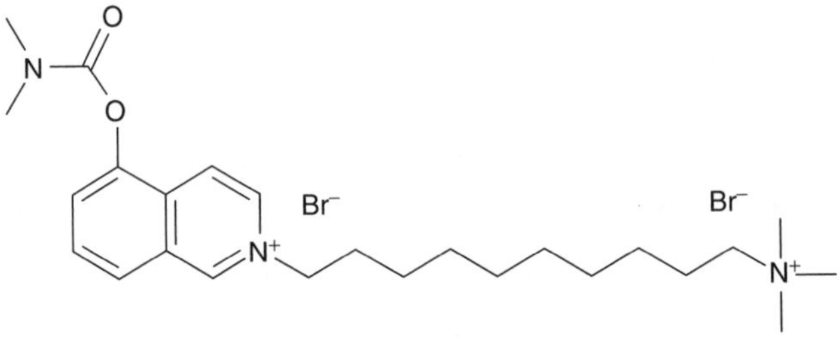

$C_{25}H_{41}N_3O_2 \cdot Br_2$

Deliquescent crystalline solid. Various other salts have been reported.

Exposure Hazards

Human toxicity values have not been established or have not been published. However, based on available information, this agent appears to be approximately one-third as toxic as VX (C01-A016).

Properties:

MW: 575.4	*VP*: Negligible	*FlP*: —
D: —	*VD*: —	*LEL*: —
MP: —	*Vlt*: —	*UEL*: —
BP: —	*H_2O*: Soluble	*RP*: Persistent
Vsc: —	*Sol*: Acetonitrile	*IP*: —

C02-A030

1-[*N*-(2-Dimethylcarbamoxybenzyl)pyrrolinio]-8-(*N*,*N*,*N*-trimethylammonio)octane dibromide

CAS: 77111-74-1
RTECS: —

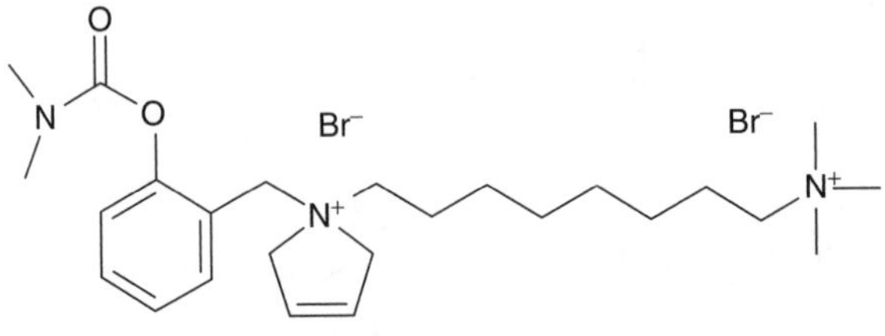

$C_{25}H_{43}N_3O_2 \cdot Br_2$

Deliquescent white solid. Various other salts have been reported.

Exposure Hazards

Human toxicity values have not been established or have not been published. However, based on available information, this agent appears to be slightly more toxic than VX (C01-A016).

Properties:

MW: 577.4	*VP*: Negligible	*FlP*: —
D: —	*VD*: —	*LEL*: —
MP: —	*Vlt*: —	*UEL*: —
BP: —	H_2O: Soluble	*RP*: Persistent
Vsc: —	*Sol*: Acetonitrile	*IP*: —

C02-A031

1-(*N*,*N*-Dimethyl-*N*-cyanomethylammonio)-10-[*N*-(3-dimethylcarbamoxy-α-picolinyl)-*N*,*N*-dimethylammonio]decane dibromide

CAS: 109973-92-4

RTECS: —

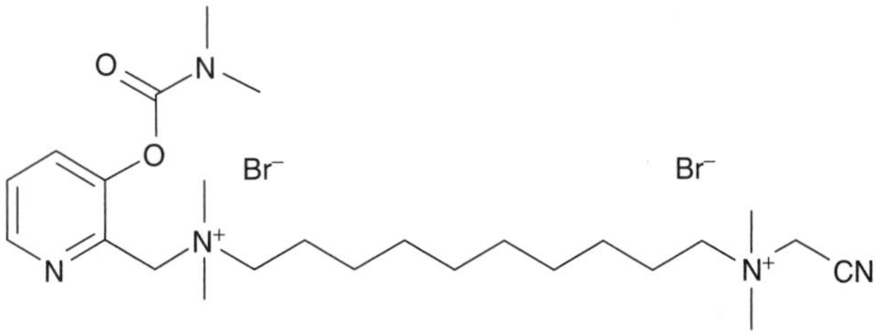

$C_{25}H_{45}N_5O_2 \cdot Br_2$

White crystalline solid. Various other salts have been reported.

Exposure Hazards

Human toxicity values have not been established or have not been published. However, based on available information, this agent appears to be slightly more toxic than VX (C01-A016).

Properties:

MW: 607.5	*VP*: Negligible	*FlP*: —
D: —	*VD*: —	*LEL*: —
MP: —	*Vlt*: —	*UEL*: —
BP: —	H_2O: Soluble	*RP*: Persistent
Vsc: —	*Sol*: Alcohols; Acetonitrile	*IP*: —

C02-A032

1-[*N*-(3-Dimethylcarbamoxy-α-picolyl)-*N*,*N*-dimethylammonio]-10-(*N*-carbamoxymethyl-*N*,*N*-dimethylammonio)decane dibromide

CAS: 78297-56-0

RTECS: —

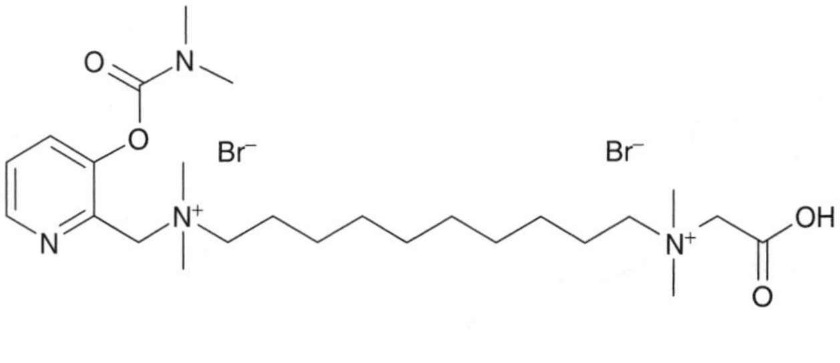

$C_{25}H_{46}N_4O_4 \cdot Br_2$
Deliquescent white crystalline solid. Various other salts have been reported.

Exposure Hazards
Human toxicity values have not been established or have not been published. However, based on available information, this agent appears to be approximately half as toxic as VX (C01-A016).

Properties:

MW: 626.5	*VP*: Negligible	*FlP*: —
D: —	*VD*: —	*LEL*: —
MP: —	*Vlt*: —	*UEL*: —
BP: —	H_2O: Soluble	*RP*: Persistent
Vsc: —	*Sol*: Acetonitrile	*IP*: —

C02-A033

1-(*N*,*N*,*N*-Trimethylammonio)-10-[*N*-(2-dimethylcarbamoxybenzyl)-*N*,*N*-dimethylammonio]decane dibromide
CAS: 77104-68-8
RTECS: —

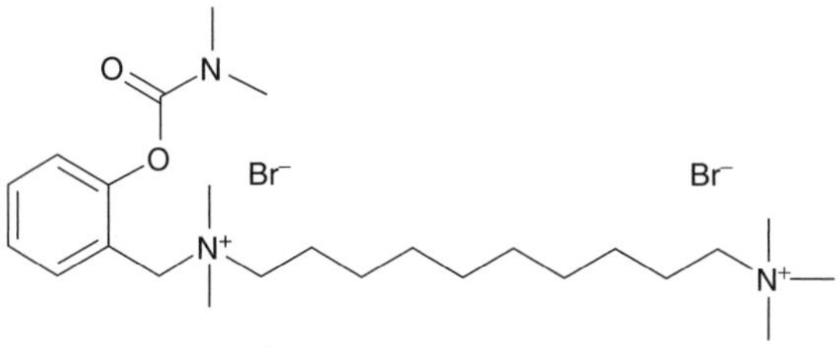

$C_{25}H_{47}N_3O_2 \cdot Br_2$
White crystalline solid. Various other salts have been reported.

Exposure Hazards
Human toxicity values have not been established or have not been published. However, based on available information, this agent appears to be just as toxic as VX (C01-A016).

Properties:

MW: 581.5	*VP*: Negligible	*FlP*: —
D: —	*VD*: —	*LEL*: —
MP: —	*Vlt*: —	*UEL*: —
BP: —	H_2O: Soluble	*RP*: Persistent
Vsc: —	*Sol*: Acetonitrile	*IP*: —

C02-A034

1-[*N*,*N*-Dimethyl-*N*-(2-hydroxy)ethylammonio]-10-[*N*-(3-dimethylcarbamoxy-α-picolinyl)-*N*,*N*-dimethylammonio]decane dibromide
CAS: 77104-62-2
RTECS: —

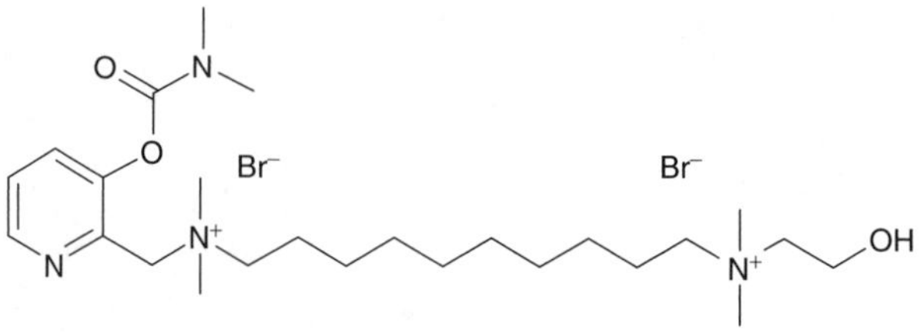

$C_{25}H_{48}N_4O_3 \cdot Br_2$
White crystalline solid. Various other salts have been reported.

Exposure Hazards
Human toxicity values have not been established or have not been published. However, based on available information, this agent appears to be approximately twice as toxic as VX (C01-A016).

Properties:

MW: 612.5	*VP*: Negligible	*FlP*: —
D: —	*VD*: —	*LEL*: —
MP: —	*Vlt*: —	*UEL*: —
BP: —	*H_2O*: Soluble	*RP*: Persistent
Vsc: —	*Sol*: Alcohols; Acetonitrile	*IP*: —

C02-A035

1-Pyridinio-10-[*N*-(3-dimethylcarbamoxy-α-picolinyl)-*N*,*N*-dimethylammonio]decane dibromide
CAS: 77223-01-9
RTECS: —

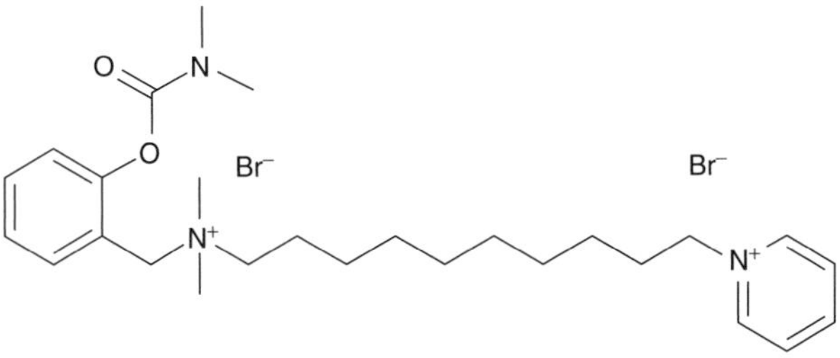

$C_{26}H_{42}N_4O_2 \cdot Br_2$
Deliquescent crystalline solid. Various other salts have been reported.

Exposure Hazards
Human toxicity values have not been established or have not been published. However, based on available information, this agent appears to be just as toxic as VX (C01-A016).

Properties:

MW: 602.5	*VP*: Negligible	*FlP*: —
D: —	*VD*: —	*LEL*: —
MP: —	*Vlt*: —	*UEL*: —
BP: —	*H_2O*: Soluble	*RP*: Persistent
Vsc: —	*Sol*: Alcohols; Pyridine	*IP*: —

C02-A036

1-(*N*-Methyl)pyrrolidinio-10-[*N*-(3-dimethylcarbamoxy-α-picolinyl)-*N*,*N*-dimethylammonio]decane dibromide
CAS: 110344-83-7
RTECS: —

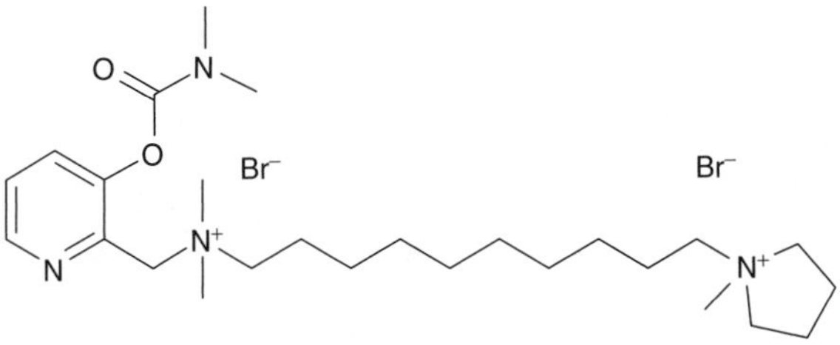

$C_{26}H_{48}N_4O_2 \cdot Br_2$
Deliquescent white solid. Various other salts have been reported.

Exposure Hazards
Human toxicity values have not been established or have not been published. However, based on available information, this agent appears to be slightly more toxic than VX (C01-A016).

Properties:

MW: 608.5	*VP*: Negligible	*FlP*: —
D: —	*VD*: —	*LEL*: —
MP: —	*Vlt*: —	*UEL*: —
BP: —	H_2O: Soluble	*RP*: Persistent
Vsc: —	*Sol*: Acetonitrile	*IP*: —

C02-A037

1-[*N*,*N*-Dimethyl-*N*-(3-hydroxy)propylammonio]-10-[*N*-(3-dimethylcarbamoxy-α-picolinyl)-*N*,*N*-dimethylammonio]decane dibromide

CAS: 77104-63-3
RTECS: —

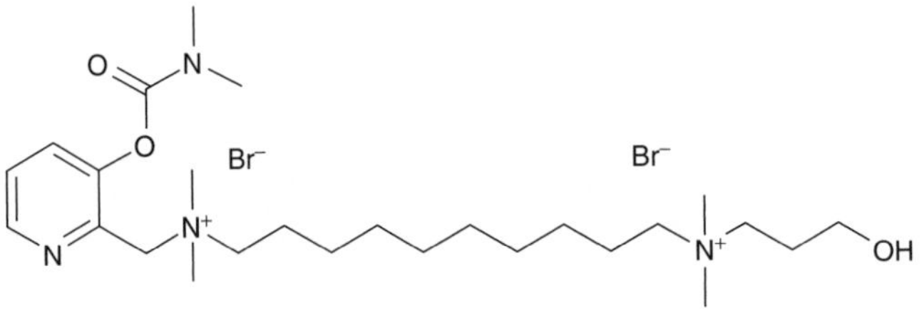

$C_{26}H_{50}N_4O_3 \cdot Br_2$
Deliquescent white solid. Various other salts have been reported.

Exposure Hazards
Human toxicity values have not been established or have not been published. However, based on available information, this agent appears to be slightly more toxic than VX (C01-A016).

Properties:

MW: 626.5	*VP*: Negligible	*FlP*: —
D: —	*VD*: —	*LEL*: —
MP: —	*Vlt*: —	*UEL*: —
BP: —	H_2O: Soluble	*RP*: Persistent
Vsc: —	*Sol*: Alcohols; Acetonitrile	*IP*: —

C02-A038

1-[*N*,*N*-Di(2-hydroxy)ethyl-*N*-methylammonio]-10-[*N*-(3-dimethylcarbamoxy-α-picolinyl)-*N*,*N*-dimethylammonio]decane dibromide

CAS: 77104-64-4
RTECS: —

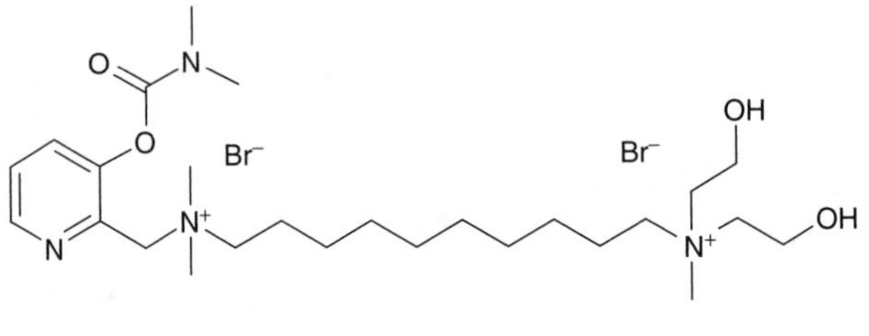

$C_{26}H_{50}N_4O_4 \cdot Br_2$
Deliquescent white solid. Various other salts have been reported.

Exposure Hazards
Human toxicity values have not been established or have not been published. However, based on available information, this agent appears to be slightly more toxic than VX (C01-A016).

Properties:

MW: 642.5	*VP*: Negligible	*FlP*: —
D: —	*VD*: —	*LEL*: —
MP: —	*Vlt*: —	*UEL*: —
BP: —	H_2O: Soluble	*RP*: Persistent
Vsc: —	*Sol*: Acetonitrile	*IP*: —

C02-A039

1-(4-Aldoximino)pyridinio-10-[*N*-(3-dimethylcarbamoxy-α-picolinyl)-*N*,*N*-dimethylammonio]decene dibromide
CAS: 77223-02-0
RTECS: —

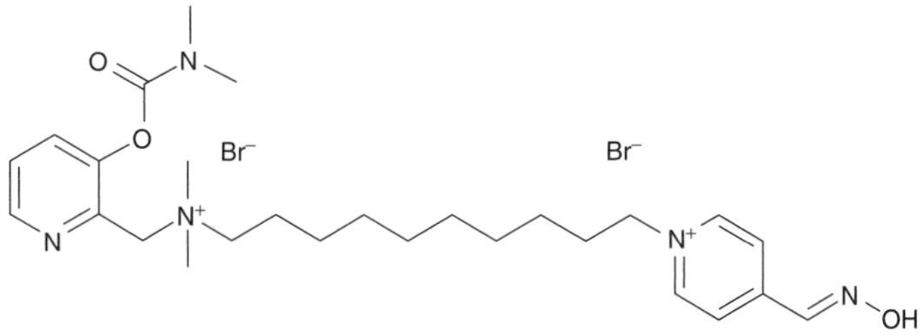

O N O Br⁻ Br⁻ N N⁺ N⁺ N OH

$C_{27}H_{43}N_5O_3 \cdot Br_2$
Deliquescent crystalline solid. Various other salts have been reported.

Exposure Hazards
Human toxicity values have not been established or have not been published. However, based on available information, this agent appears to be slightly more toxic than VX (C01-A016).

Properties:

MW: 645.5	*VP*: Negligible	*FlP*: —
D: —	*VD*: —	*LEL*: —
MP: —	*Vlt*: —	*UEL*: —
BP: —	H_2O: Soluble	*RP*: Persistent
Vsc: —	*Sol*: Acetonitrile	*IP*: —

C02-A040

1-[*N*-(2-Dimethylcarbamoxybenzyl)pyrrolinio]-10-(*N*,*N*,*N*-trimethylammonio)decane dibromide
CAS: 77111-71-8
RTECS: —

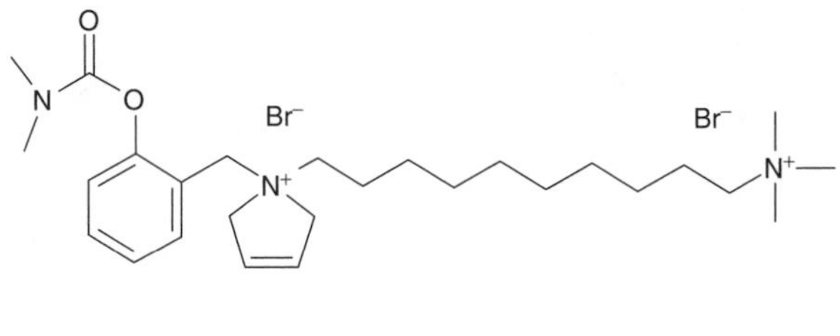

O N O Br⁻ Br⁻ N⁺ N⁺

$C_{27}H_{47}N_3O_2 \cdot Br_2$
Deliquescent white crystalline solid. Various other salts have been reported.

Exposure Hazards
Human toxicity values have not been established or have not been published. However, based on available information, this agent appears to be just as toxic as VX (C01-A016).

Properties:

MW: 605.5	*VP*: Negligible	*FlP*: —
D: —	*VD*: —	*LEL*: —
MP: —	*Vlt*: —	*UEL*: —
BP: —	*H_2O*: Soluble	*RP*: Persistent
Vsc: —	*Sol*: Acetonitrile	*IP*: —

C02-A041

1-[*N*,*N*-Dimethyl-*N*-(3-cyanopropyl)ammonio]-10-[*N*-(3-dimethylcarbamoxy-α-picolinyl)-*N*,*N*-dimethylammonio]decane dibromide

CAS: 109973-93-5
RTECS: —

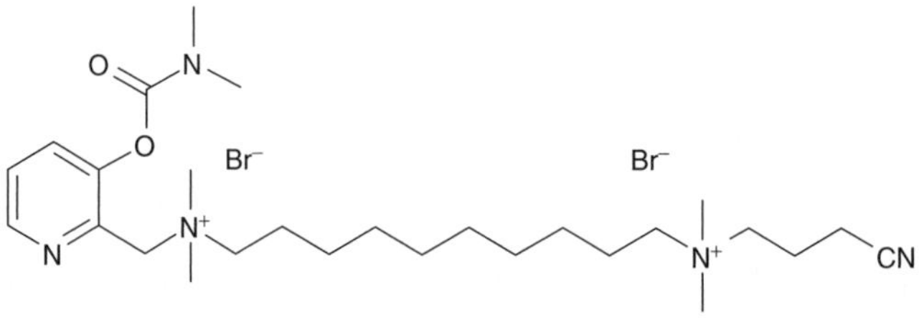

O N O Br⁻ Br⁻ N N⁺ N⁺ CN

$C_{27}H_{49}N_5O_2 \cdot Br_2$
Deliquescent white crystalline solid. Various other salts have been reported.

Exposure Hazards
Human toxicity values have not been established or have not been published. However, based on available information, this agent appears to be slightly more toxic than VX (C01-A016).

Properties:

MW: 635.5	*VP*: Negligible	*FlP*: —
D: —	*VD*: —	*LEL*: —
MP: —	*Vlt*: —	*UEL*: —
BP: —	*H_2O*: Soluble	*RP*: Persistent
Vsc: —	*Sol*: Alcohols; Acetonitrile	*IP*: —

C02-A042

1-(3-Hydroxy)quinuclidinio-10-[*N*-(3-dimethylcarbamoxy-α-picolinyl)-*N*,*N*-dimethylammonio]decane dibromide

CAS: 110344-82-6
RTECS: —

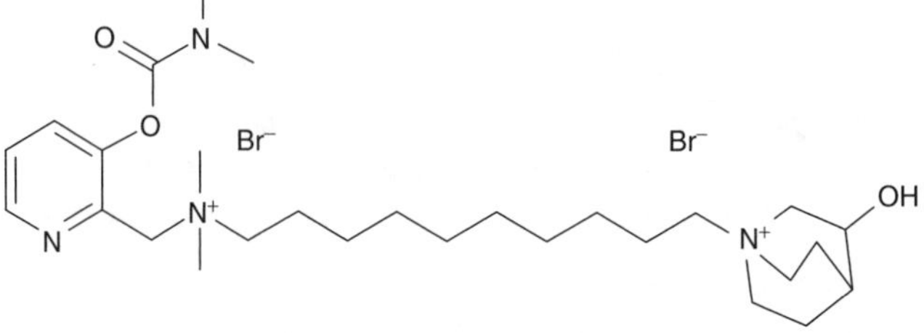

O N O Br⁻ Br⁻ OH N N⁺ N⁺

$C_{28}H_{50}N_4O_3 \cdot Br_2$
Deliquescent white solid. Various other salts have been reported.

Exposure Hazards
Human toxicity values have not been established or have not been published. However, based on available information, this agent appears to be slightly more toxic than VX (C01-A016).

Properties:

MW: 650.5	*VP*: Negligible	*FlP*: —
D: —	*VD*: —	*LEL*: —
MP: —	*Vlt*: —	*UEL*: —
BP: —	H_2O: Soluble	*RP*: Persistent
Vsc: —	*Sol*: Alcohols	*IP*: —

C02-A043

1-[*N*,*N*-Dimethyl-*N*-(2-acetoxy-2-methylethyl)ammonio]-10-[*N*-(3-dimethylcarbamoxy-α-picolinyl)-*N*,*N*-dimethylammonio]decane dibromide

CAS: 110914-01-7
RTECS: —

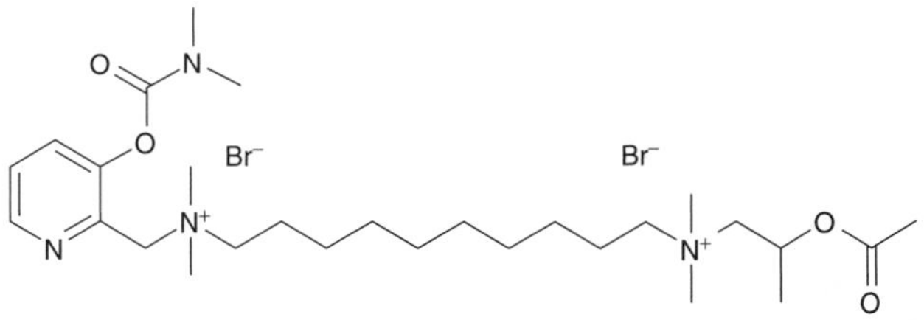

$C_{28}H_{52}N_4O_4 \cdot Br_2$
Deliquescent white solid. Various other salts have been reported.

Exposure Hazards
Human toxicity values have not been established or have not been published. However, based on available information, this agent appears to be just as toxic as VX (C01-A016).

Properties:

MW: 668.6	*VP*: Negligible	*FlP*: —
D: —	*VD*: —	*LEL*: —
MP: —	*Vlt*: —	*UEL*: —
BP: —	H_2O: Soluble	*RP*: Persistent
Vsc: —	*Sol*: Acetonitrile	*IP*: —

C02-A044

1-[*N*-(3-Hydroxy-α-picolyl)-*N*,*N*-dimethylammonio]-10-[*N*-(3-dimethylcarbamoxy-α-picolyl)-*N*,*N*-dimethylammonio]decane dibromide

CAS: 77104-09-7
RTECS: —

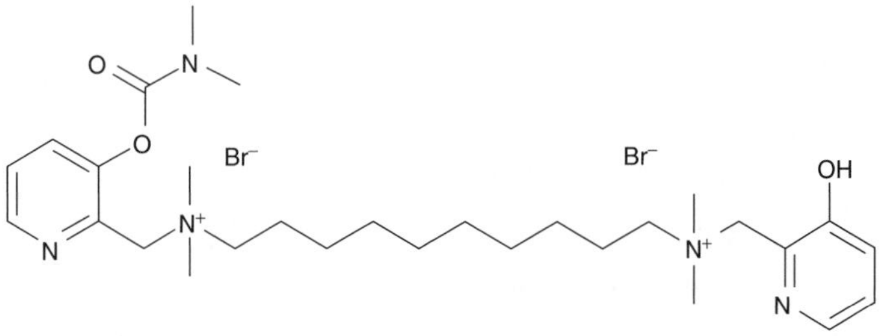

$C_{29}H_{49}N_5O_3 \cdot Br_2$
Deliquescent white crystalline solid. Various other salts have been reported.

Exposure Hazards
Human toxicity values have not been established or have not been published. However, based on available information, this agent appears to be slightly more toxic than VX (C01-A016).

Properties:

MW: 675.6	*VP*: Negligible	*FlP*: —
D: —	*VD*: —	*LEL*: —
MP: —	*Vlt*: —	*UEL*: —
BP: —	H_2O: Soluble	*RP*: Persistent
Vsc: —	*Sol*: Acetonitrile	*IP*: —

C02-A045

1-(*N*,*N*-Dimethyl-*N*-cyclohexylammonio)-10-[*N*-(3-dimethylcarbamoxy-α-picolinyl)-*N*,*N*-dimethylammonio]decane dibromide
CAS: 109973-95-7
RTECS: —

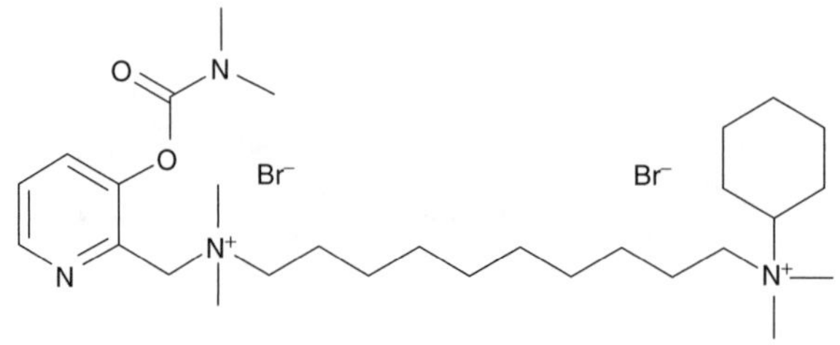

$C_{29}H_{54}N_4O_2 \cdot Br_2$
White solid. Various other salts have been reported.

Exposure Hazards
Human toxicity values have not been established or have not been published. However, based on available information, this agent appears to be approximately half as toxic as VX (C01-A016).

Properties:

MW: 650.6	*VP*: Negligible	*FlP*: —
D: —	*VD*: —	*LEL*: —
MP: —	*Vlt*: —	*UEL*: —
BP: —	H_2O: Soluble	*RP*: Persistent
Vsc: —	*Sol*: Alcohols	*IP*: —

C02-A046

1-[*N*,*N*-Dimethyl-*N*-(2-butyroxyethyl)ammonio]-10-[*N*-(3-dimethylcarbamoxy-α-picolinyl)-*N*,*N*-dimethylammonio]decane dibromide
CAS: 110914-02-8
RTECS: —

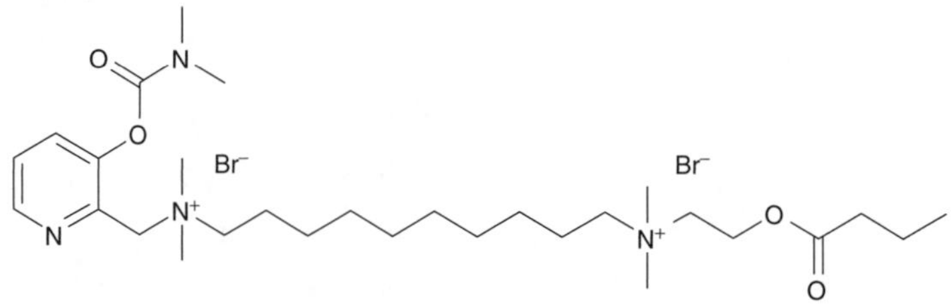

$C_{29}H_{54}N_4O_4 \cdot Br_2$
Hygroscopic white solid. Various other salts have been reported.

Exposure Hazards

Human toxicity values have not been established or have not been published. However, based on available information, this agent appears to be slightly more toxic than VX (C01-A016).

Properties:

MW: 682.6	*VP*: Negligible	*FlP*: —
D: —	*VD*: —	*LEL*: —
MP: —	*Vlt*: —	*UEL*: —
BP: —	H_2O: Soluble	*RP*: Persistent
Vsc: —	*Sol*: Acetonitrile; Chloroform	*IP*: —

C02-A047

1-(*N*,*N*,*N*-Tributylammonio)-10-[*N*-(3-dimethylcarbamoxy-α-picolinyl)-*N*,*N*-dimethylammonio]decane dibromide

CAS: 109973-94-6
RTECS: —

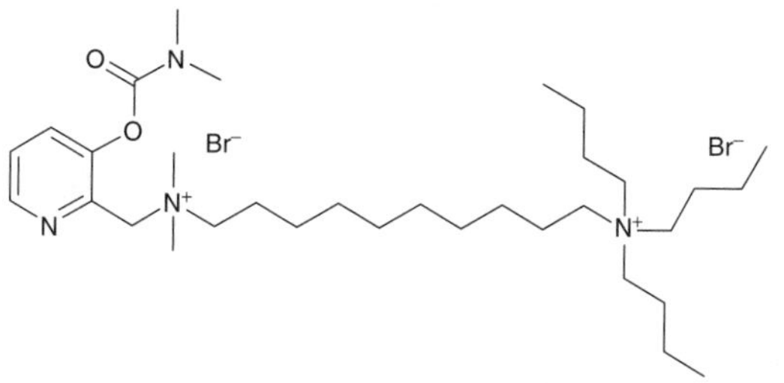

O N O N N⁺ Br⁻ N⁺ Br⁻

$C_{33}H_{64}N_4O_2 \cdot Br_2$
Deliquescent white solid. Various other salts have been reported.

Exposure Hazards

Human toxicity values have not been established or have not been published. However, based on available information, this agent appears to be slightly more toxic than VX (C01-A016).

Properties:

MW: 708.7	*VP*: Negligible	*FlP*: —
D: —	*VD*: —	*LEL*: —
MP: —	*Vlt*: —	*UEL*: —
BP: —	H_2O: Soluble	*RP*: Persistent
Vsc: —	*Sol*: Alcohols; Acetonitrile; Chloroform	*IP*: —

C02-A048

1-[*N*-(4-Dimethylcarbamoxymethyl)benzyl-*N*,*N*-dimethylammonio]-10-[*N*-(3-dimethylcarbamoxy-α-picolinyl)-*N*,*N*-dimethylammonio]decane dibromide

CAS: 77104-59-7
RTECS: —

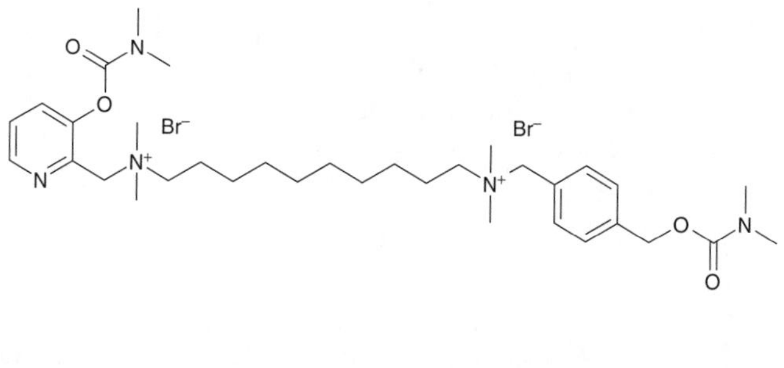

O N O N N⁺ Br⁻ Br⁻ N⁺ O N O

$C_{34}H_{57}N_5O_4 \cdot Br_2$
Deliquescent white crystalline solid. Various other salts have been reported.

Exposure Hazards
Human toxicity values have not been established or have not been published. However, based on available information, this agent appears to be just as toxic as VX (C01-A016).

Properties:

MW: 759.7	*VP*: Negligible	*FlP*: —
D: —	*VD*: —	*LEL*: —
MP: —	*Vlt*: —	*UEL*: —
BP: —	H_2O: Soluble	*RP*: Persistent
Vsc: —	*Sol*: Acetonitrile	*IP*: —

C02-A049

1,10-Bis{[10-(3-dimethylcarbamoxy-α-picolinyl)methylaminodecyl]methylamino}-decane tetramethodbromide
CAS: 110255-20-4
RTECS: —

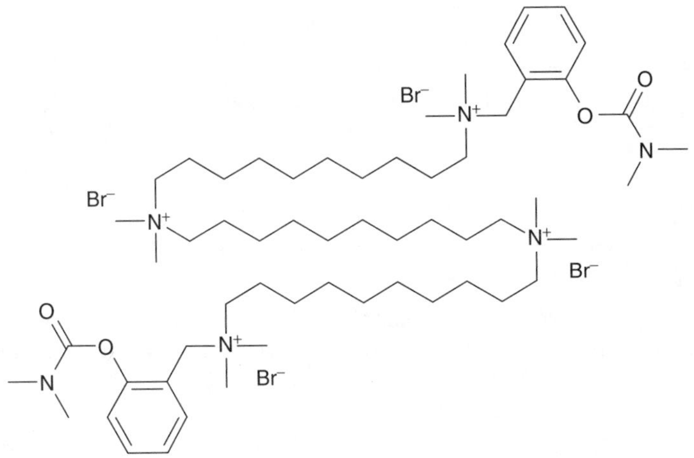

$C_{56}H_{106}N_8O_4 \cdot Br_4$
Hygroscopic white solid. Various other salts have been reported.

Exposure Hazards
Human toxicity values have not been established or have not been published. However, based on available information, this agent appears to be just as toxic as VX (C01-A016).

Properties:

MW: 1275.1	*VP*: Negligible	*FlP*: —
D: —	*VD*: —	*LEL*: —
MP: —	*Vlt*: —	*UEL*: —
BP: —	H_2O: Soluble	*RP*: Persistent
Vsc: —	*Sol*: Acetonitrile	*IP*: —

References

Sommer, Harold Z. "Method for Methylating and Quaternizing," United States Patent 3,903,135, September 2, 1975.
———. "Carbamates," United States Patent 3,919,289, November 11, 1975.
———. "Chemical Agents," United States Patent 4,692,530, September 8, 1987.
Sommer, Harold Z., and John Krenzer. "Quaternary Carbamates," United States Patent 3,901,937, August 26, 1975.
Sommer, Harold Z., John Krenzer, Omer O. Owens, and Jacob I. Miller. "Chemical Agents," United States Patent 4,677,204, June 30, 1987.
Sommer, Harold Z., and Jacob I. Miller. "Chemical Agents," United States Patent 4,675,411, June 23, 1987.
———. "Haloalkyl-carbamoxyalkyl Derivatives," United States Patent 3,956,365, May 11, 1976.
———. "Hydroxyquinuclidine Derivatives," United States Patent 3,919,241, November 11, 1975.
Sommer, Harold Z., and Omer O. Owens. "Chemical Agents," United States Patent 4,686,293, August 11, 1987.
———. "Chemical Agents," United States Patent H443, March 1, 1988.
Sommer, Harold Z., Omer O. Owens, and Jacob I. Miller. "Isoquinilinium Chemical Agents," United States Patent 4,673,745, June 16, 1987.
Sommer, Harold Z., and George E. Wicks, Jr. "Chemical Agents," United States Patent 4,241,209, December 23, 1980.
———. "Chemical Agents," United States Patent 4,241,210, December 23, 1980.
———. "Chemical Agents," United States Patent 4,241,211, December 23, 1980.
———. "Chemical Agents," United States Patent 4,241,212, December 23, 1980.
———. "Chemical Agents," United States Patent 4,241,218, December 23, 1980.
———. "Chemical Agents," United States Patent 4,672,119, June 9, 1987.
———. "Chemical Agents," United States Patent 4,672,122, June 9, 1987.
———. "Chemical Agents," United States Patent 4,672,123, June 9, 1987.
———. "Chemical Agents," United States Patent 4,672,124, June 9, 1987.
———. "Chemical Agents," United States Patent 4,672,069, June 9, 1987.
———. "Chemical Agents," United States Patent 4,677,205, June 3, 1987.
———. "Picolyl Unsymmetrical Bis-quaternary Carbamates," United States Patent 4,246,415, January 20, 1981.
———. "Unsymmetrical Bis-quaternary Amino Acids," United States Patent 4,246,418, January 20, 1981.
———. "Unsymmetrical Pyrrolino Benzyl Quaternary Compounds," United States Patent 4,240,965, December 23, 1980.
Sommer, Harold Z., George E. Wicks, Jr., and Omer O. Owens. "Chemical Agents," United States Patent 4,246,416, January 20, 1981.
———. "Ketobenzylcarbamates," United States Patent 4,677,222, June 30, 1987.
Sommer, Harold Z., George E. Wicks, Jr., and Benjamin Witten. "Chemical Agents," United States Patent 4,672,120, June 9, 1987.
United States Army Headquarters. *Chemical Agent Data Sheets Volume II, Edgewood Arsenal Special Report No. EO-SR-74002*. Washington, DC: Government Printing Office, December 1974.

Section II

Vesicant/Urticant Agents

3

Sulfur and Nitrogen Vesicants

3.1 General Information

The agents in this class are beta-halogenated thioethers, beta-halogenated alkylamines, and alkylating sulfates. The thioether agents are listed in Schedule 1 of the Chemical Weapons Convention (CWC). Only three beta-halogenated alkylamine agents, HN1 (C03-A011), HN2 (C03-A012), and HN3 (C03-A013), are specifically listed in Schedule 1. However, because of the toxicity of the agents and limited commercial application, the remaining alkylamines would be prohibited based on the Guidelines for Schedules of Chemicals.

Sulfur vesicants are first generation chemical warfare agents employed in World War I. Mustard gas (dichloroethylsulfide) was discovered in 1822. It was first employed by the Germans in 1917 at the third battle of Ypres and has been considered a major chemical agent ever since. Nitrogen vesicants are second generation chemical warfare agents developed just prior to World War II. In addition to their vesicant properties, nitrogen agents were studied as a means of poisoning an enemies' water supply because dilute aqueous solutions will rapidly decompose and form neurotoxic products. Several of these agents, including HN3 (C03-A013), were stockpiled by Nazi Germany during World War II but were never used. Modern weapons researchers have isolated and evaluated numerous other variations of the basic thiol and amine structures.

Both sulfur and nitrogen vesicants are easy to synthesize and disperse. For information on some of the chemicals used to manufacture vesicants, see the Component Section (C03-C) following information on the individual agents.

In addition to the agents detailed in this handbook, the Organisation for the Prohibition of Chemical Weapons (OPCW) identifies in its *Declaration Handbook 2002 for the Convention on the Prohibition of the Development, Production, Stockpiling, and Use of Chemical Weapons and on their Destruction* another five agents in this class. However, there is no information available in the unclassified literature concerning the physical, chemical, or toxicological properties of these additional agents.

Sulfur vesicants have been stockpiled by all countries that have pursued a chemical weapons program and have been used numerous times on the battlefield. In contrast, although nitrogen vesicants have been investigated by the United States and many other countries, concern over agent stability and a lack of a clear strategic, tactical, or production advantage over sulfur vesicants has prevented further stockpiling of these agents.

3.2 Toxicology

3.2.1 Effects

Sulfur and nitrogen vesicants produce their toxic effects by forming a highly reactive intermediate cyclic structure that alkylates nucleophiles in the cell structure. This disrupts the normal function of the affected biochemical. Although the actual damage caused by these vesicants occur within minutes of exposure, most clinical effects have a latent period ranging from hours to days. Vesicants affect both exterior and interior parts of the body causing inflammation, blisters, and general destruction of tissues. They have a greater impact on moist areas of the body. Healed burns may be hypersensitive to mechanical trauma.

Eyes are especially susceptible to vesicants. In addition to the immediate corrosive effects, the cornea of the eye can become inflamed (keratitis) after a latency of 6–10 years. This condition can progress to blindness. Corneal lesions may reoccur even after receiving a corneal transplantation.

Inhalation of vesicants can cause lung membranes to swell and become filled with liquid (pulmonary edema). Death may result from lack of oxygen.

Vesicants are also systemic agents and readily pass through the skin to affect susceptible tissue including those that produce blood. For this reason they are often described as radiomimetic poisons. In severe cases, systemic effects can include cardiovascular shock and multiorgan failure. Nitrogen vesicants can also cause central nervous system depression and cardiovascular shock. Both sulfur and nitrogen vesicants are carcinogenic.

3.2.2 Pathways and Routes of Exposure

Vesicants are hazardous through any route of exposure including inhalation, skin and eye exposure, ingestion, and broken, abraded, or lacerated skin (e.g., penetration of skin by debris). Thickened agents primarily pose a hazard through skin absorption. Dusty agents are primarily an inhalation hazard but may also cause minor skin/eye impacts. Contact with bulk dusty vesicants can produce more classical blistering and system effects.

3.2.3 General Exposure Hazards

Pure vesicants have little or no odor. However impurities can give them an easily detectable and identifiable smell. The odor of sulfur vesicants has been described as similar to garlic, horseradish, onions, or mustard. The odor of nitrogen vesicants has been described as fishy, fruity, musty, or even soap-like.

Agent vapors of both series cause eye irritation. However, there is no significant difference in the concentration that will irritate the eyes and the one that will produce eye injury. Although impacts from exposure to vesicants occur almost at once, contact with vapors or the liquid agent neither irritates the skin nor produces visible dermal injuries until after a substantial latency period. In contrast, HL (C03-A010), sulfur mustard mixed with lewisite, produces immediate pain due to the arsenic mustard component.

3.2.3.1 Sulfur Series

Lethal concentrations (LC_{50}s) for inhalation of these agents are as low as 11 ppm for a 2-minutes exposure.

Lethal percutaneous exposures (LD_{50}s) to liquid are as low as 1.4 grams per individual.

Incapacitating concentrations (ICt_{50}) for dermal exposure to these agents at moderate temperatures (i.e., between 65 and 85°F) are as low as 2 ppm for a 30-minutes exposure. Temperatures above 85°F reduce the concentration necessary to produce similar effects.

Eye irritation from exposure to agent vapors occurs at concentrations as low as 2 ppm after a 2-minutes exposure; an incapacitating concentration (ICt_{50}) for exposure of the eyes is as low as 5 ppm for a 2-minutes exposure.

3.2.3.2 Nitrogen Series

Lethal concentrations (LC_{50}s) for inhalation of these agents are as low as 60 ppm for a 2-minutes exposure.

Lethal percutaneous exposures (LD_{50}s) to liquid are as low as 1.4 grams per individual.

Incapacitating concentrations (ICt_{50}) for dermal exposure to these agents at moderate temperatures (i.e., between 65 and 85°F) are as low as 2 ppm for a 30-minutes exposure. Temperatures above 85°F reduce the concentration necessary to produce similar effects.

Eye irritation from exposure to agent vapors occurs at concentrations as low as 2 ppm after a 2-minutes exposure; an incapacitating concentration (ICt_{50}) for exposure of the eyes is as low as 4 ppm for a 2-minutes exposure.

The rate of detoxification of vesicants by the body is very low. Exposures are essentially cumulative.

3.2.4 Latency Period

3.2.4.1 Vapor/Aerosols (Mists or Dusts)

Eye irritation may become noticeable in a matter of minutes. Other signs and symptoms of exposure, including reddening of the skin (erythema), blistering (vesication), and accumulation of fluid in the lungs (pulmonary edema), do not occur until after a substantial latency period. Mixtures such as HL (C03-A010) contain lewisite (C04-A002) and will produce an immediate burning sensation on contact with the skin or eyes.

3.2.4.2 Liquids

Tissue damage occurs within minutes of exposure to vesicants, but clinical effects may not appear for up to 24 hours. Mixtures such as HL (C03-A010) contain lewisite (C04-A002) and will produce an immediate burning sensation on contact with the skin or eyes. Some agents are rapidly absorbed through the skin and extensive skin contamination may cause systemic damage.

3.2.4.3 Solids (Nonaerosol)

Tissue damage occurs within minutes of exposure to vesicants, but clinical effects may not appear for up to 24 hours. Some agents are rapidly absorbed through the skin and extensive skin contamination may cause systemic damage.

A key factor affecting the length of time before the onset of symptoms as well as the severity of the symptoms is the part of the body that is exposed to the agent. Apart from mucous membranes, the regions of the body that are the most sensitive to vesicants are warm, moist areas, and areas with thin skin such as the face, armpits, inside of the elbow, genitalia, neck, skin between the fingers, and the nail beds. The least susceptible body regions are the palms of the hands, soles of the feet, front of the knee, and outside of the elbow.

3.3 Characteristics

3.3.1 Physical Appearance/Odor

3.3.1.1 Laboratory Grade

Laboratory grade agents are typically colorless oily liquids or solids. They have little or no odor. Salts of nitrogen vesicants are typically white odorless solids. High concentrations of vesicants can cause eye irritation. Because of the lewisite (C04-A002) component, HL (C03-A010) vapors cause immediate irritation to the eyes, nose, throat, and skin.

3.3.1.2 Munition Grade

Munition grade agents are typically amber to dark brown liquids or solids. As the agent ages and decomposes, it continues to discolor until it may appear black. Production impurities and decomposition products in these agents may give them an odor. The odor of munition grade sulfur vesicants has been described as similar to garlic, horseradish, onions, or mustard; while the odor of munition grade nitrogen vesicants has been described as fishy, fruity, or soapy. Odors may become more pronounced during storage. Nitrogen vesicant agents tend to form crystalline decomposition products that precipitate out of solution on prolonged storage.

3.3.1.3 Modified Agents

Solvents have been added to vesicants to facilitate handling, to stabilize the agents, or to increase the ease of percutaneous penetration by the agents. Percutaneous enhancement solvents include dimethyl sulfoxide, *N*, *N*-dimethylformamide, *N*, *N*-dimethylpalmitamide, *N*, *N*-dimethyldecanamide, and saponin. Color and other properties of these solutions may vary from the pure agent. Odors will vary depending on the characteristics of the solvent(s) used and concentration of vesicants in the solution.

Conversely, vesicants have also been thickened with various substances to enhance deployment, increase their persistency, and increase the risk of percutaneous exposure. Thickeners include polyalkyl methacrylates (methyl, ethyl, butyl, isobutyl), poly(vinyl acetate), polystyrene, plexiglas, alloprene, polychlorinated isoprene, nitrocellulose, as well as bleached montan and lignite waxes. Military thickener K125 is a mixture of methyl, ethyl, and butyl polymethacrylates. When thickened, agents become sticky with a consistency similar to honey. Typically, not enough thickener is added to affect either the color or odor of the agent.

Vesicants have also been converted to a "dusty" form by adsorbing the liquid agent onto a solid carrier. Dusty carriers include aerogel, talc, alumina, silica gel, diatomite, kaolinite, fuller's earth, and pumice. Dusty agents appear as finely ground, free-flowing powders with individual particles in the range of 10 μm or less and are dispersed as a particulate cloud. Particles in this range can penetrate clothing and breathable protective gear, such as United States military mission-oriented protective posture (MOPP) garments. Dusty agents pose both an inhalation and contact hazard. Color and other physical properties of dusty agents will depend upon the characteristics of the carrier. Odors may vary from the unmodified agent.

3.3.1.4 Mixtures with Other Agents

In addition to mixtures containing both sulfur and nitrogen vesicants, individual members of this class have been mixed with other agents such as lewisite (C04-A002) and bis(chloroethyl)ether to prevent them from freezing in the munition as well as to enhance their toxicity. Details on HL, the standardized sulfur mustard/lewisite mixture, are reported under listing C03-A010.

3.3.2 Stability

Crude sulfur vesicants are relatively stable and stability increases with purity. Distilled materials show very little decomposition on storage. Solvents such as carbon tetrachloride and chlorobenzene have been added to enhance stability of crude material. Agents can be stored in glass or steel containers, although pressure may develop in steel containers. Sulfur vesicants rapidly corrode brass and cast iron, and permeate into ordinary rubber.

Nitrogen vesicants are relatively unstable and tend to dimerize or polymerize on storage. Polymerization is accelerated by both heat (as low as 122°F) and light. Polymerization can be self-accelerating through production of heat and may even generate enough heat to cause an explosion. Polymerization is also accelerated by the presence of polar solvents. Stabilizers, such as carbon disulfide and triphenylcarbinol, may be added to inhibit polymerization. Stabilized agents can be stored in glass or steel containers.

3.3.3 Persistency

For military purposes, unmodified vesicants are classified as persistent. Under proper conditions, agents can remain hazardous in soil and even in water for several years. Limited solubility slows the hydrolysis of liquid agents. Some hydrolysis products are highly toxic and extremely persistent. Evaporation rates range from near that of light machine oil to that of heavy motor oil.

Agents modified with thickeners last significantly longer. Dusty agents can be very persistent depending on the carrier employed and can be reaerosolized after deployment by ground traffic or strong winds.

3.3.4 Environmental Fate

Vesicant vapors have a density greater than air and tend to collect in low places. Porous material, including painted surfaces, will absorb both liquid and gaseous agent. After the initial surface contamination has been removed, agent that has been absorbed into porous material can migrate back to the surface posing both a contact and vapor hazard. Clothing may emit trapped agent vapor for up to 30 minutes after contact with a vapor cloud.

Most of these agents are insoluble in water and this limited solubility slows their hydrolysis. On standing, however, aqueous solutions of nitrogen vesicants will decompose forming neurotoxic products. Hydrolysis of agents may be further reduced if they are thickened such that a protective layer forms at the agent/water interface. Conversely, agent solubility may be increased by solvents. The specific gravities of unmodified liquid agents are much greater than that of water. These agents are soluble in most organic solvents including gasoline, oils, acetone, and alcohols.

3.4 Additional Hazards

3.4.1 Exposure

All foodstuffs in the area of a release should be considered contaminated. Unopened items packaged in glass, metal, or heavy duty plastic and exposed only to agent vapors may be used after decontamination of the container. Unopened items exposed to liquid or solid agents should be decontaminated within a few hours postexposure or destroyed. Opened or unpackaged items, or those packaged only in paper or cardboard, should be destroyed.

Meat, milk, and animal products, including hides, from animals affected or killed by vesicants should be destroyed or quarantined until tested and determined to be safe to use or consume.

3.4.2 Livestock/Pets

Although vesicants do not produce the same type of dermal damage in animals as they do in humans, they are still susceptible to the cytotoxic and systemic toxicities of these agents.

Animals can be decontaminated with shampoo/soap and water, or a 0.5% household bleach solution (see Section 3.5.3). If the animals' eyes have been exposed to agent, they should be irrigated with water or saline solution for a minimum of 30 minutes.

Unprotected feedstock (e.g., hay or grain) should be destroyed. Leaves of forage vegetation could still retain sufficient vesicant agent to produce effects for several weeks post release, depending on the level of contamination and the weather conditions.

3.4.3 Fire

Heat from a fire will increase the amount of agent vapor in the area. A significant amount of the agent could be volatilized and escape into the surrounding environment before it is consumed by the fire. Actions taken to extinguish the fire can also spread the agent. Vesicants may react with steam or water during a fire to produce toxic and/or corrosive vapors. In addition, nitrogen vesicants tend to polymerize during storage and the polymerization products may present an explosion hazard. HL (C03-A010) contains an arsenic component and combustion or hydrolysis will also produce toxic arsenical decomposition products.

3.4.4 Reactivity

Vesicants are incompatible with strong oxidizers, such as dry high-test hypochlorite (HTH) pool bleach, and will spontaneously ignite. Although these agents will decompose if dissolved in water, a lack of solubility inhibits this process. Nitrogen vesicants tend to polymerize on storage. Polymerization may generate enough heat to cause an explosion. In addition, polymerized components may present an explosion hazard.

3.4.5 Hazardous Decomposition Products

For information on individual impurities and decomposition products, see the Decomposition Products and Impurities section (C03-D) at the end of this chapter.

3.4.5.1 Hydrolysis

Vesicants produce acidic products including hydrogen chloride (HCl), hydrogen bromide (HBr), or hydrogen fluoride (HF), and ethanolamines, thioglycols, or thioethers when hydrolyzed. Arsenous oxide decomposition products from HL (C03-A010) are toxic and may also have vesicant properties. HL will also produce acetylene at higher pH.

3.4.5.2 Combustion

Volatile decomposition products may include HCl, HBr, HF, and nitrogen oxides (NO_x) or sulfur oxides (SO_x). Decomposition vapors from nitrogen vesicants may form explosive mixtures in air. In addition, a corrosive and toxic residue may remain. HL (C03-A010) will also produce toxic arsenic oxides.

3.5 Protection

3.5.1 Evacuation Recommendations

Isolation and protective action distances listed below are taken from Argonne National Laboratory Report No. ANL/DIS-00-1, *Development of the Table of Initial Isolation and Protective Action Distances for the 2000 Emergency Response Guidebook*, which is still the basis for the "when used as a weapon" scenarios in the 2004 Emergency Response Guidebook (ERG). For vesicants, these recommendations are based on a release scenario involving either a spray or explosively generated mist of vesicant that quickly settles to the ground and soaks into a depth of no more than 0.25 millimeters. A secondary cloud will then be generated by evaporation of this deposited material. Under these conditions, the difference between a small and a large release of vesicant is not based on the standard 200 liters spill used for commercial hazardous materials listed in the ERG. A small release involves 2 kilograms (approximately 1.5 liters of liquid sulfur vesicant and 1.7 liters of liquid nitrogen vesicant) and a large release involves 100 kilograms (approximately 20 gallons of liquid sulfur vesicant and 23 gallons of liquid nitrogen vesicant).

Because of uncertainties in defining the composition of HL, a mixture of sulfur mustard HD (C03-A001) and lewisite (C04-A002), the evacuation recommendations were based strictly on the lewisite component. A small release of HL involves 2 kilograms (approximately 1.3 L of liquid agent) and a large release involves 100 kilograms (approximately 17 gallons of liquid agent).

	Initial isolation (feet)	Downwind day (miles)	Downwind night (miles)
HD (sulfur mustard) *C03-A001*			
Small device (2 kilograms)	100	0.1	0.1
Large device (100 kilograms)	200	0.4	0.7
HN1 *C03-A011*			
Small device (2 kilograms)	100	0.1	0.1
Large device (100 kilograms)	200	0.4	0.8
HN2 *C03-A012*			
Small device (2 kilograms)	100	0.1	0.1
Large device (100 kilograms)	200	0.3	0.7
HN3 *C03-A013*			
Small device (2 kilograms)	100	0.1	0.1
Large device (100 kilograms)	100	0.1	0.2
HL (mustard/lewisite mixture) *C03-A010*			
Small device (2 kilograms)	100	0.1	0.2
Large device (100 kilograms)	300	0.6	1.1

3.5.2 Personal Protective Requirements

3.5.2.1 Structural Firefighters' Gear

Structural firefighters' protective clothing is recommended for fire situations only; it is not effective in spill situations or release events and should never be used as the primary chemical protective garment to enter an area contaminated with vesicants.

3.5.2.2 Respiratory Protection

Self-contained breathing apparatuses (SCBAs) or air purifying respirators (APRs) should have a National Institute for Occupational Safety and Health (NIOSH) and Chemical/Biological/Radiological/Nuclear (CBRN) certification since vesicants can be absorbed into or degrade the materials used to make some respirators. However, during emergency operations, other NIOSH approved SCBAs or APRs that have been specifically tested by the manufacturer against chemical warfare agents may be used if deemed necessary by the Incident Commander. APRs should be equipped with a NIOSH approved CBRN filter or a combination organic vapor/acid gas/particulate cartridge.

Immediately dangerous to life or health (IDLH) levels are the ceiling limit for respirators other than SCBAs. Any exposures approaching the IDLH level should be regarded with extreme caution and the use of SCBAs for respiratory protection should be considered.

3.5.2.3 Chemical Protective Clothing

Use only chemical protective clothing that has undergone material and construction performance testing against sulfur or nitrogen vesicant agents or both. Reported permeation rates may be affected by solvents, components, or impurities in munition grade or modified agents.

Because of the extreme dermal hazard posed by vesicants, responders should wear a Level A protective ensemble whenever there is a potential for exposure to any solid or liquid agent, or to an elevated or unknown concentration of agent vapor.

3.5.3 Decontamination

3.5.3.1 General

Sulfur vesicants: These agents are insoluble in water and form self-protecting decomposition products at the water/agent interface that prevent hydrolysis of the agent. These decomposition products are stable and some of them have toxic and/or vesicant properties.

Solubility of an agent will be decreased further if it is dissolved in an immiscible organic solvent or if it is thickened such that it forms a protective layer at the agent/water interface. Addition of solvents or mechanical mixing may be required to overcome insolubility problems.

Dissolved vesicants are readily destroyed by high pH (i.e., basic solutions), especially when used in combination with a strong oxidizing agent. For this reason, undiluted household bleach is an excellent agent for decontamination of these agents. Ensure that the bleach solution remains in contact with the agent for a minimum of 5 minutes. However, a large excess will be needed to ensure complete destruction of the agents. The lewisite component of HL (C03-A010) will react with basic solutions, including household bleach, to produce acetylene gas.

Sulfur vesicants are easily oxidized. However, over oxidation produces the sulfone (e.g., C03-D050), which is nearly as powerful a vesicant as the parent compound. These sulfones are also less reactive than the parent compounds and much more slowly hydrolyzed. A slurry that is 10% by weight of either super tropical bleach (STB) or HTH in water is an effective agent for oxidizing sulfur vesicants. Never use dry STB or HTH since they will react violently with vesicants and may spontaneously ignite. Basic peroxides (e.g., a solution of baking soda, 30–50% hydrogen peroxide, and an alcohol) also rapidly detoxify sulfur vesicants. Nonreactive, nonporous materials such as glass can be decontaminated with concentrated nitric acid.

Reactive oximes and their salts, such as potassium 2,3-butanedione monoximate found in commercially available Reactive Skin Decontaminant Lotion (RSDL), are extremely effective at rapidly detoxifying sulfur vesicants. Some chloroisocyanurates, similar to those found in the Canadian Aqueous System for Chemical-Biological Agent Decontamination (CASCAD), are effective at detoxifying sulfur vesicants, and so is oxone, a peroxymonosulfate triple salt.

Nitrogen vesicants: These agents are essentially insoluble in water. They will decompose on standing in water but form neurotoxic products. Some of these toxic decomposition products can last for extended periods.

Solubility of an agent will be decreased further if it is dissolved in an immiscible organic solvent or if it is thickened such that it forms a protective layer at the agent/water interface. Addition of solvents or mechanical mixing may be required to overcome insolubility problems.

Dissolved vesicants are readily destroyed by high pH (i.e., basic solutions), especially when used in combination with a strong oxidizing agent. For this reason, undiluted household bleach is an excellent agent for decontamination of these agents. Ensure that the bleach solution remains in contact with the agent for a minimum of 5 minutes. However, a large excess will be needed to ensure complete destruction of the agents.

Nitrogen vesicants are easily oxidized. However, incomplete oxidation produces the *N*-oxide, which is still highly toxic and relatively stable. A slurry that is 10% by weight of either STB or HTH in water is also an effective agent for oxidizing nitrogen vesicants. Never use dry STB or HTH since they will react violently with vesicants and may spontaneously ignite. Basic peroxides (e.g., a solution of baking soda, 30–50% hydrogen peroxide, and an alcohol) also rapidly detoxify nitrogen vesicants.

Although specific data have not been published in the unclassified literature, preliminary studies indicate that reactive oximes and their salts, such as potassium 2,3-butanedione monoximate found in commercially available RSDL, are extremely effective at rapidly detoxifying nitrogen vesicants.

3.5.3.2 Vapors

Casualties/personnel: Speed in decontamination is absolutely essential. To be effective, decontamination must be completed within 2 minutes after postexposure. However, decontamination after the initial 2 minutes should still be undertaken in order to prevent additional percutaneous absorption of the agent leading to systemic toxicity. Remove all clothing as it may continue to emit "trapped" agent vapor after contact with the vapor cloud has ceased. Shower using copious amounts of soap and water. Ensure that the hair has been washed and rinsed to remove potentially trapped vapor. To be effective, decontamination must be completed within 2 minutes of exposure. If there is a potential that the eyes have been exposed to vesicants, irrigate with water or 0.9% saline solution for a minimum of 15 minutes.

Small areas: Ventilate to remove the vapors. If condensation is present, decontaminate with copious amounts of undiluted household bleach (see Section 3.5.3.1). Allow it to stand for a minimum of 5 minutes before rinsing with water. Collect and place into containers lined with high-density polyethylene. Removal of porous material, including painted surfaces, may be required because vesicants that have been absorbed into these materials can migrate back to the surface posing both a contact and vapor hazard.

3.5.3.3 Liquids, Solutions, or Liquid Aerosols

Casualties/personnel: Speed in decontamination is absolutely essential. To be effective, decontamination must be completed within 2 minutes postexposure. However, decontamination after the initial 2 minutes should still be undertaken in order to prevent additional percutaneous absorption of the agent leading to systemic toxicity. Remove all clothing immediately. Even clothing that has not come into direct contact with the agent may contain "trapped" vapor. To avoid further exposure of the head, neck, and face to the agent, cut off potentially contaminated clothing that must be pulled over the head. Remove as much of the vesicant from the skin as fast as possible. If water is not immediately available, the agent can be absorbed with any convenient material such as paper towels, toilet paper, flour, and/or talc. To minimize both spreading the agent and abrading the skin, do not rub the agent with the absorbent. Blot the contaminated skin with the absorbent.

Use a sponge or cloth with liquid soap and copious amounts of water to wash the skin surface and hair at least three times. Do not delay decontamination to find warm or hot water if it is not readily available. Avoid rough scrubbing as this could abrade the skin and increase percutaneous absorption of residual agent. Rinse with copious amounts of water. If there is a potential that the eyes have been exposed to vesicants, irrigate with water or 0.9% saline solution for a minimum of 15 minutes.

Alternatively, a household bleach solution can be used instead of soap and water. The bleach solution should be no more than one part household bleach in nine parts water (i.e., 0.5% sodium hypochlorite) to avoid damaging the skin. Avoid any contact with sensitive areas such as the eyes. Ensure that the bleach solution remains in contact with the agent for a minimum of 5 minutes. Rinse with copious amounts of water.

Small areas: Puddles of liquid can be contained by covering with absorbent material such as vermiculite, diatomaceous earth, clay, sponges, or towels. Place the absorbed material into containers lined with high-density polyethylene. Before sealing the container, cover the contents with undiluted household bleach or an HTH/water slurry (see Section 3.5.3.1). If HL (C03-A010) is involved, then flammable acetylene gas will be generated during the neutralization process. Take appropriate actions to disperse the vapors. Decontaminate the area with copious amounts of the neutralizing agent. Ensure that it remains in contact with the agent for a minimum of 5 minutes before rinsing with water. Collect and containerize the rinseate. Ventilate the area to remove vapors. Removal of porous material, including painted surfaces, may be required because vesicants that have been absorbed into these materials can migrate back to the surface posing both a contact and vapor hazard.

3.5.3.4 Solids, Dusty Agents, or Particulate Aerosols

Casualties/personnel: Speed in decontamination is absolutely essential. To be effective, decontamination must be completed within 2 minutes after postexposure. Do not attempt to brush the agent off the individual or their clothing as this can aerosolize the agent. If possible, dampen the agent with a water mist to help prevent aerosolization. Remove all clothing immediately. To avoid further exposure of the head, neck, and face to the agent, cut off potentially contaminated clothing that must be pulled over the head. Wash the skin surface and hair at least three times with copious amounts of soap and water. Do not delay decontamination to find warm or hot water if it is not readily available. Rinse with copious amounts of water. If there is a potential that the eyes have been exposed to vesicants, irrigate with water or 0.9% saline solution for a minimum of 15 minutes.

Small areas: If indoors, close windows and doors in the area and turn off anything that could create air currents (e.g., fans, air conditioners, etc.). Avoid actions that could aerosolize the agent such as sweeping or brushing. Collect the agents using a vacuum cleaner equipped with a high-efficiency particulate air (HEPA) filter. Do not use a standard home or industrial vacuum. Do not allow the vacuum exhaust to stir the air in the affected area. Vacuum all surfaces with extreme care in a very slow and controlled manner to minimize aerosolizing the agent. Place the collected material into containers lined with high-density polyethylene. Before sealing the container, cover the contents with undiluted household bleach or the HTH/water slurry (see Section 3.5.3.1). Decontaminate the area with copious amounts of the neutralizing agent. Insure that it remains in contact with the agent for a minimum of 5 minutes before rinsing with water. Collect and containerize the rinseate.

3.6 Medical

3.6.1 CDC Case Definition

A case in which a vesicant is detected in biologic samples. The case can be confirmed if laboratory testing is not performed because either a predominant amount of clinical and nonspecific laboratory evidence is present or an absolute certainty of the etiology of the agent is known.

3.6.2 Differential Diagnosis

The following factors have been suggested as alternatives to consider when presented with a potential case of exposure to vesicant agents: chemical burns from contact with strong acids or bases; barbiturates, chemotherapeutic agents, carbon monoxide; reactions to drugs producing Stevens–Johnson syndrome and/or toxic epidermal necrolysis; autoimmune diseases such as bullous pemphigoid and pemphigus vulgaris; and staphylococcus scalded skin syndrome.

3.6.3 Signs and Symptoms

3.6.3.1 Vapors/Aerosols

Feeling of irritation and grittiness in the eyes are generally the first indications of exposure. These symptoms usually occur 30 minutes to 3 hours postexposure and, depending on the amount of agent involved, progress to soreness with a bloodshot appearance. These symptoms can be followed by swelling, pain, tearing, involuntary blinking (blepharospasm), and sensitivity to light (photophobia). Eye symptoms may be accompanied by an increase in nasal secretion, sneezing, sore throat, coughing, and hoarseness. Individuals who have been exposed to a large amount of agent may experience nausea, retching, and vomiting.

There is an asymptomatic latent period, usually 4–24 hours before skin impacts appear. With minimal exposure, skin impacts may be limited to reddening of the skin (erythema). Otherwise, it progresses to large blisters typically filled with a clear yellow fluid. Blisters do not contain active vesicant agent. These blisters usually break leaving the skin open to secondary infection. Blistering from a vapor-only exposure is generally comparable to a first-degree or second-degree burn. The skin may darken on and around the burned area.

If HL (C03-A010) is involved, then exposure of the eyes to even small amounts of vapor will produce immediate tearing, pain, and involuntary blinking (blepharospasm). Eye symptoms are rapidly followed by coughing, sneezing, and vomiting. Exposure of the skin to HL produces an immediate burning sensation. Reddening of the skin (erythema) may appear in as short a time as 5 minutes although full progression to blisters may not develop for up to 18 hours.

3.6.3.2 Liquids/Solids

There is an asymptomatic latent period, usually 4–24 hours before skin impacts appear. With minimal exposure, skin impacts may be limited to reddening of the skin (erythema). Otherwise, erythema progresses to large blisters typically filled with a clear yellow fluid or further to lesions with a central zone of localized death of cells or tissues (necrosis) and peripheral blisters. Blisters do not contain active vesicant agent. Blisters are easy to break leaving the skin open to secondary infection. Blistering from a liquid exposure can produce deep damage comparable to a third-degree burn. The skin may darken on and around the burned area.

If HL (C03-A010) is involved, then exposure of the skin will produce an immediate burning sensation, which may be quickly followed by reddening of the skin (erythema). In addition to other latent effects, casualties exposed to HL may also develop signs of systemic arsenic toxicity including diarrhea, damage to the liver, kidneys, nervous system, and the brain.

3.6.4 Mass-Casualty Triage Recommendations

3.6.4.1 Priority 1

A casualty with mild to moderate pulmonary effects less than 6 hours postexposure, or a casualty with moderately severe or severe pulmonary signs and symptoms after 6 hours postexposure.

3.6.4.2 Priority 2 (Majority of Cases)

A casualty with skin lesions covering between 5 and 50% of the body surface area (BSA), or a casualty with mild to moderate pulmonary effects after 6 hours postexposure, or a casualty with eye injuries.

3.6.4.3 Priority 3

A casualty with skin lesions covering less than 5% of the BSA, or a casualty with eye irritation or reddening, and/or slight upper respiratory complaints such as hacking cough or irritated throat 12 hours or more hours postexposure.

3.6.4.4 Priority 4

A casualty with skin lesions from *liquid exposure* to more than 50% of the BSA, or a casualty with severe pulmonary effects less than 6 hours postexposure.

3.6.5 Casualty Management

Decontaminate the casualty ensuring that all the vesicants have been removed. Rapid decontamination of any exposure is essential. If vesicants have gotten into the eyes, irrigate the eyes with water or 0.9% saline solution for at least 15 minutes. Irrigate open wounds with water or 0.9% saline solution for at least 10 minutes.

Once the casualty has been decontaminated, including the removal of foreign matter from wounds, medical personnel do not need to wear a chemical-protective mask.

There is no antidote for exposure to these agents. Treatment consists of symptomatic management of lesions. If a casualty is known to have inhaled vesicant vapors but does not display any signs or symptoms of an impacted airway, it may still be appropriate to intubate the casualty since laryngeal spasms or edema may make it difficult or impossible later. Eye lesions should be treated by saline irrigation and coating the follicular margins with petroleum jelly to prevent sticking.

If HL (C03-A010) is involved, a BAL (British-anti-Lewisite, dimercaprol) solution or ophthalmic ointment may be beneficial if administered promptly. It may also decrease the severity of skin and eye lesions if applied topically within minutes after decontamination is complete (i.e., within 2–5 minutes postexposure).

Asymptomatic individuals suspected of exposure to vesicants should be kept under observation for at least 8 hours.

Burns from *liquid exposure* to over 50% of the body surface suggest that the individual has absorbed twice a lethal dose of vesicant and the prognosis for survival of the individual is poor.

3.7 Fatality Management

Remove all clothing and personal effects segregating them as either durable or nondurable items. While it may be possible to decontaminate durable items, it may be safer and more efficient to destroy nondurable items rather than attempt to decontaminate them. Items that will be retained for further processing should be double sealed in impermeable containers, ensuring that the inner container is decontaminated before placing it in the outer one.

Vesicants that have entered the body are metabolized, hydrolyzed, or bound to tissue and pose little threat of off-gassing. To remove agents on the outside of the body, wash the remains with a 2% sodium hypochlorite bleach solution (i.e., 2 gallons of water for every gallon of household bleach) ensuring the solution is introduced into the ears, nostrils, mouth, and any wounds. This concentration of bleach will not affect remains but will neutralize vesicant agents. Higher concentrations of bleach can harm remains. Pay particular attention to areas where agent may get trapped, such as hair, scalp, pubic areas, fingernails, folds of skin, and wounds. The bleach solution should remain on the cadaver for a minimum of 5 minutes. Wash with soap and water. Ensure that all the bleach solution is removed prior to embalming as it will react with embalming fluid. All wash and rinse waste must be contained for proper disposal. Screen the remains for agent vapors and residual liquid at the conclusion of the decontamination process. If the remains must be stored before embalming, then place them inside body bags designed to contain contaminated bodies or in double body bags. If double body bags are used, seal the inner bag with duct tape, rinse, and then place in the second bag. After embalming is complete, place the remains in body bags designed to contain contaminated bodies or in double body bags. Body fluids removed during the embalming process do not pose any additional risks and should be contained and handled according to established procedures.

Standard burials are acceptable when contamination levels are low enough to allow bodies to be handled without wearing additional protective equipment. Cremation may be required if remains cannot be completely decontaminated. Although vesicant agents are destroyed at the operating temperature of a commercial crematorium (i.e., above 1000°F), the initial heating phase may volatilize some of the agents and allow vapors to escape. Additionally, if HL (C03-A010) is involved then combustion will produce toxic and and potentially volatile arsenic oxides.

C03-A

SULFUR VESICANTS

C03-A001

Mustard gas (Agent HD)

CAS: 505-60-2
RTECS: WQ090000

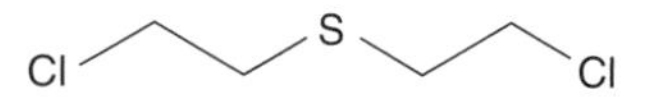

$C_4H_8SCl_2$

Oily, colorless to amber to dark brown liquid. Yellow solid below 58°F. Pure material is odorless; otherwise has an odor similar to garlic or horseradish that is detectable at approximately 0.1 ppm.

Also reported as a mixture with Sesquimustard (C03-A002); O-Mustard (C03-A003); Lewisite (C04-A002); HN1 (C03-A011); HN3 (C03-A013); Phenyldichloroarsine (C04-A005).

Exposure Hazards

Conversion Factor: 1 ppm = 6.51 mg/m^3 at 77°F
$LCt_{50(Inh)}$: 1000 mg-min/m^3 (80 ppm for a 2-min exposure)
$LCt_{50(Per)}$: $<$ 85°F: 10,000 mg-min/m^3 (50 ppm for a 30-min exposure)
$LCt_{50(Per)}$: $>$ 85°F: 5000 mg-min/m^3 (30 ppm for a 30-min exposure)
LD_{50}: 1.4 g
$ICt_{50(Skin)}$: $<$ 85°F: 500 mg-min/m^3 (3 ppm for a 30-min exposure)
$ICt_{50(Skin)}$: $>$85°F: 200 mg-min/m^3 (1 ppm for a 30-min exposure)
$ICt_{50(Eyes)}$: 75 mg-min/m^3 (6 ppm for a 2-min exposure)
Eye Irritation: 2 ppm for a 2-min exposure
$MEG_{(1\,h)}$ *Min*: 0.01 ppm; *Sig*: 0.02 ppm; *Sev*: 0.32 ppm
WPL AEL: 0.00006 ppm
STEL: 0.0005 ppm
IDLH: 0.11 ppm

Properties:

MW: 159.1	*VP*: 0.11 mmHg (77°F)	*FlP*: 221°F
D: 1.27 g/mL (77°F)	*VD*: 5.5 (calculated)	*LEL*: —
MP: 58°F	*Vlt*: 141 ppm (77°F)	*UEL*: —
BP: Decomposes	H_2O: 0.068% (77°F)	*RP*: 73
Vsc: 3.95 cS (77°F)	*Sol*: Hydrocarbons; Acetone; Alcohols	*IP*: <9 eV

Solid	**Thickened**
D: 1.33 g/cm^3 (50°F)	*Vsc*: 1000–1200 cS

Final AEGLs

AEGL-1: 1 h, 0.010 ppm	4 h, 0.003 ppm	8 h, 0.001 ppm
AEGL-2: 1 h, 0.020 ppm	4 h, 0.004 ppm	8 h, 0.002 ppm
AEGL-3: 1 h, 0.320 ppm	4 h, 0.080 ppm	8 h, 0.040 ppm

C03-A002

Sesquimustard (Agent Q)

CAS: 3563-36-8
RTECS: —

$C_6H_{12}Cl_2S_2$

Pale amber solid. Pure material has little or no odor; otherwise has an odor similar to garlic or horseradish. Also found as an impurity in Sulfur mustard (C03-A001).

Also reported as a mixture with Sulfur mustard (C03-A001).

Exposure Hazards

Conversion Factor: 1 ppm = 8.97 mg/m^3 at 77°F
$LCt_{50(Inh)}$: 200 mg-min/m^3 (11 ppm for a 2-min exposure)
$ICt_{50(Skin)}$: 300 mg-min/m^3 (1 ppm for a 30-min exposure)

These values are from older sources (ca. 1956). Reevaluation and updates of toxicity values for similar agents suggest Q may be more toxic than the values reported here.

Properties:

MW: 219.2
D: 1.272 g/cm^3 (77°F)
MP: 133°F
BP: Decomposes
BP: 358°F (15 mmHg)
Vsc: —
VP: 0.000035 mmHg (77°F)
VD: 7.6 (calculated)
Vlt: 0.5 ppm (77°F)
H_2O: 0.0025%
Sol: Hydrocarbons; Acetone; Alcohols
FlP: —
LEL: —
UEL: —
RP: 200,000
IP: —

C03-A003

***O*-Mustard** (Agent T)

CAS: 63918-89-8
RTECS: WQ3250000

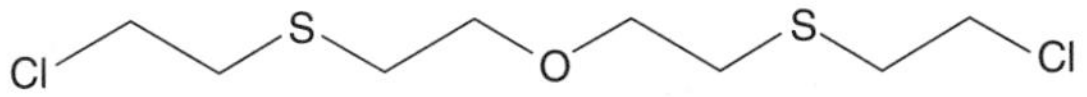

$C_8H_{16}Cl_2OS_2$

Yellow liquid. Pure material has little or no odor; otherwise has an odor similar to garlic or horseradish.

Also reported as a mixture Sulfur mustard (C03-A001). In addition, it is also found as a natural aging impurity in sulfur mustard.

Exposure Hazards

Conversion Factor: 1 ppm = 10.77 mg/m^3 at 77°F
$ICt_{50(Skin)}$: 400 mg-min/m^3 (1 ppm for a 30-min exposure). This value is from older sources (ca. 1956). Reevaluation and updates of toxicity values for similar agents suggest T may be more toxic than the values reported here.

Properties:

MW: 263.3	*VP*: 0.00003 mmHg (77°F)	*FlP*: —
D: 1.236 g/mL (77°F)	*VD*: 9.1 (calculated)	*LEL*: —
MP: 49°F	*Vlt*: 0.04 ppm (77°F)	*UEL*: —
BP: Decomposes	H_2O: Insoluble	*RP*: 100,000
Vsc: 14.7 cS (77°F)	*Sol*: —	*IP*: —

C03-A004

Mustard/sesquimustard mixture (Agent HQ)

CAS: —
RTECS: —

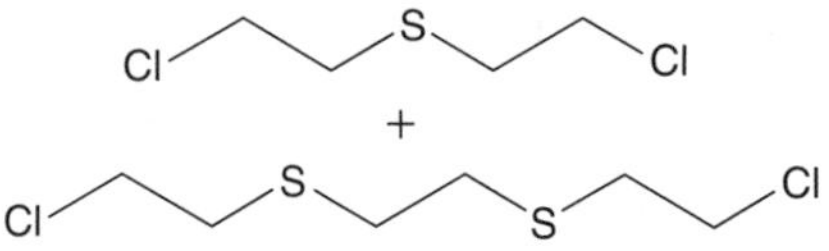

Oily, colorless to amber to dark brown liquid. Pure material is odorless; otherwise has an odor similar to garlic or horseradish. Typical mixture is 75% Sulfur mustard (C03-A001) and 25% Sesquimustard (C03-A002).

Exposure Hazards

Human toxicity values have not been established or have not been published. However, the agents in this mixture are extremely powerful vesicants and highly toxic by inhalation.

Properties:

MW: Mixture	*VP*: 0.09 mmHg (77°F) (calculated)	*FlP*: —
D: —	*VD*: 6.0 (calculated)	*LEL*: —
MP: —	*Vlt*: —	*UEL*: —
BP: Decomposes	H_2O: Insoluble	*RP*: 200,000
Vsc: —	*Sol*: Hydrocarbons; Acetone; Alcohols	*IP*: —

C03-A005

Sulfur mustard/*O*-mustard mixture (Agent HT)

CAS: 172672-28-5
RTECS: —

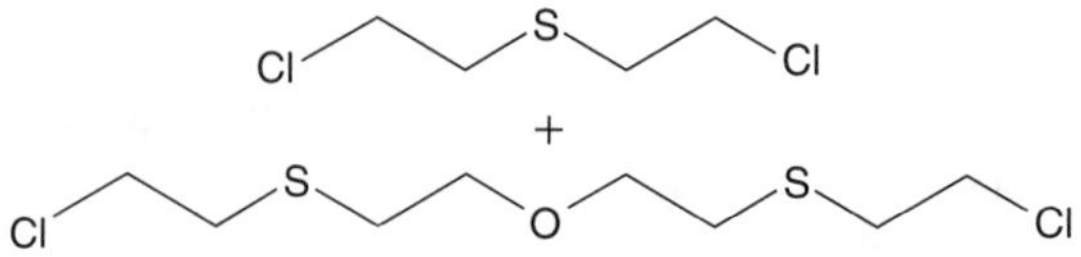

Clear, yellow to brown highly viscous liquid. Pure material is odorless; otherwise has an odor similar to garlic or horseradish. Odor is less pronounced than with Sulfur mustard alone. Typical mixture is 60% Sulfur mustard (C03-A001) and 40% *O*-Mustard (C03-A003).

Exposure Hazards

$LCt_{50(Inh)}$: 1000 mg-min/m^3 (60 ppm for a 2-min exposure)
$LCt_{50(Per)}$: 10,000 mg-min/m^3 (40 ppm for a 30-min exposure)
LD_{50}: 1.4 g
$ICt_{50(Skin)}$: <85°F: 500 mg-min/m^3 (2 ppm for a 30-min exposure)
$ICt_{50(Skin)}$: >85°: 200 mg-min/m^3 (0.9 ppm for a 30-min exposure)
$ICt_{50(Eyes)}$: 75 mg-min/m^3 (5 ppm for a 2-min exposure)
Eye Irritation: 2 ppm for a 2-min exposure

Properties:

MW: Mixture	*VP*: 0.077 mmHg (77°F)	*FlP*: 228°F
D: 1.269 g/mL (77°F)	*VD*: 6.5 (calculated)	*LEL*: —
MP: 32°F	*Vlt*: —	*UEL*: —
BP: Decomposes	*H_2O*: "Slight"	*RP*: —
Vsc: —	*Sol*: Most organic solvents	*IP*: —

C03-A006

Dibromoethyl sulfide (Bromlost)

CAS: 7617-64-3
RTECS: —

Br S Br

$C_4H_8Br_2S$

Solid. More susceptible to hydrolysis than Sulfur mustard (C03-A001).

Exposure Hazards

Conversion Factor: 1 ppm = 10.14 mg/m^3 at 77°F
Human toxicity values have not been established or have not been published.

Properties:

MW: 248.0	*VP*: —	*FlP*: —
D: 2.05 g/cm^3	*VD*: 8.6 (calculated)	*LEL*: —
MP: 70°F	*Vlt*: 39 ppm	*UEL*: —
BP: Decomposes	*H_2O*: —	*RP*: —
Vsc: —	*Sol*: —	*IP*: —

C03-A007

2,2′-Difluorodiethyl sulfide (Fluorlost)

CAS: 373-25-1
RTECS: —

F S F

$C_4H_8F_2S$

Specific information on physical appearance is not available for this agent.

Exposure Hazards

Conversion Factor: 1 ppm = 5.16 mg/m^3 at 77°F
Human toxicity values have not been established or have not been published.

Properties:

MW: 126.2	*VP*: —	*FlP*: —
D: —	*VD*: 4.4 (calculated)	*LEL*: —
MP: —	*Vlt*: —	*UEL*: —
BP: 203°F (30 mmHg)	H_2O: —	*RP*: —
Vsc: —	*Sol*: —	*IP*: —

C03-A008

2-Chloroethyl-chloromethylsulfide

CAS: 2625-76-5
RTECS: —

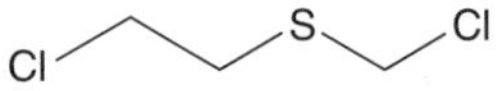

$C_3H_6Cl_2S$

Specific information on physical appearance is not available for this agent.

Exposure Hazards

Conversion Factor: 1 ppm = 5.93 mg/m^3 at 77°F
Human toxicity values have not been established or have not been published.

Properties:

MW: 145.1	*VP*: —	*FlP*: —
D: —	*VD*: 5.0 (calculated)	*LEL*: —
MP: —	*Vlt*: —	*UEL*: —
BP: 165°F (18 mmHg)	H_2O: —	*RP*: —
Vsc: —	*Sol*: —	*IP*: <9 eV

C03-A009

Dimethyl sulfate (D-Stoff)

CAS: 77-78-1
RTECS: WS8225000
UN: 1595
ERG: 156

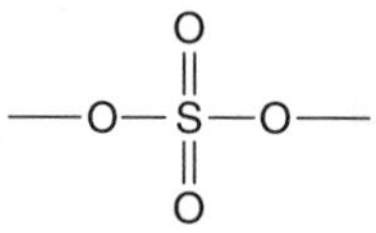

$C_2H_6O_4S$

Colorless, oily liquid with a faint odor like onions. Vapors rapidly react with moisture in the air to produce methanol and sulfuric acid.

Also reported as a mixture with Chlorosulfonic acid; Methyl chlorosulfonate (C10-A009).

Exposure Hazards

Conversion Factor: 1 ppm = 5.16 mg/m^3 at 77°F
$LCt_{50\,(Inh)}$: 500 mg/m^3 (97 ppm) for a 10-min exposure

Eye Irritation: 1 ppm; exposure duration unavailable
$MEG_{(1\,h)}$ *Min*: 0.3 ppm; *Sig*: 1 ppm; *Sev*: 7 ppm
OSHA PEL: 1 ppm [Skin]
ACGIH TLV: 0.1 ppm [Skin]
IDLH: 7 ppm

Properties:

MW: 126.1	*VP*: 0.1 mmHg	*FlP*: 182°F
D: 1.3322 g/mL	*VP*: 0.677 mmHg (77°F)	*LEL*: —
MP: −25°F	*VD*: 4.3 (calculated)	*UEL*: —
BP: Decomposes	*Vlt*: 890 ppm (77°F)	*RP*: 88
BP: 169°F (15 mmHg)	H_2O: 3%; decomposes above 64°F	*IP*: —
Vsc: —	*Sol*: Ethanol; Ether; Acetone; Aromatic hydrocarbons	

Proposed AEGLs

AEGL-1: 1 h, 0.024 ppm	4 h, 0.012 ppm	8 h, 0.0087 ppm
AEGL-2: 1 h, 0.12 ppm	4 h, 0.061 ppm	8 h, 0.043 ppm
AEGL-3: 1 h, 1.6 ppm	4 h, 0.82 ppm	8 h, 0.58 ppm

C03-A010

Mustard Lewisite mixture (Agent HL)

CAS: 378791-32-3
RTECS: —

Cl S Cl
+ Cl As Cl Cl

Oily, colorless to brownish liquid with an odor like garlic. Mixture consists of 37 to 50% Sulfur mustard (C03-A001) and +63 to 50% Lewisite (C04-A002). The eutectic mixture is 37% Sulfur mustard and 63% lewisite.

Exposure Hazards

Provisional exposure values have been published for this mixture. However, they are based solely on, and are identical to, the Sulfur mustard component (see C03-A001). These updates have not been formally adopted as of 2005.

Properties for eutectic mixture:

MW: Mixture	*VP*: 0.25 mmHg	*FlP*: —
D: —	*VD*: 6.5 (calculated)	*LEL*: —
MP: −14°F	*Vlt*: —	*UEL*: —
BP: <374°F	H_2O: —	*RP*: 30
Vsc: 1.6 cS (calculated)	*Sol*: —	*IP*: —

NITROGEN VESICANTS

C03-A011

Nitrogen mustard 1 (Agent HN1)

CAS: 538-07-8
RTECS: YE1225000

$C_8H_{13}Cl_2N$

Oily, colorless to amber liquid with a faint fishy or musty odor. Salts are odorless white solids.

Also reported as a mixture with Sulfur mustard (C03-A001).

Exposure Hazards

Conversion Factor: 1 ppm = 6.96 mg/m^3 at 77°F
$LCt_{50\,(Inh)}$: 1000 mg-min/m^3 (72 ppm for a 2-min exposure)
$LCt_{50\,(Per)}$: <85°F: 10,000 mg-min/m^3 (48 ppm for a 30-min exposure)
$LCt_{50\,(Per)}$: >85°F: 5000 mg-min/m^3 (24 ppm for a 30-min exposure)
LD_{50}: 1.4 g
$ICt_{50\,(Skin)}$: <85°F: 500 mg-min/m^3 (2.4 ppm for a 30-min exposure)
$ICt_{50\,(Skin)}$: >85°F: 200 mg-min/m^3 (1 ppm for a 30-min exposure)
$ICt_{50\,(Eyes)}$: 75 mg-min/m^3 (5.4 ppm for a 2-min exposure)
Eye Irritation: 3.5 ppm for a 2-min exposure
Skin Irritation: 0.2 ppm for a 2-min exposure

Properties:

MW: 170.1	*VP*: 0.244 mmHg (77°F)	*FlP*: —
D: 1.086 g/mL (77°F)	*VD*: 5.9 (calculated)	*LEL*: —
MP: −30°F	*Vlt*: 320 ppm	*UEL*: —
BP: Decomposes	H_2O: 0.4%	*RP*: 32
BP: 191°F (12 mmHg)	*Sol*: Most organic solvents	*IP*: —
Vsc: —		

Proposed AEGLs

AEGL-1: Not Developed		
AEGL-2: 1 h, 0.0031 ppm	4 h, 0.00080 ppm	8 h, 0.00040 ppm
AEGL-3: 1 h, 0.053 ppm	4 h, 0.013 ppm	8 h, 0.0068 ppm

C03-A012

Nitrogen mustard 2 (Agent HN2)

CAS: 51-75-2; 55-86-7 (Hydrochloride salt)
RTECS: IA1750000

$C_5H_{11}Cl_2N$

Colorless to dark liquid that has a fruity odor in high concentrations. The odor has been described as "soapy" in low concentrations. Salts are white solids.

Exposure Hazards

Conversion Factor: 1 ppm = 6.38 mg/m^3 at 77°F
$LCt_{50(Inh)}$: 1000 mg-min/m^3 (78 ppm for a 2-min exposure)
$LCt_{50(Per)}$: <85° *F*: 10,000 mg-min/m^3 (52 ppm for a 30-min exposure)
$LCt_{50(Per)}$: >85° *F*: 5,000 mg-min/m^3 (26 ppm for a 30-min exposure)
LD_{50}: 1.4 g
$ICt_{50(Skin)}$: <85° *F*: 500 mg-min/m^3 (2.6 ppm for a 30-min exposure)
$ICt_{50(Skin)}$: >85° *F*: 200 mg-min/m^3 (1 ppm for a 30-min exposure)
$ICt_{50(Eyes)}$: 75 mg-min/m^3 (5.9 ppm for a 2-min exposure)
Eye Irritation: 2 ppm for a 2-min exposure

Properties:

MW: 156.1	*VP*: 0.416 mmHg (77°F)	*FlP*: —
D: 1.118 g/mL (77°F)	*VD*: 5.4 (calculated)	*LEL*: —
MP: –94°F	*Vlt*: 550 ppm	*UEL*: —
BP: Decomposes	H_2O: 1.3%	*RP*: 19
BP: 167°F (15 mmHg)	*Sol*: Most organic solvents	*IP*: —
Vsc: —		

Hydrochloride Salt

MW: 192.6	H_2O: "Very soluble"
MP: 228°F	*Sol*: Ethanol

Proposed AEGLs

AEGL-1: Not Developed

AEGL-2: 1 h, 0.0034 ppm	4 h, 0.00088 ppm	8 h, 0.00044 ppm
AEGL-3: 1 h, 0.058 ppm	4 h, 0.015 ppm	8 h, 0.0074 ppm

C03-A013

Nitrogen mustard 3 (Agent HN3)

CAS: 555-77-1; 817-09-4 (Hydrochloride salt); 6138-32-5 (Picrate salt)
RTECS: YE2625000

$C_6H_{12}Cl_3N$

Colorless to dark liquid that is odorless when pure. Salts are white solids.

Also reported as a mixture with Sulfur mustard (C03-A001).

Exposure Hazards

Conversion Factor: 1 ppm = 8.36 mg/m^3 at 77°F
$LCt_{50(Inh)}$: 1000 mg-min/m^3 (60 ppm for a 2-min exposure)

$LCt_{50(Per)}$: <85°F: 10,000 mg-min/m^3 (40 ppm for a 30-min exposure)
$LCt_{50(Per)}$: >85°F: 5000 mg-min/m^3 (20 ppm for a 30-min exposure)
LD_{50}: 1.4 g
$ICt_{50(Skin)}$: <85°F: 500 mg-min/m^3 (2 ppm for a 30-min exposure)
$ICt_{50(Skin)}$: >85°F: 200 mg-min/m^3 (0.8 ppm for a 30-min exposure)
$ICt_{50(Eyes)}$: 75 mg-min/m^3 (4.4 ppm for a 2-min exposure)
Eye Irritation: 1.5 ppm for a 2-min exposure

Properties:

MW: 204.5	*VP*: 0.0109 mmHg (77°F)	*FlP*: —
D: 1.235 g/mL (77°F)	*VD*: 7.1 (calculated)	*LEL*: —
MP: 25°F	*Vlt*: 14 ppm (77°F)	*UEL*: —
BP: Decomposes	H_2O: 0.008%	*RP*: 650
Vsc: 5.9 cS (77°F)	*Sol*: Most organic solvents	*IP*: —

Hydrochloride salt	**Picrate salt**
MW: 241.0	*MW*: 321.6
MP: 266°F	*MP*: 278°F

Proposed AEGLs

AEGL-1: Not Developed

AEGL-2: 1 h, 0.0026 ppm	4 h, 0.00067 ppm	8 h, 0.00033 ppm
AEGL-3: 1 h, 0.044 ppm	4 h, 0.011 ppm	8 h, 0.0056 ppm

C03-A014

2,2′-Difluoro-*N*-methyldiethylamine

CAS: 352-26-1; 6089-42-5 (Hydrobromide Salt); 1598-80-7 (Hydrochloride salt)
RTECS: —

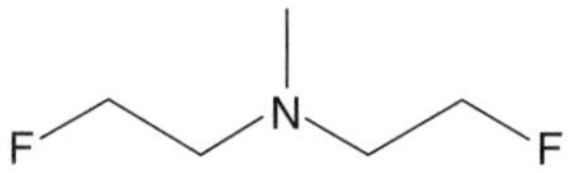

$C_5H_{11}F_2N$
Colorless liquid.

Exposure Hazards

Conversion Factor: 1 ppm = 5.03 mg/m^3 at 77°F
Human toxicity values have not been established or have not been published.

Properties:

MW: 123.1	*VP*: —	*FlP*: —
D: —	*VD*: 4.2 (calculated)	*LEL*: —
MP: —	*Vlt*: —	*UEL*: —
BP: 253°F	H_2O: —	*RP*: —
Vsc: —	*Sol*: —	*IP*: —

Hydrobromide salt	**Hydrochloride salt**
MW: 204.0	*MW*: 159.6
MP: 216°F	*MP*: 208°F

C03-A015

2-Chloroethyl methyl 1-chloroacetamide

CAS: 858817-94-4
RTECS: —

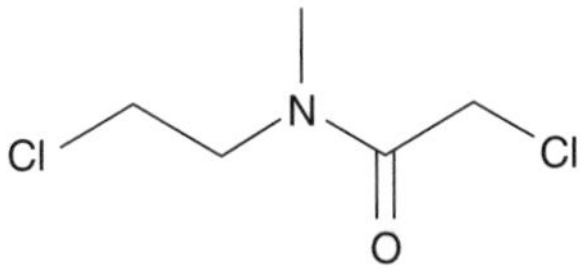

$C_5H_9Cl_2NO$

Specific information on physical appearance is not available for this agent.

Exposure Hazards

Conversion Factor: 1 ppm = 6.95 mg/m^3 at 77°F
Human toxicity values have not been established or have not been published.

Properties:

MW: 170.0	*VP*: —	*FlP*: —
D: —	*VD*: 5.9 (calculated)	*LEL*: —
MP: —	*Vlt*: —	*UEL*: —
BP: 230°F (0.8 mmHg)	H_2O: —	*RP*: —
Vsc: —	*Sol*: —	*IP*: —

C03-A016

2,2′,2″-Tribromo-triethylamine

CAS: 36647-05-9; 36647-06-0 (Hydrobromide salt)
RTECS: —

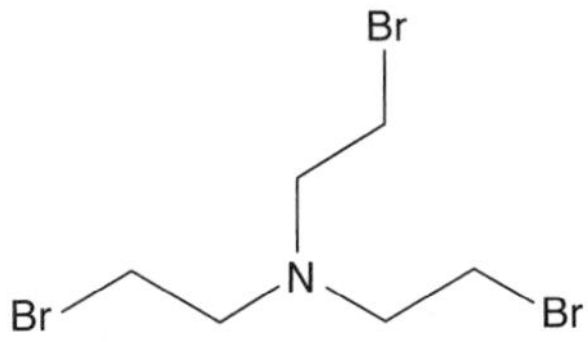

$C_6H_{12}Br_3N$

Solid.

Exposure Hazards

Conversion Factor: 1 ppm = 13.82 mg/m^3 at 77°F
Human toxicity values have not been established or have not been published.

Properties:

MW: 337.9	*VP*: —	*FlP*: —
D: —	*VD*: 12 (calculated)	*LEL*: —
MP: 86°F	*Vlt*: —	*UEL*: —
BP: 293°F (0.05 mmHg)	H_2O: —	*RP*: —
Vsc: —	*Sol*: —	*IP*: —

Hydrobromide salt

MW: 418.8
MP: 307°F

C03-A017

Bis(2-chloroethyl) 2-fluoroethylamine

CAS: 370-66-1
RTECS: —

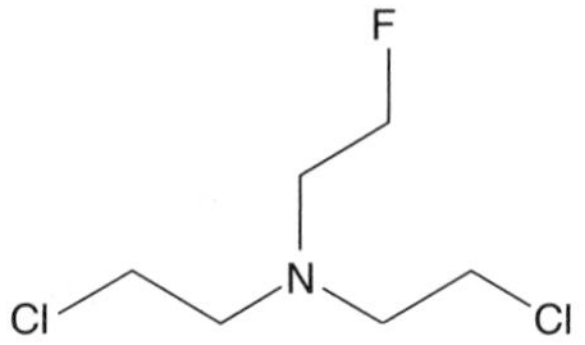

$C_6H_{12}Cl_2FN$

Specific information on physical appearance is not available for this agent.

Exposure Hazards

Conversion Factor: 1 ppm = 7.69 mg/m^3 at 77°F
Human toxicity values have not been established or have not been published.

Properties:

MW: 188.1	*VP*: —	*FlP*: —
D: —	*VD*: 6.5 (calculated)	*LEL*: —
MP: —	*Vlt*: —	*UEL*: —
BP: 239°F (13 mmHg)	H_2O: —	*RP*: —
Vsc: —	*Sol*: —	*IP*: —

C03-A018

Bis(2-fluoroethyl) 2-chloroethylamine

CAS: 370-67-2; 445-59-0 (Picrate salt)
RTECS: —

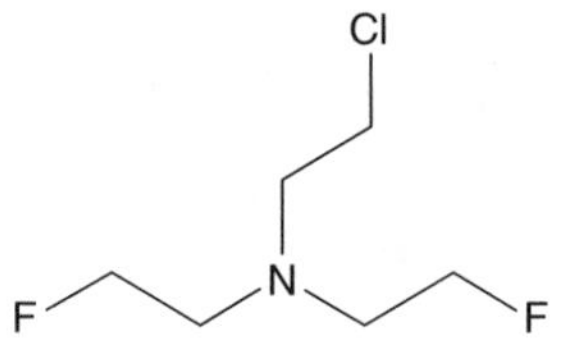

$C_6H_{12}ClF_2$

Specific information on physical appearance is not available for this agent.

Exposure Hazards

Conversion Factor: 1 ppm = 7.02 mg/m^3 at 77°F
Human toxicity values have not been established or have not been published.

Properties:

MW: 171.6	*VP*: —	*FlP*: —
D: —	*VD*: 5.9 (calculated)	*LEL*: —
MP: —	*Vlt*: —	*UEL*: —
BP: 203°F (19 mmHg)	H_2O: —	*RP*: —
Vsc: —	*Sol*: —	*IP*: —

Picrate Salt

MW: 400.7
MP: 250°F

C03-A019

2,2′,2″-Trifluoro-triethylamine

CAS: 370-68-3
RTECS: —

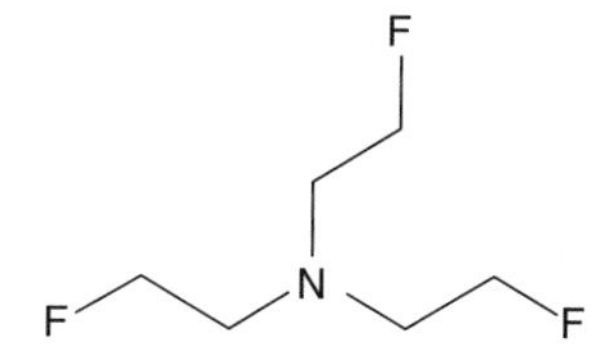

$C_6H_{12}F_3N$

Specific information on physical appearance is not available for this agent.

Exposure Hazards

Conversion Factor: 1 ppm = 6.35 mg/m^3 at 77°F
Human toxicity values have not been established or have not been published.

Properties:

MW: 155.2	*VP*: —	*FlP*: —
D: —	*VD*: 5.4 (calculated)	*LEL*: —
MP: —	*Vlt*: —	*UEL*: —
BP: 163°F (25 mmHg)	H_2O: —	*RP*: —
Vsc: —	*Sol*: —	*IP*: —

C03-A020

2-Chloropropyl 2-chloroethyl methylamine

CAS: 139881-58-6; 859060-27-8 (Picrate salt)
RTECS: —

$C_6H_{13}Cl_2N$

Specific information on physical appearance is not available for this agent.

Exposure Hazards

Conversion Factor: 1 ppm = 6.96 mg/m^3 at 77°F
Human toxicity values have not been established or have not been published.

Properties:

MW: 170.1	*VP*: —	*FlP*: —
D: —	*VD*: 5.9 (calculated)	*LEL*: —
MP: —	*Vlt*: —	*UEL*: —
BP: 201°F (21 mmHg)	H_2O: —	*RP*: —
Vsc: —	*Sol*: —	*IP*: —

Picrate salt

MW: 399.2
MP: 252°F

C03-A021

3-Chloropropyl 2-chloroethyl methylamine

CAS: 89980-59-6
RTECS: —

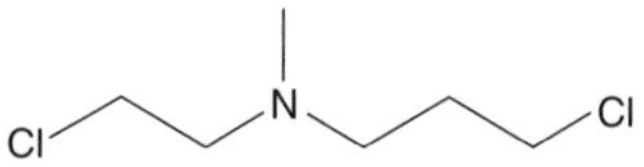

$C_6H_{13}Cl_2N$

Specific information on physical appearance is not available for this agent.

Exposure Hazards

Conversion Factor: 1 ppm = 6.96 mg/m^3 at 77°F
Human toxicity values have not been established or have not been published.

Properties:

MW: 170.1	*VP*: —	*FlP*: —
D: —	*VD*: 5.9 (calculated)	*LEL*: —
MP: —	*Vlt*: —	*UEL*: —
BP: 219°F (23 mmHg)	H_2O: —	*RP*: —
Vsc: —	*Sol*: —	*IP*: —

Picrate salt

MW: 399.2
MP: 167°F

C03-A022

Bis(2-chloroethyl) 1-propene amine

CAS: 63905-36-2
RTECS: —

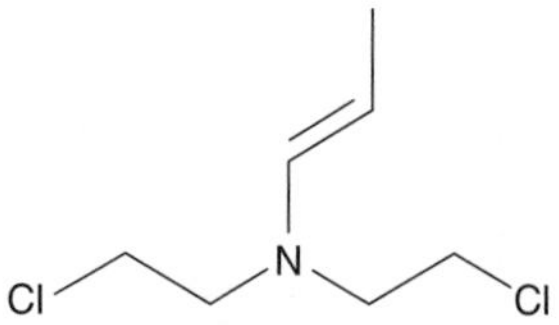

$C_7H_{13}Cl_2N$

Specific information on physical appearance is not available for this agent.

Exposure Hazards

Conversion Factor: 1ppm = 7.45mg/m^3 at 77°F
Human toxicity values have not been established or have not been published.

Properties:

MW: 182.1	*VP*: —	*FlP*: —
D: —	*VD*: 6.3 (calculated)	*LEL*: —
MP: —	*Vlt*: —	*UEL*: —
BP: 176°F (3 mmHg)	H_2O: —	*RP*: —
Vsc: —	*Sol*: —	*IP*: —

C03-A023

Bis(2-chloroethyl) 3-chloropropylamine

CAS: 61134-73-4; 100608-14-8 (Picrate salt)
RTECS: —

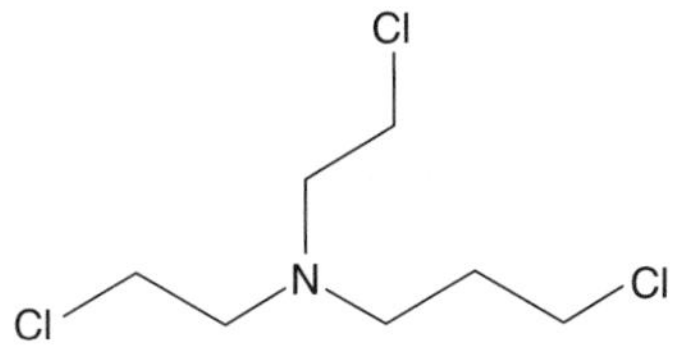

$C_7H_{14}Cl_3N$

Specific information on physical appearance is not available for this agent.

Exposure Hazards
Conversion Factor: 1 ppm = 8.94 mg/m^3 at 77°F
Human toxicity values have not been established or have not been published.

Properties:

MW: 218.6	*VP*: —	*FlP*: —
D: —	*VD*: 7.5 (calculated)	*LEL*: —
MP: —	*Vlt*: —	*UEL*: —
BP: 241°F (1.3 mmHg)	H_2O: —	*RP*: —
Vsc: —	*Sol*: —	*IP*: —

Picrate salt
MW: 447.7
MP: 199°F

C03-A024

Bis(2-chloroethyl) isopropylamine

CAS: 619-34-1
RTECS: —

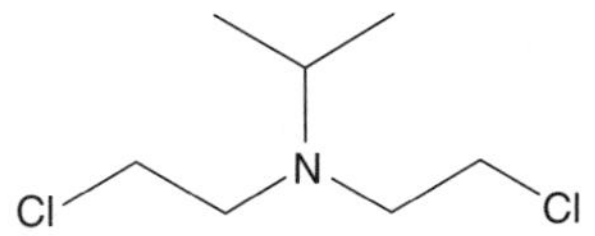

$C_7H_{15}Cl_2N$

Specific information on physical appearance is not available for this agent.

Exposure Hazards
Conversion Factor: 1 ppm = 7.53 mg/m^3 at 77°F
Human toxicity values have not been established or have not been published.

Properties:

MW: 184.1	*VP*: —	*FlP*: —
D: —	*VD*: 6.4 (calculated)	*LEL*: —
MP: —	*Vlt*: —	*UEL*: —
BP: 194°F (8 mmHg)	H_2O: —	*RP*: —
Vsc: —	*Sol*: —	*IP*: —

Hydrochloride salt

MW: 255.1
MP: 421°F

C03-A025

Bis(2-chloroethyl) propylamine

CAS: 621-68-1; 38521-66-3 (Hydrochloride salt)
RTECS: —

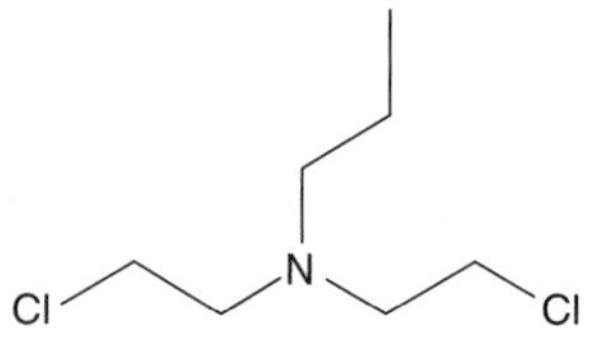

$C_7H_{15}Cl_2N$

Specific information on physical appearance is not available for this agent.

Exposure Hazards

Conversion Factor: 1 ppm = 7.53 mg/m^3 at 77°F
Human toxicity values have not been established or have not been published.

Properties:

MW: 184.1	*VP*: —	*FlP*: —
D: 1.059 g/mL (74°F)	*VD*: 6.4 (calculated)	*LEL*: —
MP: —	*Vlt*: —	*UEL*: —
BP: 205°F (10 mmHg)	*H_2O*: —	*RP*: —
Vsc: —	*Sol*: —	*IP*: —

Hydrochloride salt

MW: 220.6
MP: 221°F

C03-A026

Bis(2-chloropropyl) methylamine

CAS: 52802-03-6
RTECS: —

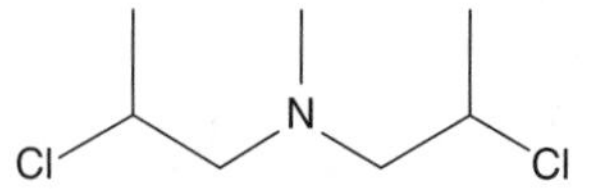

$C_7H_{15}Cl_2N$

Specific information on physical appearance is not available for this agent.

Exposure Hazards

Conversion Factor: 1 ppm = 7.53 mg/m^3 at 77°F
Human toxicity values have not been established or have not been published.

Properties:

MW: 184.1	*VP*: —	*FlP*: —
D: —	*VD*: 6.4 (calculated)	*LEL*: —
MP: —	*Vlt*: —	*UEL*: —
BP: 243°F (3 mmHg)	H_2O: —	*RP*: —
Vsc: —	*Sol*: —	*IP*: —

C03-A027

Bis(2-chloropropyl) chloroethylamine

CAS: 854873-81-7; 854873-82-8 (Picrate salt)
RTECS: —

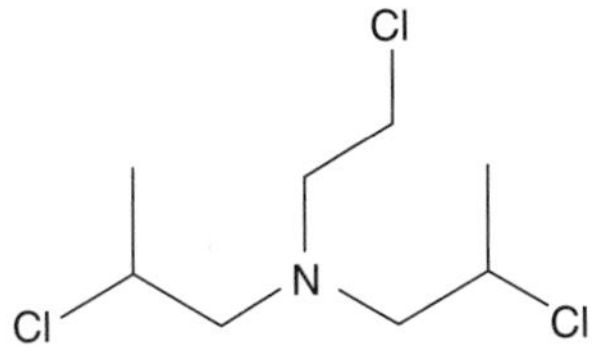

$C_8H_{16}Cl_3N$

Specific information on physical appearance is not available for this agent.

Exposure Hazards

Conversion Factor: 1 ppm = 9.51 mg/m^3 at 77°F
Human toxicity values have not been established or have not been published.

Properties:

MW: 232.6	*VP*: —	*FlP*: —
D: —	*VD*: 8.0 (calculated)	*LEL*: —
MP: —	*Vlt*: —	*UEL*: —
BP: 243°F (3 mmHg)	H_2O: —	*RP*: —
Vsc: —	*Sol*: —	*IP*: —

Picrate salt

MW: 461.7
MP: 244°F

C03-A028

Bis(2-chloroethyl) *t*-butylamine

CAS: 10125-86-7
RTECS: —

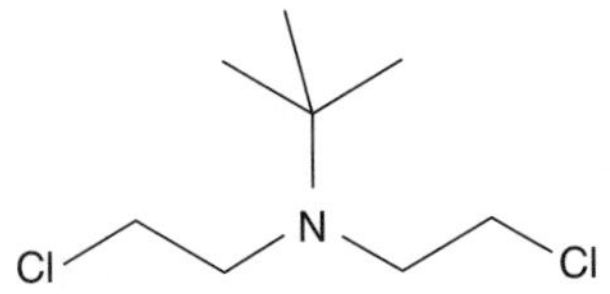

$C_8H_{17}Cl_2N$

Specific information on physical appearance is not available for this agent.

Exposure Hazards

Conversion Factor: 1 ppm = 8.10 mg/m^3 at 77°F
Human toxicity values have not been established or have not been published.

Properties:

MW: 198.1	*VP*: —	*FlP*: —
D: 1.048 g/mL (72°F)	*VD*: 6.8 (calculated)	*LEL*: —
MP: —	*Vlt*: —	*UEL*: —
BP: 160°F (2 mmHg)	*H_2O*: —	*RP*: —
Vsc: —	*Sol*: —	*IP*: —

C03-A029

Bis(2-chloroethyl) *s*-butylamine

CAS: 777799-85-6; 195210-51-6 (Isomer); 195210-50-5 (Isomer)
RTECS: —

$C_8H_{17}Cl_2N$

Specific information on physical appearance is not available for this agent.

Exposure Hazards

Conversion Factor: 1 ppm = 8.10 mg/m^3 at 77°F
Human toxicity values have not been established or have not been published.

Properties:

MW: 198.1	*VP*: —	*FlP*: —
D: 1.046 g/mL (77°F)	*VD*: 6.8 (calculated)	*LEL*: —
MP: —	*Vlt*: —	*UEL*: —
BP: 212°F (7.5 mmHg)	*H_2O*: —	*RP*: —
Vsc: —	*Sol*: —	*IP*: —

C03-A030

Bis(2-chloroethyl) isobutylamine

CAS: 87289-70-1
RTECS: —

$C_8H_{17}Cl_2N$

Specific information on physical appearance is not available for this agent.

Exposure Hazards

Conversion Factor: 1 ppm = 8.10 mg/m^3 at 77°F

Human toxicity values have not been established or have not been published.

Properties:

MW: 198.1	*VP*: —	*FlP*: —
D: 1.033 g/mL (68°F)	*VD*: 6.8 (calculated)	*LEL*: —
MP: —	*Vlt*: —	*UEL*: —
BP: 174°F (2 mmHg)	*H_2O*: —	*RP*: —
Vsc: —	*Sol*: —	*IP*: —

C03-A031

Bis(2-chloroethyl) butylamine

CAS: 42520-97-8; 55112-89-5 (Hydrochloride salt)

RTECS: —

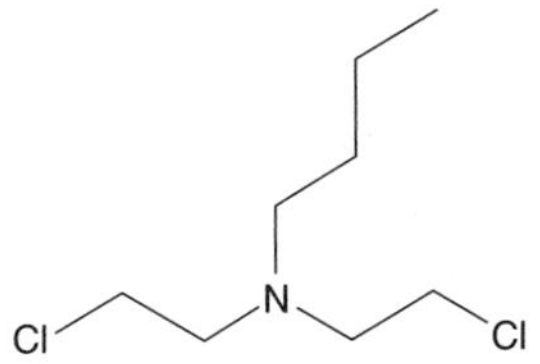

$C_8H_{17}Cl_2N$

Specific information on physical appearance is not available for this agent.

Exposure Hazards

Conversion Factor: 1 ppm = 8.10 mg/m^3 at 77°F

Human toxicity values have not been established or have not been published.

Properties:

MW: 198.1	*VP*: —	*FlP*: —
D: 1.0365 g/mL (77°F)	*VD*: 6.8 (calculated)	*LEL*: —
MP: —	*Vlt*: —	*UEL*: —
BP: 241°F (13 mmHg)	*H_2O*: —	*RP*: —
Vsc: —	*Sol*: —	*IP*: —

Hydrochloride salt

MW: 234.6

MP: 203°F

C03-C

COMPONENTS AND PRECURSORS

C03-C032

Thiodiglycol (TDG)

CAS: 111-48-8

RTECS: KM2975000

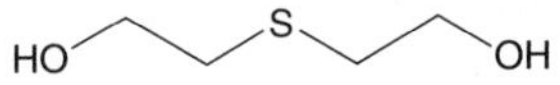

$C_4H_{10}O_2S$

Colorless to light yellow liquid with a foul odor. This material produces local skin/eye impacts.

Used industrially as an antioxidant, lubricant additive, printing-ink solvent, accelerator in the synthesis of rubber, and in the manufacture of plastics and pesticides.

This material is on the Australia Group Export Control list and Schedule 2 of the CWC.

This material is a precursor for Sulfur mustard (C03-A001) and is also commonly found as an impurity and degradation product formed during its hydrolysis.

Exposure Hazards

Conversion Factor: 1 ppm = 5.00 mg/m^3 at 77°F

Human toxicity values have not been established or have not been published.

Properties:

MW: 122.2	*VP*: 0.00002 mmHg	*FlP*: 320°F
D: 1.18 g/mL	*VP*: 0.00323 mmHg (77°F)	*LEL*: 1.2%
MP: 14°F	*VD*: 4.2 (calculated)	*UEL*: 5.2%
BP: 542°F (decomposes)	*Vlt*: 0.03 ppm	*RP*: 500,000
BP: 329°F (14 mmHg)	*H_2O*: Miscible	*IP*: —
Vsc: —	*Sol*: Acetone; Alcohols; Chloroform	

C03-C033

Sulfur monochloride

CAS: 10025-67-9
RTECS: WS4300000
UN: 1828
ERG: 137

Cl—S—S—Cl

S_2Cl_2

Light-amber to yellow-red, oily fuming liquid with a suffocating, pungent, and nauseating odor. This material is hazardous through inhalation and ingestion, and produces local skin/eye impacts. It is noncorrosive to carbon steel and iron when dry. However, when wet it will attack steel, cast iron, aluminum, stainless steel, copper and copper alloys, and many nickel based materials.

Used industrially for gold extraction and wood treatment; and to manufacture insecticides, pharmaceuticals, and dyes.

This material is on the Australia Group Export Control list and Schedule 3 of the CWC.

This material is a general precursor for sulfur based vesicants.

Exposure Hazards

Conversion Factor: 1 ppm = 5.52 mg/m^3 at 77°F

Eye Irritation: 2 ppm; exposure duration unspecified

Lethal human toxicity values have not been established or have not been published. However, based on available information, this material appears to have a considerably greater acute toxicity than HCl gas (C11-A063).

OSHA PEL: 1 ppm
ACGIH TLV: 1 ppm
NIOSH Ceiling: 1 ppm
IDLH: 5 ppm

Properties:

MW: 135.0	*VP*: 6.8 mmHg	*FlP*: 266°F
D: 1.68 g/mL	*VD*: 4.7 (calculated)	*LEL*: —
MP: −107°F	*Vlt*: 9100 ppm	*UEL*: —
BP: 276°F	H_2O: Decomposes violently	*RP*: 1.3
Vsc: —	*Sol*: Oils; Ether	*IP*: 9.40 eV

Proposed AEGLs

AEGL-1: 1 h, 0.53 ppm	4 h, 0.33 ppm	8 h, 0.17 ppm
AEGL-2: 1 h, 6.4 ppm	4 h, 4.0 ppm	8 h, 2.0 ppm
AEGL-3: 1 h, 15 ppm	4 h, 9.6 ppm	8 h, 4.8 ppm

C03-C034

Sulfur dichloride

CAS: 10545-99-0
RTECS: —

SCl_2

Fuming dark red to reddish-brown liquid with a pungent odor like chlorine. This material is hazardous through inhalation and ingestion, and produces local skin/eye impacts. It is noncorrosive to carbon steel and iron when dry. However, when wet it will attack steel, cast iron, aluminum, stainless steel, copper and copper alloys, and many nickel based materials.

Used industrially for vulcanizing oils and rubber, purifying sugar juices, as a chloridizing agent in metallurgy, and in the manufacture of insecticides.

This material is on the Australia Group Export Control list and Schedule 3 of the CWC.

This material is a general precursor for sulfur based vesicants.

Exposure Hazards

Conversion Factor: 1 ppm = 4.21 mg/m^3 at 77°F

Human toxicity values have not been established or have not been published. However, based on available information, this material appears to have a considerably greater acute toxicity than HCl gas (C11-A063).

Properties:

MW: 103.0	*VP*: 7.6 mmHg (−9°F)	*FlP*: 245°F
D: 1.621 g/mL (59°F)	*VD*: 3.6 (calculated)	*LEL*: —
MP: −108°F	*Vlt*: 12,000 ppm (−9°F)	*UEL*: —
BP: 138°F (decomposes)	H_2O: Decomposes violently	*RP*: 1.1
Vsc: —	*Sol*: Carbon tetrachloride; Benzene	*IP*: —

C03-C035

Thionyl chloride

CAS: 7719-09-7
RTECS: XM5150000

UN: 1836
ERG: 137

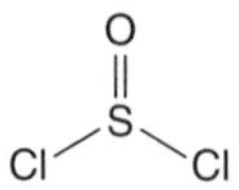

SCl_2O

Colorless to yellow to reddish fuming liquid with a pungent, suffocating odor like sulfur dioxide. This material is hazardous through inhalation, skin absorption, penetration through broken skin, and ingestion, and produces local skin/eye impacts.

Used in the manufacture of plastics, pesticides, pharmaceuticals, and dyes; and as a solvent in high energy density lithium batteries.

This material is on the Australia Group Export Control list and Schedule 3 of the CWC.

This material is used as a chlorinating agent for both nitrogen and sulfur based vesicants. Required in the manufacture of methylphosphonic dichloride (C01-C046).

Exposure Hazards

Conversion Factor: 1 ppm = 4.87 mg/m^3 at 77°F
ACGIH TLV: 1 ppm
NIOSH Ceiling: 1 ppm

Properties:

MW: 119.0	*VP*: 100 mmHg (70°F)	*FlP*: None
D: 1.638 g/mL	*VP*: 119 mmHg (77°F)	*LEL*: None
D: 1.676 g/mL (32°F)	*VD*: 4.1 (calculated)	*UEL*: None
MP: −156°F	*Vlt*: 160,000 ppm	*RP*: 0.78
BP: 169°F	*H_2O*: Decomposes violently	*IP*: —
Vsc: —	*Sol*: Benzene; Chloroform	

C03-C036

2-Chloroethanol

CAS: 107-07-3
RTECS: KK0875000
UN: 1135
ERG: 131

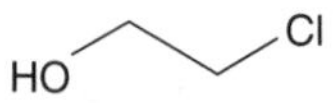

C_2H_5ClO

Colorless liquid with a sweet, pleasant odor similar to ether. This material is hazardous through inhalation, skin absorption, penetration through broken skin, and ingestion, and produces local eye impacts. It does not cause immediate irritation to warn of skin exposure.

Used industrially as a solvent and cleaning agent; used to manufacture insecticides, dyes, pharmaceuticals, thiodiethylene glycol, ethylene oxide, and ethylene glycol. Used in agriculture to treat sweet potatoes before planting.

This material is on the Australia Group Export Control list.

This material is a precursor for both nitrogen and sulfur based vesicants.

Exposure Hazards

Conversion Factor: 1 ppm = 3.29 mg/m^3 at 77°F
OSHA PEL: 5 ppm [Skin]
ACGIH Ceiling: 1 ppm [Skin]
NIOSH Ceiling: 1 ppm [Skin]
IDLH: 7 ppm

Lethal human toxicity values have not been established or have not been published. However, based on available information, the vapor of this material is more toxic than ethylene dichloride (C11-A041).

Properties:

MW: 80.5	*VP*: 4.9 mmHg	*FlP*: 140°F
D: 1.197 g/mL	*VP*: 7.18 mmHg (77°F)	*LEL*: 4.9%
MP: –90°F	*VD*: 2.8 (calculated)	*UEL*: 15.9%
BP: 262°F	*Vlt*: 6600 ppm	*RP*: 2.2
BP: 126°F (25 mmHg)	H_2O: Miscible (decomposes)	*IP*: 10.90 eV
Vsc: 2.87 cS	*Sol*: Most organic solvents	

C03-C037

Sodium sulfide

CAS: 1313-82-2
RTECS: —
UN: 1385
ERG: 135

Na_2S

Clear white to yellow-pink deliquescent crystals with an odor like rotten eggs due to formation of hydrogen sulfide. Commercial material may be yellow or brick-red lumps or flakes. It is unstable and discolors upon exposure to air. It undergoes autoxidation to form polysulfur, thiosulfate, and sulfate. It absorbs carbon dioxide from the air to form sodium carbonate. Moist sodium sulfide is spontaneously flammable upon drying in air. This material is hazardous through ingestion and produces local skin/eye impacts.

Used industrially for dehairing hides, wool pulling, ore flotation, metal refining, engraving, cotton printing, in the manufacture of paper, pharmaceuticals, rubber, and sulfur dyes. It is used in production of heavy water for nuclear reactors.

This material is on the Australia Group Export Control list.

This material is a precursor for sulfur based vesicants.

Exposure Hazards

Human toxicity values have not been established or have not been published.

Properties:

MW: 78.0	*VP*: —	*FlP*: —
D: 1.856 g/cm^3 (57°F)	*VD*: —	*LEL*: —
MP: 2156°F	*Vlt*: —	*UEL*: —
BP: —	H_2O: 18.6%	*RP*: —
Vsc: —	*Sol*: Alcohols	*IP*: —

C03-C038

Methyldiethanolamine

CAS: 105-59-9
RTECS: KL7525000

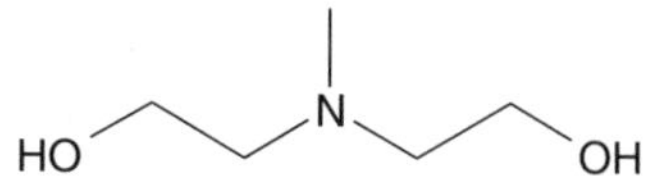

$C_5H_{13}NO_2$

Clear, colorless to light-yellow liquid with an odor like ammonia. Various salts (solids) have been reported. This material is hazardous through inhalation and ingestion, and produces local skin/eye impacts.

This material is on Schedule 3 of the CWC.

This material is a precursor for HN2 (C03-A012). It is also commonly found as a degradation product from hydrolysis of HN2.

Exposure Hazards

Conversion Factor: 1 ppm = 4.88 mg/m^3 at 77°F
Human toxicity values have not been established or have not been published.

Properties:

MW: 119.2	*VP*: 0.019 mmHg (104°F)	*FlP*: 259°F
D: 1.038 g/mL	*VD*: (calculated)	*LEL*: 0.9%
MP: −6°F	*Vlt*: 24 ppm (104°F)	*UEL*: 8.4%
BP: 477°F	*H_2O*: Miscible	*RP*: 4.1
BP: 239°F (10 mmHg)	*Sol*: —	*IP*: —
Vsc: 97.3 cS		

C03-C039

Ethyldiethanolamine

CAS: 139-87-7
RTECS: KK9800000

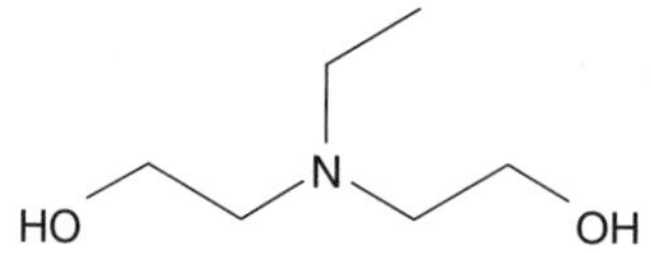

$C_6H_{15}NO_2$

Clear, colorless to light-yellow liquid with an odor like ammonia. Various salts (solids) have been reported. This material is hazardous through inhalation and ingestion, and produces local skin/eye impacts.

This material is on Schedule 3 of the CWC.

This material is a precursor for HN1 (C03-A011). It is also commonly found as a degradation product from hydrolysis of HN1.

Exposure Hazards

Conversion Factor: 1 ppm = 4.30 mg/m^3 at 77°F

Human toxicity values have not been established or have not been published.

Properties:

MW: 105.1	*VP*: —	*FlP*: 253°F
D: 1.014 g/mL	*VD*: —	*LEL*: —
MP: −58°F	*Vlt*: —	*UEL*: —
BP: 480°F	H_2O: —	*RP*: —
Vsc: —	*Sol*: —	*IP*: —

C03-C040

Triethanolamine

CAS: 102-71-6

RTECS: KL9275000

$C_6H_{15}NO_3$

Colorless to pale yellow viscous hydroscopic liquid with a slight odor like ammonia. Turns brown on exposure to air and light. Various salts (solids) have been reported. This material produces local skin/eye impacts.

Used industrially as a corrosion inhibitor for lubricants and in the manufacture of cosmetics, detergents, emulsifiers, and surfactants.

This material is on the Australia Group Export Control list and Schedule 3 of the CWC.

This material is a precursor for HN3 (C03-A013). It is also commonly found as a degradation product from hydrolysis of HN3.

Exposure Hazards

Conversion Factor: 1 ppm = 6.10 mg/m^3 at 77°F

ACGIH TLV: 0.82 ppm

Properties:

MW: 149.2	*VP*: 0.00000359 mmHg (77°F)	*FlP*: 374°F
D: 1.1242 g/mL	*VD*: 5.1 (calculated)	*LEL*: —
D: 1.0985 g/mL (140°F)	*Vlt*: 0.005 ppm (77°F)	*UEL*: —
MP: 69°F	H_2O: Miscible	*RP*: 2,000,000
BP: 636°F	*Sol*: Alcohols; Acetone; Chloroform; Benzene	*IP*: —
Vsc: 527 cS (77°F)		
Vsc: 60 cS (140°F)		

C03-D

DECOMPOSITION PRODUCTS AND IMPURITIES

C03-D041

1,2-Bis(2-hydroxyethylthio)ethane (Q-diol)

CAS: 5244-34-8
RTECS: —

$C_6H_{14}O_2S_2$

White crystalline solid. This material is hazardous through inhalation and ingestion, and produces local skin/eye impacts.

This material is commonly found as a degradation product from hydrolysis of Sulfur mustard (C03-A001).

Exposure Hazards

Conversion Factor: 1 ppm = 7.46 mg/m^3 at 77°F
Human toxicity values have not been established or have not been published.

Properties:

MW: 182.3	*VP*: —	*FlP*: —
D: —	*VD*: —	*LEL*: —
MP: 145°F	*Vlt*: —	*UEL*: —
BP: 338°F (0.5 mmHg)	H_2O: —	*RP*: —
Vsc: —	*Sol*: —	*IP*: —

C03-D042

Ethylene sulfide

CAS: 420-12-2
RTECS: KX3500000

C_2H_4S

Colorless liquid that gradually polymerizes. It may be stabilized with 0.5% butylmercaptan. Ethylene sulfide is hazardous through inhalation and ingestion, and produces local skin/eye impacts.

Used industrially as a chemical intermediate.

This material is commonly found as an impurity in Sulfur mustard (C03-A001).

Exposure Hazards

Conversion Factor: 1 ppm = 2.46 mg/m^3 at 77°F

Human toxicity values have not been established or have not been published. However, based on available information, this material appears to be approximately six times as toxic as propylene oxide.

Properties:

MW: 60.1	*VP*: 375 mmHg (77°F)	*FlP*: 50°F
D: 1.0368 g/mL (32°F)	*VD*: 2.1 (calculated)	*LEL*: —
MP: —	*Vlt*: 490,000 ppm (77°F)	*UEL*: —
BP: 131°F	H_2O: Insoluble	*RP*: 0.035
Vsc: —	*Sol*: Alcohol; Ether; Acetone; Chloroform	*IP*: —

C03-D043

1,2-Ethanedithiol

CAS: 540-63-6

RTECS: —

$C_2H_6S_2$

Liquid with a foul odor detectable at 31 ppb and easily noticeable at 5.6 ppm. This material is hazardous through inhalation and ingestion, and produces local skin/eye impacts.

Used industrially as a metal-complexing agent.

This material is a degradation product in Sulfur mustard (C03-A001).

Exposure Hazards

Conversion Factor: 1 ppm = 3.85 mg/m^3 at 77°F

Human toxicity values have not been established or have not been published.

Properties:

MW: 94.2	*VP*: 5.61 mmHg (77°F) (estimate)	*FlP*: —
D: 1.123 g/mL (74°F)	*VD*: 3.3 (calculated)	*LEL*: —
MP: –42°F	*Vlt*: 7500 ppm (77°F)	*UEL*: —
BP: 295°F	H_2O: 1.1% (77°F) (estimate)	*RP*: 1.8
BP: 169°F (150 mmHg)	*Sol*: Ethanol; Ether; Acetone; Benzene	*IP*: 9.35 eV
BP: 145°F (46 mmHg)		
Vsc: —		

C03-D044

Divinyl sulfone

CAS: 77-77-0

RTECS: KM7175000

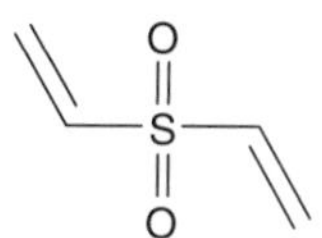

$C_4H_6O_2S$

Specific information on physical appearance is not available for this material. This material is hazardous through inhalation and ingestion, and produces local skin/eye impacts.

Used industrially in synthesis of fiber-reactive dyestuffs.

This material is a degradation product from Sulfur mustard (C03-A001).

Exposure Hazards

Conversion Factor: 1 ppm = 4.83 mg/m^3 at 77°F

Human toxicity values have not been established or have not been published. However, it can cause burns of the skin and eyes similar to those from Sulfur mustard (C03-A001), but it does not liberate chlorine or acid.

Properties:

MW: 118.2	*VP*: 0.09 mmHg	*FlP*: —
D: 1.177 g/mL (77°F)	*VD*: 4.1 (calculated)	*LEL*: —
MP: −15°F	*Vlt*: 120 ppm	*UEL*: —
BP: 226°F (15 mmHg)	H_2O: 14%	*RP*: 100
Vsc: —	*Sol*: —	*IP*: —

C03-D045

Vinyl sulfoxide

CAS: 1115-15-7
RTECS: —

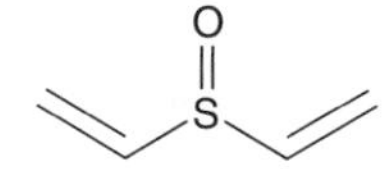

C_4H_6OS

Specific information on physical appearance is not available for this material.

This material is a degradation product from Sulfur mustard (C03-A001).

Exposure Hazards

Conversion Factor: 1 ppm = 4.18 mg/m^3 at 77°F

Human toxicity values have not been established or have not been published.

Properties:

MW: 102.2	*VP*: 0.92 mmHg	*FlP*: —
D: —	*VD*: 3.5 (calculated)	*LEL*: —
MP: —	*Vlt*: 1200 ppm	*UEL*: —
BP: 115°F (1 mmHg)	H_2O: 28%	*RP*: 11
Vsc: —	*Sol*: —	*IP*: —

C03-D046

Divinyl sulfide

CAS: 627-51-0
RTECS: —

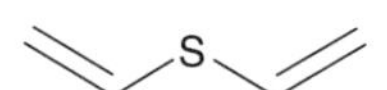

C_4H_6S

Specific information on physical appearance is not available for this material.

This material is a degradation product from Sulfur mustard (C03-A001).

Exposure Hazards

Conversion Factor: 1 ppm = 3.53 mg/m^3 at 77°F

Human toxicity values have not been established or have not been published.

Properties:

MW: 86.2	*VP*: 6.0 mmHg	*FlP*: —
D: 0.9098 g/mL	*VD*: 3.0 (calculated)	*LEL*: —
MP: —	*Vlt*: 8000 ppm	*UEL*: —
BP: 187°F	H_2O: 0.25%	*RP*: 1.8
Vsc: —	*Sol*: —	*IP*: 8.25 eV

C03-D047

2-Chloroethyl vinylsulfone

CAS: 7327-58-4
RTECS: —

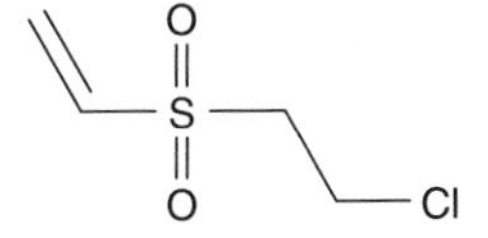

$C_4H_7ClO_2S$

Specific information on physical appearance is not available for this material.

This material is a degradation product from Sulfur mustard (C03-A001).

Exposure Hazards
Conversion Factor: 1 ppm = 6.32 mg/m^3 at 77°F
Human toxicity values have not been established or have not been published.

Properties:

MW: 154.6	*VP*: 0.023 mmHg	*FlP*: —
D: —	*VD*: 5.3 (calculated)	*LEL*: —
MP: —	*Vlt*: 31 ppm	*UEL*: —
BP: —	H_2O: 7.8%	*RP*: 350
Vsc: —	*Sol*: —	*IP*: —

C03-D048

2-Chloroethyl vinyl sulfoxide

CAS: 40709-82-8
RTECS: —

C_4H_7ClOS

Specific information on physical appearance is not available for this material.

This material is a degradation product from Sulfur mustard (C03-A001).

Exposure Hazards
Conversion Factor: 1 ppm = 5.67 mg/m^3 at 77°F
Human toxicity values have not been established or have not been published.

Properties:

MW: 138.6	*VP*: 0.064 mmHg	*FlP*: —
D: —	*VD*: 4.8 (calculated)	*LEL*: —
MP: —	*Vlt*: 86 ppm	*UEL*: —
BP: —	H_2O: 16%	*RP*: 130
Vsc: —	*Sol*: —	*IP*: —

C03-D049

2-Chloroethyl vinyl sulfide

CAS: 81142-02-1
RTECS: —

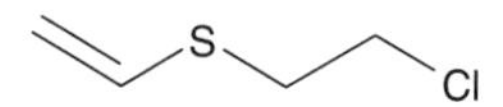

C_4H_7ClS

Specific information on physical appearance is not available for this material.

This material is a degradation product from Sulfur mustard (C03-A001).

Exposure Hazards

Conversion Factor: 1 ppm = 5.01 mg/m^3 at 77°F
Human toxicity values have not been established or have not been published.

Properties:

MW: 122.6	*VP*: 5.8 mmHg	*FlP*: —
D: —	*VD*: 4.2 (calculated)	*LEL*: —
MP: —	*Vlt*: 7800 ppm	*UEL*: —
BP: —	H_2O: 0.14%	*RP*: 1.5
Vsc: —	*Sol*: —	*IP*: —

C03-D050

Bis(2-chloroethyl) sulfone

CAS: 471-03-4
RTECS: —

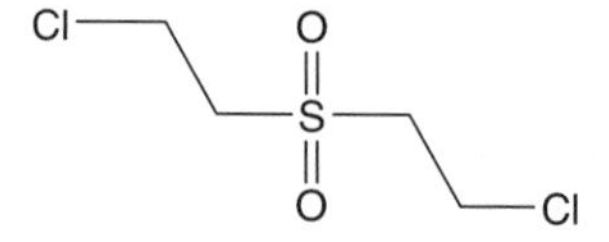

$C_4H_8Cl_2O_2S$

Specific information on physical appearance is not available for this material.

This material is a degradation product from Sulfur mustard (C03-A001).

Exposure Hazards

Conversion Factor: 1 ppm = 7.82 mg/m^3 at 77°F
Human toxicity values have not been established or have not been published. However, based on available information, this material is nearly as powerful a vesicant as the parent compound.

Properties:

MW: 191.1	*VP*: 0.96 mmHg	*FlP*: —
D: —	*VD*: 6.6 (calculated)	*LEL*: —
MP: —	*Vlt*: 1300 ppm	*UEL*: —
BP: —	H_2O: 1.1%	*RP*: 7.5
Vsc: —	*Sol*: —	*IP*: —

C03-D051

Bis(2-chloroethyl) sulfoxide

CAS: 5819-08-9
RTECS: —

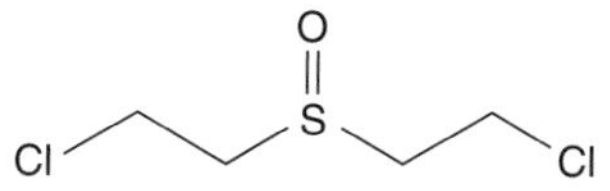

$C_4H_8Cl_2OS$

Specific information on physical appearance is not available for this material.

This material is a degradation product from Sulfur mustard (C03-A001).

Exposure Hazards

Conversion Factor: 1 ppm = 7.16 mg/m^3 at 77°F
Human toxicity values have not been established or have not been published. However, unlike the sulfone (C03-D050) this material is not a vesicant.

Properties:

MW: 175.1	*VP*: 0.65 mmHg	*FlP*: —
D: —	*VD*: 6.0 (calculated)	*LEL*: —
MP: —	*Vlt*: 870 ppm	*UEL*: —
BP: —	H_2O: 9.3%	*RP*: 12
Vsc: —	*Sol*: —	*IP*: —

C03-D052

Bis(2-chloroethyl)trisulfide

CAS: 19149-77-0
RTECS: KN3339000

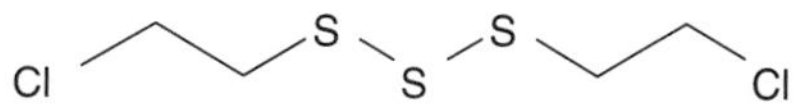

$C_4H_8Cl_2S_3$

Clear, colorless hygroscopic liquid.

This material is commonly found as an impurity in Sulfur mustard (C03-A001).

Exposure Hazards

Conversion Factor: 1 ppm = 9.13 mg/m^3 at 77°F
Human toxicity values have not been established or have not been published.

Properties:

MW: 223.2	*VP*: 1.7 mmHg	*FlP*: 135°F
D: 0.94 g/mL	*VP*: 2.96 mmHg (77°F)	*LEL*: —
MP: −90°F	*VD*: 4.6 (calculated)	*UEL*: —
BP: 324°F	*Vlt*: 2300 ppm	*RP*: 3.9
Vsc: —	H_2O: Miscible	*IP*: 9.8 eV
	Sol: —	

C03-D053

1,4-Oxathiane

CAS: 15980-15-1
RTECS: RP4200000

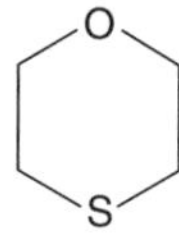

C_4H_8OS

Clear, colorless to light yellow liquid.

This material is commonly found as an impurity in Sulfur mustard (C03-A001).

Exposure Hazards

Conversion Factor: 1 ppm = 4.26 mg/m^3 at 77°F
Human toxicity values have not been established or have not been published.

Properties:

MW: 104.2	*VP*: 3.9 mmHg	*FlP*: 107°F
D: 1.105 g/mL	*VD*: 3.6 (calculated)	*LEL*: —
MP: 32°F	*Vlt*: 5200 ppm	*UEL*: —
BP: 297°F	H_2O: 16.7%	*RP*: 2.5
Vsc: —	*Sol*: —	*IP*: 8.8 eV

C03-D054

1,4-Dithiane

CAS: 505-29-3
RTECS: —

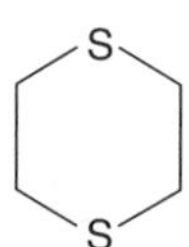

$C_4H_8S_2$

Specific information on physical appearance is not available for this material.

This material is commonly found as an impurity and degradation product from Sulfur mustard (C03-A001).

Exposure Hazards

Conversion Factor: 1 ppm = 4.92 mg/m^3 at 77°F
Human toxicity values have not been established or have not been published.

Properties:

MW: 120.2	*VP*: 0.80 mmHg	*FlP*: —
D: —	*VD*: 4.1 (calculated)	*LEL*: —
MP: —	*Vlt*: 1100 ppm	*UEL*: —
BP: 392°F	H_2O: 0.30%	*RP*: 11
Vsc: —	*Sol*: —	*IP*: 8.8 eV

C03-D055

Diethanolamine

CAS: 111-42-2
RTECS: KL2975000

HO N OH

$C_4H_{11}NO_2$

Viscous, syrupy liquid or crystalline deliquescent solid with a mild odor like ammonia. Various salts (solids) have been reported. This material is hazardous through inhalation and ingestion, and produces local skin/eye impacts.

Used in the manufacture of emulsifiers, dispersing agents, cutting oils, shampoos, cleaners, and polishes. Chemical intermediate for morpholine.

This material is a degradation product from HN1 (C03-A011).

Exposure Hazards

Conversion Factor: 1 ppm = 4.30 mg/m^3 at 77°F
$LD_{50(Ing)}$: 20 g (estimate) *ACGIH TLV*: 0.47 ppm [Skin]

Properties:

MW: 105.1	*VP*: 0.00014 mmHg (77°F)	*FlP*: 342°F
D: 1.0966 g/cm^3	*VD*: 3.7 (calculated)	*LEL*: —
MP: 82°F	*Vlt*: 0.18 ppm (77°F)	*UEL*: —
BP: 516°F	H_2O: Miscible	*RP*: 700,000
Vsc: 320 cS (86°F)	*Sol*: Alcohols	*IP*: —
Vsc: 49 cS (140°F)		

References

Agency for Toxic Substances and Disease Registry. "Blister Agents Lewisite (L) and Mustard-Lewisite Mixture (HL) ToxFAQs." April 2002.

———. "Blister Agents Sulfur Mustard Agent H/HD and Sulfur Mustard Agent HT ToxFAQs." April 2002.

———. *Managing Hazardous Materials Incidents Volume III—Medical Management Guidelines for Acute Chemical Exposures*. Rev. Edn. Washington, DC: Government Printing Office, 2000.

———. "Mustard Gas ToxFAQs." September 2001.

———. *Toxicological Profile for Mustard "Gas" (UPDATE)*. Washington, DC: Government Printing Office, September 2003.

Centers for Disease Control and Prevention. "Case Definition: Vesicant (Mustards, Dimethyl Sulfate, and Lewisite)." March 15, 2005.

———. "Counter Terrorism Card for Mustard." 2000.
———. "Facts About Sulfur Mustard." February 25, 2003.
———. "Facts About Sulfur Mustard." February 25, 2003.
Compton, James A.F. *Military Chemical and Biological Agents: Chemical and Toxicological Properties.* Caldwell, NJ: The Telford Press, 1987.
Edgewood Research Development, and Engineering Center, Department of the Army. *Material Safety Data Sheet (MSDS) for Agent HL.* Aberdeen Proving Ground, MD: Chemical Biological Defense Command, Revised June 30, 1995.
———. *Material Safety Data Sheet (MSDS) for Agent HQ.* Aberdeen Proving Ground, MD: Chemical Biological Defense Command, Revised June 30, 1995.
———. *Material Safety Data Sheet (MSDS) for Agent HT.* Aberdeen Proving Ground, MD: Chemical Biological Defense Command, Revised June 30, 1995.
———. *Material Safety Data Sheet (MSDS) for Agent Q.* Aberdeen Proving Ground, MD: Chemical Biological Defense Command, Revised June 30, 1995.
———. *Material Safety Data Sheet (MSDS) for Agent T.* Aberdeen Proving Ground, MD: Chemical Biological Defense Command, Revised May 1, 1996.
———. *Material Safety Data Sheet (MSDS) for Distilled Mustard (HD).* Aberdeen Proving Ground, MD: Chemical Biological Defense Command, Revised February 28, 1996.
Fries, Amos A. and Clarence J. West. *Chemical Warfare.* New York: McGraw-Hill Book Company, Inc., 1921.
"Interim Recommendations for Airborne Exposure Limits for Chemical Warfare Agents H and HD (Sulfur Mustard)." Federal Register 69, No. 85 (May 3, 2004): 24164–24168.
International Association of Fire Fighters (IAFF) Division of Occupational Health, Safety and Medicine. *Position on the U.S. Army Soldier & Biological Chemical Command's (SBCCOM), Withdrawn "3/30 Rule" and the New Guide: "The Risk Assessment of Using Firefighter Protective Ensemble with Self-Contained Breathing Apparatus for Rescue Operations During a Terrorist Chemical Agent Incident."* February 13, 2004.
Jackson, Kirby E. "β,β-Dichloroethyl Sulfide (Mustard Gas)." *Chemical Reviews* 15 (1934): 425–462.
Macy, Rudolph, Benjamin L. Harris, and John W. Eastes. "Vesicant," United States Patent 2,604,428, July 22, 1952.
Marrs, Timothy C., Robert L. Maynard, and Frederick R. Sidell. *Chemical Warfare Agents: Toxicology and Treatment.* Chichester, England: John Wiley & Sons, 1997.
Munro, Nancy B., Sylvia S. Talmage, Guy D. Griffin, Larry C. Waters, Annetta P. Watson, Joseph F. King, and Veronique Hauschild. "The Sources, Fate and Toxicity of Chemical Warfare Agent Degradation Products." *Environmental Health Perspectives* 107 (1999): 933–974.
National Institute of Health. *Hazardous Substance Data Bank (HSDB).* (http://toxnet.nlm.nih.gov/cgi-bin/sis/htmlgen?HSDB/). 2005.
National Institute for Occupational Safety and Health. "Emergency Response Card for Mustard." June 28, 2004.
———. *NIOSH Pocket Guide to Chemical Hazards.* Washington, DC: Government Printing Office, September 2005.
National Research Council. *Review of Acute Human-Toxicity Estimates for Selected Chemical-Warfare Agents.* Washington, DC: National Academy Press, 1997.
Petrali, John P., Tracey A. Hamilton, Betty J. Benton, Dana R. Anderson, Wesley Holmes, Robert K. Kan, Christina P. Tompkins, and Radharaman Ray. "Dimethyl Sulfoxide Accelerates Mustard Gas-Induced Skin Pathology." *Journal of Medical CBR Defense* 3 (2005) (www.jmedcbr.org).
Prentiss, Augustin M. *Chemicals in War: A Treatise on Chemical Warfare.* New York: McGraw-Hill Book Company, Inc., 1937.
Sartori, Mario. *The War Gases: Chemistry and Analysis.* Trans. L.W. Marrison. London: J. and A. Churchill, Ltd., 1939.
———. "New Developments in the Chemistry of War Gases." *Chemical Reviews* 48 (1951): 225–257.
Sidell, Frederick R. *Medical Management of Chemical Warfare Agent Casualties: A Handbook for Emergency Medical Services.* Bel Air, MD: HB Publishing, 1995.

Sidell, Fredrick R., Ernest T. Takafuji, and David R. Franz, ed. *Medical Aspects of Chemical and Biological Warfare, Textbook of Military Medicine Series, Part 1, Warfare, Weaponry, and the Casualty.* Washington, DC: Office of the Surgeon General, Department of the Army, 1997.

Sifton, David W. ed. *PDR Guide to Biological and Chemical Warfare Response.* Montvale, NJ: Thompson/Physicians Desk Reference, 2002.

Smith, Ann, Patricia Heckelman, and Maryadele J. Oneil, ed. *The Merck Index: An Encyclopedia of Chemicals, Drugs, & Biologicals.* 13th Edn. Rahway, NJ: Merck & Co., Inc., 2001.

Somani, Satu M., ed. *Chemical Warfare Agents.* New York: Academic Press, 1992.

Somani, Satu M., and James A. Romano, Jr., ed. *Chemical Warfare Agents: Toxicity at Low Levels.* Boca Raton, FL: CRC Press, 2001.

Trapp, Ralf. *The Detoxification and Natural Degradation of Chemical Warfare Agents.* Volume 3 of *SIPRI Chemical & Biological Warfare Studies.* London: Taylor & Francis, 1985.

True, Bey-Lorraine, and Robert H. Dreisbach. *Dreisbach's Handbook of Poisoning: Prevention, Diagnosis and Treatment.* 13th Edn. London, England: The Parthenon Publishing Group, 2002.

United States Army Headquarters. *Chemical Agent Data Sheets Volume I, Edgewood Arsenal Special Report No. EO-SR-74001.* Washington, DC: Government Printing Office, December 1974.

———. *Potential Military Chemical/Biological Agents and Compounds, Field Manual No. 3-11.9.* Washington, DC: Government Printing Office, January 10, 2005.

United States Army Medical Research Institute of Chemical Defense. *Medical Management of Chemical Casualties Handbook.* 3rd Edn. Aberdeen Proving Ground, MD: United States Army Medical Research Institute of Chemical Defense, July 2000.

United States Coast Guard. *Chemical Hazards Response Information System (CHRIS) Manual, 1999 Edition.* (http://www.chrismanual.com/Default.htm). March 2004.

Wachtel, Curt. *Chemical Warfare.* Brooklyn, NY: Chemical Publishing Co., Inc., 1941.

Wagner, George W. and Yu-Chu Yang. "Universal Decontaminating Solution For Chemical Warfare Agents," United States Patent 6245957, June 12, 2001.

———. "Rapid Nucleophilic/Oxidative Decontamination of Chemical Warfare Agents." *Industrial & Engineering Chemistry Research* 41 (2002): 1925–1928.

Waitt, Alden H. *Gas Warfare: The Chemical Weapon, Its Use, and Protection Against It.* Rev Edn. New York: Duell, Sloan and Pearce, 1944.

Watson, Annetta, Dennis Opresko, and Veronique Hauschild. *Evaluation of Chemical Warfare Agent Percutaneous Vapor Toxicity: Derivation of Toxicity Guidelines for Assessing Chemical Protective Ensembles, ORNL/TM-2003/180.* July 2003.

Watson, Annetta P. and Nancy B. Munro. *Reentry Planning: The Technical Basis for Offsite Recovery Following Warfare Agent Contamination, ORNL-6628.* April 1990.

Williams, Kenneth E. *Detailed Facts About Blister Agent Mustard-Lewisite Mixture (HL).* Aberdeen Proving Ground, MD: United States Army Center for Health Promotion and Preventive Medicine, 1996.

———. *Detailed Facts About Sulfur Mustard Agents H and HD.* Aberdeen Proving Ground, MD: United States Army Center for Health Promotion and Preventive Medicine, 1996.

———. *Detailed Facts About Sulfur Mustard Agent HT.* Aberdeen Proving Ground, MD: United States Army Center for Health Promotion and Preventive Medicine, 1996.

World Health Organization. *Health Aspects of Chemical and Biological Weapons.* Geneva: World Health Organization, 1970.

———. *International Chemical Safety Cards (ICSCs)* (http://www.cdc.gov/niosh/ipcs/icstart.html). 2004.

———. *Public Health Response to Biological And Chemical Weapons: WHO Guidance.* Geneva: World Health Organization, 2004.

Yang, Yu-Chu. "Chemical Reactions for Neutralizing Chemical Warfare Agents." *Chemistry & Industry* (May 1, 1995): 334–337.

Yang, Yu-Chu, James A. Baker, and J. Richard Ward. "Decontamination of Chemical Warfare Agents." *Chemical Reviews* 92 (1992): 1729–1743.

4

Arsenic Vesicants

4.1 General Information

The agents in this class are dihalo organoarsines. Other than lewisite (C04-A002), which is listed in Schedule 1, these materials are not covered by the Chemical Weapons Convention. Some of them have even seen limited commercial applications.

The majority of these materials are first generation chemical warfare agents employed in World War I. They are moderately difficult to synthesize and can be difficult to disperse effectively. For information on some of the chemicals used to manufacture arsenic vesicants, see the Component section (C04-C) following information on the individual agents.

In addition to the agents detailed in this handbook, there are other arsenic vesicants that were employed during World War I on a limited basis. However, there is little or no published information concerning the physical, chemical, or toxicological properties of these additional agents.

Toward the end of World War I, lewisite (C04-A002) was developed, produced, and weaponized but never used. Despite this, it has supplanted all other agents as the arsenic vesicant of choice and is the only one that was stockpiled in modern arsenals.

4.2 Toxicology

4.2.1 Effects

Vesicants affect both exterior and interior parts of the body. Vesicants cause inflammation, blisters, and general destruction of tissues. These agents have a greater impact on moist areas of the body. Eyes are especially susceptible to vesicants and contact results in irritation, lacrimation, and involuntary blinking (blepharospasm).

Inhalation of vesicants can cause lung membranes to swell and become filled with liquid (pulmonary edema). Death may result from lack of oxygen.

Arsenic vesicants are also systemic agents and can pass through the skin to affect susceptible tissue including blood cells and the liver. Arsenic vesicants also act as vomiting/sternatory agents (see Chapter 15) and produce violent coughing, sneezing, and regurgitation. Some arsenic vesicants are carcinogenic.

4.2.2 Pathways and Routes of Exposure

Vesicants are hazardous through any route of exposure including inhalation, skin and eye exposure, ingestion, and broken, abraded, or lacerated skin (e.g., penetration of skin by debris). Liquid agents are much more hazardous than their vapors. Thickened agents primarily pose a hazard through skin absorption.

4.2.3 General Exposure Hazards

Arsenic vesicants cause instantaneous irritation of the eyes, nose, throat, and skin, which provides warning of their presence. Extended exposures cause violent coughing, sneezing, and regurgitation. The odor of arsenic vesicants varies with the individual compound and ranges from odorless to fruity to flowery. Odors may not be discernable due to irritation.

Lethal concentrations (LC_{50}s) for inhalation of these agents are as low as 59 ppm for a 2-minutes exposure.

Lethal percutaneous exposures (LD_{50}s) to liquid agents are as low as 1.4 grams per individual.

Incapacitating concentrations (ICt_{50}) for dermal exposure to these agents at moderate temperatures (i.e., between 65 and 85°F) are as low as 2 ppm for a 30-minutes exposure. Temperatures above 85°F reduce the concentration necessary to produce similar effects.

Eye irritation from exposure to agent vapors occurs at concentrations as low as 0.9 ppm for a 2-minutes exposure; an incapacitating concentration (ICt_{50}) for exposure of the eyes is as low as 4.4 ppm for a 2-minutes exposure. Permanent eye damage may occur at concentrations as low as 90 ppm for a 2-minutes exposure.

Although sublethal doses of some arsenic vesicants are rapidly detoxified by the body, many agents are not detoxified and exposures are cumulative.

4.2.4 Latency Period

4.2.4.1 Vapor/Aerosols (Mists or Dusts)

Inhalation of high concentrations may be fatal in as short a time as 10 minutes. Otherwise, arsenic vesicants produce immediate irritation of the eyes and involuntary blinking (blepharospasm), followed by lacrimation. Nasal irritation is coupled with coughing, sneezing, and vomiting. There is an immediate burning sensation when the agent comes into contact with the skin. Other signs and symptoms of exposure, including reddening of the skin (erythema), blistering (vesication), and accumulation of fluid in the lungs (pulmonary edema), do not occur until after a substantial latency period.

4.2.4.2 Liquids

Arsenic vesicants produce immediate pain. Tissue damage occurs within minutes of exposure but clinical effects may not appear for up to 24 hours. Some agents are rapidly absorbed through the skin. Extensive skin contamination may cause systemic damage to the liver, kidneys, nervous system, red blood cells, and the brain.

4.3 Characteristics

4.3.1 Physical Appearance/Odor

4.3.1.1 Laboratory Grade

Laboratory grade agents are typically colorless oily liquids. They have little or no odor. Vapors are extremely irritating to the eyes, nose, throat, and skin.

4.3.1.2 Munition Grade

Munition grade agents are typically yellow to dark brown liquids or solids. As the agent ages and decomposes, it continues to discolor until it may appear black. Production impurities and decomposition products in these agents may give them an odor. Odors range from fruity and biting, to flowery and similar to geraniums. Odors may become more pronounced during storage.

4.3.1.3 Modified Agents

Arsenic vesicants have been thickened with various substances to enhance deployment, increase their persistency, and increase the risk of percutaneous exposure. Thickeners include polyalkyl methacrylates (methyl, ethyl, butyl, isobutyl), poly(vinyl acetate), polystyrene, plexiglas, alloprene, polychlorinated isoprene, nitrocellulose, as well as bleached montan and lignite waxes. Military thickener K125 is a mixture of methyl, ethyl, and butyl polymethacrylates. When thickened, agents become sticky with a consistency similar to honey. Typically, not enough thickener is added to affect either the color or odor of the agent.

4.3.1.4 Mixtures with Other Agents

Lewisite (C04-A002) has been mixed with sulfur mustard (C03-A001) to prevent the sulfur mustard from freezing in the shell as well as to enhance its toxicity.

4.3.2 Stability

Arsenic vesicants are stable when pure and kept dry. Stabilizers are not required. Agents can be stored in glass or steel containers, although they may attack steel at elevated temperature or if moisture is present. Arsenic vesicants corrode aluminum and brass, and will attack some rubbers and plastics.

4.3.3 Persistency

For military purposes, unmodified arsenic vesicants are classified as persistent. However, agent vapors rapidly react with high humidity to lose most of their vesicant properties. Limited solubility slows the hydrolysis of liquid agents. Some hydrolysis products are highly toxic and extremely persistent (see Section 4.4.5). Evaporation rates range from near that of water down to that of light machine oil.

Agents modified with thickeners last significantly longer.

4.3.4 Environmental Fate

Arsenic vesicant vapors have a density greater than air and tend to collect in low places. Agent vapor is rapidly decomposed by moisture in the air. Porous material, including painted surfaces, will absorb both liquid and gaseous agent. After the initial surface contamination has been removed, the agent that has been absorbed into porous material can migrate back to the surface posing both a contact and vapor hazard. Clothing may emit trapped agent vapor for up to 30 minutes after contact with a vapor cloud.

Liquid agents that do dissolve in water are rapidly decomposed. However, most of these agents are insoluble in water and this limited solubility slows their hydrolysis. Hydrolysis of agents may be further reduced if they are thickened such that a protective layer forms at the agent/water interface. The specific gravities of unmodified liquid agents are much greater than that of water. These agents are soluble in most organic solvents including gasoline, oils, acetone, and alcohols.

4.4 Additional Hazards

4.4.1 Exposure

All foodstuffs in the area of a release should be considered contaminated. Unopened items packaged in glass, metal, or heavy duty plastic and exposed only to agent vapors may be used after decontamination of the container. Unopened items exposed to liquid agents should be decontaminated within a few hours postexposure or destroyed. Opened or unpackaged items, or those packaged only in paper or cardboard, should be destroyed.

Meat and milk from animals affected by arsenic vesicants should be destroyed.

4.4.2 Livestock/Pets

Although vesicants do not produce the same type of dermal damage in animals as they do in humans, they are still susceptible to the cytotoxic and systemic toxicities of these agents.

Animals can be decontaminated with shampoo/soap and water, or a 0.5% household bleach solution (see Section 4.5.3). If the animals' eyes have been exposed to agent, they should be irrigated with water or saline solution for a minimum of 30 minutes.

Unprotected feedstock (e.g., hay or grain) should be destroyed. Leaves of forage vegetation could still retain sufficient vesicant agent, or toxic decomposition products, to produce effects for several weeks post release, depending on the level of contamination and the weather conditions.

4.4.3 Fire

Heat from a fire will increase the amount of agent vapor in the area. A significant amount of the agent could be volatilized and escape into the surrounding environment before it is consumed by the fire. Actions taken to extinguish the fire can also spread the agent. Combustion or hydrolysis of arsenic vesicants will produce volatile, toxic decomposition products (see Section 4.4.5).

4.4.4 Reactivity

Although these agents will decompose if dissolved in water, a lack of solubility inhibits this process. Vapors are decomposed rapidly by high humidity.

4.4.5 Hazardous Decomposition Products

4.4.5.1 Hydrolysis

Arsenic vesicants produce hydrogen chloride (HCl) or hydrogen bromide (HBr), and arsenous oxides or arsenic salts when hydrolyzed. Some arsenous oxide decomposition products are toxic and may also have vesicant properties. Some agents may produce acetylene at higher pHs.

4.4.5.2 Combustion

Volatile decomposition products may include HCl, HBr, and arsenic oxides. In addition, a corrosive or toxic residue or both may remain.

4.5 Protection

4.5.1 Evacuation Recommendations

Isolation and protective action distances listed below are taken from Argonne National Laboratory Report No. ANL/DIS-00-1, *Development of the Table of Initial Isolation and Protective Action Distances for the 2000 Emergency Response Guidebook*, which is still the basis for the "when used as a weapon" scenarios in the 2004 Emergency Response Guidebook (ERG). These recommendations are based on different release mechanisms (direct aerosolization of liquid agents, a spray, or an explosively generated mist) as well as various quantities of materials depending on the agent used in the release.

For lewisite (C04-A002), the release scenario involves either a spray or an explosively generated mist that quickly settles to the ground and soaks into a depth of no more than 0.25 mm. A secondary cloud will then be generated by evaporation of this deposited material. Under these conditions, the difference between a small and a large release of vesicant is not based on the standard 200 liters spill used for commercial hazardous materials listed in the ERG. A small release of lewisite involves 2 kilograms (approximately 1.1 liters) of liquid agent and a large release involves 100 kilograms (approximately 14 gallons) of liquid agent.

For the less effective agents, such as methyl, ethyl, and phenyl dichloroarsines, the release involves direct aerosolization of the liquid agents with a particle size between 2 and 5 μm. For these agents, a small release involves 30 kilograms (approximately 5 gallons) of liquid agent and a large release involves 500 kilograms (approximately 80 gallons) of liquid agent.

	Initial isolation (feet)	Downwind day (miles)	Downwind night (miles)
ED (Ethyldichloroarsine) C04 – *A001*			
Small device (30 kilograms)	100	0.2	0.5
Large device (500 kilograms)	400	0.8	1.6
L (Lewisite) C04 – *A002*			
Small device (2 kilograms)	100	0.1	0.2
Large device (100 kilograms)	300	0.6	1.1
MD (Methyldichloroarsine) C04 – *A003*			
Small device (30 kilograms)	100	0.2	0.5
Large device (500 kilograms)	400	0.8	2.2
PD (Phenyldichloroarsine) C04 – *A004*			
Small device (30 kilograms)	100	0.1	0.1
Large device (500 kilograms)	100	0.1	0.2

4.5.2 Personal Protective Requirements

4.5.2.1 Structural Firefighters' Gear

Structural firefighters' protective clothing is recommended for fire situations only; it is not effective in spill situations or release events and should never be used as the primary chemical protective garment to enter an area contaminated with arsenic vesicants.

4.5.2.2 Respiratory Protection

Self-contained breathing apparatuses (SCBAs) or air purifying respirators (APRs) should have a National Institute for Occupational Safety and Health (NIOSH) and Chemical/Biological/Radiological/Nuclear (CBRN) certification since vesicants can be absorbed into or degrade the materials used to make some respirators. However, during emergency operations, other NIOSH approved SCBAs or APRs that have been specifically tested by the manufacturer against chemical warfare agents may be used if deemed necessary by the Incident Commander. APRs should be equipped with a NIOSH approved CBRN filter or a combination organic vapor/acid gas/particulate cartridge.

Immediately dangerous to life or health (IDLH) levels are the ceiling limit for respirators other than SCBAs. However, IDLH levels have not been established for arsenical vesicants. Therefore, any potential exposure to these agents should be regarded with extreme caution and the use of SCBAs for respiratory protection should be considered.

4.5.2.3 Chemical Protective Clothing

Use only chemical protective clothing that has undergone material and construction performance testing against arsenic vesicant agents. Reported permeation rates may be affected by solvents, components, or impurities in munition grade or modified agents.

Because of the extreme dermal hazard posed by arsenic vesicants, responders should wear a Level A protective ensemble whenever there is a potential for exposure to any liquid agent, or to an elevated or unknown concentration of agent vapor.

4.5.3 Decontamination

4.5.3.1 General

Although the vesicant properties of arsenical agents can be eliminated during decontamination, arsenic is an element and cannot be destroyed. Residual arsenical compounds may still possess significant toxicity if they enter the body through ingestion, or broken, abraded, or lacerated skin (e.g., penetration of skin by debris).

Arsenic vesicants are rapidly hydrolyzed in water but a lack of solubility may slow the rate of reaction. Arsenic decomposition products are stable and some of them have toxic and/or vesicant properties that nearly equal the original agent.

Solubility of an agent will be decreased further if it is dissolved in an immiscible organic solvent or if it is thickened such that it forms a protective layer at the agent/water interface. Addition of solvents or mechanical mixing may be required to overcome insolubility problems.

These agents are readily destroyed by high pH (i.e., basic solutions), especially when used in combination with a strong oxidizing agent. For this reason, undiluted household bleach is an excellent agent for decontamination of these agents. Ensure that the bleach solution remains in contact with the agent for a minimum of 5 minutes. However, a large excess will be needed to ensure complete destruction of the agents. Lewisite (C04-A002) reacts with basic solutions, including household bleach, to produce acetylene gas.

Steam is an effective method of destroying arsenic vesicants. However, care must be taken to limit the spread of the agent and to guard against production of the toxic and potentially vesicating hydrolysis products (see Section 4.4.5).

Reactive oximes and their salts, such as potassium 2,3-butanedione monoximate found in commercially available Reactive Skin Decontaminant Lotion (RSDL), are extremely effective at rapidly detoxifying arsenic vesicants.

4.5.3.2 Vapors

Casualties/personnel: Speed in decontamination is absolutely essential. To be effective, decontamination must be completed within 2 minutes after exposure. Remove all clothing as it may continue to emit "trapped" agent vapor after contact with the vapor cloud has ceased. Shower using copious amounts of soap and water. Ensure that the hair has been washed and rinsed to remove potentially trapped vapor. If there is a potential that the eyes have been exposed to vesicants, irrigate with water or 0.9% saline solution for a minimum of 15 minutes.

Small areas: Ventilate to remove the vapors. If condensation is present, decontaminate with copious amounts of undiluted household bleach (see Section 4.5.3.1). Allow it to stand for a minimum of 5 minutes before rinsing with water. Collect and place into containers lined with high-density polyethylene. Wash the area with copious amounts of soap and water. Collect and containerize the rinseate. Removal of porous material, including painted surfaces, may be required because vesicants that have been absorbed into these materials can migrate back to the surface posing both a contact and vapor hazard.

4.5.3.3 Liquids, Solutions, or Liquid Aerosols

Casualties/personnel: Speed in decontamination is absolutely essential. To be effective, decontamination must be completed within 2 minutes after exposure. Remove all clothing immediately. Even clothing that has not come into direct contact with the agent may contain "trapped" vapor. To avoid further exposure of the head, neck, and face to the agent, cut off potentially contaminated clothing that must be pulled over the head. Remove as much of the vesicant from the skin as fast as possible. If water is not immediately available, the agent can be absorbed with any convenient material such as paper towels, toilet paper, flour, or talc. To minimize both spreading the agent and abrading the skin, do not rub the agent with the absorbent. Blot the contaminated skin with the absorbent.

Use a sponge or cloth with liquid soap and copious amounts of water to wash the skin surface and hair at least three times. Do not delay decontamination to find warm or hot water if it is not readily available. Avoid rough scrubbing as this could abrade the skin and increase percutaneous absorption of residual agent. Rinse with copious amounts of water. If there is a potential that the eyes have been exposed to vesicants, irrigate with water or 0.9% saline solution for a minimum of 15 minutes.

Alternatively, a household bleach solution can be used instead of soap and water. The bleach solution should be no more than one part household bleach in nine parts water (i.e., 0.5% sodium hypochlorite) to avoid damaging the skin. Avoid any contact with sensitive areas such as the eyes. Allow the bleach solution to remain in contact with the agent for a minimum of 5 minutes. Rinse with copious amounts of water.

Small areas: Puddles of liquid can be contained by covering with absorbent material such as vermiculite, diatomaceous earth, clay, sponges, or towels. Place the absorbed material into containers lined with high-density polyethylene. Before sealing the container, cover the contents with undiluted household bleach (see Section 4.5.3.1). If lewisite (C04-A002) is present, flammable acetylene gas will be generated during the neutralization process. Take appropriate actions to disperse the vapors. Decontaminate the area with copious amounts of the neutralizing agent. Allow it to stand for a minimum of 5 minutes before rinsing with water. Collect and containerize the rinseate. Ventilate the area to remove vapors. Removal of porous material, including painted surfaces, may be required because vesicants that have been absorbed into these material can migrate back to the surface posing both a contact and vapor hazard.

4.6 Medical

4.6.1 CDC Case Definition

A case in which an arsenic vesicant is detected in biologic samples. The case can be confirmed if laboratory testing is not performed because either a predominant amount of clinical and nonspecific laboratory evidence is present or an absolute certainty of the etiology of the agent is known.

4.6.2 Differential Diagnosis

The following factors have been suggested as alternatives to consider when presented with a potential case of exposure to vesicant agents: chemical burns from contact with strong acids or bases; barbiturates, chemotherapeutic agents, carbon monoxide; reactions to drugs producing Stevens–Johnson syndrome, or toxic epidermal necrolysis or both; autoimmune diseases such as bullous pemphigoid and pemphigus vulgaris; and staphylococcus scalded skin syndrome.

4.6.3 Signs and Symptoms

4.6.3.1 Vapors/Aerosols

Exposure of the eyes to even small amounts of arsenic vesicant vapor produces immediate tearing, pain, and involuntary blinking (blepharospasm). Eye symptoms are rapidly followed by coughing, sneezing, and vomiting. Other upper respiratory signs vary with the amount of exposure and include scratchy throat, laryngitis, and a feeling of shortness of breath. High exposures may result in low blood pressure (hypotension).

Exposure of the skin to arsenic vesicants produces an immediate burning sensation. Reddening of the skin (erythema) may appear in as short a time as 5 minutes although full progression to blisters may not develop for up to 18 hours. Although blisters tend to be deeper and more painful than produced by sulfur vesicants (Chapter 3), they heal more readily.

4.6.3.2 Liquids

Arsenic vesicants produce an immediate burning sensation. Reddening of the skin (erythema) may appear in as short a time as 5 minutes although full progression to blisters may not develop for up to 18 hours. Although blisters tend to be deeper and more painful than produced by sulfur vesicants (Chapter 3), they heal more readily.

These agents rapidly permeate through the skin. Extensive skin contamination can damage susceptible tissue including blood cells and the liver. Casualties may develop signs of systemic arsenic toxicity including diarrhea, damage to the liver, kidneys, nervous system, red blood cells, and the brain.

4.6.4 Mass-Casualty Triage Recommendations

4.6.4.1 Priority 1

A casualty with mild to moderate pulmonary effects less than 6 hours postexposure, or a casualty with moderately severe or severe pulmonary signs and symptoms after 6 hours postexposure.

4.6.4.2 *Priority 2 (Majority of Cases)*

A casualty with skin lesions covering between 5 and 50% of the body surface area (BSA), or a casualty with mild to moderate pulmonary effects after 6 hours postexposure, or a casualty with eye injuries.

4.6.4.3 *Priority 3*

A casualty with skin lesions covering less than 5% of the BSA, or a casualty with eye irritation or reddening, and/or slight upper respiratory complaints such as hacking cough or irritated throat 12 hours or more postexposure.

4.6.4.4 *Priority 4*

A casualty with skin lesions from *liquid exposure* to more than 50% of the BSA, or a casualty with severe pulmonary effects less than 6 hours postexposure.

4.6.5 Casualty Management

Decontaminate the casualty ensuring that all the vesicants have been removed. Rapid decontamination of any exposure is essential. If vesicants have gotten into the eyes, irrigate the eyes with water or 0.9% saline solution for at least 15 minutes. BAL (British-anti-Lewisite, dimercaprol) solution or ophthalmic ointment may be beneficial if administered promptly. Irrigate open wounds with water or 0.9% saline solution for at least 10 minutes.

Once the casualty has been decontaminated, including the removal of foreign matter from wounds, medical personnel do not need to wear a chemical-protective mask.

If a casualty is known to have inhaled vesicant vapors but does not display any signs or symptoms of an impacted airway, it may still be appropriate to intubate the casualty since laryngeal spasms or edema may make it difficult or impossible later. Reddening of the skin (erythema) and lesions are treated symptomatically.

BAL is the standard treatment for poisoning by arsenic compounds and will alleviate some effects from exposure to arsenic vesicants. It may also decrease the severity of skin and eye lesions if applied topically within minutes after decontamination is complete (i.e., within 2–5 minutes postexposure). Additional chelating agents for the treatment of systemic arsenic toxicity include *meso*-2,3-dimercaptosuccinic acid (DMSA) and 2,3-dimercapto-1-propanesulfonic acid (DMPS).

Asymptomatic individuals suspected of exposure to vesicants should be kept under observation for at least 8 hours.

Burns from *liquid exposure* to over 50% of the body surface suggest that the individual has received/absorbed more than a lethal dose and the prognosis is poor.

4.7 Fatality Management

Remove all clothing and personal effects segregating them as either durable or nondurable items. While it may be possible to decontaminate durable items, it may be safer and more efficient to destroy nondurable items rather than attempt to decontaminate them. Items that will be retained for further processing should be double sealed in impermeable containers, ensuring that the inner container is decontaminated before placing it in the outer one.

Vesicants that have entered the body are metabolized, hydrolyzed, or bound to tissue and pose little threat of off-gassing. To remove agents on the outside of the body, wash the remains with a 2% sodium hypochlorite bleach solution (i.e., 2 gallons of water for every gallon of household bleach), ensuring that the solution is introduced into the ears, nostrils, mouth, and any wounds. This concentration of bleach will not affect remains but will neutralize arsenic vesicant agents. Higher concentrations of bleach can harm remains. Pay particular attention to areas where agent may get trapped, such as hair, scalp, pubic areas, fingernails, folds of skin, and wounds. The bleach solution should remain on the cadaver for a minimum of 5 minutes. Wash with soap and water. Ensure that the bleach solution is removed prior to embalming, as it will react with the embalming fluid. All wash and rinse waste must be contained for proper disposal. Screen the remains for agent vapors and residual liquid at the conclusion of the decontamination process. If the remains must be stored before embalming, then place them inside body bags designed to contain contaminated bodies or in double body bags. If double body bags are used, seal the inner bag with duct tape, rinse, and then place it in the second bag. After embalming is complete, place the remains in body bags designed to contain contaminated bodies or in double body bags. Body fluids removed during the embalming process do not pose any additional risks and should be contained and handled according to established procedures.

Standard burials are acceptable when contamination levels are low enough to allow bodies to be handled without wearing additional protective equipment. Cremation may be required if remains cannot be completely decontaminated. Although arsenic vesicant agents are destroyed at the operating temperature of a commercial crematorium (i.e., above 1000°F), the initial heating phase may volatilize some of the agents and allow vapors to escape. Additionally, combustion will produce toxic and potentially volatile arsenic oxides.

C04-A

AGENTS

C04-A001

Ethyl dichloroarsine (Agent ED)
CAS: 598-14-1
RTECS: CH3500000

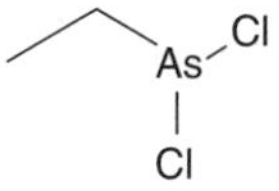

$C_2H_5AsCl_2$

Clear, colorless to yellowish, somewhat oily liquid that has a fruity but biting and irritating odor that is detectable at approximately 0.14 ppm. Vapor is irritating to both eyes and skin. Vapor concentrations that will produce blisters are hard to obtain in an open area.

Predominately acts as a vomiting agent. Blistering is usually caused by contact with liquid agent.

Also reported as a mixture with Bis(chloromethyl) ether (C10-A011).

Exposure Hazards

Conversion Factor: 1 ppm = 7.15 mg/m^3 at 77°F
$LCt_{50\ (Inh)}$: 3000 mg-min/m^3 (210 ppm for a 2-min exposure)
$ICt_{50\ (Inh)}$: 5–10 mg-min/m^3 (0.4–0.7 ppm for a 2-min exposure)

These values are from older sources (ca. 1952) and are not supported by modern data. No updated toxicity estimates have been proposed.

Properties:

MW: 174.9
D: 1.742 g/mL (57°F)
MP: –85°F
BP: Decomposes
Vsc: —
VP: 2.29 mmHg
VD: 6.0 (calculated)
Vlt: 3060 ppm
H_2O: "Slight" (slowly decomposes)
Sol: Acetone; Alcohols; Hydrocarbons
FlP: —
LEL: —
UEL: —
RP: 4
IP: <10 eV

Proposed AEGLs

AEGL-1: Not Developed
AEGL-2: 1 h, 0.0041 ppm other levels Not Developed
AEGL-3: 1 h, 0.012 ppm other levels Not Developed

C04-A002

Lewisite (Agent L)
CAS: 541-25-3; 34461-56-8 (Isomer)
RTECS: CH2975000

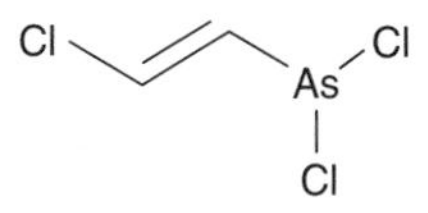

$C_2H_2AsCl_3$

Colorless to brown oily liquid that is odorless when pure. Crude material may have a violet to purple color. Impurities give it an odor similar to geranium that is detectable at approximately 0.9 ppm. There are two configurational isomers of this agent that have been studied.

Undergoes considerable decomposition when explosively disseminated.

Also reported as a mixture with Sulfur mustard (C03-A001).

Exposure Hazards

Conversion Factor: 1 ppm = 8.48 mg/m^3 at 77°F
$LCt_{50\ (Inh)}$: 1000 mg-min/m^3 (59 ppm for a 2-min exposure)
$LCt_{50(Per)}$: < 85°F: 5,000–10,000 mg-min/m^3 (20–40 ppm for a 30-min exposure)
$LCt_{50(Per)}$: > 85°F: 2500–5000 mg-min/m^3 (10–20 ppm for a 30-min exposure)
LD_{50}: 1.4 g
$ICt_{50(Skin)}$: < 85°F: 500 mg-min/m^3 (2 ppm for a 30-min exposure)
$ICt_{50(Skin)}$: > 85°F: 200 mg-min/m^3 (0.8 ppm for a 30-min exposure)
$ICt_{50(Eyes)}$: 75 mg-min/m^3 (4.4 ppm for a 2-min exposure)

These are provisional updates from older values that have not been formally adopted as of 2005.

Eye Irritation: 1.5 ppm for a 2-min exposure
$MEG_{(1\ h)}$ *Min*: 0.00035 ppm; *Sig*: —; *Sev*: —
WPL AEL: 0.00035 ppm

Properties:

MW: 207.3	*VP*: 0.35 mmHg (77°F)	*FlP*: None
D: 1.879 g/mL (77°F)	*VD*: 7.1 (calculated)	*LEL*: None
MP: 29°F	*Vlt*: 460 ppm (calculated)	*UEL*: None
MP: –49°F (munitions grade)	H_2O: Insoluble (slowly decompose)	*RP*: 20 (77°F)
BP: Decomposes	*Sol*: Organic solvents; Oils	*IP*: <10.6 eV
Vsc: 1.09 cS (77°F)		

Proposed AEGLs

AEGL-1: Not Developed
AEGL-2: 1 h, 0.014 ppm 4 h, 0.0041 ppm 8 h, 0.0021 ppm
AEGL-3: 1 h, 0.087 ppm 4 h, 0.025 ppm 8 h, 0.013 ppm

C04-A003

Methyl dichloroarsine (Agent MD)
CAS: 593-89-5
RTECS: CH4375000

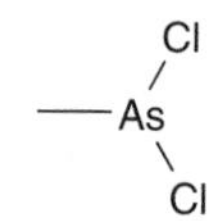

CH_3AsCl_2
Odorless liquid. Vapor is irritating to both eyes and skin, and is detectable at approximately 0.1 ppm.

Exposure Hazards

Conversion Factor: 1 ppm = 6.58 mg/m^3 at 77°F
$LCt_{50(Inh)}$: 3000–5000 mg-min/m^3 (230–380 ppm for a 2-min exposure). This toxicity estimate is from an older source (ca. 1952) and has not been updated. Current sources do not provide an estimate on a lethal concentration.
$ICt_{50(Inh)}$: 25 mg-min/m^3 (1.9 ppm for a 2-min exposure) This is a provisional update from an older value (ca. 1956) that has not been formally adopted as of 2005.

Properties:

MW: 160.9	*VP*: 7.593 mmHg	*FlP*: —
D: 1.839 g/mL	*VD*: 5.5 (calculated)	*LEL*: —
MP: −67°F	*Vlt*: 10,000 ppm	*UEL*: —
BP: 271°F	H_2O: Decomposes	*RP*: 0.8
Vsc: —	*Sol*: Most organic solvents	*IP*: 10.4 eV

Proposed AEGLs

AEGL-1: Not Developed
AEGL-2: 1 h, 0.008 ppm 4 h, 0.0023 ppm 8 h, 0.00096 ppm
AEGL-3: 1 h, 0.024 ppm 4 h, 0.0067 ppm 8 h, 0.0029 ppm

C04-A004

Phenyl dichloroarsine (Agent PD)
CAS: 696-28-6
RTECS: CH5425000
UN: 1556

$C_6H_5AsCl_2$

Colorless to yellow odorless liquid. Vapor is irritating to the nose and throat, and is detectable at approximately 0.1 ppm.

Also reported as a mixture with Sulfur mustard (C03-A001); Diphenylchloroarsine (C14-A001); *N*-Ethylcarbazole (C13-A022); Arsenic trichloride (C04-C006).

Exposure Hazards

Conversion Factor: 1 ppm = 9.12 mg/m^3 at 77°F
$LCt_{50(Inh)}$: 2600 mg-min/m^3 (140 ppm for a 2-min exposure)
$ICt_{50(Skin)}$: 200–500 mg-min/m^3 (0.7–1.8 ppm for a 30-min exposure)
"Intolerable" Irritation: 0.9 ppm for a 2-min exposure

These are provisional updates from older values (ca. 1952) that have not been formally adopted as of 2005.

Properties:

MW: 222.9	*VP*: 0.033 mmHg (77°F)	*FlP*: —
D: 1.652 g/mL	*VD*: 7.7 (calculated)	*LEL*: —
MP: −9°F	*Vlt*: 43 ppm	*UEL*: —
BP: 451°F	*H_2O*: 0.6% (rapidly decomposes)	*RP*: 200
Vsc: 1.95 cS (77°F)	*Sol*: Alcohols; Gasoline; Acetone; Ether	*IP*: <9 eV

Proposed AEGLs

AEGL-1: Not Developed
AEGL-2: 1 h, 0.016 mg/m^3 other levels Not Developed
AEGL-3: 1 h, 0.180 mg/m^3 other levels Not Developed

C04-A005

Phenyl dibromoarsine

CAS: 696-24-2
RTECS: —

$C_6H_5AsBr_2$

Colorless to faintly yellow liquid.

Exposure Hazards

Conversion Factor: 1 ppm = 12.75 mg/m^3 at 77°F
$LC_{50(Inh)}$: 200 mg/m^3 (16 ppm) for a 10-min exposure. This toxicity estimate is from an older source (ca. 1937) and has not been updated.

Properties:

MW: 311.8	*VP*: —	*FlP*: —
D: 2.1 g/mL (59°F)	*VD*: 11 (calculated)	*LEL*: —
MP: —	*Vlt*: —	*UEL*: —
BP: Decomposes	*H_2O*: —	*RP*: —
Vsc: —	*Sol*: —	*IP*: —

C04-C

COMPONENTS AND PRECURSORS

C04-C006

Arsenic trichloride
CAS: 7784-34-1
RTECS: CG1750000
UN: 1560
ERG: 157

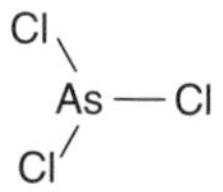

$AsCl_3$

Colorless oily liquid with an acrid odor. This material is hazardous through inhalation, skin absorption, penetration through broken skin, and ingestion, and produces local skin/eye impacts. It causes severe irritation and burns to the eyes, mucous membranes, and skin; cough, chest pain, and accumulation of fluid in the lungs (pulmonary edema).

Used in the ceramics industry and as a chemical intermediate for arsenic pharmaceuticals and arsenic insecticides.

This material is on the ITF-25 low threat list, the Australia Group Export Control list and Schedule 2 of the CWC.

This material is a general precursor for arsenical vesicants and many vomiting/sternatory agents (Chapter 14).

Exposure Hazards
Conversion Factor: 1 ppm = 7.41 mg/m^3 at 77°F
ACGIH TLV: 0.01 mg/m^3 as arsenic

Properties:

MW: 181.3	*VP*: 10 mmHg (74°F)	*FlP*: None
D: 2.163 g/mL	*VD*: 6.3 (calculated)	*LEL*: None
D: 2.1450 g/mL (77°F)	*Vlt*: 13,000 ppm	*UEL*: None
MP: 17°F	*H_2O*: Decomposes	*RP*: 0.75
BP: 267°F	*Sol*: Chloroform; Oils; Fats; Ether;	*IP*: —
Vsc: —	aqueous Hydrochloric acid	

C04-C007

Lewisite 3 (Agent L3)
CAS: 40334-70-1
RTECS: —

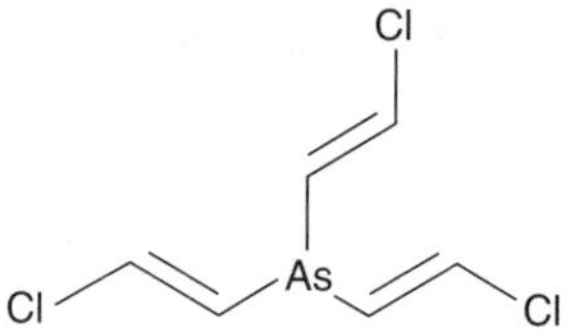

$C_6H_6AsCl_3$

Specific information on physical appearance is not available for this material.
This material is on Schedule 1 of the CWC.

This material is easily converted to Lewisite (C04-A002) and Lewisite 2 (C14-A004). It is also commonly found as an impurity and degradation product in Lewisite.

Exposure Hazards

Conversion Factor: 1 ppm = 10.61 mg/m^3 at 77°F

Human toxicity values have not been established or have not been published. However, based on available information, this material appears to have few, if any, acute vesicant or vomiting/sternatory properties.

Properties:

MW: 259.4	*VP*: —	*FlP*: —
D: 1.572 g/cm^3	*VD*: —	*LEL*: —
MP: 71°F	*Vlt*: —	*UEL*: —
BP: 500°F (decomposes)	H_2O: Insoluble	*RP*: —
BP: 280°F (12 mmHg)	*Sol*: Common organic solvents except alcohol	*IP*: —
Vsc: —		

Proposed AEGLs

AEGLs have been proposed only for mixtures of this compound with Lewisite (C04-A002).

References

Agency for Toxic Substances and Disease Registry. "Blister Agents Lewisite (L) and Mustard-Lewisite Mixture (HL) ToxFAQs." April 2002.

———. "Blister Agents: Lewisite (L) and Mustard-Lewisite Mixture (HL)." In *Managing Hazardous Materials Incidents Volume III—Medical Management Guidelines for Acute Chemical Exposures*. Rev. ed. Washington, DC: Government Printing Office, 2000.

Bartlett, Paul D., Hyp Joseph Dauben, Jr., and Leonard J. Rosen. "Preparation of Lewisite." US Patent 2,465,834, March 29, 1949.

Centers for Disease Control and Prevention. "Case Definition: Vesicant (Mustards, Dimethyl sulfate, and Lewisite)." March 15, 2005.

———. "Counter Terrorism Card for Lewisite." 2000.

———. "Facts About Lewisite." March 14, 2003.

Compton, James A.F. *Military Chemical and Biological Agents: Chemical and Toxicological Properties*. Caldwell, NJ: The Telford Press, 1987.

Edgewood Research Development, and Engineering Center, Department of the Army. *Material Safety Data Sheet (MSDS) for Lewisite (L)*. Aberdeen Proving Ground, MD: Chemical Biological Defense Command, rev. March 27, 1996.

Fries, Amos A., and Clarence J. West. *Chemical Warfare*. New York: McGraw-Hill Book Company, Inc., 1921.

Jackson, Kirby E., and Margaret A. Jackson. "The Chlorovinylarsines." *Chemical Reviews* 16 (1935): 439–52.

Lewis, W. Lee, and G.A. Perkins. "The Beta-Chlorovinyl Chloroarsines." *Industrial & Engineering Chemistry* 15 (March 1923): 290–95.

Marrs, Timothy C., Robert L. Maynard, and Frederick R. Sidell. *Chemical Warfare Agents: Toxicology and Treatment*. Chichester, England: John Wiley & Sons, 1997.

Munro, Nancy B., Sylvia S. Talmage, Guy D. Griffin, Larry C. Waters, Annetta P. Watson, Joseph F. King, and Veronique Hauschild. "The Sources, Fate and Toxicity of Chemical Warfare Agent Degradation Products." *Environmental Health Perspectives* 107 (1999): 933–74.

National Institute of Health. "Hazardous Substance Data Bank (HSDB)." http://toxnet.nlm.nih.gov/cgi-bin/sis/htmlgen?HSDB/. 2005.

Prentiss, Augustin M. *Chemicals in War: A Treatise on Chemical Warfare*. New York: McGraw-Hill Book Company, Inc., 1937.

Sartori, Mario. *The War Gases: Chemistry and Analysis.* Translated by L.W. Marrison. London: J. & A. Churchill, Ltd, 1939.

Sidell, Frederick R. *Medical Management of Chemical Warfare Agent Casualties: A Handbook for Emergency Medical Services*. Bel Air, MD: HB Publishing, 1995.

Sidell, Fredrick R., Ernest T. Takafuji, and David R. Franz, eds. *Medical Aspects of Chemical and Biological Warfare, Textbook of Military Medicine Series, Part 1, Warfare, Weaponry, and the Casualty*. Washington, DC: Office of the Surgeon General, Department of the Army, 1997.

Sifton, David W., ed. *PDR Guide to Biological and Chemical Warfare Response*. Montvale, NJ: Thompson/Physicians Desk Reference, 2002.

Smith, Ann, Patricia Heckelman, and Maryadele J. Oneil, eds. *The Merck Index: An Encyclopedia of Chemicals, Drugs, & Biologicals.* 13th ed. Rahway, NJ: Merck & Co., Inc., 2001.

Somani, Satu M., ed. *Chemical Warfare Agents.* New York: Academic Press, 1992.

United States Army Headquarters. *Chemical Agent Data Sheets Volume I, Edgewood Arsenal Special Report No. EO-SR-74001*. Washington, DC: Government Printing Office, December 1974.

———. *Potential Military Chemical/Biological Agents and Compounds, Field Manual No. 3-11.9*. Washington, DC: Government Printing Office, January 10, 2005.

United States Army Medical Research Institute of Chemical Defense. *Medical Management of Chemical Casualties Handbook.* 3rd ed. Aberdeen Proving Ground, MD: United States Army Medical Research Institute of Chemical Defense, July 2000.

United States Coast Guard. *Chemical Hazards Response Information System (CHRIS) Manual*, 1999 ed. http://www.chrismanual.com/Default.htm. 2004.

Wachtel, Curt. *Chemical Warfare*. Brooklyn, NY: Chemical Publishing Co., Inc., 1941.

Waitt, Alden H. *Gas Warfare: The Chemical Weapon, Its Use, and Protection Against It*. Rev. ed. New York: Duell, Sloan and Pearce, 1944.

Watson, Annetta, Dennis Opresko, and Veronique Hauschild. *Evaluation of Chemical Warfare Agent Percutaneous Vapor Toxicity: Derivation of Toxicity Guidelines for Assessing Chemical Protective Ensembles*. ORNL/TM-2003/180. July 2003.

Williams, Kenneth E. *Detailed Facts about Blister Agent Lewisite (L).* Aberdeen Proving Ground, MD: United States Army Center for Health Promotion and Preventive Medicine, 1996.

World Health Organization. *Public Health Response to Biological and Chemical Weapons: WHO Guidance*. Geneva: World Health Organization, 2004.

5

Urticants

5.1 General Information

The agents in this class are halogenated oximes. This class of agents is not specifically covered by the Chemical Weapons Convention. Because of the toxicity of the agents and lack of commercial application outside of limited scientific research, urticants would be prohibited based on the Guidelines for Schedules of Chemicals.

These materials are second generation chemical warfare agents developed shortly after World War I. They are moderately difficult to synthesize and disperse.

Chloroformoxime, the first urticant, was first synthesized in 1894. The more effective dihaloformoximes—phosgene oxime (C05-A001), dibromoformoxime (C05-A002), and diiodoformoxime (C05-A003)—were prepared in the late 1920s and early 1930s. Phosgene oxime was stockpiled by Nazi Germany during World War II but was never used. Since the end of World War II, urticants have been evaluated by numerous countries but stockpiled by few because of production, weaponization, and storage issues. The former Soviet Union overcame these issues and stockpiled phosgene oxime. Urticants have never been used on the battlefield.

5.2 Toxicology

5.2.1 Effects

Urticants produce instant, almost intolerable pain and cause immediate local destruction of skin and mucous membranes. No other chemical agent produces such an immediate painful onset followed by rapid tissue death. Sensations caused by exposure to these agents range from mild prickling to almost intolerable pain resembling a severe bee sting. Effects depend on the concentration of the agent and the length of the exposure. Tissue damage is more severe than that produced by vesicants (C03, C04). Direct contact of the agent with the skin produces a corrosive type lesion similar to those produced by a strong acid. Skin lesions may not fully heal for 1–3 months.

Eyes are especially susceptible to urticants. In addition to immediate pain, urticants produce lesions and inflammation to the cornea (keratitis), which may progress to blindness.

Inhalation of urticants can cause lung membranes to swell and become filled with liquid (pulmonary edema). Death may result from lack of oxygen.

Urticants are also systemic agents and rapidly pass through the skin to affect susceptible tissue. Percutaneous absorption of liquids or solids can also produce pulmonary edema and blood clots in the lungs.

5.2.2 Pathways and Routes of Exposure

Urticants are hazardous through any route of exposure including inhalation, skin and eye exposure, ingestion, and broken, abraded, or lacerated skin (e.g., penetration of skin by debris).

5.2.3 General Exposure Hazards

Urticants have a penetrating and disagreeable odor detectable at very low concentrations. Even minimal exposure causes immediate irritation and pain of the eyes, nose, mucous membranes, respiratory system, and skin. Urticants pose a significant percutaneous hazard and are absorbed through the skin within seconds.

Lethal concentrations (LC_{50}s) for inhalation of these agents are as low as 340 ppm for a 2-minute exposure.

Lethal percutaneous exposures (LD_{50}s) to liquid are estimated to be less than 2 grams per individual.

Incapacitating concentrations (ICt_{50}) for dermal exposure to these agents are as low as 0.3 ppm for a 2-minute exposure.

The rate of detoxification of these agents by the body is not known. However, because of the cellular damage caused by these agents, exposures have a cumulative risk.

5.2.4 Latency Period

Urticants produce immediate irritation and pain of the eyes, respiratory tract, and skin. Blanching, reddening of the skin (erythema), and hives develop within minutes of exposure. Blisters, localized tissue death (necrosis), and formation of scabs may be delayed for 24 hours or more. Systemic effects, including pulmonary edema, from either inhalation or percutaneous absorption of the agent, do not occur until after a substantial latency period.

5.3 Characteristics

5.3.1 Physical Appearance/Odor

5.3.1.1 Laboratory Grade

Laboratory grade agents are typically colorless crystalline solids. They have intense, penetrating, and disagreeable odors detectable at very low levels.

5.3.1.2 Munition Grade

Munition grade agents are typically amber to dark brown liquids. As the agent ages and decomposes, it continues to discolor until it may appear black. Odors are intense, penetrating, disagreeable, and violently irritating.

5.3.2 Stability

Urticants are relatively unstable and tend to decompose spontaneously unless stored at low temperatures. Below −4°F, they can be kept for extended periods. Solvents including 1, 2-dimethoxybenzene, ether, dioxane, nitromethane, and glycine act as stabilizers and may be added to help prevent decomposition of agents during storage. Agents can be stored in glass or enamel-lined containers. Urticants rapidly attack rubber and metals, especially iron.

5.3.3 Persistency

For military purposes, unmodified urticants are classified as nonpersistent. They decompose rapidly in soil, and after a short delay on most other surfaces and in water.

Urticants, even solids, have relatively high vapor pressure and evaporation or sublimation rates are nearly the same as water.

5.3.4 Environmental Fate

Urticant vapors have a density greater than air and tend to collect in low places. Porous material, including painted surfaces, will absorb both liquid and gaseous agents. After the initial surface contamination has been removed, agent that has been absorbed into porous material can migrate back to the surface posing both a contact and vapor hazard. Urticants can penetrate clothing and rubber faster than other chemical warfare agents. Clothing may emit trapped agent vapor for up to 30 minutes after contact with a vapor cloud.

Urticants are unstable and decompose rapidly in soil. Agents dissolve slowly but completely in water and may take days to decompose once in solution. The specific gravities of unmodified liquid agents are greater than that of water. These agents are also soluble in most organic solvents including gasoline and oil.

5.4 Additional Hazards

5.4.1 Exposure

The rapid skin damage caused by urticants renders the skin more susceptible to subsequent exposure of any other toxic material or agent.

All foodstuffs in the area of a release should be considered contaminated. Unopened items packaged in glass, metal, or heavy duty plastic and exposed only to agent vapors may be used after decontamination of the container. Unopened items exposed to liquid or solid agents should be decontaminated within a few hours postexposure or destroyed. Opened or unpackaged items, or those packaged only in paper or cardboard, should be destroyed.

Meat, milk, and animal products from animals exposed to or killed by urticants should be destroyed or quarantined until it has been tested and determined to be safe to consume.

Plants, fruits, vegetables, and grains should be quarantined until tested and determined to be safe to consume.

5.4.2 Livestock/Pets

Animals can be decontaminated with shampoo/soap and water (see Section 5.5.3). If the animals' eyes have been exposed to agent, they should be irrigated with water or saline solution for a minimum of 30 minutes.

Unprotected feedstock (e.g., hay or grain) should be destroyed. Although leaves of forage vegetation could still retain sufficient urticants to produce effects for a limited time, urticants are relatively nonpersistent on surfaces and should rapidly decompose depending on the level of contamination and the weather conditions.

5.4.3 Fire

Heat from a fire will increase the amount of agent vapor in the area. A significant amount of the agent could be volatilized and escape into the surrounding environment before it is consumed by the fire. Urticants are soluble in water and actions taken to extinguish the fire can also spread the agent. Runoff from firefighting efforts will pose a significant threat. A solution containing as little as 8% of active agent can cause pain within seconds of exposure. Potential decomposition products include toxic and/or corrosive gases.

5.4.4 Reactivity

Urticants rapidly react with metals, especially iron. Iron chloride, even in trace amounts, can cause explosive decomposition. These agents decompose slowly when dissolved in water. Dilute acids will retard the rate of decomposition.

Urticants react violently with strong bases. Reaction with household bleach may produce toxic gases.

5.4.5 Hazardous Decomposition Products

5.4.5.1 Hydrolysis

Urticants produce hydrogen chloride (HCl), hydrogen bromide (HBr), or hydrogen iodide (HI), and hydroxlamines when hydrolyzed.

5.4.5.2 Combustion

Volatile decomposition products may include HCl, HBr, HI, nitrogen oxides (NO_x), and toxic dimerization products.

5.5 Protection

5.5.1 Evacuation Recommendations

Isolation and protective action distances listed below are taken from Argonne National Laboratory Report No. ANL/DIS-00-1, *Development of the Table of Initial Isolation and Protective Action Distances for the 2000 Emergency Response Guidebook*, which is still the basis for the "when used as a weapon" scenarios in the 2004 Emergency Response Guidebook (ERG). CX is the only urticant addressed and recommendations are based on a release scenario involving direct aerosolization of the solid agents with a particle size between 2 and 5 μm. Under these conditions, the difference between a small and a large release of urticant is not based on the standard 200 liters spill used for commercial hazardous materials listed in the ERG. A small release of CX involves 10 kilograms (approximately 0.6 cubic feet) of bulk agent and a large release involves 500 kilograms (approximately 30 cubic feet) of bulk agent.

	Initial isolation (feet)	Downwind day (miles)	Downwind night (miles)
CX (Phosgene oxime) C05-A001			
Small device (10 kilograms)	100	0.1	0.3
Large device (500 kilograms)	300	0.6	1.9

5.5.2 Personal Protective Requirements

5.5.2.1 Structural Firefighters' Gear

Urticants can penetrate garments and rubber much faster than other chemical warfare agents. Structural firefighters' protective clothing is recommended for fire situations only; it is not effective in spill situations or release events and should never be used as the primary chemical protective garment to enter an area contaminated with urticants.

5.5.2.2 Respiratory Protection

Self-contained breathing apparatuses (SCBAs) or air purifying respirators (APRs) should have a National Institute for Occupational Safety and Health (NIOSH) and Chemical/Biological/Radiological/Nuclear (CBRN) certification since urticants can be absorbed into or degrade the materials used to make some respirators. However, during emergency operations, other NIOSH approved SCBAs or APRs that have been specifically tested by the manufacturer against chemical warfare agents may be used if deemed necessary by the Incident Commander. APRs should be equipped with a NIOSH approved CBRN filter or a combination organic vapor/acid gas/particulate cartridge.

Immediately dangerous to life or health (IDLH) levels are the ceiling limit for respirators other than SCBAs. However, IDLH levels have not been established for urticants. Therefore, any potential exposure to these agents should be regarded with extreme caution and the use of SCBAs for respiratory protection should be considered.

5.5.2.3 Chemical Protective Clothing

Currently, there is no information on performance testing of chemical protective clothing against urticants.

Because of the extreme dermal hazard posed by urticants, responders should wear a Level A protective ensemble whenever there is a potential for exposure to any solid or liquid agent, or to an elevated or unknown concentration of agent vapor.

5.5.3 Decontamination

5.5.3.1 General

Urticants are very soluble in water and can be removed from most surfaces by washing with soap and water.

Solubility of an agent will be decreased if it is dissolved in an immiscible organic solvent. Addition of solvents or mechanical mixing may be required to overcome insolubility problems.

Because of the vigorous reaction of urticants with caustics, household bleach is not an effective decontamination agent for large quantities of these materials. Reaction with hypochlorites, including household bleach, may produce toxic gases such as chlorine.

5.5.3.2 Vapors

Casualties/personnel: Speed in decontamination is absolutely essential. Because of the rapid onset of effects and the speed with which urticants are absorbed through the skin, decontamination will not be entirely effective by the time the casualty experiences pain and blanching occurs. However, decontamination must still be done as rapidly as possible postexposure. Remove all clothing as it may continue to emit "trapped" agent vapor after contact with the vapor cloud has ceased. Shower using copious amounts of soap and water. Ensure that the hair has been washed and rinsed to remove potentially trapped vapor. To be effective, decontamination must be completed within 2 minutes of exposure. If there is a potential that the eyes have been exposed to urticants, irrigate with water or 0.9% saline solution for a minimum of 15 minutes.

Small areas: Ventilate to remove the vapors. If condensation is present, decontaminate the area with copious amounts of soap and water. Collect the agent and rinseate and place into containers lined with high-density polyethylene. Although urticants rapidly break down on most surfaces, removal of porous material, including painted surfaces, may be required to prevent agents that have been absorbed into these materials from migrating back to the surface and posing an extended hazard.

5.5.3.3 Liquids, Solutions, or Liquid Aerosols

Casualties/personnel: Speed in decontamination is absolutely essential. Because of the rapid onset of effects and the speed with which urticants are absorbed through the skin, decontamination will not be entirely effective by the time the casualty experiences pain. However, decontamination must still be done as rapidly as possible postexposure. Remove all clothing immediately. Even clothing that has not come into direct contact with the agent may contain "trapped" vapor. To avoid further exposure of the head, neck, and face to the agent, cut off potentially contaminated clothing that must be pulled over the head. Remove as much of the urticant from the skin as fast as possible. If water is not immediately available, the agent can be absorbed with any convenient material such as paper towels, toilet paper, flour, or talc. To minimize both spreading the agent and abrading the skin, do not rub the agent with the absorbent. Blot the contaminated skin with the absorbent.

Use a sponge or cloth with liquid soap and copious amounts of water to wash the skin surface and hair at least three times. Do not delay decontamination to find warm or hot water if it is not readily available. Avoid rough scrubbing as this could abrade the skin and increase percutaneous absorption of residual agent. Rinse with copious amounts of water. If there is a potential that the eyes have been exposed to urticants, irrigate with water or 0.9% saline solution for a minimum of 1 hour.

Small areas: Puddles of liquid can be contained by covering with absorbent material such as vermiculite, diatomaceous earth, clay, sponges, or towels. Place the absorbed material into containers lined with high-density polyethylene. Decontaminate the area with copious amounts of soap and water. Collect and containerize the rinseate. Ventilate the area to remove vapors. Although urticants rapidly break down on most surfaces, removal of porous material, including painted surfaces, may be required to prevent agents that have been absorbed into these materials from migrating back to the surface and posing an extended hazard.

5.5.3.4 Solids or Particulate Aerosols

Casualties/personnel: Speed in decontamination is absolutely essential. Because of the rapid onset of effects and the speed with which urticants are absorbed through the skin, decontamination will not be entirely effective by the time the casualty experiences pain. However, decontamination must still be done as rapidly as possible postexposure. Do not attempt to brush the agent off of the individual or their clothing as this can aerosolize the agent. Remove all clothing immediately. To avoid further exposure of the head, neck, and face to the agent, cut off potentially contaminated clothing that must be pulled over the head. Wash the skin surface and hair at least three times with copious amounts of soap and water. Do not delay decontamination to find warm or hot water if it is not readily available. Avoid rough scrubbing as this could abrade the skin and increase percutaneous absorption of residual agent. Rinse with copious amounts of water. If there is a potential that the eyes have been exposed to vesicants, irrigate with water or 0.9% saline solution for a minimum of 1 hour.

Small areas: If indoors, close windows and doors in the area and turn off anything that could create air currents (e.g., fans, air conditioner, etc.). Dampen agent with water until thoroughly wet. Cover with absorbent materials such as vermiculite, diatomaceous earth, clay, sponges, or towels. Place the absorbed material into containers lined with high-density polyethylene. Repeat as necessary until the visible agent has been containerized. Decontaminate the area with copious amounts of soap and water. Collect and containerize the rinseate. Ventilate the area to remove vapors. Although urticants rapidly break down on most surfaces, removal of porous material, including painted surfaces, may be required to prevent agents that have been absorbed into these materials from migrating back to the surface and posing an extended hazard.

5.6 Medical

5.6.1 CDC Case Definition

The CDC has not published a specific case definition for intoxication by urticants. However, the general case definition for vesicants states: "A case in which a vesicant is detected in biologic samples. The case can be confirmed if laboratory testing is not performed because either a predominant amount of clinical and nonspecific laboratory evidence is present or an absolute certainty of the etiology of the agent is known."

5.6.2 Differential Diagnosis

The following factors have been suggested as alternatives to consider when presented with a potential case of exposure to urticants: chemical burns from contact with strong acids or bases; barbiturates, chemotherapeutic agents, and carbon monoxide; reactions to drugs producing Stevens–Johnson syndrome, toxic epidermal necrolysis or both; autoimmune diseases such as bullous pemphigoid and pemphigus vulgaris; and staphylococcus scalded skin syndrome.

5.6.3 Signs and Symptoms

5.6.3.1 Vapors/Aerosols

Urticant vapors are violently irritating to the eyes, nose, and respiratory tract and cause immediate pain. Very low concentrations can cause inflammation, lacrimation, and temporary blindness; higher concentrations can cause corneal corrosion and dimming of vision. Inhalation causes runny nose, hoarseness, and sinus pain.

Agent vapor produces pain on exposed skin within a few seconds. Within 30 seconds after contact, the exposed skin becomes white and is surrounded by an area of erythema (reddening of the skin). In about 15 minutes, the exposed skin develops hives. After 24 hours, the skin in the central blanched area becomes brown and dies. A scab is formed in a few days although pus continues to discharge from the lesion. Healing is accompanied by sloughing of the scab; itching and pain may be present throughout healing.

5.6.3.2 Liquids/Solids

Contact with the skin produces pain within a few seconds. Within 30 seconds after contact, the exposed skin becomes white and is surrounded by a red ring resembling a wagon wheel. Within an hour the area becomes swollen and within 24 hours the lesions turn yellow and form blisters. The skin then turns brown and dies. A scab is formed in a few days although pus continues to discharge from the lesion. Healing is accompanied by sloughing of the scab; itching and pain may be present throughout healing.

Urticants absorbed through the skin can cause pulmonary edema.

5.6.4 Mass-Casualty Triage Recommendations

5.6.4.1 Priority 1

A casualty with necrotic lesions and pain with early onset of airway damage.

5.6.4.2 Priority 2 (Majority of Cases)

A casualty with necrotic lesions and pain without further complications.

5.6.4.3 Priority 3

Not applicable for this class of agents.

5.6.4.4 Priority 4

Not applicable for this class of agents

5.6.5 Casualty Management

Decontaminate the casualty ensuring that all the urticants have been removed. Rapid decontamination of any exposure is essential. If urticants have gotten into the eyes, irrigate the eyes with water or 0.9% saline solution for at least 1 hour. Irrigate open wounds with water or 0.9% saline solution for at least 10 minutes.

Once the casualty has been decontaminated, including the removal of foreign matter from wounds, medical personnel do not need to wear a chemical-protective mask.

There is no antidote for exposure to these agents. Treatment consists of symptomatic management of lesions. Eye lesions should be treated by saline irrigation and coating the follicular margins with petroleum jelly to prevent sticking.

5.7 Fatality Management

Remove all clothing and personal effects segregating them as either durable or nondurable items. Although it may be possible to decontaminate durable items, it may be safer and more efficient to destroy nondurable items rather than attempt to decontaminate them. Items that will be retained for further processing should be double sealed in impermeable containers, ensuring that the inner container is decontaminated before placing it in the outer one.

Urticants that have entered the body are metabolized, hydrolyzed, or bound to tissue and pose little threat of off-gassing. To remove agents on the outside of the body, wash the remains with soap and water ensuring that the solution is introduced into the ears, nostrils, mouth, and any wounds. Pay particular attention to areas where agent may get trapped, such as hair, scalp, pubic areas, fingernails, folds of skin, and wounds. All wash and rinse waste must be contained for proper disposal.

Screen the remains for agent vapors and residual liquid at the conclusion of the decontamination process. If the remains must be stored before embalming, then place them inside body bags designed to contain contaminated bodies or in double body bags. If double body bags are used, seal the inner bag with duct tape, rinse, and then place in the second bag. After embalming is complete, place the remains in body bags designed to contain contaminated bodies or in double body bags. Body fluids removed during the embalming process do not pose any additional risks and should be contained and handled according to established procedures.

Standard burials are acceptable when contamination levels are low enough to allow bodies to be handled without wearing additional protective equipment.

C05-A

AGENTS

C05-A001

Phosgene oxime (Agent CX)
CAS: 1794-86-1
RTECS: —

OH
N
Cl Cl

$CHCl_2NO$

Colorless liquid or crystalline deliquescent solid that has an intense, penetrating, disagreeable, and violently irritating "odor" that is detectable at 0.02 ppm. CX is the most irritating of the agents in this class.

Exposure Hazards
Conversion Factor: 1 ppm = 4.66 mg/m^3 at 77°F
$LCt_{50(Inh)}$: 3200 mg-min/m^3 (340 ppm for a 2-min exposure)
LD_{50}: 1.8 g (estimate)
$ICt_{50(Inh)}$: 3 mg-min/m^3 (0.3 ppm for a 2-min exposure)

These are provisional updates from older values that have not been formally adopted as of 2005.

Properties:

MW: 113.9	*VP*: 11.2 mmHg (77°F)	*FlP*: —
D: 1.8 g/cm^3 (estimate)	*VD*: 3.9 (calculated)	*LEL*: —
MP: 95°F	*Vlt*: 390 ppm	*UEL*: —
BP: Sublimes (decomposes unless highly purified)	H_2O: 70%	*RP*: 0.8
BP: 162°F (104 mmHg)	*Sol*: Most organic solvents	*IP*: <11 eV
Vsc: —		

C05-A002

Dibromoformoxime

CAS: 74213-24-4

RTECS: —

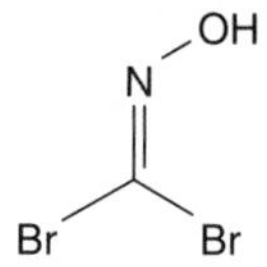

$CHBr_2NO$

Specific information on physical appearance is not available for this agent.

Exposure Hazards

Conversion Factor: 1 ppm = 8.30 mg/m^3 at 77°F

Human toxicity values have not been established or have not been published. However, this material is a much less powerful irritant than Phosgene oxime (C05-A001).

Properties:

MW: 202.8	*VP*: —	*FlP*: —
D: —	*VD*: 7.0 (calculated)	*LEL*: —
MP: 158°F	*Vlt*: —	*UEL*: —
BP: 167°F (3 mmHg)	H_2O: —	*RP*: —
Vsc: —	*Sol*: —	*IP*: —

C05-A003

Diiodoformoxime

CAS: 201205-56-3

RTECS: —

CHI_2NO

Specific information on physical appearance is not available for this agent.

Exposure Hazards

Conversion Factor: 1 ppm = 12.14 mg/m^3 at 77°F

Human toxicity values have not been established or have not been published. However, this material is a much less powerful irritant than Phosgene oxime (C05-A001).

Properties:

MW: 296.8	*VP*: —	*FlP*: —
D: —	*VD*: 10 (calculated)	*LEL*: —
MP: 156°F	*Vlt*: —	*UEL*: —
BP: —	H_2O: —	*RP*: —
Vsc: —	*Sol*: —	*IP*: —

References

Agency for Toxic Substances and Disease Registry. "Phosgene Oxime ToxFAQs." April 2002.

Centers for Disease Control and Prevention. "Case Definition of Vesicant/Blister Agent Poisoning." Interim document, December 22, 2003.

———. "Facts About Phosgene Oxime." March 18, 2003.

Compton, James A.F. *Military Chemical and Biological Agents: Chemical and Toxicological Properties*. Caldwell, NJ: The Telford Press, 1987.

Ehman, Phillip J., and Walter O. Walker. "Process for Preparing Phosgene Oxime." US Patent 2,299,742, October 27, 1942.

Hydro, William R. "Production of Dichloroformoxime." US Patent 4,558,160, December 10, 1985.

Madaus, John H., and Herman B. Urbach. "Electrolytic Production of Dichloroformoxime." US Patent 2,918,418, December 22, 1959.

Marrs, Timothy C., Robert L. Maynard, and Frederick R. Sidell. *Chemical Warfare Agents: Toxicology and Treatment*. Chichester, England: John Wiley & Sons, 1997.

Sartori, Mario. *The War Gases: Chemistry and Analysis*. Translated by L.W. Marrison. London: J. & A. Churchill, Ltd, 1939.

Sidell, Frederick R. *Medical Management of Chemical Warfare Agent Casualties: A Handbook for Emergency Medical Services*. Bel Air, MD: HB Publishing, 1995.

Sidell, Fredrick R., Ernest T. Takafuji, and David R. Franz, eds. *Medical Aspects of Chemical and Biological Warfare, Textbook of Military Medicine Series, Part 1, Warfare, Weaponry, and the Casualty*. Washington, DC: Office of the Surgeon General, Department of the Army, 1997.

Sifton, David W. ed. *PDR Guide to Biological and Chemical Warfare Response*. Montvale, NJ: Thompson/Physicians Desk Reference, 2002.

United States Army Headquarters. *Chemical Agent Data Sheets Volume I, Edgewood Arsenal Special Report No. EO-SR-74001*. Washington, DC: Government Printing Office, December 1974.

———. *NATO Handbook on the Medical Aspects of NBC Defensive Operations AmedP-6(B), Part III-Chemical, Field Manual No. 8-9*. Washington, DC: Government Printing Office, February 1996.

———. *Potential Military Chemical/Biological Agents and Compounds, Field Manual No. 3-11.9*. Washington, DC: Government Printing Office, January 10, 2005.

United States Army Medical Research Institute of Chemical Defense. *Medical Management of Chemical Casualties Handbook*. 3rd ed. Aberdeen Proving Ground, MD: United States Army Medical Research Institute of Chemical Defense, July 2000.

Williams, Kenneth E. *Detailed Facts About Blister Agent Phosgene Oxime (CX)*. Aberdeen Proving Ground, MD: United States Army Center for Health Promotion and Preventive Medicine, 1996.

Section III

Toxic Agents

6

Bicyclophosphate Convulsants

6.1 General Information

The agents in this class are bicyclophosphates and bicyclothiophosphates. This class of agents is not specifically listed in the Chemical Weapons Convention nor is it covered by the language of the general definitions in the Schedules. Some of these chemicals have been used as fire retardants, oil lubricants, and for medicinal research. They also occur as breakdown products in some synthetic turbine engine lubricants and some rigid polyurethane foams.

These materials are fourth generation chemical warfare agents that have been evaluated by various countries. They are relatively easy to synthesize. No information is available on dispersing these agents. However, because they produce negligible vapor, they will be somewhat difficult to deliver in a manner that will produce immediate casualties.

Minimal information about these agents or potential research programs concerning these agents has been published in the unclassified literature. There is no information to indicate that these agents were ever stockpiled or have ever been used on the battlefield.

6.2 Toxicology

6.2.1 Effects

These agents are potent cage convulsants that antagonize the gamma-aminobutyric acid (GABAA) inhibitory system through blockage of the chloride channel. They induce brief, muscular shock-like jerks (myoclonic seizures) that progress to generalized tonic-clonic (grand mal) convulsions and death. Subconvulsive doses may cause long-lasting central nervous system sensitization such as an increased susceptibility to audiogenic (sound induced) seizures and may lead to long-lasting changes in some social behaviors by causing regional decreases in neural substances such as serotonin, dopamine, and norepinephrine.

6.2.2 Pathways and Routes of Exposure

Bicyclophosphates are hazardous through any route of exposure including inhalation, skin exposure, ingestion, and broken, abraded, or lacerated skin (e.g., penetration of skin by debris).

6.2.3 General Exposure Hazards

Based on their chemical and physical properties, bicyclophosphates are not expected to have good warning properties. They are unlikely to have any significant odor or to cause any significant irritation of the eyes or skin.

Human toxicity data for bicyclophosphates have not been published or have not been established. However, available information indicates that under optimum conditions some of the agents are as toxic as nerve agent VX (C01-A016). Based on animal studies, bicyclophosphates are not cumulative and are rapidly eliminated from the body.

6.2.4 Latency Period

Information on the latency of this class of agents is unavailable. However, based on animal studies, there does not appear to be a significant latent period for convulsive effects. However, effects from subconvulsive doses may not be immediately obvious.

6.3 Characteristics

6.3.1 Physical Appearance/Odor

6.3.1.1 Laboratory Grade

Laboratory grade agents are typically colorless crystalline solids. Odors have not been reported.

6.3.1.2 Modified Agents

Bicyclophosphates can be dissolved in organic solvents to facilitate handling, to stabilize the agents, or to increase the ease of percutaneous penetration by the agents. Percutaneous enhancement solvents include dimethyl sulfoxide, *N,N*-dimethylformamide, *N,N*-dimethylpalmitamide, *N,N*-dimethyldecanamide, and saponin. Color and other properties of these solutions may vary from the pure agent. Odors will vary depending on the characteristics of the solvent(s) used.

6.3.2 Stability

Information on the stability of bicyclophosphates has not been published. However, based on the use of similar compounds as flame retardants and lubricant additives, these agents should be highly stable.

6.3.3 Persistency

Information on the persistency of bicyclophosphates has not been published. However, based on similar compounds these agents are likely to be persistent in the environment.

6.3.4 Environmental Fate

Bicyclophosphates are nonvolatile and do not pose a vapor hazard. Although these agents may be dissolved in volatile solvents, evaporation of the solvent does not increase the evaporation of the agent itself. Porous material, including painted surfaces, may absorb solutions of agents. After the initial surface contamination has been removed, the agent that has been absorbed into porous material can migrate back to the surface posing a contact hazard.

With the exception of methylbicyclophosphate (C06-A001), which is soluble in water, most bicyclophosphates are only slightly soluble or insoluble in water. Most of these agents are also only slightly soluble or insoluble in common organic solvents including gasoline, alcohols, and oils.

6.4 Additional Hazards

6.4.1 Exposure

Although bicyclophosphates do not inhibit acetylcholinesterase, they exhibit a synergistic toxic effect with materials that do. Individuals who have had previous exposure to cholinesterase inhibitors such as nerve agents and commercial organophosphate or carbamate pesticides may be at a greater risk from exposure to bicyclophosphates.

Exposure to doses below the level that will produce seizures can result in long-lasting central nervous system sensitization such as an increased susceptibility to sound induced (audiogenic) seizures.

All foodstuffs in the area of a release should be considered contaminated. Unopened items packaged in glass, metal, or high-density plastics and exposed only to agent aerosols may be used after decontamination of the container. Unopened items exposed to solid agents or solutions of agents should be decontaminated within a few hours postexposure or destroyed. Opened or unpackaged items, or those packaged only in paper or cardboard, should be destroyed.

Meat, milk, and animal products from animals exposed to or killed by bicyclophosphates should be destroyed or quarantined until they are tested and determined to be safe to consume.

Plants, fruits, vegetables, and grains should be quarantined until tested and determined to be safe to consume.

6.4.2 Livestock/Pets

Animals can be decontaminated with shampoo or soap and water. If the animals' eyes have been exposed to the agent, they should be irrigated with water or saline solution for a minimum of 30 minutes.

The topmost layer of unprotected feedstock (e.g., hay or grain) should be destroyed. The remaining material should be quarantined until tested. Leaves of forage vegetation could still retain sufficient agents to produce effects for several weeks post release, depending on the level of contamination and the weather conditions.

6.4.3 Fire

Some members of this class of agents have been employed as fire retardants and, therefore, are not expected to be greatly affected by a fire. However, actions taken to extinguish the fire could spread the agent. Runoff from firefighting efforts could pose a significant contact threat. Under the appropriate conditions, these agents could decompose and potentially produce toxic and/or corrosive gases.

6.4.4 Reactivity

Information on the reactivity of bicyclophosphates has not been published.

6.4.5 Hazardous Decomposition Products

Information on the decomposition products of bicyclophosphates, through either hydrolysis or combustion, has not been published.

6.5 Protection

6.5.1 Evacuation Recommendations

There are no published recommendations for isolation or protective action distances for bicyclophosphate agents released in mass casualty situations.

6.5.2 Personal Protective Requirements

6.5.2.1 Structural Firefighters' Gear

Structural firefighters' protective clothing is recommended for fire situations only; it is not effective in spill situations or release events. However, since bicyclophosphates have negligible vapor pressure, they do not pose a vapor hazard. The primary risk of exposure is through contact with aerosolized agents, solids, or solutions of agents. If chemical protective clothing is not available and it is necessary to rescue casualties from a contaminated area, then structural firefighters' gear will provide some skin protection against agent aerosols. Contact with solids and solutions should be avoided.

Even though it is likely that bicyclophosphates will be rapidly detoxified or eliminated from the body, subconvulsive insidious exposures may cause long-lasting central nervous system sensitization, placing all responders who entered the hot zone without appropriate chemical protective clothing at increased risk.

6.5.2.2 Respiratory Protection

Self-contained breathing apparatuses (SCBAs) or air purifying respirators (APRs) should have a National Institute for Occupational Safety and Health (NIOSH) and Chemical/Biological/Radiological/Nuclear (CBRN). However, during emergency operations, other NIOSH approved SCBAs or APRs that have been specifically tested by the manufacturer against chemical warfare agents may be used if deemed necessary by the incident commander. APRs should be equipped with a NIOSH approved CBRN filter or a combination organic vapor/acid gas/particulate cartridge.

Immediately dangerous to life or health (IDLH) levels are the ceiling limit for respirators other than SCBAs. However, IDLH levels have not been established for bicyclophosphates. Therefore, any potential exposure to these agents should be regarded with extreme caution and the use of SCBAs for respiratory protection should be considered.

6.5.2.3 Chemical Protective Clothing

Currently, there is no information on performance testing of chemical protective clothing against bicyclophosphates.

Because these agents do not produce any significant concentration of vapor, the primary risk of exposure is through inhalation of aerosols or by the percutaneous migration of agent following dermal exposure to solid agents or solutions containing these agents. However, it is possible that aerosolized bicyclophosphates can also enter the body and produce a toxic effect by permeating through the skin or penetrating through broken, abraded, or lacerated skin. If there is any possibility of contact with aerosolized agent, then responders should consider wearing a Level A protective ensemble.

6.5.3 Decontamination

6.5.3.1 General

Specific information on decontaminating bicyclophosphates has not been published. Apply universal decontamination procedures using soap and water.

6.5.3.2 Solutions or Liquid Aerosols

Casualties/personnel: Remove all clothing immediately. To avoid further exposure of the head, neck, and face to the agent, cut off potentially contaminated clothing that must be pulled over the head. Remove as much of the agent from the skin as fast as possible. If water is not immediately available, the agent can be absorbed with any convenient material such as paper towels, toilet paper, flour, and/or talc. To minimize both spreading the agent and abrading the skin, do not rub the agent with the absorbent. Blot the contaminated skin with the absorbent.

Use a sponge or cloth with liquid soap and copious amounts of water to wash the skin surface and hair at least three times. Do not delay decontamination to find warm or hot water if it is not readily available. Avoid rough scrubbing as this could abrade the skin and increase percutaneous absorption of residual agent. Rinse with copious amounts of water. If there is a potential that the eyes have been exposed to bicyclophosphates, irrigate with water or 0.9% saline solution for a minimum of 15 minutes.

Small areas: Puddles of liquid can be contained by covering with absorbent material such as vermiculite, diatomaceous earth, clay, sponges, or towels. Place the absorbed material into containers lined with high-density polyethylene. Wash the area and the exterior of the container with copious amounts of the soap and water. Collect and containerize the rinseate.

6.5.3.3 Solids or Particulate Aerosols

Casualties/personnel: Do not attempt to brush the agent off the individual or their clothing as this can aerosolize the agent. Remove all clothing immediately. To avoid further exposure of the head, neck, and face to the agent, cut off potentially contaminated clothing that must be pulled over the head. Wash the skin surface and hair at least three times with copious amounts of soap and water. Do not delay decontamination to find warm or hot water if it is not readily available. Rinse with copious amounts of water. If there is a potential that the eyes have been exposed to bicyclophosphates, irrigate with water or 0.9% saline solution for a minimum of 15 minutes.

Small areas: If indoors, close windows and doors in the area and turn off anything that could create air currents (e.g., fans, air conditioner, etc.). Avoid actions that could aerosolize the agent, such as sweeping or brushing. Collect the agent using a vacuum cleaner equipped with a high-efficiency particulate air (HEPA) filter. Do not use a standard home or industrial vacuum. Do not allow the vacuum exhaust to stir the air in the affected area. Vacuum all surfaces with extreme care in a very slow and controlled manner to minimize aerosolizing the agent. Place the collected material into containers lined with high-density polyethylene.

Wash the area and the exterior of the container with copious amounts of soap and water. Collect and containerize the rinseate.

6.6 Medical

6.6.1 CDC Case Definition

The Centers for Disease Control and Prevention (CDC) has not published a specific case definition for intoxication by bicyclophosphates.

6.6.2 Differential Diagnosis

The following factors have been suggested as alternatives to consider when presented with a potential case of exposure to bicyclophosphates: history of epilepsy; exposure to alcohol, cocaine, lead, camphor, strychnine, and/or carbon monoxide; medicinals such as phenothiazines; head trauma, heatstroke; encephalitis, meningitis, and tetanus.

6.6.3 Signs and Symptoms

Sudden loss of consciousness followed by tonic, then clonic contractions of the muscles. In some cases the gastrointestinal tract may be affected producing urination, defecation or both.

6.6.4 Mass-Casualty Triage Recommendations

There are no recommendations for triaging casualties exposed to bicyclophosphates. However, in general, anyone who has been exposed should be transported to a medical facility for evaluation. Individuals who are asymptomatic and have not been directly exposed to the agent can be discharged after their names, addresses, and telephone numbers have been recorded. They should be told to seek medical care immediately if symptoms develop.

6.6.5 Casualty Management

Decontaminate the casualty ensuring that all bicyclophosphates have been removed. If any bicyclophosphates have gotten into the eyes, irrigate the eyes with water or 0.9% saline solution for at least 15 minutes. Irrigate open wounds with water or 0.9% saline solution for at least 10 minutes.

Once the casualty has been decontaminated, including the removal of foreign matter from wounds, medical personnel do not need to wear a chemical-protective mask.

Ventilate the patient. If breathing is difficult, administer oxygen. It may be appropriate to intubate the casualty since seizures may make it difficult or impossible later. Otherwise, treatment consists of symptomatic management of seizures. Do not attempt to immobilize or protect the tongue. Loosen clothing around the neck and place a pillow under the head to prevent injury. Roll the casualty onto their side to prevent aspiration of fluids. Avoid unnecessary disturbances including loud or sudden noises. Treatment with anticonvulsants, including diazepam, phenobarbital, and other traditional antiepileptic drugs, has been successful in some animal studies.

6.7 Fatality Management

Remove all clothing and personal effects segregating them as either durable or nondurable items. While it may be possible to decontaminate durable items, it may be safer and more efficient to destroy nondurable items rather than attempt to decontaminate them. Items that will be retained for further processing should be double sealed in impermeable containers, ensuring that the inner container is decontaminated before placing it in the outer one.

Wash the remains with soap and water. Pay particular attention to areas where agent may get trapped, such as hair, scalp, pubic areas, fingernails, folds of skin, and wounds. All wash and rinse waste must be contained for proper disposal. Body fluids removed during the embalming process do not pose any additional risks and should be contained and handled according to established procedures.

Cadaver poses no significant secondary hazards after decontamination. Use standard burial practices.

C06-A

AGENTS

C06-A001

Methylbicyclophosphate

CAS: 1449-89-4
RTECS: —

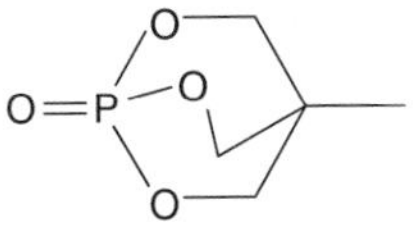

$C_7H_{13}O_4P$

Crystalline solid.

Exposure Hazards

Conversion Factor: 1 ppm = 7.86 mg/m^3 at 77°F

Human toxicity values have not been established or have not been published. However, based on available information, this agent appears to be approximately 0.1% as toxic as VX (C01-A016).

Properties:

MW: 192.2	*VP*: Negligible	*FlP*: —
D: —	*VD*: —	*LEL*: —
MP: 473°F	*Vlt*: —	*UEL*: —
BP: —	H_2O: "Soluble"	*RP*: —
Vsc: —	*Sol*: —	*IP*: —

C06-A002

Ethylbicyclophosphate

CAS: 1005-93-2
RTECS: TY6475000

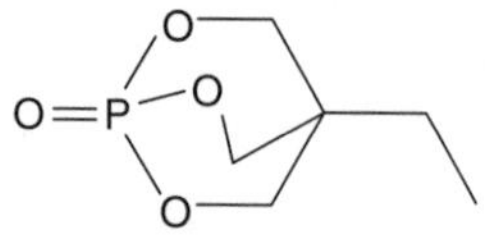

$C_6H_{11}O_4P$

Crystalline solid.

Exposure Hazards

Conversion Factor: 1 ppm = 7.29 mg/m^3 at 77°F

Human toxicity values have not been established or have not been published.

Properties:

MW: 178.1	*VP*: Negligible	*FlP*: —
D: —	*VD*: —	*LEL*: —
MP: 396°F	*Vlt*: —	*UEL*: —
BP: —	H_2O: "Slightly soluble"	*RP*: —
Vsc: —	*Sol*: —	*IP*: —

C06-A003

Isopropylbicyclophosphate

CAS: 51052-72-3

RTECS: —

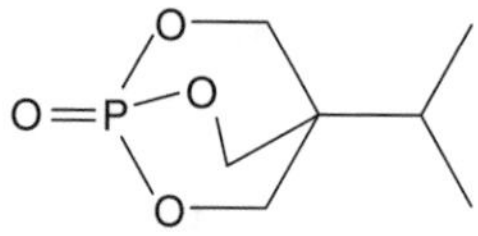

$C_7H_{13}O_4P$

Specific information on physical appearance is not available for this agent.

Exposure Hazards

Conversion Factor: 1 ppm = 7.86 mg/m^3 at 77°F

Human toxicity values have not been established or have not been published. However, based on available information, this agent appears to be less than half as toxic as VX (C01-A016).

Properties:

MW: 192.2	*VP*: —	*FlP*: —
D: —	*VD*: —	*LEL*: —
MP: —	*Vlt*: —	*UEL*: —
BP: —	H_2O: —	*RP*: —
Vsc: —	*Sol*: —	*IP*: —

C06-A004

***t*-Butylbicyclophosphate**

CAS: 61481-19-4

RTECS: —

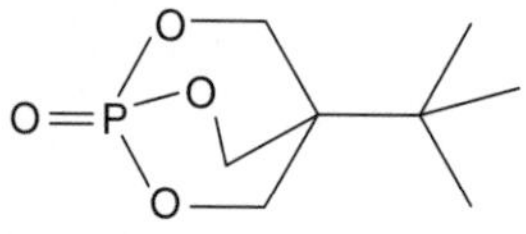

$C_8H_{15}O_4P$

Specific information on physical appearance is not available for this agent.

Exposure Hazards

Conversion Factor: 1 ppm = 8.43 mg/m^3 at 77°F

Human toxicity values have not been established or have not been published. However, based on available information, this agent appears to be just as toxic as VX (C01-A016).

Properties:

MW: 206.2	*VP*: —	*FlP*: —
D: —	*VD*: —	*LEL*: —
MP: —	*Vlt*: —	*UEL*: —
BP: —	H_2O: "Slightly soluble"	*RP*: —
Vsc: —	*Sol*: —	*IP*: —

C06-A005

1-Thio-4-ethyl-2,6,7-trioxa-1-phosphabicyclo-[2.2.2]-octane

CAS: 935-52-4

RTECS: —

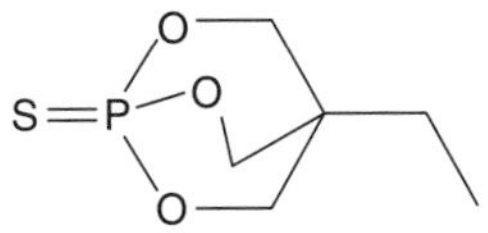

$C_6H_{11}O_3PS$

Specific information on physical appearance is not available for this agent.

Exposure Hazards

Conversion Factor: 1 ppm = 7.94 mg/m^3 at 77°F

Human toxicity values have not been established or have not been published.

Properties:

MW: 194.2	*VP*: Negligible	*FlP*: —
D: —	*VD*: —	*LEL*: —
MP: 345°F	*Vlt*: —	*UEL*: —
BP: —	H_2O: —	*RP*: —
Vsc: —	*Sol*: —	*IP*: —

References

Bellet, E.M. and J.E. Casida. "Bicyclic Phosphorus Esters. High Toxicity without Cholinesterase Inhibition." *Science* 182 (1973): 1135–36.

Casida, J.E. "Hazardous Caged Phosphorus Compounds." *Chemical and Engineering News* 52 (1974): 56.

Casida, J.E., M. Eto, A.D. Moscioni, J.L. Engel, D.S. Milbrath, and J.G. Verkade. "Structure-Toxicity Relations of 2,6,7-Trioxabicyclo[2.2.2]octanes and Related Compounds." *Toxicology and Applied Pharmacology* 36 (1976): 261–79.

Cooper, G.H., I.W. Lawston, R.L. Rickard, and T.D. Inch. "Structure-Activity Relations in 2,6,7-Trioxa-1-phosphabicyclo[2.2.2]octanes and Related Compounds." *European Journal of Medicinal Chemistry* 13 (1978): 207–12.

Eto, M., Y. Ozoe, T. Fujita, and J.E. Casida. "Significance of Branched Bridge-Head Substituent in Toxicity of Bicyclic Phosphate Esters." *Agricultural and Biological Chemistry* 40 (1976): 2113–15.

Kao, W.Y., Q.-Y. Liu, W. Ma, G.D. Ritchie, J. Lin, A.F. Nordholm, J. Rossi, III, J.L. Barker, D.A. Stenger, and J.J. Pancrazio. "Inhibition of Spontaneous GABAergic Transmission by Trimethylolpropane Phosphate." *Neurotoxicology* 20 (1999): 843–60.

Kimmerle, G., A. Een, P. Groning, and J. Thyssen. "Acute Toxicity of Bicyclic Phosphorus Esters." *Archiv fuer Toxikologie* 35 (1976): 149–52.

Lindsey, J.W., A.E. Jung, T.K. Narayanan, and G.D. Ritchie. "Acute Effects of a Bicyclophosphate Neuroconvulsant on Monoamine Neurotransitter and Metabolite Levels in the Rat Brain." *Journal of Toxicology and Environmental Health, Part A* 54 (1998): 421–29.

Milbrath, D.S., J.L. Engel, J.G. Verkade, and J.E. Casida. "Structure-Toxicity Relationships of 1-Substituted-4-alkyl-2,6,7-trioxabicyclo[2.2.2]octanes." *Toxicology and Applied Pharmacology* 47 (1979): 287–93.

Milbrath, D.S., M. Eto, and J.E. Casida. "Distribution and Metabolic Fate in Mammals of the Potent Convulsant and GABA Antagonist *tert*-Butylbicyclophosphate and its Methyl Analog." *Toxicology and Applied Pharmacology* 46 (1978): 411–20.

MP Biomedicals, Inc. *Material Safety Data Sheet (MSDS) for ETBICYPHAT*. Aurora, OH, February 17, 2004.

Rossi, J., III, G.D. Ritchie, S. McInturf, and A.F. Nordholm. "Reduction of Motor Seizures in Rats Induced by the Ethyl Bicyclophosphate Trimethylolpropane Phosphate (TMPP)." *Progress in Neuro-Psychopharmacology & Biological Psychiatry* 25 (2001): 1323–40.

7

COX Inhibiting Blood Agents

7.1 General Information

These agents inhibit the enzyme cytochrome oxidase (COX) preventing the transfer of oxygen from blood to the cells. Materials include cyanides, halogenated cyanides, and hydrogen sulfide. They are first generation warfare agents that were used during World War I. They were the first systemic agents introduced in that war. They are well-known industrial materials that were readily available at that time; most still have commercial value. Hydrogen cyanide and cyanogen chloride are listed in Schedule 3 of the Chemical Weapons Convention and are the only COX inhibiting blood agents specifically included in the convention. COX inhibiting blood agents are relatively easy to acquire or manufacture and to disperse. For information on some of the chemicals used to manufacture these blood agents, see the Component section (C07-C) following information on the individual agents.

Blood agents have been stockpiled by most countries that have pursued a chemical weapons program, and have been used a number of times on the battlefield. Although this class of agents is considered obsolete on the modern battlefield, several of these agents are still considered a significant threat as potential improvised weapons that could be utilized in urban warfare. COX inhibiting blood agents have also been used by terrorists.

7.2 Toxicology

7.2.1 Effects

COX inhibiting blood agents are compounds that produce respiratory paralysis and seizures or stop the transfer of oxygen from blood to the rest of the body by inhibiting the enzyme cytochrome oxidase. The lack of oxygen rapidly affects all body tissues, especially the central nervous system, producing headache, dizziness, confusion, stupor, nausea, and vomiting. The cyanogen halide agents will also cause lung membranes to swell and become filled with liquid (pulmonary edema).

7.2.2 Pathways and Routes of Exposure

COX inhibiting blood agents are primarily an inhalation hazard. However, liquid agents, or solutions containing these agents, are hazardous through skin and eye exposure, ingestion, and abraded skin (e.g., breaks in the skin or penetration of skin by debris). At high concentrations, agent vapor may pose a skin absorption hazard.

7.2.3 General Exposure Hazards

COX inhibiting blood agents generally do not have good warning properties. Although many cyanides have distinct odors that are evident at low levels, there is a significant portion of the population that is genetically incapable of detecting them. Hydrogen sulfide rapidly causes olfactory fatigue preventing further detection of the agent. Cyanogen halides cause eye irritation and lacrimation at low concentrations.

Lethal concentrations (LC_{50}s) for inhalation of these agents are as low as 1300 ppm for a 2-minute exposure.

Lethal percutaneous exposures (LD_{50}s) to liquefied agents are as low as 7 grams per individual.

The body detoxifies these agents relatively quickly and recovery from nonfatal acute effects is generally rapid.

7.2.4 Latency Period

7.2.4.1 *Vapor*

Effects from vapor exposure begin to appear 1–2 minutes after exposure. Pulmonary edema, caused by inhalation of cyanogen halides, does not occur until after a substantial latency period.

7.2.4.2 *Liquid*

Effects from exposure to liquid agents may be delayed from several minutes up to 2 hours after exposure. Some factors affecting the length of time before the onset of symptoms are the amount of agent involved, the amount of skin surface in contact with the agent, previous exposure to materials that chap or dry the skin (e.g., organic solvents such as gasoline or alcohols), and the addition of additives designed to enhance the rate of percutaneous penetration by the agents.

Another key factor is the part of the body that is exposed to the agent. It takes the agent longer to penetrate areas of the body that are covered by thicker and tougher skin. The regions of the body that allows the fastest percutaneous penetration are the groin, head, and neck. The least susceptible body regions are the hands, feet, front of the knee, and outside of the elbow.

7.3 Characteristics

7.3.1 Physical Appearance/Odor

7.3.1.1 *Laboratory Grade*

Blood agents are either gases or volatile liquids. Most agents are colorless. Odors vary from mildly pleasant to harsh and irritating. The ability to detect the odor of some agents is transient and may give the impression that agents are no longer present. Some agents, especially in high concentration, may cause eye irritation and tearing.

7.3.1.2 *Munition Grade*

Munition grade agents are typically colorless but can be yellow to brown and may contain crystallized decomposition products. Odors vary from mildly pleasant to harsh and irritating.

7.3.1.3 Binary/Reactive Agents

Terrorists have used binary versions created by mixing acids with reactive materials such as cyanide salts. Since the agent is formed as a result of mixing, it will be crude and consist of the agent, the individual components, and any by-products formed during the reaction. The color, odor, and consistency of the resulting agent will vary depending on the quality of the components and the degree of mixing.

The components, by-products of the reaction, or solvents used to facilitate mixing the components may have their own toxic properties and could present additional hazards. Residual components may react with common materials, such as metals, to produce other hazardous materials.

7.3.1.4 Modified Agents

Solvents, such as carbon disulfide (CS_2) and chloroform ($CHCl_3$), have been added to COX inhibiting blood agents to increase their stability in storage and increase their persistency after their release. The color, odor, and consistency of these mixtures will vary depending on the characteristics of the solvent(s) used and concentration of blood agent in the solution.

Blood agents have also been adsorbed onto a solid carrier, such as pumice, to facilitate dispersal of the agent and increase their persistency after their release. These compositions may appear as free-flowing powders or as a coarse grit. Odors may vary from the unmodified agent.

7.3.1.5 Mixtures with Other Agents

COX inhibiting blood agents have been mixed with other agents such as bromoacetone (C13-A003), chloropicrin (C10-A006), arsenic trichloride (C04-C006), and stannic chloride ($SnCl_4$) to increase their stability in storage, to enhance their toxicity, or to act as a marker during deployment.

7.3.2 Stability

Cyanide blood agents are relatively unstable and tend to polymerize on standing. Polymers can be explosive. Stabilizers or solvents can be added to inhibit decomposition. Stabilizers include phosphoric acid, sulfuric acid, powdered sodium pyrophosphate, and sulfur dioxide. Although cyanide blood agents react with metals, they can be stored in steel or other common containers if stabilized.

Hydrogen sulfide is stable and stored as a compressed, liquefied gas in aluminum or stainless steel cylinders.

7.3.3 Persistency

For military purposes, unmodified COX inhibiting blood agents are classified as nonpersistent. These agents are gases or highly volatile liquids at normal temperatures. Evaporation rates of liquid agents are more than ten times that of water. Cold weather may decrease the rate of evaporation of liquid or solid agents.

7.3.4 Environmental Fate

Other than hydrogen cyanide (C07-A001), agent vapors have a density greater than air and tend to collect in low places. However, due to the volatile nature of these materials, there is a minimal extended risk except in an enclosed or confined space.

With the exception of hydrogen cyanide, which is miscible with water, most of these agents are only slightly soluble or insoluble in water. However, agent solubility may be modified (either increased or decreased) by solvents or impurities. These agents are also soluble in many organic solvents including gasoline, alcohols, and oils.

7.4 Additional Hazards

7.4.1 Exposure

All foodstuffs in the area of a release should be considered contaminated. Unopened items packaged in glass, metal, or heavy duty plastic and exposed only to agent vapors may be used after decontamination of the container. Unopened items exposed to liquid agents or solutions of agents should be decontaminated within a few hours postexposure or destroyed. Opened or unpackaged items, or those packaged only in paper or cardboard, should be destroyed.

Meat from animals that have survived exposure to COX inhibiting blood agents should be safe to consume after a short quarantine period. Milk should be quarantined until tested.

Plants, fruits, vegetables, and grains should be quarantined until tested and determined to be safe to consume. Aeration or decontamination with soap and water may be sufficient in many cases.

7.4.2 Livestock/Pets

Animals can be decontaminated with shampoo/soap and water. If the animals' eyes have been exposed to agent, they should be irrigated with water or saline solution for a minimum of 30 minutes.

The topmost layer of unprotected feedstock (e.g., hay or grain) should be destroyed. The remaining material should be quarantined until tested. Leaves of forage vegetation are unlikely to retain sufficient agent to produce significant effects for more than a few hours post release.

7.4.3 Fire

Some COX inhibiting blood agents are flammable and can form explosive mixtures with air. Some of these agents can polymerize in their containers and explode when heated.

Most of these blood agents are gases and will be quickly dissipated or be consumed in a fire. If liquid or solid agents are present, then heat from a fire will increase the amount of agent vapor in the area. A significant amount of the agent could be volatilized and escape into the surrounding environment before it is consumed by the fire. Actions taken to extinguish the fire can also spread the agent. With the exception of hydrogen cyanide, most COX inhibiting blood agents are only slightly soluble or insoluble in water. However runoff from firefighting efforts may still pose a threat. Some of the decomposition products resulting from hydrolysis or combustion of blood agents are water soluble and highly toxic. Other potential decomposition products include toxic and/or corrosive gases. Solvents mixed with some agents are flammable and may pose an additional fire hazard. Added components may react with steam or water during a fire to produce toxic, flammable, and/or corrosive vapors.

7.4.4 Reactivity

With the exception of hydrogen sulfide, COX inhibiting blood agents must be stabilized or they will polymerize during storage. Some agents are slowly hydrolyzed by water to produce corrosive and toxic gases. They are incompatible with strong oxidizers; many are incompatible with strong corrosives. Added components may react with water to produce toxic, flammable, and/or corrosive vapors.

7.4.5 Hazardous Decomposition Products

7.4.5.1 Hydrolysis

Hydrogen cyanide is highly soluble and stable in water. Many others decompose into hydrogen cyanide and/or corrosives such as hydrogen chloride (HCl), hydrogen bromide (HBr), hydrogen fluoride (HF), or hydrogen iodide (HI). Some components may produce arsenous oxides or arsenic salts when hydrolyzed.

7.4.5.2 Combustion

Volatile decomposition products may include nitrogen oxides (NO_x), sulfur oxides (SO_x), and HCl, HBr, HF, or HI. Some components may produce arsenic oxides.

7.5 Protection

7.5.1 Evacuation Recommendations

Isolation and protective action distances listed below are taken from Argonne National Laboratory Report No. ANL/DIS-00-1, *Development of the Table of Initial Isolation and Protective Action Distances for the 2000 Emergency Response Guidebook*, which is still the basis for the "when used as a weapon" scenarios in the 2004 Emergency Response Guidebook (ERG). For COX inhibiting blood agents, these recommendations are based on two different release mechanisms (direct aerosolization of liquid agents or, for gaseous agents, sabotage of commercial containers) as well as on various quantities of material depending on the agent used in the release.

For hydrogen cyanide, the release scenario involves sabotage of a commercial container with either 60 kilograms (approximately 23 gallons) of liquefied agent or 30,000 kilograms (approximately 12,000 gallons) of liquefied agent.

For cyanogen chloride, the release involves direct aerosolization of the liquid agents with a particle size between 2 and 5 μm. In this scenario, the difference between a small and a large release is not based on the standard 200 liters spill used for commercial hazardous materials listed in the ERG. A small release involves 30 kilograms (approximately 6.5 gallons) of liquefied agent and a large release involves 500 kilograms (approximately 110 gallons) of liquefied agent.

	Initial isolation (feet)	Downwind day (miles)	Downwind night (miles)
AC (Hydrogen cyanide) *C07-A001*			
Small device (60 kilograms)	200	0.1	0.3
Large device (30,000 kilograms)	1500	1.0	2.4
CK (Cyanogen chloride) *C07-A003*			
Small device (30 kilograms)	200	0.4	1.5
Large device (500 kilograms)	1300	2.5	5.0

7.5.2 Personal Protective Requirements

7.5.2.1 Structural Firefighters' Gear

Blood agents are primarily a respiratory hazard; however, solids, liquids, or high vapor concentrations may pose a percutaneous hazard. Structural firefighters' protective clothing is recommended for fire situations only; it is not effective in spill situations or release events. If chemical protective clothing is not available and it is necessary to rescue casualties from a contaminated area, then structural firefighters' gear will provide limited skin protection against low concentrations of blood agent vapors. Contact with liquids, solids, and solutions should be avoided.

7.5.2.2 Respiratory Protection

Self-contained breathing apparatuses (SCBAs) or air purifying respirators (APRs) should have a National Institute of Occupational Safety and Health (NIOSH) and Chemical/Biological/Radiological/Nuclear (CBRN) certification. However, during emergency operations, other NIOSH approved SCBAs or APRs that have been specifically tested by the manufacturer against chemical warfare agents may be used if deemed necessary by the incident commander. APRs should be equipped with a NIOSH approved CBRN filter or a combination organic vapor/acid gas/particulate cartridge.

High concentrations of cyanides and cyanogen halides can rapidly degrade elements of some air purification filters. Care should be taken whenever using APRs in the presence of these agents.

Immediately dangerous to life or health (IDLH) levels are the ceiling limit for respirators other than SCBAs. Any exposures approaching the IDLH level should be regarded with extreme caution and the use of SCBAs for respiratory protection should be considered.

7.5.2.3 Chemical Protective Clothing

Use only chemical protective clothing that has undergone material and construction performance testing against cyanides, cyanogen halides, and hydrogen sulfide. Reported permeation rates may be affected by solvents, components, or impurities in munition grade or modified agents.

In addition to the risk of percutaneous migration of agent following dermal exposure to liquid agents, some cyanide agent vapors can also penetrate the skin and produce a toxic effect. However, these concentrations are significantly greater than concentrations that will produce similar effects if inhaled. If the concentration of vapor exceeds the level necessary to produce effects through dermal exposure, then responders should wear a Level A protective ensemble.

7.5.3 Decontamination

7.5.3.1 General

Apply universal decontamination procedures using soap and water.

7.5.3.2 Vapors

Casualties/personnel: Remove all clothing as it may continue to emit "trapped" agent vapor after contact with the vapor cloud has ceased. Shower using copious amounts of soap and water. Ensure that the hair has been washed and rinsed to remove potentially trapped vapor. If eye irritation occurs, irrigate with water or 0.9% saline solution for a minimum of 15 minutes.

Small areas: Ventilate to remove vapor. Because the boiling point of some cyanide agents is near normal room temperature (70°F), agent vapors may condense on cooler surfaces and pose a percutaneous hazard. Liquids can then revolatilize when the temperature rises. If deemed necessary, wash the area with copious amounts of soap and water. Collect and place the rinseate and place in containers lined with high-density polyethylene.

7.5.3.3 Liquids, Solutions, or Liquid Aerosols

Casualties/personnel: Remove all clothing immediately. Even clothing that has not come into direct contact with the agent may contain "trapped" vapor. To avoid further exposure of the head, neck, and face to the agent, cut off potentially contaminated clothing that must be pulled over the head. Use a sponge or cloth with liquid soap and copious amounts of water to wash the skin surface and hair at least three times. Avoid rough scrubbing. Rinse with copious amounts of water. If water is not immediately available, the agent can be absorbed with any convenient material such as paper towels, toilet paper, flour, or talc. To minimize both spreading the agent and abrading the skin, do not rub the agent with the absorbent. Blot the contaminated skin with the absorbent. If there is a potential that the eyes have been exposed, irrigate with water or 0.9% saline solution for a minimum of 15 minutes.

Small areas: Puddles of liquid can be contained by covering with absorbent material such as vermiculite, diatomaceous earth, clay, sponges, or towels. Place the absorbed material into containers lined with high-density polyethylene. Wash the area with copious amounts of the soap and water. Collect and containerize the rinseate. Ventilate the area to remove vapors.

7.5.3.4 Solids or Particulate Aerosols

Casualties/personnel: Do not attempt to brush the agent off the individual or their clothing as this can aerosolize the agent. Remove all clothing immediately. To avoid further exposure of the head, neck, and face to the agent, cut off potentially contaminated clothing that must be pulled over the head. Wash the skin surface and hair at least three times with copious amounts of soap and water. Do not delay decontamination to find warm or hot water if it is not readily available. Rinse with copious amounts of water. If there is a potential that any of the agent has gotten into the eyes, irrigate with water or 0.9% saline solution for a minimum of 15 minutes.

Small areas: Collect the agent using a vacuum cleaner equipped with a high-efficiency particulate air (HEPA) filter. Do not use a standard home or industrial vacuum. Place the collected material into containers lined with high-density polyethylene. Wash the area with copious amounts of soap and water. Collect and containerize the rinseate in containers lined with high-density polyethylene. Ventilate the area to remove any agent vapors generated during the decontamination.

7.6 Medical

7.6.1 CDC Case Definition (for Cyanides)

A case in which cyanide concentration is higher than the normal reference range (0.02–0.05 μg/mL) in whole blood, or cyanide is detected in environmental samples. The case can be confirmed if laboratory testing is not performed because either a predominant amount of clinical and nonspecific laboratory evidence is present or an absolute certainty of the etiology of the agent is known.

There is no case definition for exposure to hydrogen sulfide.

7.6.2 Differential Diagnosis

The following factors have been suggested as alternatives to consider when presented with a potential case of exposure to COX inhibiting blood agents: smoke or carbon monoxide inhalation; sedatives; stroke; methemoglobinemia; meningitis and encephalitis; and ingestion of plants or seeds containing cyanoglycosides.

7.6.3 Signs and Symptoms

7.6.3.1 Vapors/Aerosols

Most indications of cyanide and sulfide blood agent poisoning are nonspecific. Inhalation of high concentration of these agents may produce temporary rapid and deep breathing followed by convulsions and unconsciousness. Under these circumstances, the casualty will stop breathing within 2–4 minutes after exposure. Death will occur 4–8 minutes later. For most cyanides, casualties experience few effects when exposed to less than lethal doses. These may include temporary increase in breathing rate, dizziness, nausea, vomiting, and headache. Classic "cherry-red" skin and lips attributed to cyanide poisoning are not always present. In addition, cyanogen halides produce immediate eye, nose, and throat irritation similar to tear agents (Chapter 13) and vomiting agents (Chapter 14). Pulmonary edema caused by cyanogen halides may be delayed for several hours.

7.6.3.2 Liquids/Solids

Exposure to solid or liquid cyanogen halides can cause skin and eye irritation. Otherwise, casualties exposed to cyanides experience few effects at sublethal doses. Percutaneous absorption of a lethal dose may produce temporary rapid and deep breathing followed by convulsions and unconsciousness. Under these circumstances, the casualty will stop breathing within 2–4 minutes after exposure. Death will occur 4–8 minutes later.

7.6.4 Mass-Casualty Triage Recommendations

7.6.4.1 Priority 1

A casualty who is convulsing or has is post seizure, who has a heartbeat and a palpable blood pressure. The casualty may or may not be conscious and/or breathing.

7.6.4.2 Priority 2

Recovering from mild effects or from successful therapy. Generally it takes hours for full recovery.

7.6.4.3 Priority 3

Casualty who has been in a clean environment for more than 5 minutes postexposure and is conscious and talking, or a casualty in a clean environment who is unconscious but has a pulse and is breathing normally. Unconscious casualties should be monitored in case of a sudden change in status.

7.6.4.4 Priority 4

A casualty who is not breathing and does not have a heartbeat or palpable blood pressure.

7.6.5 Casualty Management

Remove casualty to fresh air. If breathing is difficult, administer oxygen. If the casualty has been exposed to either solid or liquid agent, decontaminate the casualty ensuring that all agents have been removed. However, do not delay treatment if thorough decontamination cannot be undertaken immediately. If solid or liquid agents have gotten into the eyes, or if the casualty complains of significant eye irritation, irrigate the eyes with water or 0.9% saline solution for at least 15 minutes. Irrigate open wounds with water or 0.9% saline solution for at least 10 minutes.

Once the casualty has been decontaminated, including the removal of foreign matter from wounds, medical personnel do not need to wear a chemical-protective mask.

For cyanide and cyanogen, antidote should be administered as soon as possible. The Lilly Cyanide Antidote Kit contains amyl nitrite, sodium nitrite, and sodium thiosulfate. Cobalt edentate or 4-dimethylaminophenol are alternative antidotes for cyanide poisoning. Benzodiazepines or barbiturates may be required to control severe seizures.

If cyanogen halides are suspected, asymptomatic individuals should be monitored for possible complications caused by pulmonary edema.

7.7 Fatality Management

Remove all clothing and personal effects. Heavily splashed items should be disposed of or decontaminated with soap and water. Items that will be retained for further processing should be double sealed in impermeable containers, ensuring that the inner container is decontaminated before placing it in the outer one.

COX inhibiting blood agents are highly volatile and there is little risk of extensive residual contamination unless the remains were heavily splashed with liquid agent. Agents that have entered the body pose little threat through off-gassing. Some agents, such as hydrogen sulfide, may still be detectable because of the low odor threshold. Wash the remains with soap and water. Pay particular attention to areas where agent may get trapped, such as hair, scalp, pubic areas, fingernails, folds of skin, and wounds. If remains were heavily splashed with agent, wash and rinse waste should be contained for proper disposal. Screen the remains for agent vapors at the conclusion of the decontamination process. Once the remains have been thoroughly decontaminated, no further protective action is necessary. Body fluids removed during the embalming process do not pose any additional risks and should be contained and handled according to established procedures. Use standard burial procedures.

C07-A

AGENTS

C07-A001

Hydrogen cyanide (Agent AC)
CAS: 74-90-8
RTECS: MW6825000
UN/NA: 1051
ERG: 117

HCN

Colorless gas or liquid with an odor like bitter almond or peach kernels. Odor is detectable at 0.8 ppm, but some individuals are unable to detect odor at all. Used as an industrial fumigant. It is also used in electroplating, mining; and in producing synthetic fibers, plastics, and dyes. Industrially, it can be found mixed with a variety of gases including carbon monoxide, cyanogen, and phosphine.

Also reported as a mixture with Cyanogen chloride (C07-A003); Arsenic trichloride (C04-C006).

This material is on the ITF-25 high threat, the FBI threat list, and Schedule 3 of the CWC.

Ignites 50% of the time when disseminated from an artillery shell.

Exposure Hazards

Conversion Factor: 1 ppm = 1.10 mg/m^3 at 77°F

LCt$_{50(Inh)}$: 2860 mg-min/m^3 (1300 ppm for a 2-min exposure). This is a provisional update from an older value that has not been formally adopted as of 2005. LCt$_{50}$ varies greatly due to ability of the body to detoxify the agent over time.

LC$_{50}$: 546 ppm for a 10-min exposure

MEG$_{(1h)}$ *Min*: 2 ppm; *Sig*: 7.1 ppm; *Sev*: 15 ppm

OSHA PEL: 10 ppm [Skin]

NIOSH STEL: 4.7 ppm [Skin]

ACGIH Ceiling: 4.7 ppm [Skin]

IDLH: 50 ppm

Properties:

MW: 27.0	*VP*: 630 mmHg	*FlP*: 0°F
D: 0.69 g/mL (77°F)	*VD*: 0.99 (calculated)	*LEL*: 5.6%
MP: 8°F	*Vlt*: 860,000 ppm	*UEL*: 40%
BP: 78°F	H_2O: Miscible	*RP*: 0.03
Vsc: 0.28 cS (77°F)	*Sol*: Most organic solvents	*IP*: 13.60 eV

Final AEGLs

AEGL-1: 1 h, 2.0 ppm	4 h, 1.3 ppm	8 h, 1.0 ppm
AEGL-2: 1 h, 7.1 ppm	4 h, 3.5 ppm	8 h, 2.5 ppm
AEGL-3: 1 h, 15 ppm	4 h, 8.6 ppm	8 h, 6.6 ppm

C07-A002

Cyanogen bromide (Agent CB)

CAS: 506-68-3
RTECS: —
UN: 1889
ERG: 157

BrCN

Colorless or white crystalline solid with a penetrating, stinging, or biting odor.

Also reported as a mixture with Bromoacetone (C13-A003).

Exposure Hazards

Conversion Factor: 1 ppm = 4.33 mg/m^3 at 77°F

LC$_{50(Inh)}$: 400 mg/m^3 (92 ppm for a 10-min exposure)

$IC_{50(Inh)}$: 35 mg/m^3 (8.1 ppm); exposure duration unavailable
Eye Irritation: 1.4 ppm; exposure duration unavailable

These values are from older sources (ca. 1937) and are not supported by modern data. No updated toxicity estimates have been proposed.

Properties:

MW: 105.9	*VP*: 92 mmHg	*FlP*: None
D: 2.015 g/cm^3	*VP*: 122 mmHg (77°F)	*LEL*: None
MP: 120–124°F	*VD*: 3.6 (calculated)	*UEL*: None
BP: Sublimes	*Vlt*: 120,000 ppm	*RP*: 0.1
Vsc: —	*H_2O*: Soluble	*IP*: 11.84 eV
	Sol: Alcohols; Ether	

C07-A003

Cyanogen chloride (Agent CK)
CAS: 506-77-4
RTECS: GT2275000
UN: 1589
ERG: 125

ClCN

Colorless liquid or gas with a pungent, acrid, choking odor detectable at 1 ppm. Odor can go unnoticed because of discomfort. Used as an industrial fumigant and warning agent in odorless fumigant gases. It is used in metal cleaning, ore refining, and to produce malononitrile, *s*-triazines, and synthetic rubber.

Also reported as a mixture with Hydrogen cyanide (C07-A001); Arsenic trichloride (C04-C006).

This material is on the ITF-25 low threat list and Schedule 3 of the CWC.

Will polymerize during extended storage. Polymerization may be explosive.

Exposure Hazards
Conversion Factor: 1 ppm = 2.52 mg/m^3 at 77°F
$LCt_{50(Inh)}$: 11,000 mg-min/m^3 (2200 ppm for a 2-min exposure)
$ICt_{50(Inh)}$: 7,000 mg-min/m^3 (1400 ppm for a 2-min exposure)
Eye Irritation: 4.8 ppm for a "few seconds" exposure
Skin Irritation: 4.8 ppm for a 2-min exposure
ACGIH STEL: 0.3 ppm
NIOSH STEL: 0.3 ppm

Properties:

MW: 61.5	*VP*: 010 mmHg	*FlP*: None
D: 1.18 g/mL	*VD*: 2.2 (calculated)	*LEL*: None
MP: 20°F	*Vlt*: 1,400,000 ppm	*UEL*: None
BP: 55°F	*H_2O*: 6.9%	*RP*: 0.01
Vsc: 0.337 cS (liq. gas, 77°F)	*Sol*: Most organic solvents	*IP*: 12.49 eV

C07-A004

Cyanogen iodide
CAS: 506-78-5
RTECS: NN1750000

ICN

Colorless or white needle-shaped crystals with a pungent odor. Used as a preservative in taxidermy.

Exposure Hazards

Conversion Factor: 1 ppm = mg/m^3 at 77°F
Respiratory/Eye Irritation: 16 ppm; exposure duration unavailable

Lethal human toxicity values have not been established or have not been published.

Properties:

MW: 152.9	*VP*: 1 mmHg (77°F)	*FlP*: —
D: 2.84 g/cm^3 (64°F)	*VD*: 5.3 (calculated)	*LEL*: —
MP: 298°F	*Vlt*: 1300 ppm	*UEL*: —
BP: Sublimes	H_2O: Soluble	*RP*: 8.2
Vsc: —	*Sol*: Ethanol; Ether; Volatile oils	*IP*: 10.87 eV

C07-A005

Cyanogen
CAS: 460-19-5
RTECS: GT1925000
UN: 1026
ERG: 119

NCCN

Colorless gas with a pungent odor like almonds that is detectable at 230 ppm. Vapors are irritating at 15 ppm.

Exposure Hazards

Conversion Factor: 1 ppm = 2.13 mg/m^3 at 77°F
$MEG_{(1h)}$ *Min*: 20 ppm; *Sig*: 71 ppm; *Sev*: 150 ppm
ACGIH TLV: 10 ppm

Lethal human toxicity values have not been established or have not been published. However, based on available information, this agent appears to be approximately half as toxic as Hydrogen cyanide (C07-A001).

Properties:

MW: 52.0	*VP*: 3800 mmHg	*FlP*: —
D: 0.866 g/mL (liq. gas, 77°F)	*VD*: 1.8 (calculated)	*LEL*: 6.6%
MP: –18°F	*Vlt*: 5,200,000 ppm	*UEL*: 32%
BP: –6°F	H_2O: 1%	*RP*: 0.004
Vsc: 0.633 cS (liq. gas, 77°F)	*Sol*: Alcohols; Ether	*IP*: 13.57 eV

C07-A006

Hydrogen sulfide (Agent NG)
CAS: 7783-06-4
RTECS: MX1225000
UN: 1053
ERG: 117

H_2S

Colorless gas with a strong odor of rotten eggs detectable at 0.005 ppm. However, it can cause olfactory fatigue and the sense of smell is not reliable. Used industrially to produce elemental sulfur, sulfuric acid, and heavy water for nuclear reactors.

Also reported as a mixture with Chloropicrin (C10-A006); Carbon disulfide (C11-A039).

This material is on the ITF-25 high threat list.

Exposure Hazards

Conversion Factor: 1 ppm = 1.40 mg/m^3 at 77°F
Eye Irritation: "Low concentrations"
Respiratory Irritation: 50 ppm; exposure duration unavailable
$MEG_{(1h)}$ *Min*: 0.50 ppm; *Sig*: 27 ppm; *Sev*: 50 ppm
OSHA Ceiling: 20 ppm
ACGIH TLV: 10 ppm
ACGIH STEL: 15 ppm
IDLH: 100 ppm

Lethal human toxicity values have not been established or have not been published. However, based on available information, exposure to concentrations above 500 ppm can rapidly incapacitate and kill exposed individuals.

Properties:

MW: 34.1	*VP*: 13,376 mmHg	*FlP*: —
D: 0.777 g/mL (liq. gas, 77°F)	*VD*: 1.2 (calculated)	*LEL*: 4.0%
MP: –122°F	*Vlt*: —	*UEL*: 44%
BP: –77°F	H_2O: 0.4%	*RP*: 0.001
Vsc: 0.0978 cS (liq. gas, 77°F)	*Sol*: Gasoline; Kerosene	*IP*: 10.46 eV

Interim AEGLs

AEGL-1: 1 h, 0.51 ppm	4 h, 0.36 ppm	8 h, 0.33 ppm
AEGL-2: 1 h, 27 ppm	4 h, 20 ppm	8 h, 17 ppm
AEGL-3: 1 h, 50 ppm	4 h, 37 ppm	8 h, 31 ppm

C07-C

COMPONENTS AND PRECURSORS

C07-C007

Sodium cyanide

CAS: 143-33-9
RTECS: VZ7525000
UN: 1689
ERG: 157

NaCN

White, granular, or crystalline deliquescent solid that is odorless when dry but has a faint odor like almond when wet. This material is hazardous through inhalation, skin absorption, penetration through broken skin, and ingestion, and produces local skin/eye impacts. It causes irritation of the eyes and skin, asphyxia, lassitude, headache, confusion, nausea, vomiting, increased respiratory rate, slow gasping respiration, thyroid, and blood changes.

Used industrially for electroplating, precious metal extraction; and in the synthesis of dyes, pigments, pharmaceuticals, and pesticides.

This material is on the FBI threat list and the Australia Group Export Control list.

This material is a precursor blood agent containing cyanide, many G-series nerve agents, and other cyanide containing agents.

Exposure Hazards

$LD_{50(Ing)}$: 0.2–0.3 g
OSHA PEL: 5 mg/m^3
ACGIH Ceiling: 5 mg/m^3 [10 min] [Skin]
NIOSH Ceiling: 5 mg/m^3
IDLH: 25 mg/m^3 (expressed as CN)

Other human toxicity values have not been established or have not been published.

Properties:

MW: 49.0	*VP*: Negligible	*FlP*: None
D: 1.595 g/cm^3	*VD*: —	*LEL*: None
MP: 1047°F	*Vlt*: Negligible	*UEL*: None
BP: 2725°F	*H_2O*: 58% (77°F)	*RP*: —
Vsc: —	*Sol*: —	*IP*: —

C07-C008

Potassium cyanide

CAS: 151-50-8
RTECS: TS8750000
UN: 1680
ERG: 157

KCN

White, granular, or crystalline deliquescent solid that is odorless when dry but has a faint odor like almond when wet. This material is hazardous through inhalation, skin absorption, penetration through broken skin, and ingestion, and produces local skin/eye impacts. It causes irritation of the eyes, skin, and upper respiratory system, asphyxia, lassitude, headache, confusion, nausea, vomiting, increased respiratory rate or slow gasping respiration, as well as thyroid and blood changes.

Used industrially for electroplating, precious metal extraction; and in the synthesis of dyes, pigments, pharmaceuticals, and pesticides.

This material is on the Australia Group Export Control list.

This material is a precursor blood agent containing cyanide, many G-series nerve agents, and other cyanide containing agents.

Exposure Hazards

$LD_{50(Ing)}$: 0.2–0.3 g
OSHA PEL: 5 mg/m^3
ACGIH Ceiling: 5 mg/m^3 [10 min] [Skin]
NIOSH Ceiling: 5 mg/m^3
IDLH: 25 mg/m^3 (expressed as CN)

Other human toxicity values have not been established or have not been published.

Properties:

MW: 65.1	*VP*: Negligible	*FlP*: None
D: 1.55 g/cm^3	*VD*: —	*LEL*: None
MP: 1173°F	*Vlt*: Negligible	*UEL*: None
BP: 2957°F	*H_2O*: 72% (77°F)	*RP*: —
Vsc: —	*Sol*: Glycerol; Formamide; Hydroxylamine	*IP*: —

References

Agency for Toxic Substances and Disease Registry. "Cyanide ToxFAQs." September 1997.

———. "Hydrogen Sulfide ToxFAQs." September 2004.

———. "Hydrogen Cyanide." In *Managing Hazardous Materials Incidents Volume III—Medical Management Guidelines for Acute Chemical Exposures*. Rev. ed. Washington, DC: Government Printing Office, 2000.

———. *Toxicological Profile for Cyanide*. Washington, DC: Government Printing Office, 1997.

———. *Toxicological Profile for Cyanide (DRAFT)*. Washington, DC: Government Printing Office, September 2004.

———. *Toxicological Profile for Hydrogen Sulfide*. Washington, DC: Government Printing Office, 1997.

———. *Toxicological Profile for Hydrogen Sulfide (DRAFT)*. Washington, DC: Government Printing Office, September 2004.

Brophy, Leo P., Wyndham D. Miles, and Rexmond C. Cohrane. *The Chemical Warfare Service: From Laboratory to Field*. Washington, DC: Government Printing Office, 1968, pp. 55–61.

Centers for Disease Control and Prevention. "Case Definition: Cyanide." March 9, 2005.

———. "Facts About Cyanide." February 26, 2003.

Compton, James A.F. *Military Chemical and Biological Agents: Chemical and Toxicological Properties*. Caldwell, NJ: The Telford Press, 1987.

Fries, Amos A., and Clarence J. West. *Chemical Warfare*. New York: McGraw-Hill Book Company, Inc., 1921.

Marrs, Timothy C., Robert L. Maynard, and Frederick R. Sidell. *Chemical Warfare Agents: Toxicology and Treatment*. Chichester, England: John Wiley & Sons, 1997.

Munro, Nancy B., Sylvia S. Talmage, Guy D. Griffin, Larry C. Waters, Annetta P. Watson, Joseph F. King, and Veronique Hauschild. "The Sources, Fate and Toxicity of Chemical Warfare Agent Degradation Products." *Environmental Health Perspectives* 107 (1999): 933–74.

National Institute of Health. *Hazardous Substance Data Bank (HSDB)*. http://toxnet.nlm.nih.gov/cgi-bin/sis/htmlgen?HSDB/. 2005.

National Institute for Occupational Safety and Health. "Emergency Response Card for Cyanogen Chloride." Interim Document, March 20, 2003.

———. "Emergency Response Card for Hydrogen Cyanide." Interim Document, March 20, 2003.

———. "Emergency Response Card for Potassium Cyanide." Interim Document, March 20, 2003.

———. *NIOSH Pocket Guide to Chemical Hazards*. Washington, DC: Government Printing Office, September 2005.

Olson, Kent R., ed. *Poisoning & Drug Overdose*. 4th ed. New York: Lange Medical Books/McGraw-Hill, 2004.

Prentiss, Augustin M. *Chemicals in War: A Treatise on Chemical Warfare*. New York: McGraw-Hill Book Company, Inc., 1937.

Sartori, Mario. *The War Gases: Chemistry and Analysis*. Translated by L.W. Marrison. London: J. & A. Churchill, Ltd, 1939.

Sidell, Frederick R. *Medical Management of Chemical Warfare Agent Casualties: A Handbook for Emergency Medical Services*. Bel Air, MD: HB Publishing, 1995.

Sidell, Fredrick R., Ernest T. Takafuji, and David R. Franz, eds. *Medical Aspects of Chemical and Biological Warfare, Textbook of Military Medicine Series, Part 1, Warfare, Weaponry, and the Casualty*. Washington, DC: Office of the Surgeon General, Department of the Army, 1997.

Sifton, David W., ed. *PDR Guide to Biological and Chemical Warfare Response*. Montvale, NJ: Thompson/Physicians Desk Reference, 2002.

Smith, Ann, Patricia Heckelman, Maryadele J. Oneil, eds. *Merck Index: An Encyclopedia of Chemicals, Drugs, & Biologicals*. 13th ed. Rahway, NJ: Merck & Co., Inc., 2001.

Somani, Satu M., ed. *Chemical Warfare Agents*. New York: Academic Press, 1992.

Somani, Satu M., and James A. Romano, Jr., eds. *Chemical Warfare Agents: Toxicity at Low Levels*. Boca Raton, FL: CRC Press, 2001.

Swearengen, Thomas F. *Tear Gas Munitions: An Analysis of Commercial Riot Gas Guns, Tear Gas Projectiles, Grenades, Small Arms Ammunition, and Related Tear Gas Devices*. Springfield, IL: Charles C Thomas Publisher, 1966.

True, Bey-Lorraine, and Robert H. Dreisbach. *Dreisbach's Handbook of Poisoning: Prevention, Diagnosis and Treatment*. 13th ed. London, England: The Parthenon Publishing Group, 2002.

United States Army Headquarters. *Chemical Agent Data Sheets Volume I, Edgewood Arsenal Special Report No. EO-SR-74001*. Washington, DC: Government Printing Office, December 1974.

———. *Potential Military Chemical/Biological Agents and Compounds, Field Manual No. 3-11.9*. Washington, DC: Government Printing Office, January 10, 2005.

United States Army Medical Research Institute of Chemical Defense. *Medical Management of Chemical Casualties Handbook*. 3rd ed. Aberdeen Proving Ground, MD: United States Army Medical Research Institute of Chemical Defense, July 2000.

United States Coast Guard. *Chemical Hazards Response Information System (CHRIS) Manual, 1999 Edition*. http://www.chrismanual.com/Default.htm. 2004.

Wachtel, Curt. *Chemical Warfare*. Brooklyn, NY: Chemical Publishing Co., Inc., 1941.

Waitt, Alden H. *Gas Warfare: The Chemical Weapon, Its Use, and Protection Against It*. Rev. ed. New York: Duell, Sloan and Pearce, 1944.

Williams, Kenneth E. *Detailed Facts About Blood Agent Cyanogen Chloride (CK)*. Aberdeen Proving Ground, MD: United States Army Center for Health Promotion and Preventive Medicine, 1996.

———. *Detailed Facts About Blood Agent Hydrogen Cyanide (AC)*. Aberdeen Proving Ground, MD: United States Army Center for Health Promotion and Preventive Medicine, 1996.

World Health Organization. *Health Aspects of Chemical and Biological Weapons*. Geneva: World Health Organization, 1970.

———. *International Chemical Safety Cards (ICSCs)*. http://www.cdc.gov/niosh/ipcs/icstart.html. 2004.

———. *Public Health Response to Biological and Chemical Weapons: WHO Guidance*. Geneva: World Health Organization, 2004.

Yaws, Carl L. *Matheson Gas Data Book*. 7th ed. Parsippany, NJ: Matheson Tri-Gas, 2001.

8

Arsine Blood Agents

8.1 General Information

These materials include arsine and reactive materials such as arsenide salts and alloys of arsenic that decompose to produce arsine. They are first generation warfare agents that were evaluated during World War I but have never been utilized on the battlefield. These materials are commercially available but are relatively difficult to disperse.

Although this class of agents is considered obsolete on the modern battlefield, arsine is still considered a significant threat as a potential improvised weapon that could be utilized in urban warfare.

8.2 Toxicology

8.2.1 Effects

Arsine affects the ability of the blood system to carry oxygen by destroying red blood cells. The lack of oxygen rapidly affects all body tissues, especially the central nervous system. Arsine may also affect the kidneys, liver, and heart. Most deaths related to arsine exposure are believed to be secondary to acute renal failure. Arsine is carcinogenic.

8.2.2 Pathways and Routes of Exposure

Arsine poses an inhalation hazard. However, reactive agents may also be hazardous through ingestion and abraded skin (e.g., breaks in the skin or penetration of skin by debris).

8.2.3 General Exposure Hazards

Arsine does not have good warning properties. Although it has a distinct odor, it is only detectable at levels higher than acceptable exposure limits. It does not irritate the eyes, respiratory system, or the skin.

The rate of detoxification of arsine by the body is very low. Exposures are essentially cumulative.

8.2.4 Latency Period

Effects from exposure to arsine can be delayed from 20 minutes to 36 hours depending on the level of exposure.

8.3 Characteristics

8.3.1 Physical Appearance/Odor

8.3.1.1 Laboratory Grade

Arsine is a colorless gas with a mild garlic-like odor. Effects from exposures are cumulative and may occur at levels below the odor threshold.

Reactive materials are solids that produce arsine on contact with moisture or acids. Binary munitions that mix these reactive materials with water or acid upon delivery have been developed.

8.3.2 Stability

Arsine has a high vapor pressure and is difficult to store as a liquefied gas. It is extremely flammable and is also decomposed by light, heat, and contact with various metals. It can explode on contact with warm, dry air.

Reactive solids are stable when dry.

8.3.3 Persistency

Arsine is nonpersistent and quickly dissipates or decomposes in the open environment. Solid agents will retain the potential to produce arsine (AsH_3) until they react with water.

8.3.4 Environmental Fate

Due to the volatile nature of arsine, there is minimal extended risk except in an enclosed or confined space. Arsine vapor has a density greater than air and tends to collect in low places. It is soluble in water as well as aromatic and halogenated organic solvents.

Solid agents pose an extended risk because they retain the potential to produce arsine gas until they react with water or acids.

8.4 Additional Hazards

8.4.1 Exposure

All foodstuffs in the area of release should be considered contaminated with residual arsenic. Unopened items packaged in glass, metal, or heavy duty plastic and exposed only to agent vapors may be used after decontamination of the container. Opened or unpackaged items, or those packaged only in paper or cardboard, should be destroyed.

Meat and milk from animals affected by arsine should be quarantined until tested. Exposed animals may not be suitable for consumption if they retain a sufficient body burden of arsenic.

Arsenic compounds can translocate from roots to other areas (e.g. leaves, fruits, etc.) of some plants. Plants, fruits, vegetables, and grains should be quarantined until tested.

8.4.2 Livestock/Pets

Animals can be decontaminated with shampoo/soap and water (see Section 8.5.3). If the animals' eyes have been exposed to the agent, they should be irrigated with water or 0.9% saline solution for a minimum of 30 minutes.

The topmost layer of unprotected feedstock (e.g., hay or grain) should be destroyed. The remaining material should be quarantined until tested. In the event of a large release, arsenic from decomposed arsine could remain on leaves of forage vegetation until removed by precipitation. Arsenic compounds can also translocate from roots to leaves of some plants and could pose an ingestion hazard to livestock.

8.4.3 Fire

Arsine is a highly flammable gas that can form explosive mixtures with air. Hydrogen gas produced by photolytic decomposition of the agents may also be present. Actions taken to extinguish the fire can also spread the agent. Arsine is soluble in water and runoff from firefighting efforts could pose a significant threat. Solid agents will react with water to form arsine gas. Decomposition of the agents during a fire will produce poisonous arsenic oxides that may be present in smoke from the fire. Arsenic will also be deposited in the area downwind of the fire.

8.4.4 Reactivity

Arsine is incompatible with strong oxidizers and various metals including aluminum, copper, brass, and nickel. It may be decompose on exposure to light to produce hydrogen gas and arsenic metal.

Solid agents will react with water to form arsine gas.

8.4.5 Hazardous Decomposition Products

8.4.5.1 Hydrolysis

Arsine produces arsenic acids and other arsenic products. Solid agents will react with water to form arsine gas.

8.4.5.2 Combustion

Volatile decomposition products include arsenic oxides. Arsine may decompose to hydrogen gas and arsenic metal if heated in a sealed container.

8.5 Protection

8.5.1 Evacuation Recommendations

Isolation and protective action distances listed below are taken from Argonne National Laboratory Report No. ANL/DIS-00-1, *Development of the Table of Initial Isolation and Protective Action Distances for the 2000 Emergency Response Guidebook*, which is still the basis for the "when used as a weapon" scenarios in the 2004 Emergency Response Guidebook (ERG). These recommendations are based on a release scenario involving sabotage of a commercial container. A small release involves 60 kilograms (a standard gas cylinder) and a large release involves 1500 kilograms (a large bulk container such as a rail car).

	Initial isolation (feet)	Downwind day (miles)	Downwind night (miles)
SA (Arsine) *C08-A001*			
Small device (60 kilograms)	200	0.5	1.5
Large device (1500 kilograms)	1300	2.5	5.0

8.5.2 Personal Protective Requirements

8.5.2.1 Structural Firefighters' Gear

Structural firefighters' protective clothing is recommended for fire situations only; it is not effective in spill situations or release events. Although arsine is primarily an inhalation hazard, decomposition products may pose a percutaneous risk. If chemical protective clothing is not available and it is necessary to rescue casualties from a contaminated area, then structural firefighters' gear will provide limited skin protection. Contact with solids and solutions should be avoided.

8.5.2.2 Respiratory Protection

Use a self-contained breathing apparatus (SCBA). Air purifying respirators (APRs) are not recommended for use in an arsine atmosphere because of poor warning properties and the unknown effectiveness of sorbents used in the filters.

If APRs are used, then immediately dangerous to life or health (IDLH) levels are the ceiling limit for respirators other than SCBAs. Any exposures approaching the IDLH level should be regarded with extreme caution and the use of SCBAs for respiratory protection should be considered.

8.5.2.3 Chemical Protective Clothing

Use only chemical protective clothing that has undergone material and construction performance testing against arsine or against the reactive materials used to produce arsine.

8.5.3 Decontamination

8.5.3.1 General

Apply universal decontamination procedures using soap and water.

If reactive materials have been released, then care must be taken to compensate for the heat generated by their reaction with water. Take appropriate actions to disperse the arsine vapors that will be generated.

8.5.3.2 Vapors

Casualties/personnel: Remove all clothing as it may continue to emit "trapped" agent vapor after contact with the vapor cloud has ceased. Shower using copious amounts of soap and water. Ensure that the hair has been washed and rinsed to remove potentially trapped vapor. If eye irritation occurs, irrigate with water or 0.9% saline solution for a minimum of 15 minutes.

Small areas: Ventilate to remove the vapors. If decomposition to arsenic metal or arsenic oxides has occurred, wash the area with copious amounts of soap and water. Collect and containerize the rinseate in containers lined with high-density polyethylene.

8.5.3.3 Reactive Solids

Casualties/personnel: Remove all clothing as it may contain residual solid. To avoid further exposure of the head, neck, and face to the agent, cut off potentially contaminated clothing that must be pulled over the head. Shower using copious amounts of soap and water. Care must be taken to avoid breathing any arsine generated during the decontamination process. Ensure that the hair has been washed and rinsed to remove potentially trapped vapor. If eye irritation occurs, irrigate with water or 0.9% saline solution for a minimum of 15 minutes.

Small areas: Collect the agent using a vacuum cleaner. Place the collected material into containers lined with high-density polyethylene. Wash the area with copious amounts of soap and water. Collect and containerize the rinseate in containers lined with high-density polyethylene. Ventilate the area to remove any agent vapors generated during the decontamination.

8.6 Medical

8.6.1 CDC Case Definition

No specific biologic marker/test is available for arsine exposure; however, exposure might be indicated by the detection of elevated arsenic levels in urine (greater than 50 μg/L for a spot or greater than 50 μg for a 24-hours urine) and signs of hemolysis (e.g., hemoglobinuria, anemia, or low haptoglobin), or arsine is detected in environmental samples. The case can be confirmed if laboratory testing is not performed because either a predominant amount of clinical and nonspecific laboratory evidence is present or an absolute certainty of the etiology of the agent is known.

8.6.2 Differential Diagnosis

The following factors have been suggested as alternatives to consider when presented with a potential case of exposure to arsine: abdominal disorders such as gallbladder inflammation (cholecystitis), gallstones (cholelithiasis), biliary colic, kidney stones, and acute renal failure; blood disorders including acute anemia, methemoglobinemia, and hyperkalemia; diseases such as hepatitis, leptospirosis, malaria, and urinary tract infections; rhabdomyolysis and hemolytic uremic syndrome; cold agglutinin disease and paroxysmal nocturnal hemoglobinuria; and also toxicity from smoke inhalation or exposure to toxic chemicals such as pyrogallic acid and stibine gas.

8.6.3 Signs and Symptoms

Unless high-dose exposure produces immediate fatality, exposure typically results in abdominal pain, blood in the urine (hematuria), and jaundice. Signs and symptoms due to massive hemolysis occur anywhere from 30 minutes to 24 hours postexposure. Other potential symptoms include headache, a vague feeling of bodily discomfort (malaise), a feeling of weariness or diminished energy (lassitude), difficulty in breathing (dyspnea), nausea, vomiting, bronze skin, jaundice, discoloration of the conjunctivae, numbness, weakness, burning pain, loss of reflexes, delirium, memory loss, irritability, confusion, and dizziness.

8.6.4 Mass-Casualty Triage Recommendations

8.6.4.1 Priority 1

A casualty with abdominal pain, jaundice, red or discolored conjunctivae, breath with an odor of garlic, headache, severe thirst, fever, chills, and/or numb or cold extremities.

8.6.4.2 Priority 2

A casualty with a known exposure to arsine or who reports smelling a garlic or fishy odor.

8.6.4.3 Priority 3

Anyone with a potential exposure to arsine.

8.6.4.4 Priority 4

Unlikely classification for this agent.

8.6.5 Casualty Management

Remove casualty to fresh air and decontaminate with soap and water. If the eyes of the casuality complains of eye irritation, then irrigate the eyes with water or 0.9% saline solution for at least 15 minutes. If reactive materials were used, irrigate open wounds with water or 0.9% saline solution for at least 10 minutes.

Once the casualty has been decontaminated, medical personnel do not need to wear a chemical-protective mask.

Provide oxygen for respiratory distress. In severe cases, exchange blood transfusions may be required. There are no specific antidotes for arsine poisoning. The value of chelating agents to remove arsine has only been demonstrated in limited studies. However, chelating agents do not prevent hemolysis and may not provide any significant benefit in acute arsine toxicity.

8.7 Fatality Management

Remove all clothing and personal effects and decontaminate with soap and water. If reactive materials have been released, then arsine may be generated. Take appropriate actions to disperse the vapors.

Arsine is highly volatile and there is little risk of direct residual contamination. However, potential persistent decomposition products include arsenic and arsenic oxides. Wash the remains with soap and water. Pay particular attention to areas where agent may get trapped, such as hair, scalp, pubic areas, fingernails, folds of skin, and wounds. If remains are heavily contaminated with residue, then wash and rinse waste should be contained for proper disposal. Once the remains have been thoroughly decontaminated, no further protective action is necessary. Body fluids removed during the embalming process do not pose any additional risks and should be contained and handled according to established procedures. Use standard burial procedures.

C08-A

AGENTS

C08-A001

Arsine (Agent SA)
CAS: 7784-42-1
RTECS: CG6475000
UN: 2188
ERG: 119

AsH_3

Colorless gas with a mild odor described as fishy or like garlic. It is detectable at 0.5 ppm. Used as a doping agent for solid-state electronic components, in the preparation of gallium arsenide and glass dyes, and in the manufacture of light-emitting diodes.

Decomposed by light, heat, and contact with various metals. Can explode on contact with warm, dry air. Ignites so easily that it cannot be disseminated by explosive shells.

This material is on the ITF-25 high threat list.

Exposure Hazards
Conversion Factor: 1 ppm = 3.19 mg/m^3 at 77°F
$LCt_{50(Inh)}$: 7500 mg-min/m^3 (1200 ppm for a 2-min exposure). This a provisional update from an older value that has not been formally adopted as of 2005.
$LC_{50(Inh)}$: 250 ppm for a 30-min exposure.
$MEG_{(1h)}$ *Min*: —; *Sig*: 0.17 ppm; *Sev*: 0.5 ppm
OSHA PEL: 0.05 ppm
ACGIH TLV: 0.05 ppm
NIOSH Ceiling: 0.0006 ppm
IDLH: 3 ppm

Properties:
MW: 77.9
D: 1.321 g/mL (liq. gas, 77°F)
MP: −179°F
BP: −81°F
Vsc: 0.0553 cS (liq. gas, 77°F)
VP: 11,100 mmHg
VD: 2.7 (calculated)
Vlt: —
H_2O: 20%
Sol: Benzene; Chloroform
FlP: —
LEL: 5.1%
UEL: 78%
RP: 0.001
IP: 9.89 eV

Final AEGLs
AEGL-1: Not Developed
AEGL-2: 1 h, 0.17 ppm 4 h, 0.040 ppm 8 h,0.020 ppm
AEGL-3: 1 h, 0.50 ppm 4 h, 0.13 ppm 8 h, 0.063 ppm

References

Agency for Toxic Substances and Disease Registry. "Arsine." In *Managing Hazardous Materials Incidents Volume III—Medical Management Guidelines for Acute Chemical Exposures*. Rev. ed. Washington, DC: Government Printing Office, 2000.

Centers for Disease Control and Prevention. "Case Definition: Arsine or Stibine Poisoning." March 4, 2005.

Compton, James A.F. *Military Chemical and Biological Agents: Chemical and Toxicological Properties.* Caldwell, NJ: The Telford Press, 1987.

Foulkes, Charles H. *"Gas!": The Story of the Special Brigade.* Edinburgh, England: William Blackwood & Sons Ltd, 1934.

National Institute for Occupational Safety and Health. *NIOSH Pocket Guide to Chemical Hazards.* Washington, DC: Government Printing Office, September 2005.

National Institutes of Health. *Hazardous Substance Data Bank (HSDB).* http://toxnet.nlm.nih.gov/cgi-bin/sis/htmlgen?HSDB/. 2005.

Olson, Kent R., ed. *Poisoning & Drug Overdose.* 4th ed. New York: Lange Medical Books/McGraw-Hill, 2004.

Sifton, David W., ed. *PDR Guide to Biological and Chemical Warfare Response.* Montvale, NJ: Thompson/Physicians Desk Reference, 2002.

Smith, Ann, Patricia Heckelman, and Maryadele J. Oneil, eds. *The Merck Index: An Encyclopedia of Chemicals, Drugs, & Biologicals.* 13th ed. Rahway, NJ: Merck & Co., Inc., 2001.

True, Bey-Lorraine, and Robert H. Dreisbach. *Dreisbach's Handbook of Poisoning: Prevention, Diagnosis and Treatment.* 13th ed. London, England: The Parthenon Publishing Group, 2002.

United States Army Headquarters. *Potential Military Chemical/Biological Agents and Compounds, Field Manual No. 3-11.9.* Washington, DC: Government Printing Office, January 10, 2005.

World Health Organization. *International Chemical Safety Cards (ICSCs).* http://www.cdc.gov/niosh/ipcs/icstart.html. 2004.

9

Carbon Monoxide Blood Agents

9.1 General Information

These materials include carbon monoxide and metal carbonyls that readily decompose to produce carbon monoxide (C09-A001). They are first generation warfare agents that were evaluated during World War I but largely abandoned because of the difficulty delivering an effective concentration of the agent. Between World War I and II, metal carbonyls were added to the fuels of flame weapons (i.e., flamethrowers) as a means of producing high levels of toxic carbon monoxide to augment their incendiary capabilities in situations involving soldiers occupying confined spaces such as caves and bunkers.

Although this class of agents is considered obsolete on the modern battlefield, they are still considered a significant threat as potential improvised weapons that could be utilized in urban warfare. They are commercially available but are relatively difficult to disperse over a large area.

9.2 Toxicology

9.2.1 Effects

Carbon monoxide affects the ability of the blood system to carry oxygen by binding to red blood cells preventing the absorption and transport of oxygen by the blood. The lack of oxygen rapidly affects all body tissues, especially the central nervous system. In addition, metal carbonyls can cause pulmonary edema and may damage the liver and kidneys. Metal carbonyls are carcinogenic or are suspected human carcinogens.

9.2.2 Pathways and Routes of Exposure

Carbon monoxide poses only an inhalation hazard. Metal carbonyls, however, are also hazardous through exposure of the skin and eye to either liquid or vapor, ingestion, and broken, abraded, or lacerated skin (e.g., penetration of skin by debris).

9.2.3 General Exposure Hazards

Carbon monoxide blood agents do not have good warning properties. Carbon monoxide is odorless and colorless. Although exposure to metal carbonyl vapors can cause eye irritation, these concentrations are generally not low enough to prevent potentially hazardous exposure.

Lethal concentrations (LC_{50}s) for inhalation of carbon monoxide blood agents have not been fully established. However, immediately dangerous to life or health levels (IDLHs) for inhalation of these agents are as low as 2 ppm.

9.2.4 Latency Period

9.2.4.1 Vapor/Aerosols (Mists)

Depending on the concentration of agent vapor, the effects begin to appear 1–2 minutes after initial exposure. Pulmonary edema caused by inhalation of metal carbonyls may be delayed for several hours.

9.2.4.2 Liquid

Metal carbonyls cause immediate irritation and burning of the skin, eyes, mucous membrane, and respiratory system.

9.3 Characteristics

9.3.1 Physical Appearance/Odor

9.3.1.1 Laboratory Grade

Carbon monoxide is a colorless and odorless gas. Metal carbonyls are typically colorless liquids that may be odorless or have a faint musty smell.

9.3.1.2 Munition Grade

Munition grade metal carbonyls are typically yellow to dark-red liquids. Production impurities and decomposition products may cause the agents to darken and give them additional odors.

9.3.1.3 Mixtures with Other Agents

Metal carbonyls have been mixed with incendiary fuels to increase the production of carbon monoxide when the fuel is ignited and thereby increase the lethality of flame weapons when used against soldiers hiding in confined spaces such as caves and bunkers.

9.3.2 Stability

Carbon monoxide is a stable gas. Metal carbonyls are relatively unstable and sensitive to light and moderately high temperatures. They may spontaneously ignite on contact with air. Volatile agents are stored in steel cylinders; otherwise, agents are stored in steel or glass containers. Metal carbonyls may be stored under an inert gas blanket, such as nitrogen, to prevent contact with the air.

9.3.3 Persistency

Carbon monoxide is nonpersistent and quickly dissipates in the open environment. Metal carbonyls are unstable and rapidly decompose to produce carbon monoxide and various metal oxides.

9.3.4 Environmental Fate

Due to the volatile nature of carbon monoxide, there is minimal extended risk except in an enclosed or confined space. Carbon monoxide vapor has a density only slightly less than air and does not tend to stratify significantly in confined spaces. It is insoluble in water but soluble in many organic solvents.

Metal carbonyls are unstable and are rapidly decomposed by air, light, and heat to produce carbon monoxide and various metal oxides. The vapors of unreacted carbonyls have a density greater than air and tend to collect in confined spaces. These materials are insoluble in water but soluble in many organic solvents.

9.4 Additional Hazards

9.4.1 Exposure

Combined intoxications of carbon monoxide and cyanide should not be treated with the nitrites found in cyanide antidote kits. These nitrites are used to create methemoglobinemia, which will exacerbate carbon monoxide poisoning by further reducing the ability of the blood to deliver oxygen to body tissue.

In addition to tobacco smokers, individuals who have had previous exposure to materials containing methylene chloride, such as degreasers, solvents, paint removers, and furniture strippers, are at greater risk because of an existing body burden of carbon monoxide. Approximately one-fourth to one-third of inhaled methylene chloride vapor is metabolized in the liver to carbon monoxide. In addition, methylene chloride is readily stored in body tissue. This stored material is released over time and results in elevated levels of carbon monoxide for extended periods, in some cases more than twice as long as compared with direct carbon monoxide inhalation.

Carbon monoxide poses no significant residual risk. If metal carbonyls have been released, all foodstuffs in the area should be considered contaminated with residual metal decomposition products. Unopened items packaged in glass, metal, or heavy duty plastic and exposed only to agent vapors may be used after decontamination of the container. Opened or unpackaged items, or those packaged only in paper or cardboard, should be destroyed.

Meat and milk from animals affected by metal carbonyls should be quarantined until tested for residual high levels of metals.

If metal carbonyls have been released, plants, fruits, vegetables, and grains should be quarantined until tested and determined to be safe to consume.

9.4.2 Livestock/Pets

Animals exposed to carbon monoxide do not require decontamination. If metal carbonyls have been released, animals can be decontaminated with shampoo/soap and water (see Section 9.5.3). If the animals' eyes have been exposed to agent, they should be irrigated with water or saline solution for a minimum of 30 minutes.

Carbon monoxide poses no significant residual risk to feedstock (e.g., hay or grain). If metal carbonyls have been released, the topmost layer of unprotected feedstock should be destroyed. The remaining material should be quarantined until tested for metal residue. It is unlikely that sufficient residue would remain on leaves of forage vegetation to pose a significant threat.

9.4.3 Fire

All carbon monoxide blood agents are flammable and can form explosive mixtures with air. Metal carbonyls are highly flammable and may even ignite spontaneously. Metal carbonyls decompose on heating to produce carbon monoxide. Although liquid agents are insoluble in water, actions taken to extinguish the fire can also spread the agent.

9.4.4 Reactivity

Metal carbonyls decompose in light to produce carbon monoxide. They are incompatible with strong oxidizers and readily form explosive mixtures with air. Some decompose at ordinary temperatures in contact with porous materials [e.g., activated carbon used in air purifying respirator (APR) filters] and produce carbon monoxide.

9.4.5 Hazardous Decomposition Products

9.4.5.1 Hydrolysis

Metal carbonyls produce metal oxides when hydrolyzed.

9.4.5.2 Combustion

Metal carbonyl volatile decomposition products include carbon monoxide and metal oxides.

9.5 Protection

9.5.1 Evacuation Recommendations

There are no published recommendations for isolation or protective action distances for carbon monoxide blood agents released in mass casualty situations.

9.5.2 Personal Protective Requirements

9.5.2.1 Structural Firefighters' Gear

Structural firefighters' protective clothing is recommended for fire situations only; it is not effective in spill situations or release events. Although carbon monoxide is primarily an inhalation hazard, metal carbonyls may pose a percutaneous risk. If chemical protective clothing is not available and it is necessary to rescue casualties from a contaminated area, then structural firefighters' gear will provide limited skin protection. Contact with liquids and solutions should be avoided.

9.5.2.2 Respiratory Protection

Use a self-contained breathing apparatus (SCBA). Air purifying respirators (APRs) are not recommended for use in an atmosphere containing carbon monoxide or metal carbonyls because of poor warning properties and the unknown effectiveness of sorbents used in the filters.

If APRs are used immediately dangerous to life or health levels (IDLHs) are the ceiling limit for respirators other than SCBAs. Any exposures approaching the IDLH level should be regarded with extreme caution and the use of SCBAs for respiratory protection should be considered.

9.5.2.3 Chemical Protective Clothing

Use only chemical protective clothing that has undergone material and construction performance testing against carbon monoxide and metal carbonyls. If the concentration of vapor exceeds the level necessary to produce effects through dermal exposure, then responders should wear a Level A protective ensemble.

In addition to the toxicological risk, high concentrations of metal carbonyls pose a significant fire/explosion hazard that may prevent safe entry even wearing the appropriate chemical protective apparel.

9.5.3 Decontamination

9.5.3.1 General

Apply universal decontamination procedures using soap and water.

9.5.3.2 Vapors

Casualties/personnel: Remove all clothing as it may continue to emit "trapped" agent vapor after contact with the vapor cloud has ceased. Shower using copious amounts of soap and water. Ensure that the hair has been washed and rinsed to remove potentially trapped vapor. If eye irritation occurs, irrigate with water or 0.9% saline solution for a minimum of 15 minutes.

Small areas: Ventilate to remove the vapors. If metal carbonyls have been released, wash the area with copious amounts of soap and water. Collect and containerize the rinseate in containers lined with high-density polyethylene.

9.5.3.3 Liquids or Solutions (Metal Carbonyls)

Casualties/personnel: Remove all clothing immediately. Even clothing that has not come into direct contact with the agent may contain "trapped" vapor. To avoid further exposure of the head, neck, and face to the agent, cut off potentially contaminated clothing that must be pulled over the head. Use a sponge or cloth with liquid soap and copious amounts of water to wash the skin surface and hair at least three times. Avoid rough scrubbing. Rinse with copious amounts of water. If water is not immediately available, the agent can be absorbed with any convenient material such as paper towels, toilet paper, flour, or talc. To minimize both spreading the agent and abrading the skin, do not rub the agent with the absorbent. Blot the contaminated skin with the absorbent. If there is a potential that the eyes have been exposed, irrigate with water or 0.9% saline solution for a minimum of 15 minutes.

Small areas: Puddles of liquid can be contained by covering with absorbent material such as vermiculite, diatomaceous earth, clay, sponges, or towels. Place the absorbed material into containers lined with high-density polyethylene. Wash the area with copious amounts of soap and water. Collect and containerize the rinseate. Ventilate the area to remove vapors.

9.6 Medical

9.6.1 CDC Case Definition

A case in which the carboxyhemoglobin concentration is higher than the normal reference range (greater than 5% for nonsmokers and 10% for smokers) in venous or arterial blood.

The case can be confirmed if laboratory testing is not performed because either a predominant amount of clinical and nonspecific laboratory evidence is present or an absolute certainty of the etiology of the agent is known.

9.6.2 Differential Diagnosis

The following factors have been suggested as alternatives to consider when presented with a potential case of exposure to carbon monoxide: diabetic ketoacidosis, hypothyroidism and myxedema coma, labyrinthitis, and lactic acidosis; toxic exposures resulting in methemoglobinemia; ingestion of alcohols or narcotics; and diseases that cause gastroenteritis, encephalitis, meningitis, and acute respiratory distress syndrome.

9.6.3 Signs and Symptoms

9.6.3.1 Vapors/Aerosols

Headache, tachypnea, dizziness, confusion, and chest pain. The casualty may also experience palpitations, dyspnea on exertion, drowsiness, lethargy, hallucination, agitation, nausea, vomiting, diarrhea, and coma. If metal carbonyls have been released, there may be complaints of irritation of the eyes, mucous membrane, and respiratory system. Inflammation of lung tissue (pneumonitis) caused by metal carbonyls can may be delayed 12–36 hours. They may also cause injury to the liver, kidneys, and lungs as well as degenerative changes in the central nervous system.

9.6.3.2 Liquids

Irritation and burning of the skin, eyes, mucous membrane, and respiratory system.

9.6.4 Mass-Casualty Triage Recommendations

9.6.4.1 Priority 1

A casualty with a heartbeat but is unconscious, not breathing, and has low blood pressure.

9.6.4.2 Priority 2

A casualty with known exposure to carbon monoxide blood agents, who was initially unconscious but has regained consciousness; or a casualty who shows neurological abnormalities such as dizziness, confusion, or hallucinations, has cardiac arrhythmias, bronchospasm; or complains of severe headache, difficulty in breathing or chest pain. If available, breath measurement indicates that the blood carbon monoxide level exceeds 20%.

9.6.4.3 Priority 3

A casualty with known or potential exposure to carbon monoxide blood agents but who shows no signs of neurological or cardiac abnormalities, and does not complain of discomfort (e.g., headache, difficulty breathing, etc.). If available, breath measurement indicates that the blood carbon monoxide level is less than 10%. Anyone with potential exposure to metal carbonyls should be transported to a medical facility for evaluation because of the risk of latent chemical pneumonitis from inhalation of these agents.

9.6.4.4 Priority 4

A casualty with known exposure to carbon monoxide blood agents and in cardiac arrest.

9.6.5 Casualty Management

Remove casualty to fresh air and decontaminate with soap and water. Provide oxygen for respiratory distress. If the eyes of the casualty complains of eye irritation, irrigate with water or 0.9% saline solution for at least 15 minutes. If metal carbonyls were used, irrigate open wounds with water or 0.9% saline solution for at least 10 minutes.

Once the casualty has been decontaminated, medical personnel do not need to wear a chemical-protective mask.

Continue oxygen therapy until patient is asymptomatic and blood carbon monoxide levels are below 10%. For individuals with blood carbon monoxide levels above 40%, consider transfer to a hyperbaric facility.

If metal carbonyls were used, asymptomatic individuals should be monitored for possible complications caused by pulmonary edema.

9.7 Fatality Management

Remove all clothing and personal effects and decontaminate with soap and water.

Carbon monoxide blood agents are highly volatile and there is little risk of direct residual contamination. Wash the remains with soap and water. Pay particular attention to areas where agent may get trapped, such as hair, scalp, pubic areas, fingernails, folds of skin, and wounds. Once the remains have been thoroughly decontaminated, no further protective action is necessary. Body fluids removed during the embalming process do not pose any additional risks and should be contained and handled according to established procedures. Use standard burial procedures.

C09-A

AGENTS

C09-A001

Carbon monoxide

CAS: 630-08-0
RTECS: FG3500000
UN: 1016
UN: 9202 (Cryogenic)
ERG: 119

CO

Colorless odorless gas. Used as a reducing agent in metallurgical operations and to manufacture metal carbonyls.

This material is on the ITF-25 medium threat list.

Exposure Hazards

Conversion Factor: 1 ppm = 1.15 mg/m^3 at 77°F

$LC_{50(Inh)}$: 6400 ppm. Headache and dizziness in 1–2 min followed by unconsciousness. Death in 10–15 min with continued exposure.

$LC_{50(Inh)}$: 12,800 ppm. Immediate unconsciousness. Death in 1–3 min with continued exposure.

$MEG_{(1h)}$ Min: —; *Sig*: 83 ppm; *Sev*: 330 ppm
OSHA PEL: 50 ppm
ACGIH TLV: 25 ppm
NIOSH Ceiling: 200 ppm
IDLH: 1200 ppm

Properties:

MW: 28.0	*VP*: >26,600 mmHg	*FlP*: –312°F
D: 0.624 g/mL (liq. gas, –256°F)	*VD*: 0.97 (calculated)	*LEL*: 12.5%
MP: –337°F	*Vlt*: —	*UEL*: 74%
BP: –313°F	*H_2O*: 2%	*RP*: —
Vsc: 0.097 cS (liq. gas, –238°F)	*Sol*: Alcohol; Benzene; Chloroform	*IP*: 14.01 eV

Interim AEGLs

AEGL-1: Not Developed		
AEGL-2: 1 h, 83 ppm	4 h, 33 ppm	8 h, 27 ppm
AEGL-3: 1 h, 330 ppm	4 h, 150 ppm	8 h, 130 ppm

C09-A002

Iron pentacarbonyl
CAS: 13463-40-6
RTECS: NO4900000
UN: 1994
ERG: 131

$Fe(CO)_5$

Colorless to yellow to dark-red, odorless oily liquid. Used to manufacture finely divided iron for high frequency radio and television coils.

May spontaneously ignite on contact with air (>122°F). Decomposes on heating or in light producing Carbon monoxide (C09-A001).

This material is on the ITF-25 low threat list.

Exposure Hazards
Conversion Factor: 1 ppm = 8.01 mg/m^3 at 77°F
$MEG_{(1h)}$ Min: —; *Sig*: 0.060 ppm; *Sev*: 0.18 ppm
ACGIH TLV: 0.1 ppm
ACGIH STEL: 0.2 ppm
NIOSH STEL: 0.2 ppm

Properties:

MW: 195.9	*VP*: 40 mmHg (87°F)	*FlP*: 5°F
D: 1.490 g/mL	*VD*: 6.8 (calculated)	*LEL*: 3.7%
MP: –6°F	*Vlt*: 52,000 ppm	*UEL*: 12.5%
BP: 217°F (749 mmHg)	*H_2O*: Insoluble	*RP*: 0.2
Vsc: —	*Sol*: Most organic solvents	*IP*: —

Interim AEGLs

AEGL-1: Not Developed		
AEGL-2: 1 h, 0.060 ppm	4 h, 0.037 ppm	8 h, 0.025 ppm
AEGL-3: 1 h, 0.18 ppm	4 h, 0.11 ppm	8 h, 0.075 ppm

C09-A003

Nickel carbonyl
CAS: 13463-39-3
RTECS: QR6300000
UN: 1259
ERG: 131

$Ni(CO)_4$

Colorless to yellow liquid with a musty or sooty odor. Intermediate in nickel refining. Used to produce high-purity nickel powder and to coat metals with nickel.

May spontaneously ignite or explode on contact with air (>140° F). Decomposes on heating or on contact with acids to produce Carbon monoxide (C09-A001).

Exposure Hazards
Conversion Factor: 1 ppm = 6.98 mg/m^3 at 77°F
$MEG_{(1h)}$ *Min*: —; *Sig*: 0.03 ppm; *Sev*: 0.16 ppm
OSHA PEL: 0.001 ppm
ACGIH TLV: 0.05 ppm
IDLH: 2 ppm

Properties:

MW: 170.7	*VP*: 315 mmHg	*FlP*: –11°F
D: 1.306 g/mL (77°F)	*VD*: 5.9 (calculated)	*LEL*: 2%
MP: −13°F	*Vlt*: 420,000 ppm	*UEL*: —
BP: 110°F	H_2O: 0.05%	*RP*: 0.02
Vsc: 0.378 cS (77°F)	*Sol*: Alcohol; Benzene; Acetone	*IP*: 8.28 eV

Interim AEGLs

AEGL-1: Not Developed		
AEGL-2: 1 h, 0.036 ppm	4 h, 0.0090 ppm	8 h, 0.0045 ppm
AEGL-3: 1 h, 0.16 ppm	4 h, 0.040 ppm	8 h, 0.020 ppm

References

Centers for Disease Control and Prevention. "Case Definition: Carbon Monoxide." March 25, 2005.

Fries, Amos A., and Clarence J. West. *Chemical Warfare*. New York: McGraw-Hill Book Company, Inc., 1921.

Lumsden, Malvern. *Incendiary Weapons*. Cambridge, MA: The MIT Press, 1975.

National Institute for Occupational Safety and Health. *NIOSH Pocket Guide to Chemical Hazards*. Washington, DC: Government Printing Office, September 2005.

National Institutes of Health. *Hazardous Substance Data Bank (HSDB)*. http://toxnet.nlm.nih.gov/cgi-bin/sis/htmlgen?HSDB/. 2005.

Olson, Kent R., ed. *Poisoning & Drug Overdose*. 4th ed. New York: Lange Medical Books/McGraw-Hill, 2004.

Sartori, Mario F. *The War Gases: Chemistry and Analysis*. Translated by L.W. Marrison. London: J. & A. Churchill, Ltd, 1939.

Smith, Ann, Patricia Heckelman, and Maryadele J. Oneil, eds. *The Merck Index: An Encyclopedia of Chemicals, Drugs, & Biologicals*. 13th ed. Rahway, NJ: Merck & Co., Inc., 2001.

True, Bey-Lorraine, and Robert H. Dreisbach. *Dreisbach's Handbook of Poisoning: Prevention, Diagnosis and Treatment.* 13th ed. London, England: The Parthenon Publishing Group, 2002.

United States Coast Guard. *Chemical Hazards Response Information System (CHRIS) Manual,* 1999 edition. http://www.chrismanual.com/Default.htm. 2004.

Wachtel, Curt. *Chemical Warfare.* Brooklyn, NY: Chemical Publishing Co., Inc., 1941.

Waitt, Alden H. *Gas Warfare: The Chemical Weapon, Its Use, and Protection against It.* Rev. ed. New York: Duell, Sloan and Pearce, 1944.

World Health Organization. *International Chemical Safety Cards (ICSCs).* http://www.cdc.gov/niosh/ipcs/icstart.html. 2004.

Yaws, Carl L. *Matheson Gas Data Book.* 7th ed. Parsippany, NJ: Matheson Tri-Gas, 2001.

10

Pulmonary Agents

10.1 General Information

The agents in this class cover a wide variety of chemical structures including halogens, acyl halides, various alkylating agents, as well as metallic and nonmetallic oxides. The majority of these agents are first generation warfare agents that were evaluated and used during World War I. Research into more effective pulmonary agents continued through World War II. Various metals (e.g., cadmium, selenium) have added to the fuels of flame weapons (i.e., flamethrowers) as a means of producing high levels of toxic metal fume to augment their incendiary capabilities in situations involving soldiers occupying confined spaces such as caves and bunkers. Perfluoroisobutylene (C10-A008) is listed in Schedule 2 of the Chemical Weapons Convention (CWC). Chloropicrin (C10-A006) and phosgene (C10-A003) are listed in Schedule 3. Pulmonary agents are relatively easy to acquire and disperse.

Pulmonary agents have been stockpiled by most countries that have pursued a chemical weapons program, and have been used a number of times on the battlefield. Although this class of agents is considered obsolete on the modern battlefield, several of these agents are still considered a significant threat as potential improvised weapons that could be utilized in urban warfare.

10.2 Toxicology

10.2.1 Effects

Pulmonary agents are primarily respiratory irritants. In extreme cases, membranes swell, lungs become filled with liquid (pulmonary edema), and death results from lack of oxygen. Some agents can also pass through the skin to induce systemic intoxication. In high concentrations, chlorine (C10-A001), bromine (C10-A002), and chloropicrin (C10-A006) pose a dermal hazard and can produce local blistering and corrosion. Several agents in this class are carcinogenic.

10.2.2 Pathways and Routes of Exposure

Pulmonary agents are primarily an inhalation hazard. However, at high concentration, agents and decomposition products may exhibit some corrosive properties on the skin.

Exposure of the eyes and skin to halogens can produce local blistering and corrosion. Some agents also pass through the skin to induce systemic intoxication.

10.2.3 General Exposure Hazards

As a class, pulmonary agents do not have good warning properties. However, halogens and some alkylating agents will produce eye and skin irritation at low levels.

Lethal concentrations (LC_{50}s) for inhalation of these agents are as low as 93 ppm for a 2-minutes exposure.

10.2.4 Latency Period

10.2.4.1 Vapor/Aerosols

Pulmonary effects are usually delayed from 2 to 24 hours. Exposure to very high concentrations may produce immediate symptoms. Generally, the more rapid the onset of symptoms, the more grave the prognosis.

10.3 Characteristics

10.3.1 Physical Appearance/Odor

10.3.1.1 Laboratory Grade

Most pulmonary agents are either volatile liquids or gases. These agents are typically colorless. Odors, if present, vary from mildly pleasant to harsh and irritating. Some agents, especially in high concentration, may cause eye irritation and tearing.

Metal fume agents are odorless solids dispersed as aerosols from incendiary devices. Depending on various factors, the aerosol may or may not be visible.

10.3.1.2 Modified Agents

Pulmonary agents have been absorbed into porous powders (e.g., pumice) and disseminated as dust clouds. The agents are slowly released by the dust particles thereby greatly increasing the persistency of the agents.

Some pulmonary agents are stored and shipped as concentrated solutions to facilitate handling and stabilize the agents. Odors will vary depending on the characteristics of the solvent(s) used and concentration of pulmonary agent in the solution.

10.3.1.3 Mixtures with Other Agents

During World War I, pulmonary agents were sometimes mixed with various metal chlorides to produce a visible cloud. Agents were also sometimes mixed with arsenical vesicants (Agent Index C04) to increase their lethality.

10.3.2 Stability

Pulmonary agents are stable when stored dry. If moisture is present, many agents will decompose to acid products that will corrode iron, steel, and other metals. Many commercially available gases are shipped in a compressed, liquefied form. Some of these agents, such as phosgene (C10-A003), can be found commercially as highly concentrated or saturated solutions of the gases in organic solvents.

10.3.3 Persistency

For military purposes, pulmonary agents are classified as nonpersistent. Most pulmonary agents are either volatile liquids or gases. Aerosols of nonvolatile agents are not persistent. Cold weather may decrease the rate of volatilization of liquid and solid agents and increase their persistency. Agents absorbed into porous powders may be significantly more persistent than normal. These powders can be reaerosolized after deployment by ground traffic or strong winds. Decomposition products from the breakdown of some agents can pose a persistent hazard.

10.3.4 Environmental Fate

Due to the volatile nature of most pulmonary agents, there is minimal extended risk except in an enclosed or confined space. Vapors have a density greater than air and tend to collect in low places. Solids that are dispersed as aerosols have little or no vapor pressure. Once the aerosols settle, there is minimal extended hazard from the agents unless the dust is resuspended.

Pulmonary agents have limited solubility in water and many decompose rapidly in contact with moisture (e.g., high humidity) or with water.

10.4 Additional Hazards

10.4.1 Exposure

If low volatility agent aerosols or metal fumes have been released, all foodstuffs in the area should be considered contaminated. Unopened items packaged in glass, metal, or heavy duty plastic may be used after decontamination of the container. Opened or unpackaged items, or those packaged only in paper or cardboard, should be destroyed.

Meat and milk from animals affected by metal fumes should be quarantined until tested for residual high levels of metals. Plants, fruits, vegetables, and grains should be quarantined until tested and determined to be safe to consume.

10.4.2 Livestock/Pets

Animals exposed to volatile pulmonary agents do not require decontamination. If low volatility agent aerosols have been released, animals can be decontaminated with shampoo/soap and water (see Section 10.5.3). If the animals' eyes have been exposed to the agent, they should be irrigated with water or saline solution for a minimum of 30 minutes.

If low volatility agent aerosols or metal fumes have been released, the topmost layer of unprotected feedstock (e.g., hay or grain) should be destroyed. The remaining material should be quarantined until tested for residue. It is unlikely that sufficient residue would remain on leaves of forage vegetation to pose a significant threat.

10.4.3 Fire

Most pulmonary agents are either nonflammable or difficult to ignite. Some are strong oxidizers and will support combustion. Agents may be decomposed by heat to produce other toxic and/or corrosive gases. They may react with steam or water during a fire to produce toxic or corrosive vapors or both. Although pulmonary agents have limited solubility in water, actions taken to extinguish the fire can spread the agent.

10.4.4 Reactivity

Many of these agents are incompatible with acids, bases, reducing agents, and other flammable materials. Some agents, such as chloropicrin (C10-A006) and chlorine trifluoride (C10-A015), are incompatible with oxidizers. Most pulmonary agents react with water to form acidic products.

10.4.5 Hazardous Decomposition Products

10.4.5.1 Hydrolysis

Most pulmonary agents produce corrosive decomposition products that may include hydrogen chloride (HCl), hydrogen bromide (HBr), hydrogen fluoride (HF), and/or hydrogen cyanide (HCN). Agents with metal halide additives will also form potentially toxic metallic oxides.

10.4.5.2 Combustion

Volatile decomposition products may include HCl, HBr, HF, and/or HCN. Agents containing metals will produce metallic oxides.

10.5 Protection

10.5.1 Evacuation Recommendations

Isolation and protective action distances listed below are taken from Argonne National Laboratory Report No. ANL/DIS-00-1, *Development of the Table of Initial Isolation and Protective Action Distances for the 2000 Emergency Response Guidebook*, which is still the basis for the "when used as a weapon" scenarios in the 2004 Emergency Response Guidebook (ERG). For pulmonary agents, these recommendations are based on two different release mechanisms (direct aerosolization of liquid agents or, for gaseous agents, sabotage of commercial containers) as well as various quantities of material depending on the agent used in the release.

For phosgene, the release scenario involves sabotage of a commercial container with either 60 kilograms (approximately 12 gallons) of liquefied agent or 1500 kilograms (approximately 290 gallons) of liquefied agent.

For diphosgene, the release involves direct aerosolization of the liquid agent with a particle size between 2 and 5 μm. In this scenario, the difference between a small and a large release is not based on the standard 200 liters spill used for commercial hazardous materials listed in the ERG. A small release involves 30 kilograms (approximately 5 gallons) of liquid agent and a large release involves 500 kilograms (approximately 80 gallons) of liquid agent.

	Initial isolation (feet)	**Downwind day (miles)**	**Downwind night (miles)**
CG (Phosgene) C10-A003			
Small device (60 kilograms)	500	0.8	2.0
Large device (1500 kilograms)	2500	4.5	7+
DP (Diphosgene) C10-A004			
Small device (30 kilograms)	200	0.2	0.6
Large device (500 kilograms)	600	1.0	2.8

10.5.2 Personal Protective Requirements

10.5.2.1 Structural Firefighters' Gear

Pulmonary agents are primarily a respiratory hazard; however, solids, liquids, or high vapor concentrations of some agents may pose a eye/skin hazard. Structural firefighters' protective clothing is recommended for fire situations only; it is not effective in spill situations or release events. If chemical protective clothing is not available and it is necessary to rescue casualties from a contaminated area, then structural firefighters' gear will provide limited skin protection against agent vapors. Contact with liquids, solids, and solutions should be avoided.

10.5.2.2 Respiratory Protection

Self-contained breathing apparatuses (SCBAs) or air purifying respirators (APRs) should have a National Institute of Occupational Safety and Health (NIOSH) and Chemical/Biological/Radiological/Nuclear (CBRN) certification. However, during emergency operations, other NIOSH approved SCBAs or APRs that have been specifically tested by the manufacturer against chemical warfare agents may be used if deemed necessary by the Incident Commander. APRs should be equipped with a NIOSH approved CBRN filter or a combination organic vapor/acid gas/particulate cartridge.

APRs are not recommended for use in an atmosphere containing perfluoroisobutylene (C10-A008) because of poor warning properties of the agent and a short life span of respirator cartridges.

Immediately dangerous to life or health (IDLH) levels are the ceiling limit for respirators other than SCBAs. Any exposures approaching the IDLH level should be regarded with extreme caution and the use of SCBAs for respiratory protection should be considered.

10.5.2.3 Chemical Protective Clothing

Use only chemical protective clothing that has undergone material and construction performance testing against pulmonary agents, particularly those that pose a dermal hazard such as halogens. If the concentration of vapor exceeds the level necessary to produce effects through dermal exposure, then responders should wear a Level A protective ensemble. Since chemical protective clothing is tested against laboratory grade agents, reported permeation rates may be affected by solvents if solutions of agents have been released.

10.5.3 Decontamination

10.5.3.1 General

Apply universal decontamination procedures using soap and water.

10.5.3.2 Vapors

Casualties/personnel: Remove all clothing as it may continue to emit "trapped" agent vapor after contact with the vapor cloud has ceased. Shower using copious amounts of soap and water. Ensure that the hair has been washed and rinsed to remove potentially trapped vapor. If eye irritation occurs, irrigate with water or 0.9% saline solution for a minimum of 15 minutes.

Small areas: Ventilate to remove the vapors.

10.5.3.3 Liquids, Solutions, or Liquid Aerosols

Casualties/personnel: Remove all clothing immediately. Even clothing that has not come into direct contact with the agent may contain "trapped" vapor. To avoid further exposure of the head, neck, and face to the agent, cut off potentially contaminated clothing that must be pulled over the head. Use a sponge or cloth with liquid soap and copious amounts of water to wash the skin surface and hair at least three times. Avoid rough scrubbing. Rinse with copious amounts of water. If water is not immediately available, the agent can be absorbed with any convenient material such as paper towels, toilet paper, flour, or talc. To minimize both spreading the agent and abrading the skin, do not rub the agent with the absorbent. Blot the contaminated skin with the absorbent. If there is a potential that the eyes have been exposed, irrigate with water or 0.9% saline solution for a minimum of 15 minutes.

Small areas: Puddles of liquid can be contained by covering with absorbent material such as vermiculite, diatomaceous earth, clay, sponges, or towels. Place the absorbed material into containers lined with high-density polyethylene. Wash the area with copious amounts of soap and water. Collect and containerize the rinseate. Ventilate the area to remove vapors.

10.5.3.4 Solids or Particulate Aerosols

Casualties/personnel: Do not attempt to brush the agent off the individual or their clothing as this can aerosolize the agent. Remove all clothing immediately. To avoid further exposure of the head, neck, and face to the agent, cut off potentially contaminated clothing that must be pulled over the head. Wash the skin surface and hair at least three times with copious amounts of soap and water. Do not delay decontamination to find warm or hot water if it is not readily available. Rinse with copious amounts of water. If there is a potential that the eyes have been exposed, irrigate with water or 0.9% saline solution for a minimum of 15 minutes.

Small areas: Collect the agent by using a vacuum cleaner. Place the collected material into containers lined with high-density polyethylene. Wash the area with copious amounts of soap and water. Collect and containerize the rinseate in containers lined with high-density polyethylene. Ventilate the area to remove any agent vapors generated during the decontamination.

10.6 Medical

10.6.1 CDC Case Definition

No specific biologic marker/test is available for pulmonary agents as a class; however, exposure to bromine might be indicated by detection of elevated bromide levels in serum (reference level is 50–100 mg/L), or if chlorine or bromine is released and they are detected in environmental samples. The case can be confirmed if laboratory testing is not performed because either a predominant amount of clinical and nonspecific laboratory evidence is present or an absolute certainty of the etiology of the agent is known.

10.6.2 Differential Diagnosis

The following factors have been suggested as alternatives to consider when presented with a potential case of exposure to pulmonary agents: bronchitis; asthma; bacterial or viral pneumonia; congestive heart failure with pulmonary edema; acute coronary syndrome; acute respiratory distress syndrome; pulmonary embolism; smoke inhalation; bronchial/pulmonary burns; toxic exposures resulting in methemoglobinemia; salicylate overdose; exposure to carbamate or organophosphate pesticides; inhalation of corrosive gases such as ammonia, hydrogen chloride (HCl), and hydrogen fluoride (HF) or corrosive oxides such as those produced by fires involving sulfur or phosphorus; and inhalation of hydrocarbon insecticides, carbon monoxide, or hydrogen sulfide.

10.6.3 Signs and Symptoms

10.6.3.1 Vapors

Exposure to low concentrations of pulmonary agents may not produce immediate effects. However, the severity of poisoning is not related to the presentation or magnitude of immediate symptoms. Symptoms may include headache, nausea, vomiting, eye and airway irritation, tearing, shortness of breath, coughing, wheezing, chest tightness, and delayed pulmonary edema. If halogens, chlorine trifluoride, or chloropicrin have been released, there may be redness of the skin or chemical burns.

10.6.3.2 Liquids

Irritation and burning of the skin, eyes, mucous membrane, and respiratory system. Direct contact may result in chemical burns.

10.6.4 Mass-Casualty Triage Recommendations

10.6.4.1 Priority 1

A casualty with signs of pulmonary edema within 4–6 hours postexposure but no further complications.

10.6.4.2 Priority 2

A casualty with onset of symptoms more than 4 hours post exposure, or with difficulty in breathing without objective signs. Observe closely and retriage hourly. After 12 hours, retriage every 2 hours.

10.6.4.3 Priority 3

An asymptomatic casualty with known or potential exposure to pulmonary agents. Observe closely and retriage every 2 hours for at least 6 hours before discharge.

10.6.4.4 Priority 4

A casualty with onset of symptoms (pulmonary edema, cyanosis, and hypotension; or persistent hypotension despite intensive medical care) less than 4 hours postexposure. This triage classification is resource dependent.

10.6.5 Casualty Management

Remove casualty to fresh air and decontaminate with soap and water. Provide oxygen for respiratory distress. If the eyes of the casualty complains of eye irritation, irrigate the eyes with water or 0.9% saline solution for at least 15 minutes. If solid or liquid agents were used, irrigate open wounds with water or 0.9% saline solution for at least 10 minutes.

Once the casualty has been decontaminated, medical personnel do not need to wear a chemical-protective mask.

There is no antidote for exposure to these agents. Enforce rest as even minimal physical exertion may shorten the clinical latent period. Asymptomatic individuals suspected of exposure to pulmonary agents should be monitored for possible complications caused by pulmonary edema for at least 6 h. Chelation therapy may be appropriate to minimize systemic toxicity if metal fumes have been released.

10.7 Fatality Management

Remove all clothing and personal effects and decontaminate with soap and water.

Pulmonary agents pose little risk of direct residual contamination. Wash the remains with soap and water. Pay particular attention to areas where agent may get trapped, such as hair, scalp, pubic areas, fingernails, folds of skin, and wounds. Once the remains have been thoroughly decontaminated, no further protective action is necessary. Body fluids removed during the embalming process do not pose any additional risks and should be contained and handled according to established procedures. Use standard burial procedures.

C10-A

AGENTS

C10-A001

Chlorine (Agent CL)
CAS: 7782-50-5
RTECS: FO2100000
UN: 1017
ERG: 124

Cl_2

Yellow-green gas with a pungent, irritating odor detectable at 0.01 ppm. Cloud is invisible below a concentration of approximately 1% .

Also reported as a mixture with Sulfur chloride (C03-C033); Phosgene (C10-A003); Chloropicrin (C10-A006).

This material is on the ITF-25 high threat list.

Exposure Hazards
Conversion Factor: 1 ppm = 2.90 mg/m^3 at 77° F
$LCt_{50(Inh)}$: 19,000 mg-min/m^3 (3200 ppm for a 2-min exposure)
$ICt_{50(Inh)}$: 1800 mg-min/m^3 (310 ppm for a 2-min exposure)
$MEG_{(1h)}$ *Min*: 0.50 ppm; *Sig*: 2.0 ppm; *Sev*: 20 ppm
ACGIH TLV: 0.5 ppm
ACGIH STEL: 1 ppm
OSHA Ceiling: 1 ppm
IDLH: 10 ppm

Properties:
MW: 70.9
D: 1.41 g/mL (liq. gas, 68° F)
D: 1.398 g/mL (liq. gas, 77° F)
MP: −150° F
BP: −29° F
Vsc: 0.245 cS (liq. gas, 77° F)
VP: 4,992 mmHg
VD: 2.5 (calculated)
Vlt: —
H_2O: 0.7%
Sol: Alcohols
FlP: None
LEL: None
UEL: None
RP: 0.002
IP: 11.48 eV

Final AEGLs

AEGL-1: 1 h, 0.50 ppm	4 h, 0.50 ppm	8 h, 0.50 ppm
AEGL-2: 1 h, 2.0 ppm	4 h, 1.0 ppm	8 h, 0.7 ppm
AEGL-3: 1 h, 20 ppm	4 h, 10 ppm	8 h, 7.1 ppm

C10-A002

Bromine
CAS: 7726-95-6
RTECS: EF9100000
UN: 1744
ERG: 154

Br_2

Dark reddish-brown, fuming liquid with a suffocating, irritating odor detectable at 0.05–3.5 ppm.

This material is on the ITF-25 low threat list.

Exposure Hazards
Conversion Factor: 1 ppm = 6.54 mg/m^3 at 77° F
$LC_{50(Inh)}$: 200 mg/m^3 (31 ppm) is fatal "in a very short time"
Eye Irritation: 0.32 ppm; exposure duration unavailable
Respiratory Irritation: 1.7–3.5 ppm produce severe choking
$MEG_{(1h)}$ *Min*: 0.024 ppm; *Sig*: 0.24 ppm; *Sev*: 8.5 ppm
OSHA PEL: 0.1 ppm
ACGIH TLV: 0.1 ppm
ACGIH STEL: 0.2 ppm
IDLH: 3 ppm

Properties:

MW: 159.8	*VP*: 172 mmHg	*FlP*: None
D: 3.12 g/mL	*VD*: 5.5 (calculated)	*LEL*: None
MP: 19° F	*Vlt*: 230,000 ppm	*UEL*: None
BP: 139° F	H_2O: 3.4% (77° F)	*RP*: 0.05
Vsc: 0.38 cS (68° F)	*Sol*: Common organic solvents	*IP*: 10.55 eV

Interim AEGLs

AEGL-1: 1 h, 0.033 ppm	4 h, 0.033 ppm	8 h, 0.033 ppm
AEGL-2: 1 h, 0.24 ppm	4 h, 0.13 ppm	8 h, 0.095 ppm
AEGL-3: 1 h, 8.5 ppm	4 h, 4.5 ppm	8 h, 3.2 ppm

C10-A003

Phosgene (Agent CG)
CAS: 75-44-5
RTECS: SY5600000
UN: 1076
ERG: 125

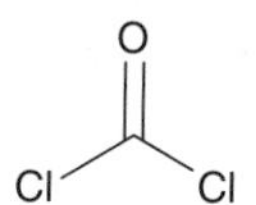

CCl_2O

Colorless gas with an odor like new mown grass or green corn detectable at 0.4–1.5 ppm. At high concentration, the odor may be strong, stifling, and unpleasant. Eyes, nose, and throat become irritated at 3–4 ppm.

Also reported as a mixture with Chlorine (C10-A001).

This material is on the ITF-25 high threat and Schedule 3 of the CWC.

Exposure Hazards

Conversion Factor: 1 ppm = 4.04 mg/m^3 at 77° F
$LCt_{50(Inh)}$: 1500 mg-min/m^3 (190 ppm for a 2-min exposure).
Exposures to 2 ppm for 80 min will not cause irritation but will produce pulmonary edema in 12–16 h.
Eye Irritation: 0.1 ppm; exposure duration unavailable
$MEG_{(1h)}$ *Min*: 0.1 ppm; *Sig*: 0.3 ppm; *Sev*: 0.75 ppm
OSHA PEL: 0.1 ppm
ACGIH TLV: 0.1 ppm
OSHA Ceiling: 0.2 ppm
IDLH: 2.0 ppm

Properties:

MW: 98.9	*VP*: 1173 mmHg	*FlP*: None
D: 1.363 g/mL (liq. gas, 77° F)	*VD*: 3.4 (calculated)	*LEL*: None
MP: −198° F	*Vlt*: —	*UEL*: None
BP: 46° F	H_2O: Very slight (decomposes)	*RP*: 0.01
Vsc: 0.284 cS (liq. gas, 77° F)	*Sol*: Most organic solvents	*IP*: 11.55 eV

Status AEGLs

AEGL-1: Not Developed		
AEGL-2: 1 h, 0.30 ppm	4 h, 0.080 ppm	8 h, 0.040 ppm
AEGL-3: 1 h, 0.75 ppm	4 h, 0.20 ppm	8 h, 0.090 ppm

C10-A004

Diphosgene (Agent DP)
CAS: 503-38-8
RTECS: LQ7350000
UN: 1076
ERG: 125

$C_2Cl_4O_2$

Colorless liquid with an odor like new mown grass or green corn detectable at 0.5 ppm.

Decomposes in contact with porous materials, including activated charcoal, or when heated above 572° F to produce two molecules of Phosgene (C10-A003). Moisture may accelerate decomposition to phosgene at normal temperatures.

Exposure Hazards

Conversion Factor: 1 ppm = 8.09 mg/m^3 at 77° F
$LCt_{50(Inh)}$: 1500 mg-min/m^3 (93 ppm for a 2-min exposure)

This is a provisional update from an older value that has not been formally adopted as of 2005. See Phosgene (C10-A003) for additional exposure hazards.

Properties:

MW: 197.8	*VP*: 4.2 mmHg	*FlP*: None
D: 1.653 g/mL	*VD*: 6.8 (calculated)	*LEL*: None
MP: −71° F	*Vlt*: 1500 ppm	*UEL*: None
BP: 261° F	*H_2O*: Insoluble	*RP*: 1.7
Vsc: —	*Sol*: Most organic solvents	*IP*: —

C10-A005

Triphosgene
CAS: 32315-10-9
RTECS: —

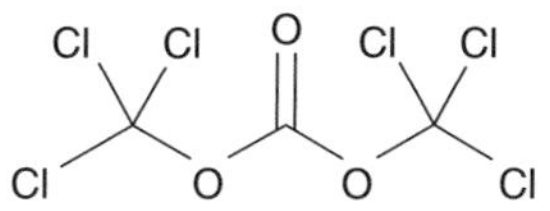

$C_2Cl_4O_2$

White to off-white crystalline material that is odorless.

Pure triphosgene decomposes when heated above 266° F to produce three molecules of Phosgene (C10-A003). Impure material decomposes at lower temperatures. Moisture may accelerate decomposition to phosgene at normal temperatures.

Exposure Hazards

Conversion Factor: 1 ppm = 12.14 mg/m^3 at 77° F

Human toxicity values have not been established or have not been published. See Phosgene (C10-A003) for additional exposure hazards.

Properties:

MW: 296.7	*VP*: —	*FlP*: —
D: —	*VD*: —	*LEL*: —
MP: 174° F	*Vlt*: —	*UEL*: —
BP: 397° F	*H_2O*: Insoluble (decomposes slowly)	*RP*: —
Vsc: —	*Sol*: —	*IP*: —

C10-A006

Chloropicrin (Agent PS)
CAS: 76-06-2
RTECS: PB6300000
UN: 1580
ERG: 154

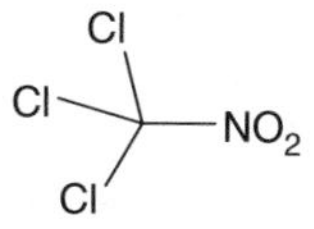

CCl_3NO_2

Colorless to faint yellow oily liquid with a stinging, pungent, intensely irritating odor that is detectable at 1.1 ppm. Concentrations between 0.3 and 1.35 ppm cause painful eye irritation in less than 30 s. Contact with the liquid produces skin lesions.

Decomposes on exposure to light. Large volumes of this chemical may be shock sensitive. When heated, this compound may decompose violently.

Also reported as a mixture with Chlorine (C10-A001); Chloroacetophenone (C13-A008); Diphosgene (C10-A004); Ethyl iodoacetate (C13-A012); Hydrogen sulfide (C07-A006); Phosgene (C10-A003); Stannic chloride; Sulfuryl chloride (C11-A088); and commercially with Methyl bromide (C11-A042).

This material is on Schedule 3 of the CWC.

Exposure Hazards

Conversion Factor: 1 ppm = 6.72 mg/m^3 at 77° F
$LCt_{50(Inh)}$: 2000 mg-min/m^3 (149 ppm for a 2-min exposure)
Eye Irritation: 0.3 ppm for a 30 s exposure
Lacrimation: 1–3 ppm; exposure duration unavailable
"Intolerable" Irritation: 15 ppm for a 1-min exposure
OSHA PEL: 0.1 ppm
ACGIH TLV: 0.1 ppm
IDLH: 2 ppm

Properties:

MW: 164.4	*VP*: 18.3 mmHg	*FlP*: None
D: 1.66 g/mL	*VD*: 5.6 (calculated)	*LEL*: None
MP: −93° F	*Vlt*: 25,000 ppm	*UEL*: None
BP: 234° F	H_2O: 0.19%	*RP*: 0.42
Vsc: —	*Sol*: Most organic solvents	*IP*: —

C10-A007

Bromopicrin
CAS: 464-10-8
RTECS: PB0100000

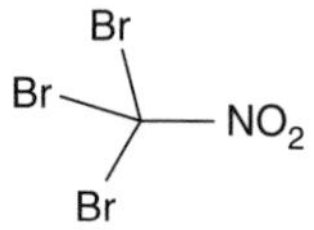

CBr_3NO_2

Prismatic crystals or oily, colorless liquid.

Exposure Hazards

Conversion Factor: 1 ppm = 12.18 mg/m^3 at 77° F
Human toxicity values have not been established or have not been published.

Properties:

MW: 297.7	*VP*: —	*FlP*: —
D: 2.79 g/mL	*VD*: 10 (calculated)	*LEL*: —
MP: 50° F	*Vlt*: —	*UEL*: —
BP: 192° F	H_2O: <0.1% (64° F)	*RP*: —
Vsc: —	*Sol*: —	*IP*: —

C10-A008

Perfluoroisobutylene (PFIB)
CAS: 382-21-8
RTECS: UD1800000

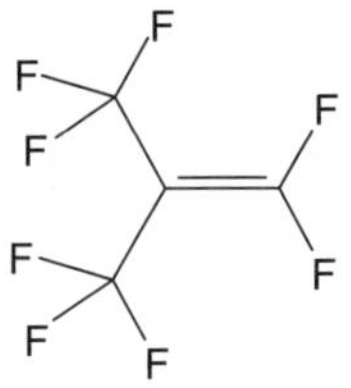

C_4F_8

Colorless, odorless gas. Produced by decomposition of Teflon when heated above 887° F.

A few hours after exposure, there is a gradual increase in temperature (up to 104° F), pulse rate (up to 120 bpm), and sometimes in respiration rate. There is no known postexposure medical or chemical treatment that slows or reverses the effects of PFIB.

Exposure Hazards

Conversion Factor: 1 ppm = 8.18 mg/m^3 at 77° F
ACGIH Ceiling: 0.01 ppm

Lethal human toxicity values have not been established or have not been published. However, based on available information, this agent appears to be approximately four times more toxic than Phosgene (C10-A003).

Properties:

MW: 200.0	*VP*: —	*FlP*: —
D: —	*VD*: 6.9 (calculated)	*LEL*: —
MP: −249° F	*Vlt*: —	*UEL*: —
BP: −20° F	H_2O: Insoluble	*RP*: —
Vsc: —	*Sol*: —	*IP*: 10.70 eV

C10-A009

Methyl chlorosulfonate (C-Stoff)
CAS: 812-01-1
RTECS: —

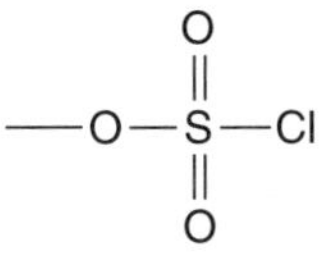

CH_3ClO_3S

Clear, viscous liquid. Vapors are highly irritating to the eyes and respiratory system. This material is hazardous through inhalation and ingestion, and produces local skin/eye impacts.

Also reported as a mixture with Dimethyl sulfate (C03-A009).

Exposure Hazards

Conversion Factor: 1 ppm = 5.34 mg/m^3 at 77° F
$LC_{50(Inh)}$: 2000 mg/m^3 (375 ppm) for a 10-min exposure
Eye Irritation: 1.3 ppm; exposure duration unavailable

These values are from older sources (ca. 1937) and are not supported by modern data. No updated toxicity estimates have been proposed.

Properties:

MW: 130.6	*VP*: 8.4 mmHg	*FlP*: —
D: 1.51 g/mL	*VD*: 4.5 (calculated)	*LEL*: —
MP: −94° F	*Vlt*: 11,000 ppm	*UEL*: —
BP: 232° F	H_2O: Reacts	*RP*: 1
Vsc: —	*Sol*: Chloroform; Carbon tetrachloride	*IP*: —

C10-A010

Ethyl chlorosulfonate (Sulvinite)
CAS: 625-01-4
RTECS: —

$C_2H_5ClO_3S$

Colorless liquid. This material is hazardous through inhalation and ingestion, and produces local skin/eye impacts.

Exposure Hazards
Conversion Factor: 1 ppm = 5.91 mg/m^3 at 77° F
$LC_{50(Inh)}$: 1000 mg/m^3 (170 ppm); exposure duration unavailable
Eye Irritation: 1 ppm; exposure duration unavailable

These values are from older sources (ca. 1937) and are not supported by modern data. No updated toxicity estimates have been proposed.

Properties:

MW: 144.6	*VP*: 2.3 mmHg	*FlP*: —
D: 1.379 g/mL	*VD*: 5.0 (calculated)	*LEL*: —
MP: —	*Vlt*: 3000 ppm	*UEL*: —
BP: 275° F (decomposes)	H_2O: Reacts	*RP*: 3.6
Vsc: —	*Sol*: Ether; Chloroform	*IP*: —

C10-A011

Bis(chloromethyl) ether
CAS: 542-88-1
RTECS: KN1575000
UN: 2249
ERG: 131

Cl O Cl

$C_2H_4Cl_2O$

Colorless liquid with a strong, unpleasant, and suffocating odor. This material is hazardous through inhalation, skin absorption, and ingestion, and produces local skin/eye impacts.

Reacts with moisture in the air to form Formaldehyde (C11-A042) and Hydrogen chloride (C11-A030).

Also reported as a mixture with Ethyldibromoarsine; Ethyldichloroarsine (C04-A001); Sulfur mustard (C03-A001).

Exposure Hazards
Conversion Factor: 1 ppm = 4.70 mg/m^3 at 77° F

Affects the equilibrium. Casualties stagger and are unable to maintain balance.

$LCt_{50(Inh)}$: 4700 mg-min/m^3 (500 ppm for a 2-min exposure)

Respiratory/Eye Irritation: 3 ppm; exposure duration unavailable
"Intolerable" Irritation: 8.5 ppm; exposure duration unavailable

These values are from older sources (ca. 1937) and are not supported by modern data. No updated toxicity estimates have been proposed.

ACGIH TLV: 0.001 ppm

Properties:

MW: 115.0	*VP*: 29.4 mmHg (77° F)	*FlP*: −0.4° F
D: 1.32 g/mL	*VD*: 4.0 (calculated)	*LEL*: —
MP: −43° F	*Vlt*: 40,000 ppm	*UEL*: —
BP: 223° F	H_2O: 2.2% (decomposes)	*RP*: 0.3
Vsc: —	*Sol*: Most organic solvents	*IP*: —

Proposed AEGLs

AEGL-1: Not Developed
AEGL-2: 1 h, 0.044 ppm   4 h, 0.028 ppm   8 h, 0.020 ppm
AEGL-3: 1 h, 0.18 ppm   4 h, 0.11 ppm   8 h, 0.075 ppm

C10-A012

Bis(bromomethyl) ether
CAS: 4497-29-4
RTECS: —

$C_2H_4Br_2O$

Colorless liquid. This material is hazardous through inhalation, skin absorption, and ingestion, and produces local skin/eye impacts.

Reacts with moisture in the air to form Formaldehyde (C11-A042) and Hydrogen bromide (C11-A029).

Also reported as a mixture with Ethyldichloroarsine (C04-A001).

Exposure Hazards

Conversion Factor: 1 ppm = 8.34 mg/m^3 at 77° F
Affects the equilibrium. Casualties stagger and are unable to maintain balance.
$LCt_{50(Inh)}$: 4000 mg-min/m^3 (240 ppm for a 2-min exposure)
Respiratory/Eye Irritation: 2.4 ppm; exposure duration unavailable
"Intolerable" Irritation: 6.0 ppm; exposure duration unavailable

These values are from older sources (ca. 1937) and are not supported by modern data. No updated toxicity estimates have been proposed.

Properties:

MW: 203.9	*VP*: 1.9 mmHg	*FlP*: —
D: 2.20 g/mL	*VD*: 7.0 (calculated)	*LEL*: —
MP: —	*Vlt*: 2.53 ppm	*UEL*: —
BP: 311° F	H_2O: Insoluble	*RP*: 3.7
Vsc: —	*Sol*: Ether; Benzene; Acetone	*IP*: —

C10-A013

Phenylcarbylamine chloride (K-Stoff)
CAS: 622-44-6
RTECS: —

UN: 1672
ERG: 151

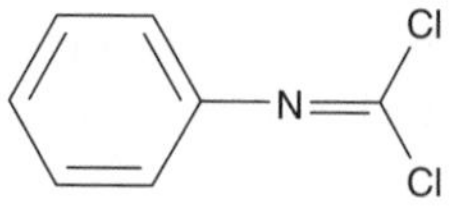

$C_7H_5Cl_2N$

Pale yellow oily liquid with an odor like onions. This material is hazardous through inhalation and ingestion, and produces local skin/eye impacts.

Exposure Hazards

Conversion Factor: 1 ppm = 7.12 mg/m^3 at 77° F
$LCt_{50(Inh)}$: 5000 mg-min/m^3 (350 ppm for a 2 min exposure)
Eye Irritation: 0.42 ppm; exposure duration unavailable
"Intolerable" Irritation: 3.5 ppm for a 1 min exposure

These values are from older sources (ca. 1937) and are not supported by modern data. No updated toxicity estimates have been proposed.

Properties:

MW: 174.0	*VP*: 0.178 mmHg (77° F)	*FlP*: —
D: 1.35 g/mL	*VD*: 6.0 (calculated)	*LEL*: —
MP: 67° F	*Vlt*: 295 ppm	*UEL*: —
BP: 410° F	H_2O: 0.04% (77° F)	*RP*: 43
Vsc: —	*Sol*: Alcohol; Ether; Chloroform	*IP*: —

C10-A014

Perchloromethyl mercaptan (Clairsite)
CAS: 594-42-3
RTECS: PB0370000
UN: 1670
ERG: 157

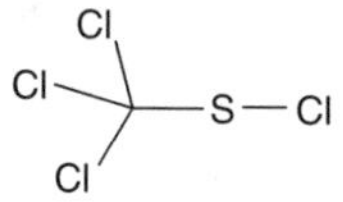

$C\ Cl_4\ S$

Pale yellow to red oily liquid with a foul, unbearable, acrid odor that is detectable at 0.016 ppm. This material is hazardous through inhalation, skin absorption, and ingestion, and produces local skin/eye impacts.

Also reported as a mixture with Sulfur dichloride (C03-C034).

Exposure Hazards

Conversion Factor: 1 ppm = 7.60 mg/m^3 at 77° F
$LCt_{50(Inh)}$: 30,000 mg-min/m^3 (2000 ppm for a 2-min exposure)
Eye Irritation: 1.3 ppm for a 2-min exposure
"Intolerable" Irritation: 9.2 ppm; exposure duration unavailable

These values are from older sources (circa 1937) and are not supported by modern data. No updated toxicity estimates have been proposed.

$MEG_{(1h)}$ *Min*: 0.014 ppm; *Sig*: 0.035 ppm; *Sev*: 0.30 ppm
OSHA PEL: 0.1 ppm
ACGIH TLV: 0.1 ppm
IDLH: 10 ppm

Properties:

MW: 185.9	*VP*: 3 mmHg	*FlP*: None
D: 1.722 g/mL	*VD*: 6.5 (calculated)	*LEL*: None
MP: —	*Vlt*: 2,400 ppm	*UEL*: None
BP: 297° F (decomposes)	H_2O: Insoluble	*RP*: 2.4
BP: 123° F (25 mmHg)	*Sol*: Ether	*IP*: —
Vsc: —		

Status AEGLs

AEGL-1: 1 h, 0.012 ppm	4 h, 0.0074 ppm	8 h, 0.0049 ppm
AEGL-2: 1 h, 0.035 ppm	4 h, 0.022 ppm	8 h, 0.014 ppm
AEGL-3: 1 h, 0.30 ppm	4 h, 0.075 ppm	8 h, 0.038 ppm

C10-A015

Chlorine trifluoride (N-Stoff)
CAS: 7790-91-2
RTECS: FO2800000
UN: 1749
ERG: 124

ClF_3

Greenish-yellow fuming liquid or colorless gas with a sweet, irritating, and suffocating odor. This material is hazardous through inhalation and ingestion, and produces local skin/eye impacts.

Can spontaneously ignite when in contact with organic matter. Reacts with moist air to produce Chlorine (C10-A001), Hydrogen fluoride (C11-A031), and Chlorine dioxide.

Exposure Hazards
Conversion Factor: 1 ppm = 3.78 mg/m^3 at 77° F
$MEG_{(1h)}$ *Min*: 0.12 ppm; *Sig*: 3.1 ppm; *Sev*: 15 ppm
OSHA Ceiling: 0.1 ppm
ACGIH Ceiling: 0.1 ppm
NIOSH Ceiling: 0.01 ppm
IDLH: 20 ppm

Properties:

MW: 92.4	*VP*: 1060 mmHg	*FlP*: —
D: 1.785 g/mL (liq. gas, 77° F)	*VD*: 3.2 (calculated)	*LEL*: —
MP: −117° F	*Vlt*: —	*UEL*: —
BP: 53° F	H_2O: Decomposes	*RP*: 0.01
Vsc: 0.231 cS (77° F)	*Sol*: —	*IP*: —

Interim AEGLs

AEGL-1: 1 h, 0.12 ppm	4 h, 0.12 ppm	8 h, 0.12 ppm
AEGL-2: 1 h, 2.0 ppm	4 h, 0.7 ppm	8 h, 0.41 ppm
AEGL-3: 1 h, 21 ppm	4 h, 7.3 ppm	8 h, 7.3 ppm

C10-A016

Disulfur decafluoride (Agent Z)
CAS: 5714-22-7
RTECS: WS4480000

F F F F
F —S———S—F
F F F F

S_2F_{10}

Colorless liquid or gas that is odorless. Sulfur dioxide is a common impurity that will give the material a characteristic odor. This material is hazardous through inhalation and ingestion, and produces local skin/eye impacts.

Exposure Hazards

Conversion Factor: 1 ppm = 10.39 mg/m^3 at 77° F
OSHA PEL: 0.025 ppm
ACGIH Ceiling: 0.01 ppm
NIOSH Ceiling: 0.01 ppm
IDLH: 1.0 ppm

Properties:

MW: 254.1	*VP*: 561 mmHg	*FlP*: None
D: 2.08 g/mL (32° F)	*VD*: 8.8 (calculated)	*LEL*: None
MP: −134° F	*Vlt*: 750,000 ppm	*UEL*: None
BP: 84° F	H_2O: Insoluble	*RP*: 0.01
Vsc: —	*Sol*: —	*IP*: —

References

Agency for Toxic Substances and Disease Registry. "Bis(chloromethyl) Ether ToxFAQs." July 1999.

———. "Chlorine ToxFAQs." April 2002.

———. "Nitrogen Oxides (Nitric Oxide, Nitrogen Dioxide, etc.) ToxFAQs." April 2002.

———. "Phosgene ToxFAQs." April 2002.

———. "Chlorine." In *Managing Hazardous Materials Incidents Volume III—Medical Management Guidelines for Acute Chemical Exposures.* Rev. ed. Washington, DC: Government Printing Office, 2000.

———."Phosgene." In *Managing Hazardous Materials Incidents Volume III—Medical Management Guidelines for Acute Chemical Exposures.* Rev ed. Washington, DC: Government Printing Office, 2000.

———. *Toxicological Profile for Bis(chloromethyl) Ether.* Washington, DC: Government Printing Office, December 1989.

———. *Toxicological Profile for Cadmium.* Washington, DC: Government Printing Office, July 1999.

———. *Toxicological Profile for Chloroform.* Washington, DC: Government Printing Office, September 1997.

Brophy, Leo P., Wyndham D. Miles, and Rexmond C. Cohrane. *The Chemical Warfare Service: From Laboratory to Field.* Washington, DC: Government Printing Office, 1968.

Centers for Disease Control and Prevention. "Case Definition: Bromine." March 4, 2005.

———. "Case Definition: Chlorine." March 9, 2005.

——— "Case Definition: Phosgene." March 10, 2005.

——— "Facts About Phosgene." March 17, 2003.

Compton, James A.F. *Military Chemical and Biological Agents: Chemical and Toxicological Properties.* Caldwell, NJ: The Telford Press, 1987.

Damle, Suresh B. "Safe Handling of Diphosgene, Triphosgene." *Chemical & Engineering News* 71 (8 Feb 1993): 4.

Fries, Amos A., and Clarence J. West. *Chemical Warfare.* New York: McGraw-Hill Book Company, Inc., 1921.
Jackson, Kirby E., and Margaret A. Jackson. "Lachrymators." *Chemical Reviews* 16 (1935): 195–242.
Lumsden, Malvern. *Incendiary Weapons.* Cambridge, MA: The MIT Press, 1975.
Marrs, Timothy C., Robert L. Maynard, and Frederick R. Sidell. *Chemical Warfare Agents: Toxicology and Treatment.* Chichester, England: John Wiley & Sons, 1997.
National Institute for Occupational Safety and Health. *NIOSH Pocket Guide to Chemical Hazards.* Washington, DC: Government Printing Office, September 2005.
National Institutes of Health. *Hazardous Substance Data Bank (HSDB).* http://toxnet.nlm.nih.gov/cgi-bin/sis/htmlgen?HSDB/. 2005.
Olson, Kent R., ed. *Poisoning & Drug Overdose.* 4th ed. New York: Lange Medical Books/McGraw-Hill, 2004.
Prentiss, Augustin M. *Chemicals in War: A Treatise on Chemical Warfare.* New York: McGraw-Hill Book Company, Inc., 1937.
Sartori, Mario F. *The War Gases: Chemistry and Analysis.* Translated by L.W. Marrison. London: J. & A. Churchill, Ltd, 1939.
Sidell, Frederick R. *Medical Management of Chemical Warfare Agent Casualties: A Handbook for Emergency Medical Services.* Bel Air, MD: HB Publishing, 1995.
Sidell, Fredrick R., Ernest T. Takafuji, and David R. Franz, eds. *Medical Aspects of Chemical and Biological Warfare, Textbook of Military Medicine Series, Part 1, Warfare, Weaponry, and the Casualty.* Washington, DC: Office of the Surgeon General, Department of the Army, 1997.
Sifton, David W., ed. *PDR Guide to Biological and Chemical Warfare Response.* Montvale, NJ: Thompson/Physicians Desk Reference, 2002.
Smith, Ann, Patricia Heckelman, and Maryadele J. Oneil, eds. *The Merck Index: An Encyclopedia of Chemicals, Drugs, & Biologicals.* 13th ed. Rahway, NJ: Merck & Co., Inc., 2001.
Somani, Satu M., ed. *Chemical Warfare Agents.* New York: Academic Press, 1992.
Somani, Satu M., and James A. Romano, Jr., eds. *Chemical Warfare Agents: Toxicity at Low Levels.* Boca Raton, FL: CRC Press, 2001.
Swearengen, Thomas F. *Tear Gas Munitions: An Analysis of Commercial Riot Gas Guns, Tear Gas Projectiles, Grenades, Small Arms Ammunition, and Related Tear Gas Devices.* Springfield, IL: Charles C Thomas Publisher, 1966.
True, Bey-Lorraine, and Robert H. Dreisbach. *Dreisbach's Handbook of Poisoning: Prevention, Diagnosis and Treatment.* 13th ed. London, England: The Parthenon Publishing Group, 2002.
United States Army Headquarters. *Chemical Agent Data Sheets Volume I, Edgewood Arsenal Special Report No. EO-SR-74001.* Washington, DC: Government Printing Office, December 1974.
———. *Potential Military Chemical/Biological Agents and Compounds, Field Manual No. 3-11.9.* Washington, DC: Government Printing Office, January 10, 2005.
United States Army Medical Research Institute of Chemical Defense. *Medical Management of Chemical Casualties Handbook.* 3rd ed. Aberdeen Proving Ground, MD: United States Army Medical Research Institute of Chemical Defense, July 2000.
United States Coast Guard. *Chemical Hazards Response Information System (CHRIS) Manual, 1999 Edition.* http://www.chrismanual.com/Default.htm. March 2004.
Wachtel, Curt. *Chemical Warfare.* Brooklyn, NY: Chemical Publishing Co., Inc., 1941.
Waitt, Alden H. *Gas Warfare: The Chemical Weapon, Its Use, and Protection against It.* Rev. ed. New York: Duell, Sloan and Pearce, 1944.
Williams, Kenneth E. *Detailed Facts About Choking Agent Phosgene (CG).* Aberdeen Proving Ground, MD: United States Army Center for Health Promotion and Preventive Medicine, 1996.
World Health Organization. *Health Aspects of Chemical and Biological Weapons: Report of a WHO Group of Consultants.* Geneva: World Health Organization, 1970.
———. *International Chemical Safety Cards (ICSCs).* Geneva: World Health Organization, 2004. http://www.cdc.gov/niosh/ipcs/icstart.html.
———. *Public Health Response to Biological and Chemical Weapons: WHO Guidance.* Geneva: World Health Organization, 2004.
Yaws, Carl L. *Matheson Gas Data Book.* 7th ed. Parsippany, NJ: Matheson Tri-Gas, 2001.

11

Toxic Industrial Agents

11.1 General Information

Chemicals in this class include agricultural and industrial chemicals that are readily available and possess appropriate chemical and toxicological properties to create a mass impact if deliberately released. These materials were selected by the Federal Bureau of Investigation (FBI), Centers for Disease Control and Prevention (CDC), and Department of Defense.

All of these chemicals pose an inhalation hazard but a toxic dose could also be obtained through skin absorption or ingestion. Factors that were considered when selecting potential candidate chemicals include global production, physical state of the material (i.e., gas, liquid, or solid), chemicals likely to cause major morbidity or mortality, potential to cause public panic and social disruption, chemicals that require special action for public health preparedness, history of previous use by the military, and/or involvement in a significant industrial accident.

In addition to the chemicals included on the other lists, the CDC also included heavy metals such as arsenic, lead, and mercury; volatile solvents such as benzene, chloroform, and bromoform; decomposition products such as dioxins and furans; polychlorinated biphenyls (PCBs); flammable industrial gases and liquids such as gasoline and propane; explosives and oxidizers; and all persistent and nonpersistent pesticides. Agents included in this volume are limited to those that are most likely to pose an acute toxicity hazard.

11.2 Toxicology

11.2.1 Effects

Many of these chemicals are similar to, but generally less toxic than, military chemical agents identified in other chapters of this book (e.g., nerve agents, vesicant agents, etc.). However, some of them are systemic poisons that act in ways that do not fit into one of the standard military classifications.

11.2.2 Pathways and Routes of Exposure

All of these materials pose an inhalation hazard. In addition, many liquid and solid agents are hazardous through ingestion, or when introduced through broken, abraded, or lacerated skin (e.g., penetration of skin by debris). Some of these materials readily pass through the skin to induce systemic intoxication. Some materials may also cause irritation to the skin and eye.

11.3 Protection

11.3.1 Evacuation Recommendations

There are no published recommendations for isolation or protective action distances for these materials deliberately released in mass casualty situations. However, traditional isolation and protective action distances for most of these materials can be found in the *Department of Transportation 2004 Emergency Response Guide* (ERG). These recommendations are based on an accidental release during transportation of the material and involving a small spill (i.e., a commercial gas cylinder or 200 liters or less of liquid material), or a large spill (i.e., more than one gas cylinder, a large gas container such as a railcar, or more than 200 liters of liquid material).

A simplified downwind hazard assessment can be developed by plotting these protective action distances in the form of a map overlay. The initial distance is the radius of a circle immediately surrounding the point of release where people may potentially be exposed to dangerous or life threatening concentrations of vapor. The downwind distance indicates the area of potential threat posed by vapors carried by the wind. These distances were not developed to account for additional contamination from a dispersal device. In such cases, the initial isolation and downwind evacuation distances should begin at the edge of any contamination caused by the release. Figure 11.1 is an example of a simplified downwind hazard assessment for an irregular release that should be used during a deliberate release scenario.

The amount of contamination and the distance an agent cloud will travel is based primarily on the physical/chemical properties of the specific agent, method of release, local terrain, and weather conditions.

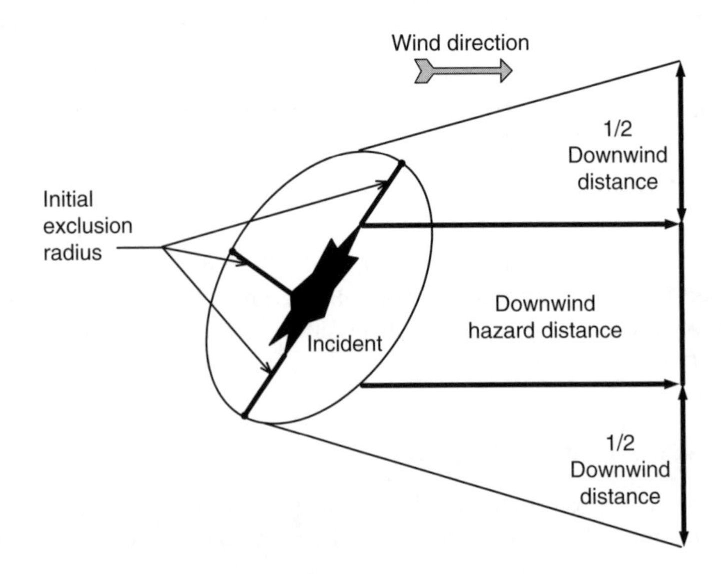

FIGURE 11.1 Irregular release simplified downwind hazard diagram

11.3.2 Personal Protective Requirements

11.3.2.1 Structural Firefighters' Gear

Structural firefighters' protective clothing is recommended for fire situations only; it is not effective in spill situations or release events. If chemical protective clothing is not available and it is necessary to rescue casualties from a contaminated area, then structural firefighters' gear will provide limited skin protection against agent vapors and aerosols. Contact with solid and liquid agents should be avoided. Depending on the material, firefighters who enter a contaminated area wearing only structural firefighters' gear may experience mild to moderate effects. Some materials, such as organophosphate pesticides, can accumulate in the body and responders who enter the hot zone without appropriate chemical protective clothing may be at increased risk during the remainder of the emergency.

11.3.2.2 Respiratory Protection

National Institute for Occupational Safety and Health (NIOSH) approved self-contained breathing apparatuses (SCBAs) or air purifying respirators (APRs) that have been tested by the manufacturer against the individual chemicals may be used. APRs should be equipped with a NIOSH approved Chemical/Biological/Radiological/Nuclear (CBRN) filter or a combination organic vapor/acid gas/particulate cartridge. Ensure that the cartridge has been tested against the specific chemical and that the concentration of the agent does not exceed the saturation level of the cartridge. APRs are not recommended for use in atmospheres containing some of these chemicals because of poor warning properties of the agent and a short life span of respirator cartridges.

Immediately dangerous to life or health (IDLH) levels are the ceiling limit for respirators other than SCBAs. Any exposures approaching the IDLH level should be regarded with extreme caution and the use of SCBAs for respiratory protection should be considered.

11.3.2.3 Chemical Protective Clothing

Use only chemical protective clothing that has undergone appropriate material and construction performance testing. If the concentration of vapor exceeds the level necessary to produce effects through dermal exposure, then responders should wear a Level A protective ensemble. Reported permeation rates may be affected by solvents if solutions of agents have been released.

11.3.3 Decontamination

11.3.3.1 General

Apply universal decontamination procedures using soap and water.

11.3.3.2 Vapors

Casualties/personnel: Remove all clothing as it may continue to emit "trapped" agent vapor after contact with the vapor cloud has ceased. Shower using copious amounts of soap and water. Ensure that the hair has been washed and rinsed to remove potentially trapped vapor. If eye irritation occurs, irrigate with water or 0.9% saline solution for a minimum of 15 minutes.

Small areas: Ventilate to remove the vapors. If deemed necessary, wash the area with copious amounts of soap and water. Collect and place into containers lined with high-density polyethylene. Removal of porous material, including painted surfaces, may be required because agents that have been absorbed into these materials may migrate back to the surface and pose a contact hazard.

11.3.3.3 Liquids, Solutions, or Liquid Aerosols

Casualties/personnel: Remove all clothing immediately. Even clothing that has not come into direct contact with the agent may contain "trapped" vapor. To avoid further exposure of the head, neck, and face to the agent, cut off potentially contaminated clothing that must be pulled over the head. Use a sponge or cloth with liquid soap and copious amounts of water to wash the skin surface and hair at least three times. Avoid rough scrubbing. Rinse with copious amounts of water. If water is not immediately available, the agent can be absorbed with any convenient material such as paper towels, toilet paper, flour, or talc. To minimize both spreading the agent and abrading the skin, do not rub the agent with the absorbent. Blot the contaminated skin with the absorbent. If there is a potential that the eyes have been exposed, irrigate with water or 0.9% saline solution for a minimum of 15 minutes.

Small areas: Small puddles of liquid can be contained by covering with absorbent materials such as vermiculite, diatomaceous earth, clay, sponges, or towels. Place the absorbed material into containers lined with high-density polyethylene. Larger puddles can be collected using vacuum equipment made of materials inert to the released material and equipped with a high-efficiency particulate air (HEPA) filter and appropriate vapor filters. If there is a potential for corrosive residue, the area should be treated with an appropriate neutralizing agent. Wash the area with copious amounts of soap and water. Collect and containerize the rinseate. Ventilate the area to remove vapors.

11.3.3.4 Solids or Particulate Aerosols

Casualties/personnel: Do not attempt to brush the agent off the individual or their clothing as this can aerosolize the agent. Remove all clothing immediately. To avoid further exposure of the head, neck, and face to the agent, cut off potentially contaminated clothing that must be pulled over the head. Wash the skin surface and hair at least three times with copious amounts of soap and water. Do not delay decontamination to find warm or hot water if it is not readily available. Rinse with copious amounts of water. If there is a potential that the eyes have been exposed, irrigate with water or 0.9% saline solution for a minimum of 15 minutes.

Small areas: If indoors, close windows and doors in the area and turn off anything that could create air currents (e.g., fans, air conditioner, etc.). Avoid actions that could aerosolize the agent such as sweeping or brushing. Collect the agent using a vacuum cleaner made of materials inert to the released material and equipped with a HEPA filter and appropriate vapor filters. Do not use a standard home or industrial vacuum. Do not allow the vacuum exhaust to stir the air in the affected area. Vacuum all surfaces with extreme care in a very slow and controlled manner to minimize aerosolizing the agent. Place the collected material into containers lined with high-density polyethylene. If there is a potential for corrosive residue, the area should be treated with an appropriate neutralizing agent. Wash the area with copious amounts of the soap and water. Collect and containerize the rinseate in containers lined with high-density polyethylene. Ventilate the area to remove any agent vapors generated during the decontamination.

11.4 Medical

11.4.1 CDC Case Definition

Case definitions exist for the following agents: anhydrous ammonia (C11-A061), arsenic trioxide (C11-A047), calcium arsenate (C11-A048), copper acetoarsenite (C11-A049), hydrogen fluoride (C11-A064), lead arsenate (C11-A050), methoxyethyl mercury acetate (C11-A052), methyl bromide (C11-A042), methyl isocyanate (C11-A127), methyl mercury dicyandiamide (C11-A053), methyl mercury hydroxide (C11-A054), nitric acid (C11-A066), paraquat (C11-A057), phenyl mercury acetate (C11-A056), phosphine (C11-A044), sodium arsenite (C11-A051), sodium fluoroacetate (C11-A058), stibine (C11-A080), and sulfuric acid (C11-A067).

11.4.2 Signs and Symptoms

11.4.2.1 Vapors

Vary depending on the specific chemical involved, concentration of the cloud, and length of exposure. Common symptoms of acute exposure (in no specific order) may include dizziness, loss of coordination, disorientation, problems in breathing (e.g., rapid breathing, chest tightness, or shortness of breath), chest pains, heart palpitations, problems seeing (e.g., blurred vision, eye irritation, pinpointing, or dilation of the pupils), heavy sweating, headache, ringing in the ears, nose or skin irritation, coughing, nausea, vomiting, convulsions, and loss of consciousness. Exposure to low concentrations of chemical may not produce immediate effects.

11.4.2.2 Liquids/Solids

Vary depending on the specific chemical involved, concentration of the agent, and length of exposure. Common symptoms of acute exposure (in no specific order) may include irritation and burning of the skin, eyes, mucous membrane, and respiratory system. Contact may also result in chemical burns to the skin and eyes.

11.4.3 Mass-Casualty Triage Recommendations

The base station physician or regional poison control center should be consulted for advice on specific situations. However, in general, anyone who has been exposed should be transported to a medical facility for evaluation. Individuals who are asymptomatic and have not been directly exposed to the chemical can be discharged after their names, addresses, and telephone numbers have been recorded. They should be instructed to seek medical care immediately if symptoms develop.

11.4.4 Casualty Management

Remove casualty to fresh air and decontaminate with soap and water. Provide oxygen for respiratory distress. If the eyes of the casualty complains of eye irritation, irrigate the eyes with water or 0.9% saline solution for at least 15 minutes. Irrigate open wounds with water or 0.9% saline solution for at least 10 minutes.

Once the casualty has been decontaminated, medical personnel do not need to wear a chemical-protective mask.

Prior to administering antidotes or other drugs, ensure that the signs and symptoms (e.g., coma, seizures, etc.) are due to chemical exposure and not the result of head trauma or other physical injury.

Consider tracheal intubation in cases of respiratory compromise. Treat patients who have bronchospasm with aerosolized bronchodilators. Use these and all catecholamines with caution because of the enhanced risk of cardiac dysrhythmias after exposure to some chemicals. When bronchodilators are needed, the lowest effective dose should be given and cardiac rhythm should be monitored. After decontamination, patients who are comatose, hypotensive, have cardiac dysrhythmias, or are having seizures should be treated according to established advanced life support (ALS) protocols.

Consult the base station physician or regional poison control center for chemical specific information.

11.5 Fatality Management

Remove all clothing and personal effects and decontaminate with soap and water. While it may be possible to decontaminate durable items, it may be safer and more efficient to destroy nondurable items rather than attempt to decontaminate them. Items that will be retained for further processing should be double sealed in impermeable containers, ensuring that the inner container is decontaminated before placing it in the outer one.

Wash the remains with soap and water. Pay particular attention to areas where agent may get trapped, such as hair, scalp, pubic areas, fingernails, folds of skin, and wounds. If remains are heavily contaminated with residue, wash and rinse waste should be contained for proper disposal.

If organophosphorus pesticides are involved, then it may be necessary to wash the remains with a 2% sodium hypochlorite bleach solution (i.e., 2 gallons of water for every gallon of household bleach). This concentration of bleach will not affect remains but will neutralize organophosphorus pesticides. Higher concentrations of bleach can harm remains. The bleach solution should remain on the cadaver for a minimum of 5 minutes before rinsing with water.

Once the remains have been thoroughly decontaminated, no further protective action is necessary. Body fluids removed during the embalming process do not pose any additional risks and should be contained and handled according to established procedures. Use standard burial procedures.

C11-A

AGENTS

C11-A001

Azinphos-methyl

CAS: 86-50-0
RTECS: TE1925000

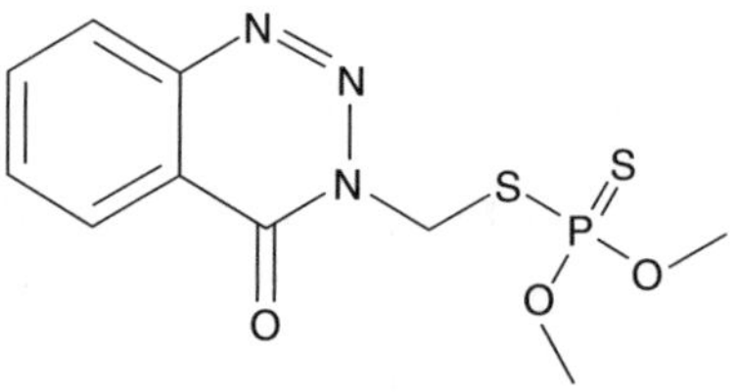

$C_{10}H_{12}N_3O_3PS_2$

Colorless to white or yellow crystalline material that has no odor. Technical grade is a cream to yellow-brown granular or waxy solid. This material is hazardous through inhalation, skin absorption, and ingestion, and produces local skin/eye impacts.

Used industrially as an insecticide, acaricide, and molluscicide.

This material is on the CDC and FBI threat lists.

Exposure Hazards

Conversion Factor: 1 ppm = 12.98 mg/m^3 at 77°4 F
OSHA PEL: 0.02 ppm [Skin]
ACGIH TLV: 0.015 ppm [Skin]
IDLH: 0.77 ppm

Properties:

MW: 317.3	*VP*: 0.0000016 mmHg	*FlP*: —
D: 1.44 g/cm^3	*VD*: 11 (calculated)	*LEL*: —
MP: 162°F	*Vlt*: 0.0023 ppm	*UEL*: —
MP: 153°F (commercial grade)	H_2O: 0.00209%	*RP*: 3,000,000
BP: Decomposes	*Sol*: Most organic solvents	*IP*: —
Vsc: —		

C11-A002

Bomyl

CAS: 122-10-1
RTECS: —

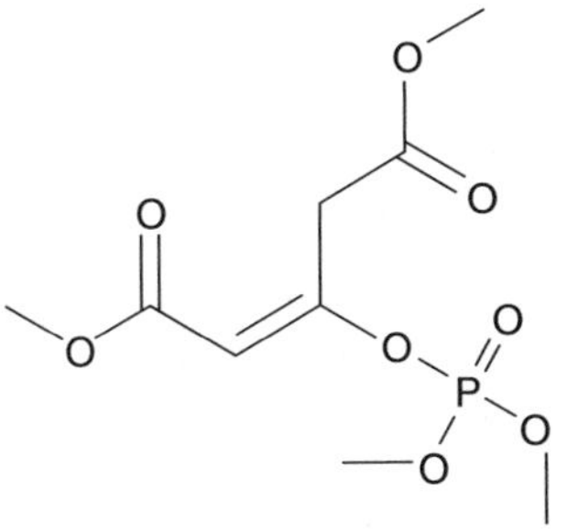

$C_9H_{15}O_8P$

Yellow oily liquid. Commercial grade is 80–90% agent. Both geometric isomers have similar toxicity. This material is hazardous through inhalation, skin absorption, penetration through broken skin, and ingestion, and produces local skin/eye impacts. This material is absorbed through the skin very quickly.

Used as an agricultural pesticide.

Exposure Hazards

Conversion Factor: 1 ppm = 11.54mg/m^3 at 77°F

Human toxicity values have not been established or have not been published.

Properties (commercial grade):

MW: 282.2	*VP*: 25 mmHg (77°F)	*FlP*: None
D: 1.2 g/mL	*VD*: 9.7 (calculated)	*LEL*: None
MP: —	*Vlt*: —	*UEL*: None
BP: 311–329°F (17 mmHg)	H_2O: "Practically insoluble"	*RP*: 0.24
Vsc: —	*Sol*: Alcohols; Acetone; Xylene; Propylene Glycol	*IP*: —

C11-A003

Bromophos-ethyl

CAS: 4824-78-6
RTECS: —

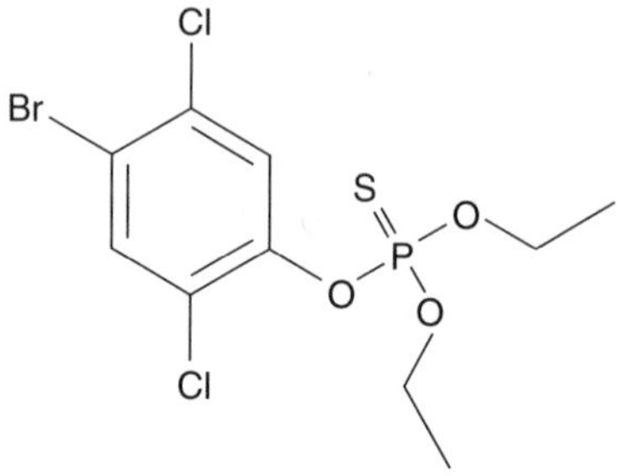

$C_{10}H_{12}BrCl_2O_3PS$

Colorless to pale-yellow liquid. This material is hazardous through inhalation, skin absorption, penetration through broken skin, and ingestion, and produces local skin/eye impacts.

Used as an agricultural pesticide.

Exposure Hazards

Conversion Factor: 1 ppm = 16.12 mg/m^3 at 77°F

Human toxicity values have not been established or have not been published.

Properties (commercial grade):

MW: 394.1	*VP*: 0.000046 mmHg (86° F)	*FlP*: —
D: 1.52–1.55 g/mL	*VD*: 14 (calculated)	*LEL*: —
MP: —	*Vlt*: 0.06 ppm (86°F)	*UEL*: —
BP: 252–271°F (0.001 mmHg)	*H_2O*: 0.000014–0.000044%	*RP*: 110,000
Vsc: —	*Sol*: Most common solvents	*IP*: —

C11-A004

Carbophenothion

CAS: 786-19-6
RTECS: TD5250000

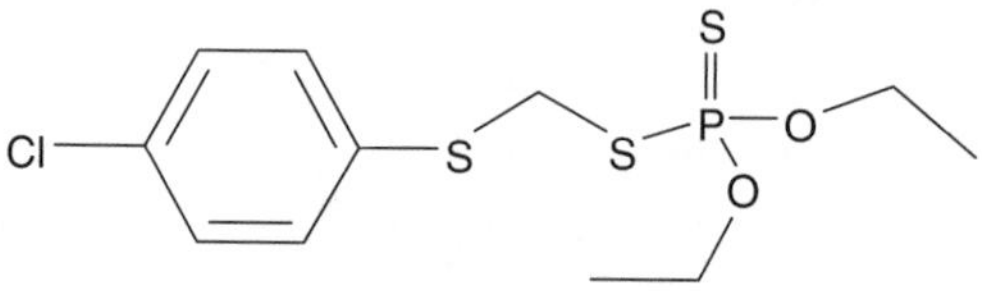

$C_{11}H_{16}ClO_2PS_3$

Colorless to light amber liquid with a foul odor. Commercial grade is 95% agent. This material is hazardous through inhalation, skin absorption, penetration through broken skin, and ingestion, and produces local skin/eye impacts. This material is absorbed through the skin very quickly.

Used as an agricultural pesticide.

Exposure Hazards

Conversion Factor: 1 ppm = 14.02 mg/m^3 at 77°F

$LD_{50(Ing)}$: 0.6 g (estimate)

Other human toxicity values have not been established or have not been published.

Properties (commercial grade):

MW: 342.9	*VP*: 0.0000003 mmHg	*FlP*: —
D: 1.285 g/mL	*VD*: 12 (calculated)	*LEL*: —
D: 1.271 g/mL (77°F)	*Vlt*: —	*UEL*: —
MP: —	H_2O: 0.0000063%	*RP*: 18,000,000
BP: 180°F (0.01 mmHg)	*Sol*: Most organic solvents	*IP*: —
Vsc: —		

C11-A005

Chlorfenvinphos

CAS: 470-90-6

RTECS: TB8750000

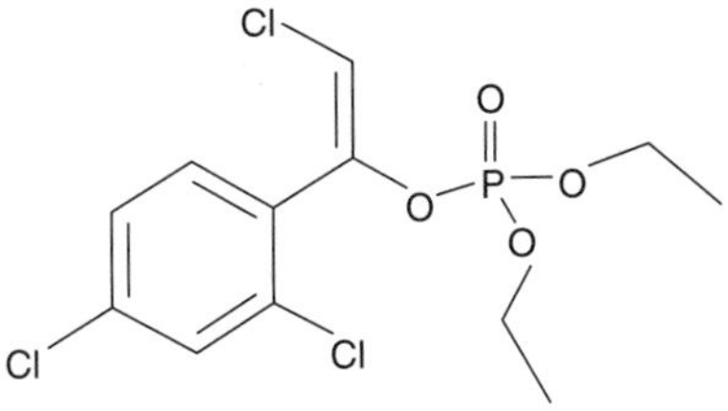

$C_{12}H_{14}Cl_3O_4P$

Colorless to yellow to amber liquid with a mild odor. Commercial grade is 90% agent. This material is hazardous through inhalation, skin absorption, penetration through broken skin, and ingestion, and produces local skin/eye impacts.

Used as an agricultural pesticide.

Exposure Hazards

Conversion Factor: 1 ppm = 14.71 mg/m^3 at 77°F

Human toxicity values have not been established or have not been published.

Properties (commercial grade):

MW: 359.6	*VP*: 0.0000075 mmHg (77°F)	*FlP*: None
D: 1.36 g/mL	*VD*: 12 (calculated)	*LEL*: None
MP: −9 to −2°F	*Vlt*: 0.01 ppm	*UEL*: None
BP: 333°F (0.5 mmHg)	H_2O: 0.012% (slowly decomposes)	*RP*: 710,000
BP: 255°F (0.008 mmHg)	*Sol*: Most organic solvents	*IP*: —
Vsc: —		

C11-A006

Demeton

CAS: 8065-48-3

RTECS: TF3150000

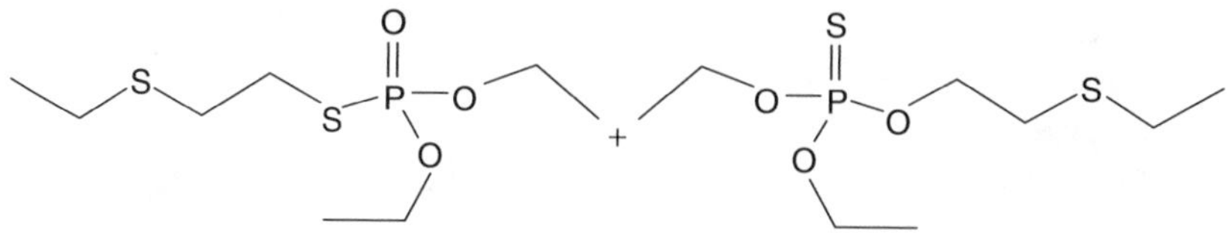

Mixture

Amber, oily liquid with an odor like sulfur. This material is hazardous through inhalation, skin absorption, penetration through broken skin, and ingestion, and produces local skin/eye impacts.

Used as an agricultural pesticide.

Exposure Hazards

Conversion Factor: 1 ppm = 10.57 mg/m^3 at 77°F
OSHA PEL: 0.009 ppm [Skin]
ACGIH TLV: 0.005 ppm [Skin]
IDLH: 0.94 ppm

Properties (commercial grade):

MW: Mixture	*VP*: 0.0003 mmHg	*FlP*: 113°F
D: 1.12 g/mL	*VD*: 8.9 (calculated)	*LEL*: —
MP: <−13°F	*Vlt*: 0.40 ppm	*UEL*: —
BP: Decomposes	*H_2O*: 0.01%	*RP*: 21,000
Vsc: —	*Sol*: —	*IP*: —

C11-A007

Dicrotophos

CAS: 141-66-2
RTECS: TC3850000

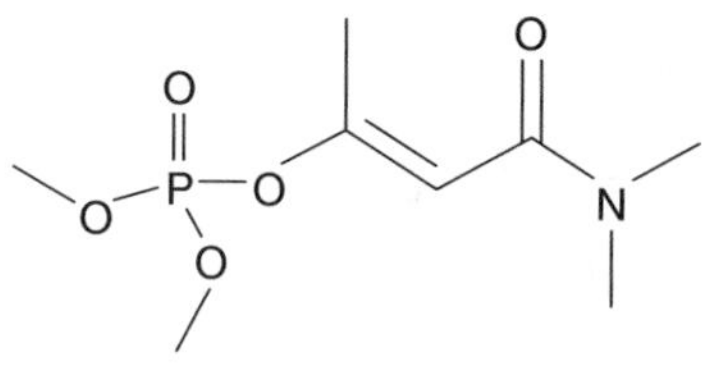

$C_8H_{16}NO_5P$

Yellow-brown liquid with a mild ester odor. Commercial material is brown. Mixture of the E and Z geometric isomers. Commercial material containing 85% E isomer, which is the active isomer. This material is hazardous through inhalation, skin absorption, penetration through broken skin, and ingestion, and produces local skin/eye impacts. It is corrosive to cast iron, mild steel, brass, and stainless steel.

Used as an agricultural pesticide.

Exposure Hazards

Conversion Factor: 1 ppm = 9.70 mg/m^3 at 77°F
ACGIH TLV: 0.005 ppm [Skin]

Properties:

MW: 237.2	*VP*: 0.0001 mmHg	*FlP*: >200°F
D: 1.216 g/mL (59°F)	*VP*: 0.00016 mmHg (77°F)	*LEL*: —
MP: —	*VD*: 8.2 (calculated)	*UEL*: —
BP: 752°F	*Vlt*: 0.13 ppm	*RP*: 64,000
BP: 266°F (0.1 mmHg)	H_2O: Miscible	*IP*: —
Vsc: —	*Sol*: Acetone; Alcohols; Chloroform; Xylene	

C11-A008

Dimefox

CAS: 115-26-4
RTECS: —

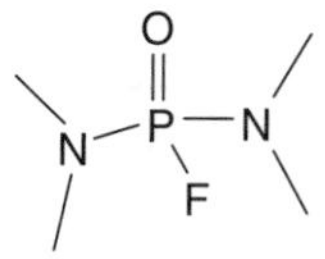

$C_4H_{12}FN_2OP$

Colorless liquid with a fishy odor. This material is hazardous through inhalation, skin absorption, penetration through broken skin, and ingestion, and produces local skin/eye impacts.

Used as an agricultural pesticide.

Exposure Hazards

Conversion Factor: 1 ppm = 6.30 mg/m^3 at 77°F
Human toxicity values have not been established or have not been published.

Properties:

MW: 154.1	*VP*: 0.11 mmHg	*FlP*: —
D: 1.1151 g/mL	*VP*: 0.36 (77°F)	*LEL*: —
MP: —	*VD*: 5.3 (calculated)	*UEL*: —
BP: 187°F (15 mmHg)	*Vlt*: 150 ppm	*RP*: 73
BP: 153°F (4 mmHg)	H_2O: Miscible	*IP*: —
Vsc: —	*Sol*: Most organic solvents	

C11-A009

Dioxathion

CAS: 78-34-2
RTECS: TE3350000

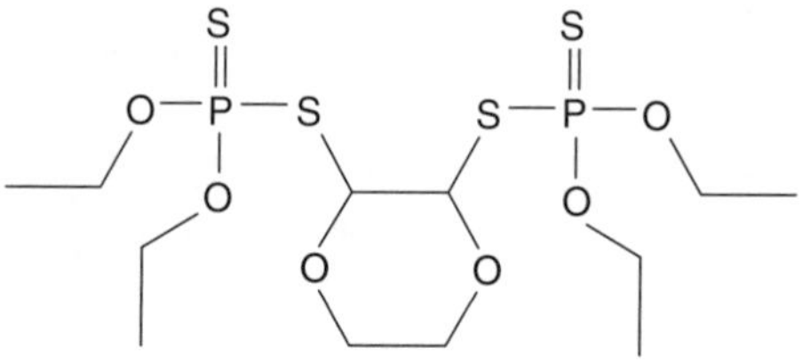

$C_{12}H_{26}O_6P_2S_4$

Viscous, brown, tan, or dark-amber liquid. Commercial product is a mixture of *cis* and *trans* isomers. This material is hazardous through inhalation, skin absorption, penetration through broken skin, and ingestion, and produces local skin/eye impacts.

Used as an agricultural pesticide.

Exposure Hazards

Conversion Factor: 1 ppm = 18.67 mg/m^3 at 77°F
ACGIH TLV: 0.005 ppm [Skin]

Properties:

MW: 456.5	*VP*: —	*FlP*: None
D: 1.257 g/mL (79°F)	*VD*: —	*LEL*: None
MP: –4°F	*Vlt*: —	*UEL*: None
BP: —	H_2O: Insoluble	*RP*: —
Vsc: 93.1 cS (77°F)	*Sol*: Alcohols; Ketones; Aromatic hydrocarbons	*IP*: —

C11-A010

Disulfoton

CAS: 298-04-4
RTECS: TD9275000

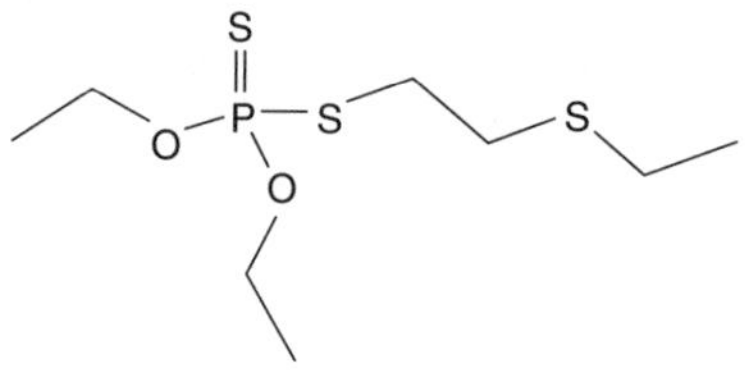

$C_8H_{19}O_2PS_3$

Oily, colorless to yellow liquid with a characteristic aromatic sulfur odor. Commercial product is a yellow to brown liquid. This material is hazardous through inhalation, skin absorption, and ingestion, and produces local skin/eye impacts.

Used industrially as an insecticide and acaricide.

This material is on the CDC and FBI threat lists.

Exposure Hazards

Conversion Factor: 1 ppm = 11.22 mg/m^3 at 77°F
ACGIH TLV: 0.004 ppm [Skin]

Properties:

MW: 274.4	*VP*: 0.00018 mmHg	*FlP*: >180°F
D: 1.144 g/mL	*VD*: 9.5 (calculated)	*LEL*: —
MP: <–13°F	*Vlt*: 0.24 ppm	*UEL*: —
BP: 235°F (0.4 mmHg)	H_2O: 0.00163%	*RP*: 33,000
Vsc: —	*Sol*: Most organic solvents	*IP*: —

C11-A011

EPN

CAS: 2104-64-5
RTECS: TB1925000

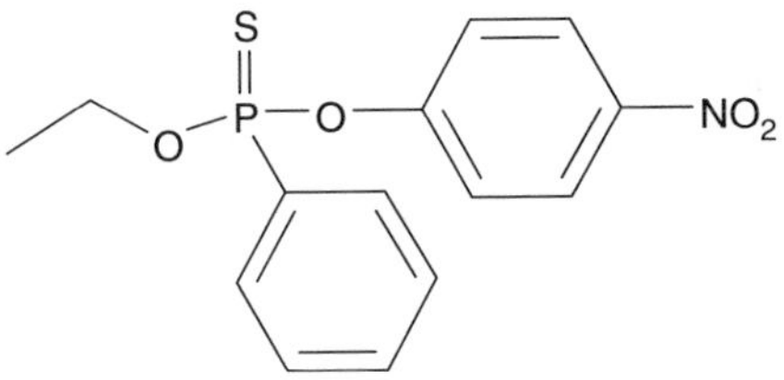

$C_{14}H_{14}NO_4PS$

White to light yellow crystalline solid or brown liquid with an aromatic odor. This material is hazardous through inhalation, skin absorption, penetration through broken skin, and ingestion, and produces local skin/eye impacts.

Used as an agricultural pesticide.

Exposure Hazards

Conversion Factor: 1 ppm = 13.22 mg/m^3 at 77°F
OSHA PEL: 0.04 ppm [Skin]
ACGIH TLV: 0.008 ppm [Skin]
IDLH: 0.38 ppm

Properties:

MW: 323.3
D: 1.270 g/cm^3 (77°F)
MP: 97°F
BP: 419°F (5 mmHg)
Vsc: —
VP: Negligible
VP: 0.0003 mmHg (212°F)
VD: —
Vlt: —
H_2O: 0.000311%
Sol: Alcohols; Ether; Acetone; Aromatic solvents
FlP: None
LEL: None
UEL: None
RP: 6,000,000
IP: —

C11-A012

Fenamiphos

CAS: 22224-92-6
RTECS: TB3675000

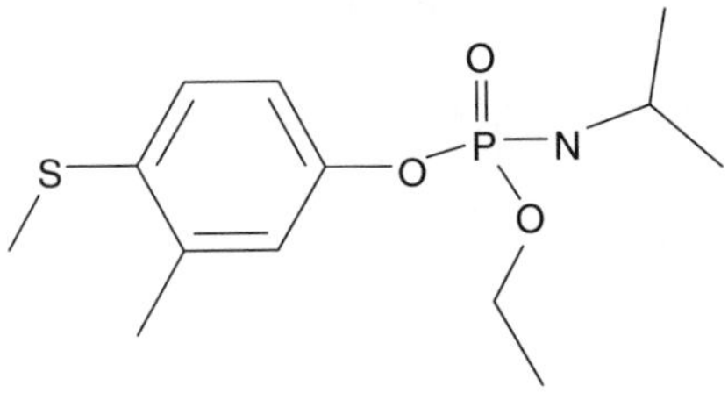

$C_{13}H_{22}NO_3PS$

Colorless crystals or off-white to tan, waxy solid. This material is hazardous through inhalation, skin absorption, penetration through broken skin, and ingestion, and produces local skin/eye impacts.

Used as an agricultural pesticide.

Exposure Hazards

Conversion Factor: 1 ppm = 12.41 mg/m^3 at 77°F
ACGIH TLV: 0.05 mg/m^3 [Skin]

Properties:

MW: 303.4	*VP*: 0.000001 mmHg (77°F)	*FlP*: —
D: 1.191 g/cm^3 (73°F)	*VP*: 0.000047 mmHg (commercial grade)	*LEL*: —
MP: 121°F	*VD*: 10 (calculated)	*UEL*: —
BP: —	*Vlt*: —	*RP*: 6,000,000
Vsc: —	H_2O: 0.0329%	*IP*: —
	Sol: Organic solvents	

C11-A013

Fenophosphon

CAS: 327-98-0
RTECS: —

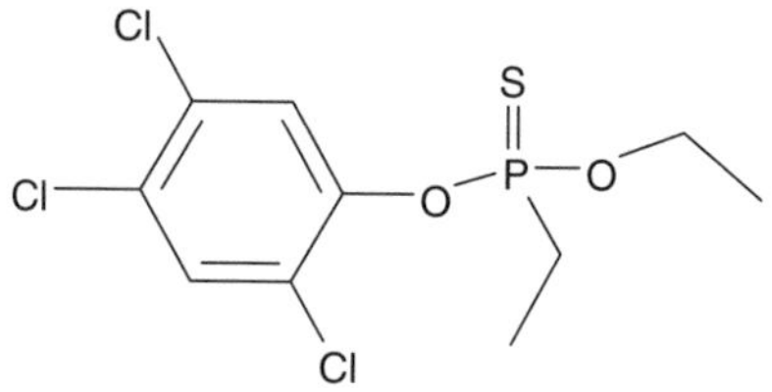

$C_{10}H_{12}Cl_3O_2PS$

Amber colored liquid. This material is hazardous through inhalation, skin absorption, penetration through broken skin, and ingestion, and produces local skin/eye impacts.

Used as an agricultural pesticide.

Exposure Hazards

Conversion Factor: 1 ppm = 13.64 mg/m^3 at 77°F
Human toxicity values have not been established or have not been published.

Properties:

MW: 333.6	*VP*: 0.000015 mmHg	*FlP*: —
D: 1.365 g/mL	*VD*: 12 (calculated)	*LEL*: —
MP: —	*Vlt*: 0.02 ppm	*UEL*: —
BP: 226.4°F (0.01 mmHg)	H_2O: 0.000059%	*RP*: 400,000
Vsc: —	*Sol*: Acetone; Alcohols; Kerosene	*IP*: —

C11-A014

Fonofos

CAS: 944-22-9

RTECS: TA5950000

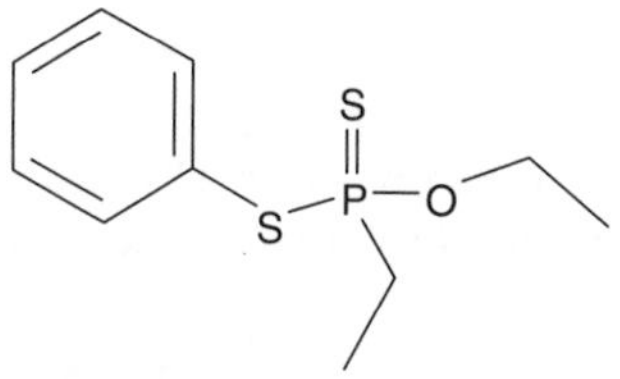

$C_{10}H_{15}OPS_2$

Clear, colorless to light-yellow liquid with an aromatic; pungent skunk-like odor. This material is hazardous through inhalation, skin absorption, and ingestion, and produces local skin/eye impacts.

Used industrially as a soil fumigant.

This material is on the CDC and FBI threat lists.

Exposure Hazards

Conversion Factor: 1 ppm = 10.07 mg/m^3 at 77°F

ACGIH TLV: 0.01 ppm [Skin]

Properties:

MW: 246.3	*VP*: 0.0002 mmHg (77°F)	*FlP*: >201°F
D: 1.16 g/cm^3	*VD*: 8.5 (calculated)	*LEL*: —
MP: 90°F	*Vlt*: 0.44 ppm	*UEL*: —
BP: 266°F (0.1 mmHg)	*H_2O*: 0.00157%	*RP*: 19,000
Vsc: —	*Sol*: Most organic solvents	*IP*: —

C11-A015

Isophenphos

CAS: 25311-71-1

RTECS: DH2255000

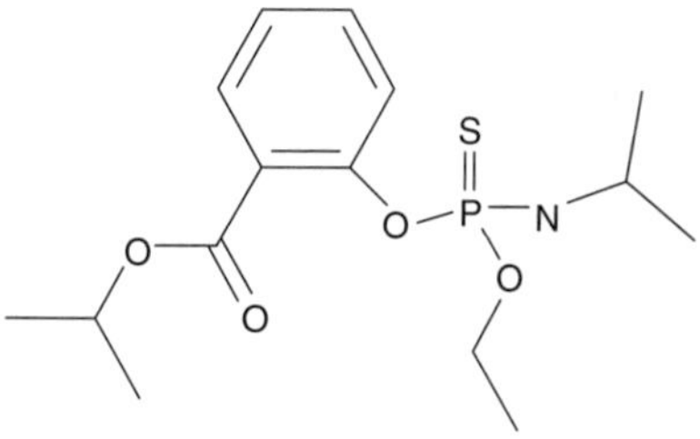

$C_{15}H_{24}NO_4PS$

Yellow-brown liquid. Commercial material is a colorless oil. This material is hazardous through inhalation, skin absorption, penetration through broken skin, and ingestion, and produces local skin/eye impacts.

Used as an agricultural pesticide.

Exposure Hazards

Conversion Factor: 1 ppm = 14.13 mg/m^3 at 77°F

Human toxicity values have not been established or have not been published.

Properties:

MW: 345.4	*VP*: 0.000003 mmHg (77°F)	*FlP*: —
D: 1.131 g/mL	*VD*: 12 (calculated)	*LEL*: —
MP: 10°F	*Vlt*: —	*UEL*: —
BP: 248°F (0.01 mmHg)	*H_2O*: 0.0022%	*RP*: 2,000,000
Vsc: —	*Sol*: Acetone; Kerosene; *n*-Hexane	*IP*: —

C11-A016

Mephosfolan

CAS: 950-10-7

RTECS: —

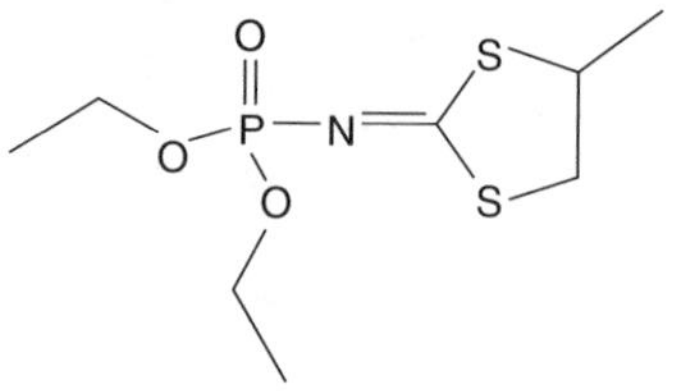

$C_8H_{16}NO_3PS_2$

Colorless to yellow to amber liquid. This material is hazardous through inhalation, skin absorption, penetration through broken skin, and ingestion, and produces local skin/eye impacts.

Used as an agricultural insecticide and acaricide.

Exposure Hazards

Conversion Factor: 1 ppm = 11.02 mg/m^3 at 77°F

Human toxicity values have not been established or have not been published.

Properties:

MW: 269.3	*VP*: 0.0000318 mmHg (77°F)	*FlP*: —
D: 1.539 g/mL (79°F)	*VD*: 9.3 (calculated)	*LEL*: —
MP: —	*Vlt*: 0.04 ppm (77°F)	*UEL*: —
BP: 248°F (0.003 mmHg)	*H_2O*: 5.7% (77°F)	*RP*: 200,000
Vsc: —	*Sol*: Acetone; Ethanol; Toluene	*IP*: —

C11-A017

Methamidophos

CAS: 10265-92-6

RTECS: TB4970000

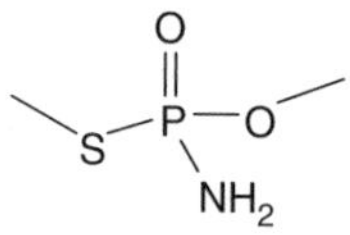

$C_2H_8NO_2PS$

Colorless crystalline solid with a mercaptan odor. This material is hazardous through inhalation, skin absorption, and ingestion, and produces local skin/eye impacts.

Used industrially as an insecticide and acaricide.

This material is on the CDC and FBI threat lists.

Exposure Hazards

Conversion Factor: 1 ppm = 5.77mg/m^3 at 77°F

Human toxicity values have not been established or have not been published.

Properties:

MW: 141.1	*VP*: 0.000035 mmHg (77°F)	*FlP*: —
D: 1.27g/cm^3	*VD*: 4.9 (calculated)	*LEL*: —
MP: 129°F	*Vlt*: 0.05 ppm	*UEL*: —
MP: 68–77°F (commercial grade)	*H_2O*: Miscible	*RP*: 240,000
BP: Decomposes	*Sol*: Alcohols; Hexane; Dichloromethane; Toluene	*IP*: —
Vsc: —		

C11-A018

Methidathion

CAS: 950-37-8
RTECS: —

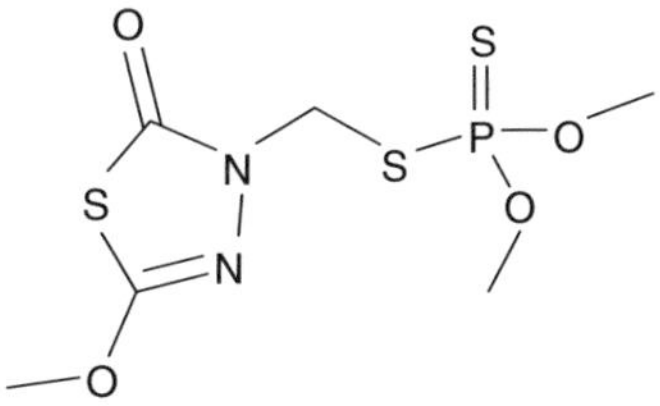

$C_6H_{11}N_2O_4PS_3$

Colorless to white crystalline solid with a characteristic odor. This material is hazardous through inhalation, skin absorption, and ingestion, and produces local skin/eye impacts.

Used industrially as an insecticide and acaricide.

This material is on the CDC and FBI threat lists.

Exposure Hazards

Conversion Factor: 1 ppm = 12.37 mg/m^3 at 77°F

Human toxicity values have not been established or have not been published.

Properties:

MW: 302.3	*VP*: 0.00000337 mmHg (77°F)	*FlP*: —
D: 1.51 g/cm^3	*VD*: 10 (calculated)	*LEL*: —
MP: 102°F	*Vlt*: 0.004 ppm	*UEL*: —
BP: —	*H_2O*: 0.0187%	*RP*: 2,000,000
Vsc: —	*Sol*: Most organic solvents	*IP*: —

C11-A019

Methyl parathion

CAS: 298-00-0
RTECS: TG0175000
UN: 2783 (solid); 3018 (liquid)
ERG: 152

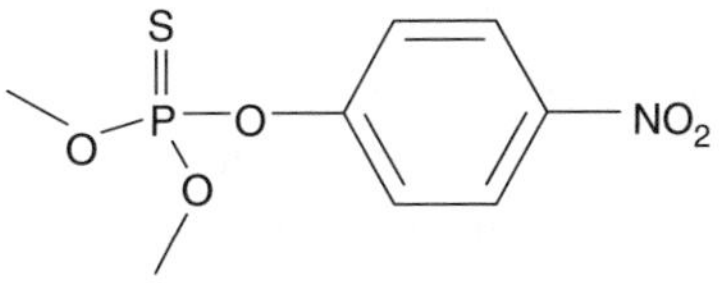

$C_8H_{10}NO_5PS$

White to tan, crystalline solid or powder. Commercial product (80% solution in xylene) is tan colored. The pure material is odorless; otherwise, it has a pungent odor like garlic or rotten eggs. This material is hazardous through inhalation, skin absorption, and ingestion, and produces local skin/eye impacts.

This material poses an explosive risk when heated above 122°F.

Used industrially as an insecticide and acaricide.

This material is on the CDC and FBI threat lists.

Exposure Hazards

Conversion Factor: 1 ppm = 10.77 mg/m^3 at 77°F
ACGIH TLV: 0.019 ppm [Skin]

Properties:

MW: 263.2	*VP*: 0.0000097 mmHg	*FlP*: 115°F
D: 1.358 g/cm^3	*VD*: 9.1 (calculated)	*LEL*: —
MP: 95°F	*Vlt*: 0.013 ppm	*UEL*: —
BP: 289°F	H_2O: 0.0055%	*RP*: 630,000
Vsc: —	*Sol*: Most organic solvents	*IP*: —

C11-A020

Mevinphos

CAS: 7786-34-7
RTECS: GQ5250000

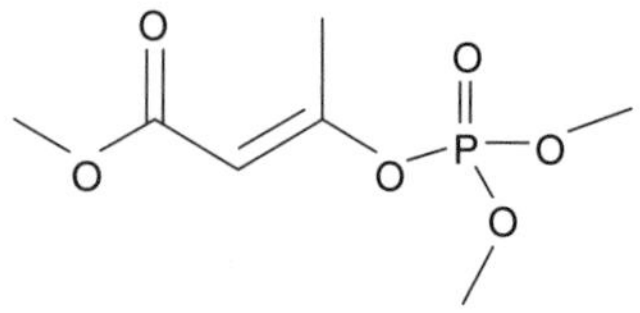

$C_7H_{13}O_6P$

Pale-yellow to orange liquid with a mild or "weak" odor. This material is hazardous through inhalation, skin absorption, and ingestion, and produces local skin/eye impacts.

Used industrially as an insecticide and acaricide. The "*cis*" isomer is about 100 times more toxic than the "*trans*" isomer.

This material is on the CDC and FBI threat lists.

Exposure Hazards

Conversion Factor: 1 ppm = mg/m^3 at 77°F
OSHA PEL: 0.01 ppm [Skin]
ACGIH TLV: 0.001 ppm [Skin]
NIOSH STEL: 0.03 ppm
IDLH: 4 ppm

Properties:

MW: 224.1	*VP*: 0.003 mmHg	*FlP*: 175°F
D: 1.25 g/mL (mixture)	*VD*: 7.7 (calculated)	*LEL*: —
D: 1.2345 g/mL (*cis* isomer)	*Vlt*: 0.17 ppm	*UEL*: —
D: 1.245 g/mL (*trans* isomer)	H_2O: Miscible (decomposes rapidly)	*RP*: 52,000
MP: 70°F (*cis* isomer)	*Sol*: Most organic solvents	*IP*: —
MP: 44°F (*trans* isomer)		
BP: Decomposes		
BP: 223°F (1 mmHg)		
Vsc: 6.6 cS (77°F)		

C11-A021

Monocrotophos

CAS: 6923-22-4
RTECS: TC4375000

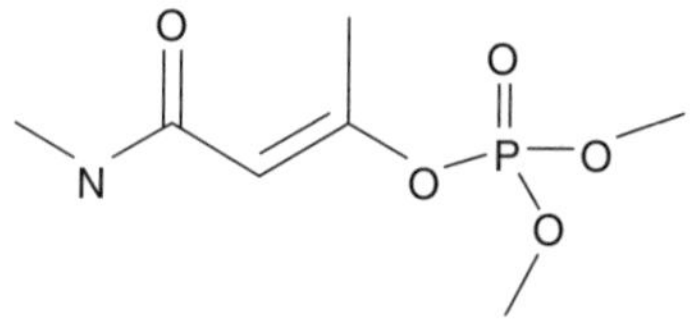

$C_7H_{14}NO_5P$

Colorless to reddish-brown solid with a mild odor. This material is hazardous through inhalation, skin absorption, and ingestion, and produces local skin/eye impacts.

Used industrially as an insecticide and acaricide.

This material is on the CDC and FBI threat lists.

Exposure Hazards

Conversion Factor: 1 ppm = 9.13 mg/m^3 at 77°F
ACGIH TLV: 0.005 ppm [Skin]

Properties:

MW: 223.2	*VP*: 0.000007 mmHg	*FlP*: >200°F
D: —	*VD*: 7.7 (calculated)	*LEL*: —
MP: 129°F	*Vlt*: 0.01 ppm	*UEL*: —
MP: 77–86°F (commercial grade)	H_2O: Miscible	*RP*: 950,000
BP: 257°F	*Sol*: Acetone; Alcohols	*IP*: —
Vsc: —		

C11-A022

Parathion

CAS: 56-38-2
RTECS: TF4550000

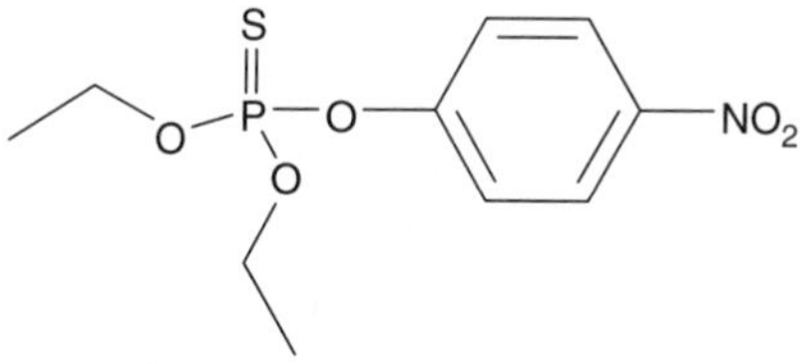

$C_{10}H_{14}NO_5PS$

Pale-yellow to dark-brown liquid with a faint odor like garlic. Odor becomes more pronounced during storage. This material is hazardous through inhalation, skin absorption (liquid), penetration through broken skin, and ingestion, and produces local skin/eye impacts.

Used industrially as an insecticide and acaricide.

This material is on the FBI and CDC threat lists.

Exposure Hazards

Conversion Factor: 1 ppm = 11.91 mg/m^3 at 77°F
$LD_{50\,(Ing)}$: 0.210 g
$MEG_{(1\,h)}$ *Min*: 0.024 ppm; *Sig*: 0.16 ppm; *Sev*: 0.8 ppm
OSHA PEL: 0.008 ppm [Skin]
ACGIH TLV: 0.004 ppm [Skin]
IDLH: 0.8 ppm

Properties:

MW: 291.3	*VP*: 0.00004 mmHg	*FlP*: 392°F
D: 1.26 g/mL (77°F)	*VD*: 10 (calculated)	*LEL*: —
MP: 43°F	*Vlt*: 0.05 ppm	*UEL*: —
BP: 707°F	H_2O: 0.0011%	*RP*: 150,000
Vsc: 12.14 cS (77°F)	*Sol*: Most organic solvents	*IP*: —

C11-A023

Phorate

CAS: 298-02-2
RTECS: TD9450000

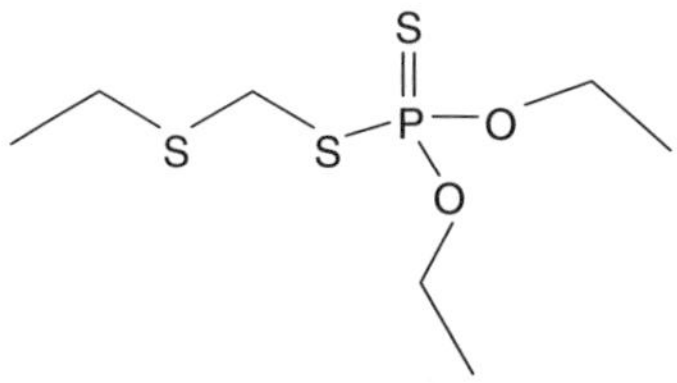

$C_7H_{17}O_2PS_3$

Colorless to light yellow liquid with a skunk-like odor. Commercial material may be light brown. This material is hazardous through inhalation, skin absorption, and ingestion, and produces local skin/eye impacts.

Used industrially as an insecticide and acaricide.

This material is on the CDC and FBI threat lists.

Exposure Hazards

Conversion Factor: 1 ppm = mg/m^3 at 77°F
ACGIH TLV: 0.005 ppm [Skin]
NIOSH STEL: 0.019 ppm [Skin]

Properties:

MW: 260.4	*VP*: 0.000638 mmHg	*FlP*: 320°F
D: 1.156 g/mL (77°F)	*VD*: 9.0 (calculated)	*LEL*: —
D: 1.167 g/mL (77°F; commercial grade)	*Vlt*: 0.85 ppm	*UEL*: —
MP: −45°F	H_2O: 0.005% (decomposes slowly)	*RP*: 9600
BP: 167°F (0.1 mmHg)	*Sol*: Most organic solvents	*IP*: —
Vsc: —		

C11-A024

Phosphamidon

CAS: 13171-21-6
RTECS: TC2800000

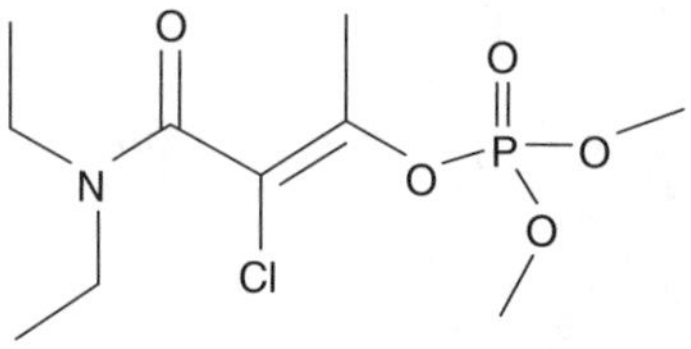

$C_{10}H_{19}ClNO_5P$

Colorless to pale-yellow, oily liquid with a "faint" odor. Commercial product is a mixture of *cis* and *trans* isomers. This material is hazardous through inhalation, skin absorption, penetration through broken skin, and ingestion, and produces local skin/eye impacts.

Used as an agricultural insecticide.

Exposure Hazards

Conversion Factor: 1 ppm = 12.26 mg/m^3 at 77°F

$LD_{50(Ing)}$: 0.49 g (estimate)

Properties:

MW: 299.7	*VP*: 0.0000165 (77°F)	*FlP*: —
D: 1.2132 g/mL (77°F)	*VD*: 10 (calculated)	*LEL*: —
MP: –49°F	*Vlt*: 0.02 ppm 77°F)	*UEL*: —
BP: 324°F (1.5 mmHg)	H_2O: Miscible	*RP*: 350,000
BP: 248°F (0.001 mmHg)	*Sol*: Most organic solvents	*IP*: —
Vsc: 58 cS (77°F; Commercial material)		

C11-A025

Schradan

CAS: 152-16-9

RTECS: —

$C_8H_{24}N_4O_3P_2$

Commercial material is a dark brown viscous liquid. This material is hazardous through inhalation, skin absorption, penetration through broken skin, and ingestion, and produces local skin/eye impacts.

Used as an agricultural systemic insecticide and acaricide.

Exposure Hazards

Conversion Factor: 1 ppm = 11.71 mg/m^3 at 77°F

Human toxicity values have not been established or have not been published.

Properties:

MW: 286.3	*VP*: 0.001 mmHg (77°F)	*FlP*: —
D: 1.1343 g/mL (77°F)	*VD*: 9.9 (calculated)	*LEL*: —
MP: 63°F	*Vlt*: 1.3 ppm (77°F)	*UEL*: —
BP: 248°F (0.5mmHg)	H_2O: Miscible	*RP*: 6000
Vsc: —	*Sol*: Alcohols; Ketones; Aromatic hydrocarbons	*IP*: —

C11-A026

Sulfotepp

CAS: 3689-24-5

RTECS: XN4375000

UN: 1704
ERG: 153

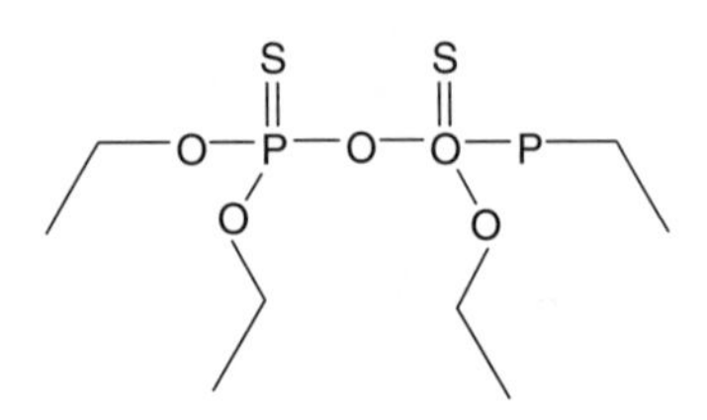

$C_8H_{20}O_5P_2S_2$

Colorless to pale-yellow liquid with an odor like garlic. This material is hazardous through inhalation, skin absorption, and ingestion, and produces local skin/eye impacts.

Used industrially as an insecticide and acaricide.

This material is on the CDC and FBI threat lists.

Exposure Hazards

Conversion Factor: 1 ppm = mg/m^3 at 77°F

Human toxicity values have not been established or have not been published.

OSHA PEL: 0.015 ppm [Skin]
ACGIH TLV: 0.0076 ppm [Skin]
IDLH: 0.76 ppm

Properties:

MW: 322.3	*VP*: 0.000105 mmHg	*FlP*: None
D: 1.196 g/mL (77°F)	*VD*: 11 (calculated)	*LEL*: None
MP: —	*Vlt*: 0.14 ppm	*UEL*: None
BP: Decomposes	H_2O: 0.0025%	*RP*: 53,000
BP: 277°F (2 mmHg)	*Sol*: Most organic solvents	*IP*: —
Vsc: —		

C11-A027

Terbufos

CAS: 13071-79-9
RTECS: —

$C_9H_{21}O_2PS_3$

Clear, colorless to pale-yellow liquid. Commercial material may be reddish to brown. This material is hazardous through inhalation, skin absorption, and ingestion, and produces local skin/eye impacts.

Used industrially as a soil insecticide.

This material is on the CDC and FBI threat lists.

Exposure Hazards

Conversion Factor: 1 ppm = mg/m^3 at 77°F
ACGIH TLV: 0.0008 ppm [Skin]

Properties:

MW: 288.4	*VP*: 0.00032 mmHg	*FlP*: 190°F
D: 1.105 g/mL (75°F)	*VD*: 9.9 (calculated)	*LEL*: —
MP: –21°F	*Vlt*: 0.43 ppm	*UEL*: —
BP: 156°F (0.01 mmHg)	H_2O: 0.0005%	*RP*: 18,000
Vsc: —	*Sol*: Most organic solvents	*IP*: —

C11-A028

Tetraethyl pyrophosphate (Agent EA 1285)

CAS: 107-49-3
RTECS: UX6825000
UN: 3018
ERG: 152

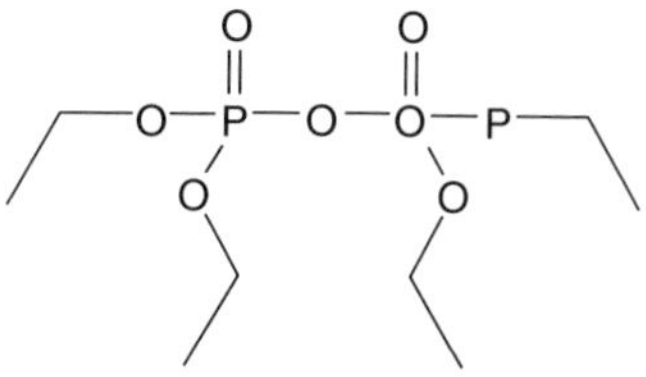

$C_8H_{20}O_7P_2$

Colorless to amber liquid with a faint fruity odor. This material is hazardous through inhalation, skin absorption, and ingestion, and produces local skin/eye impacts.

Used industrially as an insecticide and acaricide.

This material is on the CDC and FBI threat lists, and on the ITF-25 low threat list.

Exposure Hazards

Conversion Factor: 1 ppm = 11.87 mg/m^3 at 77°F
OSHA PEL: 0.004 ppm [Skin]
ACGIH TLV: 0.004 ppm [Skin]
IDLH: 0.42 ppm

Properties:

MW: 290.2	*VP*: 0.0002 mmHg	*FlP*: None
D: 1.1998 g/mL	*VP*: 0.00047 mmHg (86°F)	*LEL*: None
MP: 32°F	*VD*: 10 (calculated)	*UEL*: None
BP: Decomposes	*Vlt*: 0.27 ppm	*RP*: 29,000
BP: 255°F (1 mmHg)	H_2O: Miscible (decomposes)	*IP*: —
Vsc: —	*Sol*: Acetone; Alcohols; Aromatic and Halogenated solvents	

C11-A029

Aldicarb

CAS: 116-06-3
RTECS: UE2275000

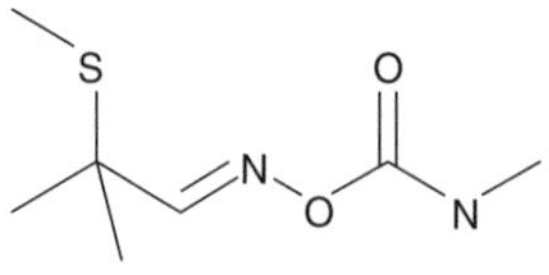

$C_7H_{14}N_2O_2S$

Colorless crystalline material with a faint sulfur odor. This material is hazardous through inhalation, skin absorption, and ingestion, and produces local skin/eye impacts.

Used industrially as a soil insecticide, acaricide, nematocide, and aviacide.

This material is on the CDC and FBI threat lists.

Exposure Hazards

Conversion Factor: 1 ppm = 7.78 mg/m^3 at 77°F
AIHA WEEL: 0.07 mg/m^3 [Skin]

Properties:

MW: 190.3	*VP*: 0.0000975 mmHg	*FlP*: —
D: 1.195 g/cm^3 (77°F)	*VD*: 6.6 (calculated)	*LEL*: —
MP: 210°F	*Vlt*: Negligible	*UEL*: —
BP: Decomposes	*H_2O*: 0.493%	*RP*: 74,000
Vsc: —	*Sol*: Acetone; Ethanol	*IP*: —

C11-A030

Bendiocarb

CAS: 22781-23-3
RTECS: FC1140000

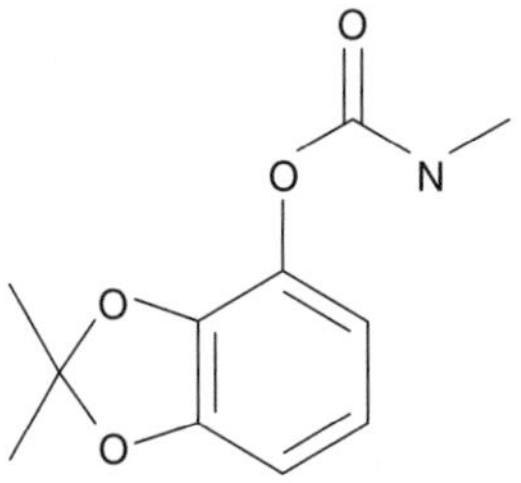

$C_{11}H_{13}NO_4$

White solid. This material is hazardous through inhalation, penetration through broken skin, and ingestion, and produces local skin/eye impacts without irritation.

Used as a household insecticide to control cockroaches, other household pests, and soil insects.

Exposure Hazards

Conversion Factor: 1 ppm = 9.13 mg/m^3 at 77°F

Human toxicity values have not been established or have not been published.

Properties:

MW: 223.2	*VP*: 0.0000345 mmHg (77°F)	*FlP*: —
D: 1.25 g/cm^3	*VD*: 7.7 (calculated)	*LEL*: —
MP: 264°F	*Vlt*: 0.05 ppm (77°F)	*UEL*: —
BP: —	*H_2O*: 0.026% (77°F)	*RP*: 200,000
Vsc: —	*Sol*: Alcohols; Acetone; Dichloromethane	*IP*: —

C11-A031

Carbofuran

CAS: 1563-66-2
RTECS: FB9450000

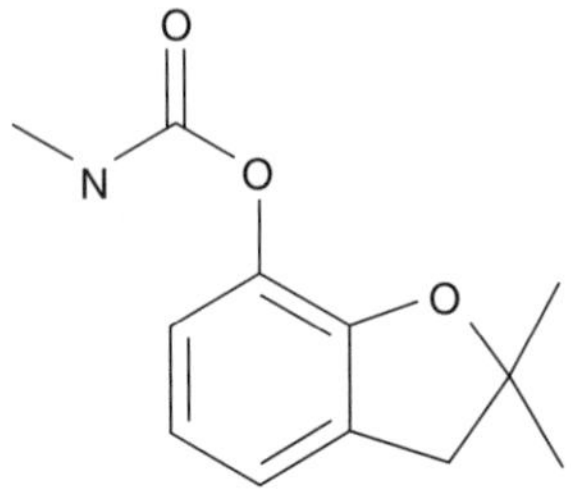

$C_{12}H_{15}NO_3$

White to grayish crystalline solid. It is odorless when pure but impurities may give it an odor similar to phenol. This material is hazardous through inhalation and ingestion, and produces local skin/eye impacts.

Used industrially as a soil insecticide, acaricide, and nematocide.

This material is on the CDC and FBI threat lists.

Exposure Hazards

Conversion Factor: 1 ppm = 9.05 mg/m^3 at 77°F
ACGIH TLV: 0.1 mg/m^3

Properties:

MW: 221.3	*VP*: 0.0000034 mmHg (77°F)	*FlP*: None
D: 1.180 g/cm^3	*VD*: 7.6 (calculated)	*LEL*: None
MP: 304°F	*Vlt*: Negligible	*UEL*: None
BP: —	*H_2O*: 0.07% (77°F)	*RP*: 2,000,000
Vsc: —	*Sol*: Acetone; Acetonitrile; Dimethylformamide; Dimethyl sulfoxide	*IP*: —

C11-A032

Methiocarb

CAS: 2032-65-7
RTECS: —

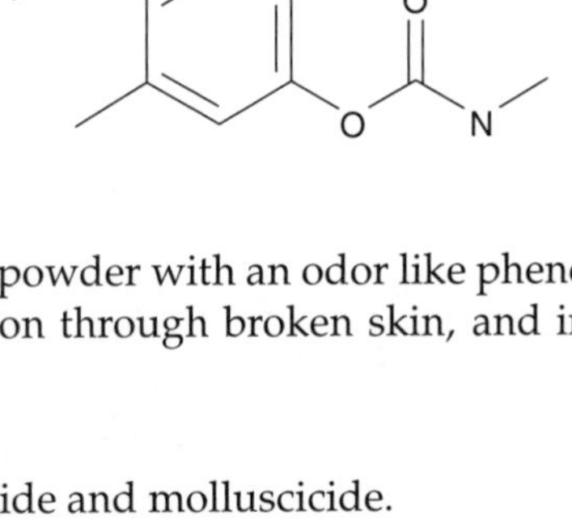

$C_{11}H_{15}NO_2S$

Colorless to white crystalline powder with an odor like phenol. This material is hazardous through inhalation, penetration through broken skin, and ingestion, and produces local skin/eye impacts.

Dust may be explosive.

Used industrially as an acaricide and molluscicide.

Exposure Hazards

Human toxicity values have not been established or have not been published.

Properties:

MW: 225.3	*VP*: Negligible	*FlP*: —
D: —	*VD*: —	*LEL*: —
MP: 246°F	*Vlt*: Negligible	*UEL*: —
BP: —	H_2O: 0.0027%	*RP*: 25,000,000
Vsc: —	*Sol*: Most organic solvents	*IP*: —

C11-A033

Dazomet

CAS: 533-74-4
RTECS: XI2800000

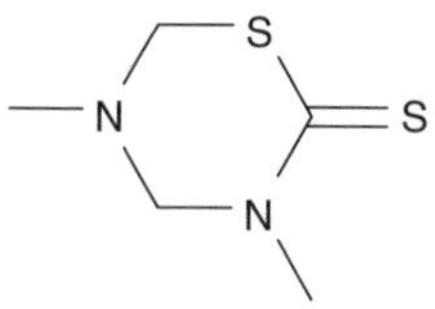

$C_5H_{10}N_2S_2$

White crystalline solid that is nearly odorless. This material is hazardous through inhalation, penetration through broken skin, and ingestion, and produces local skin/eye impacts. Even dilute solution can cause skin irritation and sensitization.

Used industrially as a soil sterilant, biocide in industrial wastewater, slimicide in pulp and paper manufacture, and as a preservative in adhesives and glues.

Exposure Hazards

Conversion Factor: 1 ppm = 6.64 mg/m^3 at 77°F

Human toxicity values have not been established or have not been published.

Properties:

MW: 162.3	*VP*: 0.00000277 mmHg	*FlP*: —
D: 1.30 g/cm^3	*VD*: 5.6 (calculated)	*LEL*: —
MP: 223°F (decomposes)	*Vlt*: —	*UEL*: —
BP: —	H_2O: 0.12% (77°F)	*RP*: 2,800,000
Vsc: —	*Sol*: Acids; Acetone; Benzene; Chloroform	*IP*: —

C11-A034

Methomyl

CAS: 16752-77-5
RTECS: AK2975000

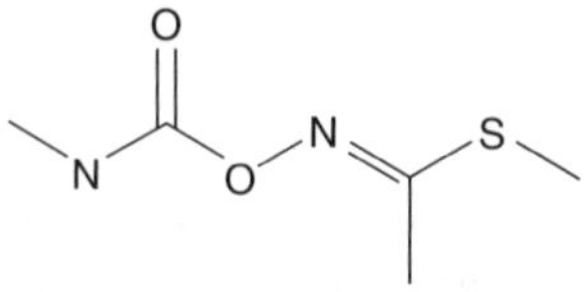

$C_5H_{10}N_2O_2S$

White, crystalline solid with a faint sulfur odor. This material is hazardous through inhalation and ingestion, and produces local skin/eye impacts.

Used industrially as an insecticide and acaricide.

This material is on the CDC and FBI threat lists.

Exposure Hazards

Conversion Factor: 1 ppm = 6.63 mg/m^3 at 77°F
ACGIH TLV: 2.5 mg/m^3

Properties:

MW: 162.2	*VP*: 0.0000054 mmHg (77°F)	*FlP*: None
D: 1.295 g/cm^3 (75°F)	*VD*: 5.6 (calculated)	*LEL*: None
MP: 172°F	*Vlt*: Negligible	*UEL*: None
BP: —	H_2O: 5.8% (77°F)	*RP*: 1,500,000
Vsc: —	*Sol*: Most organic solvents	*IP*: —

C11-A035

1,2-Dibromo-3-chloropropane

CAS: 96-12-8
RTECS: TX8750000
UN: 2872
ERG: 159

$C_3H_5Br_2Cl$

Colorless dense liquid that may become yellow or amber on standing. Commercial material is amber to dark brown. It has a pungent odor at high concentrations. This material is hazardous through inhalation, skin absorption (liquid), penetration through broken skin, and ingestion, and produces local skin/eye impacts.

Used industrially as an intermediate in organic synthesis and for production of retardants and in agriculture as a soil fumigant.

Exposure Hazards

Conversion Factor: 1 ppm = 9.67 mg/m^3 at 77°F
Irritation: 0.2 ppm; exposure duration unspecified
OSHA PEL: 0.001 ppm

Properties:

MW: 236.4	*VP*: 0.58 mmHg	*FlP*: 170°F
D: 2.08 g/mL	*VD*: 8.2 (calculated)	*LEL*: —
MP: 41°F	*Vlt*: 780 ppm	*UEL*: —
BP: 384°F (decomposes)	H_2O: 0.123%	*RP*: 11
BP: 328°F (300 mmHg)	*Sol*: Acetone; Isopropyl alcohol; Hydrocarbons	*IP*: —
Vsc: —		

C11-A036

Dibromomethane

CAS: 74-95-3
RTECS: PA7350000
UN: 2664
ERG: 160

CH_2Br_2

Clear, colorless liquid. This material is hazardous through inhalation, skin absorption (liquid), penetration through broken skin, and ingestion, and produces local skin/eye impacts.

Used industrially as a solvent and in chemical synthesis; as a fire suppressant and a gage fluid.

Exposure Hazards

Conversion Factor: 1 ppm = 7.11 mg/m^3 at 77°F

Human toxicity values have not been established or have not been published. However, based on available information, this material appears to be more toxic than either methylene chloride or methylene chlorobromide. It is carbon monoxide in the body.

Properties:

MW: 173.8	*VP*: 44.4 mmHg (77°F)	*FlP*: None
D: 2.4969 g/mL	*VD*: 6.1 (calculated)	*LEL*: None
MP: −63°F	*Vlt*: 58,000 ppm (77°F)	*UEL*: None
BP: 207°F	*H_2O*: 1.193%	*RP*: 0.12
Vsc: 0.39 cS (77°F)	*Sol*: Chloroform; Alcohol; Ether; Acetone	*IP*: 10.41 eV

C11-A037

Dichloronitroethane

CAS: 594-72-9
RTECS: KI1050000
UN: 2650
ERG: 153

NO2
Cl
Cl

$C_2H_3NO_2Cl_2$

Colorless liquid with an unpleasant odor. This material is hazardous through inhalation and ingestion, and produces local skin/eye impacts.

Used industrially as a solvent and for organic synthesis; as a fumigant for stored produce and grain.

Exposure Hazards

Conversion Factor: 1 ppm = 5.89 mg/m^3 at 77°F
OSHA Ceiling: 10 ppm
ACGIH TLV: 2 ppm
IDLH: 25 ppm

Properties:

MW: 143.9	*VP*: 16 mmHg (77°F)	*FlP*: 168°F
D: 1.4271 g/mL	*VD*: 5.0 (calculated)	*LEL*: —
MP: —	*Vlt*: 21,000 ppm (77°F)	*UEL*: —
BP: 255°F	*H_2O*: 0.25%	*RP*: 0.6
Vsc: —	*Sol*: —	*IP*: —

C11-A038

1,2-Dichloropropane

CAS: 78-87-5
RTECS: TX9625000
UN: 1279
ERG: 130

Cl Cl

$C_3H_6Cl_2$

Colorless liquid with a sweet odor similar to chloroform detectable at 130–190 ppm. This material is hazardous through inhalation, skin absorption (liquid), penetration through broken skin, and ingestion, and produces local skin/eye impacts.

Used industrially as a chemical intermediate, solvent, dry cleaning fluid, degreaser, and lead scavenger for antiknock fluids; used in agriculture as a fumigant.

Exposure Hazards

Conversion Factor: 1 ppm = 4.62 mg/m^3 at 77°F
OSHA PEL: 75 ppm
ACGIH TLV: 10 ppm
IDLH: 400 ppm

Properties:

MW: 113.0	*VP*: 53.3 mmHg (77°F)	*FlP*: 60°F
D: 1.159 g/mL	*VD*: 3.9 (calculated)	*LEL*: 3.4%
MP: −149°F	*Vlt*: 70,000 ppm (77°F)	*UEL*: 14.5%
BP: 206°F	*H_2O*: 0.28%	*RP*: 0.18
BP: 26°F (10 mmHg)	*Sol*: Miscible with organic solvents	*IP*: 10.87 eV
Vsc: —		

C11-A039

Ethyl formate

CAS: 109-94-4
RTECS: LQ8400000
UN: 1190
ERG: 129

O O H

$C_3H_6O_2$

Colorless mobile liquid with a fruity odor and slightly bitter taste. This material is hazardous through inhalation and ingestion, and produces local skin/eye impacts.

Used industrially as an intermediate in drug synthesis, in the manufacture of safety glasses, as a flavor for lemonade and for essences, in the manufacture of artificial rum and arrack, and in agriculture as a fumigant for dried fruits.

Exposure Hazards

Conversion Factor: 1 ppm = 3.03 mg/m^3 at 77°F
Eye and Nasal Irritation: 330 ppm; exposure duration unspecified
OSHA PEL: 100 ppm
ACGIH TLV: 100 ppm
IDLH: 1500 ppm

Properties:

MW: 74.1	*VP*: 200 mmHg	*FlP*: −4°F
D: 0.9168 g/mL	*VP*: 245 mmHg (77°F)	*LEL*: 2.8%
MP: −113°F	*VD*: 2.6 (calculated)	*UEL*: 16.0%
BP: 130°F	*Vlt*: 270,000 ppm	*RP*: 0.06
Vsc: 0.41 cS	H_2O: 9% (64°F); Decomposes slowly	*IP*: 10.61 eV
	Sol: Most organic solvents	

C11-A040

Ethylene bromide

CAS: 106-93-4
RTECS: KH9275000
UN: 1605
ERG: 154

Br Br

$C_2H_4Br_2$

Colorless heavy liquid with a mild, sweet odor like chloroform that is detectible at 8–10 ppm. This material is hazardous through inhalation, skin absorption, penetration through broken skin, and ingestion, and produces local skin/eye impacts.

Used industrially as a chemical intermediate in the manufacture of dyes, pharmaceuticals, and polymers; as a special solvent, as an exhaust system scavenger for leaded fuels, and in gauge fluids; used in agriculture as a fumigant.

This material is on the ITF-25 medium threat list.

Exposure Hazards

Conversion Factor: 1 ppm = 7.68 mg/m^3 at 77°F
OSHA PEL: 20 ppm
OSHA Ceiling: 30 ppm
IDLH: 770 ppm

Properties:

MW: 187.9	*VP*: 8.5 mmHg	*FlP*: None
D: 2.172 g/mL	*VP*: 11.2 mmHg (77°F)	*LEL*: None
MP: 50°F	*VD*: 6.5 (calculated)	*UEL*: None
BP: 268°F	*Vlt*: 15,000 ppm	*RP*: 0.66
Vsc: 0.795 cS	H_2O: 0.4%	*IP*: 9.45 eV
	Sol: Most organic solvents	

Proposed AEGLs

AEGL-1: Not Developed
AEGL-2: Not Developed
AEGL-3: 1 h, 26 ppm   4 h, 13 ppm   8 h, 10 ppm

C11-A041

Ethylene chloride

CAS: 107-06-2
RTECS: KI0525000
UN: 1184
ERG: 131

Cl Cl

$C_2H_4Cl_2$

Clear, colorless oily liquid that darkens on storage due to decomposition. It has a pleasant, sweet odor similar to chloroform. This material is hazardous through inhalation, skin absorption, penetration through broken skin, and ingestion, and produces local skin/eye impacts.

Used industrially as a solvent, for the production of various halogenated materials including vinyl chloride, trichloroethylene, and trichloroethane, in the manufacture of soaps, scouring compounds, wetting agents, penetrating agents, for ore flotation, and in organic synthesis. Also used as a fumigant.

Exposure Hazards

Conversion Factor: 1 ppm = 4.05 mg/m^3 at 77°F
ACGIH TLV: 10 ppm
IDLH: 50 ppm

Properties:

MW: 99.0	*VP*: 64 mmHg	*FlP*: 56°F
D: 1.2351 g/mL	*VP*: 78.9 mmHg (77°F)	*LEL*: 6.2%
MP: −32°F	*VD*: 3.4 (calculated)	*UEL*: 16%
BP: 182°F	*Vlt*: 86,000 ppm	*RP*: 0.16
Vsc: 0.68 cS	*H_2O*: 0.869%	*IP*: 11.05 eV
	Sol: Most organic solvents	

C11-A042

Methyl bromide

CAS: 74-83-9
RTECS: PA4900000
UN: 1062
ERG: 123

CH_3Br

Colorless gas that is odorless except at high concentrations; then it has a sweetish odor like chloroform. This material is hazardous through inhalation, skin absorption of the liquid, and ingestion, and the liquid produces local skin/eye impacts.

Used industrially as a fumigant, herbicide and in organic synthesis.

This material is on the CDC and FBI threat lists, and on the ITF-25 medium threat list.

Exposure Hazards

Conversion Factor: 1 ppm = 3.88 mg/m^3 at 77°F

Lethal human toxicity values have not been fully established. However, deaths have been reported for exposures of 6% for 1 h, and 1600–8200 for 4–6 h.

$MEG_{(1h)}$ *Min*: 15 ppm; *Sig*: 50 ppm; *Sev*: 200 ppm
ACGIH TLV: 1 ppm [Skin]
OSHA Ceiling: 20 ppm [Skin]
IDLH: 250 ppm

Properties:

MW: 94.9	*VP*: 1600 mmHg (77°F)	*FlP*: —
D: 1.732 g/mL (32°F)	*VD*: 3.3 (calculated)	*LEL*: 10%
MP: −135°F	*Vlt*: 2,100,000 ppm	*UEL*: 16%
BP: 38°F	H_2O: 1.34% (77°F)	*RP*: 0.006
Vsc: 0.23 cS (32°F)	*Sol*: Ethanol; Chloroform; Ether	*IP*: 10.54 eV

Proposed AEGLs

AEGL-1: Not Developed		
AEGL-2: 1 h, 210 ppm	4 h, 67 ppm	8 h, 67 ppm
AEGL-3: 1 h, 740 ppm	4 h, 230 ppm	8 h, 130 ppm

C11-A043

β-Methallyl chloride

CAS: 563-47-3
RTECS: UC8050000
UN: 2554
ERG: 130P

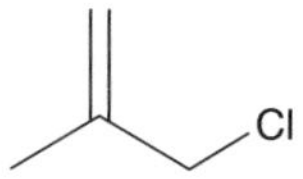

C_4H_7Cl

Colorless to pale-yellow liquid with a sharp, penetrating, disagreeable odor. This material is hazardous through inhalation and ingestion, and produces local skin/eye impacts.

Used industrially as a chemical intermediate for insecticides, plastics, and pharmaceuticals; and as a fumigant.

Exposure Hazards

Conversion Factor: 1 ppm = 3.70 mg/m^3 at 77°F

Human toxicity values have not been established or have not been published. However, concentrations above 1500 ppm cause gasping, refusal to breathe, coughing, substernal pain, and extreme respiratory distress.

Properties:

MW: 90.6	*VP*: 101.7 mmHg	*FlP*: 10°F
D: 0.9165 g/mL	*VD*: 3.1 (calculated)	*LEL*: 3.2%
D: 0.93 g/mL (Commercial material)	*Vlt*: 140,000 ppm	*UEL*: 8.1%
MP: −112°F	H_2O: 0.14% (decomposes slowly)	*RP*: 0.10
BP: 160°F	*Sol*: Alcohol; Ether; Chloroform	*IP*: —
Vsc: 4.58 cS		

C11-A044

Phosphine

CAS: 7803-51-2
RTECS: SY7525000
UN: 2199
ERG: 119

PH_3

Colorless gas that is odorless when pure; otherwise, like decaying fish or garlic that is detectable at 0.03 ppm. This material is hazardous through inhalation.

May spontaneously ignite in air if traces of diphosphorous hydride (P_2H_2) are present.

Used industrially as a fumigant, doping agent for semiconductors, and intermediate for preparation of some flame retardants.

Phosphine is generated when phosphide salts (i.e., Aluminum phosphide; Calcium phosphide; Magnesium phosphide; Magnesium aluminum phosphide; Potassium phosphide; Sodium phosphide; Stannic phosphide; Strontium phosphide; Zinc phosphide) come into contact with water.

This material is on the CDC and FBI threat lists, and on the ITF-25 medium threat list.

Exposure Hazards

Conversion Factor: 1 ppm = 1.39 mg/m^3 at 77°F
$LC_{50\,(Inh)}$: 500 ppm for a 30-min exposure
$LC_{50\,(Inh)}$: 1000 ppm "after a few breaths".
$MEG_{(1\,h)}$ *Min*: —; *Sig*: 0.3 ppm;
Sev: 1.1 ppm
OSHA PEL: 0.3 ppm
ACGIH TLV: 0.3 ppm
ACGIH STEL: 1 ppm
NIOSH STEL: 1 ppm
IDLH: 50 ppm

Properties:

MW: 34.0	*VP*: 31,400 mmHg	*FlP*: —
D: 0.491 g/mL (liq. gas, 77°F)	*VD*: 1.2 (calculated)	*LEL*: 1.79%
MP: −207°F	*Vlt*: —	*UEL*: —
BP: −126°F	H_2O: "Slight"	*RP*: 0.001
Vsc: 0.0713 cS (liq. gas, 77°F)	*Sol*: Alcohols; Ether	*IP*: 9.96 eV

Interim AEGLs

AEGL-1: Not Developed

AEGL-2: 1 h, 2 ppm	4 h, 0.5 ppm	8 h, 0.25 ppm
AEGL-3: 1 h, 3.6 ppm	4 h, 0.90 ppm	8 h, 0.45 ppm

C11-A045

Sulfonyl fluoride

CAS: 2699-79-8
RTECS: WT5075000

UN: 2191
ERG: 123

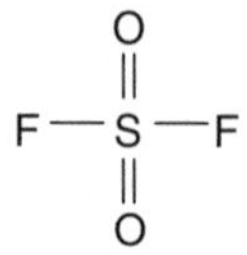

SF_2O_2

Colorless odorless gas. This material is hazardous through inhalation.

Used industrially as a fumigant and in organic synthesis.

This material is on the CDC and FBI threat lists, and on the ITF-25 medium threat list.

Exposure Hazards

Conversion Factor: 1 ppm = 4.18 mg/m^3 at 77°F
MEG$_{(1h)}$ Min: —; *Sig*: —; *Sev*: 200 ppm
OSHA PEL: 5 ppm
ACGIH TLV: 5 ppm
ACGIH STEL: 10 ppm
NIOSH STEL: 10 ppm
IDLH: 200 ppm

Properties:

MW: 102.1	*VP*: 12,750 mmHg (70°F)	*FlP*: None
D: 1.386 g/mL (liq. gas, 77°F)	*VD*: 3.5 (calculated)	*LEL*: None
MP: −213°F	*Vlt*: —	*UEL*: None
BP: −68°F	*H_2O*: 0.2% (32°F)	*RP*: 0.001
Vsc: 0.1097 cS (liq. gas, 77°F)	*Sol*: Sparingly soluble in most organic solvents	*IP*: 13.04 eV

C11-A046

Trichloroacetonitrile

CAS: 545-06-2
RTECS: AM2450000

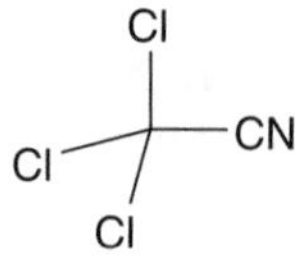

C_2Cl_3N

Clear, pale-yellow liquid. This material is hazardous through inhalation and ingestion, and produces local skin/eye impacts.

Used industrially as an insecticide.

Exposure Hazards

Conversion Factor: 1 ppm = 5.91 mg/m^3 at 77°F

Human toxicity values have not been established or have not been published.

Properties:

MW: 144.4	*VP*: 74.1 mmHg (77°F)	*FlP*: 166°F
D: 1.4403 g/mL (77°F)	*VD*: 5.0 (calculated)	*LEL*: —
MP: −44°F	*Vlt*: 97,000 ppm (77°F)	*UEL*: —
BP: 186°F	*H_2O*: 0.0715% (77°F)	*RP*: 0.11
Vsc: —	*Sol*: —	*IP*: 11.9 eV

C11-A047

Arsenic trioxide

CAS: 1327-53-3
RTECS: CG3325000
UN: 1561
ERG: 151

As_2O_3

Colorless crystalline solid that is odorless and tasteless. This material is hazardous through inhalation, and ingestion, and produces local skin/eye impacts.

Used industrially as a rodenticide, raw material for glass and ceramic production, and intermediate for preparation of other arsenical products.

Produces Arsine (C08-A001) and Arsenic oxide fumes when burned.

This material is on the CDC and FBI threat lists.

Exposure Hazards

Conversion Factor: 1 ppm = 8.09 mg/m^3 at 77°F
OSHA PEL: 0.010 mg/m^3 as arsenic
ACGIH TLV: 0.01 mg/m^3 as arsenic
NIOSH Ceiling: 0.002 mg/m^3 as arsenic (15 min)
IDLH: 5 mg/m^3 as arsenic

Properties:

MW: —	*VP*: 0.000247 mmHg (77°F)	*FlP*: None
D: 3.87–4.15 g/cm^3	*VD*: 6.8 (calculated)	*LEL*: None
MP: 525°F (arsenolite)	*Vlt*: 0.32 ppm	*UEL*: None
MP: 595°F (claudetite)	*H_2O*: 1.6% (61°F)	*RP*: 29,000
BP: Sublimes	*Sol*: Glycerol	*IP*: —
Vsc: —		

Proposed AEGLs

AEGL-1: Not Developed
AEGL-2: 1 h, 3.0 mg/m^3 4 h, 1.9 mg/m^3 8 h, 1.2 mg/m^3
AEGL-3: 1 h, 9.1 mg/m^3 4 h, 5.7 mg/m^3 8 h, 3.7 mg/m^3

C11-A048

Calcium arsenate

CAS: 7778-44-1
RTECS: CG0830000
UN: 1573
ERG: 151

$Ca_3(AsO_4)_2$

Colorless to white odorless solid. This material is hazardous through inhalation, skin absorption, penetration through broken skin, and ingestion, and produces local skin/eye impacts.

Used industrially as an insecticide, molluscicide, and preemergence herbicide.

Exposure Hazards

OSHA PEL: 0.01 mg/m^3 as arsenic
IDLH: 5 mg/m^3 as arsenic

Properties:

MW: 398.1	*VP*: Negligible	*FlP*: None
D: 3.620 g/cm^3	*VD*: —	*LEL*: None
MP: Decomposes	*Vlt*: —	*UEL*: None
BP: —	H_2O: 0.013% (77°F)	*RP*: —
Vsc: —	*Sol*: Dilute mineral acids	*IP*: —

C11-A049

Copper acetoarsenite

CAS: 12002-03-8
RTECS: GL6475000
UN: 1585
ERG: 151

$Cu_4(AsO_2)_6(CH_3CO_2)_2$

$C_4H_6As_6Cu_4O_{16}$

Brilliant bluish-green or deep green, odorless powder. This material is hazardous through inhalation and ingestion, and produces local skin/eye impacts.

Used industrially as an insecticide, fungicide, wood preservative, as paint pigment, and veterinary medication.

This material is on the FBI threat list.

Exposure Hazards

Human toxicity values have not been established or have not been published.

Properties:

MW: 1013.8	*VP*: Negligible	*FlP*: None
D: —	*VD*: —	*LEL*: None
MP: —	*Vlt*: Negligible	*UEL*: None
BP: —	H_2O: <0.1%	*RP*: —
Vsc: —	*Sol*: —	*IP*: —

C11-A050

Lead arsenate

CAS: 7784-40-9
RTECS: CG0990000
UN: 1617
ERG: 151

$HPbAsO_4$

Colorless to white odorless solid. This material is hazardous through inhalation and ingestion, and produces local skin/eye impacts.

Used industrially as an insecticide, herbicide and veterinary medication.

Exposure Hazards

ACGIH TLV: 0.15 mg/m^3 as lead

Properties:

MW: 347.1	*VP*: Negligible	*FlP*: —
D: 5.943 g/cm^3	*VD*: —	*LEL*: —
MP: 536°F (decomposes)	*Vlt*: Negligible	*UEL*: —
BP: —	H_2O: Insoluble	*RP*: —
Vsc: —	*Sol*: Ammonium hydroxide; Nitric acid	*IP*: —

C11-A051

Sodium arsenite

CAS: 7784-46-5
RTECS: CG3675000
UN: 2027
ERG: 151

$NaAsO_2$

White or grayish-white powder that is somewhat hygroscopic. It absorbs carbon dioxide from the air. Technical grade is 90–95% pure. This material is hazardous through inhalation and ingestion, and produces local skin/eye impacts.

Used industrially as a fungicide, insecticide, herbicide, wood preservation, textile drying agent, corrosion inhibitor; as a veterinary medication; and in the manufacture of specialty soaps.

This material is on the FBI threat list.

Exposure Hazards

Human toxicity values have not been established or have not been published.

Properties:

MW: 129.9	*VP*: Negligible	*FlP*: —
D: 1.87 g/cm^3	*VD*: —	*LEL*: —
MP: 1139°F	*Vlt*: Negligible	*UEL*: —
BP: —	H_2O: "Freely soluble"	*RP*: —
Vsc: —	*Sol*: Slightly in alcohol	*IP*: —

C11-A052

Methoxyethyl mercury acetate

CAS: 151-38-2
RTECS: —

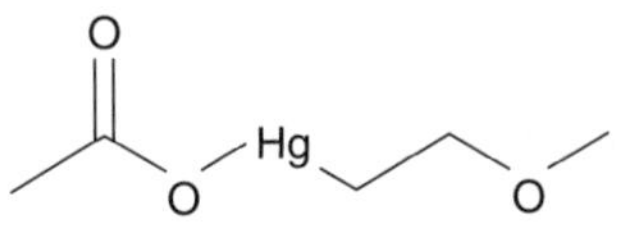

O
Hg
O O

$C_5H_{10}HgO_3$

White crystalline material. This material is hazardous through inhalation and ingestion, and produces local skin/eye impacts.

Used industrially as a fungicide and disinfectant.

This material is on the FBI threat list.

Exposure Hazards

Human toxicity values have not been established or have not been published.

Properties:

MW: 318.7	*VP*: Negligible	*FlP*: —
D: —	*VD*: —	*LEL*: —
MP: 104°F	*Vlt*: Negligible	*UEL*: —
BP: —	H_2O: "Readily soluble"	*RP*: —
Vsc: —	*Sol*: Polar organic solvents	*IP*: —

C11-A053

Methyl mercury dicyandiamide

CAS: 502-39-6
RTECS: —

$C_3H_6HgN_4$

Colorless crystalline solid. This material is hazardous through inhalation and ingestion, and produces local skin/eye impacts.

Used industrially as a seed fungicide and disinfectant, and as a timber preservative.

This material is on the FBI threat list.

Exposure Hazards

Human toxicity values have not been established or have not been published.

Properties:

MW: 298.7	*VP*: 0.000065 mmHg (95°F)	*FlP*: —
D: —	*VD*: 10 (calculated)	*LEL*: —
MP: 313°F	*Vlt*: 0.008 ppm (95°F)	*UEL*: —
BP: —	H_2O: 2.17%	*RP*: 93,000
Vsc: —	*Sol*: Acetone; Ethanol; Ethylene glycol	*IP*: —

C11-A054

Methyl mercury hydroxide

CAS: 1184-57-2
RTECS: OW4900000

CH_3HgOH

Specific information on physical appearance is not available for this material.

Used industrially as a pesticide.

Exposure Hazards

Human toxicity values have not been established or have not been published.

Properties:

MW: 232.6	*VP*: —	*FlP*: —
D: —	*VD*: —	*LEL*: —
MP: 279°F	*Vlt*: —	*UEL*: —
BP: —	H_2O: 0.1–1% (70°F)	*RP*: —
Vsc: —	*Sol*: —	*IP*: —

C11-A055

Mercuric chloride

CAS: 7487-94-7
RTECS: OV9100000
UN: 1624
ERG: 154

$HgCl_2$

White powder or crystalline material that is odorless. This material is hazardous through inhalation, skin absorption, and ingestion, and produces local skin/eye impacts.

Used industrially as an insecticide and fungicide, for leather and fur production, and as a topical antiseptic and disinfectant.

This material is on the CDC and FBI threat lists.

Exposure Hazards

Conversion Factor: 1 ppm = 11.10 mg/m^3 at 77°F

Human toxicity values have not been established or have not been published.

Properties:

MW: 271.5	*VP*: 1 mmHg (277°F)	*FlP*: None
D: 5.6 g/cm^3	*VD*: 9.4 (calculated)	*LEL*: None
MP: 531°F	*Vlt*: —	*UEL*: None
BP: 576°F	H_2O: 6.5%	*RP*: 8.4
Vsc: —	H_2O: 33% (77°F)	*IP*: —
	Sol: Polar organic solvents	

C11-A056

Phenyl mercury acetate

CAS: 62-38-4
RTECS: OV6475000
UN: 1674
ERG: 151

$C_8H_8HgO_2$

White to cream-white powder or crystalline material. Odorless when pure, impurities may give it a slight vinegar odor. This material is hazardous through inhalation, skin absorption, and ingestion, and produces local skin/eye impacts.

Used industrially as a disinfectant, paint preservative, mildewcide, slimicide, and herbicide.

This material is on the CDC and FBI threat lists.

Exposure Hazards

Conversion Factor: 1 ppm = 13.77 mg/m^3 at 77°F

Human toxicity values have not been established or have not been published.

Properties:

MW: 336.7	*VP*: 0.000006 mmHg	*FlP*: 284°F
D: 0.24 g/cm^3 (estimate)	*VD*: 12 (calculated)	*LEL*: —
MP: 300°F	*Vlt*: 0.01 ppm	*UEL*: —
BP: —	H_2O: 0.6%	*RP*: 900,000
Vsc: —	*Sol*: Acetone; Alcohol; Acetic acid	*IP*: —

C11-A057

Paraquat

CAS: 4685-14-7; 1910-42-5 (Dichloride)
RTECS: DW2275000

Cl⁻ Cl⁻
N⁺ N⁺

$C_{12}H_{14}N_2$

Salts are colorless, white, or pale-yellow hygroscopic crystalline solids. Commercial grade material is a dark red solution. This material is hazardous through inhalation, skin absorption, and ingestion, and produces local skin/eye impacts.

Used industrially as an herbicide.

This material is on the CDC and FBI threat lists.

Exposure Hazards

Conversion Factor: 1 ppm = 7.61 mg/m^3 at 77°F
MEG$_{(1h)}$ *Min*: 0.15 mg/m^3; *Sig*: 1.0 mg/m^3; *Sev*: —
OSHA PEL: 0.5 mg/m^3 (Dichloride salt)
ACGIH TLV: 0.5 mg/m^3 (Dichloride salt)
IDLH: 1 mg/m^3 (Dichloride salt)

Dichloride salt:

MW: 257.2	*VP*: Negligible	*FlP*: —
D: 1.25 g/cm^3	*VD*: —	*LEL*: —
MP: Decomposes	*Vlt*: —	*UEL*: —
BP: —	H_2O: 70%	*RP*: —
Vsc: —	*Sol*: Practically insoluble in organic solvents	*IP*: —

C11-A058

Sodium fluoroacetate

CAS: 62-74-8
RTECS: AH9100000

UN: 2629
ERG: 151

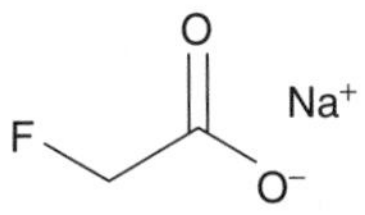

$C_2H_2FO_2{\cdot}Na$

Fluffy, colorless to white powder that is odorless. Commercial material is sometimes dyed black. This material is hazardous through inhalation, skin absorption, and ingestion, and produces local skin/eye impacts.

Used industrially in poison baits to control rodents, wild pigs, and predators.

This material is on the CDC and FBI threat lists.

Exposure Hazards

OSHA PEL: 0.05 mg/m^3 [Skin]
ACGIH TLV: 0.05 mg/m^3 [Skin]
NIOSH STEL: 0.15 mg/m^3 [Skin]
IDLH: 2.5 mg/m^3

Properties:

MW: 100.0	*VP*: Negligible	*FlP*: None
D: —	*VD*: —	*LEL*: None
MP: 392°F	*Vlt*: —	*UEL*: None
BP: Decomposes	*H_2O*: Miscible	*RP*: —
Vsc: —	*Sol*: —	*IP*: —

C11-A059

Strychnine

CAS: 57-24-9
RTECS: WL2275000

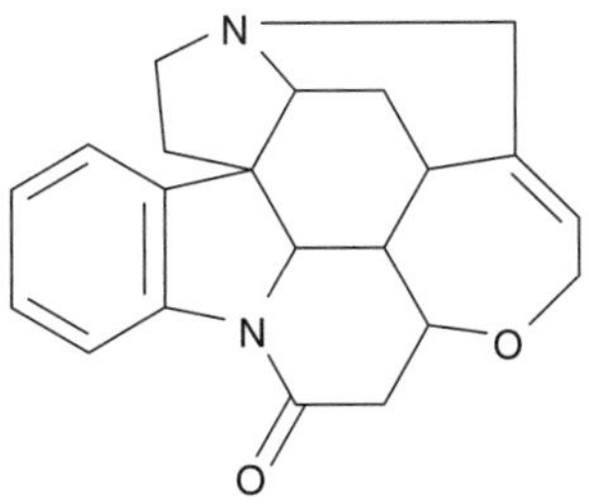

$C_{21}H_{22}N_2O_2$

Colorless and odorless solid. This material is hazardous through inhalation, skin absorption, and ingestion.

Used industrially in poison baits to control rodents, wild pigs, and predators.

This material is on the CDC and FBI threat lists.

Exposure Hazards

OSHA PEL: 0.15 mg/m^3
ACGIH TLV: 0.15 mg/m^3

Properties:

MW: 334.4	*VP*: Negligible	*FlP*: —
D: 1.36 g/cm^3	*VD*: —	*LEL*: —
MP: 549°F	*Vlt*: —	*UEL*: —
BP: Decomposes	H_2O: 0.016%	*RP*: —
BP: 518°F (5 mmHg)	*Sol*: Chloroform	*IP*: —
Vsc: —		

C11-A060

Thallium sulfate

CAS: 7446-18-6
RTECS: XG6800000

Tl_2SO_4

Colorless to white odorless crystalline solid. This material is hazardous through inhalation, skin absorption, and ingestion.

Used industrially in poison baits to control rodents and ants, and as an analytical chemistry reagent.

This material is on the CDC and FBI threat lists.

Exposure Hazards

Human toxicity values have not been established or have not been published.

Properties:

MW: 504.8	*VP*: Negligible	*FlP*: —
D: 6.77 g/cm^3	*VD*: —	*LEL*: —
MP: 1170°F	*Vlt*: —	*UEL*: —
BP: —	H_2O: 4.87% (59°F)	*RP*: —
Vsc: —	H_2O: 19.14% (212°F)	*IP*: —
	Sol: —	

C11-A061

Anhydrous ammonia

CAS: 7664-41-7
RTECS: BO0875000
UN: 1005
ERG: 125

NH_3

Colorless gas with a pungent, suffocating odor detectable at 47 ppm. Shipped as a liquefied compressed gas. This material is hazardous through inhalation and produces local skin/eye impacts.

Used in agriculture as a fertilizer and defoliant; in the manufacture of nitric acid, hydrazine, hydrogen cyanide, urethanes, acrylonitrile, nitrocellulose, nitroparaffins, melamine, ethylene diamine, and sodium carbonate; as an intermediate in producing explosives, synthetic fibers and dyes; and used industrially as a refrigerant gas, neutralizing agent in the petroleum industry, latex preservative, and the production of fuel cells.

This material is on the ITF-25 high threat list.

Exposure Hazards

Conversion Factor: 1 ppm = 0.70 mg/m^3 at 77°F
$LC_{(Inh)}$: 7100 mg/m^3 (5000 ppm); "short" exposure
Eye Irritation: 698 ppm; immediate upon exposure
Respiratory Irritation: 408 ppm; exposure duration unspecified
Skin Burns: 2–3% on wet skin
$MEG_{(1h)}$ *Min*: 25 ppm; *Sig*: 110 ppm; *Sev*: 1100 ppm
OSHA PEL: 50 ppm
ACGIH TLV: 25 ppm
ACGIH STEL: 35 ppm
NIOSH STEL: 35 ppm
IDLH: 300 ppm

Properties:

MW: 17.0
D: 0.682 g/mL (–28°F)
D: 0.602 g/mL (liq. gas, 77°F)
MP: –108°F
BP: –28°F
Vsc: 0.2243 cS (liq. gas, 77°F)
VP: 6500 mmHg
VD: 0.59 (calculated)
Vlt: —
H_2O: 34%
Sol: Alcohol; Chloroform; Ether
FlP: —
LEL: 15%
UEL: 28%
RP: 0.004
IP: 10.18 eV

Interim AEGLs

AEGL-1: 1 h, 30 ppm	4 h, 30 ppm	8 h, 30 ppm
AEGL-2: 1 h, 160 ppm	4 h, 110 ppm	8 h, 110 ppm
AEGL-3: 1 h, 1100 ppm	4 h, 550 ppm	8 h, 390 ppm

C11-A062

Hydrogen bromide

CAS: 10035-10-6
RTECS: MW3850000
UN: 1048
ERG: 125

HBr

Colorless to faintly yellow gas with a sharp, irritating acrid odor. Shipped as a liquefied compressed gas. This material is hazardous through inhalation, and ingestion, and produces local skin/eye impacts.

Used industrially as a brominating agent, reducing agent, and catalyst.

This material is on the ITF-25 high threat list.

Exposure Hazards

Conversion Factor: 1 ppm = 3.31 mg/m^3 at 77°F
$MEG_{(1h)}$ *Min*: 3 ppm (ceiling); *Sig*: 6 ppm; *Sev*: 30 ppm
OSHA PEL: 3 ppm
ACGIH Ceiling: 2 ppm
NIOSH Ceiling: 3 ppm
IDLH: 30 ppm

Properties:

MW: 80.9
D: 1.728 g/mL (liq. gas, 77°F)
VP: 15,000 mmHg
VP: 18,400 mmHg (77°F)
FlP: None
LEL: None

MP: −124°F	*VD*: 2.8 (calculated)	*UEL*: None
BP: −88°F	*Vlt*: —	*RP*: 0.001
Vsc: 0.1163 cS (liq. gas, 77°F)	H_2O: 49%	*IP*: 11.62 eV
	Sol: Most organic solvents	

Proposed AEGLs

AEGL-1: 1 h, 1.0 ppm	4 h, 1.0 ppm	8 h, 1.0 ppm
AEGL-2: 1 h, 22 ppm	4 h, 11 ppm	8 h, 11 ppm
AEGL-3: 1 h, 120 ppm	4 h, 31 ppm	8 h, 31 ppm

C11-A063

Hydrogen chloride

CAS: 7647-01-0
RTECS: MW4025000
UN: 1050
ERG: 125

HCl

Colorless to faintly yellow gas with a sharp, irritating acrid odor. Shipped as a liquefied compressed gas. This material is hazardous through inhalation, and ingestion, and produces local skin/eye impacts.

Used in metal treating/cleaning operations, petroleum well activation, refining ore in the production of tin and tantalum, hydrolyzing of starch and proteins in production of various foods, in the production synthetic rubber, vinyl chloride and alkyl chlorides; and in the manufacture of fertilizers, dyes and dyestuffs, artificial silk and pigments for paints.

This material is on the ITF-25 high threat list.

Exposure Hazards

Conversion Factor: 1 ppm = 1.49 mg/m^3 at 77°F
$MEG_{(1h)}$ *Min*: 1.8 ppm; *Sig*: 22 ppm; *Sev*: 100 ppm
OSHA Ceiling: 5 ppm
ACGIH Ceiling: 2 ppm
NIOSH Ceiling: 5 ppm
IDLH: 50 ppm

Properties:

MW: 36.5	*VP*: 30,800 mmHg	*FlP*: None
D: 0.796 g/mL (liq. gas, 77°F)	*VD*: 1.3 (calculated)	*LEL*: None
MP: −174°F	*Vlt*: —	*UEL*: None
BP: −121°F	H_2O: 67% (86°F)	*RP*: 0.001
Vsc: 0.0842 cS (liq. gas, 77°F)	*Sol*: Alcohol; Benzene; Ether	*IP*: 12.74 eV

Final AEGLs

AEGL-1: 1 h, 1.8 ppm	4 h, 1.8 ppm	8 h, 1.8 ppm
AEGL-2: 1 h, 22 ppm	4 h, 11 ppm	8 h, 11 ppm
AEGL-3: 1 h, 100 ppm	4 h, 26 ppm	8 h, 26 ppm

C11-A064

Hydrogen fluoride

CAS: 7664-39-3
RTECS: MW7875000

UN: 1052
ERG: 125

HF

Colorless fuming liquid or gas with a strong, pungent, irritating "odor". This material is hazardous through inhalation, skin absorption, and ingestion, and produces local skin/eye impacts.

Used as an etchant in the manufacture of semiconductors, frosting/polishing agent (glass/enamel), aluminum brightening agent, metal cleaning, electropolishing bath ingredient, stainless steel pickling agent, to separate uranium isotopes, and as a chemical reagent in the dye chemistry.

This material is on the ITF-25 high threat list and on the Australia Group Export Control List.

Exposure Hazards

Conversion Factor: 1 ppm = 0.82 mg/m^3 at 77°F
$MEG_{(1h)}$ *Min*: 1.0 ppm; *Sig*: 24 ppm; *Sev*: 44 ppm
OSHA PEL: 3 ppm
ACGIH TLV: 0.5 ppm
ACGIH Ceiling: 2 ppm
NIOSH Ceiling: 3 ppm
IDLH: 30 ppm

Properties:

MW: 20.0	*VP*: 783 mmHg	*FlP*: None
D: 0.941 g/mL (liq. gas, 77°F)	*VP*: 400 mmHg (37°F)	*LEL*: None
MP: −118°F	*VD*: 0.69 (calculated)	*UEL*: None
BP: 67°F	*Vlt*: —	*RP*: 0.028
Vsc: 0.2168 cS (liq. gas, 77°F)	H_2O: Miscible	*IP*: 15.98 eV
	Sol: Alcohol	

Final AEGLs

AEGL-1: 1 h, 1.0 ppm	4 h, 1.0 ppm	8 h, 1.0 ppm
AEGL-2: 1 h, 24 ppm	4 h, 12 ppm	8 h, 12 ppm
AEGL-3: 1 h, 44 ppm	4 h, 22 ppm	8 h, 22 ppm

C11-A065

Hydrogen iodide

CAS: 10034-85-2
RTECS: MW3760000
UN: 2197
ERG: 125

HI

Colorless gas with a pungent, suffocating odor that rapidly turns yellow or brown on exposure to light and air. This material is hazardous through inhalation and produces local skin/eye impacts.

Used industrially in the preparation of iodine salts, as a reducing agent, chemical analytical reagent, disinfectant, and in expectorant formulation; used in the manufacture of pharmaceuticals and other organic materials.

This material is on the ITF-25 low threat list.

Exposure Hazards

Conversion Factor: 1 ppm = 5.23 mg/m^3 at 77°F

Human toxicity values have not been established or have not been published.

Properties:

MW: 127.9	*VP*: 5637 mmHg	*FlP*: None
D: 2.54 g/mL (liq. gas, 77°F)	*VD*: 4.4 (calculated)	*LEL*: None
MP: −59°F	*Vlt*: —	*UEL*: None
BP: −32°F	H_2O: Miscible	*RP*: 0.002
Vsc: 0.24 cS (liq. gas, 77°F)	*Sol*: Alcohols	*IP*: —

Proposed AEGLs (Values based on anhydrous hydrogen bromide)

AEGL-1: 1 h, 1.0 ppm	4 h, 1.0 ppm	8 h, 1.0 ppm
AEGL-2: 1 h, 22 ppm	4 h, 11 ppm	8 h, 11 ppm
AEGL-3: 1 h, 120 ppm	4 h, 31 ppm	8 h, 31 ppm

C11-A066

Fuming nitric acid

CAS: 8007-58-7
RTECS: QU5900000
UN: 2032
ERG: 157

$HNO_3 + N_2O_4$

Fuming nitric acid is concentrated acid that contains dissolved nitrogen dioxide. It is a colorless, yellow, or red fuming liquid with an acrid, suffocating odor. Fumes are primarily nitrogen dioxide. This material is hazardous through inhalation, and ingestion, and produces local skin/eye impacts.

Used in the manufacture of pharmaceuticals, dye intermediates, explosives, various inorganic and organic nitrates, nitro compounds; and used industrially for ore flotation, metallurgy, photoengraving, and reprocessing spent nuclear fuel.

This material is on the CDC threat list, and on the ITF-25 high threat list.

Exposure Hazards

Conversion Factor: 1 ppm = 2.58 mg/m^3 at 77°F

Human toxicity values have not been established or have not been published.

The following exposure limits have been developed for nitric acid.

$MEG_{(1h)}$ *Min*: 0.5 ppm; *Sig*: 4 ppm; *Sev*: 22 ppm
OSHA PEL: 2 ppm
ACGIH TLV: 2 ppm
ACGIH STEL: 4 ppm
NIOSH STEL: 4 ppm
IDLH: 25 ppm

Properties:

MW: Mixture	*VP*: 760 mmHg (45% and 55% acid)	*FlP*: None
D: 1.64 g/mL (77°F)	*VD*: >2 (calculated)	*LEL*: None
MP: —	*Vlt*: —	*UEL*: None
BP: —	H_2O: Miscible	*RP*: 0.26
Vsc: —	*Sol*: Ether	*IP*: 11.95 eV (nitric acid)
		IP: 9.59 eV (nitrogen dioxide)

Interim AEGLs (for nitric acid)

AEGL-1: 1 h, 0.53 ppm 4 h, 0.53 ppm 8 h, 0.53 ppm
AEGL-2: 1 h, 24 ppm 4 h, 6.0 ppm 8 h, 3.0 ppm
AEGL-3: 1 h, 92 ppm 4 h, 23 ppm 8 h, 11 ppm

Interim AEGLs (for nitrogen dioxide)

AEGL-1: 1 h, 0.50 ppm 4 h, 0. 50 ppm 8 h, 0.50 ppm
AEGL-2: 1 h, 12 ppm 4 h, 8.2 ppm 8 h, 6.7 ppm
AEGL-3: 1 h, 20 ppm 4 h, 14 ppm 8 h, 11 ppm

C11-A067

Fuming sulfuric acid

CAS: 8014-95-7
RTECS: WS5605000
UN: 1831
ERG: 137

$H_2SO_4 + SO_3$

Fuming colorless to dark-brown, oily liquid with a choking, irritating "odor." This material is hazardous through inhalation, and ingestion, and produces local skin/eye impacts.

Used industrially as a chemical intermediate for sulfonate surfactants, dyes, explosives, and nitrocellulose; for petroleum refining, and as a drying agent in production of chlorine and nitric acid.

This material is on the CDC threat list and on the ITF-25 high threat list.

Exposure Hazards

Conversion Factor: 1 ppm = 7.28 mg/m^3 at 77°F

Human toxicity values have not been established or have not been published.

The following exposure limits have been developed for sulfuric acid.
MEG$_{(1h)}$ *Min*: 2 mg/m^3; *Sig*: 10 mg/m^3; *Sev*: 30 mg/m^3
OSHA PEL: 1 mg/m^3
ACGIH TLV: 0.2 mg/m^3
IDLH: 15 mg/m^3

Properties:

MW: Mixture	*VP*: 2 mmHg	*FlP*: None
D: 1.902 g/mL	*VD*: >2.8 (calculated)	*LEL*: None
MP: —	*Vlt*: —	*UEL*: None
BP: —	*H_2O*: Miscible	*RP*: 5000
Vsc: —	*Sol*: —	*IP*: —

Proposed AEGLs

AEGL-1: 1 h, 0.05 ppm 4 h, 0.05 ppm 8 h, 0.05 ppm
AEGL-2: 1 h, 2.2 ppm 4 h, 2.2 ppm 8 h, 2.2 ppm
AEGL-3: 1 h, 40 ppm 4 h, 27 ppm 8 h, 23 ppm

C11-A068

Bromine pentafluoride

CAS: 7789-30-2
RTECS: EF9350000

UN: 1745
ERG: 144

BrF_5

Colorless to pale-yellow, fuming liquid with a pungent odor. This material is hazardous through inhalation and ingestion, and produces local skin/eye impacts. It reacts violently with organic material.

Used industrially as a fluorinating agent and as an oxidizer in rocket propellant systems.

This material is on the ITF-25 low threat list.

Exposure Hazards

Conversion Factor: 1 ppm = 7.15 mg/m^3 at 77°F
ACGIH TLV: 0.1 ppm

Properties:

MW: 174.9	*VP*: 328 mmHg	*FlP*: None
D: 2.48 g/mL	*VD*: 6.1 (calculated)	*LEL*: None
D: 2.466 g/mL (77°F)	*Vlt*: 440,000 ppm	*UEL*: None
MP: −79°F	*H_2O*: Reacts violently	*RP*: 0.02
BP: 105°F	*Sol*: —	*IP*: —
Vsc: 0.193 cS (77°F)		

Proposed AEGLs

AEGL-1: Not Developed

AEGL-2: 1 h, 1.0 ppm	4 h, 0.5 ppm	8 h, 0.36 ppm
AEGL-3: 1 h, 33 ppm	4 h, 8.3 ppm	8 h, 4.2 ppm

C11-A069

Bromine trifluoride

CAS: 7787-71-5
RTECS: —
UN: 1746
ERG: 144

BrF_3

Colorless to pale-yellow liquid that fumes and smokes in air with an extremely irritating odor. Solid is crystalline material. This material is hazardous through inhalation and ingestion, and produces local skin/eye impacts. It reacts violently with organic material.

Used industrially as a fluorinating agent, electrolytic solvent; used in forming uranium fluorides for isotopic enrichment and fuel element reprocessing.

This material is on the ITF-25 low threat list.

Exposure Hazards

Conversion Factor: 1 ppm = 5.60 mg/m^3 at 77°F
IDLH: 20 ppm

Properties:

MW: 136.9	*VP*: 18 mmHg (102°F)	*FlP*: None
D: 2.803 g/mL (77°F)	*VD*: 4.7 (calculated)	*LEL*: None
MP: 48°F	*Vlt*: 23,000 ppm (102°F)	*UEL*: None
BP: 258°F	*H_2O*: Decomposes	*RP*: 0.5
Vsc: 0.457 cS (77°F)	*Sol*: —	*IP*: —

Proposed AEGLs

AEGL-1: 1 h, 0.12 ppm 4 h, 0.12 ppm 8 h, 0.12 ppm
AEGL-2: 1 h, 2.0 ppm 4 h, 0.70 ppm 8 h, 0.41 ppm
AEGL-3: 1 h, 21 ppm 4 h, 7.3 ppm 8 h, 7.3 ppm

C11-A070

Chlorine pentafluoride

CAS: 13637-63-3
RTECS: —
UN: 2548
ERG: 124

ClF_5

Colorless fuming gas with a pungent, sweet odor. This material is hazardous through inhalation and produces local skin/eye impacts.

Used industrially as a fluorinating agent and in wood pulp bleaching.

This material is on the ITF-25 low threat list.

Exposure Hazards

Conversion Factor: 1 ppm = 5.34 mg/m^3 at 77°F
$LC_{50(Inh)}$: 150 ppm for a 1 h exposure

Properties:

MW: 130.4	*VP*: 2550 mmHg	*FlP*: None
D: 1.9 g/mL (liq. gas, 68°F)	*VD*: 4.5 (calculated)	*LEL*: None
MP: −152°F	*Vlt*: —	*UEL*: None
BP: 9.1°F	H_2O: Decomposes	*RP*: —
Vsc: —	*Sol*: —	*IP*: 13.54

Proposed AEGLs

AEGL-1: 1 h, 0.30 ppm 4 h, 0.30 ppm 8 h, 0.30 ppm
AEGL-2: 1 h, 1.0 ppm 4 h, 0.50 ppm 8 h, 0.36 ppm
AEGL-3: 1 h, 8.0 ppm 4 h, 4.0 ppm 8 h, 2.8 ppm

C11-A071

Fluorine

CAS: 7782-41-4
RTECS: LM6475000
UN: 1045
ERG: 124

F_2

Pale-yellow to greenish gas with a pungent, irritating "odor." This material is hazardous through inhalation, and produces local skin/eye impacts.

Used industrially in the manufacture of fluorocarbons; as a chemical intermediate in the manufacture of sulfur hexafluoride, chlorine trifluoride, bromine trifluoride uranium hexafluoride, molybdenum hexafluoride, perchloryl fluoride, and oxygen difluoride; and as a rocket propellant.

This material is on the ITF-25 high threat list.

Exposure Hazards

Conversion Factor: 1 ppm = 1.55 mg/m^3 at 77°F
MEG$_{(1h)}$ Min: 1.7 ppm; *Sig*: 5.0 ppm; *Sev*: 13 ppm
OSHA PEL: 0.1 ppm
ACGIH TLV: 1.0 ppm
ACGIH STEL: 2.0 ppm
IDLH: 25 ppm

Properties:

MW: 38.0	*VP*: —	*FlP*: None
D: 1.28 g/mL (liq. gas, −156°F)	*VD*: 1.3	*LEL*: None
MP: −363°F	*Vlt*: —	*UEL*: None
BP: −307°F	H_2O: 0.000169% (reacts)	*RP*: —
Vsc: 0.084 cS (liq. gas, −238°F)	*Sol*: —	*IP*: 15.70 eV

Interim AEGLs

AEGL-1: 1 h, 1.7 ppm	4 h, 1.7 ppm	8 h, 1.7 ppm
AEGL-2: 1 h, 5.0 ppm	4 h, 2.3 ppm	8 h, 2.3 ppm
AEGL-3: 1 h, 13 ppm	4 h, 5.7 ppm	8 h, 5.7 ppm

C11-A072

Nitric oxide

CAS: 10102-43-9
RTECS: QX0525000
UN: 1660
ERG: 124

NO

Colorless gas that becomes reddish-brown in high concentrations. It has a sharp, suffocating, sweet odor. This material is hazardous through inhalation and produces local skin/eye impacts. It is rapidly converted in air to nitrogen dioxide.

Used industrially for the manufacture of nitric acid and nitrosyl carbonyls, in the bleaching of rayon, as a stabilizer for propylene and methyl ether, and as a medication.

This material is on the ITF-25 low threat list.

Exposure Hazards

Conversion Factor: 1 ppm = 1.23 mg/m^3 at 77°F
MEG$_{(1h)}$ Min: 0.61 ppm; *Sig*: 15 ppm; *Sev*: 25 ppm
OSHA PEL: 25 ppm
ACGIH TLV: 25 ppm
IDLH: 100 ppm

Properties:

MW: 30.0	*VP*: 26,000 mmHg	*FlP*: None
D: 1.012 g/mL (liq. gas, −166°F)	*VD*: 1.0 (calculated)	*LEL*: None
MP: −258°F	*Vlt*: —	*UEL*: None
BP: −241°F	H_2O: 4.6%	*RP*: 0.001
Vsc: 0.05 cS (liq. gas, −148°F)	H_2O: 7.38% (32°F)	*IP*: 9.27 eV
	Sol: Alcohols; Carbon disulfide	

Interim AEGLs

AEGL values for nitrogen dioxide should be used for emergency planning. Short-term exposures below 80 ppm of nitric oxide should not constitute a health hazard.

C11-A073

Nitrogen dioxide

CAS: 10102-44-0
RTECS: QW9800000
UN: 1067
ERG: 124

NO_2

Yellow-brown liquid to reddish-brown gas with a pungent, acrid, suffocating odor detectable at 1.1 ppm. This material is hazardous through inhalation, and ingestion, and produces local skin/eye impacts. Strong oxidizing agent that accelerates combustion.

Used industrially for the production of nitric acid and as a nitrating agent, oxidizing agent; polymerization inhibitor for acrylates; and as an oxidizer for rocket fuels.

This material is on the ITF-25 medium threat list.

Exposure Hazards

Conversion Factor: 1 ppm = 1.88 mg/m^3 at 77°F
Eye Irritation: 11 ppm; exposure duration unspecified
MEG$_{(1h)}$ Min: 0.5 ppm; *Sig*: 12 ppm; *Sev*: 20 ppm
OSHA Ceiling: 5 ppm
ACGIH TLV: 3 ppm
ACGIH STEL: 5 ppm
NIOSH STEL: 1 ppm
IDLH: 20 ppm

Properties:

MW: 46.0	*VP*: 720 mmHg	*FlP*: None
D: 1.448 g/mL (liq. gas, 68°F)	*VD*: 1.6 (calculated)	*LEL*: None
MP: 12°F	*Vlt*: 960,000 ppm	*UEL*: None
BP: 70°F	*H_2O*: Decomposes	*RP*: 0.02
Vsc: 0.29 cS (liq. gas, 68°F)	*Sol*: Concentrated nitric acid	*IP*: 9.75 eV

Interim AEGLs

AEGL-1: 1 h, 0.50 ppm	4 h, 0.50 ppm	8 h, 0.50 ppm
AEGL-2: 1 h, 12 ppm	4 h, 8.2 ppm	8 h, 6.7 ppm
AEGL-3: 1 h, 20 ppm	4 h, 14 ppm	8 h, 11 ppm

C11-A074

Osmium tetroxide

CAS: 20816-12-0
RTECS: RN1140000
UN: 2471
ERG: 154

OsO_4

Colorless, crystalline solid or pale-yellow mass with an unpleasant, acrid odor like chlorine detectable at 0.0019 ppm. This material is hazardous through inhalation, and ingestion, and produces local skin/eye impacts.

Used industrially as an oxidizing agent especially in conversion of olefins to glycols, and in the preparation of chlorates, peroxides, and periodates; as a biological stain for adipose tissues; and agriculturally as a micronutrient in soil for optimum microbial fixation of nitrogen.

This material is on the FBI threat list.

Exposure Hazards

Conversion Factor: 1 ppm = 10.40 mg/m^3 at 77°F
OSHA PEL: 0.0002 ppm
ACGIH TLV: 0.0002 ppm
ACGIH STEL: 0.0006 ppm
NIOSH STEL: 0.0006 ppm
IDLH: 0.1 ppm

Properties:

MW: 254.2	*VP*: 6.98 mmHg	*FlP*: None
D: 5.10 g/cm^3	*VP*: 11 mmHg (81°F)	*LEL*: None
MP: 105°F	*VD*: 8.8 (calculated)	*UEL*: None
BP: Sublimes	*Vlt*: 12,000 ppm (77°F)	*RP*: 8.9
Vsc: —	H_2O: 7.24% (77°F)	*IP*: 12.60 eV
	Sol: Benzene; Alcohol; Ether	

C11-A075

Sulfur dioxide

CAS: 7446-09-5
RTECS: WS4550000
UN: 1079
ERG: 125

SO_2

Colorless gas with a characteristic suffocating, irritating, and pungent odor detectable at 0.45–4.8 ppm. Shipped as a liquefied compressed gas. This material is hazardous through inhalation, and produces local skin/eye impacts.

Used industrially for the bleaching of flour, fruit, grain, oil straw, wood pulp, wool; in the tanning of leather; in manufacture of glass; as a reducing agent in minerals processing; as an oxidizing agent in lithium primary batteries; as a preservative for fruit and wine; as a disinfectant in breweries and food factories; as a refrigerant gas; as warning agent for liquid fumigants; and as a chemical intermediate in the production of sodium hydrosulfite, chlorine dioxide, sodium sulfate, sulfuryl chloride, thionyl chloride, and organic sulfonates.

This material is on the ITF-25 high threat list.

Exposure Hazards

Conversion Factor: 1 ppm = 2.62 mg/m^3 at 77°F
$LC_{(Inh)}$: 1000 ppm; exposure duration "10 min to several hours"
Eye Irritation: 10 ppm; exposure duration unspecified
Respiratory Irritation: 5 ppm; exposure duration unspecified
$MEG_{(1h)}$ *Min*: —; *Sig*: 3 ppm; *Sev*: 15 ppm
OSHA PEL: 5 ppm
ACGIH TLV: 2 ppm
ACGIH STEL: 5 ppm
NIOSH STEL: 5 ppm
IDLH: 100 ppm

Properties:

MW: 64.1	*VP*: 2400 mmHg	*FlP*: None
D: 1.366 g/mL (liq. gas, 77°F)	*VD*: 2.3 (calculated)	*LEL*: None
MP: −104°F	*Vlt*: —	*UEL*: None
BP: 14°F	H_2O: 7.8% (77°F)	*RP*: 0.001
Vsc: 0.188 cS (liq. gas, 77°F)	H_2O: 17.7% (32°F)	*IP*: 12.30 eV
	Sol: Acetic acid; Alcohol; Ether; Chloroform	

Interim AEGLs

AEGL-1: 1 h, 0.20 ppm	4 h, 0.20 ppm	8 h, 0.20 ppm
AEGL-2: 1 h, 0.75 ppm	4 h, 0.75 ppm	8 h, 0.75 ppm
AEGL-3: 1 h, 27 ppm	4 h, 19 ppm	8 h, 16 ppm

C11-A076

Sulfur trioxide

CAS: 7446-11-9
RTECS: WT4830000

SO_3

Colorless fuming liquid. Solid is colorless to white crystalline solid. Liquefied material may also contain a low-melting solid polymer. This material is hazardous through inhalation and ingestion, and produces local skin/eye impacts.

Used industrially as an oxidizing agent; to manufacture sulfuric acid, sulfonated oils, detergents, and explosives; used in solar energy collectors.

This material is on the ITF-25 medium threat list.

Exposure Hazards

Conversion Factor: 1 ppm = 3.27 mg/m^3 at 77°F
Eye Irritation: 1 ppm; exposure duration unspecified
Respiratory Irritation: 10 ppm; exposure duration unspecified

Other human toxicity values have not been established or have not been published.

Properties:

MW: 80.1	*VP*: 433 (77°F)	*FlP*: None
D: 1.9224 g/mL	*VD*: 2.8 (calculated)	*LEL*: None
D: 2.29 g/cm^3 (solid, 14°F)	*Vlt*: 570,000 ppm (77°F)	*UEL*: None
MP: 62°F	H_2O: Forms sulfuric acid	*RP*: 0.03
BP: 113°F	*Sol*: —	*IP*: 12.8 eV
Vsc: —		

Proposed AEGLs

AEGL-1: 1 h, 0.061 ppm	4 h, 0.061 ppm	8 h, 0.061 ppm
AEGL-2: 1 h, 2.7 ppm	4 h, 2.7 ppm	8 h, 2.7 ppm
AEGL-3: 1 h, 49 ppm	4 h, 34 ppm	8 h, 28 ppm

C11-A077

Hydrogen selenide

CAS: 7783-07-5
RTECS: MX1050000

UN: 2202
ERG: 117

H_2Se

Colorless gas with a "very offensive" odor that resembles decayed horse radish detectable at 0.3 ppm. However, can cause olfactory fatigue and sense of smell is not reliable. This material is hazardous through inhalation and produces local skin/eye impacts. It is highly flammable.

Used industrially for preparation of metallic selenides, organoselenium compounds, and preparation of semiconductor materials.

This material is on the ITF-25, medium threat list.

Exposure Hazards

Conversion Factor: 1 ppm = 3.31 mg/m^3 at 77°F
"Intolerable" Eye and Respiratory Irritation: 1.5 ppm; exposure duration unspecified
$MEG_{(1h)}$ *Min*: —; *Sig*: 0.20 ppm; *Sev*: 2.0 ppm
OSHA PEL: 0.05 ppm
ACGIH TLV: 0.05 ppm
IDLH: 1 ppm

Properties:

MW: 81.0	*VP*: 7220 mmHg (70°F)	*FlP*: —
D: 1.766 g/mL (liq. gas, 77°F)	*VD*: 2.8 (calculated)	*LEL*: —
D: 2.12 g/mL (liq. gas, -44°F)	*Vlt*: —	*UEL*: —
MP: –83.2°F	H_2O: 0.64% (77°F)	*RP*: 0.002
BP: –42°F	*Sol*: Phosgene; Carbon disulfide	*IP*: 9.88 eV
Vsc: 0.074 cS (liq. gas, 77°F)		

Proposed AEGLs

AEGL-1: Not Developed

AEGL-2: 1 h, 0.73 ppm	4 h, 0.37 ppm	8 h, 0.26 ppm
AEGL-3: 1 h, 2.2 ppm	4 h, 1.1 ppm	8 h, 0.78 ppm

C11-A078

Selenium hexafluoride

CAS: 7783-79-1
RTECS: VS9450000
UN: 2194
ERG: 125

SeF_6

Colorless gas. This material is hazardous through inhalation and produces local skin/eye impacts.

Used industrially as an electric insulator.

This material is on the ITF-25 medium threat list.

Exposure Hazards

Conversion Factor: 1 ppm = 7.89 mg/m^3 at 77°F
$MEG_{(1h)}$ *Min*: 0.15 ppm; *Sig*: 0.25 ppm; *Sev*: 2 ppm
OSHA PEL: 0.05 ppm
ACGIH TLV: 0.05 ppm
IDLH: 2 ppm

Properties:

MW: 192.9	*VP*: 651.2 mmHg (−56°F)	*FlP*: None
D: —	*VD*: 6.7 (calculated)	*LEL*: None
MP: −59°F	*Vlt*: —	*UEL*: None
BP: −30°F	H_2O: Insoluble (slowly decomposed)	*RP*: —
Vsc: —	*Sol*: —	*IP*: —

C11-A079

Silicon tetrafluoride

CAS: 7783-61-1
RTECS: VW2327000
UN: 1859
ERG: 125

SiF_4

Colorless gas that forms heavy clouds in moist air. It has a sharp, irritating and suffocating odor similar to hydrogen chloride. This material is hazardous through inhalation and produces local skin/eye impacts.

Used industrially for the manufacture of pure silicon, silane, and fluosilicic acid; used to seal water out of oil wells during drillings.

This material is on the ITF-25 medium threat list.

Exposure Hazards

Conversion Factor: 1 ppm = 4.26 mg/m^3 at 77°F
OSHA PEL: 2.5 mg/m^3 as fluorine
ACGIH TLV: 2.5 mg/m^3 as fluorine

Properties:

MW: 104.1	*VP*: —	*FlP*: None
D: 1.517 g/mL (−94°F)	*VD*: 3.6 (calculated)	*LEL*: None
D: 2.145 g/cm^3 (solid, −319°F)	*Vlt*: —	*UEL*: None
MP: −130°F (under pressure)	H_2O: Decomposes	*RP*: —
BP: Sublimes (−123°F)	*Sol*: Absolute alcohol	*IP*: —
Vsc: 0.4 cS (−58°F)		

C11-A080

Stibine

CAS: 7803-52-3
RTECS: WJ0700000
UN: 2676
ERG: 119

H_3Sb

Colorless gas with a disagreeable odor like hydrogen sulfide. It decomposes slowly on standing depositing elemental antimony. This material is hazardous through inhalation. It is highly flammable.

Used industrially in the manufacture of semiconductors, and as a fumigating agent. It may be a by-product formed during charging of some lead storage batteries.

This material is on the ITF-25 medium threat list.

Exposure Hazards

Conversion Factor: 1 ppm = 5.10 mg/m^3 at 77°F
MEG$_{(1h)}$ Min: —; *Sig*: 0.5 ppm; *Sev*: 1.5 ppm
OSHA PEL: 0.1 ppm
ACGIH TLV: 0.1 ppm
IDLH: 5 ppm

Properties:

MW: 124.8	*VP*: —	*FlP*: —
D: 2.204 g/mL (−1°F)	*VD*: 4.3 (calculated)	*LEL*: —
MP: −126°F	*Vlt*: —	*UEL*: —
BP: −1°F	*H_2O*: 0.41%	*RP*: —
Vsc: —	*Sol*: Most organic solvents	*IP*: 9.51 eV

C11-A081

Tellurium hexafluoride

CAS: 7783-80-4
RTECS: WY2800000
UN: 2195
ERG: 125

TeF_6

Colorless gas with a "repulsive" odor. This material is hazardous through inhalation.

It is a by-product of ore refining.

This material is on the ITF-25 medium threat list.

Exposure Hazards

Conversion Factor: 1 ppm = 9.88 mg/m^3 at 77°F
MEG$_{(1h)}$ Min: 0.06 ppm; *Sig*: 1 ppm; *Sev*: >1 ppm
OSHA PEL: 0.02 ppm
ACGIH TLV: 0.02 ppm
IDLH: 1 ppm

Properties:

MW: 241.6	*VP*: —	*FlP*: None
D: 2.499 g/mL (14°F)	*VD*: 8.3 (calculated)	*LEL*: None
MP: −36°F	*Vlt*: —	*UEL*: None
BP: Sublimes	*H_2O*: Decomposes	*RP*: —
Vsc: —	*Sol*: —	*IP*: —

C11-A082

Titanium tetrachloride (Agent FM)

CAS: 7550-45-0
RTECS: XR1925000
UN: 1838
ERG: 137

$TiCl_4$

Colorless to light-yellow liquid that fumes in moist air. It has a penetrating acid odor. This material is hazardous through inhalation and ingestion, and produces local skin/eye impacts.

Used industrially to produce pure titanium and titanium salts, pigments, as an additive in decorative glass, and as a polymerization catalyst.

This material is on the ITF-25 medium threat list.

Exposure Hazards

Conversion Factor: 1 ppm = 7.76 mg/m^3 at 77°F

$MEG_{(1h)}$ *Min*: 0.64 ppm; *Sig*: 2.6 ppm; *Sev*: 12.9 ppm

AIHA WEEL: 0.06 ppm

Properties:

MW: 189.7	*VP*: 9.96 mmHg	*FlP*: None
D: 1.726 g/mL	*VD*: 6.5 (calculated)	*LEL*: None
MP: −11°F	*Vlt*: 13,000 ppm	*UEL*: None
BP: 277°F	H_2O: Decomposed	*RP*: 0.72
Vsc: 0.0458 cS	*Sol*: Most organic solvents	*IP*: 11.5 eV

Proposed AEGLs

AEGL-1: 1 h, 0.070 ppm	4 h, 0.070 ppm	8 h, 0.070 ppm
AEGL-2: 1 h, 1.0 ppm	4 h, 0.21 ppm	8 h, 0.094 ppm
AEGL-3: 1 h, 5.7 ppm	4 h, 2.0 ppm	8 h, 0.91 ppm

C11-A083

Tungsten hexafluoride

CAS: 7783-82-6

RTECS: YO7720000

UN: 2196

ERG: 125

WF_6

Colorless gas or pale-yellow liquid. This material is hazardous through inhalation and produces local skin/eye impacts.

Used industrially to apply tungsten coatings to other surfaces; as a chemical intermediate and in the manufacture of electronics.

This material is on the ITF-25 high threat list.

Exposure Hazards

Conversion Factor: 1 ppm = 12.18 mg/m^3 at 77°F

Lethal human toxicity values have not been established or have not been published. However, based on available information, this material appears to be approximately one-sixth as toxic as Hydrogen fluoride (C11-A031).

Properties:

MW: 297.5	*VP*: —	*FlP*: —
D: 3.387 g/mL (liq. gas, 77°F)	*VD*: 10 (calculated)	*LEL*: —
MP: 31°F	*Vlt*: —	*UEL*: —
BP: 63°F	H_2O: Decomposes	*RP*: —
Vsc: 0.1332 cS (liq. gas, 77°F)	*Sol*: Benzene; Dioxane; Cyclohexane	*IP*: 15.2 eV

C11-A084

Boron tribromide

CAS: 10294-33-4

RTECS: ED7400000

UN: 2692
ERG: 157

BBr_3

Colorless, fuming liquid with a pungent, sharp, irritating odor. This material is hazardous through inhalation and ingestion, and produces local skin/eye impacts.

Used industrially in the manufacture of diborane, ultra high purity boron, and semiconductors; used as a catalyst in polymerizations, alkylations, and acylations.

This material is on the ITF-25 medium threat list.

Exposure Hazards

Conversion Factor: 1 ppm = 10.25 mg/m^3 at 77°F
MEG$_{(1h)}$ Min: 1 ppm; *Sig*: —; *Sev*: —
ACGIH Ceiling: 1 ppm
NIOSH Ceiling: 1 ppm

Properties:

MW: 250.5
D: 2.64 g/mL (65°F)
D: 2.698 g/mL (32°F)
MP: −51°F
BP: 194°F
Vsc: 0.28 cS (75°F)
VP: 40 mmHg (57°F)
VP: 100 mmHg (92°F)
VD: 8.6 (calculated)
Vlt: 55,000 ppm (57°F)
H_2O: Reacts violently
Sol: Carbon tetrachloride; Carbon disulfide
FlP: None
LEL: None
UEL: None
RP: 0.1
IP: 9.70 eV

C11-A085

Boron trichloride

CAS: 10294-34-5
RTECS: ED1925000
UN: 1741
ERG: 125

BCl_3

Colorless fuming liquid with a sharp, irritating, pungent odor. This material is hazardous through inhalation and produces local skin/eye impacts.

Used industrially for the production and purification of boron and boron compounds; in purification of metal alloys to remove oxides, nitrides, and carbides; as a soldering flux for alloys of aluminum, iron, zinc, tungsten, and monel; as an extinguishing agent for magnesium fires in heat treating furnaces; as a stabilizer for liquid sulfur trioxide; for the manufacture of boron filaments for advanced composites; and in semiconductors. It has been used in the field of high energy fuels and rocket propellants as a source of boron for raising the British Thermal Unit (BTU) value.

This material is on the ITF-25 high threat list.

Exposure Hazards

Conversion Factor: 1 ppm = 4.79 mg/m^3 at 77°F

Human toxicity values have not been established or have not been published.

Properties:

MW: 117.2	*VP*: 440 mmHg (30°F)	*FlP*: None
D: 1.318 g/mL (liq. gas, 77°F)	*VD*: 4.0 (calculated)	*LEL*: None
MP: −161°F	*Vlt*: —	*UEL*: None
BP: 54°F	H_2O: Decomposes	*RP*: —
Vsc: 0.170 cS (liq. gas, 77°F)	*Sol*: —	*IP*: 11.62 eV

C11-A086

Boron trifluoride

CAS: 7637-07-2
RTECS: ED2275000
UN: 1008
ERG: 125

BF_3

Colorless gas with a pungent, suffocating odor. This material is hazardous through inhalation and produces local skin/eye impacts.

Used industrially as a catalyst in organic synthesis, to protect molten magnesium and its alloys from oxidation, as a flux for soldering magnesium; manufacture of specialty electronics; as a fumigant; and in ionization chambers for detection of weak neutrons.

This material is on the ITF-25 high threat list.

Exposure Hazards

Conversion Factor: 1 ppm = 2.77 mg/m^3 at 77°F
$MEG_{(1h)}$ *Min*: 0.22 ppm; *Sig*: 5.8 ppm; *Sev*: 14 ppm
OSHA Ceiling: 1 ppm
ACGIH Ceiling: 1 ppm
IDLH: 25 ppm

Properties:

MW: 67.8	*VP*: 36,560 mmHg (8°F)	*FlP*: None
D: 1.433 g/mL (liq. gas, −94°F)	*VD*: 2.4 (calculated)	*LEL*: None
MP: −196°F	*Vlt*: —	*UEL*: None
BP: −150°F	H_2O: 106% (cold water, decomposes)	*RP*: —
Vsc: 0.05 cS (liq. gas, −58°F)	*Sol*: Most saturated hydrocarbons; Aromatic compounds	*IP*: 15.50 eV

Interim AEGLs

AEGL-1: 1 h, 2.5 ppm	4 h, 2.5 ppm	8 h, 2.5 ppm
AEGL-2: 1 h, 37 ppm	4 h, 24 ppm	8 h, 12 ppm
AEGL-3: 1 h, 110 ppm	4 h, 72 ppm	8 h, 36 ppm

C11-A087

Phosphorus pentafluoride

CAS: 7647-19-0
RTECS: TH4070000
UN: 2198
ERG: 125

PF_5

Colorless gas with an "unpleasant" odor that fumes in contact with air. This material is hazardous through inhalation and produces local skin/eye impacts.

Used industrially as a catalyst in ionic polymerizations, mild fluorinating agent; used in the manufacture of semiconductors.

This material is on the ITF-25 medium threat list.

Exposure Hazards

Conversion Factor: 1 ppm = 5.15 mg/m^3 at 77°F

Human toxicity values have not been established or have not been published. However, based on available information, the Compressed Gas Association has established an LC_{50} of 260 ppm with an exposure of 1 h, or approximately one-fifth as toxic as hydrogen fluoride.

Properties:

MW: 126.0	*VP*: 21,432 mmHg (70°F)	*FlP*: None
D: 1.13 g/mL (liq. gas, 32°F)	*VD*: 4.3 (calculated)	*LEL*: None
MP: −137°F	*Vlt*: —	*UEL*: None
BP: −120°F	*H_2O*: Reacts violently	*RP*: 0
Vsc: 0.16 cS (liq. gas, −4°F)	*Sol*: —	*IP*: 15.54 eV

C11-A088

Sulfuryl chloride

CAS: 7791-25-5

RTECS: WT4870000

UN: 1834

ERG: 137

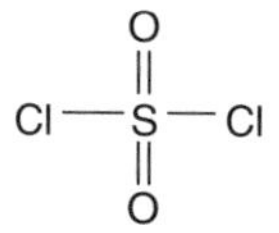

Cl_2O_2S

Clear, colorless to light yellow-green mobile liquid with a very pungent odor. This material is hazardous through inhalation and produces local skin/eye impacts.

Used industrially to treat wool to prevent shrinking; in the manufacture of rayon, rubber-base plastics, chlorophenol disinfectants, phosphate insecticides, heterocyclic fungicides and herbicides, and pharmaceuticals.

This material is on the ITF-25 medium threat list.

Exposure Hazards

Conversion Factor: 1 ppm = 5.52 mg/m^3 at 77°F

Human toxicity values have not been established or have not been published.

Properties:

MW: 135.0	*VP*: 105 mmHg	*FlP*: None
D: 1.665 g/mL	*VP*: 140 mmHg (77°F)	*LEL*: None
MP: −65°F	*VD*: 4.7 (calculated)	*UEL*: None
BP: 156°F	*Vlt*: 140,000 ppm	*RP*: 0.08
Vsc: —	*H_2O*: Reacts	*IP*: —
	Sol: Toluene; Ether	

C11-A089

Acetone cyanohydrin

CAS: 75-86-5
RTECS: OD9275000
UN: 1541
ERG: 155

OH
CN

C_4H_7NO

Clear, colorless to light yellow liquid. Pure material is practically odorless but usually has an odor of bitter almond due to residual hydrogen cyanide. This material is hazardous through inhalation, skin absorption, penetration through broken skin, and ingestion, and produces local skin/eye impacts. Forms cyanide in the body.

Used industrially in organic synthesis, in the manufacture of insecticides and pharmaceuticals, as an intermediate in the synthesis of methacrylates; used as a complexing agent for metals refining and separation.

This material is on the ITF-25 medium threat list.

Exposure Hazards

Conversion Factor: 1 ppm = 3.48 mg/m^3 at 77°F
$MEG_{(1h)}$ *Min*: 4.7 ppm; *Sig*: —; *Sev*: —
ACGIH Ceiling: 1.4 ppm as cyanide [Skin]
NIOSH Ceiling: 1 ppm [15 min limit]
AIHA WEEL: 2 ppm [Skin]
AIHA STEL: 5 ppm [Skin]

Properties:

MW: 85.1	*VP*: 0.75 mmHg	*FlP*: 165°F
D: 0.9267 g/mL (77°F)	*VD*: 2.9 (calculated)	*LEL*: 2.2%
MP: −2°F	*Vlt*: 1000 ppm	*UEL*: 12%
BP: 203°F	H_2O: Miscible	*RP*: 14
BP: 154°F (11 mmHg)	*Sol*: Most organic solvents	*IP*: —
Vsc: —		

Interim AEGLs

AEGL-1: 1 h, 2.0 ppm	4 h, 1.3 ppm	8 h, 1.0 ppm
AEGL-2: 1 h, 7.1 ppm	4 h, 3.5 ppm	8 h, 2.5 ppm
AEGL-3: 1 h, 15 ppm	4 h, 8.6 ppm	8 h, 6.6 ppm

C11-A090

Acrylonitrile

CAS: 107-13-1
RTECS: AT5250000
UN: 1093
ERG: 131P

CN

C_3H_3N

Colorless to pale-yellow liquid with an odor that is pungent, irritating, and resembles onions or garlic. It is detectable at 21.4 ppm. Typically stabilized with 35–50 ppm of methylhydroquinone. This material is hazardous through inhalation, skin absorption (liquid), penetration through broken skin, and ingestion, and produces local skin/eye impacts.

Used industrially for the manufacture of plastics, surface coatings, and adhesives; pharmaceuticals, dyes, ion exchange resins, corrosion inhibitors, water treatment resins; used agriculturally as a pesticide fumigant.

This material is on the ITF-25 medium threat list.

Exposure Hazards

Conversion Factor: 1 ppm = 2.17 mg/m^3 at 77°F
$MEG_{(1h)}$ *Min*: 10 ppm; *Sig*: 35 ppm; *Sev*: 75 ppm
OSHA PEL: 2 ppm
ACGIH TLV: 2 ppm [Skin]
OSHA Ceiling: 10 ppm [15 min] [Skin]
IDLH: 86 ppm

Properties:

MW: 53.1	*VP*: 83 mmHg	*FlP*: 30°F
D: 0.81 g/mL	*VP*: 109 mmHg (77°F)	*LEL*: 3.1%
D: 0.8004 g/mL (77°F)	*VD*: 1.8 (calculated)	*UEL*: 17%
MP: −116°F	*Vlt*: 120,000 ppm	*RP*: 0.16
BP: 171°F	*Vlt*: 180,000 ppm (86°F)	*IP*: 10.75 eV
Vsc: 0.0425 cS (77°F)	H_2O: 7.45%	
	Sol: Alcohols; Ethyl acetate; Toluene	

C11-A091

Allyl alcohol

CAS: 107-18-6
RTECS: BA5075000
UN: 1098
ERG: 131

C_3H_6O

Mobile colorless liquid with a pungent odor like mustard detectable at 0.8 ppm. This material is hazardous through inhalation, skin absorption, penetration through broken skin, and ingestion, and produces local skin/eye impacts. Acts synergistically with alcohol in blood.

Used industrially for the manufacture of flavorings, perfumes, acrolein, diallyl phthalate, diallyl isophthalate, and pharmaceuticals; used to denature alcohol. Used agriculturally as herbicide and fungicide.

This material is on the ITF-25 medium threat list.

Exposure Hazards

Conversion Factor: 1 ppm = 2.38 mg/m^3 at 77°F
Minor Eye Irritation: 5 ppm; exposure duration unspecified
Severe Eye Irritation: 25 ppm; exposure duration unspecified
Nasal Irritation: 10 ppm; exposure duration unspecified
$MEG_{(1h)}$ *Min*: 2.1 ppm; *Sig*: 4.2 ppm; *Sev*: 67 ppm
OSHA PEL: 2 ppm [Skin]
ACGIH TLV: 0.5 ppm [Skin]
NIOSH STEL: 4 ppm [Skin]
IDLH: 20 ppm

Properties:

MW: 58.1	*VP*: 17 mmHg	*FlP*: 70°F
D: 0.8540 g/mL	*VP*: 26.1 mmHg (77°F)	*LEL*: 2.5%
MP: −200°F	*VD*: 2.0 (calculated)	*UEL*: 18.0%
BP: 207°F	*Vlt*: 24,000 ppm	*RP*: 0.77
Vsc: 1.31 cS (77°F)	*Vlt*: 41,000 ppm (86°F)	*IP*: 9.63 eV
	H_2O: Miscible	
	Sol: Alcohol; Chloroform; Ether	

Interim AEGLs

AEGL-1: 1 h, 2.1 ppm	4 h, 2.1 ppm	8 h, 2.1 ppm
AEGL-2: 1 h, 14.2 ppm	4 h, 4.2 ppm	8 h, 4.2 ppm
AEGL-3: 1 h, 20 ppm	4 h, 10 ppm	8 h, 10 ppm

C11-A092

Allylamine

CAS: 107-11-9
RTECS: BA5425000
UN: 2334
ERG: 131

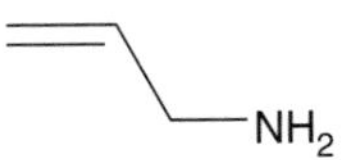

C_3H_7N

Colorless to yellow liquid with a strong, pungent ammonia odor. This material is hazardous through inhalation, skin absorption, penetration through broken skin, and ingestion, and produces local skin/eye impacts.

Used industrially for synthesis of ion-exchange resins and pharmaceutical intermediates.

This material is on the ITF-25 medium threat list.

Exposure Hazards

Conversion Factor: 1 ppm = 2.34 mg/m^3 at 77°F
Irritation: 2.5 ppm; exposure duration unspecified
"Intolerable" Irritation: 14 ppm; exposure duration unspecified

Other human toxicity values have not been established or have not been published.

Properties:

MW: 57.1	*VP*: 193 mmHg	*FlP*: −18°F
D: 0.761 g/mL	*VP*: 242 mmHg (77°F)	*LEL*: 2.2%
MP: −126°F	*VD*: 2.0 (calculated)	*UEL*: 22%
BP: 127°F	*Vlt*: 260,000 ppm	*RP*: 0.06
Vsc: —	*H_2O*: Miscible	*IP*: 8.8 eV
	Sol: Alcohol; Chloroform; Ether	

Interim AEGLs

AEGL-1: 1 h, 0.42 ppm	4 h, 0.42 ppm	8 h, 0.42 ppm
AEGL-2: 1 h, 3.3 ppm	4 h, 1.8 ppm	8 h, 1.2 ppm
AEGL-3: 1 h, 18 ppm	4 h, 3.5 ppm	8 h, 2.3 ppm

C11-A093

Carbon disulfide

CAS: 75-15-0
RTECS: FF6650000
UN: 1131
ERG: 131

CS_2

Colorless to faint-yellow mobile liquid with a sweet, ethereal odor. On standing, the odor becomes a foul stench. This material is hazardous through inhalation, skin absorption of the liquid, and ingestion, and produces local skin/eye impacts.

Carbon disulfide is highly flammable and has an autoignition temperature of 212°F.

Used industrially as a chemical intermediate in the production of rayon, carbon tetrachloride, xanthogenates, flotation agents, and pesticides; used in the cold vulcanization of vulcanized rubber, in adhesive compositions for food packaging; as a solvent for phosphorus, sulfur, selenium, bromine, iodine, fats, resins, rubbers, waxes, lacquers, camphor, resins; and in the production of optical glass, paints, enamels, varnishes, paint removers, tallow, putty preservatives, rubber cement, soil disinfectants, explosives, rocket fuel, and electronic vacuum tubes.

This material is on the ITF-25 high threat list.

Exposure Hazards

Conversion Factor: 1 ppm = 3.11 mg/m^3 at 77°F
$MEG_{(1h)}$ Min: 1.0 ppm; *Sig*: 50 ppm; *Sev*: 500 ppm
OSHA PEL: 20 ppm
OSHA Ceiling: 30 ppm
ACGIH TLV: 1 ppm [Skin]
NIOSH STEL: 10 ppm [Skin]
IDLH: 500 ppm

Properties:

MW: 76.1	*VP*: 297 mmHg	*FlP*: −22°F
D: 1.2632 g/mL	*VP*: 359 mmHg (77°F)	*LEL*: 1.3%
MP: −169°F	*VD*: 2.7 (calculated)	*UEL*: 50%
BP: 116°F	*Vlt*: 400,000 ppm	*RP*: 0.038
Vsc: 0.288 cS	*H_2O*: 0.286%	*IP*: 10.08 eV
	Sol: Alcohols; Ether; Benzene; Oils	

Interim AEGLs

AEGL-1: 1 h, 4.0 ppm	4 h, 2.5 ppm	8 h, 2.0 ppm
AEGL-2: 1 h, 160 ppm	4 h, 100 ppm	8 h, 50 ppm
AEGL-3: 1 h, 480 ppm	4 h, 300 ppm	8 h, 150 ppm

C11-A094

Chloroacetonitrile

CAS: 107-14-2
RTECS: AL8225000
UN: 2668
ERG: 131

C_2H_2ClN

Clear, colorless liquid with a pungent odor. This material is hazardous through inhalation and ingestion, and produces local skin/eye impacts.

Lacrimation usually gives adequate warning of vapor exposure.

Used to manufacture insecticides and pharmaceuticals; used agriculturally as a fumigant.

This material is on the ITF-25 medium threat list.

Exposure Hazards

Conversion Factor: 1 ppm = 3.09 mg/m^3 at 77°F

Human toxicity values have not been established or have not been published.

Properties:

MW: 75.5	*VP*: 15 mmHg (86°F)	*FlP*: —
D: 1.1930 g/mL	*VD*: 2.6 (calculated)	*LEL*: —
MP: —	*Vlt*: 19,000 ppm (86°F)	*UEL*: —
BP: 257°F	*H_2O*: 10%	*RP*: 0.79
Vsc: —	*Sol*: Hydrocarbons; Alcohols; Ether	*IP*: —

Proposed AEGLs

AEGL-1: Not Developed

AEGL-2: 1 h, 23 ppm	4 h, 13 ppm	8 h, 10 ppm
AEGL-3: 1 h, 49 ppm	4 h, 28 ppm	8 h, 21 ppm

C11-A095

Diborane

CAS: 19287-45-7
RTECS: —
UN: 1911
ERG: 119

B_2H_6

Colorless gas with a repulsive, sickly sweet odor detectable at 1.8–3.5 ppm. Industrially, it can be found diluted with a variety of gases including hydrogen, argon, nitrogen, or helium.

This material is hazardous through inhalation and produces local skin/eye impacts.

Diborane is highly flammable and has an autoignition temperature of 125°F. It is also moisture sensitive and will ignite spontaneously in moist air at room temperature.

Used industrially as a catalyst for ethylene, styrene, butadiene, acrylic, and vinyl polymerizations; reducing agent in the synthesis of organic chemicals; for production of hard boron coatings on metals and ceramics; as a component or additive for high-energy fuels; and as a chemical intermediate in the synthesis of organic boron compounds and metal borohydrides. Used in the electronics industry to improve crystal growth, to impart electrical properties in pure crystals, and as a doping agent for p-type semiconductors.

This material is on the ITF-25 high threat list.

Exposure Hazards

Conversion Factor: 1 ppm = 1.13 mg/m^3 at 77°F
$LC_{(Inh)}$: 180 mg/m^3 (159 ppm) for a 15-min exposure
$MEG_{(1h)}$ *Min*: 0.3 ppm; *Sig*: 1.0 ppm; *Sev*: 3.7 ppm
OSHA PEL: 0.1 ppm
ACGIH TLV: 0.1 ppm
IDLH: 15 ppm

Properties:

MW: 27.7	*VP*: 30,000 mmHg (62°F)	*FlP*: −130°F
D: 0.308 g/mL (liq. gas, 14°F)	*VD*: 0.96 (calculated)	*LEL*: 0.9%
MP: −265°F	*Vlt*: —	*UEL*: 98%
BP: −135°F	H_2O: Decomposes	*RP*: 0.001
Vsc: 0.0812 cS (liq. gas, 14°F)	*Sol*: Carbon disulfide; Ammonium hydroxide	*IP*: 11.38 eV

Final AEGLs

AEGL-1: Not Developed

AEGL-2: 1 h, 1.0 ppm	4 h, 0.25 ppm	8 h, 0.13 ppm
AEGL-3: 1 h, 3.7 ppm	4 h, 0.92 ppm	8 h, 0.46 ppm

C11-A096

Diketene

CAS: 674-82-8
RTECS: —
UN: 2521
ERG: 131P

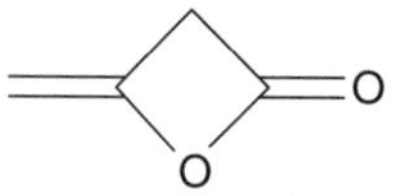

$C_4H_4O_2$

Clear, colorless to orange liquid with a pungent odor. This material is hazardous through inhalation and produces local skin/eye impacts. Violent polymerization can occur in presence of acids, bases, or heat.

Used industrially as a chemical intermediate for pharmaceuticals, dyes, pigments and toners, food preservatives, insecticides, and fungicides.

This material is on the ITF-25 medium threat list.

Exposure Hazards

Conversion Factor: 1 ppm = 3.44 mg/m^3 at 77°F
$MEG_{(1h)}$ *Min*: 1.0 ppm; *Sig*: 5.0 ppm; *Sev*: 20 ppm

Properties:

MW: 84.1	*VP*: 8.03 mmHg	*FlP*: 93°F
D: 1.0897 g/mL	*VD*: 2.9 (calculated)	*LEL*: —
MP: 19.7°F	*Vlt*: 11,000 ppm	*UEL*: —
BP: 262°F	H_2O: Decomposes	*RP*: 1.3
BP: 158°F (99 mmHg)	*Sol*: Common organic solvents	*IP*: 9.6 eV
Vsc: 0.808 cS		

C11-A097

Ethylene oxide

CAS: 75-21-8
RTECS: KX2450000
UN: 1040
ERG: 119P

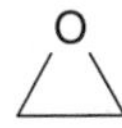

C_2H_4O

Colorless gas or liquid with an odor like ether. The odor has also been described as sweet and reminiscent of bruised apples. Industrially, it can be found diluted with a variety of gases including carbon dioxide, fluorocarbon 12, and nitrogen. This material is hazardous through inhalation, and ingestion, and produces local skin/eye impacts. Exposure to high vapor concentrations or direct contact with liquid may cause burns to the eyes and skin.

Used industrially as a fumigant for foodstuffs and textiles, sterilizing agent for surgical instruments and other heat sensitive material, agricultural fungicide, ripening agent for fruits, rocket propellant; in the manufacture of acrylonitrile, ethylene glycol, dioxane, ethylene chlorohydrin, ethanolamines, glycol ethers, carbowax, the cellosolves and carbitols, polyethylene terephthalate polyester fiber, and nonionic surfactants.

This material is on the ITF-25 high threat list.

Exposure Hazards

Conversion Factor: 1 ppm = 1.80 mg/m^3 at 77°F

Lethal human toxicity values have not been established or have not been published. However, "symptoms of illness" occur at 100 ppm (exposure duration unspecified) and "severe toxic effects" occur at 250 ppm for a 60 min exposure.

$MEG_{(1h)}$ *Min*: 7.5 ppm; *Sig*: 45 ppm; *Sev*: 200 ppm
OSHA PEL: 1 ppm
ACGIH TLV: 1 ppm
NIOSH Ceiling: 5 ppm
IDLH: 800 ppm

Properties:

MW: 44.1	*VP*: 1100 mmHg	*FlP*: −67°F
D: 0.882 g/mL (50°F)	*VP*: 1314 mmHg (77°F)	*LEL*: 3%
D: 0.862 g/mL (liq. gas, 77°F)	*VD*: 1.5 (calculated)	*UEL*: 100%
MP: −168°F	*Vlt*: —	*RP*: 0.014
BP: 51°F	H_2O: Miscible	*IP*: 10.56 eV
Vsc: 0.3016 cS (liq. gas, 77°F)	*Sol*: Most organic solvents	

Interim AEGLs

AEGL-1: Not Developed
AEGL-2: 1 h, 45 ppm 4 h, 14 ppm 8 h, 7.9 ppm
AEGL-3: 1 h, 200 ppm 4 h, 63 ppm 8 h, 35 ppm

C11-A098

Ethyleneimine

CAS: 151-56-4
RTECS: KX5075000
UN: 1185
ERG: 131P

C_2H_5N

Mobile, colorless liquid with an odor like ammonia that is detectable at 2 ppm. The odor becomes annoying at 11 ppm. This material is hazardous through inhalation, skin absorption, penetration through broken skin, and ingestion, and produces local skin/eye impacts. Dermal exposure causes skin sensitization.

Used industrially in the manufacture of pharmaceuticals, textile chemicals, adhesives, binders, petroleum refining chemicals, fuels, lubricants, coating resins, varnishes, polymerizations, lacquers, agricultural chemicals, cosmetics, ion exchange resins, photographic chemicals, surfactants; used in the paper industry and as a flocculation aid.

This material is on the ITF-25 low threat list.

Exposure Hazards

Conversion Factor: 1 ppm = 1.76 mg/m^3 at 77°F
Irritation: 100 ppm; exposure duration unspecified
MEG$_{(1h)}$ *Min*: 1.5 ppm; *Sig*: 4.6 ppm; *Sev*: 9.9 ppm
ACGIH TLV: 0.5 ppm [Skin]
IDLH: 100 ppm

Properties:

MW: 43.1	*VP*: 160 mmHg	*FlP*: 12°F
D: 0.8321 g/mL (75°F)	*VP*: 213 mmHg (77°F)	*LEL*: 3.3%
MP: −97°F	*VD*: 1.5 (calculated)	*UEL*: 54.8%
BP: 133°F	*Vlt*: 210,000 ppm	*RP*: 0.095
Vsc: —	*Vlt*: 280,000 ppm (77°F)	*IP*: 9.20 eV
	H_2O: Miscible	
	Sol: Alcohol; Ether; Benzene	

Interim AEGLs

AEGL-1: Not Developed
AEGL-2: 1 h, 4.6 ppm 4 h, 1.0 ppm 8 h, 0.47 ppm
AEGL-3: 1 h, 9.9 ppm 4 h, 2.8 ppm 8 h, 1.5 ppm

C11-A099

Formaldehyde

CAS: 50-00-0; 30525-89-4 (Paraformaldehyde)
RTECS: LP8925000; RV0540000 (Paraformaldehyde)

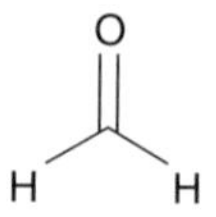

CH_2O

Nearly colorless gas with a pungent, suffocating odor detectable at 0.005–1 ppm. Solutions are colorless to light yellow oils and the solid paraformaldehyde is a white powder, granules, or flakes. This material is hazardous through inhalation, ingestion of solutions or solids, and produces local skin/eye impacts.

Used industrially in the manufacture of amino and phenolic resins, fertilizers; as a nickel-plating brightening agent, reducing agent in the recovery of gold and silver, latex coagulant, crease-proof textile finishing agent, photographic gelatin hardening agent, taxidermy preservative, cross linking agent, corrosion inhibitor, hydrogen sulfide scavenger, industrial sterilant, preservative, and biocide. Used medically as an ingredient in embalming fluids, as a fixative of histological specimens, and to alter bacterial toxins to toxoids for vaccines.

This material is on the ITF-25 high threat list.

Exposure Hazards

Conversion Factor: 1 ppm = 1.23 mg/m^3 at 77°F
Eye Irritation: 4 ppm; exposure duration unspecified
$MEG_{(1h)}$ *Min*: 1.0 ppm; *Sig*: 10 ppm; *Sev*: 25 ppm
OSHA PEL: 0.75 ppm
OSHA STEL: 2 ppm
ACGIH Ceiling: 0.3 ppm
NIOSH Ceiling: 1 ppm
IDLH: 20 ppm

Properties:

MW: 30.0	*VP*: 3890 mmHg (77°F)	*FlP*: —
D: —	*VD*: 1.1 (calculated)	*LEL*: 7.0%
MP: −134°F	*Vlt*: —	*UEL*: 73%
BP: −3.1°F	H_2O: 40%	*RP*: 0.005
Vsc: —	*Sol*: Alcohols; Ether; Acetone; Benzene	*IP*: 10.88 eV

Paraformaldehyde:

MW: Polymer	*VP*: 10.5 mmHg (77°F)	*FlP*: 158°F
D: 1.46 g/cm^3 (59°F)	*VD*: 1.1 (calculated)	*LEL*: 7%
MP: 327°F (decomposes)	*Vlt*: —	*UEL*: 73%
BP: 248°F (sublimes)	H_2O: Insoluble	*RP*: —
Vsc: —	*Sol*: Insoluble in most organic solvents	*IP*: 10.88 eV

Proposed AEGLs

AEGL-1: 1 h, 0.9 ppm	4 h, 0.9 ppm	8 h, 0.9 ppm
AEGL-2: 1 h, 14 ppm	4 h, 14 ppm	8 h, 14 ppm
AEGL-3: 1 h, 56 ppm	4 h, 35 ppm	8 h, 35 ppm

C11-A100

Hexachlorocyclopentadiene

CAS: 77-47-4
RTECS: GY1225000

UN: 2646
ERG: 135

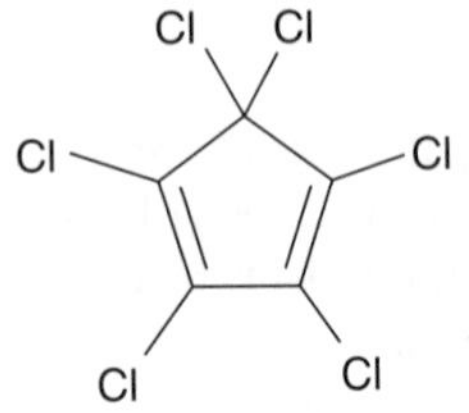

C_5Cl_6

Dense, pale-yellow to amber-colored oily liquid with an unpleasant pungent odor detectable at 0.3 ppm. This material is hazardous through inhalation, skin absorption, penetration through broken skin, and ingestion, and produces local skin/eye impacts.

Used to manufacture pesticides, shockproof plastics, acids, esters, ketones, flame retardants, fluorocarbons; used as biocide.

This material is on the ITF-25 low threat list.

Exposure Hazards

Conversion Factor: 1 ppm = 11.16 mg/m^3 at 77°F
MEG$_{(1h)}$ *Min*: 0.01 ppm; *Sig*: 0.03 ppm; *Sev*: 0.15 ppm
ACGIH TLV: 0.01 ppm

Properties:

MW: 272.8	*VP*: 0.08 mmHg (77°F)	*FlP*: None
D: 1.7019 g/mL (77°F)	*VD*: 9.4 (calculated)	*LEL*: None
MP: 16°F	*Vlt*: 110 ppm (77°F)	*UEL*: None
BP: 462°F	H_2O: 0.00018% (slowly decomposes)	*RP*: 76
Vsc: 4.56 cS (77°F)	*Sol*: Acetone; Methanol; Hexane	*IP*: —

C11-A101

Methyl mercaptan

CAS: 74-93-1
RTECS: PB4375000
UN: 1064
ERG: 117

CH_3SH

Colorless gas or water-white liquid with a disagreeable odor like garlic or rotten cabbage that is detectable at 0.0021 ppm. This material is hazardous through inhalation.

Used as a gas odorant; used to manufacture methionine, plastics, jet fuel additives, pesticides, fungicides; used as a catalyst and as a synthetic flavoring.

This material is on the ITF-25 medium threat list.

Exposure Hazards

Conversion Factor: 1 ppm = 1.97 mg/m^3 at 77°F
MEG$_{(1h)}$ *Min*: 0.5 ppm; *Sig*: 5 ppm; *Sev*: 23 ppm
ACGIH TLV: 0.5 ppm
OSHA Ceiling: 10 ppm
IDLH: 150 ppm

Properties:

MW: 48.1	*VP*: 1300 mmHg	*FlP*: 64°F
D: 0.862 g/mL (77°F)	*VP*: 1510 mmHg (77°F)	*LEL*: 3.9%
MP: −189°F	*VD*: 1.7 (calculated)	*UEL*: 21.8%
BP: 43°F	*Vlt*: —	*RP*: 0.011
Vsc: 0.291 cS (liq. gas, 77°F)	*H_2O*: 2.4% (59°F)	*IP*: 9.44 eV
	H_2O: 1.54% (77°F)	
	Sol: Alcohol; Ether; Petroleum naphtha	

Interim AEGLs

AEGL-1: Not Developed		
AEGL-2: 1 h, 47 ppm	4 h, 30 ppm	8 h, 19 ppm
AEGL-3: 1 h, 68 ppm	4 h, 43 ppm	8 h, 22 ppm

C11-A102

Octyl mercaptan

CAS: 111-88-6
RTECS: —

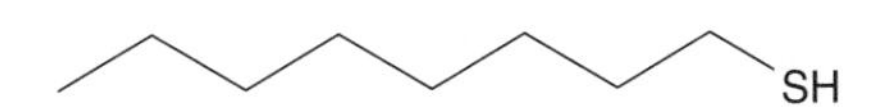

$C_8H_{18}S$

Clear, colorless to water-white liquid with a mild odor. This material is hazardous through inhalation and ingestion, and produces local skin/eye impacts.

Used industrially as a polymerization conditioner and as an intermediate in organic synthesis.

This material is on the ITF-25 medium threat list.

Exposure Hazards

Conversion Factor: 1 ppm = 5.98 mg/m^3 at 77°F
NIOSH Ceiling: 0.5 ppm [15 min]

Properties:

MW: 146.3	*VP*: 0.4245 mmHg (77°F)	*FlP*: 156°F
D: 0.8433 g/mL	*VP*: 1.55 mmHg (100°F)	*LEL*: —
MP: −57°F	*VD*: 5.0 (calculated)	*UEL*: —
BP: 390°F	*Vlt*: 560 ppm (77°F)	*RP*: 20
BP: 187°F (15 mmHg)	*H_2O*: Insoluble	*IP*: —
Vsc: 0.522 cS	*Sol*: Ethanol	

C11-A103

Tetraethyl lead

CAS: 78-00-2
RTECS: TP4550000
UN: 1649
ERG: 131

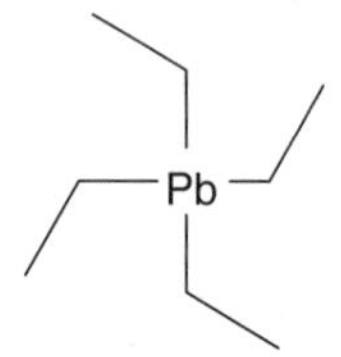

$C_8H_{20}Pb$

Colorless oily liquid with a pleasant, musty, or sweet odor. Commercial material may be dyed red, orange, or blue. This material is hazardous through inhalation, skin absorption (liquid), penetration through broken skin, and ingestion, and produces local skin/eye impacts.

Used industrially to manufacture metal alkyls, fungicides; used as a gasoline additive.

This material is on the ITF-25 low threat list.

Exposure Hazards

Conversion Factor: 1 ppm = 13.23 mg/m^3 at 77°F
Intoxication: 7.6 ppm for a 60 min exposure
$MEG_{(1h)}$ Min: 0.01 ppm as lead; *Sig*: 0.06 ppm as lead; *Sev*: 0.30 ppm as lead
OSHA PEL: 0.007 ppm as lead [Skin]
ACGIH TLV: 0.008 ppm as lead [Skin]
IDLH: 3.0 ppm as lead

Properties:

MW: 323.5	*VP*: 0.2 mmHg	*FlP*: 200°F
D: 1.653 g/mL	*VP*: 0.26 mmHg (77°F)	*LEL*: 1.8%
MP: −202°F	*VD*: 8.6	*UEL*: —
BP: 392°F (decomposes)	*Vlt*: 340 ppm (77°F)	*RP*: 22
BP: 183°F (15 mmHg)	*H_2O*: 0.00002%	*IP*: 11.10 eV
Vsc: 0.522 cS	*Sol*: Benzene; Ether; Ethanol	

C11-A104

Tetramethyl lead

CAS: 75-74-1
RTECS: TP4725000
UN: 1649
ERG: 131

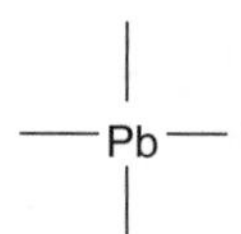

$C_4H_{12}Pb$

Colorless oily liquid with a musty or fruity odor. Commercial material may be dyed red, orange, or blue. This material is hazardous through inhalation, skin absorption (liquid), penetration through broken skin, and ingestion, and produces local skin/eye impacts.

Used industrially as an antiknock additive for gasoline.

This material is on the ITF-25 low threat list.

Exposure Hazards

Conversion Factor: 1 ppm = 10.93 mg/m^3 at 77°F
MEG$_{(1h)}$ *Min*: —; *Sig*: —; *Sev*: 3.7 ppm as lead
OSHA PEL: 0.007 ppm as lead [Skin]
ACGIH TLV: 0.014 ppm as lead [Skin]
IDLH: 3.7 ppm as lead

Other human toxicity values have not been established or have not been published. However, based on available information, this material appears to be approximately one-third as toxic as Tetraethyl lead (C11-A103).

Properties:

MW: 267.3	*VP*: 23 mmHg	*FlP*: 100°F
D: 1.995 g/mL	*VP*: 26 mmHg (77°F)	*LEL*: 1.8%
MP: −22°F	*VD*: 6.5	*UEL*: —
BP: 230°F (decomposes)	*Vlt*: 35,000 ppm	*RP*: 0.23
Vsc: 0.286 cS	H_2O: 0.0015% (77°F)	*IP*: 8.26 eV
	Sol: Fats; Oils; Gasoline	

C11-A105

Vinyl chloride

CAS: 75-01-4
RTECS: KU9625000
UN: 1086P
ERG: 116

C_2H_3Cl

Colorless gas or liquid with a pleasant, sweet odor at high concentrations (>3000 ppm). This material is prone to polymerization and is often stabilized with phenol. This material is hazardous through inhalation, skin absorption, and ingestion, and produces local skin/eye impacts.

Used industrially to produce poly(vinyl chloride) and copolymers, as an inhibitor in the production of ethylene oxide, and as a chemical intermediate for various materials including methyl chloroform, 1,1,1-trichloroethane, and chloroacetaldehyde.

This material is on the CDC threat lists.

Exposure Hazards

Conversion Factor: 1 ppm = 2.56 mg/m^3 at 77°F
OSHA PEL: 1 ppm
OSHA Ceiling: 5 ppm [15 min]
ACGIH TLV: 1 ppm

Properties:

MW: 62.5	*VP*: 2500 mmHg	*FlP*: −108°F
D: 0.969 g/mL (liq. gas, 6°F)	*VD*: 2.2 (calculated)	*LEL*: 3.6%
D: 0.9106 g/mL (liq. gas, 68°F)	*Vlt*: —	*UEL*: 33%
MP: −256°F	H_2O: 0.1% (77°F)	*RP*: 0.005
BP: 7°F	*Sol*: Most organic solvents	*IP*: 9.99 eV
Vsc: 0.193 cS (liq. gas, 77°F)		

Proposed AEGLs

AEGL-1: 1 h, 250 ppm	4 h, 140 ppm	8 h, 70 ppm
AEGL-2: 1 h, 1200 ppm	4 h, 820 ppm	8 h, 820 ppm
AEGL-3: 1 h, 4800 ppm	4 h, 3400 ppm	8 h, 3400 ppm

C11-A106

Allyl chloroformate

CAS: 2937-50-0
RTECS: LQ5775000
UN: 1722
ERG: 155

O
Cl O

$C_4H_5ClO_2$

Colorless watery liquid with a pungent odor detectable at 1.4 ppm. This material is hazardous through inhalation and ingestion, and produces local skin/eye impacts.

Monomer used in the manufacture of some polycarbonate lenses; used as a chemical intermediate for numerous compounds.

This material is on the ITF-25 medium threat list.

Exposure Hazards

Conversion Factor: 1 ppm = 4.93 mg/m^3 at 77°F

Human toxicity values have not been established or have not been published.

Properties:

MW: 120.5	*VP*: 20 mmHg (77°F)	*FlP*: 88°F
D: 1.1394 g/mL	*VD*: 4.2 (calculated)	*LEL*: —
MP: −112°F	*Vlt*: 26,000 ppm	*UEL*: —
BP: 230°F	H_2O: Insoluble (decomposes slowly)	*RP*: 0.46
Vsc: 0.622 cS	*Sol*: —	*IP*: —

C11-A107

Butyl chloroformate

CAS: 592-34-7
RTECS: LQ5890000

UN: 2743
ERG: 155

$C_5H_9ClO_2$

Clear, colorless to light yellow liquid with an unpleasant odor. This material is hazardous through inhalation and ingestion, and produces local skin/eye impacts.

Used industrially as a chemical intermediate for numerous compounds.

This material is on the ITF-25 low threat list.

Exposure Hazards

Conversion Factor: 1 ppm = 5.59 mg/m^3 at 77°F

Human toxicity values have not been established or have not been published.

Properties:

MW: 136.6	*VP*: 5.4 mmHg	*FlP*: 100°F
D: 1.057 g/mL	*VD*: 4.7 (calculated)	*LEL*: —
MP: —	*Vlt*: 7200 ppm	*UEL*: —
BP: 288°F	H_2O: Insoluble (decomposes slowly)	*RP*: 0.07
Vsc: 0.899 cS	*Sol*: —	*IP*: —

C11-A108

Carbonyl fluoride

CAS: 353-50-4
RTECS: FG6125000
UN: 2417
ERG: 125

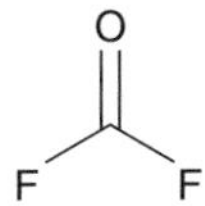

CF_2O

Hygroscopic colorless gas that fumes in moist air with a very irritating, pungent odor. This material is hazardous through inhalation and produces local skin/eye impacts. Hydrogen fluoride generated during decomposition poses a significant inhalation and dermal hazard.

Used industrially as a chemical intermediate in organic synthesis. Produced by decomposition of fluorocarbon plastics when heated between 932 and 1202°F.

This material is on the ITF-25 low threat list.

Exposure Hazards

Conversion Factor: 1 ppm = 2.70 mg/m^3 at 77°F
ACGIH TLV: 2 ppm
ACGIH STEL: 4.8 ppm
NIOSH STEL: 5.6 ppm

Properties:

MW: 66.0	*VP*: 42,000 mmHg	*FlP*: None
D: 0.987 g/mL (liq. gas, –58°F)	*VP*: 44,500 mmHg (77°F)	*LEL*: None
MP: –168°F	*VD*: 2.3 (calculated)	*UEL*: None
BP: –120°F	*Vlt*: —	*RP*: —
Vsc: —	H_2O: Reacts	*IP*: 13.02 eV
	Sol: —	

C11-A109

Carbonyl sulfide

CAS: 463-58-1
RTECS: FG6400000
UN: 2204
ERG: 119

O=S=C

COS

Colorless gas with an odor of rotten eggs. This material is hazardous through inhalation and produces local skin/eye impacts.

Used industrially as a chemical intermediate for alkyl carbonates and organic sulfur compounds.

This material is on the ITF-25 medium threat list.

Exposure Hazards

Conversion Factor: 1 ppm = 2.46 mg/m^3 at 77°F

Human toxicity values have not been established or have not been published.

Properties:

MW: 60.1	*VP*: 9412 mmHg (77°F)	*FlP*: —
D: 1.005 g/mL (liq. gas, 77°F)	*VD*: 2.1 (calculated)	*LEL*: 12%
MP: –218°F	*Vlt*: —	*UEL*: 28.5%
BP: –58°F	H_2O: 0.122% (77°F)	*RP*: 0.001
Vsc: 0.0915 cS (liq. gas, 77°F)	*Sol*: Alcohol; Toluene; Carbon disulfide	*IP*: 11.18 eV

C11-A110

Chloroacetaldehyde

CAS: 107-20-0
RTECS: AB2450000
UN: 2232
ERG: 153

O
Cl H

C_2H_3ClO

Clear, colorless liquid with an acrid, penetrating odor detectable at 0.9 ppm. Typically found as a 40% aqueous solution. This material is hazardous through inhalation, skin absorption, penetration through broken skin, and ingestion, and produces local skin/eye impacts.

Used industrially as a chemical intermediate; used agriculturally to facilitate bark removal from tree trunks and as a fungicide.

This material is on the ITF-25 low threat list.

Exposure Hazards

Conversion Factor: 1 ppm = 3.21 mg/m^3 at 77°F
MEG$_{(1h)}$ Min: 1.0 ppm; *Sig*: 22 ppm; *Sev*: 45 ppm
OSHA Ceiling: 1 ppm
ACGIH Ceiling: 1 ppm
NIOSH Ceiling: 1 ppm
IDLH: 45 ppm

Properties:

MW: 78.5
D: 1.19 g/mL (40% solution)
MP: −3°F (40% solution)
BP: 185°F
Vsc: —
VP: 100 mmHg (40% solution)
VD: 2.7 (calculated)
Vlt: —
H_2O: Miscible
Sol: Ether; Acetone; Methanol
FlP: 190°F (40% solution)
LEL: —
UEL: —
RP: —
IP: 10.61 eV

Proposed AEGLs

AEGL-1: 1 h, 1.3 ppm 4 h, 0.40 ppm 8 h, 0.23 ppm
AEGL-2: 1 h, 2.2 ppm 4 h, 0.69 ppm 8 h, 0.39 ppm
AEGL-3: 1 h, 9.9 ppm 4 h, 3.1 ppm 8 h, 1.8 ppm

C11-A111

Chloroacetyl chloride

CAS: 79-04-9
RTECS: AO6475000
UN: 1752
ERG: 156

$C_2H_2Cl_2O$

Colorless to yellowish liquid with a sharp, pungent, and irritating odor detectable at 0.140 ppm. This material is hazardous through inhalation, skin absorption, penetration through broken skin, and ingestion, and produces local skin/eye impacts.

Used industrially as an intermediate in the manufacture of chloroacetophenone, herbicides, pharmaceuticals, and organic chemical synthesis.

This material is on the ITF-25 low threat list.

Exposure Hazards

Conversion Factor: 1 ppm = 4.62 mg/m^3 at 77°F
Eye Irritation: 1 ppm; exposure duration unspecified
MEG$_{(1h)}$ Min: 0.05 ppm; *Sig*: 0.5 ppm; *Sev*: 10 ppm
ACGIH TLV: 0.05 ppm [Skin]
ACGIH STEL: 0.15 ppm [Skin]

Properties:

MW: 112.9	*VP*: 19 mmHg	*FlP*: None
D: 1.419 g/mL	*VP*: 25.2 mmHg (77°F)	*LEL*: None
MP: −7°F	*VD*: 3.9 (calculated)	*UEL*: None
BP: 223°F	*Vlt*: 25,000 ppm	*RP*: 0.49
Vsc: 0.147 cS	H_2O: Insoluble (decomposes slowly)	*IP*: 10.30 eV
	Sol: Ethyl ether; Acetone; Benzene	

Proposed AEGLs

AEGL-1: 1 h, 0.040 ppm	4 h, 0.040 ppm	8 h, 0.040 ppm
AEGL-2: 1 h, 1.6 ppm	4 h, 0.40 ppm	8 h, 0.20 ppm
AEGL-3: 1 h, 52 ppm	4 h, 13 ppm	8 h, 6.5 ppm

C11-A112

Crotonaldehyde

CAS: 4170-30-3; 123-73-9 (Trans); 15798-64-8 (Cis)
RTECS: GP9499000
UN: 1143
ERG: 131P

C_4H_6O

Water-white liquid that becomes pale yellow on contact with air. It has a tarry, extremely irritating and suffocating odor that is detectable at 0.13 ppm. It exists as two configurational isomers. This material is hazardous through inhalation and ingestion, and produces local skin/eye impacts.

Used industrially in the manufacture of butyl alcohol, butyraldehyde, quinaldine, resins, rubber antioxidants, insecticides, and other chemicals; used as a solvent, warning agent in fuel gases, as a rubber accelerator, in leather tanning, and as a denaturant in alcohol.

This material is on the ITF-25 low threat list.

Exposure Hazards

Conversion Factor: 1 ppm = 2.87 mg/m^3 at 77°F
$MEG_{(1h)}$ *Min*: 0.91 ppm; *Sig*: 4.4 ppm; *Sev*: 14 ppm
OSHA PEL: 2 ppm
ACGIH Ceiling: 0.3 ppm [Skin]
IDLH: 50 ppm

Properties:

MW: 70.1	*VP*: 19 mmHg	*FlP*: 55°F
D: 0.858 g/mL	*VP*: 30 mmHg (77°F)	*LEL*: 2.1%
MP: −105°F	*VD*: 2.4 (calculated)	*UEL*: 15.5%
BP: 219°F	*Vlt*: 25,000 ppm	*RP*: 0.62
Vsc: 0.873 cS	H_2O: 18.1%	*IP*: 9.73 eV
	H_2O: 19.2% (41°F)	
	Sol: Alcohol; Ether; Toluene	

Interim AEGLs

AEGL-1: 1 h, 0.19 ppm	4 h, 0.19 ppm	8 h, 0.19 ppm
AEGL-2: 1 h, 4.4 ppm	4 h, 1.1 ppm	8 h, 0.56 ppm
AEGL-3: 1 h, 14 ppm	4 h, 2.6 ppm	8 h, 1.5 ppm

C11-A113

Ethyl chloroformate

CAS: 541-41-3
RTECS: LQ6125000
UN: 1182
ERG: 155

$C_3H_5ClO_2$

Clear, water-white liquid with an irritating, sharp odor like hydrogen chloride. This material is hazardous through inhalation and ingestion, and produces local skin/eye impacts.

Used in the production of carbamates that are used to synthesize dyes, drugs, veterinary medicines, herbicides, and insecticides; as a solvent in the photographic industry; used as a stabilizer for PVC.

This material is on the ITF-25 low threat list.

Exposure Hazards

Conversion Factor: 1 ppm = 4.44 mg/m^3 at 77°F

Human toxicity values have not been established or have not been published.

Properties:

MW: 108.5	*VP*: 22.4 mmHg (77°F)	*FlP*: 61°F
D: 1.139 g/mL	*VD*: 3.7 (calculated)	*LEL*: 3.5%
MP: −113°F	*Vlt*: 29,000 ppm (77°F)	*UEL*: 10.2%
BP: 200°F	*H_2O*: Insoluble (decomposes slowly)	*RP*: 0.43
Vsc: 2.81 cS	*Sol*: Alcohol; Benzene; Ether	*IP*: —

C11-A114

Ethyl chlorothioformate

CAS: 2941-64-2
RTECS: LQ6950000
UN: 2826
ERG: 155

C_3H_5ClOS

Clear, colorless liquid with a foul stench. Colors of commercial material vary depending on the impurities present and range from grey to green to yellow to purple. This material is hazardous through inhalation and ingestion, and produces local skin/eye impacts.

This material is on the ITF-25 low threat list.

Exposure Hazards

Conversion Factor: 1 ppm = 5.10 mg/m^3 at 77°F

Human toxicity values have not been established or have not been published.

Properties:

MW: 124.6	*VP*: 8.3 mmHg (70°F)	*FlP*: 86°F
D: 1.195 g/mL	*VD*: 4.3 (calculated)	*LEL*: —
MP: −76°F	*Vlt*: 11,000 ppm (70°F)	*UEL*: —
BP: 270°F	H_2O: <0.1% (decomposes slowly)	*RP*: 1.1
Vsc: 2.51 cS	*Sol*: —	*IP*: —

C11-A115

Isobutyl chloroformate

CAS: 543-27-1
RTECS: —

$C_5H_9ClO_2$

Clear, colorless liquid with a pungent, unpleasant odor. Commercial material may have a very faintly brown color. This material is hazardous through inhalation and ingestion, and produces local skin/eye impacts.

This material is on the ITF-25 low threat list.

Exposure Hazards

Conversion Factor: 1 ppm = 5.59 mg/m^3 at 77°F

Human toxicity values have not been established or have not been published.

Properties:

MW: 136.6	*VP*: 6.5 mmHg	*FlP*: 87°F
D: 1.044 g/mL	*VP*: 8.35 mmHg (77°F)	*LEL*: —
MP: −112°F	*VD*: 4.7 (calculated)	*UEL*: —
BP: 264°F	*Vlt*: 8700 ppm	*RP*: 1.3
Vsc: —	H_2O: Decomposes slowly	*IP*: —
	Sol: Benzene; Chloroform; Ether	

C11-A116

Isopropyl chloroformate

CAS: 108-23-6
RTECS: LQ6475000
UN: 2407
ERG: 155

$C_4H_7ClO_2$

Colorless liquid with a strong odor. This material is hazardous through inhalation and ingestion, and produces local skin/eye impacts.

Used in the manufacture of pesticides, herbicides, veterinary medicines, polymerization initiators, blowing agents, and other chemicals.

This material is on the ITF-25 low threat list.

Exposure Hazards

Conversion Factor: 1 ppm = 5.01 mg/m^3 at 77°F

Human toxicity values have not been established or have not been published.

Properties:

MW: 122.6	*VP*: 21 mmHg	*FlP*: 82°F
D: 1.078 g/mL	*VP*: 37 mmHg (77°F)	*LEL*: 4%
MP: —	*VD*: 4.2 (calculated)	*UEL*: 15%
BP: 217°F	*Vlt*: 28,000 ppm	*RP*: 0.43
Vsc: 0.603 cS	H_2O: Decomposes	*IP*: —
	Sol: Ether	

C11-A117

Methanesulfonyl chloride

CAS: 124-63-0
RTECS: PB2790000
UN: 3246
ERG: 156

CH_3ClO_2S

Clear, colorless to slightly yellow oily liquid with a pungent odor. This material is hazardous through inhalation, skin absorption, penetration through broken skin, and ingestion, and produces local skin/eye impacts.

Used industrially as a chemical intermediate for pharmaceuticals and herbicides; used in the manufacture of flame-resistant products; used as a stabilizer for liquid sulfur trioxide.

This material is on the ITF-25 medium threat list.

Exposure Hazards

Conversion Factor: 1 ppm = 4.69 mg/m^3 at 77°F
Severe Eye Irritation: 1 ppm; exposure duration unspecified
Severe Respiratory Irritation: 10 ppm; exposure duration unspecified

Other human toxicity values have not been established or have not been published.

Properties:

MW: 114.6	*VP*: 12 mmHg (127°F)	*FlP*: 230°F
D: 1.474 g/mL	*VD*: 4.0 (calculated)	*LEL*: —
MP: −27°F	*Vlt*: 14,000 ppm (127°F)	*UEL*: —
BP: 322°F	H_2O: Insoluble (decomposes slowly)	*RP*: 0.86
BP: 144°F (18 mmHg)	*Sol*: Most organic solvents	*IP*: 11.6 eV
Vsc: 1.34 cS (77°F)		

C11-A118

Methyl chloroformate

CAS: 79-22-1
RTECS: FG3675000

UN: 1238
ERG: 155

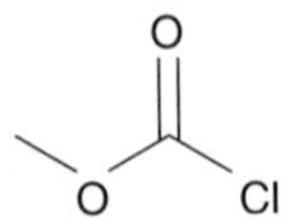

$C_2H_3ClO_2$

Clear, colorless to light yellow liquid with a sharp, unpleasant, acrid odor. This material is hazardous through inhalation, skin absorption, penetration through broken skin, and ingestion, and produces local skin/eye impacts.

Used industrially in the manufacture of carbamates, pesticides, herbicides, insecticides, and pharmaceuticals; used as a solvent in the photographic industry.

This material is on the ITF-25 medium threat list.

Exposure Hazards

Conversion Factor: 1 ppm = 3.87 mg/m^3 at 77°F
$LC_{(Inh)}$: 190 ppm for a 10 min exposure
Lacrimation: 10 ppm, exposure duration unspecified

Other human toxicity values have not been established or have not been published.

Properties:

MW: 94.5	*VP*: 108.5 mmHg (77°F)	*FlP*: 76°F
D: 1.228 g/mL	*VD*: 3.3 (calculated)	*LEL*: 6.7%
MP: −78°F	*Vlt*: 140,000 ppm (77°F)	*UEL*: 23%
BP: 159°F	H_2O: Slight (decomposes slowly)	*RP*: 0.96
Vsc: 0.393 cS	*Sol*: Alcohol; Ether; Benzene	*IP*: —

C11-A119

Propyl chloroformate

CAS: 109-61-5
RTECS: LQ6830000
UN: 2740
ERG: 155

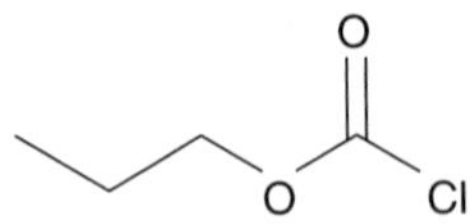

$C_4H_7ClO_2$

Colorless liquid. This material is hazardous through inhalation, skin absorption, penetration through broken skin, and ingestion, and produces local skin/eye impacts.

Used industrially as an intermediate for polymerization initiators and other chemicals.

This material is on the ITF-25 low threat list.

Exposure Hazards

Conversion Factor: 1 ppm = 5.01 mg/m^3 at 77°F

Human toxicity values have not been established or have not been published.

Properties:

MW: 122.6	*VP*: 20 mmHg (77°F)	*FlP*: 94°F
D: 1.091 g/mL	*VD*: 4.2 (calculated)	*LEL*: —
MP: —	*Vlt*: 26,000 ppm (77°F)	*UEL*: —
BP: 238°F	H_2O: Insoluble (decomposes slowly)	*RP*: 0.46
Vsc: 0.73 cS	*Sol*: Benzene; Chloroform; Ether	*IP*: —

C11-A120

Trichloroacetyl chloride

CAS: 76-02-8
RTECS: AO7140000
UN: 2442
ERG: 156

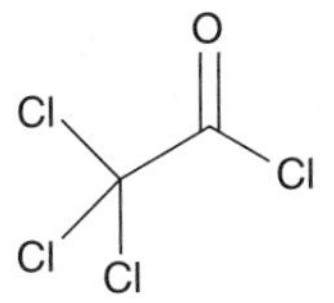

Cl, Cl, Cl, O, Cl

C_2Cl_4O

Clear, colorless to very faintly yellow liquid. This material is hazardous through inhalation and ingestion, and produces local skin/eye impacts.

Used industrially as a chemical intermediate.

This material is on the ITF-25 medium threat list.

Exposure Hazards

Conversion Factor: 1 ppm = 7.44 mg/m^3 at 77°F

Human toxicity values have not been established or have not been published.

Properties:

MW: 181.8	*VP*: 16 mmHg	*FlP*: None
D: 1.6202 g/mL	*VP*: 21.3 mmHg (77°F)	*LEL*: None
MP: −25°F	*VD*: 6.3 (calculated)	*UEL*: None
BP: 244°F	*Vlt*: 21,000 ppm	*RP*: 0.46
Vsc: —	H_2O: Decomposes	*IP*: 11.31 eV
	Sol: Ether; Alcohols	

C11-A121

Trifluoroacetyl chloride

CAS: 354-32-5
RTECS: AO7150000
UN: 3057
ERG: 125

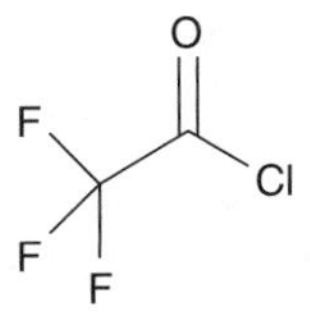

F, F, F, O, Cl

C_2ClF_3O

Colorless gas that fumes strongly in moist air. This material is hazardous through inhalation and produces local skin/eye impacts. Hydrogen fluoride generated during decomposition poses a significant inhalation and dermal hazard.

Used industrially for organic synthesis.

This material is on the ITF-25 medium threat list.

Exposure Hazards

Conversion Factor: 1 ppm = 5.42 mg/m^3 at 77°F

Human toxicity values have not been established or have not been published.

Properties:

MW: 132.5	*VP*: 3700 mmHg (77°F)	*FlP*: None
D: 1.3844 g/mL (liq. gas, 77°F)	*VD*: 4.6 (calculated)	*LEL*: None
MP: −231°F	*Vlt*: —	*UEL*: None
BP: −8°F	*H_2O*: Reacts vigorously	*RP*: 0.002
Vsc: —	*Sol*: —	*IP*: —

C11-A122

1,2-Dimethylhydrazine

CAS: 540-73-8, 306-37-6 (Dihydrochloride salt)
RTECS: —
UN: 2382
ERG: 131

N—N

$C_2H_8N_2$

Mobile, hygroscopic clear, colorless liquid that fumes in air. Slowly turns yellow after exposure to air. Has a fishy or ammonia-like odor. Various salts (solids) have been reported. This material is hazardous through inhalation, skin absorption, penetration through broken skin, and ingestion, and produces local skin/eye impacts.

Used industrially for research; limited use as a high-energy rocket fuel.

This material is on the ITF-25 medium threat list.

Exposure Hazards

Conversion Factor: 1 ppm = 2.46 mg/m^3 at 77°F

Human toxicity values have not been established or have not been published.

Properties:

MW: 60.1	*VP*: 68 mmHg (77°F)	*FlP*: 73°F
D: 0.8274 g/mL	*VD*: 2.1 (calculated)	*LEL*: —
MP: 16°F	*Vlt*: 89,000 ppm (77°F)	*UEL*: —
BP: 178°F	*H_2O*: Miscible	*RP*: 0.19
Vsc: —	*Sol*: Alcohol; Ether; Hydrocarbons	*IP*: 9.02 eV

Final AEGLs

AEGL-1: Not Developed

AEGL-2: 1 h, 3.0 ppm	4 h, 0.75 ppm	8 h, 0.38 ppm
AEGL-3: 1 h, 11 ppm	4 h, 2.7 ppm	8 h, 1.4 ppm

C11-A123

Methyl hydrazine

CAS: 60-34-4; 7339-53-9 (Hydrochloride salt)
RTECS: MV5600000
UN: 1244
ERG: 131

CH_6N_2

Hygroscopic, clear, colorless, fuming liquid with an odor like ammonia that is detectable between 0.9 and 2.8 ppm. Various salts (solids) have been reported. This material is hazardous through inhalation, skin absorption, penetration through broken skin, and ingestion, and produces local skin/eye impacts. It is a strong sensitizer. It spontaneously ignites in contact with strong oxidizers.

Used industrially to manufacture pesticides, pharmaceuticals; used as a solvent, and as a rocket fuel.

This material is on the ITF-25 medium threat list.

Exposure Hazards

Conversion Factor: 1 ppm = 1.89 mg/m^3 at 77°F
MEG$_{(1h)}$ *Min*: —; *Sig*: 1 ppm; *Sev*: 3 ppm
OSHA Ceiling: 0.2 ppm [Skin]
ACGIH TLV: 0.01 ppm [Skin]
NIOSH Ceiling: 0.04 ppm (120 min)
IDLH: 20 ppm

Properties:

MW: 46.1	*VP*: 38 mmHg	*FlP*: 17°F
D: 0.874 g/mL (77°F)	*VP*: 50 mmHg (77°F)	*LEL*: 2.5%
MP: −62°F	*VD*: 1.6 (calculated)	*UEL*: 92%
BP: 190°F	*Vlt*: 51,000 ppm	*RP*: 0.39
Vsc: 0.887 cS (77°F)	*H_2O*: Miscible	*IP*: 7.70 eV
	Sol: Ether; Alcohols; Hydrocarbons	

Final AEGLs

AEGL-1: Not Developed
AEGL-2: 1 h, 0.90 ppm   4 h, 0.23 ppm   8 h, 0.11 ppm
AEGL-3: 1 h, 2.7 ppm   4 h, 0.68 ppm   8 h, 0.34 ppm

C11-A124

Allyl isothiocyanate

CAS: 57-06-7
RTECS: NX8225000
UN: 1545
ERG: 155

N=C=S

C_4H_5NS

Clear, colorless to pale-yellow, oily liquid with an irritating, pungent odor like mustard that is detectable at 0.008 ppm. It tends to darken on storage. This material is hazardous through inhalation, skin absorption, penetration through broken skin, and ingestion, and produces local skin/eye impacts. It causes sensitization through both inhalation and skin contact.

Used industrially as a fumigant, fungicide, animal repellent, insect attractant; used in the preparation of medicinal ointments and mustard plasters; used as a flavoring agent, and denaturant for alcohol.

This material is on the ITF-25 low threat list.

Exposure Hazards

Conversion Factor: 1 ppm = 4.06 mg/m^3 at 77°F
AIHA STEL: 1 ppm [Skin]

Properties:

MW: 99.2	*VP*: 3.7 mmHg (86°F)	*FlP*: 115°F
D: 1.0175 g/mL	*VD*: 3.4 (calculated)	*LEL*: —
MP: −112°F	*Vlt*: 4800 ppm (86°F)	*UEL*: —
BP: 306°F	*H_2O*: 0.2%	*RP*: 2.8
Vsc: —	*Sol*: Most organic solvents	*IP*: —

C11-A125

Butyl isocyanate

CAS: 111-36-4
RTECS: NQ8250000
UN: 2485
ERG: 155

C_5H_9NO

Colorless liquid with a strong odor. This material is hazardous through inhalation, skin absorption, penetration through broken skin, and ingestion, and produces local skin/eye impacts.

Used industrially in the manufacture of insecticides, fungicides, and pharmaceuticals.

This material is on the ITF-25 low threat list.

Exposure Hazards

Conversion Factor: 1 ppm = 4.05 mg/m^3 at 77°F
$MEG_{(1h)}$ Min: 0.01 ppm; *Sig*: 0.05 ppm; *Sev*: 1.0 ppm

Properties:

MW: 99.1	*VP*: 10.6 mmHg	*FlP*: 66°F
D: 0.88 g/mL	*VP*: 17.6 mmHg (77°F)	*LEL*: —
MP: <–94°F	*VD*: 3.4 (calculated)	*UEL*: —
BP: 239°F	*Vlt*: 14,000 ppm	*RP*: 0.94
Vsc: —	*H_2O*: Slight (decomposes slowly)	*IP*: 10.14 eV
	Sol: —	

C11-A126

Isopropyl isocyanate

CAS: 1795-48-8
RTECS: NQ9230000
UN: 2483
ERG: 155

C_4H_7NO

Clear, colorless to faint-yellow liquid with a strong penetrating odor. This material is hazardous through inhalation, skin absorption, penetration through broken skin, and ingestion, and produces local skin/eye impacts. It causes sensitization.

Used industrially as an intermediate to manufacture pesticides and other chemicals.

This material is on the ITF-25 low threat list.

Exposure Hazards

Conversion Factor: 1 ppm = 3.48 mg/m^3 at 77°F

Human toxicity values have not been established or have not been published.

Properties:

MW: 85.1	*VP*: —	*FlP*: 27°F
D: 0.866 g/mL	*VD*: 2.9 (calculated)	*LEL*: —
MP: —	*Vlt*: —	*UEL*: —
BP: 166°F	H_2O: Decomposes	*RP*: —
Vsc: —	*Sol*: —	*IP*: —

C11-A127

Methyl isocyanate

CAS: 624-83-9
RTECS: NQ9450000
UN: 2480
ERG: 155

—N=C=O

C_2H_3NO

Colorless liquid with a strong, sharp, pungent odor. However, less than 10% of attentive persons can detect this material at the industrial exposure limits. This material is hazardous through inhalation, skin absorption, penetration through broken skin, and ingestion, and produces local skin/eye impacts.

Used industrially as an intermediate to manufacture pesticides, polyurethane foams, and plastics.

This material is on the ITF-25 medium threat list.

Exposure Hazards

Conversion Factor: 1 ppm = 2.34 mg/m^3 at 77°F
Eye Irritation: 2 ppm; exposure duration unspecified
"Unbearable" Irritation: 21 ppm; exposure duration unspecified
$MEG_{(1h)}$ *Min*: 0.025 ppm; *Sig*: 0.067 ppm; *Sev*: 0.2 ppm
OSHA PEL: 0.02 ppm [Skin]
ACGIH TLV: 0.02 ppm [Skin]
IDLH: 3 ppm

Properties:

MW: 57.1	*VP*: 348 mmHg	*FlP*: 19°F
D: 0.9599 g/mL	*VD*: 1.4	*LEL*: 5.3%
MP: −49°F	*Vlt*: 470,000 ppm	*UEL*: 26%
BP: 103°F	H_2O: 10% (59°F; decomposes slowly)	*RP*: 0.38
Vsc: —	*Sol*: —	*IP*: 10.67 eV

Final AEGLs

AEGL-1: Not Developed

AEGL-2: 1 h, 0.067 ppm	4 h, 0.017 ppm	8 h, 0.008 ppm
AEGL-3: 1 h, 0.20 ppm	4 h, 0.05 ppm	8 h, 0.025 ppm

C11-A128

Methylene bisphenyl diisocyanate

CAS: 101-68-8
RTECS: NQ9350000

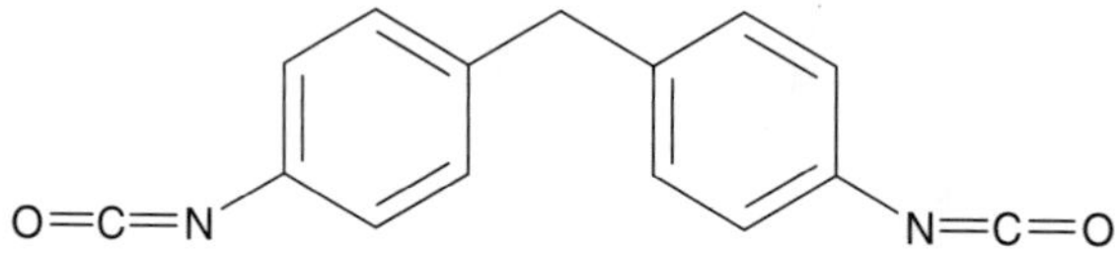

O=C=N N=C=O

$C_{15}H_{10}N_2O_2$

White to light-yellow flakes odorless flakes. This material is hazardous through inhalation and ingestion, and produces local skin/eye impacts.

Used industrially to manufacture adhesives, coatings, elastomers, and polyurethane foams. This material is on the ITF-25 low threat list.

Exposure Hazards

Conversion Factor: 1 ppm = 10.24 mg/m^3 at 77°F
$MEG_{(1h)}$ *Min*: 0.02 ppm; *Sig*: 0.20 ppm; *Sev*: 2.4 ppm
OSHA Ceiling: 0.02 ppm
ACGIH Ceiling: 0.005 ppm
NIOSH Ceiling: 0.02 ppm [10 min]
IDLH: 7 ppm

Properties:

MW: 250.3	*VP*: 0.000005 mmHg (77°F)	*FlP*: 396°F
D: 1.23 g/cm^3 (77°F)	*VD*: 8.6 (calculated)	*LEL*: —
D: 1.19 g/mL (122°F)	*Vlt*: 0.01 ppm (77°F)	*UEL*: —
MP: 99°F	*H_2O*: 0.2% (decomposes slowly)	*RP*: 1,200,000
BP: 597°F	*Sol*: Acetone; Benzene; Kerosene	*IP*: —
BP: 385°F (5 mmHg)		
Vsc: 4.22 cS (122°F)		

C11-A129

tert-Butyl isocyanate

CAS: 1609-86-5
RTECS: NQ8300000
UN: 2484
ERG: 155

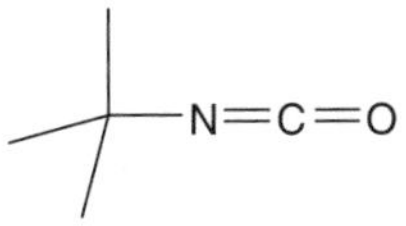

C_5H_9NO

Clear, colorless liquid. This material is hazardous through inhalation, skin absorption, penetration through broken skin, and ingestion, and produces local skin/eye impacts.

This material is on the ITF-25 low threat list.

Exposure Hazards

Conversion Factor: 1 ppm = 4.05 mg/m^3 at 77°F

Human toxicity values have not been established or have not been published.

Properties:

MW: 99.1	*VP*: 57 mmHg (77°F)	*FlP*: 24°F
D: 0.844 g/mL	*VD*: 3.4 (calculated)	*LEL*: —
MP: —	*Vlt*: 75,000 ppm (77°F)	*UEL*: —
BP: 183°F	*H_2O*: Decomposes	*RP*: 0.18
Vsc: —	*Sol*: —	*IP*: 10.14 eV

C11-A130

Toluene 2,4-diisocyanate

CAS: 584-84-9
RTECS: CZ6300000
UN: 2078
ERG: 156

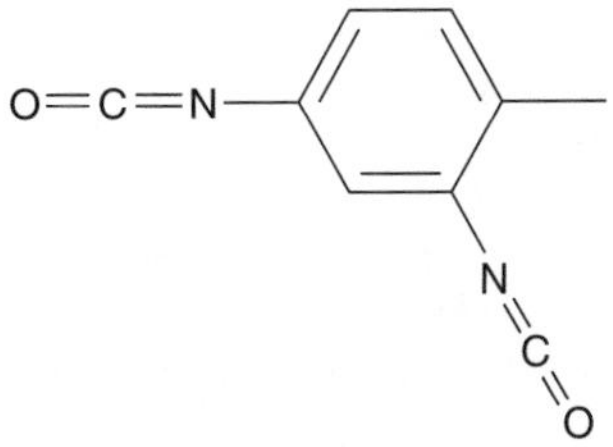

$C_9H_6N_2O_2$

Colorless solid or liquid with a sharp, pungent fruity odor that is detectable at 2.1 ppm. It turns pale yellow on exposure to air. This material is hazardous through inhalation, skin absorption, penetration through broken skin, and ingestion, and produces local skin/eye impacts.

Used industrially to manufacture polyurethane foams, elastomers, coatings; used as a cross-linking agent.

This material is on the ITF-25 low threat list.

Exposure Hazards

Conversion Factor: 1 ppm = 7.12 mg/m^3 at 77°F
$MEG_{(1h)}$ Min: 0.02 ppm; *Sig*: 0.083 ppm; *Sev*: 0.51 ppm
OSHA Ceiling: 0.02 ppm
ACGIH TLV: 0.005 ppm
ACGIH STEL: 0.02 ppm
IDLH: 2.5 ppm

Properties:

MW: 174.2	*VP*: 0.008 mmHg	*FlP*: 260°F
D: 1.22 g/mL (77°F)	*VP*: 0.01 mmHg (77°F)	*LEL*: 0.9%
MP: 70°F	*VD*: 6.0 (calculated)	*UEL*: 9.5%
BP: 484°F	*Vlt*: 11 ppm	*RP*: 940
BP: 255°F (18 mmHg)	*H_2O*: Insoluble (decomposes slowly)	*IP*: —
Vsc: 3.75 cS (77°F)	*Sol*: Ether; Acetone; Benzene	

Final AEGLs

AEGL-1: 1 h, 0.02 ppm	4 h, 0.01 ppm	8 h, 0.01 ppm
AEGL-2: 1 h, 0.083 ppm	4 h, 0.021 ppm	8 h, 0.021 ppm
AEGL-3: 1 h, 0.51 ppm	4 h, 0.32 ppm	8 h, 0.16 ppm

References

Agency for Toxic Substances and Disease Registry. "1,2-Dibromo-3-chloropropane ToxFAQs." September 1995.
———. "1,2-Dibromoethane ToxFAQs." September 1995.
———. "1,2-Dichloroethane ToxFAQs." September 2001.
———. "1,2-Dichloropropane ToxFAQs." July 1999.
———. "Acrylonitrile ToxFAQs." July 1999.
———. "Ammonia ToxFAQs." September 2004.
———. "Bromomethane ToxFAQs." September 1995.
———. "Carbon Disulfide ToxFAQs." September 1997.
———. "Chlorfenvinphos ToxFAQs." September 1997.
———. "Crotonaldehyde ToxFAQs." April 2002.
———. "Diborane ToxFAQs." April 2002.
———. "Disulfoton ToxFAQs." September 1996.
———. "Ethylene Oxide ToxFAQs." July 1999.
———. "Fluorides, Hydrogen Fluoride, and Fluorine ToxFAQs." September 2003.
———. "Formaldehyde ToxFAQs." June 1999.
———. "Hexachlorocyclopentadiene (HCCPD) ToxFAQs." June 1999.
———. "Hydrazines ToxFAQs." September 1997.

———. "Hydrogen Chloride ToxFAQs." April 2002.
———. *Managing Hazardous Materials Incidents Volume III—Medical Management Guidelines for Acute Chemical Exposures.* Rev. ed. Washington, DC: Government Printing Office, 2000.
———. "Methyl Isocyanate ToxFAQs." April 2002.
———. "Methyl Mercaptan ToxFAQs." July 1999.
———. "Methyl Parathion ToxFAQs." September 2001.
———. "Phosphine ToxFAQs." April 2002.
———. "Selenium Hexafluoride ToxFAQs." April 2002.
———. "Sulfur Dioxide ToxFAQs." June 1999.
———. "Sulfur Trioxide (SO_3) and Sulfuric Acid ToxFAQs." June 1999.
———. "Titanium Tetrachloride ToxFAQs." September 1997.
———. "Vinyl Chloride ToxFAQs." September 2004.
———. *Toxicological Profile for 1,2-Dibromo-3-chloropropane.* Washington, DC: Government Printing Office, September 1992.
———. *Toxicological Profile for 1,2-Dibromoethane.* Washington, DC: Government Printing Office, July 1992.
———. *Toxicological Profile for 1,2-Dichloroethane.* Washington, DC: Government Printing Office, August 1996.
———. *Toxicological Profile for 1,2-Dichloropropane.* Washington, DC: Government Printing Office, December 1989.
———. *Toxicological Profile for Acrylonitrile.* Washington, DC: Government Printing Office, December 1990.
———. *Toxicological Profile for Aluminum.* Washington, DC: Government Printing Office, July 1999.
———. *Toxicological Profile for Ammonia.* Washington, DC: Government Printing Office, September 2004.
———. *Toxicological Profile for Antimony and Compounds.* Washington, DC: Government Printing Office, September 1992.
———. *Toxicological Profile for Arsenic.* Washington, DC: Government Printing Office, September 2000.
———. *Toxicological Profile for Boron and Compounds.* Washington, DC: Government Printing Office, July 1992.
———. *Toxicological Profile for Bromomethane.* Washington, DC: Government Printing Office, September 1992.
———. *Toxicological Profile for Carbon Disulfide.* Washington, DC: Government Printing Office, August 1996.
———. *Toxicological Profile for Chlorfenvinphos.* Washington, DC: Government Printing Office, September 1997.
———. *Toxicological Profile for Disulfoton.* Washington, DC: Government Printing Office, August 1995.
———. *Toxicological Profile for Ethylene Oxide.* Washington, DC: Government Printing Office, 1990.
———. *Toxicological Profile for Fluorides, Hydrogen Fluoride, and Fluorine.* Washington, DC: Government Printing Office, 2003.
———. *Toxicological Profile for Formaldehyde.* Washington, DC: Government Printing Office, July 1999.
———. *Toxicological Profile for Hexachlorocyclopentadiene (HCCPD).* Washington, DC: Government Printing Office, July 1999.
———. *Toxicological Profile for Hydrazines.* Washington, DC: Government Printing Office, September 1997.
———. *Toxicological Profile for Iodine.* Washington, DC: Government Printing Office, April 2004.
———. *Toxicological Profile for Mercury.* Washington, DC: Government Printing Office, March 1999.
———. *Toxicological Profile for Methyl Mercaptan.* Washington, DC: Government Printing Office, September 1992.
———. *Toxicological Profile for Methyl Parathion.* Washington, DC: Government Printing Office, September 2001.
———. *Toxicological Profile for Selenium.* Washington, DC: Government Printing Office, September 2003.
———. *Toxicological Profile for Sulfur Dioxide.* Washington, DC: Government Printing Office, December 1998.

———. *Toxicological Profile for Sulfur Trioxide and Sulfuric Acid*. Washington, DC: Government Printing Office, December 1998.

———. *Toxicological Profile for Titanium Tetrachloride*. Washington, DC: Government Printing Office, September 1997.

———. *Toxicological Profile for Tungsten (DRAFT)*. Washington, DC: Government Printing Office, September 2003.

———. *Toxicological Profile for Vinyl Chloride (DRAFT)*. Washington, DC: Government Printing Office, September 2004.

———. *Toxicological Profile for White Phosphorus*. Washington, DC: Government Printing Office, September 1997.

———. "Titanium Tetrachloride ToxFAQs." September 1997.

Centers for Disease Control and Prevention. "Biological and Chemical Terrorism: Strategic Plan for Preparedness and Response. Recommendations of the CDC Strategic Planning Workgroup." *Morbidity and Mortality Weekly Report* 49 (RR-4) (2000): 1–14.

———. "Case Definition: Arsine or Stibine Poisoning." March 4, 2005.

Cook, H. G., B. C. Saunders, and F. E. Smith. "Esters Containing Phosphorus. VIII. Structural Requirements for High Toxicity and Miotic Action of Esters of Fluorophosphonic Acid." *Journal of the Chemical Society, Abstracts* (1949): 635–638.

Department of Health and Human Services. *National Toxicology Program, Technical Reports*. August 9, 2005. http://ntp.niehs.nih.gov/index.cfm?objectid=070E4598-9C8B-DF77-1C266EBE08732EB4. March 22, 2006.

Fries, Amos A., and Clarence J. West. *Chemical Warfare*. New York: McGraw-Hill Book Company, Inc., 1921.

National Institute for Occupational Safety and Health. *NIOSH Pocket Guide to Chemical Hazards*. Washington, DC: Government Printing Office, September 2005.

National Institutes of Health. *Hazardous Substance Data Bank (HSDB)*. http://toxnet.nlm.nih.gov/cgi-bin/sis/htmlgen?HSDB/. 2004.

Olson, Kent R., ed. *Poisoning & Drug Overdose*. 4th ed. New York: Lange Medical Books/McGraw-Hill, 2004.

Sartori, Mario F. "New Developments in the Chemistry of War Gases." *Chemical Reviews* 48 (1951): 225–257.

———. *The War Gases: Chemistry and Analysis*. Translated by L. W. Marrison. London: J. & A. Churchill, Ltd, 1939.

Sidell, Fredrick R., Ernest T. Takafuji, and David R. Franz, eds. *Medical Aspects of Chemical and Biological Warfare, Textbook of Military Medicine Series, Part 1, Warfare, Weaponry, and the Casualty*. Washington, DC: Office of the Surgeon General, Department of the Army, 1997.

Smith, Ann, Patricia Heckelman, and Maryadele J. Oneil, eds. *The Merck Index: An Encyclopedia of Chemicals, Drugs, & Biologicals*. 13th ed. Rahway, NJ: Merck & Co., Inc., 2001.

Somani, Satu M., ed. *Chemical Warfare Agents*. New York: Academic Press, 1992.

Swearengen, Thomas F. *Tear Gas Munitions: An Analysis of Commercial Riot Gas Guns, Tear Gas Projectiles, Grenades, Small Arms Ammunition, and Related Tear Gas Devices*. Springfield, IL: Charles C Thomas Publisher, 1966.

True, Bey-Lorraine, and Robert H. Dreisbach. *Dreisbach's Handbook of Poisoning: Prevention, Diagnosis and Treatment*. 13th ed. London, England: The Parthenon Publishing Group, 2002.

United States Army Headquarters. *Potential Military Chemical/Biological Agents and Compounds, Field Manual No. 3-11.9*. Washington, DC: Government Printing Office, January 10, 2005.

United States Coast Guard. *Chemical Hazards Response Information System (CHRIS) Manual, 1999 Edition*. http://www.chrismanual.com/Default.htm. March 2004.

Wachtel, Curt. *Chemical Warfare*. Brooklyn, NY: Chemical Publishing Co., Inc., 1941.

Waitt, Alden H. *Gas Warfare: The Chemical Weapon, Its Use, and Protection against It*. Rev ed. New York: Duell, Sloan and Pearce, 1944.

World Health Organization. *International Chemical Safety Cards (ICSCs)*. http://www.cdc.gov/niosh/ipcs/icstart.html. 2004.

Yaws, Carl L. *Matheson Gas Data Book*. 7th ed. Parsippany, NJ: Matheson Tri-Gas, 2001.

Section IV

Incapacitation and Riot Control Agents

12

Incapacitating Agents

12.1 General Information

Used in a military context, incapacitation means the inability of personnel to perform an assigned duty. For the purpose of this manual, incapacitation means the inability to perform any military task effectively and implies that the condition was achieved via the deliberate use of a nonlethal weapon. Incapacitating agents differ from other chemical agents in that the lethal dose is theoretically many times greater than the incapacitating dose. Under normal battlefield conditions, they do not pose a serious danger to the life of an exposed individual and do not produce any permanent injury. The military does not consider the use of lethal agents at sublethal doses as incapacitating agents.

Military incapacitating agents are third and fourth generation chemical warfare agents that became popular during the Cold War. Incapacitating agent BZ (C12-A001) and two key components necessary to synthesize BZ, 3-quinuclidinol (C12-C022) and benzilic acid (C12-C023), are listed in Schedule 2 of the Chemical Weapons Convention (CWC). In addition to military-specific agents, materials in this class encompass a wide variety of commercially available medicinal dugs that interfere with the higher functions of the brain such as attention, orientation, perception, memory, motivation, conceptual thinking, planning, and judgment. They produce their effects mainly by altering or disrupting the higher regulatory activity of the central nervous system.

Incapacitating agents are relatively easy to isolate from natural sources or to synthesize. Several agents are clandestinely synthesized and used as recreational drugs. For information on some of the chemicals used to manufacture military incapacitating agents, see the Component section (C12-C) following information on the individual agents. Although relatively easy to disperse, it is difficult to effectively control the dose received by the target population and prevent fatalities.

Incapacitating agents have been stockpiled by numerous countries and there have been unverified reports that they have been utilized on the battlefield. In addition, they have been employed by police and special forces as a way to end hostage situations (e.g., the September 2002 counter terrorism raid on the Moscow theater). These operations have met with mixed success.

12.2 Toxicology

12.2.1 Effects

Incapacitating agents produce their effects mainly by altering or disrupting the higher regulatory activity of the central nervous system. Military incapacitating agents can be separated into four fairly discrete categories: deliriants (producing confusion, hallucinations, and disorganized behavior), stimulants (essentially flooding the brain with too much information), calmatives (depressants that induce passivity or even sleep), and psychedelics (producing abnormal psychological effects resembling mental illness).

In normal usage, incapacitating agents will not cause permanent or long-lasting injury. Unlike lachrymatory agents (Chapter 13) or vomiting/sternatory agents (Chapter 14), incapacitating agents produce effects that may last for hours or even days postexposure.

12.2.2 Pathways and Routes of Exposure

Incapacitating agents are primarily a hazard through inhalation. However, exposure to liquid or solid agents may be hazardous through skin absorption (if the agent is dissolved in an appropriate solvent), ingestion, introduction through abraded skin (e.g., breaks in the skin or penetration of skin by debris), and may also produce local skin/eye effects (e.g., dilation of the pupil).

12.2.3 General Exposure Hazards

Incapacitating agents do not have good warning properties. They have little or no odor, and the vapors do not irritate the eyes. Contact with liquid or solid agents neither irritates the skin nor causes cutaneous injuries. This class of agents does not seriously endanger life except at exposures greatly exceeding an effective dose.

12.2.4 Latency Period

12.2.4.1 Vapors/Aerosols (Mists or Dusts)

Depending on the specific agent and the concentration of vapor or aerosol, the effects begin to appear in seconds or may be delayed up to several hours.

12.2.4.2 Liquids

Typically, there is a latent period with no visible effects between the time of exposure and the sudden onset of symptoms. Effects from dermal exposure may be delayed up to several days. Some factors affecting the length of time before the onset of symptoms are the amount of agent involved, the amount of skin surface in contact with the agent, previous exposure to materials that chap or dry the skin (e.g., organic solvents such as gasoline or alcohols), and addition of additives designed to enhance the rate of percutaneous penetration by the agents.

Another key factor affecting the rate of percutaneous penetration by the agent is the part of the body that is exposed. It takes the agent longer to penetrate thicker and tougher skin. The regions of the body that allow the fastest percutaneous penetration are the groin, head, and neck. The least susceptible body regions are the hands, feet, front of the knee, and outside of the elbow.

12.2.4.3 Solids (Nonaerosol)

Typically, there is a latent period with no visible effects between the time of exposure and the sudden onset of symptoms. Effects from dermal exposure may be delayed up to several days and is affected by such factors as the amount of agent involved, the amount of skin surface in contact with the agent, and the area of the body exposed (see Section 12.2.4.2). Moist, sweaty areas of the body are more susceptible to percutaneous penetration by solid agents.

12.3 Characteristics

12.3.1 Physical Appearance/Odor

12.3.1.1 Laboratory Grade

Laboratory grade agents are typically colorless liquids or solids. Depending on the specific agent, liquids may be mobile, viscous, or even waxy in nature. Many solids are salts of the free-base liquid that are colorless to white to beige crystalline materials. In either state, these materials typically have little or no odor when pure.

12.3.1.2 Munition Grade

Munition grade agents are typically white to light brown powders, waxy solids, or viscous liquids. Production impurities and decomposition products in these agents may give them an odor. Odors for all agents may become more pronounced during storage.

12.3.1.3 Modified Agents

Solvents have been added to incapacitating agents to facilitate handling, to stabilize the agents, or to increase the ease of percutaneous penetration by the agents. Percutaneous enhancement solvents include dimethyl sulfoxide, *N,N*-dimethylformamide, *N,N*-dimethylpalmitamide, *N,N*-dimethyldecanamide, and saponin. Color and other properties of these solutions may vary from the pure agent. Odors will vary depending on the characteristics of the solvent(s) used and concentration of incapacitating agent in the solution.

12.3.2 Stability

Military-specific agents are stable even under tropical conditions and can be stored in glass, aluminum, or steel. Stabilizers are not required. Some potentially dual use agents are sensitive to heat and light and may require stabilizers for long-term storage.

12.3.3 Persistency

Depending on the properties of the specific agent, unmodified incapacitating agents are classified as either nonpersistent or persistent by the military.

Addition of solvents may alter the persistency of these agents. Salts of agents have negligible vapor pressure and will not evaporate. Depending on the size of the individual particles and on any encapsulation or coatings applied to the particles, they can be reaerosolized by ground traffic or strong winds.

12.3.4 Environmental Fate

Vapors of volatile incapacitating agents have a density greater than air and tend to collect in low places. Most incapacitating agents are nonvolatile and produce negligible amounts of vapor.

Most of these agents are only slightly soluble or insoluble in water. However, the solubility of any agent may be modified (either increased or decreased) by solvents, components, or impurities. The specific gravities of unmodified liquid agents are slightly greater than that of water. Incapacitating agents are typically soluble in most organic solvents including gasoline, alcohols, and oils. Salts of agents are water soluble.

12.4 Additional Hazards

12.4.1 Exposure

All foodstuffs in the area of a release should be considered contaminated. Unopened items packaged in glass, metal, or heavy duty plastic and exposed only to agent vapors, aerosols, or to solid agents may be used after decontamination of the container. Unopened items exposed to solid agents or solutions of agents should be decontaminated within a few hours postexposure or destroyed. Opened or unpackaged items, or those packaged only in paper or cardboard, should be destroyed.

12.4.2 Livestock/Pets

Animals can be decontaminated with shampoo/soap and water (see Section 12.5.3). If the animals' eyes have been exposed to agent, they should be irrigated with water or saline solution for a minimum of 30 minutes.

The topmost layer of unprotected feedstock (e.g., hay or grain) should be destroyed. The remaining material should be quarantined until tested. Leaves of forage vegetation could still retain sufficient agent to produce effects for several weeks post release, depending on the level of contamination and the weather conditions.

12.4.3 Fire

Because of their low vapor pressures, heat from a fire will destroy incapacitating agents before generating any significant concentration of agent vapor. However, actions taken to extinguish the fire can spread the agent. Salts are water soluble and runoff from firefighting efforts will pose a significant threat. Some of the decomposition products resulting from hydrolysis or combustion of incapacitating agents are water soluble and highly toxic (see Section 12.4.5). Other potential decomposition products include toxic and/or corrosive gases.

12.4.4 Reactivity

Some incapacitating agents decompose slowly in water. Raising the pH of an aqueous solution of these agents significantly increases the rate of decomposition.

12.4.5 Hazardous Decomposition Products

12.4.5.1 Hydrolysis

Varies depending on the specific agent but may include various complex alkaloids and organic acids.

12.4.5.2 Combustion

Varies depending on the specific agent but volatile decomposition products may include hydrogen fluoride (HF), hydrogen chloride (HCl), phosgene ($COCl_2$), nitrogen oxides (NO_x), aromatic hydrocarbons such as benzene, as well as potentially toxic lower molecular weight hydrocarbons.

12.5 Protection

12.5.1 Evacuation Recommendations

Isolation and protective action distances listed below are taken from Argonne National Laboratory Report No. ANL/DIS-00-1, *Development of the Table of Initial Isolation and Protective Action Distances for the 2000 Emergency Response Guidebook*, which is still the basis for the "when used as a weapon" scenarios in the 2004 Emergency Response Guidebook (ERG). BZ is the only incapacitating agent addressed and recommendations are based on a release scenario involving direct aerosolization of the solid agents with a particle size between 2 and 5 μm. Under these conditions, the difference between a small and a large release of BZ is not based on the standard 200 liters spill used for commercial hazardous materials listed in the ERG. A small release involves 10 kilograms of powdered agent (approximately 120 cubic in.) and a large release involves 500 kilograms of powdered agent (approximately 11 cubic ft).

	Initial isolation (feet)	Downwind day (miles)	Downwind night (miles)
BZ *C12-A001*			
Small device (10 kilograms)	100	0.1	0.3
Large device (500 kilograms)	200	0.3	1.2

12.5.2 Personal Protective Requirements

12.5.2.1 Structural Firefighters' Gear

Structural firefighters' protective clothing is recommended for fire situations only; it is not effective in spill situations or release events. If chemical protective clothing is not available and it is necessary to rescue casualties from a contaminated area, then structural firefighters' gear will provide very limited skin protection against agent vapors and aerosols. Contact with solid and liquid agents should be avoided.

12.5.2.2 Respiratory Protection

Self-contained breathing apparatuses (SCBAs) or air purifying respirators (APRs) should have a National Institute for Occupational Safety and Health (NIOSH) and Chemical/Biological/Radiological/Nuclear (CBRN) certification. However, during emergency operations, other NIOSH approved SCBAs or APRs that have been specifically tested by the manufacturer against chemical warfare agents may be used if deemed necessary by the Incident Commander. APRs should be equipped with a NIOSH approved CBRN filter or a combination organic vapor/acid gas/particulate cartridge.

Immediately dangerous to life or health (IDLH) levels are the ceiling limit for respirators other than SCBAs. Any exposures approaching the IDLH level should be regarded with extreme caution and the use of SCBAs for respiratory protection should be considered.

12.5.2.3 Chemical Protective Clothing

Currently, there is no information on performance testing of chemical protective clothing against incapacitating agent.

12.5.3 Decontamination

12.5.3.1 General

Apply universal decontamination procedures using soap and water.

12.5.3.2 Vapors

Casualties/personnel: Remove all clothing as it may contain "trapped" agent. To avoid further exposure of the head, neck, and face to the agent, cut off potentially contaminated clothing that must be pulled over the head. Shower using copious amounts of soap and water. Ensure that the hair has been washed and rinsed to remove potentially trapped agent. If there is a potential that the eyes have been exposed to the agent, irrigate with water or 0.9% saline solution for a minimum of 15 minutes.

Small areas: Ventilate to remove the vapors or dissipate the aerosol. If deemed necessary, wash the area with copious amounts of soap and water. Collect and place into containers lined with high-density polyethylene. Removal of porous material, including painted surfaces, may be required because agents that have been absorbed into these materials may migrate back to the surface and pose a residual hazard.

12.5.3.3 Liquids/Solutions or Liquid Aerosols

Casualties/personnel: Remove all clothing immediately. To avoid further exposure of the head, neck, and face to the agent, cut off potentially contaminated clothing that must be pulled over the head. Use a sponge or cloth with liquid soap and copious amounts of water to wash the skin surface and hair at least three times. Do not delay decontamination to find warm or hot water if it is not readily available. Avoid rough scrubbing as this could abrade the skin and increase percutaneous absorption of residual agent. Rinse with copious amounts of water. If there is a potential that the eyes have been exposed to incapacitating agents, irrigate with water or 0.9% saline solution for a minimum of 15 minutes.

Small areas: Small puddles of liquid can be contained by covering with absorbent material such as vermiculite, diatomaceous earth, clay, sponges, or towels. Place the absorbed material into containers lined with high-density polyethylene. Larger puddles can be collected using vacuum equipment made of materials inert to the released material and equipped with a high-efficiency particulate air (HEPA) filter and appropriate vapor filters. Wash the area with copious amounts of soap and water. Collect and containerize the rinseate. Ventilate the area to remove vapors.

12.5.3.4 Solids or Particulate Aerosols

Casualties/personnel: Do not attempt to brush the agent off the individual or their clothing as this can aerosolize the agent. If possible, dampen the agent with a water mist to help prevent aerosolization. Remove all clothing immediately. To avoid further exposure of the head, neck, and face to the agent, cut off potentially contaminated clothing that must be pulled over the head. Wash the skin surface and hair at least three times with copious amounts of soap and water. Do not delay decontamination to find warm or hot water if it is not readily available. Rinse with copious amounts of water. If there is a potential that the eyes have been exposed to incapacitating agents, irrigate with water or 0.9% saline solution for a minimum of 15 minutes.

Small areas: If indoors, close windows and doors in the area and turn off anything that could create air currents (e.g., fans, air conditioner, etc.). Avoid actions that could aerosolize the agent such as sweeping or brushing. Collect the agent using a vacuum cleaner equipped with a HEPA filter. Do not use a standard home or industrial vacuum. Do not allow the vacuum exhaust to stir the air in the affected area. Vacuum all surfaces with extreme care in a very slow and controlled manner to minimize aerosolizing the agent. Place the collected material into containers lined with high-density polyethylene. Wash the area with copious amounts of soap and water. Collect and containerize the rinseate in containers lined with high-density polyethylene.

12.6 Medical

12.6.1 CDC Case Definition

A case in which incapacitating agents are detected in the urine. Remember that fentanyl derivatives and some other synthetic opioids might not be detected by routine toxicologic screens. If an analytical methodology is available, then detection of incapacitating agents in environmental samples. The case can be confirmed if laboratory testing is not performed because either a predominant amount of clinical and nonspecific laboratory evidence is present or an absolute certainty of the etiology of the agent is known.

12.6.2 Differential Diagnosis

The following factors have been suggested as alternatives to consider when presented with a potential case of exposure to incapacitating agents: conduct disorder, personality disorders, dysthymic disorder, attention deficit hyperactivity disorder, panic disorders, delirium, dementia, amnesia, anxiety, headache, migraine, brain abscess; encephalitis, CNS infection, acute respiratory distress syndrome; heat exhaustion/heatstroke, hypoxia, hypoglycemia, electrolyte abnormalities, myocardial infarction, myocarditis, diabetic ketoacidosis; substance abuse (alcohol, plant and mushroom poisoning, scopolamine-tainted heroin), withdrawal syndromes, delirium tremens, methemoglobinemia, organophosphate and carbamate pesticide exposure, carbon monoxide, and cyanides.

12.6.3 Signs and Symptoms

Varies according to the type of incapacitating agent. Care must be taken in that many signs and symptoms associated with exposure to incapacitating agents are also associated with anxiety or physical trauma. Potential indications of exposure include apprehension, restlessness, dizziness, confusion, erratic behavior, inappropriate smiling or laughing, irrational fear, difficulty in communicating (mumbling, slurred, or nonsensical speech), euphoria, lethargy, trembling, pleading, crying, perceptual distortions, hallucinations, disrobing, stumbling or staggering, blurred vision, dilated or pinpointed pupils, flushed face and skin, elevated temperature, dry mouth and skin, foul breath, stomach cramps, vomiting, difficulty in urinating, change in pulse rate (slow or elevated), change in blood pressure (lowered or elevated), changes in breathing rate, stupor, or coma.

12.6.4 Mass-Casualty Triage Recommendations

12.6.4.1 Priority 1

A casualty with cardiovascular collapse or severe hyperthermia. Immediate attention to ventilation, hemodynamic status, and temperature control could be life-saving.

12.6.4.2 Priority 2

A casualty with severe or worsening signs after exposure.

12.6.4.3 Priority 3

A casualty with mild peripheral or central nervous system effects. However these patients will not be able to manage themselves and should be controlled.

12.6.4.4 Priority 4

A casualty with severe cardiorespiratory compromise when treatment or evacuation resources are unavailable.

12.6.5 Casualty Management

Decontaminate the casualty ensuring that all incapacitating agents have been removed. If incapacitating agents have gotten into the eyes, irrigate the eyes with water or 0.9% saline solution for at least 15 minutes. Irrigate open wounds with water or 0.9% saline solution for at least 10 minutes. However, do not delay treatment if thorough decontamination cannot be undertaken immediately.

Once the casualty has been decontaminated, including the removal of foreign matter from wounds, medical personnel do not need to wear a chemical-protective mask.

Antidotes are available for some incapacitating agents. Prior to administering antidotes or other drugs, ensure that the signs and symptoms (e.g., coma, seizures, etc.) are due to chemical exposure and not the result of head trauma or other physical injury. Otherwise, general treatment consists of observation, supportive care with fluids, and possibly restraint or confinement. Casualties should be isolated in a safe area. Remove any potentially harmful material from individuals suspected of being exposed to incapacitating agents including such items as cigarettes, matches, medications, and other small items they might attempt to ingest. Observe casualties for signs of heatstroke as some incapacitating agents eliminate the ability of exposed individuals to sweat. Monitor to ensure that casualties are breathing. Casualties will usually recover from exposure to incapacitating agents without medical treatment; however, full recovery from effects may take several days.

12.7 Fatality Management

Remove all clothing and personal effects and decontaminate with soap and water. Although it may be possible to decontaminate durable items, it may be safer and more efficient to destroy nondurable items rather than attempt to decontaminate them. Items that will be retained for further processing should be double sealed in impermeable containers, ensuring that the inner container is decontaminated before placing it in the outer one.

Wash the remains with soap and water. Pay particular attention to areas where agent may get trapped, such as hair, scalp, pubic areas, fingernails, folds of skin, and wounds. If remains are heavily contaminated with residue, wash and rinse waste should be contained for proper disposal.

Once the remains have been thoroughly decontaminated, no further protective action is necessary. Body fluids removed during the embalming process do not pose any additional risks and should be contained and handled according to established procedures. Use standard burial procedures.

C12-A

AGENTS

C12-A001
3-Quinuclidinyl benzilate (Agent BZ)
CAS: 6581-06-2
RTECS: —

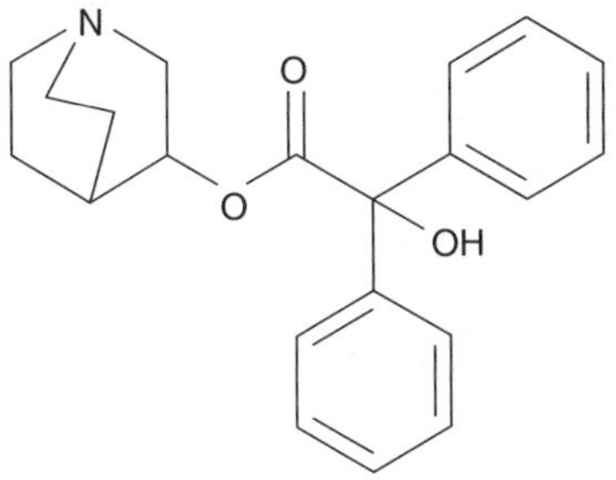

$C_{21}H_{23}NO_3$

White to beige crystalline solid that is odorless. Various salts have been reported.

Also reported as a mixture with ortho-Chlorobenzylmalononitrile (C13-A009).

Exposure Hazards

$LCt_{50(Inh)}$: 200,000 mg-min/m^3. This value is from an older source (ca. 1960) and is not supported by modern data. No updated toxicity estimates have been proposed.

$ICt_{50(Inh)}$: 100 mg-min/m^3

"Mild incapacitation": 90 mg-min/m^3 (some hallucinations)

"Severe incapacitation": 135 mg-min/m^3 (marked hallucinations)

Properties:

MW: 337.4	*VP*: Negligible	*FlP*: 475°F
D: 1.33 g/cm^3 (crystalline)	*VD*: —	*FlP*: 428°F (munition grade)
D: 0.51 g/cm^3 (powder)	*Vlt*: —	*LEL*: —
MP: 334°F	*H_2O*: 0.0012% (77°F)	*UEL*: —
BP: 774°F (estimate)	*Sol*: Common organic solvents	*RP*: 200,000
Vsc: —		*IP*: —

C12-A002

1-Methyl-4-piperidyl cyclopentyl-1-propynyl-glycolate (Agent 302196 (B))
CAS: 53034-67-6
RTECS: —

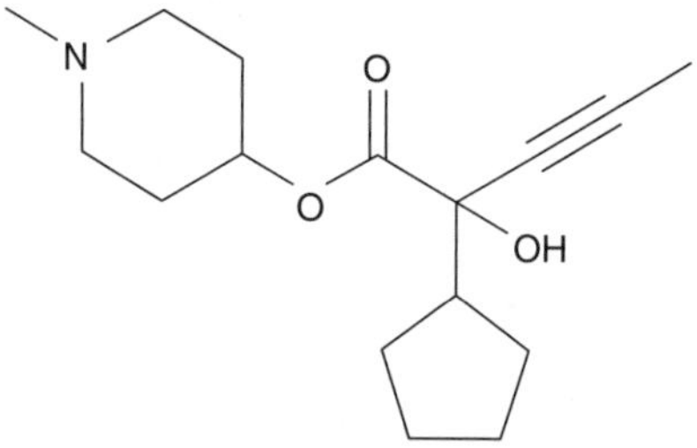

$C_{16}H_{25}NO_3$

Ivory colored powder. Various salts have been reported.

Exposure Hazards
Human toxicity values have not been established or have not been published.

Properties:

MW: 279.4	*VP*: Negligible	*FlP*: 376°F
D: 1.14 g/cm^3	*VD*: —	*LEL*: —
MP: 253°F	*Vlt*: —	*UEL*: —
BP: 671°F (estimate)	*H_2O*: 2.25% (77°F)	*RP*: 13,000,000
Vsc: —	*Sol*: Chloroform	*IP*: —

C12-A003

3-Quinuclidinyl cyclopentyl-phenylglycolate (Agent EA 3167)
CAS: 26758-53-2
RTECS: —

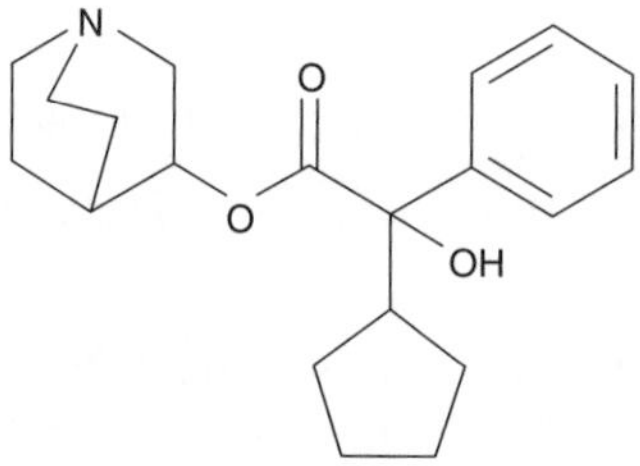

$C_{20}H_{27}NO_3$

Straw-colored, extremely viscous, and tacky liquid that is odorless. Various salts have been reported.

Exposure Hazards

Human toxicity values have not been established or have not been published.

Properties:

MW: 329.5	*VP*: Negligible	*FlP*: 448°F
D: 1.14 g/mL (77°F)	*VD*: —	*LEL*: —
MP: —	*Vlt*: —	*UEL*: —
BP: 613°F (estimate)	H_2O: <0.001%	*RP*: 11,000,000
Vsc: 15,000 cS (131°F)	*Sol*: Ethanol; Chloroform	*IP*: —

C12-A004

***N*-Methyl-4-piperidyl cyclopentyl-phenylglycolate** (Agent EA 3443)

CAS: 37803-21-0

RTECS: —

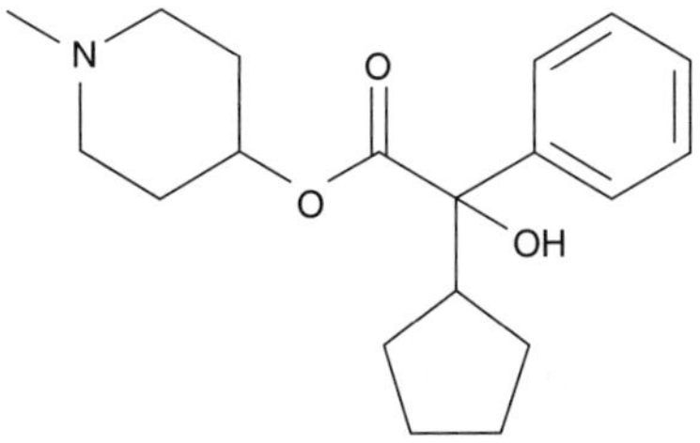

$C_{19}H_{27}NO_3$

Waxy solid or viscous liquid that is odorless. Various salts have been reported.

Exposure Hazards

$ICt_{50(Inh)}$: 54.4 mg-min/m^3

Properties:

MW: 317.5	*VP*: Negligible	*FlP*: 424°F
D: 1.1 g/mL	*VD*: —	*LEL*: —
MP: 119°F	*Vlt*: —	*UEL*: —
BP: 714°F	H_2O: 0.12% (77°F)	*RP*: 11,000,000
Vsc: 86,000 cS (86°F)	*Sol*: Most organic solvents	*IP*: —

C12-A005

1-Methyl-4-piperidyl cyclobutyl-phenylglycolate (Agent EA 3580B; Hydrochloride salt is Agent EA 3580A)

CAS: 54390-94-2

RTECS: —

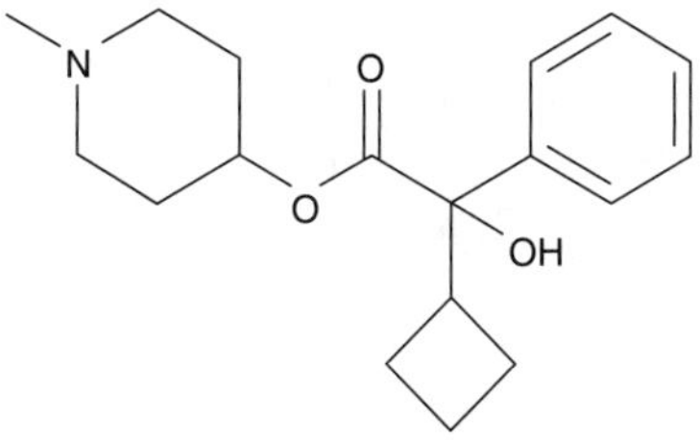

$C_{18}H_{25}NO_3$

Waxy solid to white crystalline material that is odorless. Hydrochloride salt is a white to yellow crystalline solid.

An aerosol cloud from the hydrochloride salt can be thermally generated with little decomposition of the agent.

Exposure Hazards

$ICt_{50(Inh)}$: 71 mg-min/m^3

$ICt_{50(Inh)}$: 79 mg-min/m^3 (Hydrochloride salt)

Properties:

MW: 303.4	*VP*: Negligible	*FIP*: —
D: 1.10 g/cm^3 (77°F)	*VD*: —	*LEL*: —
MP: 133°F	*Vlt*: —	*UEL*: —
BP: 710°F	H_2O: 9.42% (77°F)	*RP*: 1,200,000
Vsc: 4060 cS (77°F)	*Sol*: Most organic solvents	*IP*: —

Hydrochloride salt

MW: 339.9

D: 1.21 g/cm^3 (77°F)

MP: 387°F

H_2O: 39% (77°F)

C12-A006

1-Methyl-4-piperidyl isopropyl-phenylglycolate (Agent EA 3834B; Hydrochloride salt is Agent EA 3834A)

CAS: 75321-25-4; 137444-35-0 (Hydrochloride salt)

RTECS: —

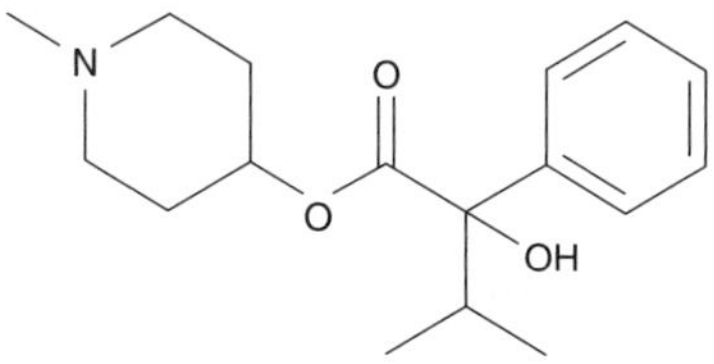

$C_{17}H_{25}NO_3$

White crystals to brown oil that is odorless. Hydrochloride salt is a white crystalline solid.

An aerosol cloud from the hydrochloride salt can be thermally generated with little decomposition of the agent. Salt has greater absorption via lungs than the free base.

Also reported as a mixture with 1-Methoxy-1,3,5-cycloheptatriene (C13-A014).

Exposure Hazards

$ICt_{50(Inh)}$: 73.4 mg-min/m^3

$ICt_{50(Inh)}$: 82.6 mg-min/m^3 (Hydrochloride salt)

Properties:

MW: 291.4	*VP*: 0.000015 mmHg (77°F)	*FIP*: 397°F
D: 1.0626 g/cm^3	*VD*: 10 (calculated)	*LEL*: —
MP: 120°F	*Vlt*: 0.02 ppm (77°F)	*UEL*: —
BP: 639°F	H_2O: <0.002%	*RP*: 390,000
Vsc: 13,525 cS (77°F)	*Sol*: Hexane; Ether	*IP*: —

Hydrochloride salt

MW: 327.9

D: 1.18 g/cm^3

MP: 399°F

H_2O: 43.9% (77°F)

C12-A007

Phencyclidine (Agent SN)
CAS: 77-10-1
RTECS: —

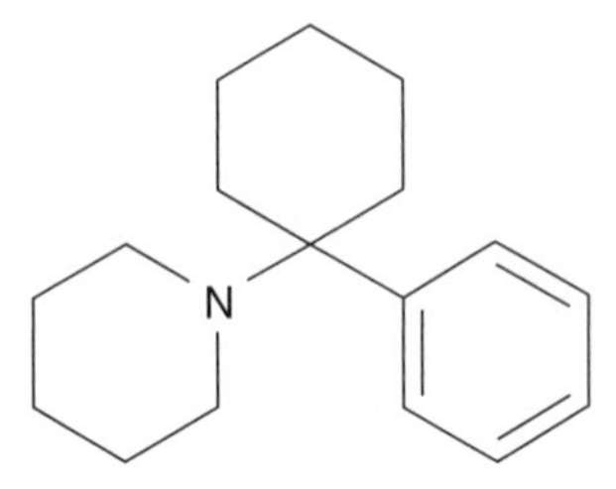

$C_{17}H_{25}N$

Colorless crystals. Various salts have been reported.

Used as a veterinary anesthetic. Also used as an illegal street drug.

Exposure Hazards

$ICt_{50(Inh)}$: 1000 mg-min/m^3. Concentration of 25–50 mg-min/m^3 produce anesthetic effects, and concentrations >100 mg-min/m^3 cause mental disturbances.

$ID_{50(Ing)}$: 0.010–0.020 g

$LD_{50(Ing)}$: 7 g

Properties:

MW: 243.4	*VP*: Negligible	*FlP*: —
D: —	*VD*: —	*LEL*: —
MP: 115°F	*Vlt*: —	*UEL*: —
BP: 275°F (1 mmHg)	H_2O: —	*RP*: —
Vsc: —	*Sol*: —	*IP*: —

Hydrochloride salt	**Hydrobromide salt**
MW: 279.9Z	*MW*: 324.3
MP: 451°F	*MP*: 417°F

C12-A008

Cocaine
CAS: 50-36-2
RTECS: YM2800000

$C_{17}H_{21}NO_4$

Colorless to white crystals or powder with no odor. Various salts have been reported.

Used medicinally and by veterinarians as a local anesthetic. Also used as an illegal street drug.

Exposure Hazards

$LD_{50(Ing)}$: 1–1.2 g (estimate)

Properties:

MW: 303.4	*VP*: Negligible	*FlP*: —
D: —	*VD*: —	*LEL*: —
MP: 208°F	*Vlt*: —	*UEL*: —
BP: 369°F (1 mmHg)	H_2O: 0.17%	*RP*: —
Vsc: —	*Sol*: Most organic solvents	*IP*: —

Hydrochloride salt	**Nitrate salt**
MW: 339.9	*MW*: 366.4
MP: 383°F	*MP*: 136°F
H_2O: 250%	H_2O: "Freely soluble"

C12-A009

Dexamphetamine

CAS: 51-64-9

RTECS: —

$C_9H_{13}N$

Colorless liquid. The sulfate salt is a white, odorless crystalline solid.

Used by veterinarians as a stimulant of the central nervous system. Also used as an illegal street drug.

Exposure Hazards

Conversion Factor: 1 ppm = 5.53 mg/m^3 at 77°F

$LD_{50(Ing)}$: 1.4–1.8 g (estimate)

Properties:

MW: 135.2	*VP*: —	*FlP*: —
D: 0.949 g/mL (59°F)	*VD*: 4.7 (calculated)	*LEL*: —
MP: —	*Vlt*: —	*UEL*: —
BP: 397°F	H_2O: "Slightly"	*RP*: —
BP: 181°F (15 mmHg)	*Sol*: Alcohol; Ether	*IP*: —
Vsc: —		

Sulfate salt

MW: 233.3

D: 1.15 g/cm^3

MP: 572°F

H_2O: 10%

C12-A010

Methamphetamine

CAS: 537-46-2

RTECS: —

$C_{10}H_{15}N$

Clear, colorless liquid with a characteristic odor resembling geranium leaves. The hydrochloride salt is a crystalline solid that is odorless.

Used medicinally as an anesthetic. Also used as an illegal street drug.

Exposure Hazards

Conversion Factor: 1 ppm = 6.10 mg/m^3 at 77°F

$LD_{50(Ing)}$: 1.4–1.8 g (estimate)

Properties:

MW: 149.2	*VP*: 0.163 mmHg (77°F)	*FlP*: —
D: —	*VD*: 5.1 (calculated)	*LEL*: —
MP: —	*Vlt*: 210 ppm (77°F)	*UEL*: —
BP: 414°F	H_2O: 1.33%	*RP*: 50
Vsc: —	*Sol*: Ethanol; Ether; Chloroform	*IP*: —

Hydrochloride salt

MW: 185.7

MP: 338°F

H_2O: 50%

C12-A011

Ecstasy

CAS: 42542-10-9

RTECS: —

$C_{11}H_{15}NO_2$

Oily liquid. The hydrochloride salt is a crystalline solid.

There is no approved medical use in the United States, but it is used as an illegal street drug.

Exposure Hazards

Conversion Factor: 1 ppm = 7.90 mg/m^3 at 77°F

$LD_{50(Ing)}$: 1.4–1.8 g (estimate)

This material is approximately three times more potent than Mescaline (C12-A014).

Properties:

MW: 193.2	*VP*: —	*FlP*: —
D: —	*VD*: 6.7 (calculated)	*LEL*: —
MP: —	*Vlt*: —	*UEL*: —
BP: 212°F (0.4 mmHg)	H_2O: —	*RP*: —
Vsc: —	*Sol*: —	*IP*: —

Hydrochloride salt
MW: 229.7
MP: 297°F

C12-A012

Mescaline
CAS: 54-04-6
RTECS: —

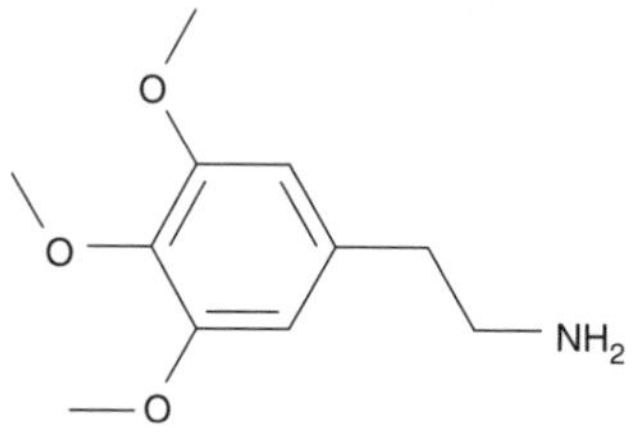

$C_{11}H_{17}NO_3$

Crystalline solid. Various salts have been reported.

Used in religious ceremonies by the North American Church of Native Americans. Also used as an illegal street drug.

Exposure Hazards
Conversion Factor: 1 ppm = 8.64 mg/m^3 at 77°F

Human toxicity values have not been established or have not been published.

Properties:

MW: 211.3	*VP*: —	*FlP*: —
D: —	*VD*: 7.3 (calculated)	*LEL*: —
MP: 97°F	*Vlt*: —	*UEL*: —
BP: 356°F (12 mmHg)	H_2O: "Moderate"	*RP*: —
Vsc: —	*Sol*: Alcohol; Chloroform; Benzene	*IP*: —

Hydrochloride salt	**Sulfate salt**
MW: 247.7	*MW*: 309.3
MP: 358°F	*MP*: 361°F

C12-A013

LSD
CAS: 50-37-3
RTECS: —

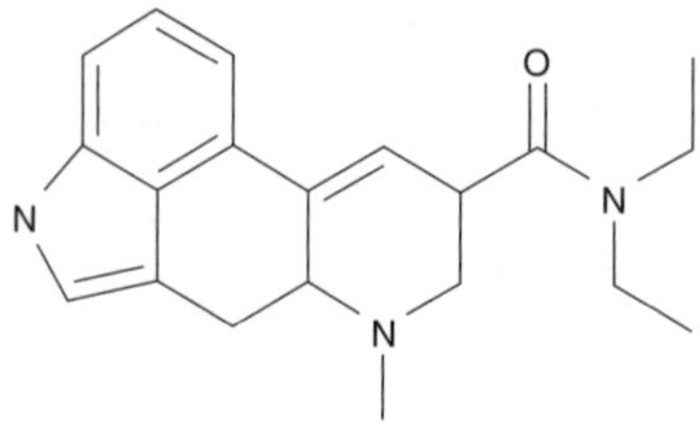

$C_{20}H_{25}N_3O$

Colorless crystalline solid that is odorless. Various salts have been reported.

Used in biochemical research as an antagonist to serotonin. Also used as an illegal street drug.

Exposure Hazards

$LD_{50(Ing)}$: 0.021–3.5 g

Properties:

MW: 323.4	*VP*: Negligible	*FlP*: —
D: —	*VD*: —	*LEL*: —
MP: 180°F	*Vlt*: —	*UEL*: —
BP: —	H_2O: 0.0002%	*RP*: —
Vsc: —	*Sol*: —	*IP*: 7.25 eV

Tartrate salt

MW: 861.0
MP: 388°F
H_2O: "Soluble"

C12-A014

Morphine

CAS: 57-27-2
RTECS: —

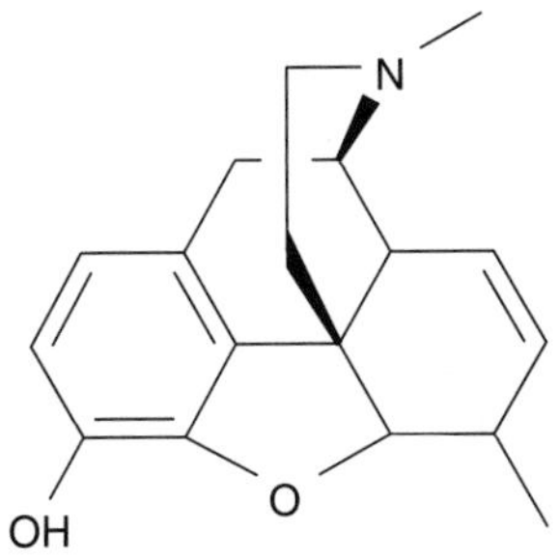

N
OH
O

$C_{17}H_{19}NO_3$

White crystalline solid that is odorless. Acetate has a slight acetic odor. Various salts have been reported.

Used medicinally as an anesthetic. Also used as an illegal street drug.

Exposure Hazards

$LD_{50(IV)}$: 0.03 g

Properties:

MW: 285.3	*VP*: —	*FlP*: —
D: 1.31 g/cm^3	*VD*: 9.8 (calculated)	*LEL*: —
MP: 387°F	*Vlt*: —	*UEL*: —
BP: Sublimes	H_2O: —	*RP*: —
Vsc: —	*Sol*: Methanol	*IP*: —

C12-A015

Fentanyl

CAS: 437-38-7
RTECS: UE5550000

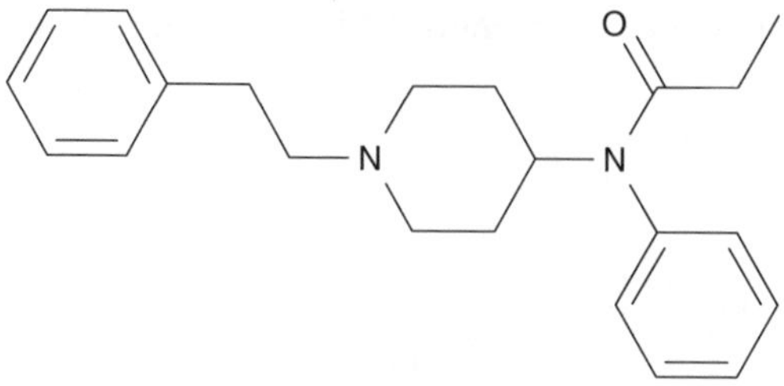

$C_{22}H_{28}N_2O$

Crystalline solid that is odorless. Various salts have been reported.

Used medicinally as an anesthetic, by veterinarians as a tranquilizer. Also used as an illegal street drug and as an illegal stimulant for racehorses.

Exposure Hazards

Human toxicity values have not been established or have not been published. However, this material is between 50 and 100 times more potent than Morphine (C12-A017).

Properties:

MW: 336.5	*VP*: Negligible	*FlP*: —
D: —	*VD*: —	*LEL*: —
MP: 181°F	*Vlt*: —	*UEL*: —
BP: —	H_2O: 0.02%	*RP*: —
Vsc: —	*Sol*: —	*IP*: —

Citrate salt

MW: 528.6
MP: 300°F
H_2O: 2.5%

C12-A016

Sufentanil

CAS: 56030-54-7
RTECS: —

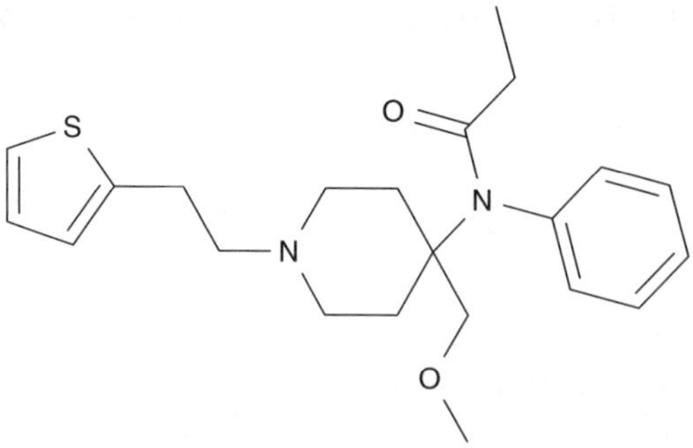

$C_{22}H_{30}N_2O_2S$

Crystalline solid. Various salts have been reported.

Used medicinally as an anesthetic.

Exposure Hazards

Human toxicity values have not been established or have not been published. However, this material is approximately 700 times more potent than Morphine (C12-A017).

Properties:

MW: 386.6	*VP*: Negligible	*FlP*: —
D: —	*VD*: —	*LEL*: —
MP: 206°F	*Vlt*: —	*UEL*: —
BP: —	H_2O: 0.0076%	*RP*: —
Vsc: —	*Sol*: —	*IP*: —

C12-A017

Alfentanil

CAS: 71195-58-9; 70879-28-6 (Hydrochloride salt)
RTECS: —

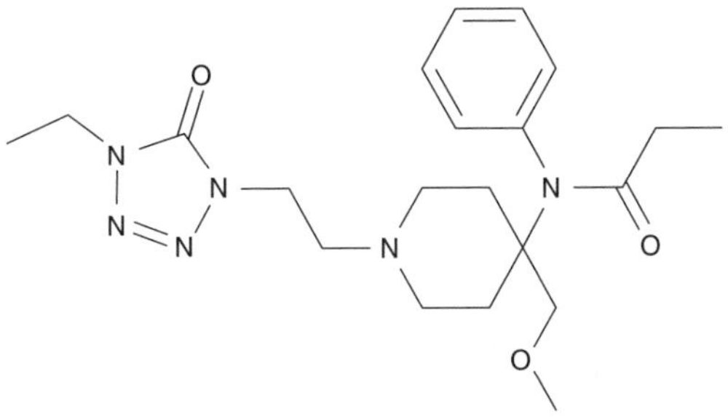

$C_{21}H_{32}N_6O_3$

Hydrochloride is a colorless solid with no odor. This material is hazardous through inhalation, penetration through broken skin, and ingestion. May cause nausea and vomiting as well as muscle rigidity that includes the chest wall causing apnea.

Used medicinally as a short-acting general anesthetic. Effects dissipate after approximately 20 min.

Exposure Hazards

Human toxicity values have not been established or have not been published.

Properties:

MW: 416.5	*VP*: Negligible	*FlP*: —
D: —	*VD*: —	*LEL*: —
MP: 285°F (Hydrochloride salt)	*Vlt*: —	*UEL*: —
BP: —	H_2O: 0.0035%	*RP*: —
Vsc: —	H_2O: Soluble (Hydrochloride salt)	*IP*: —
	Sol: —	

C12-A018

Lofentanil

CAS: 61380-40-3; 61380-41-4 (Oxalate salt)
RTECS: —

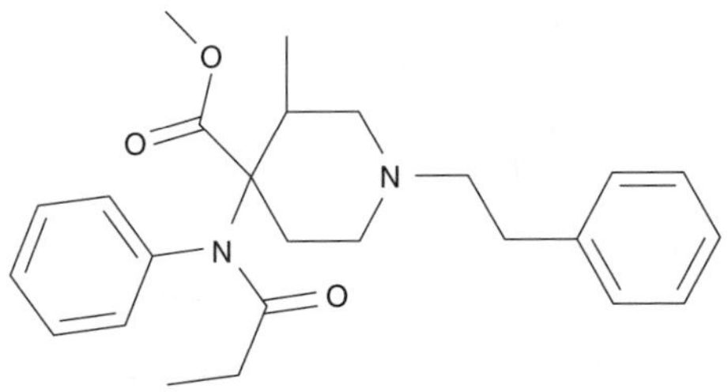

$C_{25}H_{32}N_2O_3$

Oxalate is a colorless solid with no odor. This material is hazardous through inhalation, penetration through broken skin, and ingestion.

Used medicinally as an anesthetic.

Exposure Hazards

Human toxicity values have not been established or have not been published.

Properties:

MW: 408.5	*VP*: Negligible	*FlP*: —
D: —	*VD*: —	*LEL*: —
MP: 351°F (Oxalate salt)	*Vlt*: —	*UEL*: —
BP: —	H_2O: 0.00007%	*RP*: —
Vsc: —	*Sol*: —	*IP*: —

C12-A019

Prolixin

CAS: 69-23-8; 146-56-5 (Hydrochloride salt); 5002-47-1 (Decanoate salt); 3093-66-1 (Dimaleate salt);
2746-81-8 (Enanthate salt)
RTECS: —

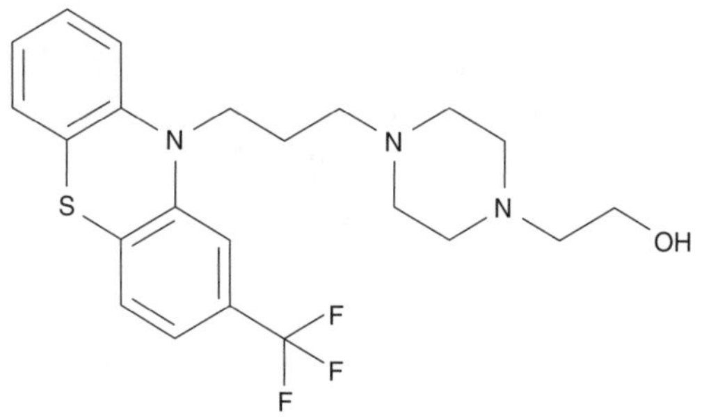

$C_{22}H_{26}F_3N_3OS$

Dark brown viscous oil. The hydrochloride salt is a white crystalline solid, while the decanoate and enanthate esters are pale yellow to yellowish orange viscous liquids or oily solids.

Used medicinally as an anesthetic.

Exposure Hazards

Human toxicity values have not been established or have not been published.

Properties:

MW: 437.5	*VP*: Negligible	*FlP*: —
D: —	*VD*: —	*LEL*: —
MP: —	*Vlt*: —	*UEL*: —
BP: 482°F (0.3 mmHg)	H_2O: 0.003% (99°F)	*RP*: —
Vsc: —	*Sol*: —	*IP*: 8.64 eV

Dihydrochloride salt

MW: 510.4
MP: 436°F
H_2O: "Soluble"

C12-A020

Halothane

CAS: 151-67-7
RTECS: KH6550000

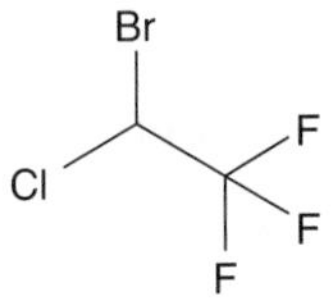

$C_2HBrClF_3$

Clear, colorless liquid with a sweetish, pleasant odor.

Used medicinally as an inhalation anesthetic.

Exposure Hazards

Conversion Factor: 1 ppm = 8.07 mg/m^3 at 77°F
ACGIH TLV: 50 ppm
NIOSH Ceiling: 2.0 ppm [60 min limit]

Properties:

MW: 197.4	*VP*: 243 mmHg	*FlP*: None
D: 1.871 g/mL	*VP*: 302 mmHg (77°F)	*LEL*: None
MP: –180°F	*VD*: 6.8 (calculated)	*UEL*: None
BP: 122°F	*Vlt*: 330,000 ppm	*RP*: 0.029
Vsc: —	*H_2O*: 0.345%	*IP*: —
	H_2O: 0.407% (77°F)	
	Sol: Hydrocarbon solvents	

C12-A021

Methoxyflurane

CAS: 76-38-0
RTECS: KN7820000

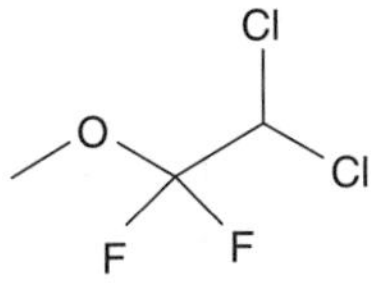

$C_3H_4Cl_2F_2O$

Clear, colorless liquid with a fruity odor.

Used medicinally as an inhalation anesthetic.

Exposure Hazards

Conversion Factor: 1 ppm = 6.75 mg/m^3 at 77°F
NIOSH Ceiling: 2.0 ppm [60 min limit]

Properties:

MW: 165.0	*VP*: 23 mmHg	*FlP*: 145°F
D: 1.4223 g/mL	*VP*: 49.1 mmHg (77°F)	*LEL*: 7%
MP: –31°F	*VD*: 5.7 (calculated)	*UEL*: —

BP: 221°F	*Vlt*: 65,000 ppm	*RP*: 0.16
BP: 124°F (100 mmHg)	H_2O: 2.8% (100°F)	*IP*: 11 eV
Vsc: —	*Sol*: —	

COMPONENTS AND PRECURSORS C12-C

C12-C022

3-Quinuclidinol
CAS: 1619-34-7
RTECS: VD6191700

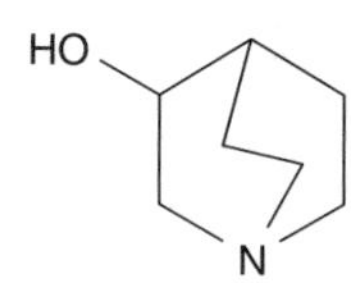

$C_7H_{13}NO$

White to beige crystalline powder. This material produces local skin/eye impacts.

Used industrially as an intermediate in the synthesis of pharmaceuticals.

This material is on the Australia Group Export Control list and Schedule 2 of the CWC.

This material is a component for numerous incapacitating agents and several organophosphorus nerve agents (Chapter 1).

Exposure Hazards
Conversion Factor: 1 ppm = 5.20 mg/m^3 at 77°F

Human toxicity values have not been established or have not been published.

Properties:

MW: 127.2	*VP*: —	*FlP*: —
D: —	*VD*: —	*LEL*: —
MP: 424°F	*Vlt*: —	*UEL*: —
BP: —	H_2O: "Soluble"	*RP*: —
Vsc: —	*Sol*: —	*IP*: ≤8.1 eV

C12-C023

Benzilic acid
CAS: 76-93-7
RTECS: DD2064000

$C_{14}H_{12}O_3$

White to cream powder. This material produces local skin/eye impacts.

Used industrially as an intermediate in the synthesis of pharmaceuticals.

This material is on the Australia Group Export Control list and Schedule 2 of the CWC.

This material is a component for numerous incapacitating agents.

Exposure Hazards

Conversion Factor: 1 ppm = 9.33 mg/m^3 at 77°F

Human toxicity values have not been established or have not been published.

Properties:

MW: 228.2	*VP*: —	*FlP*: —
D: —	*VD*: —	*LEL*: —
MP: 300°F	*Vlt*: —	*UEL*: —
BP: 356°F	H_2O: "Slightly soluble"	*RP*: —
Vsc: —	*Sol*: —	*IP*: —

C12-C024

Methyl Benzilate

CAS: 76-89-1

RTECS: —

$C_{15}H_{14}O_3$

Specific information on physical appearance is not available for this material.

Used industrially as an intermediate in the synthesis of pharmaceuticals.

This material is on the Australia Group Export Control list.

This material is a component for numerous incapacitating agents.

Exposure Hazards

Conversion Factor: 1 ppm = 9.91 mg/m^3 at 77°F

Human toxicity values have not been established or have not been published.

Properties:

MW: 242.3	*VP*: —	*FlP*: —
D: —	*VD*: —	*LEL*: —
MP: 163°F	*Vlt*: —	*UEL*: —
BP: 369°F (13 mmHg)	H_2O: —	*RP*: —
Vsc: —	*Sol*: —	*IP*: —

References

Centers for Disease Control and Prevention. "Biological and Chemical Terrorism: Strategic Plan for Preparedness and Response. Recommendations of the CDC Strategic Planning Workgroup." *Morbidity and Mortality Weekly Report* 49 (RR-4) (2000): 1–14.

———. "Case Definition: 3-Quinuclidinyl Benzilate (BZ)." March 11, 2005.

———. "Case Definition: Opioids (Fentanyl, Etorphine, or Others)." March 11, 2005.

Compton, James A.F. *Military Chemical and Biological Agents: Chemical and Toxicological Properties.* Caldwell, NJ: The Telford Press, 1987.

Lakoski, Joan M., W. Bosseau Murray, and John M. Kenny. *The Advantages and Limitations of Calmatives for Use as Non-Lethal Technique.* College of Medicine, Applied Research Laboratory, The Pennsylvania State University, October 3, 2000.

National Institute for Occupational Safety and Health. *NIOSH Pocket Guide to Chemical Hazards.* Washington, DC: Government Printing Office, September 2005.

National Institutes of Health. *Hazardous Substance Data Bank (HSDB).* http://toxnet.nlm.nih.gov/cgi-bin/sis/htmlgen?HSDB/. 2004.

Olson, Kent R., ed. *Poisoning & Drug Overdose.* 4th ed. New York: Lange Medical Books/McGraw-Hill, 2004.

Perrine, Daniel M. *The Chemistry of Mind-Altering Drugs: History, Pharmacology, and Cultural Context.* Washington, DC: American Chemical Society, 1996.

Sidell, Fredrick R., Ernest T. Takafuji, and David R. Franz, ed. *Medical Aspects of Chemical and Biological Warfare, Textbook of Military Medicine Series, Part 1, Warfare, Weaponry, and the Casualty.* Washington, DC: Office of the Surgeon General, Department of the Army, 1997.

Sifton, David W., ed. *PDR Guide to Biological and Chemical Warfare Response.* Montvale, NJ: Thompson/Physicians Desk Reference, 2002.

Smith, Ann, Patricia Heckelman, and Maryadele J. Oneil, ed. *The Merck Index: An Encyclopedia of Chemicals, Drugs, & Biologicals.* 13th ed. Rahway, NJ: Merck & Co., Inc., 2001.

Sommer, Harold Z., and Jacob I. Miller. "Quaternary Quinuclidinones." US Patent 3,919,240, November 11, 1975.

True, Bey-Lorraine, and Robert H. Dreisbach. *Dreisbach's Handbook of Poisoning: Prevention, Diagnosis and Treatment.* 13th ed. London, England: The Parthenon Publishing Group, 2002.

United States Army Headquarters. *Chemical Agent Data Sheets Volume I, Edgewood Arsenal Special Report No. EO-SR-74001.* Washington, DC: Government Printing Office, December 1974.

———. *Chemical Agent Data Sheets Volume II, Edgewood Arsenal Special Report No. EO-SR-74002.* Washington, DC: Government Printing Office, December 1974.

———. *Potential Military Chemical/Biological Agents and Compounds, Field Manual No. 3-11.9.* Washington, DC: Government Printing Office, January 10, 2005.

United States Army Medical Research Institute of Chemical Defense. *Medical Management of Chemical Casualties Handbook.* 3rd ed. Aberdeen Proving Ground, MD: United States Army Medical Research Institute of Chemical Defense, July 2000.

Williams, Kenneth E. *Detailed Facts About Psychedelic Agent 3-Quinuclidinyl Benzilate (BZ).* Aberdeen Proving Ground, MD: United States Army Center for Health Promotion and Preventive Medicine, 1996.

World Health Organization. *International Chemical Safety Cards (ICSCs).* http://www.cdc.gov/niosh/ipcs/icstart.html. 2004.

———. *Health Aspects of Chemical and Biological Weapons: Report of a WHO Group of Consultants.* Geneva: World Health Organization, 1970.

———. *Public Health Response to Biological and Chemical Weapons: WHO Guidance.* Geneva: World Health Organization, 2004.

13

Irritating and Lachrymatory Agents

13.1 General Information

The majority of these materials are alkylating agents that react with the moisture in the eyes to cause irritation. Under normal battlefield conditions, they do not pose a serious danger to the life of an exposed individual and do not produce any permanent injury. They are first generation chemical warfare agents and were the first agents deployed in World War I. Since the end of World War I, numerous new agents have been developed, typically with greater irritating power and less toxicity. Under the general purpose criterion of the Chemical Weapons Convention (CWC) the use of irritating and lachrymatory agents is banned during a war. However, they may still be used by the military during operations other than war such as when responding to incidents of civil unrest. Lachrymatory agents are also used by police forces throughout the world to control rioters and disband unruly crowds. In some countries, agents can be purchased by individuals for personal protection.

13.2 Toxicology

13.2.1 Effects

Irritating and lachrymatory agents cause intense eye pain and tears. They may also irritate the respiratory tract, causing the sensation that the casualty has difficulty breathing. In high concentrations, they are irritating to the skin and cause a temporary burning or itching sensation. High concentration can cause nausea, vomiting, and blistering on the skin. In an enclosed or confined space, very high concentration can be lethal.

13.2.2 Pathways and Routes of Exposure

Irritating and lachrymatory agents are primarily an eye-contact and inhalation hazard. Aerosols and vapors are irritating to the eyes and skin at low concentrations but are otherwise relatively nontoxic via these routes. However, exposure to bulk liquid or solid agents may be hazardous through skin absorption, ingestion, and introduction through abraded skin (e.g., breaks in the skin or penetration of skin by debris).

13.2.3 General Exposure Hazards

Irritating and lachrymatory agents have excellent warning properties. In general, they produce eye, respiratory, and/or skin irritation at concentrations well below lethal levels.

This class of agents does not seriously endanger life except at exposures greatly exceeding an effective dose, usually only achieved in a confined or enclosed space.

13.2.4 Latency Period

Exposure to irritating and lachrymatory agents produces immediate effects.

13.3 Characteristics

13.3.1 Physical Appearance/Odor

13.3.1.1 Laboratory Grade

Laboratory grade agents are typically colorless to yellow liquids or solids. They typically have little or no odor when pure. If present, odors range from sweetish to floral to pepper-like. Most simply cause a burning sensation in the nose and nasal passages.

13.3.1.2 Munition Grade

Munition grade agents are typically off-white to yellow to brown. As the agent ages and decomposes it continues to discolor. Production impurities and decomposition products in these agents may give them additional odors.

13.3.1.3 Modified Agents

Solvents have been added to these materials to increase the efficacy of the agents, to facilitate handling, to stabilize the agents, or to aid in dispersing the agents. Typical solvents include propylene glycol, benzene, carbon tetrachloride, chloroform, and/or trioctylphosphite. Solvents may pose toxic hazards themselves (e.g., chloroform, carbon tetrachloride, and benzene). Color and other properties of these solutions may vary from the pure agent. Odors will vary depending on the characteristics of the solvent(s) used and concentration of agent in the solution.

Agents have also been micropulverized, encapsulated, or treated with flowing agents such as silica aerogel to facilitate their dispersal and increase their persistency. Color and other physical properties of the agent may be affected by these additives.

13.3.2 Stability

Modern irritating and lachrymatory agents are stable and can be stored even under tropical conditions. Older agents, typically simple halogenated acyl or aryl compounds, tend to be sensitive to air, moisture, and/or light. Some older agents are also prone to polymerization during storage. Stabilizers may be added to enhance stability and increase shelf life. Stabilizers include butylphenol, butylhydroquinone, amyl nitrate, and calcium carbonate. Modern agents can typically be stored in aluminum, glass, or steel containers. Older agents typically require glass, enamel lined, or lead lined containers. Many halogenated agents tend to deteriorate in the presence of metals such as iron and aluminum.

13.3.3 Persistency

As typically deployed, unmodified irritating and lachrymatory agents are classified as nonpersistent by the military. However, bulk solid agents deployed for the purpose of area denial may persist for weeks or months. Depending on the size of the individual particles and on any encapsulation or coatings applied to the particles, they can be reaerosolized by ground traffic or strong winds.

13.3.4 Environmental Fate

Many irritating and lachrymatory agents are nonvolatile and produce negligible amounts of vapor. Vapors of volatile agents have a density greater than air and tend to collect in low places.

Most agents are insoluble in water and have specific gravities that range from near water to greater than water. Lack of solubility inhibits reaction of these agents with water. Further, solvents used to disperse irritating and lachrymatory agents are generally insoluble in water and will help prevent interaction of the agent with water. Solvents may have densities less than or greater than water and may cause agents to either float or sink in a water column. Most of these agents are soluble in organic solvents including gasoline, alcohols, and ketones.

Agents may be absorbed into porous material, including painted surfaces, and these materials may be difficult to decontaminate.

13.4 Additional Hazards

13.4.1 Exposure

All foodstuffs in the area of a release should be considered contaminated. Unopened items packaged in glass, metal, or heavy duty plastic and exposed only to agent vapors or aerosols may be used after decontamination of the container. Unopened items exposed to solid or liquid agents, or solutions of agents, should be decontaminated within a few hours postexposure or destroyed. Opened or unpackaged items, or those packaged only in paper or cardboard, should be destroyed.

Plants, fruits, and vegetables should be washed thoroughly with soap and water. Skins should be removed prior to use.

13.4.2 Livestock/Pets

Animals can be decontaminated with shampoo/soap and water (see Section 13.5.3). If the animals' eyes have been exposed to agent, they should be irrigated with water or saline solution for a minimum of 30 minutes.

The topmost layer of unprotected feedstock (e.g., hay or grain) should be destroyed. The remaining material should be quarantined until tested.

13.4.3 Fire

Heat from a fire will increase the amount of agent vapor in the area. A significant amount of the agent could be volatilized and escape into the surrounding environment before the agent is consumed by the fire. Actions taken to extinguish the fire can also spread the agent. Although many irritating and lachrymatory agents are only slightly soluble or insoluble in water, runoff from firefighting efforts will still pose a potential contact threat. Many decompose to produce toxic and/or corrosive gases such as hydrogen chloride (HCl), hydrogen cyanide (HCN), and phosgene ($COCl_2$). Some of the decomposition products resulting from hydrolysis or combustion of incapacitating agents are water soluble and highly toxic (see Section 13.4.5). In addition, solvents used in many formulation are highly flammable.

13.4.4 Reactivity

Irritating and lachrymatory agents either do not react with water or are very slowly decomposed by it. Some agents may be corrosive and react with metal. In some cases these reactions may be violent. Most of these agents are incompatible with strong oxidizers, including household bleach, and may produce toxic decomposition products. Solvents used to disperse agents may be incompatible with strong oxidizers and may also decompose to form toxic and/or corrosive decomposition products.

13.4.5 Hazardous Decomposition Products

13.4.5.1 Hydrolysis

Irritating and lachrymatory agents are generally stable or very slowly decomposed by water. Further, solvents used to disperse these agents are generally insoluble in water and will help prevent interaction of the agent with water. However, should hydrolysis occur, decomposition products may include HCl, HCN, hydrogen bromide (HBr), and/or aromatic hydrocarbons, as well as complex condensation products.

13.4.5.2 Combustion

Volatile decomposition products may include HCl, HBr, HCN, $COCl_2$, nitrogen oxides (NO_x), aromatic hydrocarbons such as benzene, and/or halogenated aromatic compounds.

13.5 Protection

13.5.1 Evacuation Recommendations

Isolation and protective action distances listed below are taken from Argonne National Laboratory Report No. ANL/DIS-00-1, *Development of the Table of Initial Isolation and Protective Action Distances for the 2000 Emergency Response Guidebook*, which is still the basis for the "when used as a weapon" scenarios in the 2004 Emergency Response Guidebook (ERG). For irritating and lachrymatory agents, these recommendations are based on a release scenario involving direct aerosolization of the solid agents with a particle size between 2 and 5 μm. Under these conditions, the difference between a small and a large release is not based on the standard 200 liters spill used for commercial hazardous materials listed in the ERG. A small release involves 10 kilograms of powdered agent (approximately 200 cubic inch) and a large release involves 500 kilograms of powdered agent (approximately 18 cubic feet).

	Initial isolation (feet)	Downwind day (miles)	Downwind night (miles)
CA (Bromobenzyl cyanide) *C13-A004*			
Small device (10 kilograms)	100	0.1	0.3
Large device (500 kilograms)	500	1.0	2.6
CN (Chloroacetophenone) *C13-A008*			
Small device (10 kilograms)	100	0.1	0.3
Large device (500 kilograms)	400	0.7	2.0
CS (o-Chlorobenzylmalononitrile) *C13-A009*			
Small device (10 kilograms)	200	0.2	0.7
Large device (500 kilograms)	800	1.6	3.5

13.5.2 Personal Protective Requirements

13.5.2.1 Structural Firefighters' Gear

Structural firefighters' protective clothing is recommended for fire situations only; it is not effective in spill situations or release events. If chemical protective clothing is not available and it is necessary to rescue casualties from a contaminated area, then structural firefighters' gear will provide very limited skin protection against agent vapors and aerosols. Contact with solid and liquid agents should be avoided.

13.5.2.2 Respiratory Protection

Self-contained breathing apparatuses (SCBAs) or air purifying respirators (APRs) should have a National Institute for Occupational Safety and Health (NIOSH) and Chemical/Biological/Radiological/Nuclear (CBRN) certification. However, during emergency operations, other NIOSH approved SCBAs or APRs that have been specifically tested by the manufacturer against chemical warfare agents may be used if deemed necessary by the incident Commander. APRs should be equipped with a NIOSH approved CBRN filter or a combination organic vapor/acid gas/particulate cartridge.

Immediately dangerous to life or health (IDLH) levels are the ceiling limit for respirators other than SCBAs. Any exposures approaching the IDLH level should be regarded with extreme caution and the use of SCBAs for respiratory protection should be considered.

13.5.2.3 Chemical Protective Clothing

Irritating and lachrymatory agents are primarily an eye and respiratory hazard; however, at elevated vapor/aerosol concentrations or in contact with bulk material, agents may also pose a dermal hazard. In addition, solvents used in agent formulations may also pose respiratory or contact hazards.

Use only chemical protective clothing that has undergone material and construction performance testing against the specific agent that has been released. Since chemical protective clothing is tested against relatively pure agents, reported permeation rates may be affected by solvents, components, or impurities in munition grade agents.

13.5.3 Decontamination

13.5.3.1 General

Apply universal decontamination procedures using soap and water. If available, an alkaline soap/detergent works best. Do not use bleach or detergents containing bleach as they may interact with agents to produce toxic decomposition products. Alternatively, an aqueous solution of sodium bicarbonate (i.e., baking soda) may be used.

Casualties should be warned that there is a mild reaction between water and the agent and that they could experience a burning sensation during the decontamination process.

13.5.3.2 Vapors

Casualties/personnel: Aeration and ventilation. If decontamination is deemed necessary, remove all clothing as it may contain "trapped" agent. Flush skin with cool water followed by showering with copious amounts of soap and warm water. Do not use hot water as it will increase the burning sensation on the skin. Ensure that the hair has been washed and rinsed to remove potentially trapped vapor. For severe eye irritation, irrigate with water or 0.9% saline solution for a minimum of 15 minutes. Do not allow casualties to rub their eyes or skin as this may exacerbate agent effects.

Small areas: Ventilate to dissipate the aerosol. If deemed necessary, wash the area with copious amounts of alkaline soap/detergent and water. Collect and place into containers lined with high-density polyethylene. Removal of porous material, including painted surfaces, may be required because these materials may be difficult to decontaminate.

13.5.3.3 Liquids, Solutions, or Liquid Aerosols

Casualties/personnel: Remove all clothing immediately. To avoid further exposure of the head, neck, and face to the agent, cut off potentially contaminated clothing that must be pulled over the head. Flush skin with cool water. After flushing, use a sponge or cloth with liquid soap and copious amounts of water to wash the skin surface and hair at least three times. Do not use hot water as it will increase the burning sensation on the skin. Avoid rough scrubbing as this could abrade the skin and increase discomfort. Rinse with copious amounts of water. For severe eye irritation, irrigate with water or 0.9% saline solution for a minimum of 15 minutes. Do not allow casualties to rub their eyes or skin as this may exacerbate agent effects.

Small areas: Small puddles of liquid can be contained by covering with absorbent material such as vermiculite, diatomaceous earth, clay, sponges, or towels. Place the absorbed material into containers lined with high-density polyethylene. Larger puddles can be collected using vacuum equipment made of materials inert to the released material and equipped with appropriate vapor filters. Wash the area with copious amounts of an alkaline soap/detergent and water. Collect and containerize the rinseate. Removal of porous material, including painted surfaces, may be required because these materials may be difficult to decontaminate. Ventilate the area to remove vapors.

13.5.3.4 Solids or Particulate Aerosols

Casualties/personnel: Do not attempt to brush the agent off the individual or their clothing as this can aerosolize the agent. If possible, dampen the agent with a water mist to help prevent aerosolization. Remove all clothing immediately. To avoid further exposure of the head, neck, and face to the agent, cut off potentially contaminated clothing that must be pulled over the head. Flush skin with cool water. After flushing, use a sponge or cloth with liquid soap and copious amounts of water to wash the skin surface and hair at least three times. Do not use hot water as it will increase the burning sensation on the skin. Avoid rough scrubbing as this could abrade the skin and increase discomfort. Rinse with copious amounts of water. For severe eye irritation, irrigate with water or 0.9% saline solution for a minimum of 15 minutes. Do not allow casualties to rub their eyes or skin as this may exacerbate agent effects.

Small areas: If indoors, close windows and doors in the area and turn off anything that could create air currents (e.g., fans, air conditioner, etc.). Avoid actions that could aerosolize the agent such as sweeping or brushing. Collect the agent using a vacuum cleaner equipped with a high-efficiency particulate air (HEPA) filter. Do not use a standard home or industrial vacuum. Do not allow the vacuum exhaust to stir the air in the affected area. Vacuum all surfaces with extreme care in a very slow and controlled manner to minimize aerosolizing the agent. Place the collected material into containers lined with high-density polyethylene. Wash the area with copious amounts of an alkaline soap/detergent and water. Collect and containerize the rinseate in containers lined with high-density polyethylene. Removal of porous material, including painted surfaces, may be required because these materials may be difficult to decontaminate.

13.6 Medical

13.6.1 CDC Case Definition

The case can be confirmed if there is either a predominant amount of clinical and nonspecific laboratory evidence or an absolute certainty of the etiology of the agent is known.

13.6.2 Differential Diagnosis

The following factors have been suggested as alternatives to consider when presented with a potential case of exposure to irritating agents: anxiety, anaphylaxis, conjunctivitis, pneumonia, ultraviolet keratitis; thermal or chemical burns; inhalation of smoke, hydrocarbons, ammonia, hydrogen sulfide, phosgene, halogens (e.g., chlorine), sulfuric acid, hydrogen chloride (HCl), or nickel carbonyl; sodium azide; acute respiratory distress syndrome, chronic obstructive pulmonary disease and emphysema, congestive heart failure, and pulmonary edema.

13.6.3 Signs and Symptoms

13.6.3.1 Vapors/Aerosols

Irritating and lachrymatory agents produce intense eye pain and tearing. They may also produce burning or stinging sensations of exposed mucous membranes (e.g., nose and mouth) and skin. Symptoms may also include rhinorrhea (runny nose), sneezing, coughing, respiratory discomfort (e.g., tightness of the chest or inability to breathe), nausea, and/or vomiting. Increases in ambient temperature and/or humidity exacerbate agent effects. Effects from solvents will be minimal in comparison to the impacts caused by the actual agents themselves.

13.6.3.2 Solids/Solutions

General signs and symptoms may include intense pain of the eyes and mucous membranes, tearing, as well as localized irritation and burning of the skin.

13.6.4 Mass-Casualty Triage Recommendations

Typically not required. Casualties will usually recover unassisted shortly after removal from the contaminated atmosphere. Consult the base station physician or regional poison control center for advice on specific situations.

13.6.5 Casualty Management

Decontaminate the casualty ensuring that all the irritating and lachrymatory agents have been removed. For severe eye irritation, irrigate with water or 0.9% saline solution for a minimum of 15 minutes. Do not allow casualties to rub their eyes or skin as this may exacerbate agent effects. Irrigate open wounds with water or 0.9% saline solution for at least 10 minutes.

Once the casualty has been decontaminated, including the removal of foreign matter from wounds, medical personnel do not need to wear a chemical-protective mask.

Casualties will usually recover unassisted from exposure to irritating agents within 15 minutes after removal from the contaminated atmosphere. Most patients can be discharged safely. Rarely a patient with significant respiratory findings may merit admission.

13.7 Fatality Management

Remove all clothing and personal effects and decontaminate with soap and water. Do not use bleach or detergents containing bleach as they may interact with agents to produce toxic decomposition products.

Wash the remains with soap and water. Pay particular attention to areas where agent may get trapped, such as hair, scalp, pubic areas, fingernails, folds of skin, and wounds. If remains are heavily contaminated with residue, wash and rinse waste should be contained for proper disposal.

Once the remains have been thoroughly decontaminated, no further protective action is necessary. Body fluids removed during the embalming process do not pose any additional risks, and should be contained and handled according to established procedures. Use standard burial procedures.

C13-A

AGENTS

C13-A001

Acrolein (Papite)
CAS: 107-02-8
RTECS: AS1050000
UN: 1092
ERG: 131P

O
H

C_3H_4O

Colorless to greenish-yellow liquid with a pungent, piercing, disagreeable odor detectable at 0.3 ppm. Unstable and prone to polymerization; often stabilized with amyl nitrate or hydroquinone. May form shock-sensitive peroxides during storage.

Used industrially as a pesticide, warning agent in refrigerants, and in the manufacturing of glycerol, polyurethane, polyester resins, and pharmaceuticals.

Exposure Hazards

Conversion Factor: 1 ppm = 2.29 mg/m^3 at 77°F
$LCt_{50(Inh)}$: 3500 mg-min/m^3 (760 ppm for a 2-min exposure)
Eye Irritation: 0.06 ppm; exposure duration unspecified
"Intolerable" Irritation: 22 ppm; exposure duration unspecified

These values are from older sources (ca. 1937). No updated toxicity estimates have been proposed.

$MEG_{(1h)}$ Min: 0.03 ppm; *Sig*: 0.1 ppm; *Sev*: 1.4 ppm
OSHA PEL: 0.10 ppm
ACGIH TLV: 0.10 ppm [Skin]
NIOSH STEL: 0.3 ppm
IDLH: 2.0 ppm

Properties:

MW: 56.1	*VP*: 210 mmHg	*FlP*: −15°F
D: 0.84 g/mL (59°F)	*VD*: 1.9 (calculated)	*LEL*: 2.8%
MP: −126°F	*Vlt*: 280,000 ppm	*UEL*: 31%
BP: 127°F	*H_2O*: 40%	*RP*: 0.063
Vsc: 0.35 cS	*Sol*: Alcohol; Ether; Acetone	*IP*: 10.13 eV

Interim AEGLs

AEGL-1: 1 h, 0.030 ppm	4 h, 0.030 ppm	8 h, 0.030 ppm
AEGL-2: 1 h, 0.10 ppm	4 h, 0.10 ppm	8 h, 0.10 ppm
AEGL-3: 1 h, 1.4 ppm	4 h, 0.48 ppm	8 h, 0.27 ppm

C13-A002

Benzyl bromide (Cyclite)
CAS: 100-39-0
RTECS: XS7965000
UN: 1737
ERG: 156

Br

C_7H_7Br

Colorless to yellow to brownish liquid with a pleasant and aromatic odor, resembling water cress.

Used industrially as a chemical intermediate.

Exposure Hazards

Conversion Factor: 1 ppm = 6.99 mg/m^3 at 77°F
$LCt_{50(Inh)}$: 45,000 mg-min/m^3 (3220 ppm for a 2-min exposure)
Eye Irritation: 0.6 ppm; exposure duration unspecified
"Intolerable" Irritation: 4.3 ppm; exposure duration unspecified

These values are from older sources (ca. 1937). No updated toxicity estimates have been proposed.

Properties:

MW: 171.0	*VP*: 0.450 mmHg	*FlP*: 174°F
D: 1.4362 g/mL	*VP*: 1.0 mmHg (90°F)	*LEL*: —

MP: 27°F
BP: 394°F
Vsc: —

VD: 5.9 (calculated)
Vlt: 600 ppm
H_2O: 0.0385% (slowly decomposes)
Sol: Alcohol; Ether; Benzene

UEL: —
RP: 17
IP: 8.99 eV

C13-A003

Bromoacetone (Agent BA)
CAS: 598-31-2
RTECS: —
UN: 1569
ERG: 131

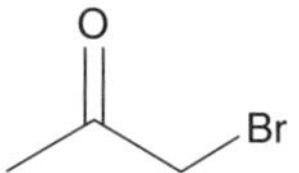

C_3H_5BrO

Colorless to violet liquid with a pungent odor. It is unstable and decomposed by heat and light.

Used industrially as a chemical intermediate.

Exposure Hazards
Conversion Factor: 1 ppm = 5.60 mg/m^3 at 77°F
$LCt_{50(Inh)}$: 32,000 mg-min/m^3 (2860 ppm for a 2-min exposure)
Eye Irritation: 0.27 ppm; exposure duration unspecified
"Intolerable" Irritation: 1.8 ppm; exposure duration unspecified

Liquid agent produces blisters on exposed skin.

These values are from older sources (ca. 1937). No updated toxicity estimates have been proposed.

Properties:

MW: 137.0
D: 1.634 g/mL (73°F)
MP: –34°F
BP: 279°F
BP: 146°F (50 mmHg)
Vsc: —

VP: 9 mmHg
VD: 4.7 (calculated)
Vlt: 13,000 ppm
H_2O: "Slight"
Sol: Alcohol; Acetone; Ether; Benzene

FlP: —
LEL: —
UEL: —
RP: 0.9
IP: 9.73 eV

C13-A004

Bromobenzyl cyanide (Agent CA)
CAS: 5798-79-8
RTECS: AL8090000
UN: 1694
ERG: 159

C_8H_6BrN

Yellow solid or yellow to brown liquid with an odor like soured or rotting fruit. Undergoes considerable decomposition when a large explosive charge is used to disseminate the agent.

Exposure Hazards

Conversion Factor: 1 ppm = 8.02 mg/m^3 at 77°F
$LCt_{50(Inh)}$: 8,000–11,000 mg-min/m^3 (500–690 ppm for a 2-min exposure)
$ICt_{50(Eyes)}$: 30 mg-min/m^3 (1.9 ppm for a 2-min exposure)
Eye Irritation: 0.04 ppm; exposure duration unspecified

Properties:

MW: 196.0	*VP*: 0.012 mmHg	*FlP*: None
D: —	*VD*: 6.8 (calculated)	*LEL*: None
MP: 84°F	*Vlt*: 14 ppm	*UEL*: None
BP: 468°F (decomposes)	*Vlt*: 34 ppm (86°F)	*RP*: 640
BP: 270°F (12 mmHg)	H_2O: "Slightly soluble"	*IP*: <10 eV
Vsc: —	*Sol*: Common organic solvents	

C13-A005

Bromomethylethyl ketone (Bn-Stoff)
CAS: 816-40-0
RTECS: —

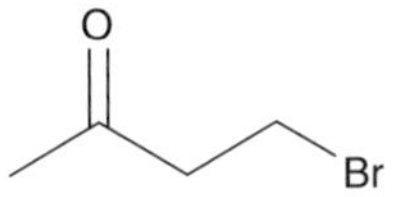

C_4H_7BrO

Light yellow liquid.

Exposure Hazards

Conversion Factor: 1 ppm = 6.18 mg/m^3 at 77°F
$LCt_{50(Inh)}$: 20,000 mg-min/m^3 (1600 ppm for a 2-min exposure)
Eye Irritation: 2 ppm; exposure duration unspecified
"Intolerable" Irritation: 2.6 ppm; exposure duration unspecified

These values are from older sources (ca. 1937). No updated toxicity estimates have been proposed.

Properties:

MW: 151.0	*VP*: 15 mmHg (57°F)	*FlP*: —
D: 1.479 g/mL	*VD*: 5.2 (calculated)	*LEL*: —
MP: —	*Vlt*: 20,000 ppm	*UEL*: —
BP: 293°F (decomposes)	H_2O: "Insoluble"	*RP*: 0.53
Vsc: —	*Sol*: —	*IP*: —

C13-A006

Capsaicin (Agent OC)
CAS: 404-86-4
RTECS: RA8530000

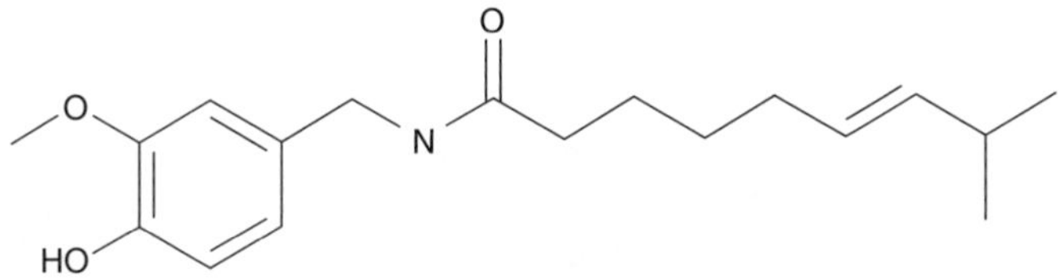

$C_{18}H_{27}NO_3$
Crystalline solid.
Used medicinally and as a food flavoring.

Exposure Hazards
Conversion Factor: 1 ppm = 12.49 mg/m^3 at 77°F
$LD_{50(Ing)}$: 350 g (estimate)

Properties:

MW: 305.4	*VP*: Negligible	*FlP*: —
D: —	*VD*: —	*LEL*: —
MP: 149°F	*Vlt*: —	*UEL*: —
BP: 410°F (0.01 mmHg)	H_2O: "Practically insoluble"	*RP*: 38,000,000
Vsc: —	*Sol*: Alcohol; Ether; Benzene; Chloroform	*IP*: —

C13-A007

Chloroacetone (Tonite)
CAS: 78-95-5
RTECS: UC0700000
UN: 1695
ERG: 131

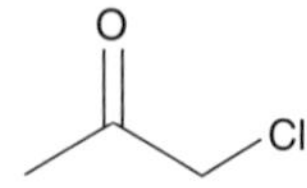

C_3H_5ClO

Clear liquid with a pungent odor similar to hydrochloric acid. Readily breaks down during storage; often stabilized with water or calcium carbonate. Turns dark and forms resins on prolonged exposure to light.

Used industrially as a chemical intermediate in the manufacture of couplers for color photography, as a photo polymerization agent for vinyl compounds, as a solvent, and as an enzyme inactivator in biological research.

Exposure Hazards
Conversion Factor: 1 ppm = 3.78 mg/m^3 at 77°F
$LCt_{50(Inh)}$: 23,000 mg-min/m^3 (3040 ppm for a 2-min exposure)
Eye Irritation: 4.8 ppm; exposure duration unspecified
"Intolerable" Irritation: 26 ppm; exposure duration unspecified

These values are from older sources (ca. 1937). No updated toxicity estimates have been proposed.

$MEG_{(1h)}$ *Min*: 1 ppm; *Sig*: —; *Sev*: —
ACGIH Ceiling: 1 ppm [Skin]

Properties:

MW: 92.5	*VP*: 12.0 mmHg (77°F)	*FlP*: 104°F
D: 1.123 g/mL (77°F)	*VD*: 3.7 (calculated)	*LEL*: —
MP: –48°F	*Vlt*: 15,800 ppm (77°F)	*UEL*: —
BP: 248°F	H_2O: 9%	*RP*: 0.88
Vsc: —	*Sol*: Alcohols; Ether; Chloroform	*IP*: 9.92 eV

Proposed AEGLs

AEGL-1: Not Developed

AEGL-2: 1 h, 4.4 ppm	4 h, 1.1 ppm	8 h, 1.1 ppm
AEGL-3: 1 h, 13 ppm	4 h, 3.3 ppm	8 h, 3.3 ppm

C13-A008

Chloroacetophenone (Agent CN)

CAS: 532-27-4
RTECS: AM6300000
UN: 1697
ERG: 153

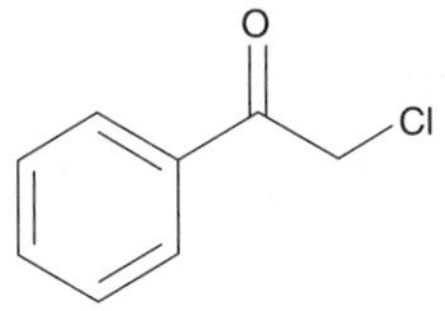

C_8H_7ClO

Colorless to white or gray crystalline solid with a sharp, fragrant odor like apple blossoms that is detectable at 0.02 ppm.

Used industrially as a chemical intermediate for pharmaceuticals and as a denaturant for alcohol.

Also reported as a mixture with various solvents including CNB (mixture of 10% chloroacetophenone, 45% benzene, and 45% carbon tetrachloride), CNC (mixture of 30% chloroacetophenone and 70% chloroform), CND (mixture of chloroacetophenone and ethylene dichloride), and CNS (C13-A015).

Exposure Hazards

Conversion Factor: 1 ppm = 6.32 mg/m^3 at 77°F

$LCt_{50(Inh)}$: 7000 mg-min/m^3 (550 ppm for a 2-min exposure) when dispersed as a solution of the agent dissolved in a solvent.

$LCt_{50(Inh)}$: 14,000 mg-min/m^3 (1100 ppm for a 2-min exposure) when dispersed as an aerosol from a thermal device.

$ICt_{50(Eyes)}$: 80 mg-min/m^3 (6.3 ppm for a 2 min exposure)

Eye Irritation: 0.02–0.06 ppm; exposure duration unspecified

Exposure to high concentrations of aerosolized agent can cause severe skin irritation and even produce blistering similar to mustard agents (C03–C05).

$MEG_{(1h)}$ *Min*: —; *Sig*: —; *Sev*: 2.4 ppm

OSHA PEL: 0.05 ppm

ACGIH TLV: 0.05 ppm

IDLH: 2.4 ppm

Properties:

MW: 154.6	*VP*: 0.0041 mmHg	*FlP*: 244°F
D: 1.318 g/cm^3	*VD*: 5.3 (calculated)	*LEL*: —
MP: 129°F	*Vlt*: 5.4 ppm	*UEL*: —
BP: 478°F	*Vlt*: 170 ppm (125°F)	*RP*: 2000
Vsc: —	H_2O: Insoluble	*IP*: 9.44 eV
	Sol: Most organic solvents	

C13-A009

***o*-Chlorobenzyl-malononitrile** (Agent CS)
CAS: 2698-41-1
RTECS: OO3675000

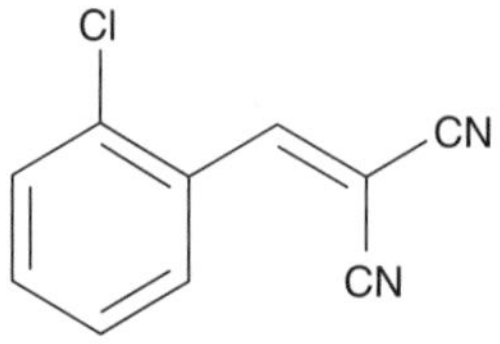

$C_{10}H_5ClN_2$

Colorless solid with a pungent, pepper-like odor.

Also reported as a 1% mixture in Trioctylphosphite (CSX).

Exposure Hazards

Conversion Factor: 1 ppm = 7.71 mg/m^3 at 77°F
$LCt_{50(Inh)}$: 52,000–61,000 mg-min/m^3 (3400–4000 ppm for a 2-min exposure). This is a provisional update from an older value that has not been formally adopted as of 2005.
"Intolerable" Irritation: 0.91 ppm; exposure duration unspecified
$MEG_{(1h)}$ *Min*: 0.05 ppm; *Sig*: —; *Sev*: 0.26 ppm
OSHA PEL: 0.05 ppm
ACGIH Ceiling: 0.05 ppm [Skin]
NIOSH Ceiling: 0.05 ppm [Skin]
IDLH: 0.26 ppm

Properties:

MW: 188.6
D: 1.04 g/cm^3
D: 0.24 g/cm^3 (bulk)
MP: 203°F
BP: 590°F (decomposes)
Vsc: —
VP: 0.000034 mmHg
VD: 6.5 (calculated)
Vlt: 0.092 ppm
H_2O: 0.008% (77°F)
Sol: Acetone; Hexane: Benzene; Methylene chloride
FlP: 387°F
MEC: 25 g/m^3
UEL: —
RP: 21,000
IP: <10 eV

C13-A010

Chloromethyl chloroformate (Palite)
CAS: 22128-62-7
RTECS: —
UN: 2745
ERG: 157

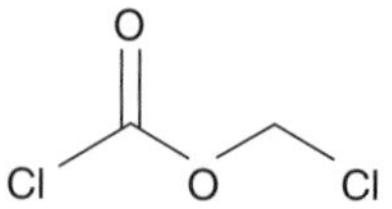

$C_2H_2Cl_2O_2$

Clear, colorless liquid with a pungent odor.

Used industrially as a chemical intermediate.

Also reported as a mixture with Dichloromethyl chloroformate (C13-A018); Stannic chloride.

Exposure Hazards

Conversion Factor: 1 ppm = 5.27 mg/m^3 at 77°F

$LCt_{50(Inh)}$: 10,000 mg-min/m^3 (950 ppm for a 2-min exposure)

$ICt_{50(Eyes)}$: 50 mg-min/m^3 (4.7 ppm for a 2-min exposure)

These values are from older sources (ca. 1937) and are not supported by modern data. No updated toxicity estimates have been proposed.

Eye Irritation: 0.38 ppm; exposure duration unspecified

Properties:

MW: 128.9	*VP*: 5.6 mmHg	*FlP*: 203°F
D: 1.449 g/mL	*VP*: 100 mmHg (127°F)	*LEL*: —
MP: —	*VD*: 4.4 (calculated)	*UEL*: —
BP: 225°F	*Vlt*: 7600 ppm	*RP*: 0.97
Vsc: —	H_2O: Decomposes	*IP*: —
	Sol: Hydrocarbon solvents	

C13-A011

Dibenz-(b,f)-1,4-oxazepine (Agent CR)

CAS: 257-07-8

RTECS: HQ3950000

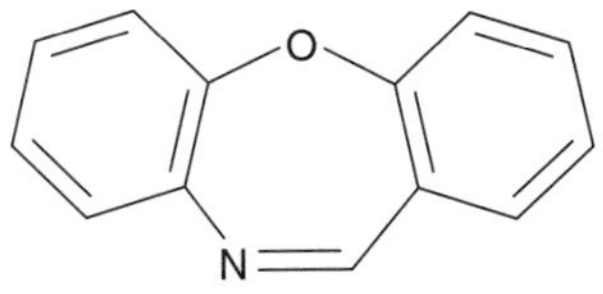

$C_{13}H_9NO$

Yellow needles or brown powder with a peppery odor and produces a burning sensation.

Exposure Hazards

Conversion Factor: 1 ppm = 7.98 mg/m^3 at 77°F

$ICt_{50(Eyes)}$: 0.15 mg-min/m^3 (0.009 ppm for a 2-min exposure)

Eye Irritation: 0.0003–0.0005 ppm for a 1-min exposure

Properties:

MW: 195.2	*VP*: 0.000059 mmHg	*FlP*: 370°F
D: 1.56 g/mL	*VD*: 6.7 (calculated)	*LEL*: —
MP: 160°F	*Vlt*: 0.079 ppm	*UEL*: —
BP: 634°F	H_2O: 0.008%	*RP*: 120,000
Vsc: —	*Sol*: Ethanol; Ether; Benzene; Chloroform	*IP*: $<$9 eV

C13-A012

Ethyl iodoacetate (Agent SK)

CAS: 623-48-3

RTECS: AI3575000

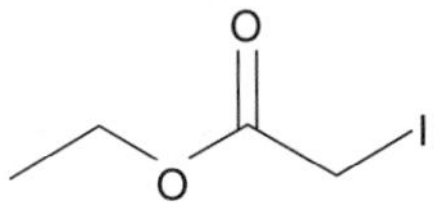

$C_4H_7IO_2$

Colorless oily liquid that is light and moisture sensitive. Becomes brown during storage due to liberation of iodine.

Also reported as a mixture with Chloropicrin (C10-A006) and as a mixture in Ethyl acetate and Ethanol (KSK).

Exposure Hazards

Conversion Factor: 1 ppm = 8.75 mg/m^3 at 77°F
$LCt_{50(Inh)}$: 15,000 mg-min/m^3 (860 ppm for a 2-min exposure)
"Intolerable" Irritation: 1.7 ppm; exposure duration unspecified

These values are from older sources (ca. 1939) and are not supported by modern data. No updated toxicity estimates have been proposed.

Properties:

MW: 214.0	*VP*: 0.54 mmHg	*FlP*: 169°F
D: 1.808 g/mL	*VD*: 7.4 (calculated)	*LEL*: —
MP: —	*Vlt*: 350 ppm	*UEL*: —
BP: 355°F	H_2O: —	*RP*: 13
BP: 163°F (16 mmHg)	*Sol*: —	*IP*: —
Vsc: —		

C13-A013

Iodoacetone (Bretonite)
CAS: 3019-04-3
RTECS: —

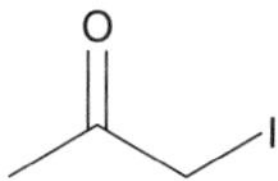

C_3H_5IO

Faintly yellow liquid. Becomes brown during storage due to liberation of iodine. Decomposes on standing to 1,3-Diiodoacetone.

Also reported as a mixture with Stannic chloride.

Exposure Hazards

Conversion Factor: 1 ppm = 7.53 mg/m^3 at 77°F
$LCt_{50(Inh)}$: 19,000 mg-min/m^3 (1300 ppm for a 2-min exposure)
$ICt_{50(Eyes)}$: 100 mg-min/m^3 (6.6 ppm for a 2-min exposure)
"Intolerable" Irritation: 13 ppm; exposure duration unspecified

These values are from older sources (ca. 1937) and are not supported by modern data. No updated toxicity estimates have been proposed.

Properties:

MW: 184.0	*VP*: —	*FlP*: —
D: 1.8 g/mL	*VD*: 6.3 (calculated)	*LEL*: —
MP: —	*Vlt*: 400 ppm	*UEL*: —
BP: 216°F	H_2O: —	*RP*: 20
BP: 122°F (11 mmHg)	*Sol*: —	*IP*: 9.3 eV
Vsc: —		

C13-A014

1-Methoxy-1,3,5-cycloheptatriene (Agent CH)
CAS: 1714-38-1; 69044-01-5 (Fluorosulfate salt); 25059-10-3 (Perchlorate salt); 25482-84-2 (Perchlorate salt); 54098-95-2 (Perchlorate salt); 29630-03-3 (Tetrafluoroborate salt)
RTECS: —

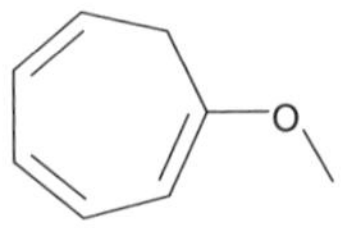

$C_8H_{10}O$

Colorless to brown liquid with a sweetish odor. Various salts have been reported.

Also reported as a mixture with 1-Methyl-4-piperidyl isopropylphenylglycolate (C12-A006).

Exposure Hazards
Conversion Factor: 1 ppm = 5.00 mg/m^3 at 77°F
Human toxicity values have not been established or have not been published.

Properties:

MW: 122.2	*VP*: 1.3 mmHg (77°F)	*FlP*: 133°F
D: 0.9717 g/mL (77°F)	*VD*: 4.2 (calculated)	*LEL*: —
MP: –153°F	*Vlt*: 1700 ppm (77°F)	*UEL*: —
BP: 345°F	H_2O: 0.072%	*RP*: 7
Vsc: 1.50 cS (77°F)	*Sol*: Aromatic organic solvents	*IP*: 7.23 eV

C13-A015

CNS
CAS: 675600-78-9
RTECS: —

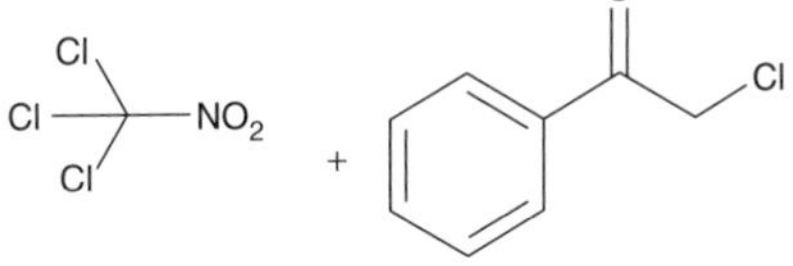

Mixture

Liquid mixture of 38.4% Chloropicrin (C10-A006), 23% Chloroacetophenone (C13-A008), and 38.4% Chloroform. Historically, the odor of this mixture has been compared to flypaper.

Exposure Hazards
Conversion Factor: 1 ppm = mg/m^3 at 77°F
$LCt_{50(Inh)}$: 11,400 mg-min/m^3
$ICt_{50(Eyes)}$: 60 mg-min/m^3

In addition to lacrimation, the chloropicrin component may cause pulmonary edema, vomiting, nausea, and diarrhea.

Properties:

MW: Mixture	*VP*: 78 mmHg	*FlP*: None
D: 1.47 g/mL	*VD*: 5 (calculated)	*LEL*: None
MP: 36°F (precipitation occurs)	*Vlt*: —	*UEL*: None
BP: 140°F	H_2O: Insoluble	*RP*: —
Vsc: —	*Sol*: —	*IP*: —

C13-A016

***o*-Nitrobenzyl chloride** (Cedenite)
CAS: 612-23-7
RTECS: XS9092000

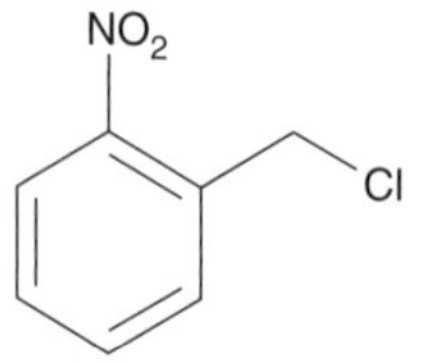

$C_7H_6ClNO_2$

Pale yellow crystals.

Used industrially for photographic imaging and in medicinal research.

Exposure Hazards
Conversion Factor: 1 ppm = 7.02 mg/m^3 at 77°F
$ICt_{50(Eyes)}$: 15 mg-min/m^3 (1.1 ppm for a 2-min exposure)
Eye Irritation: 0.26 ppm; exposure duration unspecified

These values are from older sources (ca. 1939) and are not supported by modern data. No updated toxicity estimates have been proposed.

Properties:

MW: 171.6	*VP*: —	*FlP*: 234°F
D: —	*VD*: 5.9 (calculated)	*LEL*: —
MP: 118°F	*Vlt*: —	*UEL*: —
BP: 261°F (10 mmHg)	H_2O: Insoluble	*RP*: —
Vsc: —	*Sol*: Alcohol; Ether; Benzene	*IP*: <10 eV

C13-A017

Thiophosgene (Lacrymite)
CAS: 463-71-8
RTECS: XN2450000
UN: 2474
ERG: 131

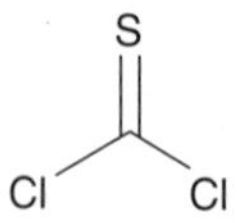

CCl_2S

Reddish-yellow liquid with a sharp, choking, foul odor. Air and moisture sensitive; decomposes above 392°F to carbon disulfide and carbon tetrachloride.

Used industrially as a chemical intermediate.

Exposure Hazards
Conversion Factor: 1 ppm = 4.70 mg/m^3 at 77°F

Human toxicity values have not been established or have not been published.

Properties:

MW: 115.0	*VP*: —	*FlP*: —
D: 1.508 g/mL (59°F)	*VD*: 4.0 (calculated)	*LEL*: —
MP: —	*Vlt*: —	*UEL*: —
BP: 163°F	H_2O: Insoluble (decomposes)	*RP*: —
Vsc: —	*Sol*: Ether	*IP*: 9.61 eV

C13-A018

Dichloromethyl chloroformate

CAS: 22128-63-8

RTECS: —

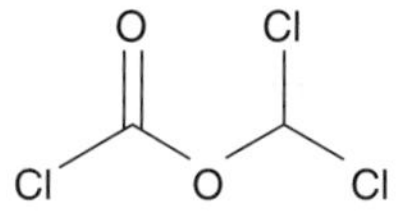

$C_2HCl_3O_2$

Colorless liquid.

Also reported as a mixture with Chloromethyl chloroformate (C13-A010).

Exposure Hazards

Conversion Factor: 1 ppm = 6.68 mg/m^3 at 77°F

"Intolerable" Irritation: 11 ppm; exposure duration unspecified

These values are from older sources (ca. 1939) and are not supported by modern data. No updated toxicity estimates have been proposed.

Properties:

MW: 163.4	*VP*: 5 mmHg	*FlP*: —
D: 1.56 g/mL (59°F)	*VD*: 5.6 (calculated)	*LEL*: —
MP: —	*Vlt*: 6700 ppm	*UEL*: —
BP: 230°F	H_2O: —	*RP*: 1.6
BP: 129°F (100 mmHg)	*Sol*: —	*IP*: —
Vsc: —		

C13-A019

Benzyl chloride

CAS: 100-44-7

RTECS: XS8925000

UN: 1738

ERG: 156

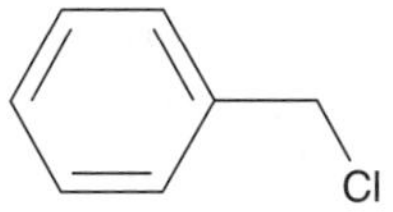

C_7H_7Cl

Colorless to slightly yellow liquid with a pungent, aromatic odor.

Used industrially as a chemical intermediate in the manufacture of pharmaceuticals, perfumes, dyes, synthetic tannins, and artificial resins.

Exposure Hazards

Conversion Factor: 1 ppm = 5.18 mg/m^3 at 77°F
"Intolerable" Irritation: 16 ppm; exposure duration unspecified

These values are from older sources (ca. 1939) and are not supported by modern data. No updated toxicity estimates have been proposed.

OSHA PEL: 1 ppm
ACGIH TLV: 1 ppm
IDLH: 10 ppm

Properties:

MW: 126.6
D: 1.1028 g/mL
MP: –49°F
BP: 354°F
BP: 210°F (62 mmHg)
Vsc: —

VP: 1.23 mmHg (77°F)
VD: 4.4 (calculated)
Vlt: 1200 ppm
H_2O: 0.0525%
Sol: Most organic solvents

FlP: 153°F
LEL: —
UEL: —
RP: 7.3
IP: 9.14 eV

C13-A020

Benzyl iodide

CAS: 620-05-3
RTECS: —
UN: 2653
ERG: 156

C_7H_7I

White crystalline solid.

Exposure Hazards

Conversion Factor: 1 ppm = mg/m^3 at 77°F
$LCt_{50(Inh)}$: 30,000 mg-min/m^3 (1700 ppm for a 2-min exposure)
Eye Irritation: 0.2 ppm; exposure duration unspecified
"Intolerable" Irritation: 3.4 ppm for 1-min exposure

These values are from older sources (ca. 1939) and are not supported by modern data. No updated toxicity estimates have been proposed.

Properties:

MW: 218.0
D: 1.7429 g/cm^3
MP: 75°F
BP: 438°F (decomposes)
BP: 200°F (10 mmHg)
Vsc: —

VP: —
VD: 7.5 (calculated)
Vlt: 130 ppm
H_2O: Insoluble
Sol: Alcohol: Ether; Benzene

FlP: —
LEL: —
UEL: —
RP: 64
IP: 8.73 eV

C13-A021

Ethyl bromoacetate

CAS: 105-36-2
RTECS: AF6000000

UN: 1603
ERG: 155

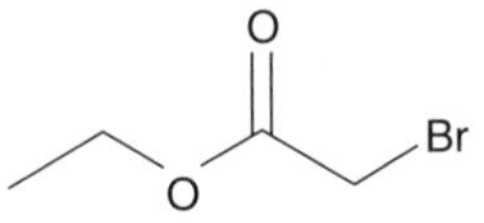

$C_4H_7BrO_2$

Clear colorless liquid with a pungent odor.

Used industrially as a chemical intermediate.

Exposure Hazards

Conversion Factor: 1 ppm = 6.83 mg/m^3 at 77°F
$LCt_{50(Inh)}$: 23,000 mg-min/m^3 (1700 ppm for a 2-min exposure)
Eye Irritation: 0.44 ppm; exposure duration unspecified
"Intolerable" Irritation: 5.9 ppm; exposure duration unspecified

These values are from older sources (ca. 1937) and are not supported by modern data. No updated toxicity estimates have been proposed.

Properties:

MW: 167.0	*VP*: 3.4 mmHg (77°F)	*FlP*: 118°F
D: 1.5032 g/mL	*VD*: 5.8 (calculated)	*LEL*: —
MP: –36°F	*Vlt*: 3100 ppm	*UEL*: —
BP: 335°F	H_2O: 0.7%	*RP*: 2.3
Vsc: —	*Sol*: Acetone; Benzene; Ethanol; Ether	*IP*: —

C13-A022

***N*-Ethylcarbazole**

CAS: 86-28-2
RTECS: FE6225700

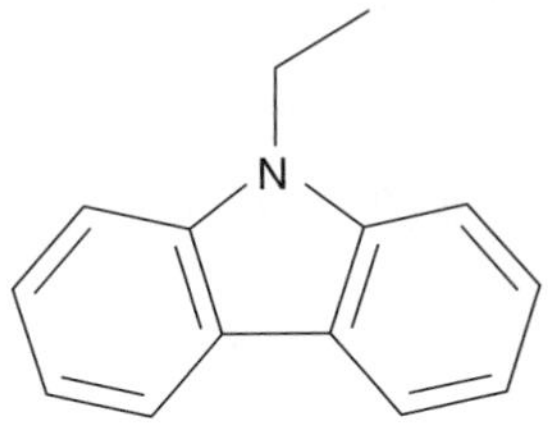

$C_{14}H_{13}N$

White to brown flaky solid.

Exposure Hazards

Conversion Factor: 1 ppm = 7.99 mg/m^3 at 77°F

Human toxicity values have not been established or have not been published.

Properties:

MW: 195.3	*VP*: 0.02 mmHg (167°F)	*FlP*: 367°F
D: 1.16 g/cm^3	*VD*: 6.7 (calculated)	*LEL*: —
MP: 153°F	*Vlt*: —	*UEL*: —
BP: 374°F	H_2O: Insoluble	*RP*: 420
Vsc: —	*Sol*: Alcohol; Ether	*IP*: —

C13-A023

Xylyl bromide (T-Stoff)
CAS: 28258-59-5 (Mixture); 620-13-3 (Meta isomer); 89-92-9 (Ortho isomer); 104-81-4 (Para isomer)
RTECS: —
UN: 1701
ERG: 152

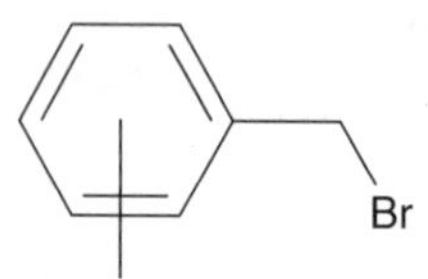

C_8H_9Br

Mixture of the ortho, meta, and para isomers. Colorless to light yellow, slightly viscous liquid with a pungent and aromatic odor, resembling lilacs or elder blossoms. Weapons grade material is a black liquid.

Used industrially as a chemical intermediate.

Exposure Hazards
Conversion Factor: 1 ppm = 7.57 mg/m^3 at 77°F
$LCt_{50(Inh)}$: 56,000 mg-min/m^3 (3700 ppm for a 2-min exposure)
Eye Irritation: 0.24 ppm; exposure duration unspecified
"Intolerable" Irritation: 2.0 ppm; exposure duration unspecified

These values are from older sources (ca. 1937) and are not supported by modern data. No updated toxicity estimates have been proposed.

Properties:

MW: 185.1	*VP*: —	*FlP*: —
D: 1.36 g/mL	*VD*: 6.4 (calculated)	*LEL*: —
MP: —	*Vlt*: 80 ppm	*UEL*: —
BP: 410–428°F	*H_2O*: Insoluble	*RP*: 120
Vsc: —	*Sol*: Most organic solvents	*IP*: —

C13-A024

Xylylene bromide
CAS: — (Mixture); 91-13-4 (Ortho isomer); 626-15-3 (Meta isomer); 623-24-5 (Para isomer)
RTECS: —

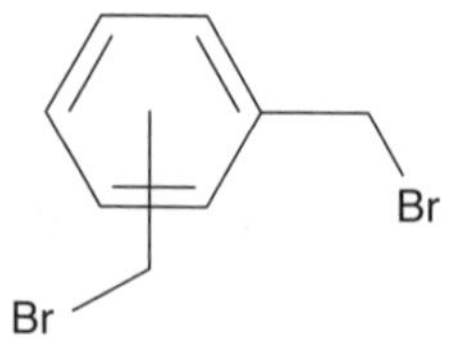

$C_8H_8Br_2$

Mixture of the ortho, meta, and para isomers. White to beige crystalline solid.

Exposure Hazards
Human toxicity values have not been established or have not been published.

Properties:

MW: 264.0	*VP*: Negligible	*FlP*: —
D: 1.96 g/cm^3 (ortho isomer)	*VD*: —	*LEL*: —
MP: 196°F (ortho isomer)	*Vlt*: —	*UEL*: —
MP: 288°F (para isomer)	H_2O: —	*RP*: —
BP: 473°F (para isomer)	*Sol*: —	*IP*: —
Vsc: —		

References

Agency for Toxic Substances and Disease Registry. *Toxicological Profile for Acrolein*. Washington, DC: Government Printing Office, December 1990.

Brophy, Leo P., Wyndham D. Miles, and Rexmond C. Cohrane. *The Chemical Warfare Service: From Laboratory to Field*. Washington, DC: Government Printing Office, 1968.

Compton, James A.F. *Military Chemical and Biological Agents: Chemical and Toxicological Properties*. Caldwell, NJ: The Telford Press, 1987.

Edgewood Research Development, and Engineering Center, Department of the Army. *Material Safety Data Sheet (MSDS) for Agent CS*. Aberdeen Proving Ground, MD: Chemical Biological Defense Command, Revised June 30, 1995.

———. *Material Safety Data Sheet (MSDS) for Riot Control Agent CR*. Aberdeen Proving Ground, MD: Chemical Biological Defense Command, Revised June 30, 1995.

Fries, Amos A., and Clarence J. West. *Chemical Warfare*. New York: McGraw-Hill Book Company, Inc., 1921.

Grant, George A. "Safe Sensory Irritant." US Patent 4,598,096, July 1, 1986.

Jackson, Kirby E., and Margaret A. Jackson. "Lachrymators." *Chemical Reviews* 16 (1935): 195–242.

Langford, Gordon E. "Scale-up and Synthesis of 1-Methoxycyglohepta-1,3,5-triene." US Patent 4,978,806, December 18, 1990.

Marrs, Timothy C., Robert L. Maynard, and Frederick R. Sidell. *Chemical Warfare Agents: Toxicology and Treatment*. Chichester, England: John Wiley & Sons, 1997.

National Institute for Occupational Safety and Health. *NIOSH Pocket Guide to Chemical Hazards*. Washington, DC: Government Printing Office, September 2005.

National Institutes of Health. *Hazardous Substance Data Bank (HSDB)*. http://toxnet.nlm.nih.gov/cgi-bin/sis/htmlgen?HSDB/. 2004.

Olson, Kent R., ed. *Poisoning & Drug Overdose*. 4th ed. New York: Lange Medical Books/McGraw-Hill, 2004.

Prentiss, Augustin M. *Chemicals in War: A Treatise on Chemical Warfare*. New York: McGraw-Hill Book Company, Inc., 1937.

Sartori, Mario F. *The War Gases: Chemistry and Analysis*. Translated by L.W. Marrison. London: J. & A. Churchill, Ltd, 1939.

Sidell, Frederick R. *Medical Management of Chemical Warfare Agent Casualties: A Handbook for Emergency Medical Services*. Bel Air, MD: HB Publishing, 1995.

Sidell, Fredrick R., Ernest T. Takafuji, and David R. Franz, eds. *Medical Aspects of Chemical and Biological Warfare, Textbook of Military Medicine Series, Part 1, Warfare, Weaponry, and the Casualty*. Washington, DC: Office of the Surgeon General, Department of the Army, 1997.

Sifton, David W., ed. *PDR Guide to Biological and Chemical Warfare Response*. Montvale, NJ: Thompson/Physicians Desk Reference, 2002.

Smith, Ann, Patricia Heckelman, and Maryadele J. Oneil, eds. *The Merck Index: An Encyclopedia of Chemicals, Drugs, & Biologicals*. 13th ed. Rahway, NJ: Merck & Co., Inc., 2001.

Somani, Satu M., ed. *Chemical Warfare Agents*. New York: Academic Press, 1992.

Somani, Satu M., and James A. Romano, Jr., eds. *Chemical Warfare Agents: Toxicity at Low Levels*. Boca Raton, FL: CRC Press, 2001.

Swearengen, Thomas F. *Tear Gas Munitions: An Analysis of Commercial Riot Gas Guns, Tear Gas Projectiles, Grenades, Small Arms Ammunition, and Related Tear Gas Devices.* Springfield, IL: Charles C Thomas Publisher, 1966.

True, Bey-Lorraine, and Robert H. Dreisbach. *Dreisbach's Handbook of Poisoning: Prevention, Diagnosis and Treatment.* 13th ed. London, England: The Parthenon Publishing Group, 2002.

United States Army Headquarters. *Chemical Agent Data Sheets Volume II, Edgewood Arsenal Special Report No. EO-SR-74002.* Washington, DC: Government Printing Office, December 1974.

———. *Potential Military Chemical/Biological Agents and Compounds, Field Manual No. 3-11.9.* Washington, DC: Government Printing Office, January 10, 2005.

United States Army Medical Research Institute of Chemical Defense. *Medical Management of Chemical Casualties Handbook.* 3rd Edn. Aberdeen Proving Ground, MD: United States Army Medical Research Institute of Chemical Defense, July 2000.

United States Coast Guard. *Chemical Hazards Response Information System (CHRIS) Manual, 1999 Edition.* http://www.chrismanual.com/Default.htm. March 2004.

Wachtel, Curt. *Chemical Warfare.* Brooklyn, NY: Chemical Publishing Co., Inc., 1941.

Waitt, Alden H. *Gas Warfare: The Chemical Weapon, Its Use, and Protection Against It.* Rev. ed. New York: Duell, Sloan and Pearce, 1944.

Williams, Kenneth E. *Detailed Facts About Tear Agent 2-Chloroacetophenone (CN).* Aberdeen Proving Ground, MD: United States Army Center for Health Promotion and Preventive Medicine, 1996.

———. *Detailed Facts About Tear Agent Bromobenzylcyanide (CA).* Aberdeen Proving Ground, MD: United States Army Center for Health Promotion and Preventive Medicine, 1996.

———. *Detailed Facts About Tear Agent Chloracetophenone in Benzene and Carbon Tetrachloride (CNB).* Aberdeen Proving Ground, MD: United States Army Center for Health Promotion and Preventive Medicine, 1996.

———. *Detailed Facts About Tear Agent Chloroacetophenone and Chloropicrin in Chloroform (CNS).* Aberdeen Proving Ground, MD: United States Army Center for Health Promotion and Preventive Medicine, 1996.

———. *Detailed Facts About Tear Agent o-Chlorobenzylidene Malononitrile (CS).* Aberdeen Proving Ground, MD: United States Army Center for Health Promotion and Preventive Medicine, 1996.

World Health Organization. *Health Aspects of Chemical and Biological Weapons: Report of a WHO Group of Consultants.* Geneva: World Health Organization, 1970.

———. *International Chemical Safety Cards (ICSCs).* http://www.cdc.gov/niosh/ipcs/icstart.html. 2004.

———. *Public Health Response to Biological and Chemical Weapons: WHO Guidance.* Geneva: World Health Organization, 2004.

14

Vomiting/Sternatory Agents

14.1 General Information

The majority of these materials are halo or cyano organoarsines. Under normal battlefield conditions, they do not pose a serious danger to the life of an exposed individual and do not produce any permanent injury. Although the only agent in this class that is specifically banned under the Chemical Weapons Convention is lewisite 2 (C14-A004), which is listed in Schedule 1 because it can be readily converted into lewisite 1 (C04-A002), the use of all the other vomiting/sternatory agents is banned during a war under the general purpose criterion of the convention. They may still be used by law enforcement and the military during operations other than war such as when responding to incidents of civil unrest or to disband unruly crowds. However, because of their toxicity, this class of agents has been abandoned for other riot control agents (see Chapter 13).

Vomiting/sternatory agents are first generation chemical warfare agents employed in World War I in an attempt to defeat the existing mask filters. Toward the end of the war, adamsite (C14-A003) was developed, produced, and weaponized but never used. It was first used when the British dropped thermal generators containing a mixture of adamsite and diphenylchloroarsine (C14-A001) from aircraft during the Russian Civil War. This was the first reported deployment of air delivered chemical munitions.

These agents are moderately difficult to synthesize. They are relatively easy to disperse as thermally generated aerosols or as aerosolized solutions.

14.2 Toxicology

14.2.1 Effects

Vomiting/sternatory agents are primarily respiratory irritants. Effects from exposure are usually delayed for several minutes and include violent, uncontrolled sneezing and coughing, pain in the nose and throat, nausea, vomiting, chills, abdominal cramps, nasal discharge, and/or tears. Severe headaches and depression often follow exposure to vomiting/sternatory agents. Effects may persist for several hours postexposure. They may produce dermatitis on exposed skin. When released in an enclosed or confined space, they can cause serious illness or death. Most vomiting/sternatory agents contain arsenic as a constituent and decomposition products may pose a serious health hazard.

14.2.2 Pathways and Routes of Exposure

Vomiting/sternatory agents are primarily a hazard through inhalation. Aerosols are very irritating to the skin and eyes at low concentrations but relatively nontoxic via these routes. However, exposure to bulk liquid or solid agents may be hazardous through skin and eye exposure, ingestion, introduction through abraded skin (e.g., breaks in the skin or penetration of skin by debris). Ingestion of some decomposition products may pose a significant hazard.

14.2.3 General Exposure Hazards

Most bulk agents have very little odor in pure form although impurities may give some agents an odor of garlic or bitter almonds. Aerosols of vomiting/sternatory agents have excellent warning properties, producing eye, respiratory, and/or skin irritation at concentrations well below lethal levels.

Lethal concentrations (LC_{50}s) for inhalation of these agents are as low as 5000 mg/m^3 for a 2-minute exposure.

Incapacitating concentrations (ICt_{50}) due to sneezing and regurgitation for inhalation of these agents are as low as 6 mg/m^3 for a 2-minute exposure.

Eye irritation from exposure to agent vapors occurs at concentrations as low as 0.3 mg/m^3.

14.2.4 Latency Period

14.2.4.1 Aerosols

Depending on dose, the effects from exposure may be delayed from 30 seconds to several minutes, and may last up to several hours. Mild effects may persist for several days.

14.3 Characteristics

14.3.1 Physical Appearance/Odor

14.3.1.1 Laboratory Grade

Laboratory grade agents are typically colorless to yellow or green liquids or solids. Pure materials are typically odorless.

14.3.1.2 Munition Grade

Munition grade agents are typically colorless to yellow, green, or brown. As the agent ages, colors may become more pronounced. Production impurities and decomposition products in these agents may give them an odor. Odors for some agents have been described as similar to garlic or bitter almonds. Odors may become more pronounced during storage.

14.3.1.3 Modified Agents

Solvents have been added to these materials to facilitate handling or to increase the ease of percutaneous penetration by the agents. Color and other properties of these solutions may vary from the pure agent. Odors will vary depending on the characteristics of the solvent(s) used and concentration of agent in the solution.

Agents have also been micropulverized, encapsulated, or treated with flowing agents to facilitate their dispersal and increase their persistency. Color and other physical properties may be affected by these additives.

14.3.1.4 Mixtures with Other Agents

Vomiting/sternatory agents have been mixed with other agents such as phosgene (C11-A003), diphosgene (C10-A004), phenyl dichloroarsine (C04-A004), *N*-ethylcarbazole (C14-A022), arsenic trichloride (C04-C006), and triphenylarsine.

14.3.2 Stability

Most vomiting/sternatory agents are stable at typical temperatures. Apomorphine (C14-A005) is stable as a salt but is air and light sensitive. It is unstable in solution.

Agents can be stored in glass, steel, or Teflon-coated containers when pure. If moisture is present, they may rapidly corrode aluminum, iron, bronze, and brass.

14.3.3 Persistency

When vomiting/sternatory agents are employed as aerosols they are not classified as persistent by the military. Normally, there is minimal secondary risk once the initial aerosol has settled. However, depending on the size of the individual particles and on any encapsulation or coatings applied to the particles, they may be reaerosolized by ground traffic or strong winds. Bulk liquid or solid agents can persist in the environment for extended periods. Decomposition products from the breakdown of these agents can pose a persistent hazard.

14.3.4 Environmental Fate

Most vomiting/sternatory agents are nonvolatile and produce negligible amounts of vapor. They are deployed as dust aerosols. Once the aerosols settle, there is minimal extended hazard from the agents unless the dusts are resuspended. Decomposition products can be persistent hazards.

Most of these agents are insoluble in water. However, the solubility of any agent may be increased by solvents, components, or impurities. The specific gravities of unmodified agents are greater than that of water. With the exception of adamsite (C14-A003), vomiting/sternatory agents are soluble in most organic solvents. Adamsite has only limited solubility in common organic solvents except acetone.

14.4 Additional Hazards

14.4.1 Exposure

All foodstuffs in the area of a release should be considered contaminated. Unopened items packaged in glass, metal, or heavy duty plastic and exposed only to agent aerosols may be used after decontamination of the container. Unopened items exposed to solid or liquid agents, or solutions of agents, should be decontaminated within a few hours postexposure or destroyed. Opened or unpackaged items, or those packaged only in paper or cardboard, should be destroyed.

Plants, fruits, vegetables, and grains should be washed thoroughly with soap and water.

14.4.2 Livestock/Pets

Animals can be decontaminated with shampoo/soap and water. If the animals' eyes have been exposed to agent, they should be irrigated with water or saline solution for a minimum of 30 minutes.

The topmost layer of unprotected feedstock (e.g., hay or grain) should be destroyed. The remaining material should be quarantined until tested.

14.4.3 Fire

Although not normally volatile, heat from a fire will increase the amount of vomiting/sternatory agent vapor in the area. Actions taken to extinguish the fire can also spread the agent and steam generated in combating the fire could create agent aerosols. Combustion of vomiting/sternatory agents will produce volatile toxic metal (i.e., arsenic, antimony, lead) decomposition products. In addition, combustion of these agents may produce toxic and/or corrosive gases such as hydrogen chloride (HCl) and hydrogen cyanide (HCN).

14.4.4 Reactivity

Vomiting/sternatory agents decompose slowly in water. Some agents are self-protecting and form an oxide coating that delays further hydrolysis. Agents may be corrosive to some metals.

14.4.5 Hazardous Decomposition Products

14.4.5.1 Hydrolysis

Varies depending on the specific agent but may include HCl and HCN. Several produce diphenylarsenious oxide. Other organic oxides of arsenic, antimony, or lead may also be present.

14.4.5.2 Combustion

Volatile decomposition products may include HCl, HCN, nitrogen oxides (NO_x), benzene, and oxides of arsenic, antimony, or lead.

14.5 Protection

14.5.1 Evacuation Recommendations

Isolation and protective action distances listed below are taken from Argonne National Laboratory Report No. ANL/DIS-00-1, *Development of the Table of Initial Isolation and Protective Action Distances for the 2000 Emergency Response Guidebook*, which is still the basis for the "when used as a weapon" scenarios in the 2004 Emergency Response Guidebook (ERG). For vomiting/sternatory agents, these recommendations are based on a release scenario involving direct aerosolization of the solid agents with a particle size between 2 and 5 μm. Under these conditions, the difference between a small and a large release is not based on the standard 200 liters spill used for commercial hazardous materials listed in the ERG. A small release involves 10 kilograms of powdered agent (approximately 200 cubic inch) and a large release involves 500 kilograms of powdered agent (approximately 18 cubic feet).

	Initial isolation (feet)	Downwind day (miles)	Downwind night (miles)
DA (Diphenylchloroarsine) *C14-A001*			
Small device (10 kilograms)	200	0.2	0.7
Large device (500 kilograms)	600	1.4	3.2
DC (Diphenylcyanoarsine) *C14-A002*			
Small device (10 kilograms)	100	0.1	0.5
Large device (500 kilograms)	800	1.4	3.3
DM (Adamsite) *C14-A003*			
Small device (10 kilograms)	200	0.2	0.7
Large device (500 kilograms)	600	1.4	3.2

14.5.2 Personal Protective Requirements

14.5.2.1 Structural Firefighters' Gear

Structural firefighters' protective clothing is recommended for fire situations only; it is not effective in spill situations or release events. If chemical protective clothing is not available and it is necessary to rescue casualties from a contaminated area, then structural firefighters' gear will provide very limited skin protection against agent vapors and aerosols. Contact with solid and liquid agents should be avoided.

14.5.2.2 Respiratory Protection

Self-contained breathing apparatuses (SCBAs) or air purifying respirators (APRs) should have a National Institute for Occupational Safety and Health (NIOSH) and Chemical/Biological/Radiological/Nuclear (CBRN) certification since nerve agents can degrade the materials used to make some respirators. However, during emergency operations, other NIOSH approved SCBAs or APRs that have been specifically tested by the manufacturer against chemical warfare agents may be used if deemed necessary by the Incident Commander. APRs should be equipped with a NIOSH approved CBRN filter or a combination organic vapor/acid gas/particulate cartridge.

Immediately dangerous to life or health (IDLH) levels are the ceiling limit for respirators other than SCBAs. However, IDLH levels have not been established for vomiting/sternatory agents. Therefore, any potential exposure to elevated concentrations of these agents should be regarded with extreme caution and the use of SCBAs for respiratory protection should be considered.

14.5.2.3 Chemical Protective Clothing

Vomiting/sternatory agents are primarily an eye and respiratory hazard; however, at elevated aerosol concentrations or in contact with bulk material, agents may also pose a dermal hazard. Currently, there is no information on performance testing of chemical protective clothing against vomiting/sternatory agent.

14.5.3 Decontamination

14.5.3.1 General

Apply universal decontamination procedures using soap and water.

14.5.3.2 Solutions or Liquid Aerosols

Casualties/personnel: Remove all clothing immediately. To avoid further exposure of the head, neck, and face to the agent, cut off potentially contaminated clothing that must be pulled over the head. Use a sponge or cloth with liquid soap and copious amounts of water to wash the skin surface and hair at least three times. Do not use hot water as it may increase skin irritation. Avoid rough scrubbing as this could abrade the skin and increase discomfort. Rinse with copious amounts of water. For severe eye irritation, irrigate with water or 0.9% saline solution for a minimum of 15 minutes. Do not allow casualties to rub their eyes or skin as this may exacerbate agent effects.

Small areas: Small puddles of liquid can be contained by covering with absorbent material such as vermiculite, diatomaceous earth, clay, sponges, or towels. Place the absorbed material into containers lined with high-density polyethylene. Larger puddles can be collected using vacuum equipment made of materials inert to the released material and equipped with appropriate vapor filters. Wash the area with copious amounts of an alkaline soap/detergent and water. Collect and containerize the rinseate. Removal of porous material, including painted surfaces, may be required because these materials may be difficult to decontaminate. Arsenic or antimony metal and/or oxides, due to decomposition of the agents, may be present and require additional decontamination.

14.5.3.3 Solids or Particulate Aerosols

Casualties and personnel: Do not attempt to brush the agent off the individual or their clothing as this can aerosolize the agent. Remove all clothing immediately. To avoid further exposure of the head, neck, and face to the agent, cut off potentially contaminated clothing that must be pulled over the head. Use a sponge or cloth with liquid soap and copious amounts of water to wash the skin surface and hair at least three times. Do not use hot water as it may increase skin irritation. Avoid rough scrubbing as this could abrade the skin and increase discomfort. Rinse with copious amounts of water. For severe eye irritation, irrigate with water or 0.9% saline solution for a minimum of 15 minutes. Do not allow casualties to rub their eyes or skin as this may exacerbate agent effects.

Small areas: If indoors, close windows and doors in the area and turn off anything that could create air currents (e.g., fans, air conditioners, etc.). Avoid actions that could aerosolize the agent such as sweeping or brushing. Collect the agent using a vacuum cleaner equipped with a high-efficiency particulate air (HEPA) filter. Do not use a standard home or industrial vacuum. Do not allow the vacuum exhaust to stir the air in the affected area. Vacuum all surfaces with extreme care in a very slow and controlled manner to minimize aerosolizing the agent. Place the collected material into containers lined with high-density polyethylene. Wash the area with copious amounts of an alkaline soap/detergent and water. Collect and containerize the rinseate in containers lined with high-density polyethylene. Removal of porous material, including painted surfaces, may be required because these materials may be difficult to decontaminate. Arsenic or antimony metal and/or oxides, due to decomposition of the agents, may be present and require additional decontamination.

14.6 Medical

14.6.1 CDC Case Definition

The case can be confirmed if there is either a predominant amount of clinical and nonspecific laboratory evidence or an absolute certainty of the etiology of the agent is known.

14.6.2 Differential Diagnosis

The following factors have been suggested as alternatives to consider when presented with a potential case of exposure to irritating agents: anxiety, anaphylaxis, conjunctivitis, pneumonia, ultraviolet keratitis; inhalation of smoke, hydrocarbons, ammonia, hydrogen sulfide, phosgene, halogens such as chlorine, sulfuric acid, hydrogen chloride (HCl), or nickel carbonyl; sodium azide, street drugs; acute respiratory distress syndrome, chronic obstructive pulmonary disease and emphysema, congestive heart failure, pulmonary edema, and anthrax.

14.6.3 Signs and Symptoms

Progression of symptoms is generally irritation of the eyes and mucous membranes, viscous discharge from the nose similar to that caused by a cold, violent uncontrollable sneezing and coughing, severe headache, acute pain and difficulty breathing (tightness of the chest), nausea, and vomiting. Mental depression may occur. Severe effects last from 30 minutes to several hours. Minor effects may persist for over 24 hours.

14.6.4 Mass-Casualty Triage Recommendations

Typically not required. Casualties will usually recover unassisted after removal from the contaminated atmosphere. Consult the base station physician or regional poison control center for advice on specific situations.

14.6.5 Casualty Management

Decontaminate the casualty ensuring that all the agents have been removed. For severe eye irritation, irrigate with water or 0.9% saline solution for a minimum of 15 minutes. Do not allow casualties to rub their eyes or skin as this may exacerbate agent effects. Irrigate open wounds with water or 0.9% saline solution for at least 10 minutes.

Once the casualty has been decontaminated, including the removal of foreign matter from wounds, medical personnel do not need to wear a chemical-protective mask.

Casualties will usually recover unassisted from exposure to vomiting/sternatory agent although it may take several hours after removal from the contaminated atmosphere. Vigorous exercise may lessen and shorten symptoms. Most patients can be discharged safely. Rarely a patient with significant respiratory findings may merit admission.

14.7 Fatality Management

Remove all clothing and personal effects. Because of the potential for hazardous residual metal content (i.e., arsenic, lead, antimony), it may be appropriate to ship nondurable items to a hazardous waste disposal facility. Otherwise, decontaminate with soap and water.

Wash the remains with soap and water. Pay particular attention to areas where agent may get trapped, such as hair, scalp, pubic areas, fingernails, folds of skin, and wounds. If remains are heavily contaminated with residue, wash and rinse waste should be contained for proper disposal.

Once the remains have been thoroughly decontaminated, no further protective action is necessary. Body fluids removed during the embalming process do not pose any additional risks and should be contained and handled according to established procedures. Use standard burial procedures.

C14-A

AGENTS

C14-A001

Diphenylchloroarsine (Agent DA)
CAS: 712-48-1
RTECS: CG9900000
UN: 1769
ERG: 156

$C_{12}H_{10}AsCl$

Colorless odorless crystals. Weapons grade material is a dark brown, thick, viscous, semisolid resembling shoe polish. It has in the past been used industrially as a wood preservative, pesticide, and herbicide for cacti.

Also reported as a mixture with Adamsite (C14-A003); Phosgene (C10-A003); Diphosgene (C10-A004); Phenyldichloroarsine (C04-A005), *N*-Ethylcarbazole (C14-A022); Arsenic Trichloride (C04-C007); and Triphenylarsine.

Exposure Hazards

Conversion Factor: 1 ppm = 10.82 mg/m^3 at 77°F

$LCt_{50(Inh)}$: 15,000 mg-min/m^3. This value is from older sources (ca. 1942) and is not supported by modern data. No updated toxicity estimate has been proposed.

$ICt_{50(Inh)}$: 12 mg-min/m^3

Eye Irritation: 0.3 mg/m^3; exposure duration unspecified

The hydrolysis product, diphenylarsenious oxide, is very poisonous if ingested.

Properties:

MW: 264.5	*VP*: 0.0036 mmHg (113°F)	*FlP*: 662°F
D: 1.387 g/mL (122°F)	*VD*: 9.1	*LEL*: —
MP: 113°F	*Vlt*: 4.4 ppm (113°F)	*UEL*: —
MP: 100°F (weapons grade)	H_2O: Insoluble	*RP*: >2000
BP: 631°F (decomposes)	*Sol*: Acetone; Ethanol; Ether; Chloroform	*IP*: <9 eV
BP: 338°F (1 mmHg)		
Vsc: —		

Proposed AEGLs

AEGL-1: Not Developed		
AEGL-2: 1 h, 0.39 mg/m^3	4 h, 0.098 mg/m^3	8 h, 0.049 mg/m^3
AEGL-3: 1 h, 1.2 mg/m^3	4 h, 0.30 mg/m^3	8 h, 0.15 mg/m^3

C14-A002
Diphenylcyanoarsine (Agent DC)
CAS: 23525-22-6
RTECS: —

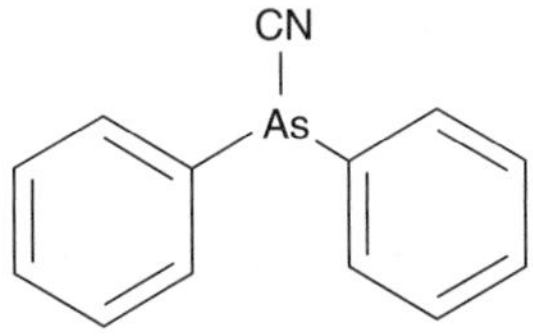

$C_{13}H_{10}AsN$

Colorless crystalline solid with an odor like garlic or bitter almonds detectable at 0.001–0.0005 ppm. Undergoes considerable decomposition when explosively disseminated.

Also reported as a mixture with Phenyldichloroarsine (C04-A005).

Exposure Hazards

Conversion Factor: 1 ppm = 10.44 mg/m^3 at 77°F
$LCt_{50(Inh)}$: 10,000 mg-min/m^3 (480 ppm for a 2-min exposure)
$IC_{50(Inh)}$: 60 mg/m^3 (5.7 ppm) for a 30-s exposure
$IC_{50(Inh)}$: 4 mg/m^3 (0.38 ppm) for a 5-min exposure

These values are from older sources (ca. 1942) and are not supported by modern data. No updated toxicity estimates have been proposed.

Properties:

MW: 255.2
D: 1.45 g/cm^3
D: 1.334 g/mL (95°F)
MP: 91°F
BP: 662°F (decomposes)
Vsc: —
VP: 0.0002 mmHg
VD: 8.8 (calculated)
Vlt: 0.27 ppm
H_2O: Insoluble
Sol: Most organic solvents
FlP: "Low"
LEL: —
UEL: —
RP: 31,000
IP: <9 eV

C14-A003

Adamsite (Agent DM)
CAS: 578-94-9
RTECS: SG0680000
UN: 1698
ERG: 154

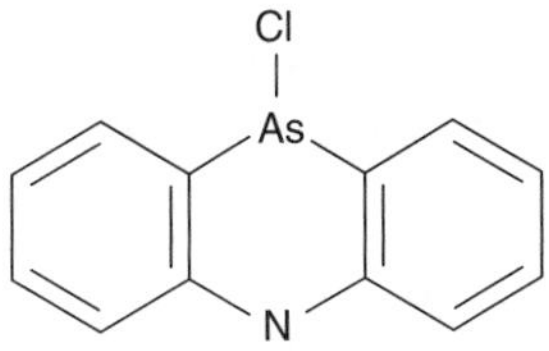

$C_{12}H_9AsClN$

Bright canary-yellow crystals that are odorless but produce irritation. Weapons grade material is a dark green to brown solid. When dispersed as a particulate cloud from a thermal munition, it has a characteristic "smoky" odor.

Also reported as a mixture with Diphenylchloroarsine (C14-A001).

Exposure Hazards

Conversion Factor: 1 ppm = 11.35 mg/m^3 at 77°F

$LCt_{50(Inh)}$: 11,000 mg-min/m^3

$ICt_{50(Inh)}$: 22–150 mg-min/m^3

Eye & Respiratory Irritation: 0.38 mg/m^3 for 1–2 min exposure

These are provisional updates from older values that have not been formally adopted as of 2005.

The hydrolysis product, diphenylaminoarsenious oxide, is very poisonous if ingested.

Properties:

MW: 277.6	*VP*: Negligible	*FlP*: None
D: 1.648 g/cm^3	*VD*: —	*LEL*: None
MP: 383°F	*Vlt*: —	*UEL*: None
BP: 770°F (decomposes)	H_2O: 0.0064%	*RP*: —
Vsc: —	*Sol*: Acetone	*IP*: —

Proposed AEGLs

AEGL-1: 1 h, 0.016 mg/m^3	4 h, 0.0022 mg/m^3	8 h, 0.00084 mg/m^3
AEGL-2: 1 h, 2.6 mg/m^3	4 h, 0.36 mg/m^3	8 h, 0.14 mg/m^3
AEGL-3: 1 h, 6.4 mg/m^3	4 h, 0.91 mg/m^3	8 h, 0.34 mg/m^3

C14-A004

Lewisite 2 (Agent L2)

CAS: 40334-69-8; 50361-06-3 (Isomer)

RTECS: —

Cl
As
Cl Cl

$C_4H_4AsCl_3$

Clear yellow to yellowish-brown liquid. In conjunction with Lewisite 3 (C04-C007), Lewisite 2 is a major synthetic by-product in the production of Lewisite (C04-A002). It is readily converted into Lewisite.

Exposure Hazards

Conversion Factor: 1 ppm = 9.55 mg/m^3 at 77°F

Human toxicity values have not been established or have not been published. Although L2 acts primarily as a vomiting/sternatory agent, it does have limited vesicant power similar to Lewisite (C04-A002).

Properties:

MW: 233.4	*VP*: —	*FlP*: —
D: 1.702 g/mL	*VD*: 8.1 (calculated)	*LEL*: —
MP: —	*Vlt*: —	*UEL*: —
BP: Decomposes	H_2O: Insoluble (slowly decomposes)	*RP*: —
BP: 235°F (11 mmHg)	*Sol*: Most organic solvents	*IP*: —
Vsc: —		

AEGLs have been proposed only for mixtures of this compound with Lewisite (C04-A002).

C14-A005

Apomorphine

CAS: 58-00-4; 314-19-2 (Hydrochloride salt); 41372-20-7 (Hydrochloride salt); 6191-56-6 (Diacetate)

RTECS: —

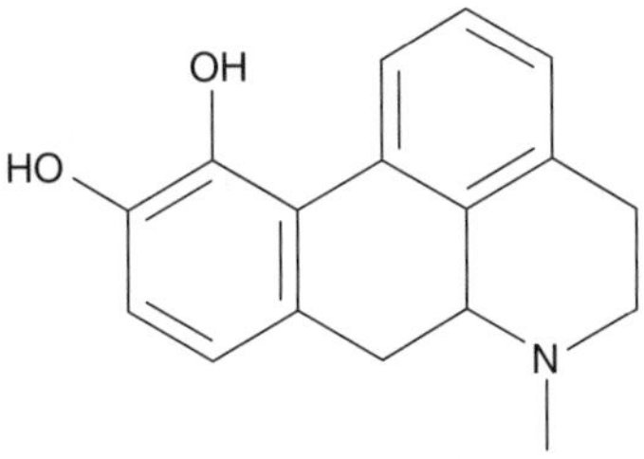

$C_{17}H_{17}NO_2$

White to grayish crystalline solid. It is air and light sensitive, and rapidly becomes green on standing. Solutions are unstable. Various salts have been reported.

Used as an emetic in veterinary medicine.

Exposure Hazards

Conversion Factor: 1 ppm = 10.93 mg/m^3 at 77°F

Human toxicity values have not been established or have not been published.

Properties:

MW: 267.3	*VP*: Negligible	*FlP*: —
D: —	*VD*: —	*LEL*: —
MP: 383°F (decomposes)	*Vlt*: —	*UEL*: —
BP: —	H_2O: 1.66% (61°F)	*RP*: —
Vsc: —	*Sol*: Alcohol; Acetone; Ether; Chloroform	*IP*: —

References

Brophy, Leo P., Wyndham D. Miles, and Rexmond C. Cohrane. *The Chemical Warfare Service: From Laboratory to Field*. Washington, DC: Government Printing Office, 1968.

Centers for Disease Control and Prevention. "Case Definition: Vesicant (Mustards, Dimethyl Sulfate, and Lewisite)." March 15, 2005.

Compton, James A.F. *Military Chemical and Biological Agents: Chemical and Toxicological Properties*. Caldwell, NJ: The Telford Press, 1987.

Fries, Amos A., and Clarence J. West. *Chemical Warfare*. New York: McGraw-Hill Book Company, Inc., 1921.

Green, Stanley Joseph, and Thomas Slater Price. "LVI. The Chlorovinylchloroarsines." *Journal of the Chemical Society* 119 (1921): 452.

Jackson, Kirby E. "Sternutators." *Chemical Reviews* 17 (1935): 251–92.

Jackson, Kirby E., and Margaret A. Jackson. "The Chlorovinylarsines." *Chemical Reviews* 16 (1935): 439–52.

Lewis, W. Lee, and G.A. Perkins. "The Beta-Chlorovinyl Chloroarsines." *Industrial & Engineering Chemistry* 15 (March 1923): 290–95.

Munro, Nancy B., Sylvia S. Talmage, Guy D. Griffin, Larry C. Waters, Annetta P. Watson, Joseph F. King, and Veronique Hauschild. "The Sources, Fate and Toxicity of Chemical Warfare Agent Degradation Products." *Environmental Health Perspectives* 107 (1999): 933–74.

National Institutes of Health. *Hazardous Substance Data Bank (HSDB).* http://toxnet.nlm.nih.gov/cgi-bin/sis/htmlgen?HSDB/. 2004.

Olson, Kent R., ed. *Poisoning & Drug Overdose.* 4th ed. New York: Lange Medical Books/McGraw-Hill, 2004.

Prentiss, Augustin M. *Chemicals in War: A Treatise on Chemical Warfare.* New York: McGraw-Hill Book Company, Inc., 1937.

Sartori, Mario F. *The War Gases: Chemistry and Analysis.* Translated by L.W. Marrison. London: J. & A. Churchill, Ltd, 1939.

Sidell, Frederick R. *Medical Management of Chemical Warfare Agent Casualties: A Handbook for Emergency Medical Services.* Bel Air, MD: HB Publishing, 1995.

Sidell, Fredrick R., Ernest T. Takafuji, and David R. Franz, eds. *Medical Aspects of Chemical and Biological Warfare, Textbook of Military Medicine Series, Part 1, Warfare, Weaponry, and the Casualty.* Washington, DC: Office of the Surgeon General, Department of the Army, 1997.

Sifton, David W., ed. *PDR Guide to Biological and Chemical Warfare Response.* Montvale, NJ: Thompson/Physicians Desk Reference, 2002.

Smith, Ann, Patricia Heckelman, and Maryadele J. Oneil, eds. *The Merck Index: An Encyclopedia of Chemicals, Drugs, & Biologicals.* 13th ed. Rahway, NJ: Merck & Co., Inc., 2001.

Somani, Satu M., ed. *Chemical Warfare Agents.* New York: Academic Press, 1992.

Somani, Satu M., and James A. Romano, Jr., eds. *Chemical Warfare Agents: Toxicity at Low Levels.* Boca Raton, FL: CRC Press, 2001.

Swearengen, Thomas F. *Tear Gas Munitions: An Analysis of Commercial Riot Gas Guns, Tear Gas Projectiles, Grenades, Small Arms Ammunition, and Related Tear Gas Devices.* Springfield, IL: Charles C Thomas Publisher, 1966.

True, Bey-Lorraine, and Robert H. Dreisbach. *Dreisbach's Handbook of Poisoning: Prevention, Diagnosis and Treatment.* 13th ed. London, England: The Parthenon Publishing Group, 2002.

United States Army Headquarters. *Chemical Agent Data Sheets Volume I, Edgewood Arsenal Special Report No. EO-SR-74001.* Washington, DC: Government Printing Office, December 1974.

———. *Potential Military Chemical/Biological Agents and Compounds, Field Manual No. 3-11.9.* Washington, DC: Government Printing Office, January 10, 2005.

United States Army Medical Research Institute of Chemical Defense. *Medical Management of Chemical Casualties Handbook.* 3rd ed. Aberdeen Proving Ground, MD: United States Army Medical Research Institute of Chemical Defense, July 2000.

United States Coast Guard. *Chemical Hazards Response Information System (CHRIS) Manual, 1999 Edition.* http://www.chrismanual.com/Default.htm. March 2004.

Wachtel, Curt. *Chemical Warfare.* Brooklyn, NY: Chemical Publishing Co., Inc., 1941.

Waitt, Alden H. *Gas Warfare: The Chemical Weapon, Its Use, and Protection Against It.* Rev. ed. New York: Duell, Sloan and Pearce, 1944.

Williams, Kenneth E. *Detailed Facts About Vomiting Agent Adamsite (DM).* Aberdeen Proving Ground, MD: United States Army Center for Health Promotion and Preventive Medicine, 1996.

World Health Organization. *International Chemical Safety Cards (ICSCs).* http://www.cdc.gov/niosh/ipcs/icstart.html. 2004.

———. *Health Aspects of Chemical and Biological Weapons: Report of a WHO Group of Consultants.* Geneva: World Health Organization, 1970.

———. *Public Health Response to Biological and Chemical Weapons: WHO Guidance.* Geneva: World Health Organization, 2004.

15

Malodorants

15.1 General Information

The majority of these materials are organic sulfur compounds that may also contain an odor intensifier. These chemicals are generally volatile liquids at room temperature with odors that are detectable at very low levels. Under normal battlefield conditions, these materials do not pose a serious danger to the life of an exposed individual and do not produce any permanent injury. Since approximately 0.2% of the population is unable to detect odors (anosmic), compositions may contain multiple malodorant components.

Obnoxious smelling agents have been used throughout history to provide camouflage and assist in breaching hardened defensive positions. Ancient Chinese armories included various "stink bombs." Research into development of new and more effective malodorants began after World War I. During World War II, an agent known as "Who Me?," which smelled of rotting food and carcasses, was developed by the Allies and tested by resistance fighters on German and Japanese soldiers in an effort to humiliate and embarrass them. However, the agent was volatile and difficult to deliver and did not always produce the expected response. As a result of this failure, malodorants were largely abandoned in the United States until the 1980s. Currently, malodorants are high on the list of nonlethal research priorities for use in riot control, to clear facilities, to deny an area, or as a taggant. Stench weapons in development include concentrates of natural odors such as those produced by skunks, rotting meat, excrement, and body odor. Since odors can provoke varying reactions in people based on their social and cultural conditioning, malodorants offer the possibility of ethnic or cultural targeting.

Natural malodorants derived from a biological entity are prohibited under the Biological Weapons Convention. Malodorants composed strictly of synthetic chemicals would be defined as riot control agents and the Chemical Weapons Convention bans the use of such agents during a war. However, they may still be used by the military during operations other than war such as when responding to incidents of civil unrest. The military has also used many of these materials as simulants for lethal chemical agents. Malodorants have also been developed for use by police to control rioters and disband unruly crowds.

15.2 Toxicology

15.2.1 Effects

Malodorants have strong, repulsive characteristics including nausea, gagging, and/or vomiting. Unpleasant odors impede cognitive performance, increase feelings of discomfort, and heighten a perception of illness. If an odor is perceived to be harmful or hazardous, they can stimulate a feeling of panic and fear, and cause exposed personnel to want to flee.

Reactions vary in different people based on the concentration of the odor and on social and cultural conditioning. Extended exposure may desensitize individuals. High concentrations may cause severe physiological trauma. In an enclosed or confined space, very high concentration can be lethal.

15.2.2 Pathways and Routes of Exposure

Malodorants are primarily an inhalation hazard. Aerosols and vapors are extremely foul smelling at low concentrations but are otherwise relatively nontoxic. However, exposure to bulk liquid or solid agents may be hazardous through skin absorption, ingestion, and introduction through abraded skin (e.g., breaks in the skin or penetration of skin by debris).

15.2.3 General Exposure Hazards

Malodorants do not seriously endanger life except at exposures greatly exceeding an effective dose, usually only achieved in a confined or enclosed space.

15.2.4 Latency Period

Exposure to malodorants produces immediate effects.

15.3 Characteristics

15.3.1 Physical Appearance/Odor

15.3.1.1 Laboratory Grade

Laboratory grade agents are typically colorless to yellow liquids or solids.

15.3.1.2 Munition Grade

Munition grade agents typically consist of at least one malodorant agent (10–90%) and an odor intensifier (0.5–5%) dissolved in a liquid carrier. Solvents include volatile hydrocarbons, plant/vegetable oils, and water. Solvents typically pose minimal toxic hazards themselves. Compositions are typically colorless to yellow liquids. As the agent ages and decomposes it may discolor and become brown.

15.3.1.3 Modified Agents

Agents have been microencapsulated to facilitate their dispersal and increase their persistency.

15.3.2 Stability

Malodorant compositions are stable during storage but some ingredients are light and air sensitive.

15.3.3 Persistency

As typically deployed, malodorants are classified as nonpersistent by the military. However, heavy aerosols or release of bulk material may cause contamination that will persist for days or weeks.

15.3.4 Environmental Fate

Malodorant vapors have a density greater than air and tend to collect in low places. The majority of these agents are insoluble in water and have a wide range of specific gravities that cause them to either float or sink in water. Further, solvents used in the formulation are often insoluble in water and can change the specific gravities of the active agents. Most of these agents are soluble in common organic solvents including oils, alcohols, and ketones.

Agents may be absorbed into porous material, including painted surfaces, and these materials may be difficult to decontaminate.

15.4 Additional Hazards

15.4.1 Exposure

All foodstuffs in the area of a release should be considered contaminated. Unopened items packaged in glass, metal, or heavy duty plastic and exposed only to agent vapors or aerosols may be used after decontamination of the container. Opened or unpackaged items, or those packaged only in paper or cardboard, should be destroyed.

Plants, fruits, vegetables, and grains should be washed thoroughly with soap and water and aerated to remove residual odor.

15.4.2 Livestock/Pets

Animals can be decontaminated with shampoo/soap and water or an aqueous solution of mild oxidants (see Section 15.5.3).

The topmost layer of unprotected feedstock (e.g., hay or grain) should be destroyed. The remaining material should be aerated to remove residual odor.

15.4.3 Fire

Heat from a fire will increase the amount of agent vapor in the area. A significant amount of the agent could be volatilized and escape into the surrounding environment before the agent is consumed by the fire. Actions taken to extinguish the fire can also spread the agent. Although many malodorants are only slightly soluble or insoluble in water, runoff from firefighting efforts will still pose a potential contact threat.

15.4.4 Reactivity

Malodorants generally do not react with water or are very slowly decomposed by water. Most of these agents are incompatible with strong oxidizers. Solvents used to disperse agents may also be incompatible with strong oxidizers.

15.4.5 Hazardous Decomposition Products

15.4.5.1 Hydrolysis

Malodorants are generally stable or very slowly decomposed by water.

15.4.5.2 Combustion

Volatile decomposition products may include sulfur oxides (SO_x), hydrogen sulfide (H_2S), and nitrogen oxides (NO_x).

15.5 Protection

15.5.1 Evacuation Recommendations

There are no published recommendations for isolation or protective action distances for malodorants released in mass casualty situations.

15.5.2 Personal Protective Requirements

15.5.2.1 Structural Firefighters' Gear

Structural firefighters' protective clothing is recommended for fire situations only; it is not effective in spill situations or release events. If chemical protective clothing is not available and it is necessary to rescue casualties from a contaminated area, then structural firefighters' gear will provide very limited skin protection against agent vapors and aerosols. Contact with solid and liquid agents should be avoided.

15.5.2.2 Respiratory Protection

Self-contained breathing apparatuses (SCBAs) or air purifying respirators (APRs) should have a National Institute for Occupational Safety and Health (NIOSH) and Chemical/Biological/Radiological/Nuclear (CBRN) certification. However, during emergency operations, other NIOSH approved SCBAs or APRs that have been specifically tested by the manufacturer against chemical warfare agents may be used if deemed necessary by the Incident Commander. APRs should be equipped with a NIOSH approved CBRN filter or a combination organic vapor/acid gas/particulate cartridge.

Immediately dangerous to life or health (IDLH) levels are the ceiling limit for respirators other than SCBAs. Any exposures approaching the IDLH level should be regarded with extreme caution and the use of SCBAs for respiratory protection should be considered.

15.5.2.3 Chemical Protective Clothing

Malodorants are primarily an inhalation hazard; however, at elevated vapor/aerosol concentrations or in contact with bulk material, agents may also pose a dermal hazard.

Use only chemical protective clothing that has undergone material and construction performance testing against the specific agent that has been released. Reported permeation rates may be affected by solvents, components, or impurities in munition grade agents.

15.5.3 Decontamination

15.5.3.1 General

Apply universal decontamination procedures using soap and water.

Some malodorants are insoluble in water and difficult to remove with soap and water. In such situations, an aqueous solution of mild oxidants may be effective in destroying the odorous ingredients of the agent. Published examples include

one quart of 3% household hydrogen peroxide with one-quarter cup of baking soda and one teaspoon of liquid dish soap, or

one cup of a sodium perborate bleach and three tablespoons of liquid dish soap in 1 gallon of water.

The mixture is applied with a sponge and allowed to remain in contact with the agent for several minutes. It is then rinsed off with water. Either of these decontamination mixtures could cause discoloration of hair or fabrics.

15.5.3.2 Vapors

Casualties/personnel: Aeration and ventilation. If decontamination is deemed necessary, remove all clothing as it may continue to emit "trapped" agent vapor after contact with the vapor cloud has ceased. Shower using copious amounts of soap and water. Ensure that the hair has been washed and rinsed to remove potentially trapped vapor. If necessary, an aqueous solution containing a mild oxidant can be used (see Section 15.5.3.1). Avoid any contact with sensitive areas such as the eyes. Rinse with copious amounts of water. If eye irritation occurs, irrigate with water or 0.9% saline solution for a minimum of 15 minutes.

Small areas: Ventilate to remove the vapors. If deemed necessary, wash the area with copious amounts of soap and water. Collect and place into containers lined with high-density polyethylene. If necessary, an aqueous solution containing a mild oxidant can be used (see Section 15.5.3.1). Removal of porous material, including painted surfaces, may be required because agents that have been absorbed into these materials may migrate back to the surface and pose a contact hazard.

15.5.3.3 Liquids, Solutions, or Liquid Aerosols

Casualties/personnel: Remove all clothing immediately. Even clothing that has not come into direct contact with the agent may contain "trapped" vapor. To avoid further exposure of the head, neck, and face to the agent, cut off potentially contaminated clothing that must be pulled over the head. Use a sponge or cloth with liquid soap and copious amounts of water to wash the skin surface and hair at least three times. Avoid rough scrubbing. Rinse with copious amounts of water. If necessary, an aqueous solution containing a mild oxidant can be used (see Section 15.5.3.1). Avoid any contact with sensitive areas such as the eyes. Rinse with copious amounts of water. If eye irritation occurs, irrigate with water or 0.9% saline solution for a minimum of 15 minutes.

Small areas: Small puddles of liquid can be contained by covering with absorbent material such as vermiculite, diatomaceous earth, clay, sponges, or towels. Place the absorbed material into containers lined with high-density polyethylene. Larger puddles can be collected using vacuum equipment made of materials inert to the released material and equipped with appropriate vapor filters. Wash the area with copious amounts of soap and water. If necessary, an aqueous solution containing a mild oxidant can be used (see Section 15.5.3.1). Collect and containerize the rinseate. Removal of porous material, including painted surfaces, may be required because these materials may be difficult to decontaminate. Ventilate the area to remove vapors.

15.6 Medical

15.6.1 CDC Case Definition

The Centers for Disease Control and Prevention (CDC) has not published a specific case definition for intoxication by malodorants.

15.6.2 Differential Diagnosis

Anyone exposed to malodorants will be immediately identifiable.

15.6.3 Signs and Symptoms

15.6.3.1 Vapors

Nausea, gagging, vomiting, dizziness, loss of coordination, disorientation, shortness of breath, and headache.

15.6.3.2 Liquids

Irritation and burning of the skin, eyes, mucous membrane, and respiratory system.

15.6.4 Mass-Casualty Triage Recommendations

Typically not required. Casualties will usually recover unassisted after removal from the contaminated atmosphere and decontamination. Consult the base station physician or regional poison control center for advice on specific situations.

15.6.5 Casualty Management

Decontaminate the casualty ensuring that all the agents have been removed. For severe eye irritation, irrigate with water or 0.9% saline solution for a minimum of 15 minutes. Irrigate open wounds with water or 0.9% saline solution for at least 10 minutes.

Once the casualty has been decontaminated, including the removal of foreign matter from wounds, medical personnel do not need to wear a chemical-protective mask.

Casualties will usually recover from exposure to malodorants shortly after removal from the contaminated atmosphere. Most patients can be discharged safely. Rarely a patient with significant respiratory findings may merit admission.

15.7 Fatality Management

Remove all clothing and personal effects and decontaminate with soap and water. Wash the remains with soap and water. If necessary, an aqueous solution containing a mild oxidant can be used (see Section 15.5.3.1). Pay particular attention to areas where agent may get trapped, such as hair, scalp, pubic areas, fingernails, folds of skin, and wounds. If remains are heavily contaminated with residue, wash and rinse waste should be contained for proper disposal.

Once the remains have been thoroughly decontaminated, no further protective action is necessary. Use standard burial procedures.

C15-A

AGENTS

C15-A001

Mercaptoethanol

CAS: 60-24-2
RTECS: KI5600000
UN: 2966
ERG: 153

HS OH

C_2H_6OS

Clear, colorless mobile liquid with a strong disagreeable odor (stench) detectable at 0.12–0.64 ppm.

Used in industry as a chemical intermediate for dyestuffs, pharmaceuticals, rubber chemicals, flotation agents, insecticides, and plasticizers; used as a water-soluble reducing agent and reagent in biochemical research.

Exposure Hazards

Conversion Factor: 1 ppm = 3.20 mg/m^3 at 77°F
AIHA WEEL: 0.2 ppm [Skin]

Properties:

MW: 78.1	*VP*: 1.756 mmHg (77°F)	*FlP*: 165°F
D: 1.1143 g/mL	*VD*: 2.7 (calculated)	*LEL*: 2.3%
MP: −148°F	*Vlt*: 2300 ppm	*UEL*: 18%
BP: Decomposes	H_2O: Miscible	*RP*: 6.5
BP: 131°F (13 mmHg)	*Sol*: Most organic solvents	*IP*: —
Vsc: 3.08 cS		

C15-A002

Butyl mercaptan

CAS: 109-79-5
RTECS: EK6300000
UN: 2347
ERG: 130

HS

$C_4H_{10}S$

Colorless mobile liquid with a strong, obnoxious odor like garlic, cabbage, or a skunk depending on the concentration.

This material has been used by the military as a simulant of a nonpersistent agent to evaluate chemical equipment. It is also used to activate detector kits and alarms. It has been used during field exercises to simulate the threat posed by toxic agent vapor and assist in training soldiers on the proper use of protective equipment.

Used industrially as a solvent and chemical intermediate for insecticides, acaricides, herbicides, and defoliants. It is used in agriculture as a deer repellant.

Exposure Hazards

Conversion Factor: 1 ppm = 3.69 mg/m^3 at 77°F
OSHA PEL: 10 ppm
ACGIH TLV: 0.5 ppm
NIOSH Ceiling: 0.5 ppm [15 min]
IDLH: 500 ppm

Properties:

MW: 90.2
D: 0.8368 g/mL (77°F)
MP: −176°F
BP: 209°F
Vsc: 0.667 cS
VP: 35 mmHg
VP: 45.5 mmHg (77°F)
VD: 3.1 (calculated)
Vlt: 60,000 ppm (77°F)
H_2O: 0.0595%
Sol: Alcohol; Ether
FlP: 35°F
LEL: —
UEL: —
RP: 0.23
IP: 9.15 eV

C15-A003

***s*-Butyl mercaptan**

CAS: 513-53-1
RTECS: —

HS

$C_4H_{10}S$

Colorless, mobile liquid with a heavy skunk-like odor.

Used industrially as a chemical intermediate for cadusafos and as an odorant for natural gas.

Exposure Hazards

Conversion Factor: 1 ppm = 3.69 mg/m^3 at 77°F

Human toxicity values have not been established or have not been published.s

Properties:

MW: 90.2
D: 0.8299 g/mL (63°F)
MP: −265°F
BP: 185°F
Vsc: —
VP: 80.7 mmHg (77°F)
VD: 3.1 (calculated)
Vlt: 110,000 ppm
H_2O: 0.132%
Sol: Alcohol; Ether; Benzene
FlP: −10°F
LEL: —
UEL: —
RP: 0.13
IP: 9.10 eV

C15-A004

Isobutyl Mercaptan

CAS: 513-44-0
RTECS: —

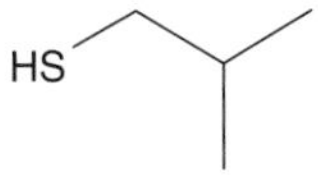

$C_4H_{10}S$

Specific information on physical appearance is not available for this agent.

Exposure Hazards

Conversion Factor: 1 ppm = 3.69 mg/m^3 at 77°F

Human toxicity values have not been established or have not been published.

Properties:

MW: 90.2	*VP*: 69.8 mmHg (77°F)	*FlP*: —
D: 1.4388 g/mL	*VD*: 3.1 (calculated)	*LEL*: —
MP: −229°F	*Vlt*: 92,000 ppm	*UEL*: —
BP: 191°F	H_2O: 0.17%	*RP*: 0.15
Vsc: —	*Sol*: —	*IP*: 9.12 eV

C15-A005

***t*-Butyl mercaptan**

CAS: 75-66-1

RTECS: TZ7660000

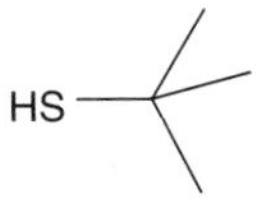

$C_4H_{10}S$

Colorless, mobile liquid with a strong, offensive, skunk-like odor.

Used industrially as an odorant for natural gas, chemical intermediate, and bacterial nutrient.

Exposure Hazards

Conversion Factor: 1 ppm = 3.69 mg/m^3 at 77°F

Human toxicity values have not been established or have not been published.

Properties:

MW: 90.2	*VP*: 181 mmHg (77°F)	*FlP*: —
D: 0.7943 g/mL (77°F)	*VD*: 3.1 (calculated)	*LEL*: —
MP: 31°F	*Vlt*: 240,000 ppm	*UEL*: —
BP: 147°F	H_2O: 0.2% (77°F)	*RP*: 0.059
Vsc: —	*Sol*: Alcohol; Acetone; Ether	*IP*: 9.03 eV

C15-A006

Mercaptoethyl sulfide

CAS: 3570-55-6

RTECS: —

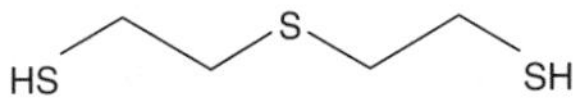

$C_4H_{10}S_3$

Colorless to yellow liquid with a strong, offensive, skunk-like odor.

Exposure Hazards

Conversion Factor: 1 ppm = 6.31 mg/m^3 at 77°F

Human toxicity values have not been established or have not been published.

Properties:

MW: 154.3	*VP*: —	*FlP*: 194°F
D: 1.183 g/mL (77°F)	*VD*: —	*LEL*: —
MP: —	*Vlt*: —	*UEL*: —
BP: 276°F (10 mmHg)	H_2O: —	*RP*: —
Vsc: —	*Sol*: —	*IP*: —

C15-A007

Butyl sulfide

CAS: 544-40-1

RTECS: ER6417000

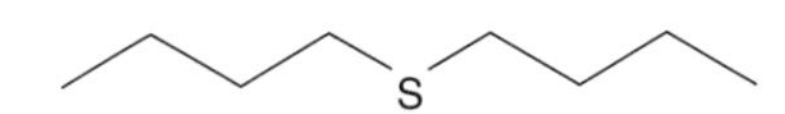

$C_8H_{18}S$

Clear, colorless to slightly yellow liquid with a skunk-like stench. This material is hazardous through inhalation and produces local skin/eye impacts.

Exposure Hazards

Conversion Factor: 1 ppm = 5.98 mg/m^3 at 77°F

Human toxicity values have not been established or have not been published.

Properties:

MW: 146.3	*VP*: 1.2 mmHg (77°F)	*FlP*: 169°F
D: 0.838 g/mL	*VD*: 5.0 (calculated)	*LEL*: —
MP: –112°F	*Vlt*: 1600 ppm (77°F)	*UEL*: —
BP: 372°F	H_2O: 0.0039% (77°F)	*RP*: 7.0
Vsc: —	*Sol*: —	*IP*: 8.40 eV

C15-A008

Amyl mercaptan

CAS: 110-66-7

RTECS: SA3150000

UN: 1111

ERG: 130

$C_5H_{12}S$

Colorless to yellow liquid with a strong, disagreeable odor like garlic detectable at 0.000094–0.070 ppm.

Used industrially as a chemical intermediate, synthetic flavoring ingredient, and odorant to locate gas leaks.

Exposure Hazards

Conversion Factor: 1 ppm = 4.26 mg/m^3 at 77°F

Human toxicity values have not been established or have not been published.

Properties:

MW: 104.2	*VP*: 13.8 mmHg (77°F)	*FlP*: 65°F
D: 0.8421 g/mL	*VD*: 3.6 (calculated)	*LEL*: —
MP: −104°F	*Vlt*: 18,000 ppm	*UEL*: —
BP: 260°F	H_2O: 0.0156%	*RP*: 0.72
Vsc: —	*Sol*: Alcohol; Ether	*IP*: —

C15-A009

***s*-Amyl mercaptan**
CAS: 2084-19-7
RTECS: —

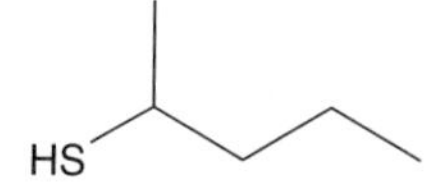

$C_5H_{12}S$

Colorless liquid with a disagreeable odor (stench).

Exposure Hazards
Conversion Factor: 1 ppm = 4.26 mg/m^3 at 77°F

Human toxicity values have not been established or have not been published.

Properties:

MW: 104.2	*VP*: —	*FlP*: —
D: 0.833 g/mL	*VD*: 3.6 (calculated)	*LEL*: —
MP: −170°F	*Vlt*: —	*UEL*: —
BP: 235°F	H_2O: "Slight"	*RP*: —
Vsc: —	*Sol*: Alcohol	*IP*: —

C15-A010

Isoamyl mercaptan
CAS: 541-31-1
RTECS: —

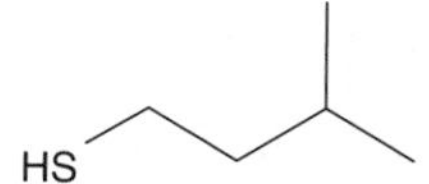

$C_5H_{12}S$

Clear, colorless to slightly yellow liquid with a strong disagreeable odor (stench).

Exposure Hazards
Conversion Factor: 1 ppm = 4.26 mg/m^3 at 77°F

Human toxicity values have not been established or have not been published.

Properties:

MW: 104.2	*VP*: 41.4 mmHg (100°F)	*FlP*: 64°F
D: 0.835 g/mL	*VD*: 3.6 (calculated)	*LEL*: —
MP: —	*Vlt*: 52,000 ppm	*UEL*: —
BP: 244°F	H_2O: —	*RP*: 2.5
Vsc: —	*Sol*: —	*IP*: —

C15-A011

t-Amyl mercaptan
CAS: 1679-09-0
RTECS: —

$C_5H_{12}S$

Specific information on physical appearance is not available for this agent.

Exposure Hazards
Conversion Factor: 1 ppm = 4.26 mg/m^3 at 77°F

Human toxicity values have not been established or have not been published.

Properties:

MW: 104.2	*VP*: —	*FlP*: 30°F
D: 0.812 g/mL	*VD*: 3.6 (calculated)	*LEL*: —
MP: −155°F	*Vlt*: —	*UEL*: —
BP: 210°F	H_2O: —	*RP*: —
Vsc: —	*Sol*: —	*IP*: —

C15-A012

Allyl sulfide
CAS: 592-88-1
RTECS: BC4900000

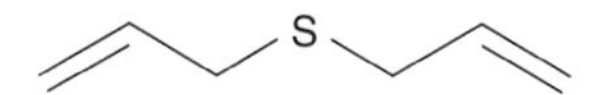

$C_6H_{10}S$

Colorless liquid with a strong odor like garlic or horseradish.

Used industrially for the manufacture of food flavors.

Exposure Hazards
Conversion Factor: 1 ppm = 4.67 mg/m^3 at 77°F

Human toxicity values have not been established or have not been published.

Properties:

MW: 114.2	*VP*: 9.22 mmHg (77°F)	*FlP*: 115°F
D: 0.888 g/mL (81°F)	*VD*: 3.9 (calculated)	*LEL*: —
MP: −117°F	*Vlt*: 12,000 ppm	*UEL*: —
BP: 282°F	H_2O: "Practically insoluble"	*RP*: 1.0
Vsc: —	*Sol*: Alcohol; Chloroform; Ether	*IP*: 8.52 eV

C15-A013

1-Hexanethiol
CAS: 111-31-9
RTECS: MO4550000

$C_6H_{14}S$

Clear liquid with a stench. This material is hazardous through inhalation, skin absorption, penetration through broken skin, and ingestion, and produces local skin/eye impacts. Reacts with air on long storage.

Exposure Hazards

Conversion Factor: 1 ppm = 4.83 mg/m^3 at 77°F
NIOSH Ceiling: 2.7 mg/m^3 [15 min]

Properties:

MW: 118.2	*VP*: —	*VP*: —
D: 0.8405 g/mL	*VD*: —	*VD*: —
MP: −113°F	*Vlt*: —	*Vlt*: —
BP: 304°F	*H_2O*: Insoluble	*H_2O*: Insoluble
BP: 178°F (10 mmHg)	*Sol*: —	*Sol*: —
Vsc: —		

C15-A014

1-Dodecanethiol

CAS: 112-55-0
RTECS: JR3155000

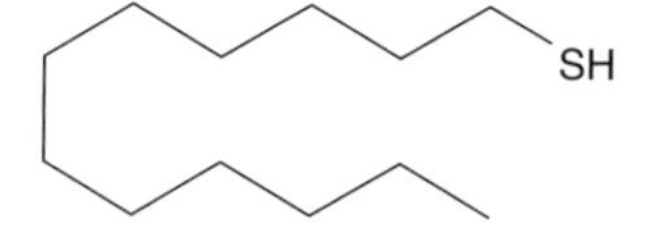

$C_{12}H_{26}S$

Colorless, water-white, or pale-yellow, oily liquid with a mild, skunk-like odor detectable at 0.5 ppm. This material is hazardous through inhalation and ingestion, and produces local skin/eye impacts. It causes irritation to the eyes, skin, respiratory system; cough; dizziness, dyspnea (breathing difficulty), lassitude (weakness, exhaustion), confusion, cyanosis; abdominal pain, nausea; skin sensitization.

Used industrially as a chemical intermediate in the manufacture of synthetic rubber, plastics, pharmaceuticals, insecticides, fungicides, nonionic detergents; and used in industry as a flotation reagent for the removal of metals from wastes.

This material has also been used as a simulant in government tests.

Exposure Hazards

Conversion Factor: 1 ppm = 8.28 mg/m^3 at 77°F
ACGIH TLV: 0.1 ppm
NIOSH Ceiling: 0.5 ppm [15 min]

Properties:

MW: 202.4	*VP*: 0.00853 mmHg	*FlP*: 262°F
D: 0.8435 g/mL	*VD*: 7.0 (calculated)	*LEL*: —
MP: 19°F	*Vlt*: 11 ppm	*UEL*: —
BP: 500–514°F	*H_2O*: 0.00002% (77°F)	*RP*: 820
BP: 297°F (15 mmHg)	*Sol*: —	*IP*: —
Vsc: —		

C15-A015

1-Octadecanethiol
CAS: 2885-00-9
RTECS: —

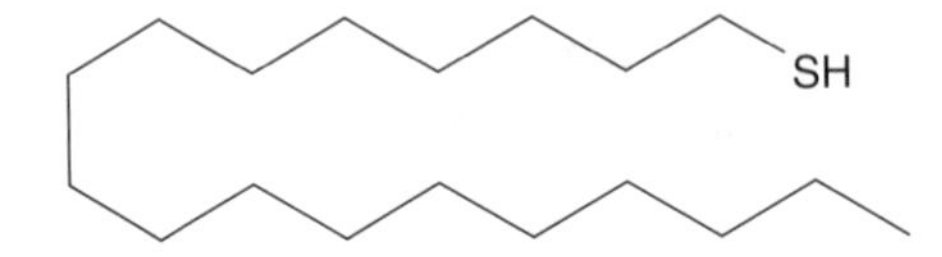

$C_{18}H_{38}S$

White solid with a foul stench. This material is hazardous through inhalation, skin absorption, penetration through broken skin, and ingestion, and produces local skin/eye impacts. It causes irritation of the eyes, skin, respiratory system; headache, dizziness, lassitude (weakness, exhaustion), cyanosis, nausea, convulsions.

This material has also been used as a simulant in government tests.

Exposure Hazards
Conversion Factor: 1 ppm = 11.72 mg/m^3 at 77°F
NIOSH Ceiling: 0.5 ppm [15 min]

Properties:

MW: 286.6	*VP*: <0.37 mmHg	*FlP*: 365°F
D: 0.85 g/cm^3	*VD*: 9.9 (calculated)	*LEL*: —
MP: 84°F	*Vlt*: —	*UEL*: —
BP: 405°F (1 mmHg)	*H_2O*: Insoluble	*RP*: >16
Vsc: —	*Sol*: —	*IP*: —

C15-A016

Cadaverine
CAS: 462-94-2; 1476-39-7 (Hydrochloride salt)
RTECS: SA0200000

H_2N NH_2

$C_5H_{14}N_2$

Clear, colorless to slightly yellow hygroscopic liquid with the odor of rotting flesh. Various salts have been reported.

Exposure Hazards
Conversion Factor: 1 ppm = mg/m^3 at 77°F

Human toxicity values have not been established or have not been published.

Properties:

MW: 102.2	*VP*: 0.666 mmHg	*FlP*: 144°F
D: 0.873 g/mL	*VD*: 3.5 (calculated)	*LEL*: —
MP: 48°F	*Vlt*: 890 ppm	*UEL*: —
BP: 352°F	*H_2O*: —	*RP*: 15
Vsc: —	*Sol*: —	*IP*: —

C15-A017

Putrescine
CAS: 110-60-1
RTECS: EJ6800000

$C_4H_{12}N_2$

White solid or liquid with a putrid odor like rotting flesh. Various salts have been reported.

Exposure Hazards
Conversion Factor: 1 ppm = 3.61 mg/m^3 at 77°F

Human toxicity values have not been established or have not been published.

Properties:

MW: 88.2	*VP*: 4.12 mmHg (77°F)	*FlP*: 145°F
D: 0.877 g/mL	*VD*: 3.0 (calculated)	*LEL*: —
MP: 81°F	*Vlt*: 5400 ppm	*UEL*: —
BP: 316°F	*H_2O*: 4%	*RP*: 2.6
Vsc: —	*Sol*: —	*IP*: —

C15-A018

Butyric acid
CAS: 107-92-6
RTECS: ES5425000
UN: 2820
ERG: 153

$C_4H_8O_2$

Colorless oily liquid with a pungent putrid odor like rancid butter or vomit detectable at 0.003–2.5 ppm. Various salts have been reported.

This material has been used by the military as a simulant of a nonpersistent agent to evaluate chemical equipment. It has been used during field exercises to simulate the threat posed by toxic agents and assist in training soldiers on the proper use of protective equipment.

A mixture of butyric acid (2%), magnesium silicate (5%), and water (93%) was once fielded as a simulant for Mustard gas (C03-A01), with the military designation AS.

Used industrially as a chemical intermediate for pharmaceuticals, emulsifiers, disinfectants, perfumes; artificial flavorings; in the manufacture of cellulose derivatives in lacquers and plastics; used as a leather tanning agent, as a sweetening agent for gasoline; in the food industry to add body to butter, cheese, butterscotch, caramel, fruit, and nut flavors; and in some countries it is used as an antifungal agent in the preservation of high moisture grains.

Exposure Hazards
Conversion Factor: 1 ppm = 3.60 mg/m^3 at 77°F

Human toxicity values have not been established or have not been published.

Properties:

MW: 88.1	*VP*: 0.43 mmHg	*FlP*: 161°F
D: 0.9577 g/mL	*VD*: 3.0 (calculated)	*LEL*: 2.0%
MP: 18°F	*Vlt*: 580 ppm	*UEL*: 10%
BP: 329°F	H_2O: Miscible	*RP*: 25
Vsc: 1.68 cS	*Sol*: Alcohol; Ether	*IP*: 10.17 eV

C15-A019

Isobutyric acid

CAS: 79-31-2
RTECS: NQ4375000
UN: 2529
ERG: 132

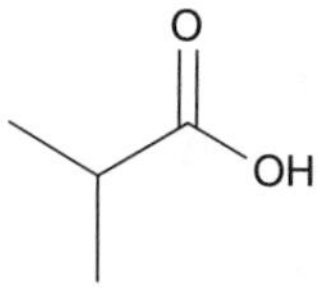

OH

$C_4H_8O_2$

Colorless liquid with a sharp odor like butter fat.

Used industrially as a chemical intermediate for the synthesis of esters, manufacture of varnish, as a tanning agent, and as a food additive (flavor).

Exposure Hazards

Conversion Factor: 1 ppm = 3.60 mg/m^3 at 77°F

Human toxicity values have not been established or have not been published.

Properties:

MW: 88.1	*VP*: 1.81 mmHg (77°F)	*FlP*: 131°F
D: 0.950 g/mL	*VD*: 3.0 (calculated)	*LEL*: 2.0%
MP: −53°F	*Vlt*: 2400 ppm	*UEL*: 9.2%
BP: 306°F (decomposes)	H_2O: 16.7%	*RP*: 5.9
Vsc: 1.516 cS	*Sol*: Alcohol; Chloroform; Ether	*IP*: 10.24 eV

C15-A020

2-Methylbutyric acid

CAS: 116-53-0
RTECS: —

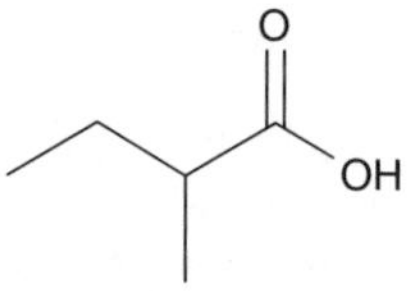

OH

$C_5H_{10}O_2$

Colorless to slightly yellow liquid with a strong disagreeable odor.

Exposure Hazards

Conversion Factor: 1 ppm = 4.18 mg/m^3 at 77°F

Human toxicity values have not been established or have not been published.

Properties:

MW: 102.1	*VP*: 0.491 mmHg (77°F)	*FlP*: 171°F
D: 0.9365 g/mL	*VD*: 3.5 (calculated)	*LEL*: 1.2%
MP: −94°F	*Vlt*: 650 ppm	*UEL*: 5.7%
BP: 352°F	H_2O: 4.5%	*RP*: 20
Vsc: 2.242 cS	*Sol*: —	*IP*: 10.27 eV

C15-A021

3-Methylindole
CAS: 83-34-1
RTECS: —

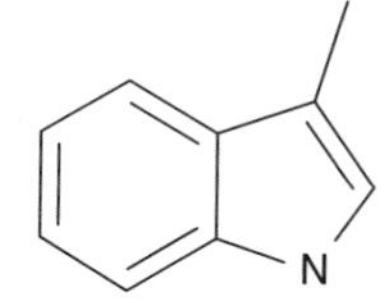

C_9H_9N

White crystalline solid that darkens and turns brown with exposure to air. At very low levels, it has a pleasant, sweet, warm odor similar to overripe fruit. Otherwise, a putrid odor like fecal matter.

This material acts as an odor intensifier.

Used industrially as a fixative for perfumes, artificial civet, food additive (flavoring), and medication.

Exposure Hazards

Conversion Factor: 1 ppm = 5.37 mg/m^3 at 77°F

Human toxicity values have not been established or have not been published.

Properties:

MW: 131.2	*VP*: 0.00555 mmHg	*FlP*: 270°F
D: —	*VD*: 4.5 (calculated)	*LEL*: —
MP: 203°F	*Vlt*: 7.3 ppm	*UEL*: —
BP: 509°F	H_2O: 0.05% (77°F)	*RP*: 1600
BP: 284°F (13 mmHg)	*Sol*: Acetone; Alcohol; Benzene; Ether	*IP*: —
Vsc: —		

C15-A022

4-Methyl indole
CAS: 16096-32-5
RTECS: —

N

C_9H_9N

Clear, yellow to brown, light sensitive liquid with a disagreeable odor.

This material acts as an odor intensifier.

Exposure Hazards

Conversion Factor: 1 ppm = 5.37 mg/m^3 at 77°F

Human toxicity values have not been established or have not been published.

Properties:

MW: 131.2	*VP*: —	*FlP*: >234°F
D: 1.062 g/mL	*VD*: 4.5 (calculated)	*LEL*: —
MP: 41°F	*Vlt*: —	*UEL*: —
BP: 513°F	*H_2O*: —	*RP*: —
BP: 248°F (20 mmHg)	*Sol*: —	*IP*: 7.6 eV
Vsc: —		

C15-A023

6-Methyl indole

CAS: 3420-02-8

RTECS: —

N

C_9H_9N

Light sensitive solid with a disagreeable odor.

This material acts as an odor intensifier.

Exposure Hazards

Conversion Factor: 1 ppm = 5.37 mg/m^3 at 77°F

Human toxicity values have not been established or have not been published.

Properties:

MW: 131.2	*VP*: —	*FlP*: >234°F
D: 1.059 g/cm^3	*VD*: 4.5 (calculated)	*LEL*: —
MP: —	*Vlt*: —	*UEL*: —
BP: 234°F (5 mmHg)	*H_2O*: —	*RP*: —
Vsc: —	*Sol*: —	*IP*: 7.54 eV

C15-A024

Standard Bathroom Malodor

CAS: —

RTECS: —

Mixture of 62.82% dipropylene glycol, 21.18% mercaptoacetic acid, 6% *n*-hexanoic acid, 6% *N*-methyl morpholine, 2.18% *p*-cresyl isovalerate, 0.91% 2-naphthalenethiol, and 0.91% skatole.

Exposure Hazards

Human toxicity values have not been established or have not been published.

References

Dalton, Pamela, and Gary K. Beauchamp. *Establishment of Odor Response Profiles: Ethnic, Racial and Cultural Influences.* Philadelphia, PA: Monell Chemical Senses Center, February 4, 1999.

National Institute for Occupational Safety and Health. *NIOSH Pocket Guide to Chemical Hazards.* Washington, DC: Government Printing Office, September 2005.

National Institutes of Health. *Hazardous Substance Data Bank (HSDB).* http://toxnet.nlm.nih.gov/cgi-bin/sis/htmlgen?HSDB/. 2004.

Pinney, Virginia Ruth. "Malodorant Compositions." US Patent 6,242,489, June 5, 2001.

———. "Non-lethal Weapon Systems." US Patent 6,386,113, May 14, 2002.

Prentice, John A. "Defensive and Offensive Projector Composition." US Patent 1,643,954, October 4, 1927.

Smith, Ann, Patricia Heckelman, and Maryadele J. Oneil, eds. *The Merck Index: An Encyclopedia of Chemicals, Drugs, & Biologicals.* 13th ed. Rahway, NJ: Merck & Co., Inc., 2001.

United States Air Force. *Development of Candidate Chemical Simulant List: The Evaluation of Candidate Chemical Simulants Which May Be Used in Chemically Hazardous Operations, Technical Report AFAMRL-TR-82-87.* Washington, DC: Government Printing Office, 1982.

United States Coast Guard. *Chemical Hazards Response Information System (CHRIS) Manual, 1999 Edition.* http://www.chrismanual.com/Default.htm. March 2004.

World Health Organization. *International Chemical Safety Cards (ICSCs).* http://www.cdc.gov/niosh/ipcs/icstart.html. 2004.

Yaws, Carl L. *Matheson Gas Data Book.* 7th ed. Parsippany, NJ: Matheson Tri-Gas, 2001.

Section V

Biological Agents

16

Toxins

16.1 General Information

Toxins are any poisonous substances that can be produced by an animal, plant, or microbe. Because of their complexity, most toxins are difficult to synthesize in large quantities by traditional chemical means. However, they may be harvested from cultured sources or produced by genetically engineered microbes. Toxins are odorless, tasteless, and nonvolatile. Ricin (C16-A036) and saxitoxin (C16-A018) are the only toxins listed in the Chemical Weapons Convention (Schedule 1).

Several countries have stockpiled a limited number of toxins. Their use on the battlefield has been alleged (e.g., Laos, Kampuchea, and Afghanistan) but not documented to the extent that it is universally accepted. Toxins have been used for political assassinations (e.g., 1978 murder of Georgi Markov with ricin) and terrorists have threatened the use of toxins, usually through contamination of food or water supplies.

16.2 Toxicology

16.2.1 Effects

Toxins present a variety of both incapacitating and lethal effect. Most toxins of military significance can be broadly classified in one of two ways. Neurotoxins disrupt the nervous system and interfere with nerve impulse transmission similar to nerve agents (Chapter 1). However, all neurotoxins do not operate through the same mechanism of action or do they produce the same symptoms. Cytotoxins are poisons that destroy cells or impair cellular activities. Symptoms may resemble those of vesicants (Chapter 3) or they may resemble food poisoning or other diseases. Toxins may also produce effects that are a combination of these general categories. The consequences of intoxication from any individual toxin can vary widely with route of exposure and dose. In addition, some toxins act as biomediators and cause the body to release excessive, and therefore harmful, amounts of chemicals that are normally produced by the body.

Although classified as biological weapons, toxins are chemicals. They are not alive and do not replicate themselves like pathogens (Chapters 17–20). They are not communicable; to be affected, an individual must come into direct contact with the toxin.

16.2.2 Pathways and Routes of Exposure

Toxins are primarily hazardous through inhalation, ingestion, and broken, abraded, or lacerated skin (e.g., penetration of skin by debris). A small number of toxins, such as the mycotoxins, produce skin lesions and systemic illness through skin and eye exposure. Toxins may also be dissolved in select solvents and delivered as dilute solutions that pose a percutaneous hazard.

16.2.3 General Exposure Hazards

In general, toxins do not have good warning properties. They are nonvolatile and do not have an odor. Although some toxins irritate the skin and eyes, in most cases they do not. Many neurotoxins will produce severe pain in contact with any abrasion or laceration.

Individual toxins may be effective through multiple pathways and the route of exposure may significantly change the signs and symptoms associated with any given toxin. In most cases, effects are most severe when the toxin is inhaled.

16.2.4 Latency Period

Effects from exposure to toxins can appear within minutes or be delayed for days. The impacts from some toxins, especially cytotoxins, may occur within minutes but symptoms may not appear for hours. The route of exposure to the toxin can significantly change the latency period.

16.3 Characteristics

16.3.1 Physical Appearance/Odor

16.3.1.1 Laboratory Grade

Pure toxins are typically colorless, white, tan, or yellow solids. Venoms—crude mixtures of toxins and other natural chemicals produced by animals such as snakes, spiders, and scorpions—are colorless to yellow or brown liquids.

16.3.1.2 Modified Agents

Toxins have been dissolved in solvents to facilitate handling, to stabilize them, or to create a percutaneous hazard. Percutaneous enhancement solvents include dimethyl sulfoxide, *N,N*-dimethylformamide, *N,N*-dimethylpalmitamide, *N,N*-dimethyldecanamide, and saponin. Color and other properties of these solutions may vary from the pure agent. Odors will vary depending on the characteristics of the solvent(s) used.

Toxins have also been micropulverized and microencapsulated to facilitate their dispersal and increase their persistency. Color and other physical properties of the toxin may be affected by these modifications.

16.3.2 Stability

Very much dependent on the specific toxin. Many are sensitive to heat or light or both. Freeze drying or isolation as salts increases their stability and shelf life.

16.3.3 Persistency

For military purposes, toxins are generally nonpersistent. In cases where toxins have been micropulverized, microencapsulated, or otherwise modified to facilitate their dispersal, reaerosolization by ground traffic or strong winds may be a concern.

16.3.4 Environmental Fate

All toxins are nonvolatile. Once the initial aerosol has settled, there is minimal inhalation hazard unless the toxin is released as an aerosolized powder that has been modified to increase the potential of reaerosolization. Solubility in water depends on the specific toxin, presence of solvents, and isolation as salts.

16.4 Additional Hazards

16.4.1 Exposure

All foodstuffs in the area of a release should be considered contaminated. Unopened items packaged in glass, metal, or heavy plastic and exposed only to aerosols may be used after decontamination of the container. All unopened items exposed to bulk agents should be decontaminated within a few hours postexposure or destroyed. Opened or unpackaged items, or those packaged only in paper or cardboard, should be destroyed.

Meat, milk, and animal products from animals affected or killed by toxins should be destroyed.

16.4.2 Livestock/Pets

Animals can be decontaminated with shampoo/soap and water, or a 0.5% household bleach solution (see Section 16.5.3). If the animals' eyes have been exposed to agent, they should be irrigated with water or saline solution for a minimum of 30 minutes.

Unprotected feedstock (e.g., hay or grain) should be destroyed. Depending on the specific toxin released, the level of contamination, and the weather conditions, leaves of forage vegetation could still retain sufficient agents to produce effects for several days post release.

16.4.3 Fire

Toxins are not volatile and the heat from a fire will destroy these agents. However, actions taken to extinguish the fire may spread the agent before it is destroyed. Runoff from firefighting efforts may pose a potential contact threat through exposure of broken, abraded, or lacerated skin, or through accidental ingestion. Smoke from a fire may contain acrid, irritating, and/or toxic decomposition products.

16.4.4 Reactivity

Varies depending on the specific toxin. Some decompose rapidly after they become wet. Many react with strong acids, bases, or oxidizing agents.

16.5 Protection

16.5.1 Evacuation Recommendations

There are no published recommendations for isolation or protective action distances for toxins released in mass casualty situations.

16.5.2 Personal Protective Requirements

16.5.2.1 Structural Firefighters' Gear

Structural firefighters' protective clothing is recommended for fire situations only; it is not effective in spill situations or release events. However, toxins have negligible vapor pressure and do not pose a vapor hazard. The primary risk of exposure is through contact with aerosolized agents, bulk agents (e.g., spilled liquids or solids), or solutions of agents. If chemical protective clothing is not available and it is necessary to rescue casualties from a contaminated area, then structural firefighters' gear will provide some skin protection against most toxin aerosols. Contact with bulk material and solutions should be avoided. However, any responder with areas of cut or lacerated skin should not make entry because they place the individual at extreme risk of subcutaneous exposure.

Structural firefighters' protective clothing should never be used as the primary chemical protective garment to enter an area contaminated with dermally hazardous toxins.

There is also a significant hazard posed by injection of toxins through contact with contaminated debris. Appropriate protection to avoid any potential laceration or puncture of the skin is essential.

16.5.2.2 Respiratory Protection

Self-contained breathing apparatuses (SCBAs) or air purifying respirators (APRs) should have a National Institute for Occupational Safety and Health (NIOSH) and Chemical/Biological/Radiological/Nuclear (CBRN). However, during emergency operations, other NIOSH approved SCBAs or APRs that have been specifically tested by the manufacturer against chemical warfare agents may be used if deemed necessary by the incident commander. APRs should be equipped with a NIOSH approved CBRN filter or a combination organic vapor/acid gas/particulate cartridge.

Immediately dangerous to life or health (IDLH) levels are the ceiling limit for respirators other than SCBAs. However, IDLH levels have not been established for toxins. Therefore, any potential exposure to aerosols of these agents should be regarded with extreme caution and the use of SCBAs for respiratory protection should be considered.

16.5.2.3 Chemical Protective Clothing

Currently, there is no information available on performance testing of chemical protective clothing against toxins.

In the event that dermally hazardous toxins have been released, responders should wear a Level A protective ensemble. Also, because of the extreme hazard posed by toxin aerosols to any area of cut or lacerated skin, responders should wear a Level A protective ensemble whenever there is any potential for exposure to airborne agent.

Since there is a significant hazard posed by injection of toxins through contact with debris, appropriate protection to avoid any potential abrasion, laceration, or puncture of the skin is essential.

16.5.3 Decontamination

16.5.3.1 General

Apply universal decontamination procedures using soap and water.

Most toxins are readily destroyed by high pH (i.e., basic solutions), especially when used in combination with a strong oxidizing agent. For this reason, undiluted household bleach is an excellent agent for decontamination of these agents. Ensure that the bleach solution remains in contact with the toxin for a minimum of 10 minutes.

16.5.3.2 Liquids, Solutions, or Liquid Aerosols

Casualties/personnel: Remove all clothing immediately. To avoid further exposure of the head, neck, and face to the agent, cut off potentially contaminated clothing that must be pulled over the head. Use a sponge or cloth with liquid soap and copious amounts of water to wash the skin surface and hair at least three times. Do not delay decontamination to find warm or hot water if it is not readily available. Avoid rough scrubbing as this could abrade the skin and increase the potential for movement of any residual toxin through the skin barrier. Rinse with copious amounts of water. If there is a potential that the eyes have been exposed to the agent, irrigate with water or 0.9% saline solution for a minimum of 15 minutes.

Alternatively, a household bleach solution can be used instead of soap and water. The bleach solution should be no more than one part household bleach in nine parts water (i.e., 0.5% sodium hypochlorite) to avoid damaging the skin. Avoid any contact with sensitive areas such as the eyes. Ensure that the bleach solution remains in contact with the agent for a minimum of 10 minutes. Rinse with copious amounts of water.

Small areas: Puddles of liquid can be absorbed by covering with absorbent material such as vermiculite, diatomaceous earth, clay, sponges, or towels. Place the absorbed material into containers lined with high-density polyethylene. Wash the area with copious amounts of soap and water or undiluted household bleach (see Section 16.5.3.1). If bleach is used, then rewash the area with soap and water. Collect and containerize the rinseate in containers lined with high-density polyethylene.

16.5.3.3 Solids or Particulate Aerosols

Casualties/personnel: Do not attempt to brush the agent off the individual or their clothing as this can aerosolize the agent. Remove all clothing immediately. To avoid further exposure of the head, neck, and face to the agent, cut off potentially contaminated clothing that must be pulled over the head. Wash the skin surface and hair at least three times with copious amounts of soap and water. Do not delay decontamination to find warm or hot water if it is not readily available. Rinse with copious amounts of water. If there is a potential that the eyes have been exposed to toxins, irrigate with water or 0.9% saline solution for a minimum of 15 minutes.

Alternatively, a household bleach solution can be used instead of soap and water. The bleach solution should be no more than one part household bleach in nine parts water (i.e., 0.5% sodium hypochlorite) to avoid damaging the skin. Avoid any contact with sensitive areas such as the eyes. Ensure that the bleach solution remains in contact with the agent for a minimum of 10 minutes. Rinse with copious amounts of water.

Small areas: Extreme care must be exercised when dealing with dry or powdered agents as toxins may adhere to the skin or clothing and then be spread to other areas. Because of the minute quantities needed to produce a response in an exposed individual, cross contamination can pose a significant inhalation or puncture hazard later.

If indoors, close windows and doors in the area and turn off anything that could create air currents (e.g., fans, air conditioner, etc.). Avoid actions that could aerosolize the agent such as sweeping or brushing. Collect the agent using a vacuum cleaner equipped with a high-efficiency particulate air (HEPA) filter. Do not use a standard home or industrial vacuum. Do not allow the vacuum exhaust to stir the air in the affected area. Vacuum all surfaces with extreme care in a very slow and controlled manner to minimize aerosolizing the agent. Place the collected material into containers lined with high-density polyethylene. Wash the area with copious amounts of soap and water or undiluted household bleach (see Section 16.5.3.1). If bleach is used, then rewash the area with soap and water. Collect and containerize the rinseate in containers lined with high-density polyethylene.

16.6 Medical

16.6.1 CDC Case Definition

A case in which the toxin or appropriate metabolite is detected in urine, serum, or plasma, or detection of the specific toxin in environmental samples unless there could be a local source of the toxin (e.g., the molds that produce mycotoxins have been found in some residential and industrial settings, and the toxins have been implicated in some cases of "sick building" syndrome).

A case should not be considered due to toxin poisoning if another confirmed diagnosis exists to explain the signs and symptoms. However, the case can be confirmed if either a predominant amount of clinical and nonspecific laboratory evidence is present or an absolute certainty of the etiology of the agent is known.

16.6.2 Differential Diagnosis

Patients should be viewed epidemiologically. Save clinical and environmental samples for diagnosis. The best early diagnostic sample for most toxins is a swab of the nasal mucosa.

16.6.3 Signs and Symptoms

Highly variable depending on the specific toxin, route of exposure, and dose. Even symptoms presented by toxins with the same general classification (i.e., neurotoxin or cytotoxin) may vary depending on the specific mechanism of action within the body.

16.6.4 Mass-Casualty Triage Recommendations

Because of the wide variety of potential toxins and their symptoms, there are no universal recommendations for triaging casualties exposed to toxins as a class. However, in general, anyone who has been exposed should be transported to a medical facility for evaluation. Individuals who are asymptomatic and have not been directly exposed to the agent can be discharged after their names, addresses, and telephone numbers have been recorded. They should be told to seek medical care immediately if symptoms develop.

16.6.5 Casualty Management

Decontaminate the casualty ensuring that all the toxins have been removed. Extreme care must be exercised when dealing with dry or powdered agents as toxins may adhere to the skin or clothing and present an inhalation hazard. If any agents have gotten into the eyes, irrigate the eyes with water or 0.9% saline solution for at least 15 minutes. Irrigate open wounds with water or 0.9% saline solution for at least 10 minutes.

Once the casualty has been decontaminated, including the removal of foreign matter from wounds, medical personnel do not need to wear a chemical-protective mask.

Treatment primarily consists of supportive care. Ventilate patient if they have difficulty breathing and administer oxygen. Be prepared to treat for shock. Monitor and support cardiac and respiratory functions as necessary. If the identity of the toxin is known, administer antidote if available. Unlike chemical agents, toxins can cause an immune response. Vaccines are available for some toxins but generally require more than 4 weeks for the body to produce antibodies. Passive immunotherapy is effective for some neurotoxins but must be instituted shortly after exposure. The utility of antibody therapy drops sharply at or shortly after the onset of the first signs of disease.

16.7 Fatality Management

Remove all clothing and personal effects segregating them as either durable or nondurable items. Although it may be possible to decontaminate durable items, it may be safer and more efficient to destroy nondurable items rather than attempt to decontaminate them. Items that will be retained for further processing should be double sealed in impermeable containers, ensuring that the inner container is decontaminated before placing it in the outer one.

Extreme care must be exercised when dealing with dry or powdered agents as toxins may adhere to the skin or clothing and present an inhalation hazard.

Wash the remains with a 2% sodium hypochlorite bleach solution (i.e., 2 gallons of water for every gallon of household bleach) ensuring that the solution is introduced into the ears, nostrils, mouth, and any wounds. This concentration of bleach will not affect remains but will destroy all toxins on the skin surface, greatly reducing the risk of secondary exposure. Higher concentrations of bleach can harm remains. Pay particular attention to areas where agent may get trapped, such as hair, scalp, pubic areas, fingernails, folds of skin, and wounds. The bleach solution should remain on the cadaver for a minimum of 10 minutes. Wash with soap and water. Ensure that the bleach solution is completely removed prior to embalming as it will react with embalming fluid. All wash and rinse waste must be contained for proper disposal.

Once the remains have been thoroughly decontaminated, no further protective action is necessary. Use standard burial procedures.

C16-A

NEUROTOXINS

C16-A001

Aconitine

CAS: 302-27-2

RTECS: AR5960000

$C_{34}H_{47}NO_{11}$

Molecular Weight: 645.7

Rapid-acting channel-activating neurotoxin. It is obtained from the leaves and roots of various plants including wolfbane (*Aconitum lycoctonum*) and monkshood (*Aconitum napellus*). It is an off-white powder that is insoluble in water (0.03%) but soluble in chloroform and benzene. Various salts have been reported.

This material is hazardous through inhalation, penetration through broken skin, and ingestion. The toxin is lipid soluble and once in the body can become stored in body fat. This can result in cumulative effect. Symptoms include nausea, vomiting, diarrhea, slow heart rate (bradycardia), restlessness, incoordination (ataxia), vertigo, difficulty breathing (dyspnea), low body temperature (hypothermia), convulsions, headache, and pallor. Ingestion causes a burning or tingling sensation with subsequent numbness on the lips, tongue, mouth, and throat. It produces similar effects on other mucous membranes. Death results from respiratory or cardiac failure.

Used in medical research to produce heart arrhythmia in experimental animals.

Toxicology

$LD_{(Ing)}$: 0.02–0.06 g

Other human toxicity values have not been established or have not been published. However, based on available information, this material appears to be approximately one-tenth as toxic as the nerve agent VX (C02-004).

C16-A002

Anatoxin A

CAS: 64285-06-9; 92142-32-0 (Stereoisomer); 85514-42-7 (Racemic mixture); 92844-80-9 (*p*-Toluenesulfonate salt); 92216-03-0 (Hydrochloride salt); 70470-07-4 (Hydrochloride salt); 122564-82-3 (Butenedioate salt)
RTECS: KM5528500

$C_{10}H_{15}NO$

Molecular Weight: 165.2

Very rapid-acting paralytic neurotoxin (postsynaptic depolarizing neuromuscular blocker) that binds to the same receptor as acetylcholine (nicotinic cholinergic receptor agonist) producing neural and muscular stimulation. However acetylcholinesterase does not hydrolyze the toxin. It is obtained from blue-green algae (*Anabaena* spp.). It is water soluble, but inactivated after several days as an aqueous solution. It is "probably tolerant" to chlorine at purification concentrations. It is destroyed by heat, light, and high pH. Various salts (solids) have been reported.

This material is hazardous through inhalation, penetration through broken skin, and ingestion. Symptoms mimic nerve agent exposure and include twitching, incoordination, tremors, paralysis, and respiratory arrest. Symptoms typically begin within 5 min. Death from ingestion can be delayed up to 3 h, depending on the dose. There is no specific treatment.

Toxicology

Human toxicity values have not been established or have not been published. However, based on available information, this material appears to be approximately one-fifth as toxic as the nerve agent VX (C02-004).

C16-A003

Anatoxin-A(S)
CAS: 103170-78-1
RTECS: —

$C_7H_{17}N_4O_4P$

Molecular Weight: 252.2

Neurotoxin obtained from freshwater blue-green cyanobacteria (*Anabaena* spp., *Aphanizomenon* spp., and *Oscillatoria* spp.). It is the only known organophosphate toxin produced by cyanobacteria.

This material is hazardous through inhalation, penetration through broken skin, and ingestion. It is an irreversible inhibitor of acetylcholinesterase and produces symptoms similar to an organophosphate pesticide or military nerve agent, including difficulty breathing with a feeling of shortness of breath or tightness of the chest, sweating, nausea, vomiting, involuntary urination/defecation, and a feeling of weakness.

Toxicology
Human toxicity values have not been established or have not been published. However, based on available information, this material appears to be approximately twice as toxic as the nerve agent VX (C02-004).

C16-A004

Batrachotoxin
CAS: 23509-16-2
RTECS: CR3990000

$C_{31}H_{42}N_2O_6$

Molecular Weight: 538.7

Very rapid-acting paralytic neurotoxin that binds to sodium channels of nerve and muscle cells depolarizing neurons by increasing the sodium channel permeability. It is obtained from South American poison-dart frogs (*Phyllobates aurotaenia, Phyllobates terribilis*). It is insoluble in water but soluble in hydrocarbons and other nonpolar solvents. The dried toxin can remain active for at least a year. However, it is relatively nonpersistent in the environment.

This material is hazardous through inhalation and penetration through broken skin. Once in the body, the toxin can become stored in body fat and may have a cumulative effect. Symptoms include loss of balance and coordination, profound weakness, irregular heart rhythms, convulsions, and cyanosis. These symptoms occur quickly and are produced in rapid succession. The toxin is lipid soluble and once in the body can become store in body fat. This can result in cumulative effect. Batrachotoxin has no effect on the skin but produces a long-lasting painful stinging sensation in contact with the smallest scratch. If ingested, it is only toxic if a lesion exists in the gastrointestinal tract.

Toxicology
Human toxicity values have not been established or have not been published. However, based on available information, this material appears to be approximately 1000 times more toxic than the nerve agent VX (C02-004).

C16-A005

Botulinum toxins (Agent X)
CAS: 93384-43-1 (A); 93384-44-2 (B); 93384-46-4 (D); 93384-47-5 (E); 107231-15-2 (F)
RTECS: ED9300000

Large protein

Molecular Weight: 150,000

Delayed-action paralytic neurotoxins that block the release of acetylcholine causing a symmetric, descending flaccid paralysis of motor and autonomic nerves. Paralysis always begins with the cranial nerves. Toxins are obtained from an anaerobic bacteria (*Clostridium botulinum*). Toxin A is a white powder or crystalline solid that is readily soluble in water. It is stable for up to 7 days as an aqueous solution. All toxins are destroyed by heat and decompose when exposed to air for more than 12 h.

These materials are hazardous through inhalation, penetration through broken skin, and ingestion. Botulinum toxins are unusual in that they are more toxic when ingested than when inhaled. They are most toxic when injected. Symptoms include dizziness, difficulty swallowing and speaking, blurred or double vision, sensitivity to light, and muscular weakness progressing from the head downward. In some cases may produce nausea and profuse vomiting. Symptoms from *ingestion* usually begin within 12–72 h, but can be delayed for up to 8 days. Symptoms from *inhalation* have a more rapid onset, usually 3–6 h. Onset of symptoms from wound botulism is usually 3 days or longer. Death results from respiratory failure. Recovery is prolonged and can take months. Those who survive may have fatigue and shortness of breath for years.

Used medicinally as a muscle relaxant.

They are on the HHS Select Agents list, the USDA High Consequence list, the Australia Group Core list, and are listed as a Category A Potential Terrorism Agent by the CDC.

Toxicology
$LD_{50(Inh)}$: 0.0002 mg (estimate)
$LD_{50(Ing)}$: 0.00007 mg (estimate)

C16-A006

Brevetoxins
CAS: 98112-41-5 (A); 79580-28-2 (B); 85079-48-7 (C)
ICD-10: T61.2
RTECS: XW5886000; XW5885000

Toxin A: $C_{49}H_{70}O_{13}$ (MW 867.1)
Toxin C: $C_{49}H_{72}O_{14}$ (MW 885.1)

Paralytic neurotoxins that bind to sodium channels of nerve and muscle cells causing muscle contractions. They are obtained from the dinoflagellate that causes "red-tide" (*Gymnodinium breve*). Toxins are typically light tan crystalline solids. They are insoluble in water and very unstable.

These materials are hazardous through inhalation and ingestion, and cause eye irritation. Toxins are lipid soluble and once in the body can become stored in body fat. This can result in a cumulative effect. Symptoms from ingestion include burning, prickling, or tingling sensations in the mouth as well as reversal of temperature sensations progressing to paralysis of the lips and extremities, with dizziness, incoordination (ataxia), muscle aches, nausea, and diarrhea. Inhalation causes bronchospasm and difficulty breathing (dyspnea). Symptoms tend to resolve themselves quickly and completely. There have been no deaths reported from these toxins.

Toxicology

CDC Case Definition: *Ingestion*: a combination of gastrointestinal (e.g., abdominal pain, vomiting, diarrhea) and neurologic signs (e.g., paresthesias, reversal of hot and cold temperature sensation, vertigo, ataxia). Symptoms have a latency period ranging from 15 min to 18 h. *Inhalational*: cough, dyspnea, and bronchospasm. Laboratory criteria for diagnosis is (1) detection by an enzyme-linked immunosorbent assay (ELISA) method in biologic samples (this is not a certified method for detection of brevetoxins) or (2) any concentration of brevetoxin in environmental samples.

The case can be confirmed if laboratory testing is not performed because either a predominant amount of clinical and nonspecific laboratory evidence of a nerve agent is present or an absolute certainty of the etiology of the agent is known.

Human toxicity values have not been established or have not been published.

C16-A007

Bungarotoxins

CAS: 37209-28-2; 12687-39-7 (α); 12778-32-4 (β)
RTECS: —

Protein

Mixture of neurotoxins that block the acetylcholine receptors. The β-bungarotoxin is a presynaptic neural toxin, α-bungarotoxin is a postsynaptic neural toxin, and κ-bungarotoxin is specific to the neuronal receptors in ganglions. They are obtained from the venom of the banded krait (*Bungarus multicinctus*).

This material is hazardous through inhalation and penetration through broken skin. Symptoms include paralysis with rapid progression to respiration failure and death.

Used as a research tool in neurophysiology.

Toxicology

$LD_{50(Sub)}$: 0.0076 g

C16-A008

Ciguatoxin

CAS: 11050-21-8 (CTX-1); 142185-85-1 (CTX-2); 139341-09-6 (CTX-3); 524945-49-1 (CTX-4)
ICD-10: T61.0
RTECS: —

$C_{60}H_{86}O_{19}$

Molecular Weight: 1111.3

Complex polyether paralytic neurotoxins that bind to sodium channels of nerve and muscle cells causing muscle contractions. They are heat stable (not destroyed by cooking or freezing) white solids obtained from dinoflagellates (*Gambierdiscus toxicus*).

These materials are hazardous through ingestion. Symptoms from ingestion include burning, prickling, or tingling sensations in the mouth as well as reversal of temperature sensations, diarrhea, vomiting, and pain in the abdomen and lower extremities. There may also be paralysis of the lips and extremities. Severe cases may progress to low blood pressure (hypotension) with paradoxical slow heart rate (bradycardia), and further to coma and respiratory arrest. Recovery may be prolonged.

Toxicology

Human toxicity values have not been established or have not been published.

C16-A009

Cobrotoxin

CAS: 12584-83-7
RTECS: —

Protein

Molecular Weight: 11,000

Neurotoxin obtained from the Formosan cobra (*Naja naja atra*). It is a relatively heat stable, water soluble, crystalline solid.

This material is hazardous by penetration through broken skin; can cause localized pain, tearing, and blurred vision in contact with the eyes. Symptoms include drooping upper eyelid (ptosis), paralysis of the motor nerves of the eye (ophthalmoplegia), difficulty in swallowing (dysphagia), difficulty speaking (dysphasia), profuse salivation, nausea, vomiting, abdominal pain, muscular weakness followed by flaccid paralysis, chest pain or tightness, shortness of breath, and respiratory failure.

Toxicology

$LD_{50(Sub)}$: 0.0029 g

C16-A010

Conotoxins

CAS: 115797-06-3
RTECS: —

Proteins or small peptides

Neurotoxins obtained from mollusks (sea snails) of the genus *Conus*. α-Conotoxins bind to and inhibit the acetylcholine receptor; δ-conotoxins and μ-conotoxins block voltage-gated sodium channels; κ-conotoxins block voltage-gated potassium channels in muscles; and ω-conotoxins block voltage-gated calcium channels and also block conduction at the neuromuscular junctions of skeletal muscles. Venoms are white, gray, yellow, or black viscous liquids. Toxins are water soluble solids that are highly stable.

These materials are hazardous through inhalation, penetration through broken skin, and ingestion. Symptoms include vague feeling of bodily discomfort (malaise), impaired coordination, weakness, paralysis, numbness, nausea, difficulty in swallowing (dysphagia), vomiting, inability to speak (aphonia), absence of reflexes (areflexia), temporary cessation of breathing (apnea), severe itching (pruritus), and double vision (diplopia).

Used as a research tool in molecular biology.

They are on the HHS Select Agents list and the Australia Group Core list.

Toxicology

LD_{50}: 0.00021–0.00042 g (estimate, pathway unspecified)

C16-A011

Diamphotoxin

CAS: 87915-42-2
RTECS: —

Protein

Molecular Weight: 60,000

Neurotoxin obtained from the larva of leaf-cutting beetles (*Diamphidia nigro-ornata pupae*) that is used by bushmen of Southern Africa as an arrow poison. It blocks neuromuscular function and also attacks the heart muscles (cardiotoxic) and is destructive to red blood cells (hemolytic). Dried poison is stable for over a year.

This material is hazardous by penetration through broken skin.

Toxicology

Human toxicity values have not been established or have not been published.

C16-A012

Gonyautoxins

CAS: 60748-39-2 (GTX1); 60748-40-5 (GTX2); 60748-41-6 (GTX2′); 60748-42-7 (GTX3); 64296-26-0 (GTX4); 69866-34-8 (GTX7); 122139-78-0 (GTX-B1); 82810-44-4 (GTX-B2); 80248-94-8 (PX2); 89614-45-9 (PX3); 89674-98-6 (PX4)

RTECS: —

Natural sulfate homologues of Saxitoxin (C16-A016).

Rapid-acting paralytic neurotoxins that blocks transient sodium channels and inhibits depolarization of nerve cells. They are some of the causative agents of paralytic shellfish poisoning (PSP). They are obtained from dinoflagellates (*Gonyaulax* spp., *Alexandrium* spp.) and cyanobacteria (*Anabaena circinalis*).

These materials are hazardous through inhalation, penetration through broken skin, and ingestion. Symptoms include tingling, burning, numbness, drowsiness, incoherent speech, respiratory paralysis, and possibly cardiovascular collapse.

Toxicology

Human toxicity values have not been established or have not been published.

C16-A013

α-Latrotoxin

CAS: 65988-34-3

RTECS: OE9020000

Protein

Molecular Weight: approximately 125,000

Neurotoxin that produces a massive release of transmitters from cholinergic and adrenergic nerve endings resulting in continuous stimulation of muscles. It also induces formation of an ion channel allowing the inward flow of calcium ions into the nerve cell. It is a white powder obtained from the venom of the black widow spider.

This material is hazardous by penetration through broken skin. Symptoms include sweating, salivation, vomiting, contractions, cramps, respiration is shallow and rapid, hypertension, rapid heart rate (tachycardia), and cardiac arrhythmia. May be mistaken for heart attack. Produces intense pain that becomes excruciating over time. Skin may be hyper pain sensitive.

Used as a research tool in molecular biology.

Toxicology

Human toxicity values have not been established or have not been published.

C16-A014

Neosaxitoxin

CAS: 64296-20-4

RTECS: —

$C_{10}H_{17}N_7O_5$

Molecular Weight: 315.3

Natural analogue of Saxitoxin (C16-A016).

Rapid-acting paralytic neurotoxin that reversibly blocks the voltage-gated sodium channels at neuronal level thus stopping the propagation of the nerve impulse. One of the causative agents of PSP. It is a white solid obtained from dinoflagellates (*Gonyaulax tamarensis* and *Gonyaulax catanella*) and cyanobacteria (*Aphanizomenon flos-aquae, Anabaena circinalis, Lyngbya majuscula,* and *Oscillatoria mougeotti*).

This material is hazardous through inhalation, penetration through broken skin, and ingestion. Symptoms include tingling, burning, numbness, drowsiness, incoherent speech, respiratory paralysis, and possibly cardiovascular collapse.

It is being tested for medicinal use as a local anesthetic.

Toxicology

Human toxicity values have not been published.

C16-A015

Palytoxin

CAS: 11077-03-5 (from *Palythoa caribaerum*); 77734-92-0 (from *Palythoa toxica*); 77734-91-9 (from *Palythoa toxica*)

RTECS: RT6475000

$C_{129}H_{223}N_3O_{54}$

Molecular Weight: 2680

Rapid-acting neurotoxin that causes irreversible depolarization of neural and muscular tissue by an unknown mechanism. It has a very potent effect on coronary arteries and death may result from constriction of the blood vessels of the heart. It is a solid obtained from bacterium associated with soft corals (*Palythoa caribaerum* and *Palythoa toxica*). It is soluble in water and alcohol, and stable to heat.

This material is hazardous through inhalation, skin absorption (high doses), penetration through broken skin, and ingestion. Delayed effects from exposure to high concentrations include disintegration of red blood cells. Symptoms include drowsiness, weakness, vomiting, respiratory distress, diarrhea, convulsions, shock, low body temperature, and death.

Toxicology

Human toxicity values have not been established or have not been published. However, based on available information, this material appears to be more than 8000 times as toxic as the nerve agent VX (C02-004).

C16-A016

Saxitoxin (Agent SS)
CAS: 35523-89-8; 63038-80-2 (Racemic mixture); 220355-66-8 (Diacetate salt); 35554-08-6 (Dihydrochloride salt); 63038-81-3 (Sulfate salt)
ICD-10: T61.2
RTECS: UY8708500

$C_{10}H_{17}N_7O_4$

Molecular Weight: 299.3

Rapid-acting channel-activating neurotoxin that blocks transient sodium channels and inhibits depolarization of nerve cells. It is one of the causative agents of Paralytic Shellfish Poisoning (PSP). It is obtained from dinoflagellates (*Gonyaulax tamarensis, Gonyaulax catanella*). It is soluble and stable in water and resistant to chlorine at 10 ppm. Iodine has no effect at 16 ppm. Various salts (solids) have been reported. The dihydrochloride salt is a white hygroscopic powder that is water soluble.

This material is hazardous through inhalation, penetration through broken skin, and ingestion. Symptoms from ingestion include a tingling or burning sensation of the lips, face, and tongue, progressing to the fingertips, arms, legs, and neck. Additional symptoms include incoordination, dizziness, headache, a floating sensation, muscle weakness, vertigo, vomiting, nausea, drooling, and abdominal pain. Symptoms occur in 10 min to 4 h depending on the route of exposure. Severe flaccid paralysis can lead to death from respiratory failure in 1–24 h. If the casualty survives 18 h, recovery is usually rapid and complete.

Used as a research tool in molecular biology.

It is on Schedule 1 of the CWC, the HHS Select Agents list, and the Australia Group Core list.

Toxicology

CDC Case Definition: (1) A case in which saxitoxin in urine is detected or (2) detection of saxitoxin in ingested compounds or seafood.

The case can be confirmed if laboratory testing is not performed because either a predominant amount of clinical and nonspecific laboratory evidence is present or an absolute certainty of the etiology of the agent is known.

$LC_{50(Inh)}$: 5 mg /m^3
$LD_{50(Inh)}$: 0.00049 g
$LD_{50(Sub)}$: 0.00040 g
$LD_{50(Ing)}$: 0.0003–0.001 g

C16-A017

Taipoxin
CAS: 52019-39-3
RTECS: —

Protein

Molecular Weight: 46,000

Neurotoxin obtained from the venom of the Australian taipan snake.

This material is hazardous by penetration through broken skin. Symptoms include dizziness, nausea, vomiting, sweating, headache, abdominal pain, and swollen lymph nodes that are painful when touched. Neurologic symptoms include drooping upper eyelid (ptosis), paralysis of the motor nerves of the eye (ophthalmoplegia), slurred speech, difficulty breathing (dyspnea), and sudden unconsciousness.

Toxicology

$LD_{50(Sub)}$: 0.0074 g

C16-A018

Tetanus toxin

CAS: 676570-37-9
RTECS: XW5807000

Protein

Molecular Weight: 150,000

Delayed-action neurotoxin that blocks the release of acetylcholine that is a crystalline solid obtained from bacteria (*Clostridium tetani*). Dried material is stable for years when stored between 39 and 45°F; otherwise, it is relatively unstable and very sensitive to heat.

This material is hazardous through inhalation, penetration through broken skin, and ingestion. Symptoms include muscle spasms (frequently of the jaw muscle) progressing to rigid paralysis. Generalized spasms can be induced by sensory stimulation. Even minor stimulation may trigger these spasms. Spasms may be so severe as to cause bone fractures. Spasms affecting the larynx, diaphragm, and intercostal muscles lead to respiratory failure. Involvement of the autonomic nervous system results in an irregular heartbeat (cardiac arrhythmias), rapid heart rate (tachycardia), and high blood pressure (hypertension).

Toxicology

CDC Case Definition: Acute onset of hypertonia and/or painful muscular contractions (usually of the muscles of the jaw and neck) and generalized muscle spasms without other apparent medical cause.

$LD_{50(Ing)}$: 0.0002 g (estimate)

Human toxicity values have not been fully established or have not been published. However, based on available information, this material appears to have approximately half as toxic as Botulinum toxins (C16-A005).

C16-A019

Tetrodotoxin

CAS: 4368-28-9; 39920-04-2 (Racemic mixture); 629653-73-2 (Acetate salt); 18660-81-6 (Citrate salt); 17522-62-2 (Formate salt); 4664-41-9 (Hydrobromide salt); 4664-40-8 (Picrate salt); 129497-92-3 (Trifluoroacetate salt)
ICD-10: T61.2
RTECS: IO1450000

$C_{11}H_{17}N_3O_8$

Molecular Weight: 319.3

Rapid-acting neurotoxin that inhibits sodium-ion channels in neural and muscular tissue. It does not affect the neuromuscular junction. It is colorless crystals or a white powder that is obtained from puffer fish (*Arothron* sp.), frogs, newts, dinoflagellates (*Takifugu poecilonotus*), and bacteria (*Pseudoalteromonas tetraodonis*). It is heat stable but darkens on heating above

428°F without decomposition. Slightly soluble and "probably stable" in water but rapidly inactivated by chlorine at 50 ppm. It is soluble in dilute acetic acid but practically insoluble in most other organic solvents. Various salts (solids) have been reported.

This material is hazardous through inhalation, penetration through broken skin, and ingestion. Symptoms appear 10 min to 4 h after ingestion and include nausea, vomiting, dizziness, paleness, and a vague feeling of bodily discomfort (malaise). A sensation of tingling or prickling that progresses to numbness may be present. General weakness, dilation of the pupils, twitching, tremors, and loss of coordination follow. Death is due to respiratory arrest. For survivors, recovery is usually complete in 24 h.

Used as a research tool in molecular biology.

It is on the HHS Select Agents list and the Australia Group Core list.

Toxicology

CDC Case Definition: Rapid onset of one of the following neurologic and gastrointestinal signs or symptoms after consumption of material potentially containing tetrodotoxin: (1) oral paresthesias (might progress to include the arms and legs), (2) cranial nerve dysfunction, (3) weakness (might progress to paralysis), or (4) nausea or vomiting. There are no biological markers for exposure to tetrodotoxin.

$LD_{50(Inh)}$: 0.0001–0.0002 g
$LD_{50(Ing)}$: 0.002 g

C16-A020

Tityustoxin

CAS: 39465-37-7
RTECS: XR3030000

Polypeptide

Rapid-acting neurotoxin that binds to sodium channels in nerve tissue leading to an increase in the release of neurotransmitters. It is a solid obtained from the venom of the Brazilian scorpion *Tityus serrulatus*.

This material is hazardous through inhalation and penetration through broken skin. Causes immediate pain in contact with any break in the skin. General symptoms include hyperexcitability, restlessness, salivation, lacrimation, accelerated respiration, convulsions, contractions, and muscular twitching, followed by spastic paralysis with rigid limbs. Death is due to respiratory failure and can occur within a few minutes or be delayed for several hours.

Used as a research tool in neurobiology.

Toxicology

Human toxicity values have not been established or have not been published. However, based on available information, this material appears to be slightly more toxic than the nerve agent VX (C02-004).

C16-A021

Veratridine

CAS: 71-62-5
RTECS: YX5600000

$C_{36}H_{51}NO_{11}$

Molecular Weight: 673.8

Neurotoxin that preferentially binds to activated sodium channels and increases the intracellular calcium concentration. It prolongs the action potential duration in the heart. It is obtained from sabadilla seeds (*Schoenocaulon officinale*). Yellowish-white amorphous powder that retains water and melts at 356°F. It is insoluble in water but slightly soluble in ether. Various salts (solids) have been reported. The nitrate is sparingly soluble in water.

This material is hazardous through inhalation, penetration through broken skin, and ingestion. The toxin is lipid soluble and once in the body can become stored in body fat. This can result in cumulative effect. Symptoms from ingestion include severe nausea, vomiting, burning, prickling, itching, or tingling skin sensation (paresthesia), weakness, low blood pressure (hypotension), slow heart rate (bradycardia), and loss of consciousness (syncope).

Used industrially as a pesticide and for medicinal purposes.

Toxicology

Human toxicity values have not been established or have not been published. However, based on available information, this material appears to be approximately 100 times less toxic than the nerve agent VX (C02-004).

C16-A

CYTOTOXINS

C16-A022

Abrin

CAS: 1393-62-0
RTECS: AA5250000

Protein

Molecular Weight: 63,000–67,000

Delayed-action cytotoxins that inhibits protein synthesis (ribosomal inactivating protein). They are obtained from the seed of the Jequirity beans plant (*Abrus precatorius*). Typically yellowish-white powders that are insoluble in distilled water but soluble in salt water. They are fairly heat stable.

These materials are hazardous through inhalation, penetration through broken skin, and ingestion, and produces local skin/eye impacts (redness and pain). Exposure may cause sensitizations. Symptoms from *inhalation* may occur within 8 h postexposure. Initial symptoms include difficulty breathing, fever, cough, nausea, and tightness in the chest progressing to heavy sweating, pulmonary edema, low blood pressure, and respiratory failure. Symptoms from *ingestion* may occur in less than 6 h but usually are delayed for 1–3 days. Initial symptoms include vomiting and diarrhea that may become bloody progressing to severe dehydration and low blood pressure. Other signs or symptoms may include hallucinations, seizures, and blood in the urine. Within several days, the liver, spleen, and kidneys stop working. Death from abrin poisoning could take place within 72 h postexposure, depending on the route of exposure and the dose received. If death has not occurred in 3–5 days, the casualty usually recovers.

They are on the HHS Select Agents list and the Australia Group Core list.

Toxicology

Human toxicity values have not been established or have not been published. However, based on available information, this material appears to be approximately twice as toxic as the nerve agent VX (C02-004) when injected into the body.

C16-A023

Aflatoxins

CAS: 1402-68-2; 27261-02-5 (B1); 7220-81-7 (B2); 31223-79-7 (B2a); 1385-95-1 (G1); 7241-98-7 (G2); 24147-91-9 (G2a); 6795-23-9 (M1); 6885-57-0 (M2); 32215-02-4 (P1); 52819-96-2 (Q1)

RTECS: —

Delayed action cytotoxins that inhibit the synthesis of nucleic acids. They are obtained from various molds/fungi (*Aspergillus flavus, Aspergillus parasiticus*). They are colorless to pale-yellow crystalline materials melting above 450°F. The "B" toxins fluoresce blue in the presence of UV light while the "G" toxins fluoresce green. They are only slightly soluble in water, but are soluble in methanol, acetone, and chloroform. Aqueous solutions are "probably stable" and "probably tolerant" to chlorine at purification concentrations.

These materials are hazardous through inhalation and ingestion. Symptoms include vomiting, abdominal pain, convulsions, pulmonary edema, coma, and death. All aflatoxins are potential carcinogens. Aflatoxin B1 is possibly the most potent natural liver carcinogen known. High-level exposure to aflatoxins produces acute localized death of liver tissue (necrosis) followed by the growth of scar tissue (cirrhosis), progressing to formation of a malignant tumor (carcinoma) in the liver.

They are on the Australia Group Core list.

Toxicology

Human toxicity values have not been established or have not been published.

C16-A024

Cardiotoxins

CAS: 11061-96-4

RTECS: —

Protein

Molecular Weight: 6700 (CTX III)

Cytotoxins obtained from cobra venom (*Naja naja atra*). Hydrophobic solids that cause irreversible depolarization of cell membrane and cellular destruction as well as contraction of skeletal and smooth muscle, including the heart.

This material is hazardous by penetration through broken skin. Symptoms include heart irregularities and low blood pressure.

Toxicology

$LD_{50(Sub)}$: 0.032 g

C16-A025

Cholera toxin

CAS: 9012-63-9

RTECS: —

Protein

Cytotoxin that inactivates G proteins involved in cellular metabolism. It is obtained from bacteria (*Vibrio cholerae,* C17-A025). It is a water soluble white powder that is stable for years when stored at 39°F.

This material is hazardous by penetration through broken skin and ingestion. Symptoms include vomiting, abdominal cramps, and watery diarrhea.

Used as a research tool in molecular biology and in the production of vaccines.

It is on the Australia Group Core list.

Toxicology

Human toxicity values have not been established or have not been published. However, based on available information, this material appears to be less than one-tenth as toxic as the nerve agent VX (C02-004).

C16-A026

***Clostridium perfringens* toxins**

CAS: —

RTECS: —

Cytotoxins obtained from bacteria (*Clostridium perfringens*).

This material is hazardous through inhalation, penetration through broken skin, and ingestion. Wound contamination results in death of skeletal muscles and soft tissue. Symptoms of ingestion include nausea and diarrhea usually without vomiting. Inhalation produces pulmonary complications.

It is on the HHS Select Agents list, the USDA High Consequence list, the Australia Group Core list, and is listed as a Category B Potential Terrorism Agent by the CDC.

Toxicology

Human toxicity values have not been established or have not been published.

C16-A027

Maitotoxin

CAS: 59392-53-9

RTECS: OM5470000

Polycyclic ether

Molecular Weight: 3422.0

Cytotoxin that affects the voltage-gated calcium channels causing an increased calcium influx into the cell, ultimately resulting in breakup of the cell's nuclear envelope (blebbing). It is obtained from dinoflagellates (*Gambierdiscus toxicus*). It is a water soluble, colorless, amorphous solid. It is the largest nonbiopolymeric and the most lethal nonpeptide natural product.

This material is hazardous through inhalation, penetration through broken skin, and ingestion. This toxin, in conjunction with the neurotoxin ciguatoxin (C16-A008), is usually associated with ciguatera fish poisoning (CFP). Although maitotoxin is approximately four times more toxic than ciguatoxin, the symptoms of CFP are dominated by the neurologic presentation from ciguatoxin. The specific symptoms associated with unique exposure to maitotoxin have not been established or have not been published.

Used as a research tool in neurophysiology.

Toxicology

Human toxicity values have not been established or have not been published. However, based on available information, this material appears to be approximately 100 times more toxic than the nerve agent VX (C02-004).

C16-A028

Microcystin LR

CAS: 128657-50-1
RTECS: GT2810000

$C_{49}H_{74}N_{10}O_{12}$

Molecular Weight: 995.2

Rapid-acting cytotoxin that disrupts cell membranes in the liver (hepatoxin) causing an accumulation of blood in the liver. It is the most toxic of the Microcystins. It is a solid obtained from freshwater blue-green cyanobacteria (*Microcystis aeruginosa, Microcystis cyanea*). It is heat stable and water soluble. Aqueous solutions are "probably stable" and resistant to chlorine at 100 ppm. It is also soluble in alcohol and acetone.

This material is hazardous through inhalation, penetration through broken skin, and ingestion. Symptoms include shivering, and rapid, deep breathing, progressing to twitching, convulsions, gasping respirations, and death. Shock and death occur within a matter of hours.

It is on the Australia Group Core list.

Toxicology

ID_{50}: 0.001–0.01 g (estimate, route unspecified)

Human toxicity values have not been established or have not been published. However, based on available information, this material appears to be approximately one-third as toxic as the nerve agent VX (C02-004) when inhaled, but to have a similar toxicity when injected.

C16-A029

Modeccin

CAS: 65988-88-7
RTECS: —

Protein

Molecular Weight: 57,000

Cytotoxin that inhibits protein synthesis (ribosomal inactivating protein). It is obtained from the roots of *Adenia digitata* plant.

This material is hazardous through inhalation, penetration through broken skin, and ingestion. Specific signs and symptoms of exposure to modeccin have not been established or have not been published. However, based on its similarity to other ribosomal inactivating proteins, such as abrin (C16-A022) and ricin (C16-A031), general symptoms may include fever, fatigue, weakness, and muscle pain. Symptoms from *inhalation* may include irritation and pain in the mucous membranes, cough, chest tightness, shortness of breath, low blood oxygen (hypoxemia), pulmonary edema, and multisystem organ dysfunction. Symptoms from *ingestion* may include abdominal pain, vomiting, diarrhea (may be bloody), dehydration, and multisystem organ dysfunction. Symptoms may persist for several days before death or recovery.

It is on the Australia Group Core list.

Toxicology
Human toxicity values have not been established or have not been published. However, based on available information, this material appears to be approximately 1000 times more toxic than the nerve agent VX (C02-004) when injected.

C16-A030

Pertussis toxin
CAS: 70323-44-3
RTECS: XW5883750

Protein

Cytotoxin that inactivates G proteins involved in cellular metabolism. It is a solid obtained from bacteria (*Bordetella pertussis*).

This material is hazardous through inhalation and penetration through broken skin. Specific signs and symptoms of exposure to this toxin have not been established or have not been published. However, the major symptoms associated with whooping cough, the disease caused by *Bordetella pertussis*, are related to the effects of this toxin.

Used as a research tool in molecular biology.

Toxicology
Human toxicity values have not been established or have not been published.

C16-A031

Ricin (Agent W)
CAS: 9009-86-3
RTECS: VJ2625000

Protein

Molecular Weight: 65,000

Delayed-action cytotoxin that inhibits protein synthesis (ribosomal inactivating protein) that is obtained from castor beans (*Ricinus communis*). Waste from production of castor oil contains about 5% ricin by weight. It is a white powder that is soluble in water and relatively heat stable. Aqueous solutions are resistant to chlorine at 10 ppm. It is persistent in the environment.

This material is hazardous through inhalation, penetration through broken skin, and ingestion. General symptoms may include fever, fatigue, weakness, muscle, and joint pain. Symptoms from *inhalation* include irritation and pain in the mucous membranes, cough, chest tightness, shortness of breath, low blood oxygen (hypoxemia), pulmonary edema, and multisystem organ dysfunction. Symptoms from *ingestion* may include abdominal pain, vomiting, diarrhea (may be bloody), dehydration, and multisystem organ dysfunction. Symptoms may persist for several days before death or recovery.

Used as a research tool in molecular biology, and has been used as a rodenticide.

It is on Schedule 1 of the CWC, the HHS Select Agents list, the Australia Group Core list, and is listed as a Category B Potential Terrorism Agent by the CDC.

Toxicology

CDC Case Definition: A clinically compatible case with (1) detection of urinary ricinine, an alkaloid in the castor bean plant; or (2) detection of ricin in environmental samples. The case can be confirmed if laboratory testing is not performed because either a predominant amount of clinical and nonspecific laboratory evidence is present or an absolute certainty of the etiology of the agent is known.

$LD_{50(Inh)}$: 0.001 g
$LD_{50(Ing)}$: 0.070 g

C16-A032

Shiga toxin
CAS: 75757-64-1
RTECS: —

Protein

Molecular Weight: 55,000

Ribosome inactivating cytotoxic protein that irreversibly inhibits protein synthesis in cells, causing cell death. It is a solid obtained from bacteria (*Shigella dysenteriae*).

This material is hazardous through inhalation and ingestion. Symptoms from *ingestions* include diarrhea (may be bloody), dysentery, and hemolytic-uremic syndrome (HUS). Symptoms from *inhalation* are not fully documented but may result in breathing difficulty due to fluid accumulation in the lungs.

It is on the HHS Select Agents list, the USDA High Consequence list, and the Australia Group Core list.

Toxicology

Human toxicity values have not been established or have not been published. However, based on available information, this material appears to be several magnitudes more toxic than the nerve agent VX (C02-004).

C16-A033

Staphylococcal enterotoxin B (Agent PG)
CAS: 11100-45-1
RTECS: XW5807700

Protein

Rapid-acting cytotoxin capable of producing either incapacitating or lethal effects. It is one of seven enterotoxins obtained from bacteria (*Staphylococcus aureus*). It is a white, fluffy solid that is water soluble and heat stable (not destroyed by cooking or freezing). Aqueous solutions are "probably stable." It resists chlorine used in municipal water systems. As a freeze-dried powder, it can be stored for more than a year.

This material is hazardous through inhalation and ingestion. Symptoms from *inhalation* include fever, headache, muscle pain (myalgia), red and inflamed whites of the eyes (conjunctivitis), nonproductive cough, difficulty in breathing (dyspnea), elevated white blood count (leukocytosis), nausea, anorexia, and vomiting but not diarrhea. May progress to chest pain and pulmonary edema. Symptoms from *ingestion* usually occur in 3–4 h postexposure and include the sudden onset of vomiting, abdominal cramps, nausea, explosive watery diarrhea, and severe weakness.

It is on the HHS Select Agents list, the USDA High Consequence list, the Australia Group Core list, and is listed as a Category B Potential Terrorism Agent by the CDC.

Toxicology

$LD_{50(Inh)}$: 0.0014 g
$ID_{50(Inh)}$: 0.000028 g

C16-A034

Verotoxins

CAS: 620190-09-2 (VT-1); 153834-62-9 (VT-2)
RTECS: —

Proteins

Ribosome activating cytotoxic proteins that irreversibly inhibit protein synthesis in cells, causing cell death. They are obtained from bacteria (*Escherichia coli* serotype O157:H7). Verotoxin 1 is almost identical to shiga toxin (C16-A032) and differs only by a single amino acid. Verotoxin 2 has significant differences.

These materials are hazardous through inhalation and ingestion. Symptoms from *ingestions* include diarrhea (may be bloody), dysentery, and hemolytic-uremic syndrome (HUS). Symptoms from *inhalation* are not fully documented but may result in difficulty breathing due to fluid accumulation in the lungs.

They are on the Australia Group Core list.

Toxicology

Human toxicity values have not been established or have not been published. However, based on the similarity to shiga toxin (C16-A032), these materials are likely to be several magnitudes more toxic than the nerve agent VX (C02-004).

C16-A035

Viscumin

CAS: 83590-17-4
RTECS: —

Protein

Molecular Weight: 115,000

Ribosome inactivating cytotoxic protein that irreversibly inhibits protein synthesis in cells, causing cell death. It is a solid obtained from the mistletoe plant (*Viscum album*).

This material is hazardous through inhalation, penetration through broken skin, and ingestion. Symptoms include headache, fever and chills, seizures, slow heart rate (bradycardia), abnormal blood pressure, vomiting, and death.

It is being tested for medicinal use as an antitumor agent.

It is on the Australia Group Core list.

Toxicology

Human toxicity values have not been established or have not been published. However, based on available information, this material appears to have a comparable toxicity to ricin (C16-A031).

C16-A036

Volkensin
CAS: 91933-11-8
RTECS: —

Protein

Molecular Weight: 62,000

Ribosome inactivating cytotoxic protein that irreversibly inhibits protein synthesis in cells, causing cell death. It is a solid obtained from the roots of the kilyambiti plant (*Adenia volkensii*).

This material is hazardous through inhalation, penetration through broken skin, and ingestion. May be a sensitizer. Specific signs and symptoms of exposure to modeccin have not been established or have not been published. However, based on its similarity to other ribosomal inactivating proteins, such as abrin (C16-A022) and ricin (C16-A031), general symptoms may include fever, fatigue, weakness, and muscle pain. Symptoms from *inhalation* may include irritation and pain in the mucous membranes, cough, chest tightness, shortness of breath, low blood oxygen (hypoxemia), pulmonary edema, and multisystem organ dysfunction. Symptoms from ingestion may include abdominal pain, vomiting, diarrhea (may be bloody), dehydration, and multisystem organ dysfunction. Symptoms may persist for several days before death or recovery.

Used as a research tool in neurology.

It is on the Australia Group Core list.

Toxicology
Human toxicity values have not been established or have not been published. However, based on available information, this material appears to be several magnitudes more toxic than the nerve agent VX (C02-004).

C16-A

DERMALLY HAZARDOUS CYTOTOXINS

C16-A037

T2 Mycotoxin
CAS: 21259-20-1
RTECS: YD0100000

$C_{24}H_{34}O_9$

Molecular Weight: 466.5

Rapid-acting dermally hazardous cytotoxin that inhibits protein synthesis and affects clotting factors in the blood. It is capable of producing incapacitating or lethal effects. T2 is obtained from various molds and fungi (*Fusarium* sp.). It is a colorless crystalline solid of white powder that melts at 304°F. Impure samples may be a colorless to slightly yellow oil. It is slightly soluble in water, but soluble in ethyl acetate, acetone, ethanol, chloroform, methylene chloride, diethyl ether, and dimethyl sulfoxide (DMSO). It is heat stable and can be stored at room temperature for years.

This material is hazardous through inhalation, skin absorption, penetration through broken skin, ingestion, and produces local skin/eye impacts. Trichothecenes are radiomimetic and may cause bone marrow suppression, liver failure, and internal bleeding.

They are also immunosuppressive. T2 is the most toxic member of the trichothecene mycotoxins. Symptoms include severe skin and eye irritation possibly progressing to blistering and scabbing, cough (if inhaled), incoordination (ataxia), low blood pressure (hypotension), excessive bleeding and a lack of clotting (coagulopathy), nausea, shortness of breath, dizziness, chest pains, vomiting and diarrhea (may be bloody), abdominal pain, headache, dizziness, fever, and lack of appetite. Death may be delayed for several weeks. Exposures are cumulative and repeated exposures may be cumulative to a lethal dose.

It is on the HHS Select Agents list, the USDA High Consequence list, and the Australia Group Core list.

Toxicology

CDC Case Definition: A clinically compatible case in which laboratory tests of environmental samples have confirmed exposure. The case can be confirmed if laboratory testing is not performed because either a predominant amount of clinical and nonspecific laboratory evidence is present or an absolute certainty of the etiology of the agent is known.

$LC_{50(Inh)}$: 200–5800 mg/m^3
$LD_{50(Per)}$: 0.14–0.84 g (In a penetrant solvent)
$ID_{50(Ing)}$: 0.78 mg/L (drinking water)
Nausea and Vomiting: 0.05 mg/L (drinking water)

C16-A038

Other trichothecene mycotoxins

Dermally hazardous cytotoxins obtained from various molds and fungi (*Fusarium* sp.). They are colorless, crystalline solids that are heat stable and can be stored for long periods.

These materials are hazardous through inhalation, skin absorption, penetration through broken skin, ingestion, and produces local skin/eye impacts. Trichothecenes are radiomimetic and may cause bone marrow suppression, liver failure, and internal bleeding. They are also immunosuppressive. Specific signs and symptoms of exposure to these toxins have not been established or have not been published. However, general symptoms may include severe skin and eye irritation possibly progressing to blistering, cough (if inhaled), incoordination (ataxia), low blood pressure (hypotension), excessive bleeding and a lack of clotting (coagulopathy), nausea, vomiting, diarrhea, abdominal pain, lack of appetite, headache, dizziness, and fever.

Toxicology

Human toxicity values have not been established or have not been published.

Deoxynivalenol ($C_{15}H_{20}O_6$)
CAS: 51481-10-8
RTECS: YD0167000
Molecular Weight: 296.3
Melting point: 304°F.
Soluble in water and ethanol.

Diacetoxyscirpenol ($C_{15}H_{20}O_6$). On the HHS Select Agents list and the Australia Group Core list.
CAS: 2270-40-8
RTECS: YD0112000
Molecular Weight: 296.3
Melting point: 322°F.
Soluble in chloroform.

Fusarenon X ($C_{17}H_{22}O_8$)
CAS: 23255-69-8
RTECS: —
Molecular Weight: 354.4
Melting point: 196°F
Soluble in water, alcohols, chloroform, and ethyl acetate. It hydrolyzes to form Nivalenol.

HT-2 Toxin ($C_{22}H_{32}O_8$). On the Australia Group Core list.
CAS: 26934-87-2
RTECS: YD0050000
Molecular Weight: 424.5
Melting point: —
Soluble in acetone, ethyl acetate, diethyl ether, chloroform, and methylene chloride.

Neosolaniol ($C_{19}H_{26}O_8$)
CAS: 36519-25-2
RTECS: YD0080000
Molecular Weight: 382.4
Melting point: 325°F
Soluble in acetone, ethyl acetate, diethyl ether, chloroform, and methylene chloride.

Nivalenol ($C_{15}H_{20}O_7$
CAS: 23282-20-4
RTECS: YD0165000
Molecular Weight: 312.3
Melting point: 432°F (decomposes)
Slightly soluble in water and soluble in methanol, ethanol, and acetonitrile.

Nivalenol Diacetate ($C_{19}H_{24}O_9$)
CAS: 14287-82-2
RTECS: —
Molecular Weight: 396.4
Melting point: —
Soluble in acetone, ethyl acetate, diethyl ether, chloroform, and methylene chloride. It hydrolyzes to form Nivalenol.

C16-A039

Macrocyclic trichothecene mycotoxins

Dermally hazardous cytotoxins obtained from various molds and fungi (*Stachybotrys atra, Myrothecium* sp.). They are colorless, crystalline solids that are heat stable and can be stored for long periods.

These materials are hazardous through inhalation, skin absorption, penetration through broken skin, ingestion, and produces local skin/eye impacts. Trichothecenes are radiomimetic and may cause bone marrow suppression, liver failure, and internal bleeding. They are also immunosuppressive. Specific signs and symptoms of exposure to these toxins have not been established or have not been published. However, based on its similarity to other trichothecene mycotoxins, general symptoms may include severe skin and eye irritation possibly progressing to blistering, cough (if inhaled), incoordination (ataxia), low blood pressure (hypotension), excessive bleeding and a lack of clotting (coagulopathy), sore throat, headache, fatigue, nausea, vomiting, diarrhea, abdominal pain, lack of appetite, headache, dizziness, and fever.

Toxicology

Human toxicity values have not been established or have not been published.

Roridin A ($C_{29}H_{40}O_9$)
CAS: 14729-29-4
RTECS: VL0355000
Molecular Weight: 532.6
Melting point: 388°F
Soluble in chloroform.

Satratoxin H ($C_{29}H_{36}O_9$)
CAS: 53126-64-0
RTECS: —
Molecular Weight: 528.6
Melting point: —

Verrucarin A ($C_{27}H_{34}O_9$)
CAS: 3148-09-2
RTECS: WH1314900
Molecular Weight: 502.6
Melting point: >680°F (decomposes)
Soluble in chloroform.

C16-A040

Zearalenone
CAS: 17924-92-4
RTECS: DM2550000

$C_{18}H_{22}O_5$

Molecular Weight: 318.4

Dermally hazardous cytotoxins obtained from various molds and fungi (*Fusarium* sp.). It is a white, crystalline solid that melts at 322°F. It is "practically insoluble" in water, but soluble in benzene, acetonitrile, methylene chloride, methanol, ethanol, and acetone. It is somewhat heat stable and can be stored for long periods.

This material is hazardous through inhalation, skin absorption, penetration through broken skin, ingestion, and produces local skin/eye impacts. It is known to mimic estrogen in the body (hyperestrogenism). In animals, it has caused feminization of male animals and interfered with conception, ovulation, and fetal development in female animals. Specific signs and symptoms of acute high-dose exposure to zearalenone have not been established or have not been published.

References

Amdur, Mary O., John Doull, M.D., and Curtis D. Klaassen, eds. *Casarett and Doull's Toxicology.* 4th ed. New York: Pergamon Press, Inc., 1991.

Arnon, Stephen S., Robert Schechter, Thomas V. Inglesby, Donald A. Henderson, John G. Bartlett, Michael S. Ascher, Edward Eitzen, Anne D. Fine, Jerome Hauer, Marcelle Layton, Scott Lillibridge, Michael T. Osterholm, Tara O'Toole, Gerald Parker, Trish M. Perl, Philip K. Russell,

David L. Swerdlow, and Kevin Tonat. "Botulinum Toxin as a Biological Weapon: Medical and Public Health Management." *Journal of the American Medical Association* 285 (2001): 1059–70.

Brooks, David W. *Teaching and Research Web Site (University of Nebraska).* http://www.cci.unl.edu/Teacher/NSF/C10/C10Links/chemistry.about.com/library/blpoison.htm. 2006.

Burrows, Dickinson W., and Sara E. Renner. "Biological Warfare Agents as Threats to Potable Water." *Environmental Health Perspectives* 107 (1999): 975–84.

Canada Minister of National Health and Welfare, Laboratory Centre for Disease Control, Office of Biosafety. *Material Safety Data Sheet-Infectious Substances: Clostridium botulinum.* January 23, 2001.

Centers for Disease Control and Prevention. "Biological and Chemical Terrorism: Strategic Plan for Preparedness and Response. Recommendations of the CDC Strategic Planning Workgroup." *Morbidity and Mortality Weekly Report* 49 (RR-4) (2000): 1–14.

———. "Case Definitions for Infectious Conditions Under Public Health Surveillance." *Morbidity and Mortality Weekly Report* 46 (RR-10) (1997): 39.

———. "Facts About Botulism." September 2001.

———. "Facts About Cyanobacteria & Cyanobacterial Harmful Algal Blooms." Undated.

———. "Facts About Ricin." October 23, 2003.

———. "Ricin FAQ." February 25, 2003.

Chandrasekhar, J. "Maitotoxin—Holder of Two World Records." *Resonance* (May 1996): 68–70.

Chin, James, ed. *Control of Communicable Diseases Manual.* 17th ed. Washington, DC: American Public Health Association, 2000.

Chorus, Ingrid, and Jamie Bartram, eds. *Toxic Cyanobacteria in Water: A Guide to Their Public Health Consequences, Monitoring and Management.* London, England: E & FN Spon, 1999.

Compton, James A.F. *Military Chemical and Biological Agents: Chemical and Toxicological Properties.* Caldwell, NJ: The Telford Press, 1987.

Department of Agriculture. 9 CFR Part 121—"Possession, Use, and Transfer of Biological Agents and Toxins." 2005.

Donohue-Rolfe, Arthur, David W.K. Acheson, Anne V. Kane, and Gerald T. Keusch. "Purification of Shiga Toxin and Shiga-Like Toxins I and II by Receptor Analog Affinity Chromatography with Immobilized P1 Glycoprotein and Production of Cross-Reactive Monoclonal Antibodies." *Infection and Immunity* 57 (December 1989): 3888–893.

Estacion, Mark, and William P. Schilling. "Maitotoxin-Induced Membrane Blebbing and Cell Death in Bovine Aortic Endothelial Cells." *BMC Physiology* 1 (2001): 2.

European Mycotoxin Awareness Network. *Mycotoxin Fact Sheets.* http://www.mycotoxins.org/. 2006.

Franz, David R. *Defense Against Toxin Weapons.* Rev. ed. Fort Detrick, MD: United States Army Medical Research and Materiel Command, 1997.

Friedman, Melissa A., and Bonnie E. Levin. "Neurobehavioral Effects of Harmful Algal Bloom (HAB) Toxins: A Critical Review." *Journal of the International Neuropsychological Society* 11 (2005): 331–38.

Fry, Bryan Grieg. Australian Venom Research Unit, Department of Pharmacology, School of Medicine, University of Melbourne. http://www.kingsnake.com/toxinology. 2006.

Gasperi-Campani, Anna, Luigi Barbieri, Enzo Lorenzoni, Lucio Montanaro, Simonetta Sperti, Eugenio Bonetti, and Fiorenzo Stirpe. "Modeccin, the Toxin of Adenia digitata: Purification, Toxicity and Inhibition of Protein Synthesis *In Vitro.*" *Biochemical Journal* 174 (1978): 491–96.

Harpe, Jon de la, E. Reich, Karl A. Reich, and Eugene B. Dowdle. "Diamphotoxin: The Arrow Poison of the !Kung Bushmen." *The Journal of Biological Chemistry* 258 (1983): 11924–931.

Henderson, Donald A., Thomas V. Inglesby, and Tara O'Toole, eds. *Bioterrorism: Guidelines for Medial and Public Health Management.* Chicago, IL: AMA Press, 2002.

Herenda, D. *Manual on Meat Inspection for Developing Countries.* Reprint 2000. Rome, Italy: Food and Agriculture Organization of the United Nations, 1994.

International Programme on Chemical Safety. *Environmental Health Criteria Monographs.* http://www.inchem.org/pages/ehc.html. 2006.

———. *Joint Expert Committee on Food Additives (JECFA)—Monographs and Evaluations.* http://www.inchem.org/pages/jecfa.html. 2006.

Ishiguro, Masatsune, Takao Takahashi, Gunki Funatsu, Katsuya Hayashi, and Masaru Funatsu. "Biochemical Studies on Ricin. I. Purification of Ricin." *Journal of Biochemistry* (Tokyo, Japan) (1964): 587–92.

Kabat, Elvin E., Michael Heidelberger, and Ada E. Bezer. "A Study of the Purification and Properties of Ricin." *Journal of Biological Chemistry* 168 (1947): 629–39.

Kahn, Cynthia M., and Susan E. Aiello, eds. *The Merck Veterinary Manual.* 8th ed. Online Version. Whitehouse Station, NJ: Merck & Co., Inc., 1998.

Katircioglu, Hikmet, Beril S. Akin, and Tahir Atici. "Microalgal Toxin(s): Characteristics and Importance." *African Journal of Biotechnology* 3 (2004): 667–74.

Kishi, Yoshito. "Complete Structure of Maitotoxin." *Pure & Applied Chemistry* 70 (1998): 339–44.

Kortepeter, Mark, George Christopher, Ted Cieslak, Randall Culpepper, Robert Darling, Julie Pavlin, John Rowe, Kelly McKee, Jr., and Edward Eitzen, Jr., eds. *Medical Management of Biological Casualties Handbook.* 4th ed. Fort Detrick, MD: United States Army Medical Research Institute of Infectious Diseases, February 2001.

Lindler, Luther E., Frank J. Lebeda, and George W. Korch, eds. *Biological Weapons Defense: Infectious Diseases and Counterbioterrorism.* Totowa, NJ: Humana Press, Inc., 2005.

List Biological Laboratories, Inc. *Material Safety Data Sheets (MSDSs) for Bacterial Toxins.* http://www.listlabs.com. 2006.

Mankiewicz, Joanna, Malgorzata Tarczynska, Zofia Walter, and Maciej Zalewski. "Natural Toxins from Cyanobacteria." *Acta Biologica Cracoviensia Series Botanica* 45 (2003): 9–20.

Mebs, Dietrich. *Venomous and Poisonous Animals: A Handbook for Biologists, Toxicologists and Toxinologists, Physicians and Pharmacists.* Stuttgart: Medpharm Scientific Publishers, 2002.

Moriyama, Hideo. "Purification and Properties of Ricin." *Igaku to Seibutsugaku* (1947): 163–66.

National Institute for Occupational Safety and Health. "Emergency Response Card for Abrin." Interim Document, March 24, 2003.

———. "Emergency Response Card for Ricin." Interim Document, March 24, 2003.

National Institutes of Health. *Hazardous Substance Data Bank (HSDB).* http://toxnet.nlm.nih.gov/cgi-bin/sis/htmlgen?HSDB/. 2004.

Olsnes, Sjur, Fiorenzo Stirpe, Kirsten Sandvig, and Alexander Pihl. "Isolation and Characterization of Viscumin, a Toxic Lectin from *Viscum album* L. (Mistletoe)." *The Journal of Biological Chemistry* 257 (1982): 13263–270.

Olson, Kent R., ed. *Poisoning & Drug Overdose.* 4th ed. New York: Lange Medical Books/McGraw-Hill, 2004.

Patocka Jiri, and Ladislav Streda. "Plant Toxic Proteins and Their Current Significance for Warfare and Medicine." *Journal of Applied Biomedicine* 1 (2003): 141–47.

Public Health Service. 42 CFR Part 73—"Select Agents and Toxins." 2004: 443–58.

Schmitt, Clare K., Karen C. Meysick, and Alison D. O'Brien. "Bacterial Toxins: Friends or Foes?" *Emerging Infectious Diseases* 5 (April/June 1999): 224–34.

Sidell, Fredrick R., Ernest T. Takafuji, and David R. Franz, eds. *Medical Aspects of Chemical and Biological Warfare, Textbook of Military Medicine Series, Part 1, Warfare, Weaponry, and the Casualty.* Washington, DC: Office of the Surgeon General, Department of the Army, 1997.

Sifton, David W., ed. *PDR Guide to Biological and Chemical Warfare Response.* Montvale, NJ: Thompson/Physicians Desk Reference, 2002.

Singh, Bal Ram, and Anthony T. Tu, eds. *Natural Toxins 2: Structure, Mechanism of Action, and Detection. Volume 391 of Advances in Experimental Medicine and Biology.* New York: Plenum Press, 1996.

Smith, Ann, Patricia Heckelman, and Maryadele J. Oneil, eds. *The Merck Index: An Encyclopedia of Chemicals, Drugs, & Biologicals.* 13th ed. Rahway, NJ: Merck & Co., Inc., 2001.

Somani, S.M., ed. *Chemical Warfare Agents.* New York: Academic Press, 1992.

Somani, Satu M., and James A. Romano, Jr., eds. *Chemical Warfare Agents: Toxicity at Low Levels.* Boca Raton, FL: CRC Press, 2001.

Stirpe, Fiorenzo, Kirsten Sandvig, Sjur Olsnes, and Alexander Pihl. "Action of Viscumin, a Toxic Lectin from Mistletoe, on Cells in Culture." *The Journal of Biological Chemistry*, 257 (1982): 13271–277.

Stirpe, Fiorenzo, Luigi Barbieri, Ada Abbondanza, Anna Ida Falasca, Alex N.F. Brown, Kirsten Sandvig, Sjur Olsnes, and Alexander Pihl. "Properties of Volkensin, a Toxic Lectin from Adenia volkensii." *The Journal of Biological Chemistry* 260 (1985): 14589–595.

Sweeney, Edel C., Alexander G. Tonevitsky, Rex A. Palmer, Hidie Niwa, Uwe Pfueller, Juergen Eck, Hans Lentzen, Igor I. Agapov, and Mikhail P. Kirpichnikov. "Mistletoe Lectin I Forms a Double Trefoil Structure." *FEBS Letters* 431 (1998): 367–70.

True, Bey-Lorraine, and Robert H. Dreisbach. *Dreisbach's Handbook of Poisoning: Prevention, Diagnosis and Treatment.* 13th ed. London, England: The Parthenon Publishing Group, 2002.

United States Army Headquarters. *Potential Military Chemical/Biological Agents and Compounds, Field Manual No. 3-11.9.* Washington, DC: Government Printing Office, January 10, 2005.

United States Department of Agriculture Grain Inspection, Packers and Stockyards Administration. *Aflatoxin Handbook.* Washington, DC: Government Printing Office, 2005.

———. *DON (Vomitoxin) Handbook.* Rev. ed. Washington, DC: Government Printing Office, 2004.

———. *Grain Fungal Diseases & Mycotoxin Reference.* Rev. ed. Washington, DC: Government Printing Office, 2004.

United States Department of Health and Human Services Centers for Disease Control and Prevention and National Institutes of Health. *Biosafety in Microbiological and Biomedical Laboratories (BMBL).* 4th ed. Washington, DC: US Government Printing Office, 1999.

United States Food & Drug Administration, Center for Food Safety & Applied Nutrition. *Foodborne Pathogenic Microorganisms and Natural Toxins Handbook.* January 30, 2003. http://www.cfsan.fda.gov/~mow/preface.html. April 18, 2005.

Woods Hole Oceanographic Institution. "Human Illness Associated with Harmful Algae." *The Harmful Algae Page.* http://www.whoi.edu/redtide/illness/illness.html. 2006.

World Health Organization. *Fact Sheet #270 Botulism.* Geneva: Health Communications and Public Relations, August 2002.

———. *Health Aspects of Chemical and Biological Weapons: Report of a WHO Group of Consultants.* Geneva: World Health Organization, 1970.

———. *Public Health Response to Biological and Chemical Weapons: WHO Guidance.* Geneva: World Health Organization, 2004.

Yang, Chen-Chung. "Crystallization and Properties of the Cobrotoxin from Formosan Cobra Venom." *The Journal of Biological Chemistry* 240 (1965): 1616–18.

17

Bacterial Pathogens

17.1 General Information

Bacteria are single-celled microorganisms. They are easy to grow but production, isolation, harvesting, and storage of large quantities of these organisms is difficult. Bacteria come in various shapes including rods (bacilli), spheres (cocci), and commas or spirals (spirilla). Individual organisms range in size from less than one micron to tens of microns. They may or may not be able to move on their own (i.e., motile). Bacteria can be aerobic, that is, they can live and grow in the presence of oxygen, or they can be anaerobic and live without oxygen.

In adverse conditions, some bacteria can enter a dormant state known as a spore. Spores can remain dormant for decades and can survive under extreme temperatures and other adverse environmental conditions. Unlike fungi (Chapter 20), formation of spores is not related to reproduction and is done strictly as a protective mechanism. Upon reactivation, each spore produces a single active bacteria. Spores are normally spherical or oval and are only a fraction of the size of the active (i.e., vegetative) cell.

Pathogens employed as biological weapons can be used for both lethal and incapacitating purposes. They cause disease by invading tissues or by producing toxins (Chapter 16) that are detrimental to the infected individual. Pathogens can be selected to target a specific host (e.g., humans, cows, pigs) or they may pose a broad threat to both animals and to people.

Pathogens deployed as antianimal biological weapons are generally used to produce lethal effects in an agriculturally significant species such as cows, pigs, or chickens. Although these pathogens are selected to target a specific animal species, there is a possibility that the disease may migrate to humans. The diseases produced by these crossover pathogens may be difficult to diagnose for medical personnel not trained in exotic pathology.

Other pathogens are selected to produce lethal effects in an agriculturally significant crop species such as wheat, corn, or rice. There is little potential for migration of these pathogens to humans or animals.

A final group of biological warfare pathogens are those used as simulants to model the release of other, more hazardous agents. Pathogens employed as biological warfare simulants do not generally pose a significant risk to people, animals, or plants. However, individuals with respiratory illness or suppressed immune systems may be at risk should they be exposed to an infectious dose of the agent.

Bacteria can be stored as either liquids (e.g., organisms concentrated growth media) or powders (e.g., spores or freeze-dried mixtures of agent and growth media) and are easy to disperse. However, because they are living organisms and can be killed during the dispersal process, there are limitations to the methods that can be used. They can also be stored and dispersed via infected vectors (e.g., mosquitoes, fleas). In most cases, large-scale attacks will be clandestine and only detected through epidemiological analysis of resulting disease patterns. Localized or small-scale attacks may take the form of "anthrax" letters. Even in these cases, without the inclusion of a threat, the attack may go unnoticed until the disease appears in exposed individuals (e.g., the initial 2001 anthrax attack at American Media Inc., which claimed the life of Robert Stevens).

In general, unless a local reservoir (i.e., intermediate host that may or may not be affected by the bacteria) is established, pathogens are easily killed by unfavorable environmental factors such as fluctuations in temperature, humidity, food sources, or ultraviolet light. For this reason, their persistency is generally limited to days. However, bacterial spores are highly resistant to impacts from changes in environmental factors. Agents that can form spores can survive in this state for decades and then become active again under the proper conditions. In addition, pathogens can be freeze-dried and remain in a preserved state almost indefinitely. Freeze-dried pathogens are reactivated when exposed to moisture.

It is possible that local insects can become both a reservoir and a vector for the pathogen. Under these circumstances, the pathogen can survive well after the initial release and can rapidly spread beyond the immediately affected area. In many cases, once a vector is infected, it is capable of transmitting the disease throughout its life span. Some pathogens are transmitted directly to the young of the vector so that the next generation is born infected. Response activities must also include efforts to contain and eliminate these vectors (e.g., application of pesticides).

Incubation times for diseases resulting from infection vary depending on the specific pathogen, but are generally on the order of days to weeks. Exposures to extremely high doses of some pathogens may reduce the incubation period to as short as several hours. The pathway of exposure (e.g., inhalation, ingestion) can also cause a significant change in the incubation time required as well as the clinical presentation of the disease.

Some diseases caused by bacteria are communicable and easily transferred from an infected individual to anyone in close proximity. Typically, this occurs when the infected individual coughs or sneezes creating an infectious aerosol. These aerosols enter the body of a new host through inhalation and/or contact of the aerosol with the mucous membranes of the eyes, nose, or mouth. In addition, although intact skin is an effective barrier against most pathogens, abrasions, or lacerations circumvents this protective barrier and allows entry of the pathogen into the body.

In addition to direct aerosol exposure, some bacteria are potentially associated with exposure through ingestion of contaminated food, water or both. Pathogens that normally infect individuals through an ingestion pathway also pose a significant risk of secondary infections from the "fecal/oral" cycle. Some individuals or animals can become asymptomatic carriers and are capable of spreading the disease long after their recovery (e.g., Typhoid Mary). Mechanical vectors (e.g., flies, roaches) can also carry pathogens and spread the disease to food not directly contaminated.

17.2 Response

17.2.1 Personal Protective Requirements

17.2.1.1 Responding to the Scene of a Release

A number of conditions must be considered when selecting protective equipment for individuals at the scene of a release. For instances such as "anthrax" letters when the mechanism of release is known and it does not involve an aerosol generating device, then responders can use Level C with high-efficiency particulate air (HEPA) filters.

If an aerosol generating device is employed (e.g., sprayer), or the dissemination method is unknown and the release is ongoing, then responders should wear a Level A protective ensemble. Once the device has stopped generating the aerosol or has been rendered inoperable, and the aerosol has settled, then responders can downgrade to Level B.

In all cases, there is a significant hazard posed by contact of contaminated material with skin that has been cut or lacerated, or through injection of pathogens by contact with debris. Appropriate protection to avoid any potential abrasion, laceration, or puncture of the skin is essential. Individuals with damaged or open skin should not be allowed to enter the contaminated area.

17.2.1.2 Working with Infected Individuals

For most bacteria, use infection control guidelines standard precautions. If appropriate, or the identity of the bacteria is unknown, use additional droplet and airborne precautions. Avoid direct contact with wounds or wound drainage.

17.2.1.2.1 Infection Control Guidelines Standard Precautions

1. Wash hands after patient contact.
2. Wear gloves when handling potentially infectious items that are contaminated with blood, body fluids, or excreta.
3. Wear a mask and eye protection during procedures likely to generate splashes or sprays of infectious fluids.
4. Handle all contaminated and potentially contaminated equipment and linen in a manner that will prevent cross contamination.
5. Take care when handling sharps. When practical, use a mouthpiece or other ventilation device as an alternative to mouth-to-mouth resuscitation.

17.2.1.2.2 Droplet Precautions

Standard precautions plus

1. Place the patient in a private room or cohort them with someone with the same infection. Maintain at least 3 feet between patients.
2. Wear a mask when working within 3 feet of the patient.
3. Limit movement and transport of the patient. Place a mask on the patient if they need to be moved.

17.2.1.2.3 Airborne Precautions

Standard precautions plus

1. Place the patient in a private room with negative air pressure that has a minimum of six air changes per hour and appropriate filtration before the air is discharged from the room.
2. Wear respiratory protection when entering the room.
3. Limit movement and transport of the patient. Place a mask on the patient if they need to be moved.

17.2.2 Decontamination

17.2.2.1 Food

All foodstuffs in the area of a release should be considered contaminated. Unopened items may be used after decontamination of the container. Opened or unpackaged items should be destroyed. Fruits and vegetables should be washed thoroughly with antimicrobial soap and water. Many pathogens can survive in food containers for extended periods.

17.2.2.2 Casualties/Personnel

Infected individuals: Unless the individual is reporting directly from the scene of an attack (e.g., "anthrax" letter, aerosol release, etc.) then decontamination is not necessary. Use standard protocols for individuals that may be infected with a communicable disease transmissible via an aerosol.

Direct Exposure: In the event that an individual is at the scene of a known or suspected attack (e.g., white-powder letter, aerosol release, etc.), have them wash their hands and face thoroughly with antimicrobial soap and water as soon as possible. If antimicrobial soap is not available, use any available soap or shampoo. They should also blow their nose to remove any agent particles that may have been captured by nasal mucous. Remove all clothing and seal in a plastic bag. To avoid further exposure of the head, neck, and face to the agent, cut off potentially contaminated clothing that must be pulled over the head. Shower using copious amounts of antimicrobial soap (if available) and water. Ensure that the hair has been washed and rinsed to remove potentially trapped agent. The Centers for Disease Control and Prevention (CDC) does not recommend that individuals use bleach or other disinfectants on their skin.

17.2.2.3 Animals

Unless the animals are at the scene of an attack, then decontamination is not necessary.

Apply universal decontamination procedures using antimicrobial soap and water. If antimicrobial soap is not available, use any available soap, shampoo, or detergent. In some cases, severe infection may require euthanasia of animals or herds. Consult local/state veterinary assistance office. If the pathogen has not been identified, then wear a fitted N95 protective mask, eye protection, disposable protective coverall, disposable boot covers, and disposable gloves when dealing with infected animals.

17.2.2.4 Plants

Removal and destruction of infected plants may be required. Incineration of impacted fields may be required. Consult local/state agricultural assistance office. Wear disposable protective coverall, disposable boot covers, and disposable gloves to prevent spread of contamination. Many plant pathogens are spread via insects and response activities must also include efforts to contain and eliminate these vectors.

17.2.2.5 Property

Surface disinfectants: Compounds containing phenolics, chlorhexidine (not effective against bacteria spores), quaternary ammonium salts (additional activity if bis-n-tributyltin oxide present), hypochlorites such as household bleach, alcohols such as 70–95% ethanol and isopropyl (not effective against bacteria spores), potassium peroxymonosulfate, hydrogen peroxide, iodine/iodophores, and triclosan.

The military also identifies the following "nonstandard" decontaminants: Detrochlorite (thickened bleach mixture of diatomaceous earth, anionic wetting agent, calcium hypochlorite, and water), 3% aqueous peracetic acid solution, 1% aqueous hyamine solution, and a 10% aqueous sodium or potassium hydroxide solution.

Large area fumigants: Gases including formaldehyde, ethylene oxide, or chlorine dioxide. These materials are highly toxic to humans and animals and fumigation operations must be adequately controlled to prevent unnecessary exposure. Additional methods include vaporized hydrogen peroxide and an ionized hydrogen peroxide aerosol.

Fomites: Some pathogens may be absorbed into clothing or bedding causing these items to become infectious and capable of transmitting the disease. Others may contain vectors (e.g., lice, ticks) that pose a transmission hazard. Deposit items in an appropriate biological waste container and send to a medical waste disposal facility.

Alternatively, cotton or wool articles can be boiled in water for 30 minutes, autoclaved at 253°F for 45 minutes, or immersed in a 2% household-bleach solution (i.e., 1 gallon of bleach in 2 gallons of water) for 30 minutes followed by rinsing.

17.3 Fatality Management

Unless the cadaver is coming directly from the scene of an attack (e.g., "anthrax" letter, aerosol release), process the body according to established procedures for handling potentially infectious remains.

Because of the nature of biological warfare agents, it is highly unlikely that a contaminated cadaver will be recovered from the scene of an attack unless it is from an individual who died from trauma or other complications while the attack was ongoing. If a fatality is grossly contaminated with a biological agent, wear disposable protective clothing with integral hood and booties, disposable gloves, and an N-95 respirators and eye protection or powered air-purifying respirator (PAPRs) equipped with N-95 or HEPA filter.

Remove all clothing and personal effects. Items that will be retained for further processing should be double sealed in impermeable containers, ensuring that the inner container is decontaminated before placing it in the outer one. Otherwise, dispose of contaminated articles at an appropriate medical waste disposal facility.

Thoroughly wash the remains with antimicrobial soap and water. Pay particular attention to areas where agent may get trapped, such as hair, scalp, pubic areas, fingernails, folds of skin, and wounds. If deemed appropriate, the cadaver can be treated with a surface disinfectant listed in Section 17.2.2.

If there is a potential that vectors may be involved, care must be taken to kill any vectors (e.g., lice, fleas) remaining either on the cadaver or residing in fomites. Remove all potentially infested clothing depositing it in a container that will trap and eliminate vectors. Dispose contaminated particles at an appropriate medical waste disposal facility.

Once the remains have been thoroughly decontaminated, process the body according to established procedures for handling potentially infectious bodies. Use appropriate burial procedures.

In 2004 (*Morbidity and Mortality Weekly Report* 53), the CDC determined that the risks of occupational exposure to biological terrorism agents outweighed the advantages of embalming fatalities and recommended that, unless death was due to a bacteria that forms spores (e.g., *Bacillus anthracis*), the body should be buried without embalming. Fatalities due to spore forming bacteria should be cremated without embalming.

C17-A

AGENTS

C17-A001

Bacillus anthracis (Agent N)

Anthrax

ICD-10: A22

It is an aerobic, gram-positive, spore-forming, rod-shaped bacterium. Dry spores are stable for decades. Spores are stable in water for up to 2 years and are resistant to chlorine at purification concentrations. It is endemic in many countries of the world, particularly in tropical and subtropical areas. This is a biosafety level 2 agent.

It is on the HHS Select Agents list, the USDA High Consequence list, the Australia Group Core list, and is listed as a Category A Potential Terrorism Agent by the CDC.

In People:

CDC Case Definition: An illness with acute onset characterized by several distinct clinical forms including: *Cutaneous*: a skin lesion evolving over 2–6 days from a papule, through a vesicular stage, to a depressed black eschar. *Inhalation*: a brief prodrome resembling a viral respiratory illness followed by development of hypoxia and dyspnea, with x-ray evidence of mediastinal widening. *Intestinal*: severe abdominal distress followed by fever and signs of septicemia. *Oropharyngeal*: mucosal lesion in the oral cavity or oropharynx, cervical adenopathy and edema, and fever. Laboratory criteria for diagnosis is (1) isolation of *B. anthracis* from a clinical specimen; or (2) fourfold or greater rise in either the anthrax enzyme-linked immunosorbent assay (ELISA) or electrophoretic immunotransblot (EITB) titer between acute- and convalescent-phase serum specimens obtained ≥2 weeks apart; or (3) anthrax ELISA titer ≥64 or an EITB reaction to the protective antigen and/or lethal factor bands in one or more serum samples obtained after onset of symptoms; or (4) demonstration of *B. anthracis* in a clinical specimen by immunofluorescence.

Communicability: Direct person-to-person transmission is rare. When dealing with infected individuals, use standard contact precautions. Avoid direct contact with wounds or wound drainage.

Normal Routes of Exposure: Inhalation; Ingestion; Abraded skin; Mucous membranes; possibly Vectors (biting flies).

Infectious Dose: 8,000–55,000 spores (Inhalation).

Secondary Hazards: Spores; Blood; Body tissue; Fomites.

Incubation: Hours (high dose) to 7 days. May be prolonged up to 2 months.

Signs and Symptoms: *Inhalation*: Often biphasic, but symptoms may progress rapidly. Mild and nonspecific (i.e., flu-like) progressing to respiratory distress with fever and shock following in 3–5 days. Death occurs shortly thereafter. *Skin*: Itching followed by formation of a lesion progressing to a black eschar surrounded by swelling 7–10 days after appearance of the initial lesion. Usually little or no pain. *Ingestion*: Nausea, anorexia, vomiting, and fever progressing to severe abdominal pain, vomiting blood, and diarrhea that is almost always bloody. Acute abdomen picture with rebound tenderness may develop.

Suggested Alternatives for Differential Diagnosis: Abdominal aneurysm, aortic dissection, pleural effusion, subarachnoid hemorrhage, superior vena cava syndrome, hantavirus pulmonary syndrome, mediastinitis, fulminate mediastinal tumors; pneumonia, gastroenteritis, meningitis, ecthyma, rat bite fever, spider bite, leprosy, plague, tularemia, coccidioidomycosis, diphtheria, glanders, histoplasmosis, psittacosis, typhoid fever, and rickettsial pox.

Mortality Rate (untreated): Cutaneous: $\leq$20%; Inhalation: $\leq$100%.

In Agricultural Animals:

Target species: Cattle, sheep, goats, horses.

Communicability: Direct transmission does not occur.

Normal Routes of Exposure: Inhalation; Ingestion; Abraded Skin; Mucous Membranes; Vectors (biting flies).

Secondary Hazards: Spores; Blood; Body tissue; Animal products such as hides, hair, wool, bones, feedstuffs, or handicrafts.

Incubation: 1–2 weeks in pigs and horses.

Signs and Symptoms: Anthrax cases usually produce sudden death. When present, the following signs may be seen: fever, restlessness, edematous swelling of the throat and neck, difficulty swallowing, difficulty in breathing (dyspnea), agitation, bleeding from orifices and subcutaneous hemorrhage, and disorientation. Death due to choking or toxemia.

Postmortem findings include dark-tarry blood discharge from body orifices, absence of rigor mortis, hemorrhage of the mucous membranes, bloating, and rapid decomposition of the carcass.

Suggested Alternatives for Differential Diagnosis: May be difficult to differentiate from other causes of sudden death such as lightning strikes, peracute lead poisoning, peracute blackleg (blackquarter), acute leptospirosis, bacillary hemoglobinuria, hypomagnesaemia tetani, acute bloat, septicemic forms of other diseases.

Mortality Rate (untreated): "Very high."

C17-A002

Bacillus subtilis (Agent BG)

BW Simulant

It is an aerobic, gram-positive, spore-forming, rod-shaped bacterium. It is very common in soil, water, and air.

No significant harmful health effects to humans or animals are expected from exposure to this pathogen unless the individual has a compromised respiratory system or suppressed immune system. Direct contact with large quantities of *B. subtilis spores* may cause redness or irritation of the skin.

Some strains have been used as a medicinal product to treat dysentery and other intestinal problems, and as a pesticide to control plant diseases and fungal pathogens.

C17-A003

Bacillus thuringiensis (Agent BT)

BW Simulant

B. thuringiensis is an aerobic, gram-positive, spore-forming, rod-shaped bacterium. Vegetative cells are motile and approximately 1 μm wide and 5 μm long. It is very common in soil and on plants.

No significant harmful health effects to humans, animals, or plants are expected from exposure to this pathogen. Some strains have been used as pesticides to control crop damaging insects.

C17-A004

Brucella **species** including *B. abortus*; *B. melitensis*; *B. suis* (Agent US)

Brucellosis

ICD-10: A23; A23.0 (*B. melitensis)*; A23.1 (*B. abortus)*; A23.1 (*B. suis)*

They are aerobic, gram-negative, nonsporing, nonmotile, cocci or rod-shaped bacterium. They are stable in water for 20–72 days. The most pathogenic of the species for man is *B. melitensis*, followed by *B. suis* and then *B. abortus*. The natural reservoirs are cattle (*B. abortus*); sheep and goats (*B. melitensis*); and pigs (*B. suis*). These are biosafety level 3 agents.

They are on the HHS Select Agents list, the USDA High Consequence list, the Australia Group Core list, and are listed as Category B Potential Terrorism Agents by the CDC.

In People:

CDC Case Definition: An illness characterized by acute or insidious onset of fever, night sweats, undue fatigue, anorexia, weight loss, headache, and arthralgia. Laboratory criteria for diagnosis is (1) isolation of *Brucella* species from a clinical specimen; or (2) fourfold or greater rise in *Brucella* agglutination titer between acute- and convalescent-phase serum specimens obtained ≥2 weeks apart and studied at the same laboratory; or (3) demonstration by immunofluorescence of *Brucella* species in a clinical specimen.

Communicability: Direct person-to-person transmission does not occur. When dealing with infected individuals who have open skin lesions, use standard contact precautions. Avoid direct contact with wounds or wound drainage.

Normal Routes of Exposure: Inhalation; Ingestion; Abraded skin; Mucous membranes.

Infectious Dose: 10–100 organisms.

Secondary Hazards: Blood and body fluids; Body tissue.

Incubation: 5–60 days.

Signs and Symptoms: Are nonspecific and consist of irregular fever, headache, profound weakness and fatigue, chills and sweating, generalized severe joint and muscle pain (myalgia), anorexia, weight loss, and depression. Joint complications are common.

Suggested Alternatives for Differential Diagnosis: Influenza, infectious mononucleosis, hepatitis, leptospirosis, infective endocarditis, malaria, tuberculosis, typhoid fever, cryptococcosis, histoplasmosis, ankylosing spondylitis and undifferentiated spondyloarthropathy, collagen vascular disease, chronic fatigue syndrome, malignancy, and osteomyelitis.

Mortality Rate (untreated): ≤2%.

In Agricultural Animals:

Target species: Cattle, sheep, goats, pigs.

Communicability: Direct transmission is possible.

Normal Routes of Exposure: Ingestion; Mucous membranes; Mating.

Secondary Hazards: Blood and body fluids; Body tissue; Vectors (mechanical).

Incubation: Varies depending on route of entry and condition of host.

Signs and Symptoms: Fever, increased respiration and depression; inflammation of testes and epididymis; swelling of the last scrotum; atrophic testicles; edematous placenta and fetus; infertility; abortion in the last 3–4 months of pregnancy; and lymph-filled cystic cavities (hygromas) on the knees, stifles, hock, and angle of the haunch.

Suggested Alternatives for Differential Diagnosis: In *cattle*: Other causes of abortion including infectious bovine rhinotracheitis, vibriosis, leptospirosis, trichomoniasis, mycoplasma infections, mycosis, and nutritional and physiological causes. Brucellosis should always be suspected when there are multiple late-term abortions in a herd. In *pigs*: Other common diseases causing abortion are Aujeszky's disease, leptospirosis, salmonellosis, streptococcidiosis, classical swine fever, and parvorisosis. In *sheep, goats*: Abortions due to chlamydiosis or coxiellosis.

Mortality Rate (untreated): Not normally fatal.

C17-A005

Chlamydia psittaci (Agent SI)

Psittacosis
ICD-10: A70

It is a gram-negative, spherical (0.4–0.6 μm diameter) bacterium. Survival of the bacteria outside the host depends on the source: infected fluid from eggs—52 h; bird droppings—a few days; bird feed—2 months; glass—15 days; and straw—20 days. The natural reservoir is birds. This is a biosafety level 2 agent. It is highly communicable from infected birds to people.

It is on the Australia Group Core list and is listed as a Category B Potential Terrorism Agent by the CDC.

In People:

CDC Case Definition: An illness characterized by fever, chills, headache, sensitivity to light (photophobia), cough, and myalgia. Laboratory criteria for diagnosis is (1) isolation of *C. psittaci* from respiratory secretions; or (2) fourfold or greater increase in antibody against *C. psittaci* by complement fixation or microimmunofluorescence (MIF) to a reciprocal titer of $\geq$32 between paired acute- and convalescent-phase serum specimens; or (3) presence of immunoglobulin M antibody against *C. psittaci* by MIF to a reciprocal titer of $\geq$16.

Communicability: Direct person-to-person transmission is rare. When dealing with infected individuals, use standard contact precautions.

Normal Routes of Exposure: Inhalation; Abraded skin; Mucous membranes.

Infectious Dose: Unknown.

Secondary Hazards: Aerosols (cough, sneeze); Blood and body fluids; Fecal (from birds); Fomites (from birds).

Incubation: 7–28 days.

Signs and Symptoms: Diagnosis of psittacosis can be difficult. There is a variable clinical presentation but may include fever, headache, muscle pain (myalgia), chills and upper or lower respiratory tract disease, and dry cough. Pneumonia is often evident in chest x-rays.

Suggested Alternatives for Differential Diagnosis: Brucellosis, chlamydial pneumonias, infective endocarditis, legionnaires disease, mycoplasma infections, pneumonia, *Coxiella burnetii* infection, *Francisella tularensis* infection, Q fever, tuberculosis, tularemia, typhoid fever, and all atypical pneumonia.

Mortality Rate (untreated): $\leq$20%.

In Agricultural Animals:

Target species: Poultry, pheasants.

Communicability: Direct transmission is possible.

Normal routes of exposure: Inhalation; Ingestion; Mucous Membranes.

Secondary Hazards: Aerosols; Body fluids; Fecal matter; Fomites.

Incubation: Varies, most common 3–10 days.

Signs and Symptoms: Ruffled feathers; loss of condition; conjunctivitis; nasal and ocular discharge; and pale, greenish, or watery feces.

Postmortem findings include inflammation of the lungs, airsacs, liver, heart, spleen, kidneys, and peritoneum.

Suggested Alternatives for Differential Diagnosis: *Mycoplasma gallisepticum* infection, pasteurellosis, and salmonellosis including pullorum disease; bacillary white diarrhea.

Mortality Rate (untreated): "High," varies greatly depending on bird species.

C17-A006

Clavibacter michiganensis* subspecies *sepedonicus

Ring rot of potatoes

It is a gram-positive rod-shaped bacterium.

In wet soil, the bacterium can survive for several months; in drier soil it can survive for more than a year. Ring rot infection can pass through one or more field generations without causing symptoms in stems and tubers. Capable of surviving 2–5 years in dried slime on surfaces of crates, bins, burlap sacks, or harvesting and grading machinery, even if exposed to temperatures well below freezing.

It is on the Australia Group Core list.

In Plants:

Target species: Potatoes; can colonize roots of sugar beets.

Normal routes of transmission: Contact through wounds in tubers.

Secondary Hazards: Crop debris; Mechanical Vectors (contaminated containers; farm implements; wash water); Vectors (Colorado potato beetle; flea beetle; leafhoppers; sucking insects such as aphids).

Signs: Wilting of the leaf margins, especially on the lower leaves and often on only one side of the plant. Leaves curl and progressively lose their shine (appearing dull and greasy) with the onset of yellowing, browning, and eventual necrosis.

In the early stages, the tissue around the vascular bundles appears semitranslucent, glassy, and water-soaked. As the infection progresses, the vascular ring becomes discolored and the tissue around it degrades and develops a colorless rot with a paste-like consistency.

Squeezing a cut tuber produces a pale creamy, cheese-like ooze. Later, the discoloration becomes a more distinct brown and the necrosis can extend into the surrounding tissue. In advanced stages the bacterium can ooze from the heel end and eyes. This may result in reddish-brown, slightly sunken, and star-shaped lesions on the skin from which bacteria may ooze, causing soil particles to adhere. Yield loss may reach 50%.

C17-A007

Clostridium botulinum

Produces Botulinum Toxin (C16-A005)
ICD-10: A05.1

It is an anaerobic, gram-positive, spore-forming, rod-shaped bacterium that produces neurotoxins (see C16-A005), especially in low-acid foods. The natural reservoir is soil. It is also found in honey. This is a biosafety level 2 agent.

It is on the HHS Select Agents list, the USDA High Consequence list, and the Australia Group Core list.

In People:

CDC Case Definition: See Botulinum Toxin (C16-A005).

Communicability: Direct person-to-person transmission does not occur.

Normal Routes of Exposure: Danger is from exposure to the toxin (Inhalation; Ingestion; Abraded skin).

Infectious Dose: Not applicable.

Secondary Hazards: Spores; Residual toxin.

Incubation: Not applicable.

Signs and Symptoms: See Botulinum toxin (C16-A005).

Suggested Alternatives for Differential Diagnosis: See Botulinum toxin (C16-A005).

Mortality Rate (untreated): See Botulinum toxin (C16-A005).

C17-A008

Clostridium perfringens Epsilon toxin producing (Agent G)

Gas Gangrene
ICD-10: A05.2

It is an anaerobic, gram-positive, spore-forming, rod-shaped bacterium that produces cytotoxins (see C16-A028). It is stable in water (sewage) and resistant to chlorine at purification concentrations. It can grow at an extremely rapid rate with generation times as short as 10 min. Causes gas gangrene, enteritis necroticans, food poisoning, and nonfood-borne enterotoxemic infections. Gas gangrene results from wound contamination with spores. Only rare cases of pulmonary infections, and no apparent disease caused by inhalation of spores. It is widely distributed in the environment and frequently occurs in the intestines of humans and many animals.

It is on the Australia Group Core list.

In People:

CDC Case Definition: None established.

Communicability: Direct person-to-person transmission does not occur. When dealing with infected individuals who have open skin lesions, use standard contact precautions.

Normal Routes of Exposure: Inhalation; Ingestion; Abraded Skin.

Infectious Dose: >100,000,000 vegetative cells via ingestion.

Secondary Hazards: Fecal matter (if ingestional).

Incubation: 8–22 h (enteritis); 6 h to several days (gas gangrene).

Signs and Symptoms: *Gas Gangrene*—Rapid heart rate (tachycardia), excessive perspiration (diaphoresis), and possibly altered mental status. Crepitance or subcutaneous air, tense edema, discolored wound discharge, blisters, with necrotic tissue. Because gas gangrene affects the deep muscle tissue, the superficial skin often appears normal early in the disease course but eventually turns pale and then to a gray or purplish red color. Pain is often out of proportion to physical findings. Decreased pain or anesthesia at the site of infection can indicate that cutaneous nerve endings are being destroyed and that the disease is advanced. Progression to toxemia and shock can be rapid. *Enteritis*—Acute onset of abdominal cramps with diarrhea lasting less than 1 day. Vomiting is rare.

Suggested Alternatives for Differential Diagnosis: *Gas Gangrene*—Cellulitis, necrotizing fasciitis, other causes of necrotizing myositis, deep venous thrombosis and thrombophlebitis, cutaneous anthrax, vaccinia vaccination, acute gout, septic arthritis, familial mediterranean fever, fixed drug reaction, pyoderma gangrenosa, Sweet syndrome, Wells syndrome, carcinoma erysipeloides, pyomyositis, water-borne skin infections, and other causes of soft tissue gas (e.g., pneumomediastinum, pneumothorax, fractured larynx, fractured trachea).

Mortality Rate (untreated): >25%.

In Agricultural Animals:

Target species: All species.

Communicability: Direct transmission does not occur.

Normal routes of exposure: Ingestion; Abraded Skin.

Secondary Hazards: None.

Incubation: Varies, as short as 2 h for ingestional.

Signs and Symptoms: Causes enteritis, dysentery, and toxemia in horses, sheep, cattle, and pigs. Mortality may be high in lambs, calves, pigs, and foals. In birds, typically the only sign is a sudden increase in mortality (≤50%). However, birds with depression, ruffled feathers, and diarrhea may also be seen. Gangrenous dermatitis is characterized by gangrenous necrosis of the skin and a sharp increase in mortality (≤60%).

Suggested Alternatives for Differential Diagnosis: Salmonellosis, pasteurellosis, enterotoxemia due to *E. coli*, rabies, pregnancy toxemia, polioencephalomalatia, acute rumen impaction, louping-ill; hypocalcemia, hypomagnesemia, and acute lead poisoning.

Mortality Rate (untreated): "Variable," depending on the species and location of infection.

C17-A009

Clostridium tetani

Lockjaw

ICD-10: A35

It is an anaerobic, gram-positive, spore-forming, rod-shaped bacterium that produces neurotoxins (see C16-A020). Spores may remain viable for years if protected from light and heat. They can be destroyed by boiling water. Spores are unable to grow in normal tissue or even in wounds unless necrosis is present. The bacteria remain localized in the necrotic tissue at the original site of infection and multiply. It is found in soil and intestinal tracts of animals. This is a biosafety level 2 agent.

It is on the Australia Group Warning list.

In People:

CDC Case Definition: Acute onset of hypertonia and/or painful muscular contractions (usually of the muscles of the jaw and neck) and generalized muscle spasms without other apparent medical cause.

Communicability: Direct person-to-person transmission does not occur. When dealing with infected individuals who have open skin lesions, use standard contact precautions.

Normal Routes of Exposure: Wounds (punctures, lacerations, burns).

Infectious Dose: Unknown.

Secondary Hazards: Spores.

Incubation: 3–21 days.

Signs and Symptoms: Trismus (i.e., lockjaw), stiffness, neck rigidity, difficulty in swallowing (dysphagia), restlessness, and reflex spasms. Subsequently, muscle rigidity becomes the major manifestation. Muscle rigidity spreads from the jaw and facial muscles to the extensor muscles of the limbs. Patients with tetanus may present with abdominal tenderness and guarding, mimicking an acute abdomen. Reflex spasms develop in most patients and can be triggered by minimal external stimuli such as noise, light, or touch. Sustained contraction of facial musculature produces a sneering grin expression known as risus sardonicus. Tetanic seizures may occur and portend a poor prognosis.

Suggested Alternatives for Differential Diagnosis: Conversion disorder, mandible dislocations, encephalitis, hypocalcemia, meningitis, peritonsillar abscess, rabies, black widow spider envenomations, hemorrhagic stroke, subarachnoid hemorrhage, medication-induced dystonic reactions, intraoral disease, odontogenic infections, globus hystericus, hepatic encephalopathy, hysteria, strychnine poisoning, acute abdomen, and intracranial hemorrhage.

Mortality Rate (untreated): $\leq$90%.

In Agricultural Animals:

Target species: Horses, cattle, goats, sheep.

Communicability: Direct transmission does not occur.

Normal routes of exposure: Wounds (punctures); Birthing.

Secondary Hazards: Spores.

Incubation: 4 days to 4 months.

Signs and Symptoms: The first symptoms are the protrusion of the nictitating membrane, and the involvement of facial and jaw muscles leading to lock jaw. Usually develops first in the limbs followed by the muscles of the trunk except in horses, which present from the head down. Causes spasmodic contraction of the voluntary muscles and increased sensitivity to stimuli. Consciousness is not affected.

Suggested Alternatives for Differential Diagnosis: Strychnine poisoning, hypocalcemia of mares, cerebrospinal meningitis, lactation tetany of cattle, enzootic muscular dystrophy, enterotoxaemia of lambs, polioencephalomalacia. Clinical signs and history of recent trauma are usually adequate for a diagnosis of tetanus.

Mortality Rate (untreated): $\leq$80%.

C17-A010

Coxiella burnetii (Agent OU)

Q fever

ICD-10: A78

It is a gram-negative, sporing bacterium that used to be considered a rickettsiae. In addition to the spore form, it exists in two vegetative forms: a large one that is found in infected cells, and a small one that exists outside a host. It can survive in the environment for up to 10 months and also exposure to many standard disinfectants. Although the disease is usually subclinical in ruminants (e.g., cattle, sheep, and goats), it can cause anorexia, abortions or both. Inhalation of dust contaminated with placental tissue or birth fluids is a common way humans contract the disease. Domestic animals (e.g., cats, rabbits, birds, etc.), are also susceptible to infection and can act as a source of infection for humans. Ticks may transmit the disease among domestic ruminants, but are not thought to play an important role in transmission of the disease to humans. This is a biosafety level 3 agent.

It is on the HHS Select Agents list, the USDA High Consequence list, the Australia Group Core list, and is listed as a Category B Potential Terrorism Agent by the CDC.

In People:

CDC Case Definition: A febrile illness usually accompanied by rigors, myalgia, malaise, and retrobulbar headache. Severe disease can include acute hepatitis, pneumonia, and meningoencephalitis. Clinical laboratory findings may include elevated liver enzyme levels and abnormal chest film findings. Asymptomatic infections may also occur. Laboratory criteria for diagnosis is (1) fourfold or greater change in antibody titer to *C. burnetii* phase II or phase I antigen in paired serum specimens ideally taken 3–6 weeks apart; or (2) isolation of *C. burnetii* from a clinical specimen by culture; or (3) demonstration of *C. burnetii* in a clinical specimen by detection of antigen or nucleic acid.

Communicability: Direct person-to-person transmission is rare. It may be communicable if pneumonia is present. When dealing with infected individuals, use standard contact precautions except in cases of pneumonia, then use droplet precautions.

Normal Routes of Exposure: Inhalation; Ingestion; Mucous membranes; Vectors (ticks).

Infectious Dose: 10 organisms (Inhalation).

Secondary Hazards: Aerosols (contaminated dust; coughs, sneezes from individuals with pneumonia); Milk and urine (from ruminants); Fecal (from ruminants).

Incubation: 14–39 days (can be less depending on the dose).

Signs and Symptoms: Variable symptoms ranging from inapparent or nonspecific to severe. Initial acute symptoms include sudden onset of fever, chills, severe sweating, severe headache, weakness, a vague feeling of bodily discomfort (malaise), pain in the joints (arthralgias), confusion, sore throat, nonproductive cough, nausea, vomiting, diarrhea, abdominal pain, and chest pain. May progress to pneumonia, hepatitis, or endocarditis. Total recovery may be prolonged.

Suggested Alternatives for Differential Diagnosis: Hepatitis, Legionnaires disease, myocarditis, pericarditis, cardiac tamponade, pneumonia, ehrlichiosis, relapsing fever, Rocky Mountain spotted fever, and tularemia.

Mortality Rate (untreated): $\leq$3%.

C17-A011

Escherichia coli

BW Simulant

ICD-10: A04.0

It is a facultative anaerobic, gram-negative, motile, rod-shaped bacterium propelled by long, rapidly rotating flagella. It is part of the common bacterium that normally inhabits the intestinal tracts of humans and animals. It helps protect the intestinal tract from bacterial infection and aids in digestion.

Although used as a simulant, it can cause acute bacterial meningitis, pneumonia, intra-abdominal infections, enteric infections, urinary tract infections, septic arthritis, endophthalmitis, suppurative thyroiditis, sinusitis, osteomyelitis, endocarditis, and skin and soft tissue infections. There are also strains of *E. coli* (C17-A015) that produce lethal cytotoxins (C16-A052).

C17-A012

***Escherichia coli*, Serotype O157:H7**

Enterohemorrhagic *Escherichia coli*

ICD-10: A04.3

It is a facultative anaerobic, gram-negative, motile, rod-shaped bacterium propelled by long, rapidly rotating flagella. It can survive well outside a host for extended periods. Although *E. coli* is part of the common bacterium that normally inhabits the intestinal tracts of humans and animals, serotype O157:H7 produces a shiga-like cytotoxin (C16-A052) that causes hemorrhage in the intestines. Cattle, including calves, are one of the reservoirs for infection in humans. This is a biosafety level 2 agent.

E. coli can also cause acute bacterial meningitis, pneumonia, intra-abdominal infections, enteric infections, urinary tract infections, septic arthritis, endophthalmitis, suppurative thyroiditis, sinusitis, osteomyelits, endocarditis, and skin and soft tissue infections.

It is on the Australia Group Core list, and is listed as a Category B Potential Terrorism Agent by the CDC.

In People:

CDC Case Definition: An infection of variable severity characterized by diarrhea (often bloody) and abdominal cramps. Illness may be complicated by hemolytic uremic syndrome (HUS) or thrombotic thrombocytopenic purpura (TTP); asymptomatic infections may also occur and the organism may cause extraintestinal infections. Laboratory criteria for diagnosis is isolation of Shiga toxin-producing *E. coli* from a clinical specimen. *E. coli* O157:H7 isolates may be assumed to be Shiga toxin-producing. For all other *E. coli* isolates, Shiga toxin production or the presence of Shiga toxin genes must be determined to be considered STEC.

Communicability: Direct person-to-person transmission is possible through fecal/oral. When dealing with infected individuals, use standard contact precautions. Wash hands frequently.

Normal Routes of Exposure: Ingestion.

Infectious Dose: 10 organisms (estimate).

Secondary Hazards: Aerosols (possibly); Fecal matter; Vectors (mechanical).

Incubation: 2–8 days.

Signs and Symptoms: Severe cramping with bowel movements ranging from nonbloody diarrhea to stools that are almost pure blood. Vomiting occurs in approximately half the cases. There is generally no fever associated with the infection. Usually lasts 8 days. Can progress to kidney failure.

Suggested Alternatives for Differential Diagnosis: Enterobacter infections, enterococcal infections, klebsiella infections, proteus infections, providencia infections, *Pseudomonas aeruginosa* infections, serratia, shigellosis, and streptococcus group B infections.

Mortality Rate (untreated): ≤0.5% overall; ≤15% hemolytic uremic syndrome; ≤99% for thrombotic thrombocytopenic purpura.

In Agricultural Animals:

Target species: Cattle, pigs, sheep.

Communicability: Direct transmission is possible through fecal/oral and nursing.

Normal routes of exposure: Ingestion.

Secondary Hazards: Fecal matter; Vectors (mechanical).

Incubation: Not available.

Signs and Symptoms: Clinical signs range from peracute death with no signs of illness, to CNS involvement with incoordination (ataxia), paralysis, and recumbency. Swelling around the eyes (periocular edema), forehead and submandibular regions, difficulty breathing (dyspnea), and anorexia are common. Edema may be accompanied by hemorrhage. There may be profuse diarrhea that may contain blood and mucus.

Suggested Alternatives for Differential Diagnosis: Enterobacter infections.

Mortality Rate (untreated): 90% (Pigs).

C17-A013

Francisella tularensis (Agent UL)

Tularemia
ICD-10: A21

It is an aerobic, gram-negative, nonmotile, nonsporing, coccobacillus that requires cystine for growth. The organism can remain viable for weeks in soil, water, carcasses, and hides. It is resistant for months at temperatures of freezing or below. The natural reservoir is rodents, particularly rabbits and hares. This is a biosafety level 2 agent.

It is on the HHS Select Agents list, the USDA High Consequence list, the Australia Group Core list, and is listed as a Category A Potential Terrorism Agent by the CDC.

In People:

CDC Case Definition: An illness characterized by several distinct forms, including the following: Ulceroglandular (cutaneous ulcer with regional lymphadenopathy); glandular (regional lymphadenopathy with no ulcer); oculoglandular (conjunctivitis with preauricular lymphadenopathy); oropharyngeal (stomatitis or pharyngitis or tonsillitis and cervical lymphadenopathy); intestinal (intestinal pain, vomiting, and diarrhea); pneumonic (primary pleuropulmonary disease); typhoidal (febrile illness without early localizing signs and symptoms). Clinical diagnosis is supported by evidence or history of a tick or deerfly bite, exposure to tissues of a mammalian host of *F. tularensis*, or exposure to potentially contaminated water. Laboratory criteria for a presumptive diagnosis is (1) elevated serum antibody titer(s) to *F. tularensis* antigen (without documented fourfold or greater change) in a patient with no history of tularemia vaccination or (2) detection of *F. tularensis* in a clinical specimen by fluorescent assay. Laboratory criteria for a confirmatory diagnosis is (1) isolation of *F. tularensis* in a clinical specimen or (2) fourfold or greater change in serum antibody titer to *F. tularensis* antigen.

Communicability: Direct person-to-person transmission does not occur. When dealing with infected individuals, use standard contact precautions. Avoid direct contact with wounds or wound drainage.

Normal Routes of Exposure: Inhalation; Ingestion; Abraded Skin; Mucous Membranes; Vectors (mosquitoes; ticks; biting flies).

Infectious Dose: 5–10 Organisms (inhalation); >1,000,000 Organisms (ingestion).

Secondary Hazards: Blood and body fluids; Body tissue; Fomites (with vectors).

Incubation: 1–14 days.

Signs and Symptoms: Tularemia can be difficult to differentiate from other diseases because it can have multiple clinical manifestations. It can lead to enlarged lymph nodes similar to plague (buboes) except the buboes are more likely to ulcerate. In general, the onset is usually abrupt, with fever (100–104°F), headache, chills and rigors, generalized body aches (often prominent in the low back), an inflammation of the mucous membrane lining the nose, and sore throat. A pulse–temperature dissociation has been observed. A dry or slightly productive cough and substernal pain or tightness frequently occur with or without signs of pneumonia. Nausea, vomiting, and diarrhea may occur progressing to sweats, fever, chills, progressive weakness, malaise, anorexia, and weight loss. If inhaled, there may be hemorrhagic inflammation of the airways early in the course of illness, which may progress to bronchopneumonia or a typhoidal syndrome. If transmitted via a bite, there may be ulceration at the site of the bite with swelling of the regional lymph node.

Suggested Alternatives for Differential Diagnosis: Psittacosis, Q fever, plague, diphtheria, tick-borne diseases, mycotic infections.

Mortality Rate (untreated): $\leq$35% (Pulmonary).

In Agricultural Animals:

Target species: Affects over 250 species including mammals, birds, reptiles, and fish.

Communicability: Direct transmission does not occur.

Normal Routes of Exposure: Inhalation; Ingestion; Abraded Skin; Mucous Membranes; Vectors (mosquitoes; ticks; biting flies).

Secondary Hazards: Blood and body fluids; Body tissue; Fomites (with vectors).

Incubation: 1–10 days.

Signs and Symptoms: Characterized by sudden onset of high fever, lethargy, anorexia, stiffness, and incoordination (ataxia). Pulse and respiratory rates are increased. Coughing, diarrhea, dehydration, with frequent urination (pollakiuria) may develop. Prostration and death may occur in a few hours or days. Subclinical cases may be common.

Suggested Alternatives for Differential Diagnosis: Plague and other septicemic diseases, acute pneumonia.

Mortality Rate (untreated): Varies by species.

C17-A014

Legionella pneumophila

Legionnaire's Disease
ICD-10: A48.1; A48.2

It is an aerobic, gram-negative (poorly stained), nonsporing, rod-shaped bacterium. It can survive for months in tap or distilled water. It does not naturally occur in animals. The natural reservoir is water systems. This is a biosafety level 2 agent.

It is on the Australia Group Warning list.

In People:

CDC Case Definition: Legionellosis is associated with two clinically and epidemiologically distinct illnesses: Legionnaires' disease, which is characterized by fever, myalgia, cough, and clinical or radiographic pneumonia; and Pontiac fever, a milder illness without pneumonia. Laboratory criteria for diagnosis of a suspect case is (1) by seroconversion: fourfold or greater rise in antibody titer to specific species, or serogroups of *Legionella* other than *L. pneumophila* serogroup 1 (e.g., *L. micdadei*, *L. pneumophila* serogroup 6); or (2) by seroconversion: fourfold or greater rise in antibody titer to multiple species of *Legionella* using pooled antigen and validated reagents; or (3) by the detection of specific *Legionella* antigen or staining of the organism in respiratory secretions, lung tissue, or pleural fluid by direct fluorescent antibody (DFA) staining, immunohistochemstry (IHC), or other similar methods, using validated reagents; or (4) by detection of *Legionella* species by a validated nucleic acid assay. Laboratory criteria for diagnosis of a confirmed is (1) by culture: isolation of any *Legionella* organism from respiratory secretions, lung tissue, pleural fluid, or other normally sterile fluid; or (2) by detection of *L. pneumophila* serogroup 1 antigen in urine using validated reagents; or (3) by seroconversion: fourfold or greater rise in specific serum antibody titer to *L. pneumophila* serogroup 1 using validated reagents.

Communicability: Direct person-to-person transmission does not occur. When dealing with infected individuals, use standard contact precautions.

Normal Routes of Exposure: Inhalation.

Infectious Dose: "Low."

Secondary Hazards: None.

Incubation: 2–16 days.

Signs and Symptoms: Anorexia, malaise, myalgia, and headache progressing to fever, chills, nonproductive cough, abdominal pain, diarrhea, and pneumonia. Blood-streaked phlegm may be present. The severity of disease ranges from a mild cough to a rapidly fatal pneumonia. Death occurs through progressive pneumonia with respiratory failure and/or shock and multiorgan failure. Recovery always requires antibiotic treatment and is usually complete after several weeks or months. A variation of the disease that does not involve pneumonia is known as Pontiac fever and has a recovery within 5 days.

Suggested Alternatives for Differential Diagnosis: Bronchitis, pneumonia, meningitis, gastroenteritis, septic shock, congestive heart failure and pulmonary edema, pleural effusion, costochondritis, prostatitis, adult respiratory distress syndrome (ARDS), HIV infection and AIDS, and Q fever.

Mortality Rate (untreated): ≤25%.

C17-A015

Liberobacter africanus
Huanglongbing

This bacteria has not yet been successfully cultured outside of citrus plants or the psyllid vectors that carry it. Once infected, vectors remain capable of transmitting the disease for their entire lives, but progeny are free of the bacterium. It causes symptoms only under relatively cool conditions (68–77°F) and generally has a milder effect than *L. asiaticus* (C17-A019).

It is one of the USDA Select Agent Listed Plant Pathogens.

In Plants:

Target species: Citrus plants.

Normal routes of transmission: Vectors (Psyllids *Diaphorina citri* Kuwayama and *Trioza erytreae*).

Secondary Hazards: Fruit; Budwood; Crop debris.

Signs: Early symptoms include leaves that are yellow, or have a mottled or blotchy appearance. Initial presentation may be on a single shoot or branch but then spreads throughout the tree. Affected trees show twig dieback. Fruit are sparse, small, abnormal in appearance, and fail to color properly. The affected fruit are lopsided, small, and remain green. The fruit often contain aborted seeds.

C17-A016

Liberobacter asiaticus

Huanglongbing

This bacteria has not yet been successfully cultured outside citrus plants or the psyllid vectors that carry it. Once infected, vectors remain capable of transmitting the disease for their entire lives, but progeny are free of the bacterium. It is heat-tolerant and able to cause symptoms at temperatures above 86°F. It generally causes greater damage than *L. africanus* (C17-A018).

It is one of the USDA Select Agent Listed Plant Pathogens.

In Plants:

Target species: Citrus plants.

Normal routes of transmission: Vectors (Psyllids *Diaphorina citri* Kuwayama and *Trioza erytreae*).

Secondary Hazards: Fruit; Budwood; Crop debris.

Signs: Early symptoms include leaves that are yellow, or have a mottled or blotchy appearance. Initial presentation may be on a single shoot or branch but then spreads throughout the tree. Affected trees show twig dieback. Fruit are sparse, small, abnormal in appearance, and fail to color properly. The affected fruit are lopsided, small, and remain green. The fruit often contain aborted seeds.

C17-A017

Mycoplasma capricolum

Contagious Caprine Pleuropneumonia

It is a gram-negative, nonmotile bacterium. They are among the smallest of bacterial organisms and lack a true cell wall. This organism requires a host to survive.

It is on the USDA High Consequence list and the Australia Group Core list.

In People:

Does not occur in humans.

In Agricultural Animals:

Target species: Goats.

Communicability: Direct transmission is possible.

Normal routes of exposure: Inhalation.

Secondary Hazards: Aerosols (cough, sneeze).

Incubation: 1–8 weeks.

Signs and Symptoms: Causes severe fibrinous pleuropneumonia characterized by respiratory distress, coughing, nasal discharge, and high mortality rate. In the peracute form, the animal dies within 3 days showing minimal respiratory signs. In the acute form symptoms include weakness, anorexia, fever, mucopurulent nasal discharge, excess salivation, and diarrhea. Respiration is accelerated and painful, and may be accompanied by a grunt. Coughing is frequent, violent, and productive. In the terminal stages animals are unable to move. There is also a septicemic form of the disease that does not affect the respiratory tract.

Postmortem findings in acute cases include serofibrinous pleuritis, pneumonia with varying degrees of lung consolidation, and swollen bronchial lymph nodes. Cuts on the surface of the lung have a granular appearance and produce straw-colored exudate.

Current United Nations recommendations include slaughter of all infected and in-contact goats. Quarantine and movement controls should be implemented.

Suggested Alternatives for Differential Diagnosis: Peste des petits ruminants, contagious agalactia syndrome, caseous lymphadenitis, pasteurellosis, and other causes of pneumonia.

Mortality Rate (untreated): $\leq$100%.

C17-A018

Mycoplasma mycoides mycoides

Contagious bovine pleuropneumonia

It is a gram-negative, nonmotile bacterium. They are among the smallest of bacterial organisms and lack a true cell wall. This organism requires a host to survive.

It is on the USDA High Consequence list and the Australia Group Core list.

In People:

Does not occur in humans.

In Agricultural Animals:

Target species: Cattle.

Communicability: Direct transmission is possible.

Normal routes of exposure: Inhalation. Transplacental infection of unborn calves can also occur.

Secondary Hazards: Aerosols (cough, sneeze).

Incubation: 3–16 weeks.

Signs and Symptoms: In acute cases, signs include fever, anorexia, and depression. Respiration is accelerated and painful, and may be accompanied by a grunt. Percussion of the chest is painful; respiration is rapid, shallow, and abdominal. The animal may experience nose bleeds, and purulent or mucoid nasal discharges that progress to coughing and chest pain. In calves up to 6 months there may also be arthritis with swelling of affected joints. If the animal is going to die, it usually occurs within 3 weeks. Subacute and chronic cases are common. Clinical signs are milder and include temperature, loss of condition, and respiratory problems that may only be apparent after exercise. Symptoms last from 3 to 4 weeks and then the animal appears to recover. Subclinical cases occur and may be important as carriers. Carriers may not be detectable either clinically or serologically and constitute a serious problem in control programs.

Postmortem findings in acute cases include severe pneumonia with copious yellow exudates. One or both lungs may be completely consolidated with a characteristic "marbled" appearance. Affected areas are pink to dark red, swollen with a firm consistency. Most of the time lesions are unilateral. In chronic cases, necrotic lung tissue becomes encapsulated. Pleural adhesions are common.

In an unprecedented outbreak, current United Nations recommendations include slaughter of all infected and in-contact herds. Quarantine and movement controls should be implemented. Otherwise vaccination with quarantine and movement controls.

Suggested Alternatives for Differential Diagnosis: Pasteurellosis and other causes of pneumonia, East Coast fever, traumatic pericarditis, hydatid cyst, actinobacillosis and tuberculosis, and bovine farcy.

Mortality Rate (untreated): ≤50%.

C17-A019

Pseudomonas mallei (Agent LA)

Glanders

ICD-10: A24.0

It is an aerobic, gram-negative, nonmotile, nonsporing, rod-shaped bacterium. It can survive in the environment for up to 2 months in sheltered positions. It is normally a disease of horses. It is a rare and sporadic disease in humans, but it is transmissible by direct contact with sick animals or infected materials. The natural reservoirs are horses, donkeys, and mules. This is a biosafety level 3 agent.

It is on the HHS Select Agents list, the USDA High Consequence list, the Australia Group Core list, and is listed as a Category B Potential Terrorism Agent by the CDC.

In People:

CDC Case Definition: Has not been developed.

Communicability: Direct person-to-person transmission does not occur. When dealing with infected individuals, use standard contact precautions.

Normal Routes of Exposure: Abraded Skin; Mucous Membranes.

Infectious Dose: Unknown.

Secondary Hazards: Aerosols (cough, sneeze); Contact; Body fluids; Fomites.

Incubation: 1–14 days.

Signs and Symptoms: Pneumonia with or without bacteremia. Pulmonary abscesses, fluid (pleural effusion) and pus (empyema) in the chest cavity may occur. In acute cases pus is discharged from the nose. There are ulcers in the mucous membranes of the nose and possibly the pharynx.

Suggested Alternatives for Differential Diagnosis: Other causes of pneumonia.

Mortality Rate (untreated): 95% (acute cases).

In Agricultural Animals:

Target species: Horses.

Communicability: Direct transmission is possible.

Normal routes of exposure: Inhalation; Ingestion; Mucous membranes.

Secondary Hazards: Contact; Body fluids; Fomites.

Incubation: 3–7 days.

Signs and Symptoms: High fever; coughing; thick nasal discharge; rapidly spreading, deep ulceration of the nasal mucosa; submaxillary lymph nodes swollen and painful; nodules on the skin, abdomen, and lower limbs; death in 1–2 weeks. In the cutaneous form, the lymphatics are enlarged and nodular abscesses ("buds") of 0.5–2.5 cm develop, which ulcerate and discharge yellow oily pus.

Postmortem findings include lymph nodes that are enlarged, fibrotic, and abscessed; the nasal cavity, pharynx, larynx, and trachea show nodules, ulcers, and stellate scars; the lungs have seed-like (miliary), firm, rounded, encapsulated gray nodules resembling tubercles; as well as cutaneous lesions.

Affected animals should be destroyed and carcasses burned or buried. Premises should be thoroughly cleaned and disinfected. In contact animals should be quarantined and tested.

Suggested Alternatives for Differential Diagnosis: Epizootic lymphangitis, ulcerative lymphangitis, strangles, dourine, melioidosis, and fungal infections.

Mortality Rate (untreated): "Usually fatal."

C17-A020

Pseudomonas pseudomallei

Melioidosis
ICD-10: A24

It is an aerobic, gram-negative, motile, nonsporing, rod-shaped bacterium. It can survive for many months in surface water and up to 3 months in shaded soil. The natural reservoir is soil and water. This is a biosafety level 2 agent. Additional primary containment and personnel precautions may be indicated for activities with a high potential for aerosol or droplet production.

It is on the HHS Select Agents list, the USDA High Consequence list, the Australia Group Core list, and is listed as a Category B Potential Terrorism Agent by the CDC.

In People:

CDC Case Definition: Has not been developed.

Communicability: Direct person-to-person transmission is rare. When dealing with infected individuals, use standard contact precautions.

Normal Routes of Exposure: Inhalation; Ingestion; Abraded Skin; Mucous Membranes.

Infectious Dose: Unknown.

Secondary Hazards: Aerosols (body fluids); Blood and body fluids; Body tissue.

Incubation: 2 days to years.

Signs and Symptoms: Manifestations range from asymptomatic pulmonary consolidation to necrotizing pneumonia and/or rapidly fatal septicemia. Symptoms include fever, rigors, night sweats, muscle pain (myalgia), anorexia, and headache. Depending on the route of exposure, additional symptoms may include urticaria progressing to skin lesions, swollen lymph glands, chest pain, cough, pneumonia, pulmonary abscesses, fluid in the chest cavity (pleural effusion), sensitivity to light (photophobia), lacrimation, and diarrhea.

Suggested Alternatives for Differential Diagnosis: Other causes of pneumonia, typhoid fever, tuberculosis, plague, anthrax infection, smallpox.

Mortality Rate (untreated): 90% (Septicemia).

In Agricultural Animals:

Target species: Pigs, sheep, goats.

Communicability: Direct transmission is rare.

Normal Routes of Exposure: Inhalation; Ingestion; Abraded Skin; Mucous Membranes.

Secondary Hazards: Aerosols (body fluids); Blood and body fluids; Body tissue; Vectors (mechanical—rodents).

Incubation: Variable.

Signs and Symptoms: Disease most likely due to percutaneous inoculation often develops at distant sites without evidence of active infection at the inoculation site. The organs most commonly affected include the lungs, spleen, liver, and associated lymph nodes. Causes fever, prostration, cough, pneumonia with respiratory distress, difficulty breathing (dyspnea), nasal and ocular discharges, colic and diarrhea; nervous symptoms such as rhythmic, oscillating motions of the eyes (nystagmus), circling, incoordination, and spasms; anorexia; abscesses in superficial lymph nodes, joints, the udder, and the viscera; and arthritis. May produce partial paralysis.

Postmortem findings include multiple abscesses in most organs especially in the regional lymph nodes, spleen, and liver. Abscesses contain a thick, caseous greenish-yellow, or off-white pus. There is usually no calcification. Additional findings may include pneumonic changes in the lungs, suppurative polyarthritis with the joint capsules containing fluid and large masses of greenish-yellow pus, and meningoencephalitis.

Suggested Alternatives for Differential Diagnosis: Symptomatology is not very characteristic and the disease is difficult to diagnose. Consider tuberculosis, nonspecific purulent conditions, caseous lymphadenitis, actinobacillosis. In horses consider strangles and glanders.

Mortality Rate (untreated): Sheep $\leq$90%.

C17-A021

***Ralstonia solanacearum* race 3, biovar 2**

Bacterial wilt

It is a bacterial pathogen that causes common wilt. There are five races of *Ralstonia solanacearum* with different hosts and geographic distributions. Race 3 is found worldwide except in the United States and Canada.

It is one of the USDA Select Agent Listed Plant Pathogens and is on the Australia Group Core list.

In Plants:

Target species: Geraniums, tomatoes, eggplant, peppers, and potatoes.

Normal routes of transmission: Contaminated soil; Contaminated irrigation water.

Secondary Hazards: Mechanical vectors (planting, harvesting, track out). There is little or no risk of transmission through leaf contact or splashing of water from leaf-to-leaf.

Signs: The primary symptom is wilted and/or yellowed leaves similar to wilting symptoms caused by other pathogens. However, leaf spots are rarely present. Initially, leaves wilt during the day, but recover during the night time. Plants may become stunted and slightly yellowed (chlorotic). Vascular discoloration of the stem is common, and roots may sometimes turn brown. Bacterial streaming may be seen if stem sections are placed into water. In potatoes, bacterial ooze may collect in the tuber eyes and soil may stick to secretions. Eventually, plants fail to recover and die.

Current APHIS control action requires elimination of the infected plants and destruction of all plants within 1 m of an infected shipment.

C17-A022

Salmonella typhi

Typhoid fever

ICD-10: A01.0

It is an aerobic and facultatively anaerobic, gram-negative, motile, nonsporing, rod-shaped bacterium. It can survive outside a host on the skin for up to 20 min, in dust for up to 30 days, and in feces for up to 62 days. Does not naturally occur in animals. This is a biosafety level 2 agent.

It is on the Australia Group Core list, and is listed as a Category B Potential Terrorism Agent by the CDC.

In People:

CDC Case Definition: An illness caused by *S. typhi* that is often characterized by insidious onset of sustained fever, headache, malaise, anorexia, relative bradycardia, constipation or diarrhea, and nonproductive cough. However, many mild and atypical infections occur. Carriage of *S. typhi* may be prolonged. Laboratory criteria for diagnosis is isolation of *S. typhi* from blood, stool, or other clinical specimen.

Communicability: Direct person-to-person transmission is possible through fecal/oral. When dealing with infected individuals, use standard contact precautions. Wash hands frequently.

Normal Routes of Exposure: Ingestion.

Infectious Dose: 100,000 Organisms (Ingestion).

Secondary Hazards: Blood and body fluids; Fecal matter; Vectors (mechanical).

Incubation: 7–28 days.

Signs and Symptoms: Insidious onset of sustained fever, severe headache, a vague feeling of bodily discomfort (malaise), chills, cough, nausea, vomiting, abdominal pain, rose spots, muscle aches, loss of appetite, usually constipation (although it may cause diarrhea), and bloody stools. There are few clinical features that reliably distinguish it from a variety of other infectious diseases. Severe cases produce confusion, delirium, intestinal perforation, and death. The illness may last for 3–4 weeks but individuals may become asymptomatic carriers capable of spreading the disease (e.g., Typhoid Mary).

Suggested Alternatives for Differential Diagnosis: Campylobacteriosis, cryptosporidiosis, cyclosporiasis, *E. coli* infections, *Listeria* monocytogenes, shigellosis, *Vibrio* infections, yersiniosis, ingestion of bacterial toxins such as staphylococcal enterotoxins or botulinum toxin.

Mortality Rate (untreated): 12%.

C17-A023

Serratia marcescens (Agent SM)

BW Simulant

It is a facultatively anaerobic, gram-negative, nonsporing, rod-shaped bacterium with a blood red color. It can survive outside a host in dust for up to 35 days. The main risk factor for exposure to *S. marcescens* is hospitalization. In the hospital, it tends to cause infections of the respiratory and urinary tracts rather than the gastrointestinal tract. However, it causes opportunistic infections of the endocardium, eyes, blood, wounds, urinary, and respiratory tracts. Notorious nosocomial pathogen responsible for about 4% of hospital acquired pneumonias. Until the 1960s, medical researchers used *S. marcescens* as a biological marker for studying the transmission of microorganisms because it was considered a harmless saprophyte. It is now recognized as an opportunistic human pathogen.

In People:

CDC Case Definition: None established.

Communicability: Direct person-to-person transmission does not occur.

Normal Routes of Exposure: Inhalation; Abraded Skin; Mucous Membranes.

Infectious Dose: Unknown.

Secondary Hazards: None.

Incubation: Unknown.

Signs and Symptoms: Depend on the site of infection. Infection may produce osteomyelitis or arthritis; pneumonia [with chills, productive cough, low blood pressure (hypotension), difficulty breathing (dyspnea), or chest pain]; meningitis or cerebral abscesses (with headache, fever, vomiting, stupor, coma); or intra-abdominal infections (with biliary drainage, hepatic abscess, pancreatic abscess, peritoneal exudate).

Suggested Alternatives for Differential Diagnosis: Infections from other bacteria such as Enterobacter, *E. coli*, Klebsiella, Proteus, and Providencia, producing meningitis, pneumonia, or sepsis.

Mortality Rate (untreated): 26% (Nosocomial bloodstream infection).

C17-A024

Shigella dysenteriae (Agent Y)

Shigellosis
ICD-10: A03

It is an aerobic, gram-negative, nonsporing, nonmotile, rod-shaped bacterium. It can survive outside a host in water for up to 3 days, in feces for up to 11 days, in/on flies for up to 12 days. It rarely occurs in animals except primates. The organism is frequently found in water polluted with human feces. This is a biosafety level 2 agent.

It is on the Australia Group Core list.

In People:

CDC Case Definition: An illness of variable severity characterized by diarrhea, fever, nausea, cramps, and tenesmus. Asymptomatic infections may occur. Laboratory criteria for diagnosis is isolation of *Shigella* from a clinical specimen.

Communicability: Direct person-to-person transmission is possible through fecal/oral. When dealing with infected individuals, use standard contact precautions. Wash hands frequently.

Normal Routes of Exposure: Ingestion.

Infectious Dose: 10–180 organisms (Ingestion).

Secondary Hazards: Fecal matter; Vectors (mechanical—flies, roaches).

Incubation: 12 h to 7 days.

Signs and Symptoms: Abdominal pain, cramps, diarrhea, fever, vomiting, tenesmus, and blood, pus, or mucus in stools. Infections also cause mucosal ulceration, rectal bleeding, drastic dehydration. Serious less frequent complications include sepsis, seizures, convulsions, rectal prolapse, toxic megacolon, intestinal perforation, renal failure, and hemolytic uremic syndrome.

Suggested Alternatives for Differential Diagnosis: Amebiasis, cholera, salmonellosis, schistomosis, yersiniosis, *Clostridium* difficile colitis, colon cancer, Crohn's disease, ulcerative colitis.

Mortality Rate (untreated): ≤15%; ≤50% (Hemolytic uremic syndrome).

C17-A025

Vibrio cholerae (Agent HO)

Cholera
ICD-10: A00

It is a facultative anaerobic, gram-negative, toxin producing, rod- or curved rod-shaped bacterium. Cholera toxin is a potent protein enterotoxin elaborated by the organism in the small intestine. The El Tor strain can survive in fresh water for long periods. It is easily killed by chlorine at purification concentrations. The natural reservoir is humans, although marine shellfish and plankton can also serve as reservoirs. This is a biosafety level 2 agent.

It is on the Australia Group Core list, and is listed as a Category B Potential Terrorism Agent by the CDC.

In People:

CDC Case Definition: An illness characterized by diarrhea and/or vomiting; severity is variable. Laboratory criteria for diagnosis is (1) isolation of toxigenic (i.e., cholera toxin-producing) *V. cholerae* O1 or O139 from stool or vomitus; or (2) serologic evidence of recent infection.

Communicability: Direct person-to-person transmission is possible through fecal/oral. When dealing with infected individuals, use standard contact precautions. Wash hands frequently.

Normal Routes of Exposure: Ingestion.

Infectious Dose: 1,000,000 Organisms. Varies markedly with gastric acidity.

Secondary Hazards: Fecal matter; Vomit; Vectors (mechanical—flies).

Incubation: 6 h to 5 days.

Signs and Symptoms: Vomiting is a prominent early manifestation of the disease followed rapidly by abdominal cramps with profuse watery diarrhea (opaque white liquid that does not have a bad odor; often described as resembling rice water). Bowel movements are frequent and often uncontrolled. Stool volume is more than that from any other infectious diarrhea. Diarrhea and vomiting can lead to severe dehydration, vascular collapse, shock, and death. Dehydration can develop within hours after the onset of symptoms. This contrasts with disease produced by infection from any other enteropathogen.

Suggested Alternatives for Differential Diagnosis: Gastroenteritis.

Mortality Rate (untreated): $\leq$50%.

C17-A026

Xanthomonas albilineans

Leaf scald disease

It is an aerobic, gram-negative, nonsporing, rod-shaped bacterium. It is motile with a single polar flagellum.

It is on the Australia Group Core list.

In Plants:

Target species: Sugar cane.

Normal routes of transmission: Cross contamination; Airborne (Unknown mechanism).

Secondary Hazards: Mechanical Vectors (planting, harvesting, track out, possibly insects); Crop debris.

Signs: Initially, leaves may have a thin white line but eventually become slightly yellowed (chlorotic). The vascular tissue is damaged and leaves begin to die from the edges towards the midrib, ultimately resulting in plant death. It has the capacity to destroy whole fields in a few months.

C17-A027

Xanthomonas campestris pv. citri

Citrus canker

It is a gram-negative, nonsporing, motile, rod-shaped bacterium.

It is on the Australia Group Core list.

In Plants:

Target species: Many varieties of citrus including grapefruits, oranges, lemons, limes, and tangerines.

Normal routes of transmission: Contact (entry through leaf pores or wounds in green portions of the plant).

Secondary Hazards: Airborne (wind, rain, storm); Crop debris; Mechanical Vectors (planting, harvesting, track out).

Signs: Plants infected with citrus canker have characteristic lesions on leaves, stems, and fruit with raised, brown water-soaked margins usually with a yellow hallow effect around the lesion. Incubation is typically 14–60 days. Older lesions may fall out, creating a shot-hole effect. Fruit production declines and then stops. The disease ultimately kills the tree.

C17-A028

Xanthomonas oryzae pv. oryzae

Bacterial leaf blight

It is a gram-negative, nonsporing, motile, rod-shaped bacterium.

It is one of the USDA Select Agent Listed Plant Pathogens, the Australia Group Awareness list.

In Plants:

Target species: Rice.

Normal routes of transmission: Contact (entry through leaf pores or wounds in green portions of the plant).

Secondary Hazards: Airborne (wind, rain, storm); Mechanical vectors (irrigation); Crop debris.

Signs: Infected leaves of young plants wilt and roll up, turning grayish-green to yellow, until the whole seedling dies. Plants that survive the disease are stunted and yellowish. Mature rice plants develop lesions that look like water-soaked stripes on the leaf blades and eventually increase in length and width becoming yellow to grayish-white until the entire leaf dries up. Bacterial discharge appears on newer lesion early in the morning that looks like a milky dewdrop.

C17-A029

Xylella fastidiosa

Pierce's disease

It is a gram-negative, motile bacterium that is limited to the xylem of plants.

It is one of the USDA Select Agent Listed Plant Pathogens, the Australia Group Awareness list.

In Plants:

Target species: Grapes, Almond trees, Alfalfa, Oleander, Citrus trees.

Normal routes of transmission: Vectors (Sharpshooter *Homalodisca coagulate* and spittlebugs).

Secondary Hazards: None.

Signs: The leaves become slightly yellowed (chlorotic) along the margins before drying, or the outer leaf may dry suddenly while still green. Foliar symptoms gradually spread out toward the cane from the point of infection. The oldest leaves show severe scorching and the youngest leaves may not have any symptoms. "Scorched" leaves detach from the distal end of the petiole (leaf stem) rather than from the base of the petiole, leaving the bare petioles attached to canes, often well after normal leaf fall. Eventually the canes become affected and dieback along with the roots.

C17-A030

Yersinia pestis (Agent LE)

Plague

ICD-10: A20

It is a facultative anaerobic, gram-negative, nonsporing, intracellular, nonmotile, oval shaped bacterium. It grows slowly even at optimal temperatures. It is stable in water for up to 16 days. At or near freezing temperatures it will remain alive for months or years. It is killed by exposure to several hours of sunlight. The natural reservoir is numerous mammalian species. Cattle, horses, sheep, and pigs are not known to develop symptomatic illness from plague. It does cause illness in cats and dogs. This is a biosafety level 2 agent. Additional primary containment and personnel precautions may be indicated for activities with a high potential for aerosol or droplet production.

It is on the Australia Group Core list, and is listed as a Category A Potential Terrorism Agent by the CDC.

In People:

CDC Case Definition: Plague is transmitted to humans by fleas or by direct exposure to infected tissues or respiratory droplets; the disease is characterized by fever, chills, headache, malaise, prostration, and leukocytosis that manifests in one or more of the following principal clinical forms: Regional lymphadenitis (bubonic plague); septicemia without an evident bubo (septicemic plague); plague pneumonia, resulting from hematogenous spread in bubonic or septicemic cases (secondary pneumonic plague), or inhalation of infectious droplets (primary pneumonic plague); pharyngitis and cervical lymphadenitis resulting from exposure to larger infectious droplets or ingestion of infected tissues (pharyngeal plague). Laboratory criteria for a Presumptive diagnosis is (1) elevated serum antibody titer(s) to *Y. pestis* fraction 1 (F1) antigen (without documented fourfold or greater change) in a patient with no history of plague vaccination or (2) detection of F1 antigen in a clinical specimen by fluorescent assay. Laboratory criteria for a Confirmatory diagnosis is (1) isolation of *Y. pestis* from a clinical specimen or (2) fourfold or greater change in serum antibody titer to *Y. pestis* F1 antigen.

Communicability: Direct person-to-person transmission is possible. When dealing with infected individuals, use standard contact precautions. Avoid direct contact with wounds or wound drainage.

Normal Routes of Exposure: Inhalation; Abraded Skin; Vectors (fleas—*Xenopsylla cheopis and Pulex irritans*).

Infectious Dose: 340 organisms (Inhalation).

Secondary Hazards: Aerosols (cough, sneeze); Blood and body fluids; Body tissue; Fomites (with vectors); Vector cycle.

Incubation: 1–6 days.

Signs and Symptoms: Initial signs and symptoms may be nonspecific and include fever, chills, a vague feeling of bodily discomfort (malaise), muscle pain (myalgia), nausea, sore throat, and headache. In bubonic plague, the lymph system becomes impacted with swelling and tenderness in nodes near the site of the bite. Additional signs and symptoms include fever, chills, and prostration. Fluid from the swollen node (bubo) is infectious. If the lungs become infected, pneumonia may develop. Initial signs and symptoms include fever, chills, cough, and difficulty breathing progressing rapidly to shock and death if not treated early. Respiratory involvement poses significant hazard of person-to-person transmission through aerosol generated by coughing.

Suggested Alternatives for Differential Diagnosis: Scarlet fever, cellulitis, cat scratch disease, gas gangrene, necrotizing fasciitis, tick-borne diseases such as Rocky Mountain spotted fever, pneumonia, septic shock, acute respiratory distress syndrome (ARDS), disseminated intravascular coagulation.

Mortality Rate (untreated): ≤60% (bubonic); ≤100% (septicemic and pneumonic).

C17-A031

Yersinia pseudotuberculosis

Yersiniosis

ICD-10: A28.2

It is a gram-negative, nonsporing, oval-shaped bacterium. It is primarily a zoonotic disease of birds and mammals, with humans as incidental hosts. Often seen in hamsters, guinea pigs, and chinchillas. It has also been associated with enterocolitis and diarrhea in young sheep at pasture that are debilitated from factors such as starvation and cold weather. This is a biosafety level 2 agent.

It is on the Australia Group Warning list.

In People:

CDC Case Definition: None established.

Communicability: Direct person-to-person transmission is possible through fecal/oral. When dealing with infected individuals, use standard contact precautions. Wash hands frequently.

Normal Routes of Exposure: Ingestion.

Infectious Dose: 1,000,000 Organisms (Ingestion).

Secondary Hazards: Fecal matter; Fomites.

Incubation: 3–10 days.

Signs and Symptoms: Symptoms include pain in the lower-right abdominal area resembling appendicitis, as well as fever, headache, pharyngitis, anorexia, vomiting, and possibly watery diarrhea. May also produce arthritis, inflammation of the iris (iritis), cutaneous ulceration. Infection may also produce abscesses in the liver, bone infection (osteomyelitis), and septicemia. Carriers may be asymptomatic. May also cause infections of other sites such as wounds, joints, and the urinary tract.

Suggested Alternatives for Differential Diagnosis: Appendicitis, mesenteric lymphadenitis.

Mortality Rate (untreated): "Low."

References

Acha, Pedro N., and Boris Szyfres. *Zoonoses and Communicable Diseases Common to Man and Animals, Scientific and Technical Publication No. 580.* 3rd ed. Vol. 1, Bacterioses and Mycoses. Washington, DC: Pan American Health Organization, 2003.

Burrows, Dickinson W., and Sara E. Renner. "Biological Warfare Agents as Threats to Potable Water." *Environmental Health Perspectives* 107 (1999): 975–84.

California Department of Food and Agriculture. Animal Health and Food Safety Services. Animal Health Branch. *Biosecurity: Selection and Use of Surface Disinfectants.* Revision June 2002.

Canada Minister of National Health and Welfare. Laboratory Centre For Disease Control, Office of Biosafety. *Material Safety Data Sheet-Infectious Substances: Bacillus anthracis.* January 23, 2001.

———. *Material Safety Data Sheet-Infectious Substances: Brucella* spp. *(B. abortus, B. canis, B. melitensis, B. suis).* January 23, 2001.

———. *Material Safety Data Sheet-Infectious Substances: Burkholderia (Pseudomonas) mallei.* January 23, 2001.

———. *Material Safety Data Sheet-Infectious Substances: Burkholderia (Pseudomonas) pseudomallei.* January 23, 2001.

———. *Material Safety Data Sheet-Infectious Substances: Chlamydia psittaci.* January 23, 2001.

———. *Material Safety Data Sheet-Infectious Substances: Clostridium botulinum.* January 23, 2001.

———. *Material Safety Data Sheet-Infectious Substances: Clostridium perfringens.* July 21, 2003.

———. *Material Safety Data Sheet-Infectious Substances: Corynebacterium diphtheriae.* January 23, 2001.

———. *Material Safety Data Sheet-Infectious Substances: Escherichia coli, Enterohemorrhagic.* September 27, 2001.

———. *Material Safety Data Sheet-Infectious Substances: Francisella tularensis.* May 25, 2001.

———. *Material Safety Data Sheet-Infectious Substances: Legionella pneumophila.* April 23, 2001.

———. *Material Safety Data Sheet-Infectious Substances: Salmonella typhi.* March 2001.

———. *Material Safety Data Sheet-Infectious Substances: Serratia* spp. April 23, 2001.

———. *Material Safety Data Sheet-Infectious Substances: Shigella* spp. March 2001.

———. *Material Safety Data Sheet-Infectious Substances: Vibrio cholerae,* serovar 01, serogoup O139 (Bengal). September 27, 2001.

———. *Material Safety Data Sheet-Infectious Substances: Yersinia enterocolitica, Yersinia pseutotuberculosis.* March 5, 2001.

———. *Material Safety Data Sheet-Infectious Substances: Yersinia pestis.* March 5, 2001.

Centers for Disease Control and Prevention. "Case Definitions for Infectious Conditions Under Public Health Surveillance." *Morbidity and Mortality Weekly Report* 46 (RR-10) (1997).

———. *Case Definitions.* http://www.cdc.gov/epo/dphsi/casedef/case_definitions.htm. 2006.

———. *"Facts About Pneumonic Plague."* Undated.

———. "Biological and Chemical Terrorism: Strategic Plan for Preparedness and Response. Recommendations of the CDC Strategic Planning Workgroup." *Morbidity and Mortality Weekly Report* 49 (RR-4) (2000): 1–14.

———. "Medical Examiners, Coroners, and Biologic Terrorism: A Guidebook for Surveillance and Case Management." *Morbidity and Mortality Weekly Report* 53 (RR-8) (2004).

Chin, James, ed. *Control of Communicable Diseases Manual.* 17th ed. Washington, DC: American Public Health Association, 2000.

Chung, K.-R. and R.H. Brlansky. *Citrus Diseases Exotic to Florida: Huanglongbing (Citrus Greening), Document PP-210.* Plant Pathology Department, Florida Cooperative Extension Service, Institute of Food and Agricultural Sciences, University of Florida, June 2005.

Committee on Foreign Animal Diseases of the United States Animal Health Association. *Foreign Animal Diseases.* Revised 1998. Richmond, VA: Pat Campbell & Associates and Carter Printing Company, 1998.

Compton, James A.F. *Military Chemical and Biological Agents: Chemical and Toxicological Properties.* Caldwell, NJ: The Telford Press, 1987.

Dennis, David T., Thomas V. Inglesby, Donald A. Henderson, John G. Bartlett, Michael S. Ascher, Edward Eitzen, Anne D. Fine, Arthur M. Friedlander, Jerome Hauer, Marcelle Layton, Scott R. Lillibridge, Joseph E. McDade, Michael T. Osterholm, Tara O'Toole, Gerald Parker, Trish M. Perl, Philip K. Russell, and Kevin Tonat. "Tularemia as a Biological Weapon: Medical and Public Health Management." *Journal of the American Medical Association* 285 (2001): 2763–773.

Department for Environment, Food and Rural Affairs. *Contingency Plan—Potato Ring Rot.* http://www.defra.gov.uk/planth/ring.pdf#search=%22Ring%20Rot%20of%20Potatoes%22. September 11, 2006.

———. *Ring Rot of Potato.* November 2003. www.defra.gov.uk/planth/pestnote/ringrot.pdf. September 11, 2006.

Department of Agriculture. 9 CFR Part 121—"Possession, Use, and Transfer of Biological Agents and Toxins," 2005: 753–67.

European and Mediterranean Plant Protection Organization. *Diagnostic Protocols for Regulated Pests.* February 15, 2006. http://archives.eppo.org/EPPOStandards/diagnostics.htm. September 25, 2006.

Fediaevsky, Alexandre. *Manual for the Recognition of Exotic Diseases of Livestock: A Reference Guide for Animal Health Staff Food and Agriculture.* 2nd ed. Organization of the United Nations Secretariat of the Pacific Community. http://www.spc.int/rahs/manual/manuale.html. November 1, 2005.

Floyd, Joel, and Conrad Krass. *New Pest Response Guidelines: Huanglongbing, Citrus Greening Disease, Version 1.1.* United States Department of Agriculture, Animal and Plant Health Inspection Service, May 10, 2006.

Gottwald, Tim R., James H. Graham, and Tim S. Schubert. "Citrus Canker: The pathogen and its Impact." *Plant Health Progress Online.* August/September 2002. http://www.apsnet.org/online/feature/citruscanker/. September, 2006.

Henderson, Donald A., Thomas V. Inglesby, and Tara O'Toole, eds. *Bioterrorism: Guidelines for Medial and Public Health Management.* Chicago, IL: AMA Press, 2002.

Herenda, D. *Manual on Meat Inspection for Developing Countries.* Reprint 2000. Rome, Italy: Food and Agriculture Organization of the United Nations, 1994.

Holty, Jon-Erik C., Dena M. Bravata, Hau Liu, Richard A. Olshen, Kathryn M. McDonald, and Douglas K. Owens. "Systematic Review: A Century of Inhalational Anthrax Cases from 1900 to 2005." *Annals of Internal Medicine* 144 (February 21, 2006): 270–80.

Inglesby, Thomas V., David T. Dennis, Donald A. Henderson, John G. Bartlett, Michael S. Ascher, Edward Eitzen, Anne D. Fine, Arthur M. Friedlander, Jerome Hauer, John F. Koerner, Marcelle Layton, Joseph McDade, Michael T. Osterholm, Tara O'Toole, Gerald Parker, Trish M. Perl, Philip K. Russell, Monica Schoch-Spana, and Kevin Tonat. "Plague as a Biological Weapon: Medical and Public Health Management." *Journal of the American Medical Association* 283 (2000): 2281–290.

Inglesby, Thomas V., Donald A. Henderson, John G. Bartlett, Michael S. Ascher, Edward Eitzen, Arthur M. Friedlander, Jerome Hauer, Joseph McDade, Michael T. Osterholm, Tara O'Toole, Gerald Parker, Trish M. Perl, Philip K. Russell, and Kevin Tonat. "Anthrax as a Biological Weapon: Medical and Public Health Management." *Journal of the American Medical Association* 281 (1999): 1735–745.

Kahn, Cynthia M., and Susan E. Aiello, eds. *The Merck Veterinary Manual.* 9th ed. Online Version. Whitehouse Station, NJ: Merck & Co., Inc., 1998.

Kortepeter, Mark, George Christopher, Ted Cieslak, Randall Culpepper, Robert Darling, Julie Pavlin, John Rowe, Kelly McKee, Jr., and Edward Eitzen, Jr., eds. *Medical Management of Biological Casualties Handbook.* 4th ed. Fort Detrick, MD: United States Army Medical Research Institute of Infectious Diseases, February 2001.

Lemay, Andrea, Scott Redlin, Glenn Fowler, and Melissa Dirani. *Pest Data Sheet: Ralstonia solanacearum race 3 biovar 2.* United States Department of Agriculture, Animal and Plant Health Inspection Service, February 12, 2003.

Lindler, Luther E., Frank J. Lebeda, and George W. Korch, eds. *Biological Weapons Defense: Infectious Diseases and Counterbioterrorism.* Totowa, NJ: Humana Press, Inc., 2005.

Pan American Health Organization. *Emergency Vector Control After Natural Disaster.* Scientific Publication No. 419. Washington, DC: Pan American Health Organization, 1982.

Plant Protection and Quarantine. *New Pest Response Guidelines: Ralstonia solanacearum race 3 biovar 2, Version 4.0*. United States Department of Agriculture, Animal and Plant Health Inspection Service, January 15, 2004.

———. *Questions and Answers on Ralstonia solanacearum Race 3 Biovar 2*. United States Department of Agriculture, Animal and Plant Health Inspection Service, April 2004.

———. *Citrus Canker Eradication Program: Environmental Assessment, April 1999*. United States Department of Agriculture, Animal and Plant Health Inspection Service, April 19, 1999.

Public Health Service. 42 CFR Part 73—"*Select Agents and Toxins*," 2004: 443–58.

Richmond, Jonathan Y., and Robert W. McKinney, eds. *Biosafety in Microbiological and Biomedical Laboratories*. 4th ed. Washington, DC: US Government Printing Office, 1999.

Schmitt, Clare K., Karen C. Meysick, and Alison D. O'Brien. "Bacterial Toxins: Friends or Foes?" *Emerging Infectious Diseases 5* (April/June 1999): 224–34.

Sidell, Fredrick R., Ernest T. Takafuji, and David R. Franz, eds. *Medical Aspects of Chemical and Biological Warfare, Textbook of Military Medicine Series, Part 1, Warfare, Weaponry, and the Casualty*. Washington, DC: Office of the Surgeon General, Department of the Army, 1997.

Sifton, David W., ed. *PDR Guide to Biological and Chemical Warfare Response*. Montvale, NJ: Thompson/Physicians Desk Reference, 2002.

Somani, Satu M., ed. *Chemical Warfare Agents*. New York: Academic Press, 1992.

The Ohio State University. Plant Pathology. *Ohio State University Extension Fact Sheet: Bacterial Ring Rot of Potatoes*. http://ohioline.osu.edu/hyg-fact/3000/3103.html. September 11, 2006.

Turnbull, P.C.B. *Guidelines for the Surveillance and Control Anthrax in Humans and Animals*. 3rd ed. Document WHO/EMC/ZDI./98.6. Geneva: World Health Organization, 1998.

United States Army Headquarters. *Potential Military Chemical/Biological Agents and Compounds, Field Manual No. 3-11.9*. Washington, DC: Government Printing Office, January 10, 2005.

———. *Technical Aspects of Biological Defense, Technical Manual No. 3-216*. Washington, DC: Government Printing Office, January 12, 1971.

United States Department of Agriculture. Office of Pest Management National Plant Disease Recovery System. "Recovery Plan for Huanglongbing (HLB) or Citrus Greening Caused by 'Candidatus' *Liberibacter africanus, L. asiaticus, and L. americanus*." September 20, 2006.

———. "Recovery Plan for Ralstonia solanacearum Race 3 Biovar 2 Causing Brown Rot of Potato, Bacterial Wilt of Tomato, and Southern Wilt of Geranium." October 11, 2006.

United States Food & Drug Administration, Center for Food Safety & Applied Nutrition. *Foodborne Pathogenic Microorganisms and Natural Toxins Handbook*. January 30, 2003. http://www.cfsan.fda.gov/~mow/preface.html. April 18, 2005.

World Health Organization. *Health Aspects of Chemical and Biological Weapons: Report of a WHO Group of Consultants*. Geneva: World Health Organization, 1970.

———. *Public Health Response to Biological and Chemical Weapons: WHO Guidance*. Geneva: World Health Organization, 2004.

———. *Fact Sheet #107 Cholera*. Geneva: Health Communications and Public Relations, March 1996.

———. *Fact Sheet #108 Epidemic Dysentery*. Geneva: Health Communications and Public Relations, March 1996.

———. *Fact Sheet #124 Foodborne Diseases, Emerging*. Geneva: Health Communications and Public Relations, January 2002.

———. *Fact Sheet #125 Escherichia coli O157:H7*. Geneva: Health Communications and Public Relations, July 1996.

———. *Fact Sheet #139 Drug-Resistant Salmonella*. Geneva: Health Communications and Public Relations, April 2005.

———. *Fact Sheet #149 Typhoid Fever*. Geneva: Health Communications and Public Relations, March 1997.

———. *Fact Sheet #173 Brucellosis*. Geneva: Health Communications and Public Relations, July 1997.

———. *Fact Sheet #264 Anthrax*. Geneva: Health Communications and Public Relations, October 2001.

———. *Fact Sheet #267 Plague*. Geneva: Health Communications and Public Relations, February 2005.

———. *Fact Sheet #270 Botulism*. Geneva: Health Communications and Public Relations, August 2002.

———. *Fact Sheet #285 Legionellosis*. Geneva: Health Communications and Public Relations, February 2005.

———. *Fact Sheet #89 Diphtheria*. Geneva: Health Communications and Public Relations, May 29, 1998.

———. *Guidelines for the Control of Shigellosis, Including Epidemics Due to Shigella dysenteriae 1*. Geneva: WHO Press, 2005.

World Organization for Animal Health. "Technical Disease Card for Contagious Bovine Pleuropneumonia." April 22, 2002.

———. *Manual of Diagnostic Tests and Vaccines for Terrestrial Animals 2004*. Updated July 22, 2005. http://www.oie.int/. January 2006.

18

Viral Pathogens

18.1 General Information

Viruses are the simplest type of microorganism consisting of a protein outer coat containing genetic material (i.e., RNA, DNA). They are much smaller than bacteria and vary in size from 0.01 to 0.27 μm. They are an obligate (single set of conditions) intracellular parasite and lack a system for their own metabolism. They replicate by taking over metabolic processes of the invaded host cell and redirecting them toward virus production. Approximately 60% of infectious disease in man is caused by a virus. Most treatment for viral infections is supportive because of limited antiviral medications.

Viruses are very costly, time consuming, and difficult to grow in large quantities. They only replicate in living cells and cannot be grown in a growth media like bacteria (Chapter 17). In some cases, viruses can be grown on the chorioallantoic membrane of fertilized eggs.

Pathogens employed as biological weapons can be used for both lethal and incapacitating purposes. They cause disease by invading tissue or by producing toxins (Chapter 16) that are detrimental to the infected individual. Pathogens can be selected to target a specific host (e.g., humans, cows, pigs) or they may pose a broad threat to both animals and to people.

Pathogens deployed as antianimal biological weapons are generally used to produce lethal effects in an agriculturally significant species such as cows, pigs, or chickens. Although these pathogens are selected to target a specific animal species, there is the possibility that the disease may migrate to humans. The diseases produced by these crossover pathogens may be difficult for medical personnel not trained in exotic pathology to diagnose.

Other pathogens are selected to produce lethal effects in an agriculturally significant crop species such as wheat, corn, or rice. Symptoms in plants vary greatly and include mottling of leaves, stunting, lesions, leaf rolling, and death. Natural transmission from plant to plant is often by insects. There is little potential for migration of these pathogens to humans or animals.

A final group of biological warfare pathogens are those used as simulants to model the release of other, more hazardous agents. Pathogens employed as biological warfare simulants do not generally pose a significant risk to people, animals, or plants. However, individuals with respiratory illness or suppressed immune systems may be at risk should they be exposed to an infectious dose of the agent.

Some viruses can be crystallized, similar to chemical compounds, while others can be stored as freeze-dried powders. In these forms, viruses are easy to disperse. However, because they are living organisms and can be killed during the dispersal process there are limitations to the methods that can be used. They can also be stored and dispersed via infected vectors (e.g., mosquitoes, fleas). In most cases, large-scale attacks will be clandestine and only detected through epidemiological analysis of resulting disease patterns. Localized or small-scale attacks may take the form of "anthrax" letters. Even in these cases, without the inclusion of a threat the attack may go unnoticed until the disease appears in exposed individuals (e.g., the initial 2001 anthrax attack at American Media Inc., which claimed the life of Robert Stevens).

In general, unless a local reservoir (i.e., intermediate host that may or may not be affected by the virus) is established, pathogens are easily killed by unfavorable environmental factors such as fluctuations in temperature, humidity, food sources, or ultraviolet light. For this reason, their persistency is generally limited to days. However, freeze-dried pathogens can remain in a preserved state almost indefinitely and are reactivated when exposed to moisture.

It is possible that local insects can become both a reservoir and a vector for the pathogen. Under these circumstances, the pathogen can survive well after the initial release and can rapidly spread beyond the immediately affected area. In many cases, once a vector is infected, it is capable of transmitting the disease throughout its life span. Some pathogens are transmitted directly to the young of the vector so that the next generation is born infected. Response activities must also include efforts to contain and eliminate these vectors (e.g., application of pesticides).

Incubation times for diseases resulting from infection vary depending on the specific pathogen, but are generally on the order of days to weeks. Exposures to extremely high doses of some pathogens may reduce the incubation period to as short as several hours. The pathway of exposure (e.g., inhalation, ingestion) can also cause a significant change in the incubation time required as well as the clinical presentation of the disease.

Some diseases caused by viruses are communicable and easily transferred from an infected individual to anyone in close proximity. Typically, this occurs when the infected individual coughs or sneezes creating an infectious aerosol. These aerosols enter the body of a new host through inhalation and/or contact of the aerosol with the mucous membranes of the eyes, nose, or mouth. In addition, although intact skin is an effective barrier against most pathogens, abrasions, or lacerations circumvents this protective barrier and allows entry of the pathogen into the body.

In addition to direct aerosol exposure, some viruses are potentially associated with exposure through ingestion of contaminated food and/or water. Pathogens that normally infect individuals through an ingestion pathway also pose a significant risk of secondary infections from the "fecal/oral" cycle. Some individuals or animals can become asymptomatic carriers and are capable of spreading the disease long after their recovery. Mechanical vectors (e.g., flies, roaches) can also carry pathogens and spread the disease to food not directly contaminated.

18.2 Response

18.2.1 Personal Protective Requirements

A number of conditions must be considered when selecting protective equipment for individuals at the scene of a release. For instances such as "anthrax" letters when the mechanism of release is known and it does not involve an aerosol generating device, then responders can use Level C with high-efficiency particulate air (HEPA) filters.

If an aerosol generating device is employed (e.g., sprayer), or the dissemination method is unknown and the release is ongoing, then responders should wear a Level A protective ensemble. Once the device has stopped generating the aerosol or has been rendered inoperable, and the aerosol has settled, then responders can downgrade to Level B.

In all cases, there is a significant hazard posed by contact of contaminated material with skin that has been cut or lacerated, or through injection of pathogens by contact with debris. Appropriate protection to avoid any potential abrasion, laceration, or puncture of the skin is essential. Individuals with damaged or open skin should not be allowed to enter the contaminated area.

18.2.1.1 Working with Infected Individuals

For most viruses, use infection control guidelines standard precautions. If appropriate, or the identity of the virus is unknown, use additional droplet and airborne precautions. Avoid direct contact with wounds or wound drainage. If there are hemorrhagic symptoms, use VHF-specific barrier precautions.

18.2.1.1.1 Infection Control Guidelines Standard Precautions

1. Wash hands after patient contact.
2. Wear gloves when handling potentially infectious items that are contaminated with blood, body fluids, or excreta.
3. Wear a mask and eye protection during procedures likely to generate splashes or sprays of infectious fluids.
4. Handle all contaminated and potentially contaminated equipment and linen in a manner that will prevent cross contamination.
5. Take care when handling sharps. When practical use a mouthpiece or other ventilation device as an alternative to mouth-to-mouth resuscitation.

18.2.1.1.2 Droplet Precautions

Standard precautions plus

1. Place the patient in a private room or cohort them with someone with the same infection. Maintain at least 3 feet between patients.
2. Wear a mask when working within 3 feet of the patient.
3. Limit movement and transport of the patient. Place a mask on the patient if they need to be moved.

18.2.1.1.3 Airborne Precautions

Standard precautions plus

1. Place the patient in a private room with negative air pressure that has a minimum of six air changes per hour and appropriate filtration before the air is discharged from the room.
2. Wear respiratory protection when entering the room.
3. Limit movement and transport of the patient. Place a mask on the patient if they need to be moved.

18.2.1.1.4 VHF-Specific Barrier Precautions

Standard precautions plus

1. Consolidate patients in the same area of the hospital to minimize exposures to other patients and healthcare workers. Place the patient in a private room with negative air pressure that has a minimum of six air changes per hour and appropriate filtration before the air is discharged from the room. Restricted access of nonessential staff and visitors to patient's room.
2. Use dedicated medical equipment (e.g., stethoscopes, point-of-care monitors).
3. Wear appropriate personal protective equipment (PPE) when entering the room. Clean hands prior to donning PPE for patient contact. Wear goggles and face shield [unless wearing a full-face powered air-purifying respirator (PAPR)], N-95 mask or PAPR, double gloves, impermeable gown, and leg and shoe coverings.
4. After patient contact, remove gown, leg and shoe coverings, and gloves in a designated decontamination area. Hands should be washed prior to removal of respiratory and eye protection (i.e., mask/respirator, face shield, and goggles) to minimize potential exposure of mucous membranes. Wash hands again after removal of facial PPE.
5. Ensure environmental disinfection with appropriate commercial hospital disinfectant or a 1% household bleach solution.

18.2.2 Decontamination

18.2.2.1 Food

All foodstuffs in the area of a release should be considered contaminated. Unopened items may be used after decontamination of the container. Opened or unpackaged items should be destroyed. Fruits and vegetables should be washed thoroughly with antimicrobial soap and water. Many pathogens can survive in food containers for extended periods. Some pathogens can survive in turbid water for long periods.

18.2.2.2 Casualties/Personnel

Infected individuals: Unless the individual is reporting directly from the scene of an attack (e.g., "anthrax" letter, aerosol release, etc.) then decontamination is not necessary. Use standard protocols for individuals that may be infected with communicable diseases transmissible via an aerosol.

Direct exposure: In the event that an individual is at the scene of a known or suspected attack (e.g., white-powder letter, aerosol release, etc.), have them wash their hands and face thoroughly with antimicrobial soap and water as soon as possible. If antimicrobial soap is not available, use any available soap or shampoo. They should also blow their nose to remove any agent particles that may have been captured by nasal mucous. Remove all clothing and seal in a plastic bag. To avoid further exposure of the head, neck, and face to the agent, cut off potentially contaminated clothing that must be pulled over the head. Shower using copious amounts of antimicrobial soap (if available) and water. Ensure that the hair has been washed and rinsed to remove potentially trapped agent. The CDC does not recommend that individuals use bleach or other disinfectants on their skin.

18.2.2.3 Animals

Unless the animals are at the scene of an attack then decontamination is not necessary.

Apply universal decontamination procedures using antimicrobial soap and water. If antimicrobial soap is not available, use any available soap, shampoo, or detergent. In some cases, severe infection may require euthanasia of animals and/or herds. Consult local/state veterinary assistance office. If the pathogen has not been identified, then wear a fitted N95 protective mask, eye protection, disposable protective coverall, disposable boot covers, and disposable gloves when dealing with infected animals.

18.2.2.4 Plants

May require removal and destruction of infected plants. Incineration of impacted fields may be required. Consult local/state agricultural assistance office. Wear disposable protective coverall, disposable boot covers, and disposable gloves to prevent spread of contamination. Many plant pathogens are spread via insects and response activities must also include efforts to contain and eliminate these vectors.

18.2.2.5 Property

Surface disinfectants: Compounds containing phenolics, chlorhexidine, potassium peroxymonosulfate, hypochlorites such as household bleach, and iodine/iodophores. Other disinfectants like alcohols and quaternary ammonium salts will work but are less effective.

The military also identifies the following "nonstandard" decontaminants: Detrochlorite (thickened bleach mixture of diatomaceous earth, anionic wetting agent, calcium hypochlorite, and water), 3% aqueous peracetic acid solution, 1% aqueous hyamine solution, and a 10% aqueous sodium or potassium hydroxide solution.

Large area fumigants: Gases including formaldehyde, ethylene oxide, or chlorine dioxide. These materials are highly toxic to humans and animals, and fumigation operations must be adequately controlled to prevent unnecessary exposure. Additional methods include vaporized hydrogen peroxide and an ionized hydrogen peroxide aerosol.

Fomites: Some pathogens may be absorbed into clothing or bedding causing these items to become infectious and capable of transmitting the disease. Others may contain vectors (e.g., lice, ticks) that pose a transmission hazard. Deposit items in an appropriate biological waste container and send to a medical waste disposal facility.

Alternatively, cotton or wool articles can be boiled in water for 30 minutes, autoclaved at 253°F for 45 minutes, or immersed in a 2% household-bleach solution (i.e., 1 gallon of bleach in 2 gallons of water) for 30 minutes followed by rinsing.

18.3 Fatality Management

Unless the cadaver is coming directly from the scene of an attack (e.g., "anthrax" letter, aerosol release), process the body according to established procedures for handling potentially infectious remains.

For fatalities that are grossly contaminated with a biological agent (e.g., individuals who died from trauma or other complications at the scene of a biological attack), remove all clothing and personal effects. Items that will be retained for further processing should be double sealed in impermeable containers, ensuring that the inner container is decontaminated before placing it in the outer one. Otherwise, dispose contaminated articles at an appropriate medical waste disposal facility.

Thoroughly wash the remains with antimicrobial soap and water. Pay particular attention to areas where agent may get trapped, such as hair, scalp, pubic areas, fingernails, folds of skin, and wounds. If deemed appropriate, the cadaver can be treated with a surface disinfectant listed in Section 18.2.2.

If there is a potential that vectors may be involved, care must be taken to kill any vectors (e.g., lice, ticks) remaining either on the cadaver or residing in fomites. Remove all potentially infested clothing depositing it in a container that will trap and eliminate vectors. Dispose contaminated articles at an appropriate medical waste disposal facility.

Once the remains have been thoroughly decontaminated, no further protective action is necessary. Use standard burial procedures.

Once the remains have been thoroughly decontaminated, process the body according to established procedures for handling potentially infectious bodies. Use appropriate burial procedures.

In 2004 (*Morbidity and Mortality Weekly Report* 53), the Centers for Disease Control and Prevention determined that the risks of occupational exposure to biological terrorism agents outweighed the advantages of embalming fatalities and recommended that, unless death was due to a highly infectious virus (e.g., smallpox and hemorrhagic fever viruses), the body should be buried without embalming. Fatalities due to highly infectious viruses should be cremated without embalming.

C18-A

AGENTS

C18-A001

African horse sickness

Reoviridae

It is normally found in Sub Saharan Africa, the Middle East, India, and Pakistan. The natural reservoirs are horses, donkeys, and mules. It is relatively heat stable and can survive at room temperature for over 30 days. It is resistant to common disinfectants but is destroyed by sun light.

It is on the USDA High Consequence list and the Australia Group Core list.

In People:

Does not occur in humans.

In Agricultural Animals:

Target species: Horses, mules, donkeys.

Communicability: Direct transmission does not occur.

Normal routes of exposure: Vectors (midges—*Culicoides* species).

Secondary Hazards: Blood and body fluids; Body tissue (Viscera).

Incubation: 2–14 days.

Signs and Symptoms: Causes fever, labored breathing with increased respiratory rate, and depression. There may be paroxysmal coughing. May cause conjunctivitis, subcutaneous edema of the head, neck, brisket, thorax, and ventral abdomen. There may be edema in the supraorbital fossa above the eye.

Postmortem findings include edema in the periorbital tissue, neck muscles, ligamentum nuchae, intermuscular and lungs; hemorrhage of the tongue, intestinal serosa, kidneys, and pericardium; and general subcutaneous edema and hemorrhage.

There is no cure for this disease; affected animals are destroyed.

Suggested Alternatives for Differential Diagnosis: Colics, anthrax, equine rhinopneumonitis, equine infectious anemia, equine influenza, equine encephalosis, equine viral arteritis, and piroplasmosis. Field diagnosis may be virtually impossible.

Mortality Rate (untreated): ≤95% (Horses); ≤50% (Mules); ≤10% (Donkeys).

C18-A002

African swine fever

Asfarviridae

It is normally found in Sub Saharan Africa. The natural reservoir are ticks, wild pigs, and warthogs. Ticks remain infected for life. It is resistant to cleaning and disinfection. It can survive for 15 weeks in putrefied blood, 70 days in blood on wooden boards, and 11 days in feces at room temperature. It can survive for up to 6 months in infected meats and can even survive in smoked or partly cooked sausages and other pork products.

It is on the USDA High Consequence list and the Australia Group Core list.

In People:

Does not occur in humans.

In Agricultural Animals:

Target species: Pigs.

Communicability: Direct transmission is possible.

Normal routes of exposure: Ingestion; Mucous Membranes; Vectors (ticks).

Secondary Hazards: Contact; Blood and body fluids; Body tissue; Fecal matter; Fomites (with vectors).

Incubation: 3–15 days.

Signs and Symptoms: Fever, incoordination, labored breathing, coughing, nasal, and ocular discharge; loss of appetite with diarrhea and vomiting; cyanosis of the extremities and hemorrhages of skin.

Postmortem findings include hemorrhage of the skin, kidneys, heart, serous membranes, gastrohepatic, and renal lymph nodes; accumulation of fluid in the pericardial sac, thorax, and abdominal cavity.

Suggested Alternatives for Differential Diagnosis: Hog cholera, salmonellosis, erysipelas, Glasser's disease, erysipelas, septicemic salmonellosis, pasteurellosis, *Haemophilus suis* infection, *Streptococcus suis* infection, eperythrozoonosis, porcine dermatitis and nephropathy syndrome, porcine reproductive and respiratory syndrome, pseudorabies, coumarin poisoning, and salt poisoning.

Mortality Rate (untreated): ≤100%.

C18-A003

Akabane

Bunyaviridae

It is normally found in Africa, Asia, and Australia. The natural reservoirs are mosquitoes and midges. Insects remain infected for life. It causes abortions, stillbirths, premature births, and various dysfunctions or deformities in newborns. Adult animals are not clinically affected by the virus.

It is on the USDA High Consequence list.

In People:

Does not occur in humans.

In Agricultural Animals:

Target species: Cattle, sheep, goats.

Communicability: Direct transmission does not occur.

Normal routes of exposure: Vectors (mosquitoes—*Aedes vexans, Culex triteeniorhynchus, Anopheles funestus*; midges—*Culicoides* species).

Secondary Hazards: None.

Incubation: 1–6 days.

Signs and Symptoms: Causes abortions, stillbirths and fetal deformities including blindness, nystagmus, deafness, dullness, incoordination, and/or paralysis.

Suggested Alternatives for Differential Diagnosis: Bluetongue, Aino virus, Chuzan virus, Cache Valley virus.

Mortality Rate (untreated): Most offspring eventually die due to complications.

C18-A004

Highly pathogenic avian influenza

Orthomyxoviridae

It is normally found worldwide. The natural reservoir is wild birds. Virus can remain viable in feces and water for up to 32 days. It remains viable at moderate temperatures for long periods in the environment and can survive indefinitely in frozen material.

This form of avian influenza, due to the H5 and H7 subtypes, occurs infrequently in humans. In most cases the clinical picture is limited to conjunctivitis. However, may take a much more lethal form.

It is on the USDA High Consequence list and the Australia Group Core.

In People:

CDC Case Definition: None established.

Communicability: Direct person-to-person transmission is possible. When dealing with infected individuals, use airborne precautions.

Normal Routes of Exposure: Inhalation; Mucous membranes.

Infectious Dose: Unknown.

Secondary Hazards: Aerosols (cough, sneeze); Fomites; Fomites (from people and poultry).

Incubation: 18–72 h.

Signs and Symptoms: Abrupt onset of fever with chills, sore throat, muscle pain (myalgia), severe frontal headache, sensitivity to light (photophobia), burning sensation or pain in the eyes, weakness and severe fatigue, nonproductive cough, chest pain, and difficulty breathing (dyspnea). Nasal discharge is typically absent. May progress to acute encephalopathy with altered mental status, coma, seizures, and incoordination (ataxia).

Suggested Alternatives for Differential Diagnosis: Adenoviruses, arenaviruses, California encephalitis, coxsackieviruses, cytomegalovirus, dengue fever, eastern equine encephalitis, echoviruses, infectious mononucleosis, Japanese encephalitis, Lyme disease, meningitis, parainfluenza virus, rhinoviruses, bacterial sepsis, severe acute respiratory syndrome (SARS), St Louis encephalitis, upper respiratory infection, Venezuelan encephalitis, and West Nile encephalitis.

Mortality Rate (untreated): $\leq$70%.

In Agricultural Animals:

Target species: Poultry.

Communicability: Direct transmission is possible.

Normal routes of exposure: Inhalation; Mucous Membranes.

Secondary Hazards: Contact; Blood and body fluids; Body tissue; Fecal matter; Fomites.

Incubation: Several hours to 7 days.

Signs and Symptoms: In peracute cases, sudden death without clinical signs. Otherwise, severe respiratory distress, blood-tinged oral and nasal discharges, watery eyes and sinuses, cyanosis of the combs, wattle and shanks, swelling of the head, diarrhea that may be greenish, nervous signs, and sudden death (as soon as 24 h after the appearance of the first signs). Birds that survive may develop CNS complications.

Suggested Alternatives for Differential Diagnosis: Infectious bronchitis, infectious laryngotracheitis, Newcastle disease, mycoplasmosis, duck viral enteritis, infectious coryza, ornithobacteriosis, turkey coryza, fowl cholera, aspergillosis, heat exhaustion, and severe water deprivation.

Mortality Rate (untreated): $\leq$50%.

C18-A005

Banana bunchy top virus

Nanoviridae

It is normally found in Southeast Asia, the Pacific islands, and parts of India and Africa. It does not occur in Central or South America.

It is on the Australia Group Awareness list.

In Plants:

Target species: Bananas.

Normal routes of transmission: Vectors (aphids). They can retain the virus through their adult life.

Secondary Hazards: Mechanical vectors (planting, harvesting); Crop debris.

Signs: The first symptoms consist of darker green streaks in a pattern of "dots" and "dashes" ("Morse code" streaking) on the lower portion of the midrib. Later, the streaks appear on the secondary veins of the leaf and then on the leaf blade. Suckers that develop after infection are usually severely stunted, which causes the leaves at the top of the stem to become bunched. Leaves are usually short, stiff, erect, and more narrow than normal. Severely infected banana plants usually will not fruit. If the plant does bear fruit, it will likely be distorted and twisted.

There is no cure for the disease. Plants that are infected become reservoirs of the virus and must be destroyed.

C18-A006

Bluetongue

Reoviridae

It is an arbovirus transmitted by midges that are normally found worldwide. The natural reservoirs are cattle and midges. Midges remain infective for life. It can survive for years in blood stored at 68°F.

It is on the USDA High Consequence list and the Australia Group Core list.

In People:

Does not occur in humans.

In Agricultural Animals:

Target species: Cattle, sheep, goats, deer.

Communicability: Direct transmission does not occur.

Normal routes of exposure: Vectors (midges—*Culicoides* species).

Secondary Hazards: Blood.

Incubation: 5–20 days.

Signs and Symptoms: Disease can be quite variable. Peracute cases die as a result of severe pulmonary edema leading to difficulty breathing (dyspnea), frothing from the nostrils, and death by asphyxiation. Otherwise, characterized by fever, edema of lips, nose, face, submandibular area, eyelids, and sometimes ears; vesicles, ulcers and necrosis in and around the mouth (gums, cheeks, and tongue); congestion of the mouth, nose, nasal cavity, conjunctiva, and coronary bands; lacrimation, drooling (ptyalism), stiffness, lameness, depression; increased respiratory rate, and possible increase in sensitivity to sensory stimuli (hyperesthesia).

Suggested Alternatives for Differential Diagnosis: Foot-and-mouth disease, vesicular stomatitis, peste des petits ruminants, photosensitisation, nasal botfly infestation, pneumonia, akabane infection, epizootic hemorrhagic disease of deer, contagious ecthyma, polyarthritis, footrot, foot abscesses, plant poisonings, and coenurosis.

Mortality Rate (untreated): ≤30% (Sheep).

C18-A007

Camelpox

Poxviridae

It is normally found in northern Africa and southwestern Asia. It is a poorly characterized orthopoxvirus that causes severe disease in camels. It is genetically closely related genetically to the smallpox virus. This virus has rarely, if ever, caused disease in people.

It is on the USDA High Consequence list.

In People:

Rarely occurs in humans. Appears as lesions on the hands of individuals working with camels.

In Agricultural Animals:

Target species: Camels.

Communicability: Direct transmission is possible.

Normal routes of exposure: Not specified.

Secondary Hazards: Contact.

Incubation: Approximately 13 days.

Signs and Symptoms: Formation of papules that progress through pustules and vesicles ultimately forming crusts.

Suggested Alternatives for Differential Diagnosis: No published suggestions.

Mortality Rate (untreated): ≤25% (young camels).

C18-A008

Central European tick-borne encephalitis

Flaviviridae
ICD-10: A84.1

It is transmitted by ticks and is normally found in Europe and the western part of the former Soviet Union. The natural reservoirs are ticks, rodents, and small mammals. Ticks remain infected for life. This is a biosafety level 4 agent.

It is on the HHS Select Agents.

In People:

CDC Case Definition for arboviral encephalitis: Neuroinvasive disease requires the presence of fever and at least one of the following in the absence of a more likely clinical explanation: (1) Acutely altered mental status (e.g., disorientation, obtundation, stupor, or coma), or (2) other acute signs of central or peripheral neurologic dysfunction (e.g., paresis or paralysis, nerve palsies, sensory deficits, abnormal reflexes, generalized convulsions, or abnormal movements), or (3) increased white blood cell concentration in cerebrospinal fluid associated with illness clinically compatible with meningitis (e.g., headache or stiff neck). *Laboratory criteria* for diagnosis is (1) fourfold or greater change in virus-specific serum antibody titer, or (2) isolation of virus from or demonstration of specific viral antigen or genomic sequences in tissue, blood, CSF, or other body fluid, or (3) virus-specific immunoglobulin M (IgM) antibodies demonstrated in CSF by antibody-capture enzyme immunoassay (EIA), or (4) virus-specific IgM antibodies demonstrated in serum by antibody-capture EIA and confirmed by demonstration of virus-specific serum immunoglobulin G (IgG) antibodies in the same or a later specimen by another serologic assay (e.g., neutralization or hemagglutination inhibition). *Laboratory criteria* for a *probable case* is (1) stable (less than or equal to a twofold change) but elevated titer of virus-specific serum antibodies, or (2) virus-specific serum IgM antibodies detected by antibody-capture EIA but with no available results of a confirmatory test for virus-specific serum IgG antibodies in the same or a later specimen.

Communicability: Direct person-to-person transmission does not occur. When dealing with infected individuals, use standard contact precautions.

Normal Routes of Exposure: Ingestion (unpasteurized milk); Vectors (ticks—*Ixodes* species).

Infectious Dose: Unknown.

Secondary Hazards: Fomites (with vectors).

Incubation: 8–20 days (from ticks); 4–7 days (ingestion).

Signs and Symptoms: Typically biphasic, the first phase consists of nonspecific flu-like symptoms that last about a week. This is followed by a 1–7 day asymptomatic period. The third and final phase is an abrupt onset of encephalitis, which affects the central nervous system phase producing tremors, dizziness, and altered sensorium. Recovery is prolonged.

Suggested Alternatives for Differential Diagnosis: Meningitis, basilar artery blood clots (thrombosis), cardioembolic stroke, cavernous sinus syndromes, cerebral venous blood clots (thrombosis), confusional states and acute memory disorders, epileptic and epileptiform encephalopathies, febrile seizures, haemophilus meningitis, intracranial hemorrhage, leptomeningeal carcinomatosis, subdural pus (empyema), or bruise (hematoma).

Mortality Rate (untreated): $\leq$2%.

In Agricultural Animals:

Target species: Sheep, goats.

Communicability: Direct transmission does not occur.

Normal routes of exposure: Vectors (ticks—*Ixodes* species).

Secondary Hazards: Fomites (with vectors).

Incubation: Unknown.

Signs and Symptoms: Usually asymptomatic but may develop symptoms of meningoencephalitis including spastic paralysis of the rear legs, lockjaw, and chattering of the teeth.

Suggested Alternatives for Differential Diagnosis: Other causes of meningitis, encephalitis, or meningoencephalitis.

Mortality Rate (untreated): ≤12% (Symptomatic cases).

C18-A009

Cercopithecine herpes-1 virus

Herpesviridae
ICD-10: B00.4

It is normally found worldwide. The natural reservoir is primates. Monkeys infected with this virus usually have no symptoms or only mild ones. This is a biosafety level 2 agent.

It is on the HHS Select Agents list.

In People:

CDC Case Definition: None established.

Communicability: Direct person-to-person transmission does not occur. When dealing with infected individuals, use standard contact precautions.

Normal Routes of Exposure: Abraded skin; Mucous membranes.

Infectious Dose: Unknown.

Secondary Hazards: Blood and body fluids; Body tissue.

Incubation: 3–28 days.

Signs and Symptoms: Symptoms are acute and include fever, headache, encephalitis, vesicular skin lesions at site of the exposure, and variable neurological patterns. Involvement of the respiratory center and death usually occurs in 1–21 days after onset of symptoms. Survivors usually have considerable residual disability. May produce severe permanent neurologic impairment requiring lifelong institutionalization.

Suggested Alternatives for Differential Diagnosis: No published suggestions.

Mortality Rate (untreated): ≤80%.

C18-A010

Chikungunya

Togaviridae
ICD-10: A92.0

It is normally found in tropical areas throughout the world. The natural reservoirs are mosquitoes and primates. Mosquitoes remain infective for life. Monkeys infected with this virus usually have no symptoms or only mild ones. This is a biosafety level 3 agent.

It is on the Australia Group Core list.

In People:

CDC Case Definition: None established.

Communicability: Direct person-to-person transmission does not occur. When dealing with infected individuals, use standard contact precautions.

Normal Routes of Exposure: Vectors (mosquitoes—*Aedes aegypti*).

Infectious Dose: Unknown.

Secondary Hazards: Vector cycle; Vectors (mechanical).

Incubation: 2–12 days.

Signs and Symptoms: Symptoms include high fever, nausea, vomiting, arthritis in wrist, knee, ankle, and small joints of extremities. A rash may develop in 1–10 days. Recovery may be prolonged but immunity is long lasting.

Suggested Alternatives for Differential Diagnosis: Dengue, measles, Rocky Mountain spotted fever, rubella, tick bite fever, epidemic typhus, Q fever, typhoid, malaria, trypanosomiasis, hepatitis, infectious mononucleosis, herpes, and influenza.

Mortality Rate (untreated): There have been no documented fatalities associated with this disease.

C18-A011

Crimean-Congo hemorrhagic fever

Bunyaviridae
ICD-10: A98.0

It is normally found in southern and Eastern Europe, Central Asia, the Mediterranean, northwestern China, Africa, and the Indian subcontinent. The natural reservoirs are ticks and numerous animal species. Animals (e.g., cattle, sheep, goats, horses, pigs, hares, hedgehogs) infected with this virus usually have no clinical symptoms or suffer only a mild illness. The hemorrhagic fever is highly pathogenic and notable for aerosol transmission. This is a biosafety level 4 agent.

It is on the HHS Select Agents list and the Australia Group Core list.

In People:

CDC Case Definition: None established.

Communicability: Direct person-to-person transmission is possible. It is highly communicable in a hospital setting. When dealing with infected individuals, use VHF-specific barrier precautions.

Normal Routes of Exposure: Inhalation mucous membranes; Vectors (ticks—*Ixodes hyalomma*).

Infectious Dose: Unknown.

Secondary Hazards: Aerosols (cough, sneeze); Contact; Blood and body fluids; Body tissue; Fomites (with vectors).

Incubation: 1–13 days.

Signs and Symptoms: Sudden onset of fever, headache, pain in the muscles (myalgia) and joints (arthralgias), abdominal pain, and vomiting. Sore throat, conjunctivitis, jaundice, sensitivity to light (photophobia), and various sensory and mood alterations may develop. A rash appears on the chest and stomach, spreading to the rest of the body. May present hemorrhagic symptoms. Latter stages of the illness are characterized by vomiting, diarrhea, shock, and hemorrhagic phenomena. Recovery may be prolonged. Treatment is supportive and requires intensive care and isolation.

Suggested Alternatives for Differential Diagnosis: Malaria, typhoid fever, shigellosis, meningococcemia, salmonella infection, other tick-borne diseases, rickettsial infections, leukemia, lupus, disseminated intravascular coagulation, hemolytic uremic syndrome, leptospirosis, thrombocytopenic purpura, and idiopathic or thrombotic thrombocytopenic purpura.

Mortality Rate (untreated): ≤30%.

C18-A012

Dengue fever

Flaviviridae

ICD-10: A90, A91

It is normally found in the Caribbean, the South Pacific, Southeast Asia, the West Indies, India, and the Middle East. The natural reservoirs are humans, primates, and mosquitoes. Does not produce disease in animals. This is a biosafety level 2 (classical) or 3 (hemorrhagic) agent. Typically a fulminant, nonlethal disease; however, it may progress to a hemorrhagic form. Stabile outside a host in dried blood and exudates for up to several days at room temperature.

It is on the Australia Group Core list.

In People:

CDC Case Definition: An acute febrile illness characterized by frontal headache, retro-ocular pain, muscle (myalgia) and joint (arthralgias) pain, and rash. The principal vector is the *Aedes aegypti* mosquito and transmission usually occurs in tropical or subtropical areas. Severe manifestations (e.g., dengue hemorrhagic fever and dengue shock syndrome) are rare but may be fatal. *Laboratory criteria* for diagnosis is (1) isolation of dengue virus from serum and/or autopsy tissue samples, or (2) demonstration of a fourfold or greater rise or fall in reciprocal immunoglobulin G (IgG) or immunoglobulin M (IgM) antibody titers to one or more dengue virus antigens in paired serum samples, or (3) demonstration of dengue virus antigen in autopsy tissue or serum samples by immunohistochemistry or by viral nucleic acid detection.

Communicability: Direct person-to-person transmission does not occur. Patients are infectious and can inoculate any mosquitoes that feed on them from shortly before a fever is present until the fever passes. When dealing with infected individuals, use standard contact precautions. If hemorrhagic symptoms are present, use VHF-specific barrier precautions.

Normal Routes of Exposure: Inhalation; Abraded Skin; Vectors (mosquitoes—*Aedes aegypti*).

Infectious Dose: 1–10 organisms (Inhalation).

Secondary Hazards: Aerosols; Vector cycle.

Incubation: 3–14 days.

Signs and Symptoms: *Normal*: Sudden onset of fever with intense headache, pain behind the eyes, nausea, and vomiting. Fever may be biphasic. A rash with generalized reddening of the skin occurs. Hemorrhage may be present. Recover may be prolonged and accompanied with fatigue and depression. *Hemorrhagic*: Above symptoms suddenly worsen. A marked weakness occurs with severe restlessness and facial pallor. Skin becomes blotchy and extremities are cool. Small hemorrhages (petechia) may progress to bleeding nose and gums. Death usually occurs from shock.

Suggested Alternatives for Differential Diagnosis: Hepatitis, meningitis, Rocky Mountain spotted fever, malaria, yellow fever, leptospirosis, rickettsioses, river viruses, scrub typhus, typhoid, and other viral infections.

Mortality Rate (untreated): "Rare" (classic dengue); ≤40% (hemorrhagic).

C18-A013

Dobrava hemorrhagic fever

Bunyaviridae
ICD-10: A98.5

It is a hantavirus that is normally found in the former Yugoslavia. The natural reservoirs are small rodents and the virus is shed in their urine. Infection occurs after inhalation of dust contaminated with excreta from infected rodents or from aerosol of animal blood or fluids. Does not produce disease in animals. This is a biosafety level 3 agent.

It is on the Australia Group Core list.

In People:

CDC Case Definition for hantavirus pulmonary syndrome: An illness characterized by one or more of the following clinical features: A febrile illness (i.e., temperature greater than 101°F) characterized by bilateral diffuse interstitial edema that may radiographically resemble ARDS, with respiratory compromise requiring supplemental oxygen, developing within 72 h of hospitalization, and occurring in a previously healthy person and/or an unexplained respiratory illness resulting in death, with an autopsy examination demonstrating noncardiogenic pulmonary edema without an identifiable cause. *Laboratory criteria* for diagnosis is (1) detection of hantavirus-specific immunoglobulin M or rising titers of hantavirus-specific immunoglobulin G, or (2) detection of hantavirus-specific ribonucleic acid sequence by polymerase chain reaction in clinical specimens, or (3) detection of hantavirus antigen by immunohistochemistry.

Communicability: Direct person-to-person transmission is rare. When dealing with infected individuals, use VHF-specific barrier precautions.

Normal Routes of Exposure: Inhalation; Ingestion; Abraded skin; Mucous membranes.

Infectious Dose: Unknown.

Secondary Hazards: Aerosols (blood, fluids, contaminated dust); Blood and body fluids (from mice); Fecal (from mice); Fomites (from mouse habitation).

Incubation: 2 days to 2 months.

Signs and Symptoms: Characterized by sudden onset of high fever, headache, a vague feeling of bodily discomfort (malaise), dizziness, anorexia, abdominal and/or lower back pain, nausea and vomiting progressing to nonproductive cough, difficulty breathing (dyspnea), and rapid development of respiratory insufficiency. The next phase involves renal failure and protein in the urine (proteinuria). This is followed by excessive urination (diuresis) that may progress into dehydration and severe shock. Normal blood pressure returns and there is a dramatic drop in urine production. Recovery may be prolonged.

Suggested Alternatives for Differential Diagnosis: Drug induced noncardiac pulmonary edema, acute respiratory distress syndrome, pneumonic plague, tularemia, Q fever, and viral influenza.

Mortality Rate (untreated): ≤50%.

C18-A014

Eastern equine encephalitis

Togaviridae
ICD-10: 83.2

It is normally found in the Americas and the Caribbean. The natural reservoirs are birds and mosquitoes. Mosquitoes remain infective for life. Infection is also transferred directly to mosquito eggs. Horses may provide transient amplification of the virus. This is a biosafety level 3 agent. It does not survive outside a host.

It is on the HHS Select Agents list, the USDA High Consequence list, the Australia Group Core list, and is listed as a Category B Potential Terrorism Agent by the CDC.

In People:

CDC Case Definition: Arboviral infections may be asymptomatic or may result in illnesses of variable severity sometimes associated with central nervous system (CNS) involvement. When the CNS is affected, clinical syndromes ranging from febrile headache to aseptic meningitis to encephalitis may occur, and these are usually indistinguishable from similar syndromes caused by other viruses. Arboviral meningitis is characterized by fever, headache, stiff neck, and pleocytosis. Arboviral encephalitis is characterized by fever, headache, and altered mental status ranging from confusion to coma with or without additional signs of brain dysfunction (e.g., paresis or paralysis, cranial nerve palsies, sensory deficits, abnormal reflexes, generalized convulsions, and abnormal movements). *Laboratory criteria* for diagnosis is (1) fourfold or greater change in virus-specific serum antibody titer, or (2) isolation of virus from or demonstration of specific viral antigen or genomic sequences in tissue, blood, cerebrospinal fluid (CSF), or other body fluid, or (3) virus-specific immunoglobulin M (IgM) antibodies demonstrated in CSF by antibody-capture enzyme immunoassay (EIA), or (4) virus-specific IgM antibodies demonstrated in serum by antibody-capture EIA and confirmed by demonstration of virus-specific serum immunoglobulin G (IgG) antibodies in the same or a later specimen by another serologic assay (e.g., neutralization or hemagglutination inhibition).

Communicability: Direct person-to-person transmission does not occur. Humans can infect new mosquitoes during the fever phase. When dealing with infected individuals, use standard contact precautions and protect from vectors.

Normal Routes of Exposure: Vectors (mosquitoes—*Culex* species, *Aedes* species, *Coquillettidia* species).

Infectious Dose: Unknown.

Secondary Hazards: Vector cycle.

Incubation: 5–15 days.

Signs and Symptoms: Difficult to diagnose because of the lack of specific symptoms. Once symptoms arise, the patient often deteriorates rapidly. Symptoms include headache, nausea or vomiting, confusion, seizures, semiconsciousness (somnolence), neck stiffness, a vague feeling of bodily discomfort (malaise) and weakness, cranial nerve palsies, sensitivity to light (photophobia), autonomic disturbances such as drooling (ptyalism), fever, chills, abdominal pain, diarrhea, sore throat, severe joint pain (arthralgias) and muscle pain (myalgia), and respiratory difficulty.

Suggested Alternatives for Differential Diagnosis: Bartonellosis, brucellosis, other causes of encephalitis, coxsackieviruses, cryptococcosis, cysticercosis, cytomegalovirus, histoplasmosis, legionellosis, leptospirosis, listeria, lyme disease, malaria, rabies, tuberculosis, mumps, stroke, metabolic encephalopathy, Reye syndrome, *Bartonella* infection, *Naegleria* infection, Ebstein-Barr virus, prion disease, toxic ingestions, and AIDS.

Mortality Rate (untreated): ≤70%.

In Agricultural Animals:

Target species: Horses.

Communicability: Direct transmission does not occur.

Normal routes of exposure: Vectors (mosquitoes—*Culex* species).

Secondary Hazards: Vector cycle.

Incubation: 18–24 h.

Signs and Symptoms: Impaired vision, aimless wandering, head pressing, circling, inability to swallow, irregular ataxic gait, paralysis, convulsions, and death. Most deaths occur within 2–3 days.

Suggested Alternatives for Differential Diagnosis: Rabies, hepatoencephalopathy, leukoencephalomalacia, protozoal encephalomyelitis, equine herpes virus 1, verminous meningoencephalomyelitis, cranial trauma, botulism, and meningitis. In *birds*: Newcastle disease virus, avian encephalomyelitis virus, botulism, and listeriosis.

Mortality Rate (untreated): ≤90%.

C18-A015

Ebola hemorrhagic fever

Filoviridae
ICD-10: A98.3

It is normally found in Africa. The natural reservoir is unknown. It can survive in blood specimens for several weeks at room temperature, but does not survive for long periods after drying. Also produces disease in primates with symptoms similar to those seen in humans. This is a biosafety level 4 agent.

It is on the HHS Select Agents list, the Australia Group Core list, and is listed as a Category A Potential Terrorism Agent by the CDC.

In People:

CDC Case Definition: None established.

Communicability: Direct person-to-person transmission is possible. It is communicable up to 7 days after clinical recovery. When dealing with infected individuals, use VHF-specific barrier precautions.

Normal Routes of Exposure: Inhalation (Reston only); Ingestion; Abraded Skin; Mucous Membranes.

Infectious Dose: 1–10 organisms (Aerosol).

Secondary Hazards: Aerosols (blood, body fluids); Contact; Blood and body fluids; Body tissue; Fecal matter; Vomit; Fomites.

Incubation: 2–21 days.

Signs and Symptoms: Onset is abrupt. Symptoms include fever, headache, pain in the joints (arthralgias) and muscles (myalgia), sore throat, and weakness, followed by diarrhea, vomiting, and stomach pain progressing in some cases to rash, red eyes, hiccups, and both external and internal bleeding. Terminally ill patients often have a normal temperature, but appear to be dull, with rapid breathing, are unable to urinate, and progress into shock.

Suggested Alternatives for Differential Diagnosis: Leptospirosis, malaria, salmonella infection, lupus, typhoid fever, shigellosis, meningococcemia, rickettsial infections, Crimean-Congo hemorrhagic fever, thrombocytopenic purpura, acute surgical abdomen, acute leukemia, disseminated intravascular coagulation, hemolytic uremic syndrome.

Mortality Rate (untreated): $\leq$88%.

C18-A016

Flexal hemorrhagic fever

Arenaviridae

Is a member of the New World hemorrhagic fever viruses normally found in Brazil. The natural reservoir is the rice rat (*Oryzomys*) and the virus is shed in their urine. Infection occurs after inhalation of dust contaminated with excreta from infected rats or from aerosol of animal blood or fluids. Does not produce disease in animals. This is a biosafety level 3 agent.

It is on the HHS Select Agents list, and the Australia Group Core list.

In People:

CDC Case Definition: None established.

Communicability: Not specified.

Normal Routes of Exposure: Inhalation; Ingestion; Abraded skin; Mucous membranes.

Infectious Dose: Unknown.

Secondary Hazards: Aerosols (blood, fluids, contaminated dust); Blood and body fluids; Body tissue; Fecal matter; Fomites.

Incubation: 7–16 days.

Signs and Symptoms: Gradual onset of sustained fever accompanied by a vague feeling of bodily discomfort (malaise), headache, severe muscle pain (myalgia) and joint pain (arthralgias), dizziness, sensitivity to light (photophobia), sore throat, nausea, vomiting, and diarrhea progressing to hemorrhagic symptoms including nosebleed (epistaxis), bloody vomit, and blood in stool or urine. Patients may deteriorate due to vascular or neurologic complications. The illness lasts 2–14 days after the onset of symptoms. Recovery may be prolonged.

Suggested Alternatives for Differential Diagnosis: Disseminated intravascular coagulation, hemolytic uremic syndrome, leptospirosis, malaria, salmonella infection, lupus, thrombocytopenic purpura, typhoid fever, shigellosis, meningococcemia, rickettsial infections, acute leukemia, and idiopathic or thrombotic thrombocytopenic purpura.

Mortality Rate (untreated): Not specified.

C18-A017

Foot-and-mouth disease (Agent OO)

Picornaviridae

ICD-10: B08.8

It is normally found in parts of Europe, Africa, Asia, and South America. The natural reservoirs are wild and domestic cloven-footed animals. It is one of the most contagious diseases of these animals. Airborne spread is possible up to 40 miles over land and 180 miles over water. It can survive for up to 2 weeks in dry fecal material and for 6 months in slurry, about 1 week in urine, and over 3 months in hay.

It is on the USDA High Consequence list and the Australia Group Core list.

In People:

Does not occur in humans.

In Agricultural Animals:

Target species: Cattle, sheep, pigs.

Communicability: Direct transmission is possible.

Normal routes of exposure: Inhalation; Ingestion; Mucous membranes.

Secondary Hazards: Aerosols (breathing); Contact; Blood and body fluids; Body tissue; Fomites; Vectors (mechanical—Cats, Dogs, Birds).

Incubation: 2–14 days.

Signs and Symptoms: Characterized by fever, sudden appearance of vesicles or blisters on the mouth, nose, feet, and teats. The blisters quickly rupture to leave erosions or ulcers. Animals with mouth ulcers drool (ptyalism) and back off of feed. Due to sore feet, animals prefer to lie down. Cattle may also lose one or both horns of the foot. Animals with teat lesions are hard to milk and prone to mastitis.

Many countries slaughter all infected and in-contact animals. Quarantine and movement controls are implemented. Carcasses are either burned or buried on or close to the premises. Buildings are thoroughly washed and disinfected.

Suggested Alternatives for Differential Diagnosis: Rinderpest, infectious bovine rhinotracheitis, bovine herpes mammillitis, malignant catarrhal fever, Peste des petits ruminants, vesicular stomatitis, bluetongue, bovine viral diarrhea, and foot rot in cattle, vesicular exanthema of swine, swine vesicular disease, and foreign bodies or trauma.

Mortality Rate (untreated): "Rarely"; ≤60% (Young animal).

C18-A018

Goatpox

Poxviridae

It is normally found in Africa and parts of Europe and Asia. The natural reservoir is goats. Considered to be the most severe poxvirus diseases in domestic animals. It can persist for up to 6 months in shaded animal pens, and for at least 3 months in dry scabs on the fleece, skin, and hair from infected animals.

It is on the USDA High Consequence list and the Australia Group Core list.

In People:

Does not occur in humans.

In Agricultural Animals:

Target species: Goats.

Communicability: Direct transmission is possible.

Normal routes of exposure: Inhalation; Abraded skin; Mucous membranes.

Secondary Hazards: Aerosols (cough, sneeze, contaminated dust); Contact; Body fluids; Fecal matter; Fomites; Vectors (mechanical—stable fly).

Incubation: 5–14 days.

Signs and Symptoms: Characterized by sudden onset of fever, discharges from the nose and eyes, and drooling (ptyalism). Animals have a loss of appetite and show a reluctance to move. Pox lesions appear on the skin but are most obvious on face, eyelids and ears, perineum, and tail. Lesions begin as an area of reddening, then progressing to papules, vesicles, pustules with exudation, and finally to scabs. Lesions also appear on the mucous membranes of the nostrils, mouth, and vulva. Animals suffer acute respiratory distress. Healing is very slow. Mortality peaks about 2 weeks after the onset of the skin lesions.

Suggested Alternatives for Differential Diagnosis: Contagious ecthyma, contagious pustular dermatitis, bluetongue, mycotic dermatitis, sheep scab, mange, insect bites, parasitic pneumonia, and photosensitization.

Mortality Rate (untreated): ≤50%.

C18-A019

Guanarito hemorrhagic fever

Arenaviridae

ICD-10: A96.8

A member of the New World Hemorrhagic fever viruses normally found in Venezuela. The natural reservoirs are the short-tailed cane mouse (*Zygodontomys brevicauda*) and the cotton rat (*Sigmodon alstoni*), and the virus is shed in their urine. Infection occurs after inhalation of dust contaminated with excreta from infected mice or from aerosol of animal blood or fluids. Does not produce disease in animals. This is a biosafety level 4 agent.

It is on the HHS Select Agents list, and the Australia Group Core list.

In People:

CDC Case Definition: None established.

Communicability: Direct person-to-person transmission does not occur. When dealing with infected individuals, use standard contact precautions; droplet precautions; airborne precautions; VHF-specific barrier precautions.

Normal Routes of Exposure: Inhalation; Ingestion; Abraded skin; Mucous membranes.

Infectious Dose: Unknown.

Secondary Hazards: Aerosols (blood, fluids, contaminated dust); Blood and body fluids; Body tissue; Fecal matter; Fomites.

Incubation: 7–16 days.

Signs and Symptoms: Fever, prostration, headache, severe joint pain (arthralgias), cough, sore throat, nausea, vomiting, diarrhea, nosebleed (epistaxis), bleeding gums, and black tarry stool. Other symptoms include conjunctivitis, enlargement of the cervical lymph nodes, facial edema, and small hemorrhages (petechia), low blood platelet count (thrombocytopenia), decreased white blood cells (leukopenia), and pulmonary edema.

Suggested Alternatives for Differential Diagnosis: Disseminated intravascular coagulation, hemolytic uremic syndrome, leptospirosis, malaria, salmonella infection, lupus, thrombocytopenic purpura, typhoid fever, shigellosis, meningococcemia, rickettsial infections, acute leukemia, and idiopathic or thrombotic thrombocytopenic purpura.

Mortality Rate (untreated): ≤30%.

C18-A020

Hantaan

Bunyaviridae
ICD-10: A98.5

It is a hantavirus that produces a viral hemorrhagic fever. It is normally found in Central and Eastern Asia. The natural reservoir is the striped field mouse (*Apodemus agrarius*) and the virus is shed in their urine. Infection occurs after inhalation of dust contaminated with excreta from infected mice or from aerosol of animal blood or fluids. Does not produce disease in animals. This is a biosafety level 3 agent.

It is on the Australia Group Core list and is listed as a Category C Potential Terrorism Agent by the CDC.

In People:

CDC Case Definition for hantavirus pulmonary syndrome: An illness characterized by one or more of the following clinical features: A febrile illness (i.e., temperature greater than 101°F) characterized by bilateral diffuse interstitial edema that may radiographically resemble ARDS, with respiratory compromise requiring supplemental oxygen, developing within 72 h of hospitalization, and occurring in a previously healthy person and/or an unexplained respiratory illness resulting in death, with an autopsy examination demonstrating noncardiogenic pulmonary edema without an identifiable cause. *Laboratory criteria* for diagnosis is (1) detection of hantavirus-specific immunoglobulin M or rising titers of hantavirus-specific immunoglobulin G, or (2) detection of hantavirus-specific ribonucleic acid sequence by polymerase chain reaction in clinical specimens, or (3) detection of hantavirus antigen by immunohistochemistry.

Communicability: Direct person-to-person transmission is rare. When dealing with infected individuals, use VHF-specific barrier precautions.

Normal Routes of Exposure: Inhalation; Ingestion; Abraded skin; Mucous membranes.

Infectious Dose: Unknown.

Secondary Hazards: Aerosols (blood, fluids, contaminated dust); Blood and body fluids (from mice); Fecal (from mice); Fomites (from mouse habitation).

Incubation: 2–60 days.

Signs and Symptoms: Characterized by sudden onset of high fever, headache, a vague feeling of bodily discomfort (malaise), dizziness, anorexia, abdominal and/or lower back pain, nausea and vomiting progressing to nonproductive cough, difficulty breathing (dyspnea), and rapid development of respiratory insufficiency. The next phase involves renal failure and protein in the urine (proteinuria). This is followed by excessive urination (diuresis) which may progress into dehydration and severe shock. Normal blood pressure returns and there is a dramatic drop in urine production. Recovery may be prolonged.

Suggested Alternatives for Differential Diagnosis: Acute respiratory distress syndrome, congestive heart failure, pulmonary edema, AIDS, pneumonia, cardiogenic shock, septic shock, phosgene toxicity, phosphine toxicity, salicylate toxicity with pulmonary edema, influenza, plague, tularemia, and anthrax.

Mortality Rate (untreated): $\leq$5%.

C18-A021

Hendra virus

Paramyxoviridae
ICD-10: B33.8

It is normally found only in Australia. The natural reservoir is fruit bats (*Pteropus* species), however, there is no evidence of direct transmission from bats to humans. Special precautions should be taken when examining a horse suspected of having the disease or performing a necropsy. Although it is not considered highly communicable, this is an enhanced biosafety level 3 agent.

It is on the HHS Select Agents list, the USDA High Consequence list, the Australia Group Core list, and is listed as a Category C Potential Terrorism Agent by the CDC.

In People:

CDC Case Definition: None established.

Communicability: Direct person-to-person transmission does not occur. When dealing with infected individuals, use standard contact precautions.

Normal Routes of Exposure: Undetermined, possibly Inhalation; Mucous Membranes.

Infectious Dose: Unknown.

Secondary Hazards: Body fluids (from horses); Body tissue (from horses); Urine (from horses).

Incubation: 4 days to 3 months.

Signs and Symptoms: Flu-like symptoms that may progress to pneumonitis, respiratory failure, renal failure, arterial blood clots (thrombosis), cardiac arrest, and meningoencephalitis.

Suggested Alternatives for Differential Diagnosis: No suggestions.

Mortality Rate (untreated): "High."

In Agricultural Animals:

Target species: Horses.

Communicability: Direct transmission does not occur.

Normal routes of exposure: Undetermined, possibly inhalation; Mucous membranes.

Secondary Hazards: Body fluids.

Incubation: 8–14 days.

Signs and Symptoms: Fever, anorexia, depression, increased respiratory and heart rates, and respiratory distress. Other potential signs include facial edema and frothy nasal discharge.

Suggested Alternatives for Differential Diagnosis: African horse sickness, anthrax, botulism, pasteurellosis, equine influenza, peracute equine herpesvirus 1 infection, and ingestion of plant or agricultural poisons.

Mortality Rate (untreated): "High."

C18-A022

Hog cholera

Flaviviridae

It is normally found worldwide. The natural reservoirs are pigs and wild boar. It can survive for many months in frozen or refrigerated meat and is not affected by mild forms of curing.

It is on the USDA High Consequence list and the Australia Group Core list.

In People:

Does not occur in humans.

In Agricultural Animals:

Target species: Pigs.

Communicability: Direct transmission is possible.

Normal routes of exposure: Ingestion; Abraded Skin; Mucous Membranes.

Secondary Hazards: Blood and body fluids; Body tissue; Fecal matter; Vectors (mechanical—track out).

Incubation: 2–14 days.

Signs and Symptoms: Has a highly variable clinical picture. It has acute and chronic forms, and virulence varies from severe, with high mortality, to mild or even subclinical. The severe acute form is characterized by fever, lack of appetite, depression, constipation followed by diarrhea. May progress to incoordination or convulsions. Conjunctivitis is frequent and is manifested by encrustation of the eyelids and the presence of dirty streaks below the eyes caused by the accumulation of dust and feed particles. In the chronic form of the disease, pigs often survive more than 30 days. After an initial acute febrile phase, pigs may show apparent recovery but then relapse, with anorexia, depression, fever, and progressive loss of condition.

Postmortem findings include widespread small hemorrhages (petechia) and bruises (ecchymoses), especially in lymph nodes, kidneys, spleen, bladder, and larynx. Infarction may be seen, notably in the spleen. Most pigs show a nonsuppurative encephalitis with vascular cuffing.

No treatment should be attempted. Confirmed cases and in-contact animals should be slaughtered, and measures taken to protect other pigs.

Suggested Alternatives for Differential Diagnosis: African swine fever, salmonellosis, erysipelas, anticoagulant poisoning, and hemolytic disease of the newborn, porcine dermatitis and nephropathy syndrome and postweaning multisystemic wasting syndrome, pseudorabies, parvovirus, and border disease.

Mortality Rate (untreated): $\leq$100%.

C18-A023

Infectious porcine encephalomyelitis

Picornaviridae

It is normally found in Europe and Madagascar. The natural reservoir is pigs. It is a highly infectious disease that is similar to polio in humans. It can survive in the environment for more than 5 months and can survive even longer in liquid manure.

It is on the Australia Group Core list.

In People:

Does not occur in humans.

In Agricultural Animals:

Target species: Pigs.

Communicability: Direct transmission is possible.

Normal routes of exposure: Ingestion.

Secondary Hazards: Fecal matter; Fomites.

Incubation: 5–14 days.

Signs and Symptoms: Fever, anorexia, depression, and incoordination (ataxia). Progressive paralysis, beginning in the hindquarters and ascending toward the head, may develop. In severe cases cannot rise from a dog-sitting position. Although animals that are mildly affected may recover, death usually results from paralysis of the respiratory muscles.

Suggested Alternatives for Differential Diagnosis: Classical swine fever, African swine fever, pseudorabies, rabies, Japanese encephalitis, edema disease, hemagglutinating encephalomyelitis, bacterial meningoencephalitis, porcine reproductive and respiratory syndrome (PRRS) virus, hypoglycemia, water deprivation/salt intoxication, and other toxins.

Mortality Rate (untreated): $\leq$90%.

C18-A024

Japanese encephalitis (Agent AN)

Flaviviridae

ICD-10: A83.0

It is normally found in Japan, Southeast Asia, the Indian subcontinent, and parts of Oceania. The natural reservoirs are water birds such as herons and egrets, and mosquitoes. Swine acts as amplifying host and has very important role in epidemiology of the disease. Human and horses are dead-end hosts and disease is manifested as fatal encephalitis. This is a biosafety level 3 agent.

It is on the USDA High Consequence list and the Australia Group Core list.

In People:

CDC Case Definition: Arboviral infections may be asymptomatic or may result in illnesses of variable severity sometimes associated with central nervous system (CNS) involvement. When the CNS is affected, clinical syndromes ranging from febrile headache to aseptic meningitis to encephalitis may occur, and these are usually indistinguishable from similar syndromes caused by other viruses. Arboviral meningitis is characterized by fever, headache, stiff neck, and pleocytosis. Arboviral encephalitis is characterized by fever, headache, and altered mental status ranging from confusion to coma with or without additional signs of brain dysfunction (e.g., paresis or paralysis, cranial nerve palsies, sensory deficits, abnormal reflexes, generalized convulsions, and abnormal movements). *Laboratory criteria* for diagnosis is (1) fourfold or greater change in virus-specific serum antibody titer, or (2) isolation of virus from or demonstration of specific viral antigen or genomic sequences in tissue, blood, cerebrospinal fluid (CSF), or other body fluid, or (3) virus-specific immunoglobulin M (IgM) antibodies demonstrated in CSF by antibody-capture enzyme immunoassay (EIA), or (4) virus-specific IgM antibodies demonstrated in serum by antibody-capture EIA and confirmed by demonstration of virus-specific serum immunoglobulin G (IgG) antibodies in the same or a later specimen by another serologic assay (e.g., neutralization or hemagglutination inhibition).

Communicability: Direct person-to-person transmission does not occur. When dealing with infected individuals, use standard contact precautions.

Normal routes of exposure: Vectors (mosquitoes—*Culex* species and *Aedes* species).

Infectious Dose: Unknown.

Secondary Hazards: None.

Incubation: 5–15 days.

Signs and Symptoms: Symptoms range from mild fever and headache to high fever, chills, nausea, and vomiting, headache, sensitivity to light (photophobia) and objective neurologic signs, stupor, disorientation, tremors, coma, and spastic paralysis. Up to 80% of severe cases develop neuropsychiatric sequelae including seizure disorders, motor or cranial nerve paresis, movement disorders, and/or mental retardation.

Suggested Alternatives for Differential Diagnosis: Other forms of encephalitis (e.g., California, Eastern Equine, St Louis, West Nile, Murray Valley), malaria, dengue fever, meningitis, tuberculosis, typhoid fever, enteroviruses, herpes simplex, and Nipah virus.

Mortality Rate (untreated): ≤60%.

In Agricultural Animals:

Target species: Horses

Communicability: Direct transmission does not occur.

Normal routes of exposure: Vectors (mosquitoes—*Culex* species and *Aedes* species).

Secondary Hazards: Pigs are an amplifying host.

Incubation: 8–10 days (horses).

Signs and Symptoms: Fever, anorexia, difficulty in swallowing, jaundice, small hemorrhages (petechia) in visible mucous membranes, sluggish movement, incoordination, staggering and falling, transient neck rigidity, and radial paralysis. Abortions are common. May progress to blindness, profuse sweating, muscle trembling, aimless wandering with violent and demented behavior, followed by collapse and death.

Suggested Alternatives for Differential Diagnosis: In *pigs*: Nipah virus, Aujeszky's disease, brucellosis, porcine reproductive and respiratory syndrome (PRRS) virus, Classical swine fever, parvovirus. In *horses*: Equine encephalomyelitis (Western, Eastern, and Venezuelan), Rabies, Borna disease, Lead poisoning, Tetanus.

Mortality Rate (untreated): ≤25% (Horses); nonfatal (Pigs).

C18-A025

Junin hemorrhagic fever

Arenaviridae
ICD-10: A96.0

It is a viral hemorrhagic fever normally found in Argentina. The natural reservoir is mice (*Calomys* species) and the virus is shed in their urine. Infection occurs after inhalation of dust contaminated with excreta from infected mice or from aerosol of animal blood or fluids. Does not produce disease in animals. This is a biosafety level 3 agent.

It is on the HHS Select Agents list, the Australia Group Core list, and is listed as a Category A Potential Terrorism Agent by the CDC.

In People:

CDC Case Definition: None established.

Communicability: Direct person-to-person transmission is rare. It is communicable during the fever phase. When dealing with infected individuals, use VHF-specific barrier precautions.

Normal Routes of Exposure: Inhalation; Abraded skin.

Infectious Dose: Unknown.

Secondary Hazards: Aerosols (blood, fluids, contaminated dust); Blood and body fluids (from mice); Body tissue (from mice); Fecal (from mice); Fomites (from mouse habitation).

Incubation: 7–16 days.

Signs and Symptoms: Gradual onset of sustained fever accompanied by a vague feeling of bodily discomfort (malaise), headache, severe muscle pain (myalgia) and joint pain (arthralgias), dizziness, sensitivity to light (photophobia), sore throat, nausea, vomiting, and diarrhea progressing to hemorrhagic symptoms including nosebleed (epistaxis), bloody vomit, and blood in stool or urine. Patients may deteriorate due to vascular or neurologic complications. The illness lasts 2–14 days after the onset of symptoms. Recovery may be prolonged. A vaccine is available.

Suggested Alternatives for Differential Diagnosis: Leptospirosis, malaria, salmonella infection, lupus, typhoid fever, shigellosis, meningococcemia, rickettsial infections, thrombocytopenic purpura, acute leukemia, disseminated intravascular coagulation, hemolytic uremic syndrome.

Mortality Rate (untreated): ≤30%.

C18-A026

Kyasanur forest disease

Flaviviridae
ICD-10: A98.2

It is a viral hemorrhagic fever normally found in South Asia. The natural reservoirs are ticks, monkeys, rodents, and porcupines. Ticks remain infective for life. This is a biosafety level 4 agent.

It is on the HHS Select Agents and the Australia Group Core list.

In People:

CDC Case Definition: None established.

Communicability: Direct person-to-person transmission does not occur. When dealing with infected individuals, use VHF-specific barrier precautions.

Normal Routes of Exposure: Inhalation; Abraded Skin; Mucous Membranes; Vectors (ticks—*Haemophysalis spinigera*).

Infectious Dose: Unknown.

Secondary Hazards: Aerosols (blood, fluids, contaminated dust); Fomites (with vectors).

Incubation: 3–8 days.

Signs and Symptoms: Sudden onset of fever, headache, muscle pain (myalgia), anorexia and insomnia, followed by severe prostration with diarrhea and vomiting. Other symptoms include low white blood cell count (leukopenia), slow heart rate (bradycardia), low blood pressure (hypotension), formation of papulovesicular lesions on the palate, and swelling of the cervical and axillary lymph nodes. Coughing, abdominal pain, and hemorrhagic symptoms may be present. May follow a biphasic course with remission followed by high fever, gastrointestinal and bronchial problems, and meningoencephalitis. Recovery may be prolonged.

Suggested Alternatives for Differential Diagnosis: No suggestions.

Mortality Rate (untreated): ≤10%.

C18-A027

Lassa fever

Arenaviridae
ICD-10: A96.2

It is a viral hemorrhagic fever normally found in west Africa. The natural reservoir is the multimammate rat (*Mastomys natalensis*) and the virus is shed in their urine. Infection occurs after inhalation of dust contaminated with excreta from infected rats or from aerosol of animal blood or fluids. Does not produce disease in animals. This is a biosafety level 4 agent.

It is on the HHS Select Agents list, the Australia Group Core list, and is listed as a Category A Potential Terrorism Agent by the CDC.

In People:

CDC Case Definition: None established.

Communicability: Direct person-to-person transmission is possible. The virus is excreted in urine for 3–9 weeks and in semen for up to 3 months. When dealing with infected individuals, use VHF-specific barrier precautions.

Normal Routes of Exposure: Inhalation; Abraded Skin; Mucous Membranes.

Infectious Dose: Unknown.

Secondary Hazards: Aerosols (blood, fluids, contaminated dust); Blood and body fluids; Body tissue; Fecal matter; Body fluids (from rats); Fecal (from rats); Fomites; Fomites (from rats habitation).

Incubation: 6–21 days.

Signs and Symptoms: About 80% of infections are asymptomatic. Symptoms are varied and nonspecific, and diagnosis is often difficult. The onset of the disease is usually gradual. Initial symptoms include fever, general weakness, and a vague feeling of bodily discomfort (malaise) progressing to headache, sore throat, muscle pain (myalgia), chest pain, nausea, vomiting, diarrhea, cough, and abdominal pain. Although hemorrhagic syndrome is less common, severe cases may progress to facial swelling, fluid in the lung cavity, and bleeding from mouth, nose, vagina, or gastrointestinal tract. Further complications include low blood pressure, shock, seizures, tremor, disorientation, and coma. Deafness occurs in 25% of patients; half of these may recover some function after several months. Transient hair loss and gait disturbance may occur during recovery.

Suggested Alternatives for Differential Diagnosis: Leptospirosis, malaria, salmonella infection, lupus, typhoid fever, shigellosis, meningococcemia, yellow fever and other viral hemorrhagic fevers, rickettsial infections, thrombocytopenic purpura, acute leukemia, disseminated intravascular coagulation, hemolytic uremic syndrome.

Mortality Rate (untreated): ≤40%; ≤15% (hospitalized).

C18-A028

Louping ill

Flaviviridae
ICD-10: A84.8

It is normally found in Scotland, Wales, Northern England, and Ireland. The natural reservoirs are sheep, deer, and ticks. Ticks remain infective for life. Sheep, red grouse, and ptarmigans may act as amplifying hosts. This is a biosafety level 3 agent.

It is on the Australia Group Core list.

In People:

CDC Case Definition: None established.

Communicability: Direct person-to-person transmission does not occur. When dealing with infected individuals, use standard contact precautions.

Normal Routes of Exposure: Inhalation; Abraded Skin; Vectors (ticks—*Ixodes racinus*).

Infectious Dose: Unknown.

Secondary Hazards: Aerosols (during slaughter); Blood Body tissue (from slaughtered animals); Fomites (with vectors).

Incubation: 2–8 days.

Signs and Symptoms: Disease is biphasic. Initial symptoms are flu-like and include fever, headache, pain behind the eyes, and a vague feeling of bodily discomfort (malaise). Initial phase does not appear to involve the central nervous system. However, 4–10 days after apparent recover, there is a return of fever with meningoencephalitis or symptoms resembling paralytic poliomyelitis. Convalescence may be prolonged.

Suggested Alternatives for Differential Diagnosis: Other tick-borne encephalitis, influenza, poliomyelitis.

Mortality Rate (untreated):0% .

In Agricultural Animals:

Target species: Sheep; although cattle, goats, and horses can contract the disease.

Communicability: Direct transmission is possible in goats.

Normal routes of exposure: Ingestion; Vectors (ticks—*Ixodes racinus*).

Secondary Hazards: Ingestion (goat milk); Fomites (with vectors); Vector cycle (in sheep).

Incubation: 6–18 days.

Signs and Symptoms: Disease is usually biphasic. Initial signs are nonspecific and include fever, depression, anorexia, and possibly constipation. During the second phase the virus may invade the central nervous system. If it does not, the animal will recover rapidly and develop a durable protective immunity. Symptoms of involvement of the central nervous system include muscular tremors and incoordination (ataxia), increase in sensitivity to sensory stimuli (hyperesthesia), and development of the characteristic louping gait. Animals are often hypersensitive to noise and touch and will go into convulsive spasms if disturbed. Symptoms may progress to paralysis of the lower half of the body, convulsions, tetanic spasms, coma, and finally death. Some recovered animals may exhibit residual partial paralysis (paresis) or sideways twitching of the neck (torticollis). All recovered animals are solidly immune for life. Concurrent infection with toxoplasmosis or tick-borne fever can increase the severity of louping ill.

Suggested Alternatives for Differential Diagnosis: *Sheep*: Scrapie, pregnancy toxemia, hypocalcemia, tetanus, listeriosis, tick pyemia, hypocuprosis, rabies, hydatid disease, and various plant poisons. *Cattle*: Malignant catarrhal fever, listeriosis, pseudorabies, bovine spongiform encephalopathy, rabies, hypomagnesemia, hypocalcemia, acute lead poisoning, and certain plant poisons.

Mortality Rate (untreated): ≤60% (Sheep).

C18-A029

Lumpy skin disease

Poxviridae

It is a disease of cattle that is normally found in Africa. The natural reservoir is cattle. It is very resistant to physical and chemical agents. It can persist in necrotic skin for 40 days and remains viable in air-dried hides for over 18 days at ambient temperature.

It is on the USDA High Consequence list and the Australia Group Core list.

In People:

Does not occur in humans.

In Agricultural Animals:

Target species: Cattle.

Communicability: Direct transmission is possible.

Normal routes of exposure: Ingestion; Abraded skin; Mucous membranes; Vectors (mosquitoes; biting flies).

Secondary Hazards: Contact; Body fluids (saliva); Body tissue; Fomites (hides).

Incubation: 2–5 weeks.

Signs and Symptoms: Initial symptoms include fever, watery eyes, increased nasal secretions, drooling (ptyalism), diarrhea, loss of appetite, reduced milk production, depression, and reluctance to move. This is followed by the eruption of various sized skin nodules that may cover the whole body. They can be found on any part of the body but are most numerous on the head and neck, perineum, genitalia and udder, and the limbs. The nodules are painful and involve all layers of the skin. Skin lesions may show scab formation. Regional lymph nodes are enlarged and full of fluid. Secondary bacterial infection can complicate healing and recovery. Final resolution of lesions may take 2–6 months, and nodules can remain visible 1–2 years.

Suggested Alternatives for Differential Diagnosis: Besnoitiosis, bovine ephemeral fever, bovine farcy, bovine herpes mammillitis, bovine papular stomatitis, cattle grubs, cutaneous tuberculosis, demodicosis, dermatophilosis, *Hypoderma bovis* infection, insect bites, onchocercosis, photosensitization, pseudo-lumpyskin disease, rinderpest, ringworm, skin allergies, sporadic bovine lymphomatosis, sweating weakness of calves, urticaria, vesicular disease.

Mortality Rate (untreated): $\leq$3%; $\leq$10% (Calves).

C18-A030

Lymphocytic choriomeningitis

Arenaviridae

ICD-10: A87.2

Virus is found worldwide. The natural reservoir is the common house mouse (*Mus musculus*) and the virus is shed in their urine. Infection occurs after inhalation of dust contaminated with excreta from infected mice or from aerosol of animal blood or fluids. Disease can be passed to rodent pets such as mice and hamsters. Does not produce disease in animals. The virus normally has little effect on healthy people but can be deadly for people whose immune system has been weakened. This is a biosafety level 3 agent.

It is on the Australia Group Core.

In People:

CDC Case Definition: None established.

Communicability: Direct person-to-person transmission does not occur. It is communicable only between a mother and a fetus during pregnancy. When dealing with infected individuals, use standard contact precautions.

Normal Routes of Exposure: Inhalation; Ingestion; Abraded Skin; Mucous Membranes.

Infectious Dose: Unknown.

Secondary Hazards: Aerosols (blood, fluids, contaminated dust); Body fluids (from mice); Fecal (from mice); Fomites (from mouse habitation).

Incubation: 8–13 days.

Signs and Symptoms: Produces a biphasic febrile illness. Symptoms of the first phase are nonspecific and fever, a vague feeling of bodily discomfort (malaise), anorexia, muscle pain (myalgia), headache, nausea, and vomiting. Additional symptoms may include sore throat, cough, and pain in the joints (arthralgias), chest, testicles, and salivary gland. Symptoms of the second phase include fever, increased headache, a stiff neck, drowsiness, confusion, sensory disturbances, and/or paralysis. Recovery is generally complete but may be prolonged. Immunosuppressed individuals may develop hemorrhagic fever syndrome.

Suggested Alternatives for Differential Diagnosis: Typhoid fever, leptospirosis, influenza, mumps, enteroviruses, arboviral encephalitis.

Mortality Rate (untreated): <1%.

C18-A031

Machupo hemorrhagic fever

Arenaviridae
ICD-10: A96.1

It is a viral hemorrhagic fever normally found in Bolivia. The natural reservoir is the vesper mouse (*Calomys callosus*) and the virus is shed in their urine. Infection occurs after inhalation of dust contaminated with excreta from infected mice or from aerosol of animal blood or fluids. Does not produce disease in animals. This is a biosafety level 4 agent.

It is on the HHS Select Agents list, the Australia Group Core list, and is listed as a Category A Potential Terrorism Agent by the CDC.

In People:

CDC Case Definition: None established.

Communicability: Direct person-to-person transmission is rare. When dealing with infected individuals, use VHF-specific barrier precautions.

Normal Routes of Exposure: Inhalation; Ingestion; Abraded skin; Mucous membranes.

Infectious Dose: Unknown.

Secondary Hazards: Aerosols (blood, fluids, contaminated dust); Body fluids; Blood and body fluids (from mice); Fecal (from mice); Fomites (from mouse habitation).

Incubation: 7–16 days.

Signs and Symptoms: Gradual onset of sustained fever accompanied by a vague feeling of bodily discomfort (malaise), headache, severe muscle pain (myalgia) and joint pain (arthralgias), dizziness, sensitivity to light (photophobia), sore throat, nausea, vomiting, and diarrhea progressing to hemorrhagic symptoms including nosebleed (epistaxis), small hemorrhages (petechia) on the upper body, bloody vomit, and blood in stool or urine. Patients may deteriorate due to vascular or neurologic complications. Can cause nerve deafness. Recovery may be prolonged.

Suggested Alternatives for Differential Diagnosis: Leptospirosis, malaria, salmonella infection, lupus, typhoid fever, shigellosis, meningococcemia, rickettsial infections, thrombocytopenic purpura, acute leukemia, disseminated intravascular coagulation, hemolytic uremic syndrome.

Mortality Rate (untreated): ≤30%.

C18-A032

Malignant catarrhal fever virus

Herpesviridae

Virus is found worldwide. The natural reservoirs are sheep, goats, and wild ruminants.

It is on the USDA High Consequence list.

In People:

Does not occur in humans.

In Agricultural Animals:

Target species: Cattle; American bison.

Communicability: Direct transmission (cattle to cattle) is rare.

Normal routes of exposure: Inhalation; Ingestion.

Secondary Hazards: Aerosols; Body fluids; Fecal matter; Vectors (mechanical).

Incubation: Unknown.

Signs and Symptoms: Morbidity is low but mortality is high. High fever, depression, hemorrhagic diarrhea, bloody urine, muzzle encrustation; erosions on the lips, tongue, gums, soft and hard palate; swollen reddened eyelids, corneal opacity and conjunctivitis; sensitivity to light (photophobia); bilateral ocular and nasal discharges; difficulty breathing (dyspnea) and cyanosis; loss of appetite; reluctance to swallow because of esophageal erosions and drooling; enlarged body lymph nodes; lameness; depression, trembling, hyporesponsiveness, stupor, aggressiveness, and convulsions.

Postmortem findings in cattle include crater like erosions of the nose, mouth, conjunctiva, esophagus, and gastrointestinal tract; intestinal edema and small hemorrhages (petechia); white areas in the kidneys; "tiger striping" in the distal colon; swollen and reddened abomasal folds; and the lungs may be congested, swollen, or emphysematous.

Suggested Alternatives for Differential Diagnosis: Bovine viral diarrhea/mucosal disease, rinderpest, bluetongue, foot and mouth disease, vesicular stomatitis, pneumonic pasteurellosis, photosensitive dermatitis, infectious bovine rhinotracheitis, theileriosis, rabies, and the tick-borne encephalitides.

Mortality Rate (untreated): $\leq$100%.

C18-A033

Marburg hemorrhagic fever

Filoviridae

ICD-10: A98.4

It is a viral hemorrhagic fever normally found in Africa. The natural reservoir is unknown. Other than humans, the disease has only been documented in primates, with a similar clinical picture. This is a biosafety level 4 agent.

It is on the HHS Select Agents list, the Australia Group Core list, and is listed as a Category A Potential Terrorism Agent by the CDC.

In People:

CDC Case Definition: None established.

Communicability: Direct person-to-person transmission is possible. It is communicable up to 7 days after clinical recovery. When dealing with infected individuals, use VHF-specific barrier precautions.

Normal Routes of Exposure: Ingestion; Abraded skin; Mucous membranes.

Infectious Dose: 1–10 organisms (Aerosol).

Secondary Hazards: Aerosols (blood, body fluids); Contact; Blood and body fluids; Body tissue; Fecal matter; Vomit; Fomites.

Incubation: 3–10 days.

Signs and Symptoms: Onset of the disease is sudden. Early signs and symptoms are nonspecific and include fever, chills, headache, and diffuse and nonspecific muscular pain. After approximately 5 days, a nonpruritic, centripetal, pinhead-sized erythematous eruption develops that rapidly progresses into a maculopapular rash that may hemorrhage. The rash is most prominent on the chest, back, and stomach. Additional symptoms include nausea, vomiting, abdominal and chest pain, sore throat, and diarrhea progressing to jaundice, inflammation of the pancreas, severe weight loss, delirium, shock, liver failure, massive hemorrhaging, and multiorgan dysfunction. Death is due to a combination of hemorrhage, capillary leakage, shock, and end-organ failure. For survivors, recovery from may be prolonged.

Suggested Alternatives for Differential Diagnosis: Leptospirosis, malaria, salmonella infection, lupus, typhoid fever, shigellosis, meningococcemia, rickettsial infections, Crimean-Congo hemorrhagic fever, thrombocytopenic purpura, acute surgical abdomen, acute leukemia, disseminated intravascular coagulation, hemolytic uremic syndrome.

Mortality Rate (untreated): ≤25%.

C18-A034

Menangle virus

Paramyxoviridae

It is an emerging disease normally found in Australia. The natural reservoir is bats (*Pteropus* species). Does not produce disease in the bats. This is a biosafety level 3 agent.

It is on the USDA High Consequence list.

In People:

CDC Case Definition: None established.

Communicability: Direct person-to-person transmission is undocumented. When dealing with infected individuals, use standard contact precautions.

Normal Routes of Exposure: Unknown.

Infectious Dose: Unknown.

Secondary Hazards: Suspected as contact with blood, body fluids, and fecal matter from bats and/or pigs.

Incubation: Unknown.

Signs and Symptoms: Influenza-like illness with a rash.

Suggested Alternatives for Differential Diagnosis: No suggestions.

Mortality Rate (untreated): 0%.

In Agricultural Animals:

Target species: Pigs.

Communicability: Direct transmission is undocumented.

Normal routes of exposure: Unknown.

Secondary Hazards: Suspected as contact with blood, body fluids, and fecal matter from bats.

Incubation: Unknown.

Signs and Symptoms: Causes stillbirths with deformities.

Suggested Alternatives for Differential Diagnosis: No suggestions.

Mortality Rate (untreated): "High" (Fetus).

C18-A035

Monkeypox

Poxviridae
ICD-10: B04

Although initially a disease of nonhuman primates in central and western Africa, the disease has spread to the United States via importation of African rodents [e.g., rope squirrels (*Funiscuirus* species), tree squirrels (*Heliosciurus* species), Gambian giant pouched rats (*Cricetomys* species), brushtail porcupines (*Atherurus* species), dormice (*Graphiurus* species), and striped mice (*Hybomys* species)]. It has infected indigenous rodents in the United States such as prairie dogs with subsequent transfer to individuals who own these animals as pets or otherwise have contact with them. This is a biosafety level 2 agent.

It is on the HHS Select Agents list and the Australia Group Core list.

In People:

CDC Case Definition: Rash (macular, papular, vesicular, or pustular; generalized or localized; discrete or confluent), with fever (subjective or measured temperature of ≥99.3°F). Other signs and symptoms include chills and/or sweats, headache, backache, lymphadenopathy, sore throat, cough, and shortness of breath. *Epidemiologic criteria* for diagnosis is (1) exposure to an exotic or wild mammalian pet obtained on or after April 15, 2003, with clinical signs of illness (e.g., conjunctivitis, respiratory symptoms, and/or rash); or (2) exposure to an exotic or wild mammalian pet with or without clinical signs of illness that has been in contact with either a mammalian pet or a human with monkeypox; or (3) exposure to a suspect, probable, or confirmed human case of monkeypox. *Laboratory criteria* for diagnosis is (1) isolation of monkeypox virus in culture; or (2) demonstration of monkeypox virus DNA by polymerase chain reaction testing of a clinical specimen; or (3) demonstration of virus morphologically consistent with an orthopoxvirus by electron microscopy in the absence of exposure to another orthopoxvirus; or (4) demonstration of presence of orthopoxvirus in tissue using immunohistochemical testing methods in the absence of exposure to another orthopoxvirus.

Communicability: Direct person-to-person transmission is possible. When dealing with infected individuals, use airborne precautions.

Normal Routes of Exposure: Inhalation; Ingestion; Abraded skin and Animal bites; Mucous membranes.

Infectious Dose: Unknown.

Secondary Hazards: Aerosol; Contact; Body fluids; Fomites.

Incubation: 7–17 days.

Signs and Symptoms: Clinical symptoms in humans similar to that seen in smallpox although often milder. Symptoms include extreme fatigue, fever, muscular, and back pain, with evolution of discolored spots (maculas) progressing successively to elevated bumps (papules), blisters (vesicles), pus filled pimples (pustules), and finally scabs. In addition, infected persons experience enlarged neck and groin lymph nodes. Vaccination with the smallpox vaccine immunizes against monkeypox.

Suggested Alternatives for Differential Diagnosis: Smallpox, pseudocowpox/paravaccinia, Varicella-zoster, tularemia, plague, parapox virus, Eczema herpeticum.

Mortality Rate (untreated): ≤10%.

In Agricultural Animals:

Target species: Rodent and Lagomorph pets.

CDC Case Definition for animals: Rash (macular, papular, vesicular, or pustular; generalized or localized; discrete or confluent). Other possible signs and symptoms include conjunctivitis, coryza, cough, anorexia, and/or lethargy. *Epidemiologic criteria* for diagnosis is (1) originating from the shipment of rodents from Ghana to Texas on April 9, 2003 (i.e., Gambian giant rats, rope squirrels, tree squirrels, striped mice, brush-tailed porcupines, and dormice); or (2) originating from a pet holding facility where wild or exotic mammalian pets with suspect, probable, or confirmed monkeypox have been reported; or (3) exposure to a wild or exotic mammalian pet that has been diagnosed with suspect, probable, or confirmed monkeypox; or (4) exposure to a suspect, probable, or confirmed human case of monkeypox. *Laboratory criteria* for diagnosis is (1) isolation of monkeypox virus in culture; or (2) demonstration of monkeypox virus DNA by polymerase chain reaction testing in a clinical specimen; or (3) demonstration of virus morphologically consistent with an orthopoxvirus by electron microscopy in the absence of exposure to another orthopoxvirus; or (4) demonstration of presence of orthopox virus in tissue using immunohistochemical testing methods in the absence of exposure to another orthopoxvirus.

Communicability: Direct transmission is possible.

Normal routes of exposure: Inhalation; Ingestion; Abraded Skin; Mucous Membranes.

Secondary Hazards: Aerosols (cough, sneeze); Contact; Body fluids; Fomites.

Incubation: Unknown.

Signs and Symptoms: Depending on the species involved, symptoms may include cough, discharge from the eyes (appear cloudy or crusty) or the nostrils, swelling in the limbs from enlarged lymph nodes, a bumpy, or blister-like rash. Animals may just appear to be very tired and may not be eating or drinking. Some species may present no respiratory signs and limited dermatologic involvement.

Suggested Alternatives for Differential Diagnosis: No suggestions.

Mortality Rate (untreated): "Low."

C18-A036

Murray valley encephalitis

Flaviviridae
ICD-10: A83.4

It is normally found in Australia and New Guinea. The natural reservoirs are mosquitoes, birds, foxes, and opossums. Mosquitoes remain infective for life. Domestic animals do not manifest clinical symptoms. This is a biosafety level 3 agent.

It is on the Australia Group Core list.

In People:

CDC Case Definition: Arboviral infections may be asymptomatic or may result in illnesses of variable severity sometimes associated with central nervous system (CNS) involvement. When the CNS is affected, clinical syndromes ranging from febrile headache to aseptic meningitis to encephalitis may occur, and these are usually indistinguishable from similar syndromes caused by other viruses. Arboviral meningitis is characterized by fever, headache, stiff neck, and pleocytosis. Arboviral encephalitis is characterized by fever, headache, and altered mental status ranging from confusion to coma with or without additional signs of brain dysfunction (e.g., paresis or paralysis, cranial nerve palsies, sensory deficits, abnormal reflexes, generalized convulsions, and abnormal movements). *Laboratory criteria* for diagnosis is (1) fourfold or greater change in virus-specific serum antibody titer, or (2) isolation of virus from or demonstration of specific viral antigen or genomic sequences in tissue, blood, cerebrospinal fluid (CSF), or other body fluid, or (3) virus-specific immunoglobulin M (IgM) antibodies demonstrated in CSF by antibody-capture enzyme immunoassay (EIA), or (4) virus-specific IgM antibodies demonstrated in serum by antibody-capture EIA and confirmed by demonstration of virus-specific serum immunoglobulin G (IgG) antibodies in the same or a later specimen by another serologic assay (e.g., neutralization or hemagglutination inhibition).

Communicability: Direct person-to-person transmission does not occur. When dealing with infected individuals, use standard contact precautions.

Normal Routes of Exposure: Vectors (mosquitoes—*Culex annulirostris; Aedes* species).

Infectious Dose: Unknown.

Secondary Hazards: None.

Incubation: 5–15 days.

Signs and Symptoms: Symptoms appear in less than 1% of infected individuals. Symptoms range from mild fever and headache to high fever, headache, stupor, tremors, coma and spastic paralysis, and progressive CNS damage. There is neurologic sequelae including paraplegia, impaired gait and motor control, and decreased intellect in up to 40% of mild cases and all severe cases.

Suggested Alternatives for Differential Diagnosis: Meningitis, basilar artery blood clots (thrombosis), cardioembolic stroke, cavernous sinus syndromes, cerebral venous blood clots (thrombosis), confusional states and acute memory disorders, epileptic and epileptiform encephalopathies, febrile seizures, haemophilus meningitis, intracranial hemorrhage, leptomeningeal carcinomatosis, subdural pus (empyema), or bruise (hematoma).

Mortality Rate (untreated): $\leq$40%.

C18-A037

Newcastle disease (Agent OE)

Paramyxoviridae
ICD-10: B30.8

It is endemic in many countries of the world. The natural reservoir is birds. It is highly contagious and can survive for several weeks in a warm, humid environment, such as on birds' feathers, manure, and other materials. It can survive indefinitely in frozen material.

It is on the USDA High Consequence list and the Australia Group Core list.

In People:

CDC Case Definition: None established.

Communicability: Direct person-to-person transmission does not occur. When dealing with infected individuals, use standard contact precautions.

Normal Routes of Exposure: Inhalation; Mucous membranes.

Infectious Dose: Unknown.

Secondary Hazards: Aerosols (contaminated fomites); Fomites (from birds).

Incubation: 1–4 days.

Signs and Symptoms: Symptoms include self-limiting conjunctivitis with tearing and pain. May progress to a generalized illness with "flu-like" symptoms including elevated temperature, chills, and sore throat.

Suggested Alternatives for Differential Diagnosis: No suggestions.

Mortality Rate (untreated): 0%.

In Agricultural Animals:

Target species: Poultry.

Communicability: Direct transmission is possible.

Normal routes of exposure: Inhalation; Ingestion; Mucous Membranes.

Secondary Hazards: Aerosols (cough, sneeze, contaminated fomites); Body fluids; Fecal matter; Fomites; Vectors (mechanical—track out).

Incubation: 2–15 days.

Signs and Symptoms: Clinical signs are very variable depending on strain of virus, species and age of bird, concurrent disease and preexisting immunity. Symptoms include edema of the head with swelling of the lower eyelid, often accompanied by red, inflamed whites of the eyes (conjunctivitis) and dark ring around the eye (black eye); excessive fluids from the respiratory tract causing sneezing, gasping for air, nasal discharge, coughing, increased respiration; circling, falling, paralyzed wings, and sideways twisting of the head and neck (torticollis); depression and loss of appetite with greenish, watery diarrhea 2–3 days after onset of illness; and sudden death. In hens that have survived the disease, there is a tendency to lay misshapen eggs or develop egg yolk peritonitis.

Postmortem findings include edema of the head and neck; hemorrhage and erosions of the esophagus; hemorrhage of the tracheal mucosa; hemorrhages throughout the gastrointestinal tract that may have ulcerated and become necrotic; inflammation of the intestine with marked involvement of the caecal tonsils and Peyer's patches; hemorrhage of the mucosal lining of the proventriculus, especially at the junction with the esophagus. Peracute deaths will often show no discernible lesions in some of the first birds dying in an outbreak.

Suggested Alternatives for Differential Diagnosis: Infectious bronchitis, laryngotracheitis, fowl cholera, avian influenza, fowl pox, psittacosis, mycoplasmosis, avian encephalomyelitis, coryza, salmonellosis, Marek's disease and Pacheco's disease in parrots, vitamin E deficiency, and deprivation of water, air, or feed.

Mortality Rate (untreated): ≤100%.

C18-A038

Nipah virus

Paramyxoviridae

ICD-10: B33.8

It is normally found only in Malaysia. The natural reservoir is unknown, but believed to be fruit bats (*Pteropid* species). Special precautions should be taken when examining a pig suspected of having the disease or performing a necropsy. This is a biosafety level 4 agent.

It is on the HHS Select Agents list, the USDA High Consequence list, the Australia Group Core list, and is listed as a Category C Potential Terrorism Agent by the CDC.

In People:

CDC Case Definition: None established.

Communicability: Direct person-to-person transmission does not occur. When dealing with infected individuals, use standard contact precautions.

Normal Routes of Exposure: Undetermined, possibly Inhalation; Mucous membranes.

Infectious Dose: Unknown.

Secondary Hazards: Body fluids (from pigs); Body tissue (from pigs).

Incubation: 4–18 days.

Signs and Symptoms: Flu-like symptoms with fever, headache, drowsiness, cough, abdominal pain, nausea, vomiting, weakness, problems with swallowing, and blurred vision that may progress to encephalitis with drowsiness, disorientation, severe hypertension, rapid heart rate (tachycardia), very high temperature, convulsions, and coma. Encephalitis may be delayed up to 4 months postexposure. Nipah virus is also known to cause relapse encephalitis.

Suggested Alternatives for Differential Diagnosis: Meningitis, basilar artery blood clots (thrombosis), cardioembolic stroke, cavernous sinus syndromes, cerebral venous blood clots (thrombosis), confusional states and acute memory disorders, epileptic and epileptiform encephalopathies, febrile seizures, haemophilus meningitis, intracranial hemorrhage, leptomeningeal carcinomatosis, subdural pus (empyema), or bruise (hematoma).

Mortality Rate (untreated): $\leq$50%.

In Agricultural Animals:

Target species: Pigs.

Communicability: Direct transmission is possible.

Normal routes of exposure: Undetermined, possibly Inhalation; Ingestion; Mucous Membranes.

Secondary Hazards: Body fluids; Body tissues.

Incubation: 8–14 days.

Signs and Symptoms: Although highly contagious, morbidity and mortality are not excessive and clinical signs are not significantly different from other pig diseases. Most pigs developed a fever with respiratory involvement, often with a severe barking cough. May progress to encephalitis. Sows may abort.

Suggested Alternatives for Differential Diagnosis: Classical swine fever, porcine reproductive and respiratory syndrome, pseudorabies, swine enzootic pneumonia caused by *Mycoplasma hyopneumoniae*, porcine pleuropneumonia caused by *Actinobacillus pleuropneumonia* and *Pasteurellosis* species.

Mortality Rate (untreated): $\leq$5%.

C18-A039

Omsk hemorrhagic fever

Flaviviridae

ICD-10: A98.1

It is normally found in western Siberia. The natural reservoirs are ticks, rodents, and muskrats. Ticks remain infective for life. This is a biosafety level 4 agent.

It is on the HHS Select Agents list and the Australia Group Core list.

In People:

CDC Case Definition: None established.

Communicability: Direct person-to-person transmission does not occur. When dealing with infected individuals, use VHF specific barrier precautions.

Normal Routes of Exposure: Inhalation; Ingestion; Abraded skin; Mucous membranes; Vectors (ticks—*Dermacentor reticulates; Dermacentor marginatus; Ixodes persulcatus*).

Infectious Dose: Unknown.

Secondary Hazards: Aerosols; Body tissue (from rodents, muskrats); Blood and body fluids (from rodents, muskrats); Fomites (with vectors).

Incubation: 3–7 days.

Signs and Symptoms: Sudden onset of chills, headache, fever, pain in lower back and limbs, severe prostration, red and inflamed eyes, a papulovesicular rash on the soft palate, diarrhea, and vomiting. Central nervous system abnormalities develop after 1–2 weeks. May progress to hemorrhage from gums, nose, gastrointestinal tract, and lungs. Recovery may be prolonged but previous infection leads to immunity.

Suggested Alternatives for Differential Diagnosis: No suggestions.

Mortality Rate (untreated): ≤2%.

C18-A040

O'nyong-nyong

Togaviridae
ICD-10: A92.1

It is normally found in Africa. The natural reservoir is unknown. It is transmitted by mosquitoes, which remain infective for life. Does not produce disease in animals. This is a biosafety level 2 agent.

In People:

CDC Case Definition: None established.

Communicability: Direct person-to-person transmission does not occur. When dealing with infected individuals, use standard contact precautions.

Normal Routes of Exposure: Vectors (mosquitoes—*Anopheles funestus, Anopheles gambiae*).

Infectious Dose: Unknown.

Secondary Hazards: Aerosols (blood); Blood.

Incubation: 3–11 days.

Signs and Symptoms: Self limiting febrile disease. Symptoms include arthritis in wrist, knee, ankle, and small joints of extremities typically followed by a maculopapular rash in 1–10 days. Swelling of the cheeks and palate can occur. May present hemorrhagic symptoms. Recovery may be prolonged.

Suggested Alternatives for Differential Diagnosis: No suggestions.

Mortality Rate (untreated): 0%.

C18-A041

Oropouche virus disease

Bunyaviridae
ICD-10: A93.0

It is normally found mainly in the Amazon region, the Caribbean, and Panama. The natural reservoir is unknown. It is transmitted by ticks, which remain infective for life. Does not produce disease in animals. This is a biosafety level 3 agent.

It is on the Australia Group Core list.

In People:

CDC Case Definition: None established.

Communicability: Direct person-to-person transmission does not occur. When dealing with infected individuals, use standard contact precautions.

Normal Routes of Exposure: Vectors (mosquitoes—Ochlerotatus serratus; midge Culicoides paraensis).

Infectious Dose: Unknown.

Secondary Hazards: Aerosols (blood, fluids).

Incubation: 4–8 days.

Signs and Symptoms: Initial symptoms include abrupt onset of fever, chills, headache, a vague feeling of bodily discomfort (malaise), joint pain (arthralgias) and/or muscle pain (myalgia), inflammation of the eyes with an intolerance to light. There may also be occasional nausea and vomiting. May progress to inflammation of the meninges and/or the brain. Relapse may occur 1–2 weeks after initial symptoms disappear. The infection is usually self-limiting and complications are rare. Patients usually recover fully with no long-term ill effects.

Suggested Alternatives for Differential Diagnosis: No suggestions.

Mortality Rate (untreated): 0%.

C18-A042

Pandemic influenza

Orthomyxoviridae
ICD-10: J09 - J18

Influenza is one of the most common infectious diseases in humankind and is one of the most contagious airborne infectious diseases. The natural reservoirs are humans, horses, swine, and birds. It can survive outside a host in dried mucous for several hours. This is a biosafety level 2 agent.

In People:

CDC Case Definition: Clinically compatible illness. Laboratory criteria for diagnosis is (1) influenza virus isolation in tissue cell culture from respiratory specimens; or (2) reverse-transcriptase polymerase chain reaction (RT–PCR) testing of respiratory specimens; or (3) immunofluorescent antibody staining (direct or indirect) of respiratory specimens; or (4) rapid influenza diagnostic testing of respiratory specimens; or (5) immunohistochemical (IHC) staining for influenza viral antigens in respiratory tract tissue from autopsy specimens; or (6) fourfold rise in influenza hemagglutination inhibition (HI) antibody titer in paired acute and convalescent sera.

Communicability: Direct person-to-person transmission is possible. It is communicable for 3–5 days after the onset of symptoms. When dealing with infected individuals, use airborne precautions.

Normal Routes of Exposure: Inhalation; Mucous Membranes.

Infectious Dose: 2–790 (plaque forming units).

Secondary Hazards: Aerosols (cough, sneeze); Body fluids; Body tissue (from animals); Body fluids (from animals).

Incubation: 1–4 days.

Signs and Symptoms: Abrupt onset with fever, chills, severe sore throat, diffuse muscle pain (myalgia), severe headache, sensitivity to light (photophobia), eye pain, runny nose, weakness and severe fatigue, nonproductive cough, wheezing, rhonchi, chest pain, and difficulty breathing (dyspnea). May progress to acute encephalopathy.

Suggested Alternatives for Differential Diagnosis: Adenoviruses, arenaviruses, rhinoviruses, coxsackieviruses, cytomegalovirus, echoviruses, infectious mononucleosis, parainfluenza virus, bacterial sepsis, upper respiratory infection, Lyme disease, encephalitis, meningitis, dengue fever, acute HIV infection.

Mortality Rate (untreated): Variable.

C18-A043

Peste des petits ruminants

Paramyxoviridae

It is primarily a disease of goats and sheep that is found normally in most African countries. Cattle and pigs can develop silent infections. It is not very contagious; transmission requires close contact with infected animals. It has a long survival time in chilled and frozen tissues and can survive at 140°F for up to 60 min.

It is on the USDA High Consequence list and the Australia Group Core list.

In People:

Does not occur in humans.

In Agricultural Animals:

Target species: Goats, sheep.

Communicability: Direct transmission is possible.

Normal routes of exposure: Inhalation; Mucous membranes.

Secondary Hazards: Aerosols (cough, sneeze); Body fluids; Fecal matter; Fomites.

Incubation: 3–10 days.

Signs and Symptoms: The natural disease is usually more severe in goats than sheep. Symptoms include sudden fever, depression, dull coat and loss of appetite. Initially, animals have congested mucous membranes and a dry muzzle. This progresses to a profuse mucopurulent nasal discharge that may crust over and block the nostrils and give a putrid odor to the breath. Conjunctiva are congested and there is a purulent discharge that can encrust the eyelids, cementing them together. Lesions develop in the oral cavity causing drooling (ptyalism). There is respiratory distress and sign of bronchopneumonia, as well as erosive stomatitis and severe watery blood-stained diarrhea leading to dehydration and weight loss. May cause abortions.

Postmortem findings include necrotic lesions in the mouth and nose, congestion of the ileocecal valve, engorgement and blackening of the folds in the cecum, colon, and rectum with "zebra striping," enlarged spleen, edematous lymph nodes, and bronchopneumonia.

Suggested Alternatives for Differential Diagnosis: Rinderpest, pasteurellosis, contagious caprine pleuropneumonia, bluetongue, heartwater, contagious ecthyma, foot-and-mouth disease, Nairobi sheep disease, sheep pox, coccidiosis, salmonellosis, Arsenic poisoning, gastrointestinal helminth infestations.

Mortality Rate (untreated): $\leq$90%.

C18-A044

Plum pox

Potyvirus

It is normally found in North America, Europe, the Middle East, North Africa, India, and Chile. It is transmitted by aphids. The virus is not persistent in the aphid vector and only stays viable in its mouthparts for approximately 1 h.

It is on the USDA Select Agent Listed Plant Pathogens.

In Plants:

Target species: Stone fruit including peaches, nectarines, apricots, plums, and almonds.

Normal routes of transmission: Vectors (aphids).

Secondary Hazards: Nursery stock; Budwood; Mechanical vectors (grafting).

Signs: Symptoms of infection appear on leaves, flowers, and fruit but may considerably depending on the tree species or cultivar. Early symptoms may include color-breaking in the blossoms such as darker pink stripes on the flower petals. Later, it may be characterized by yellowing line patterns and blotches, or necrotic rings on expanded leaves. Infected fruit can develop yellow rings or blotches, or brown rings, and may be severely deformed and bumpy. The seeds of many infected apricots and some plums show rings.

There is no cure or treatment for the disease once a tree becomes infected. Current recommendations include destruction of all infected trees.

C18-A045

Potato andean latent *Tymovirus*

Tymovirus

It is normally found in South and Central America.

It is on the Australia Group Core list.

In Plants:

Target species: Potatoes.

Normal routes of transmission: Contact; Vectors (*Epithrix* beetles).

Secondary Hazards: None; Mechanical vectors (grafting); Crop debris.

Signs: Symptoms are dependent on environmental conditions and are most severe when there is a wide daily fluctuation in temperature. Symptoms include mosaic with necrotic flecking, curling, and leaf-tip necrosis.

C18-A046

Potato spindle tuber viroid

Pospiviroid

It is normally found in North and South America, China, the former Soviet Union, and Australia.

It is on the Australia Group Core list.

In Plants:

Target species: Potatoes; tomatoes.

Normal routes of transmission: Contact Vectors (aphids).

Secondary Hazards: Mechanical vectors (planting); Seeds.

Signs: Mild strains produce no obvious symptoms. Symptoms are dependent on environmental conditions and are most severe in hot conditions. Symptoms may be mild in initial infections but become progressively worse in the following generations. *Potatoes*: may include reduced plant size, uprightness and clockwise arrangement of leaves on the stem when the plant is viewed from above, as well as dark green leaves with a rough, wrinkled surface. Tubers may be reduced in size or misshapen with evenly distributed prominent eyes. *Tomatoes*: Stunted growth with a "bunchy top" caused by shortened internodes. Leaves are yellow or purple and often become curled and twisted. The top leaves are reduced in size; the veins of the middle and bottom leaves become necrotic. Tomato fruit is hard, small, and dark green. There is no form of natural resistance.

C18-A047

Powassan encephalitis

Flaviviridae
ICD-10: A84.8

It is normally found in Canada, the northern United States, and parts of Central Asia. The natural reservoirs are ticks, small mammals, and birds. Ticks remain infective for life. Does not produce disease in animals. This is a biosafety level 3 agent. It does not survive outside a host.

It is on the Australia Group Core list.

In People:

CDC Case Definition: Arboviral infections may be asymptomatic or may result in illnesses of variable severity sometimes associated with central nervous system (CNS) involvement. When the CNS is affected, clinical syndromes ranging from febrile headache to aseptic meningitis to encephalitis may occur, and these are usually indistinguishable from similar syndromes caused by other viruses. Arboviral meningitis is characterized by fever, headache, stiff neck, and pleocytosis. Arboviral encephalitis is characterized by fever, headache, and altered mental status ranging from confusion to coma with or without additional signs of brain dysfunction (e.g., paresis or paralysis, cranial nerve palsies, sensory deficits, abnormal reflexes, generalized convulsions, and abnormal movements). *Laboratory criteria* for diagnosis is (1) fourfold or greater change in virus-specific serum antibody titer, or (2) isolation of virus from or demonstration of specific viral antigen or genomic sequences in tissue, blood, cerebrospinal fluid (CSF), or other body fluid, or (3) virus-specific immunoglobulin M (IgM) antibodies demonstrated in CSF by antibody-capture enzyme immunoassay (EIA), or (4) virus-specific IgM antibodies demonstrated in serum by antibody-capture EIA and confirmed by demonstration of virus-specific serum immunoglobulin G (IgG) antibodies in the same or a later specimen by another serologic assay (e.g., neutralization or hemagglutination inhibition).

Communicability: Direct person-to-person transmission does not occur. When dealing with infected individuals, use standard contact precautions.

Normal Routes of Exposure: Inhalation; Ingestion; Abraded Skin; Vectors (ticks—*Ixodes cookie, Ixodes marxi, Ixodes spinipalpus*).

Infectious Dose: Unknown.

Secondary Hazards: Fomites (with vectors).

Incubation: 7–14 days.

Signs and Symptoms: Range from mild fever and headache to high fever, headache, stupor, disorientation, tremors, meningoencephalitis, convulsions, spastic paralysis, and coma. There is a high incidence of neurologic sequelae.

Suggested Alternatives for Differential Diagnosis: Meningitis, basilar artery blood clots (thrombosis), cardioembolic stroke, cavernous sinus syndromes, cerebral venous blood clots (thrombosis), confusional states and acute memory disorders, epileptic and epileptiform encephalopathies, febrile seizures, haemophilus meningitis, intracranial hemorrhage, leptomeningeal carcinomatosis, subdural pus (empyema), or bruise (hematoma).

Mortality Rate (untreated): ≤60%.

C18-A048
Pseudorabies virus

Herpesviridae

Found worldwide. The natural reservoir is pigs, but it can infect cattle, sheep, cats, dogs, goats, raccoons, opossums, skunks, and rodents. It can survive up to 5 weeks in pig meat. It can persist for up to 7 h in air with a relative humidity of greater than 55% and be spread by the wind for up to 2 km. It can survive for up to 7 h in nonchlorinated well water; 2 days in anaerobic lagoon effluent; 2 days in green grass, soil, feces, and shelled corn; and for 4 days in straw bedding.

It is on the Australia Group Core list.

In People:

No documented infections in humans.

In Agricultural Animals:

Target species: Pigs; cattle; sheep; goats.

Communicability: Direct transmission is possible.

Normal routes of exposure: Inhalation; Ingestion; Mucous membranes.

Secondary Hazards: Aerosols (cough, sneeze); Contact; Body fluids; Body tissue; Fecal matter.

Incubation: Varies.

Signs and Symptoms: Fever, coughing, sneezing, difficulty breathing (dyspnea), constipation, vomiting, anorexia with weight loss; incoordination (ataxia), prostration, weakness, muscular twitching, teeth grinding, involuntary eye movements, convulsions, "goose-stepping" gait, circling, and cardiac irregularities. Other symptoms may resemble rabies including self-mutilation, jaw and pharyngeal paralysis, excess salivation, general paralysis, and death.

Postmortem findings in pigs include purulent inflammation of the nasal lining, pharynx and tonsils, congestion or consolidation of the lungs, congested meninges, congested lymph nodes with small hemorrhages, small white to yellow necrotic foci in liver and spleen of affected animals and aborted fetuses, and necrotic placentitis. In other species postmortem findings are often only edema, congestion, and hemorrhage of the spinal cord.

Suggested Alternatives for Differential Diagnosis: Porcine polioencephalomyelitis, rabies, classical swine fever, African swine fever, hemagglutination encephalomyelitis virus infection, erysipela, Nipah virus, streptococcal meningoencephalitis, hypoglycemia, organic arsenic and mercury poisoning, salt poisoning, swine influenza, and congenital tremor.

Mortality Rate (untreated): ≤100% (Piglets); ≤50% (Nursery Pigs); ≤2% (Adult Pigs); ≤100% (Cattle, Sheep).

C18-A049

Puumala hemorrhagic fever

Bunyaviridae
ICD-10: A98.5

It is a hantavirus that is normally found in Europe, Russia, and Scandinavia. The natural reservoir is the bank vole (*Clethrionomys glareolus*) and the virus is shed in it's urine. Infection occurs after inhalation of dust contaminated with excreta from infected voles or from aerosol of animal blood or fluids. Does not produce disease in animals. This is a biosafety level 3 agent.

It is on the Australia Group Core list, and is listed as a Category C Potential Terrorism Agent by the CDC.

In People:

CDC Case Definition: None established.

Communicability: Direct person-to-person transmission does not occur. When dealing with infected individuals, use VHF-specific barrier precautions.

Normal Routes of Exposure: Inhalation; Ingestion; Abraded skin; Mucous membranes.

Infectious Dose: Unknown.

Secondary Hazards: Aerosols (blood, fluids, contaminated dust); Blood and body fluids (from voles); Fecal (from voles); Fomites (from vole habitation).

Incubation: 2–4 weeks.

Signs and Symptoms: Produces hemorrhagic fever with renal syndrome (HFRS); characterized by sudden onset of high fever, headache, a vague feeling of bodily discomfort (malaise), dizziness, anorexia, abdominal and/or lower back pain, nausea, and vomiting. Progresses to rapid heart rate (tachycardia), hypoxemia, renal failure, and protein in the urine (proteinuria). This is followed by excessive urination (diuresis), which may progress to dehydration and severe shock. Normal blood pressure returns and there is a dramatic drop in urine production. Recovery may be prolonged.

Suggested Alternatives for Differential Diagnosis: Acute poststreptococcal glomerulonephritis, spotted fevers, typhus, malaria, hepatitis, Colorado tick fever, septicemia, heat stroke, disseminated intravascular coagulation, leptospirosis, hemolytic uremic syndrome.

Mortality Rate (untreated): $\leq$1%.

C18-A050

Rabies

Rhabdoviridae
ICD-10: A82

It is present in most of Europe, throughout Africa, the Middle East, most of Asia, and the Americas. It is a highly lethal disease that can affect all warm-blooded animals. This is a biosafety level 2 agent unless there is a high risk of aerosol production then it should be treated as a biosafety level 3 agent.

It is on the Australia Group Core list.

In People:

CDC Case Definition: Laboratory criteria for diagnosis is (1) a positive direct fluorescent antibody test (preferably performed on central nervous system tissue); or (2) isolation of rabies virus (in cell culture or in a laboratory animal).

Communicability: Direct person-to-person transmission is possible but unlikely. When dealing with infected individuals, use standard contact precautions.

Normal Routes of Exposure: Inhalation; Abraded skin; Mucous membranes.

Infectious Dose: Unknown.

Secondary Hazards: Aerosols (Unusual); Body fluids; Body fluids (from animals).

Incubation: 20–90 days; although may last up to 1 year.

Signs and Symptoms: Headache, fever, and a vague feeling of bodily discomfort (malaise) progressing to delirium, psychosis, restlessness, thrashing, muscular fasciculations, seizures, and aphasia. Attempting to drink or having air blown in the face may produce severe laryngeal or diaphragmatic spasms and a sensation of choking. Other symptoms include rapid heart rate (tachycardia), hypertension, hyperventilation, drooling, differences in the size of the pupils, lacrimation, salivation, perspiration, and postural low blood pressure (hypotension) progressing to systemic paralysis followed by delirium, stupor, then coma and death.

Suggested Alternatives for Differential Diagnosis: Encephalitis, Herpes Simplex, tetanus, Guillain–Barre syndrome, poliomyelitis, transverse myelitis, cerebrovascular accident, psychosis, intracranial mass, epilepsy, atropine poisoning, and Creutzfeldt–Jacob disease.

Mortality Rate (untreated): >99%.

In Agricultural Animals:

Target species: All warm-blooded animals.

Communicability: Direct transmission is possible.

Normal routes of exposure: Inhalation; Ingestion; Injection; Mucous membranes.

Secondary Hazards: Aerosols (Unusual); Body fluids; Body tissue.

Incubation: Varies, typically 2 weeks to several months.

Signs and Symptoms: Symptoms include acute behavioral changes [anorexia, signs of apprehension or nervousness, irritability, hyperexcitability, incoordination (ataxia), altered phonation, aggression] and unexplained progressive paralysis (initially throat and masseter muscles, leading to profuse salivation and inability to swallow but progressing rapidly to all parts of the body).

Suggested Alternatives for Differential Diagnosis: Canine distemper, canine hepatitis, Aujeszky's disease, Borna disease, encephalomyelitis, listeriosis, cryptococcosis traumatic injuries or foreign bodies in the oropharynx or esophagus, acute psychosis in dogs and cat, Teschen's disease, erysipela, sodium fluoroacetate poisoning, lead poisoning, chlorinated hydrocarbon poisoning, organophosphate pesticide poisoning, and nitrogen trichloride poisoning.

Mortality Rate (untreated): >95%.

C18-A051

Rift Valley fever

Bunyaviridae
ICD-10: A92.4

Generally found in East Africa and the Middle East. The primary natural reservoir is not known. The disease is transmitted by mosquitoes, which remain infective for life. Infection is also transferred directly to mosquito eggs. It can survive in dried discharges for extended periods. This is a biosafety level 3 agent.

It is on the HHS Select Agents list, the USDA High Consequence list, the Australia Group Core list, and is listed as a Category C Potential Terrorism Agent by the CDC.

In People:

CDC Case Definition: None established.

Communicability: Direct person-to-person transmission does not occur. When dealing with infected individuals, use standard contact precautions unless hemorrhagic symptoms exist; then use VHF-specific barrier precautions.

Normal Routes of Exposure: Inhalation; Ingestion; Abraded skin; Mucous membranes; Vectors (mosquitoes—*Aedes* species and *Culex* species; sandflies).

Infectious Dose: 1–10 organisms (Inhalation).

Secondary Hazards: Aerosols (blood, body fluids); Blood, body tissue, and body fluids (from slaughtered animals).

Incubation: 1–6 days.

Signs and Symptoms: Initial symptoms include fever, generalized weakness, back pain, dizziness, and extreme weight loss. There may also be occasional nausea and vomiting. Fever may be biphasic. In some cases the illness can become hemorrhagic or cause inflammation of the brain leading to headaches, coma, or seizures. Can also cause inflammation of the eyes with an intolerance to light. Up to 10% of affected patients may have some permanent vision loss.

Suggested Alternatives for Differential Diagnosis: Drug eruptions, Henoch–Schönlein purpura, rubella, meningococcemia, Rocky Mountain spotted fever, rubella, tick bite fever, typhus, Q fever, septicemia typhoid, plague, leptospirosis, malaria, trypanosomiasis, hepatitis, chikungunya, herpes, influenza, mononucleosis, acute anemia, acute leukemias, disseminated intravascular coagulation, encephalitis, hemolytic uremic syndrome, thrombotic thrombocytopenic purpura, meningitis, pleural effusion, and sepsis.

Mortality Rate (untreated): $\leq$1%.

In Agricultural Animals:

Target species: Sheep, cattle, goats.

Communicability: Direct transmission does not occur.

Normal routes of exposure: Vectors (mosquitoes; sandflies).

Secondary Hazards: Vector cycle.

Incubation: 1–6 days.

Signs and Symptoms: Signs of the disease tend to be nonspecific, rendering it difficult to recognize in individual cases. Fever, salivation, inflammation in the mouth, anorexia, abdominal pain, depression, weakness, sensitivity to light (photophobia), mucopurulent nasal discharge and encrustation around the muzzle, vomiting, diarrhea (may be hemorrhagic), and jaundice. Abortions may reach 100%.

Postmortem findings include small hemorrhages present in the internal organs, carcasses may be jaundiced, fluid in body cavities that frequently is blood-stained, intestinal inflammation, edematous and hemorrhagic gall bladder, and the liver may be necrotic.

Suggested Alternatives for Differential Diagnosis: Wesselbron disease, Nairobi sheep disease, bluetongue, brucellosis, heartwater, rinderpest, peste des petits ruminants, vibriosis, trichomonosis, enterotoxemia of sheep, ephemeral fever, bacterial septicaemias, East Coast fever, fungal conditions, other viral and bacterial causes of abortion, hepatotoxins, and toxic plants.

Mortality Rate (untreated): ≤70% (Calves); ≤10% (Adult cattle); ≤90% (Lambs less than 1 week); ≤20% (Adult sheep).

C18-A052

Rinderpest

Paramyxoviridae

It is endemic in many countries of Asia and Africa. The natural reservoir is wild ruminants. It is the most lethal plague known in cattle. It remains viable for long periods in chilled or frozen tissues.

It is on the USDA High Consequence list and the Australia Group Core list.

In People:
Does not occur in humans.

In Agricultural Animals:

Target species: Cattle, buffalo.

Communicability: Direct transmission is possible.

Normal routes of exposure: Inhalation; Mucous membranes.

Secondary Hazards: Aerosols (cough, sneeze); Contact; Blood and body fluids; Body tissue; Fecal matter; Fomites.

Incubation: 3–15 days.

Signs and Symptoms: Fever, anorexia, depression, and discharge from the eyes and nose. Pinpoint necrotic lesions appear on the inside of the mouth that rapidly form a cheesy plaque. Further symptoms include severe abdominal pain, thirst, difficulty breathing (dyspnea), and watery diarrhea containing blood, mucus, and mucous membranes. Recovery is prolonged and may be complicated by concurrent infections due to immunosuppression.

Postmortem findings include dehydrated carcass with fecal staining of the legs; erosions of the mucosa in the mouth, pharynx, and esophagus; edema or emphysema of the lungs; mucopurulent nasal exudate; congestion, edema, and erosion of the abomasal mucosa; severe congestion, ulceration in the large intestine as well as "tiger striping" hemorrhages; hemorrhage in the spleen, gallbladder, and urinary bladder; and enlarged and edematous lymph nodes.

Suggested Alternatives for Differential Diagnosis: Bovine viral diarrhea, East Coast fever, foot-and-mouth disease, infectious bovine rhinotracheitis, malignant catarrhal fever, vesicular stomatitis, paratuberculosis, and arsenic poisoning.

Mortality Rate (untreated): ≤90%.

C18-A053

Rocio encephalitis

Flaviviridae
ICD-10: A83.6

It is transmitted by mosquitoes and normally found in Brazil. The natural reservoir is unknown but probably mosquitoes and birds. It is not known whether it produces disease in animals. This is a biosafety level 3 agent.

It is on the Australia Group Core list.

In People:

CDC Case Definition: None established.

Communicability: Direct person-to-person transmission does not occur. When dealing with infected individuals, use standard contact precautions.

Normal Routes of Exposure: Vectors (mosquitoes—*Aedes* species).

Infectious Dose: Unknown.

Secondary Hazards: Unknown.

Incubation: 5–15 days.

Signs and Symptoms: Sudden onset of fever, headache, vomiting and possibly abdominal pain, progressing to neck stiffness, mental confusion, motor disturbances, and difficulty with equilibrium. Survivors may suffer significant impairment of mental functions.

Suggested Alternatives for Differential Diagnosis: No suggestions.

Mortality Rate (untreated): 10%.

C18-A054

Russian spring-summer encephalitis

Flaviviridae
ICD-10: A84.0

It is normally found in Asia. The natural reservoirs are ticks, rodents, and small mammals. Ticks remain infective for life. This is a biosafety level 4 agent.

It is on the HHS Select Agents list and the Australia Group Core list.

In People:

CDC Case Definition for arboviral encephalitis: Neuroinvasive disease requires the presence of fever and at least one of the following in the absence of a more likely clinical explanation: (1) acutely altered mental status (e.g., disorientation, obtundation, stupor, or coma), or (2) other acute signs of central or peripheral neurologic dysfunction (e.g., paresis or paralysis, nerve palsies, sensory deficits, abnormal reflexes, generalized convulsions, or abnormal movements), or (3) increased white blood cell concentration in cerebrospinal fluid associated with illness clinically compatible with meningitis (e.g., headache or stiff neck). *Laboratory criteria* for diagnosis is (1) fourfold or greater change in virus-specific serum antibody titer, or (2) isolation of virus from or demonstration of specific viral antigen or genomic sequences in tissue, blood, CSF, or other body fluid, or (3) virus-specific immunoglobulin M (IgM) antibodies demonstrated in CSF by antibody-capture enzyme immunoassay (EIA), or (4) virus-specific IgM antibodies demonstrated in serum by antibody-capture EIA and confirmed by demonstration of virus-specific serum immunoglobulin G (IgG) antibodies in the same or a later specimen by another serologic assay (e.g., neutralization or hemagglutination inhibition). *Laboratory criteria* for a *probable case* is (1) stable (less than or equal to a twofold change) but elevated titer of virus-specific serum antibodies or (2) virus-specific serum IgM antibodies detected by antibody-capture EIA but with no available results of a confirmatory test for virus-specific serum IgG antibodies in the same or a later specimen.

Communicability: Direct person-to-person transmission does not occur. When dealing with infected individuals, use standard contact precautions.

Normal Routes of Exposure: Vectors (ticks—*Ixodes persulcatus; Ixodes ricinus*).

Infectious Dose: Unknown.

Secondary Hazards: Fomites (with vectors).

Incubation: 7–14 days.

Signs and Symptoms: Symptoms include sudden onset of intense headache, fever, nausea, vomiting, and sensitivity to light (photophobia) followed by central nervous system abnormalities such as stupor, tremors, delirium, focal epilepsy and flaccid paralysis (especially in the shoulder), and coma. Recovery is prolonged. Sequelae may include paralysis of the upper extremities and back.

Suggested Alternatives for Differential Diagnosis: Meningitis, basilar artery blood clots (thrombosis), cardioembolic stroke, cavernous sinus syndromes, cerebral venous blood clots (thrombosis), confusional states and acute memory disorders, epileptic and epileptiform encephalopathies, febrile seizures, haemophilus meningitis, intracranial hemorrhage, leptomeningeal carcinomatosis, subdural pus (empyema), or bruise (hematoma).

Mortality Rate (untreated): ≤20%.

In Agricultural Animals:

Target species: Sheep; goats.

Communicability: Direct transmission does not occur.

Normal routes of exposure: Vectors (ticks—*Ixodes persulcatus; Ixodes ricinus; Haemaphysalis longicornis*).

Secondary Hazards: Fomites (with vectors).

Incubation: Unknown.

Signs and Symptoms: Usually asymptomatic but may develop explosive meningoencephalitis with weakness in all the limbs, rigidity, lockjaw, chattering of the teeth, and tetanic spasms.

Suggested Alternatives for Differential Diagnosis: Other causes of meningitis, encephalitis, or meningoencephalitis.

Mortality Rate (untreated): ≤100% (Symptomatic cases).

C18-A055

Sabia hemorrhagic fever

Arenaviridae

ICD-10: A96.8

A member of the New World hemorrhagic fever viruses normally found in Brazil. The natural reservoir is unknown but believed to be rodents. It is likely that the virus is shed in their urine. Infection occurs after inhalation of dust contaminated with excreta from infected rodents or from aerosol of animal blood or fluids. It is not believed to produce disease in animals. This is a biosafety level 4 agent.

It is on the HHS Select Agents list and the Australia Group Core list.

In People:

CDC Case Definition: None established.

Communicability: Direct person-to-person transmission is not documented, but may be possible. When dealing with infected individuals, use VHF-specific barrier precautions.

Normal Routes of Exposure: Unknown, but believed to be inhalation; Abraded skin; Mucous membranes.

Infectious Dose: Unknown.

Secondary Hazards: Unknown, but believed to be Aerosols (blood, fluids, contaminated dust); Blood and body fluids (from rodents); Fecal matter (from rodents); Fomites (from rodent habitation).

Incubation: 7–14 days.

Signs and Symptoms: Symptoms include fever, chills, a vague feeling of bodily discomfort (malaise), sore throat, headache, muscle pain (myalgia), conjunctivitis, nausea, vomiting, abdominal pain, diarrhea, bleeding gums, and low white blood cell count (leukopenia). May progress to tremors, convulsions, pulmonary edema, internal hemorrhage, coma, shock, and death.

Suggested Alternatives for Differential Diagnosis: Drug eruptions, Henoch-Schönlein purpura, rubella, meningococcemia, Rocky Mountain spotted fever, rubella, tick bite fever, typhus, Q fever, septicemia typhoid, plague, leptospirosis, malaria, trypanosomiasis, hepatitis, chikungunya, herpes, influenza, mononucleosis, acute anemia, acute leukemias, disseminated intravascular coagulation, encephalitis, hemolytic uremic syndrome, thrombotic thrombocytopenic purpura, meningitis, pleural effusion, and sepsis.

Mortality Rate (untreated): Unknown.

C18-A056

Seoul hemorrhagic fever

Bunyaviridae
ICD-10: A98.5

It is a hantavirus that is found worldwide. The natural reservoir is Norway rats (*Rattus norvegicus*) and the virus is shed in their urine. Infection occurs after inhalation of dust contaminated with excreta from infected rats or from aerosol of animal blood or fluids. Does not produce disease in animals. This is a biosafety level 3 agent.

It is on the Australia Group Core list and is listed as a Category C Potential Terrorism Agent by the CDC.

In People:

CDC Case Definition: None established.

Communicability: Direct person-to-person transmission does not occur. When dealing with infected individuals, use VHF specific barrier precautions.

Normal Routes of Exposure: Inhalation; Ingestion; Abraded skin; Mucous membranes.

Infectious Dose: Unknown.

Secondary Hazards: Aerosols (blood, fluids, contaminated dust); Blood and body fluids (from rats); Fecal (from rats); Fomites (from rat habitation).

Incubation: 2–4 weeks.

Signs and Symptoms: Symptoms include high fever, fatigue, anorexia, vomiting, pain in the upper back (dorsalgia), diffuse muscle pain (myalgia), abdominal pain, small hemorrhages (petechia) on the soft palate, enlargement of the liver (hepatomegaly), protein in the urine (proteinuria), low blood platelet count (thrombocytopenia), and increased white blood cells (lymphocytosis). There may also be some mild renal and hepatic dysfunction.

Suggested Alternatives for Differential Diagnosis: Drug induced noncardiac pulmonary edema, acute respiratory distress syndrome, atypical pneumonias, and viral influenza.

Mortality Rate (untreated): ≤15%.

C18-A057

Sheeppox

Poxviridae

It is normally found in Africa, the Middle East, the Indian subcontinent, and much of Asia. It is a serious, often fatal disease of sheep. It can persist for up to 6 months in shaded animal pens, and for at least 3 months in dry scabs on the fleece, skin, and hair from infected animals.

It is on the USDA High Consequence list and the Australia Group Core list.

In People:

Does not occur in humans.

In Agricultural Animals:

Target species: Goats.

Communicability: Direct transmission is possible.

Normal routes of exposure: Inhalation; Abraded skin; Mucous membranes.

Secondary Hazards: Aerosols (cough, sneeze, contaminated dust); Contact; Body fluids; Fecal matter; Fomites; Vectors (mechanical—stable fly).

Incubation: 4–8 days.

Signs and Symptoms: Characterized by sudden onset of fever, discharges from the nose and eyes, and drooling (ptyalism). Animals have a loss of appetite and show a reluctance to move. Pox lesions appear on the skin but are most obvious on face, eyelids and ears, perineum, and tail. Lesions begin as an area of reddening, then progressing to papules, vesicle, pustule with exudation, and finally to scabs. Lesions also appear on the mucous membranes of the nostrils, mouth, and vulva. Animals suffer acute respiratory distress. Healing is very slow. Mortality peaks about 2 weeks after the onset of the skin lesions.

Suggested Alternatives for Differential Diagnosis: Contagious ecthyma, contagious pustular dermatitis, bluetongue, mycotic dermatitis, sheep scab, mange, insect bites, parasitic pneumonia, and photosensitization.

Mortality Rate (untreated): ≤50%.

C18-A058

Sin nombre

Bunyaviridae

ICD-10: J12.8

It is a hantavirus that is normally found in the southwestern United States. The natural reservoir is the deer mouse (*Peromyscus maniculatus*) and the virus is shed in their urine. Infection occurs after inhalation of dust contaminated with excreta from infected mice or from aerosol of animal blood or fluids. Does not produce disease in animals. This is a biosafety level 3 agent. Outside a host the virus can live for several days.

It is on the Australia Group Core list, and is listed as a Category C Potential Terrorism Agent by the CDC.

In People:

CDC Case Definition for hantavirus pulmonary syndrome: An illness characterized by one or more of the following clinical features: A febrile illness (i.e., temperature greater than 101°F) characterized by bilateral diffuse interstitial edema that may radiographically resemble ARDS, with respiratory compromise requiring supplemental oxygen, developing within 72 h of hospitalization, and occurring in a previously healthy person and/or an unexplained respiratory illness resulting in death, with an autopsy examination demonstrating noncardiogenic pulmonary edema without an identifiable cause. *Laboratory criteria* for diagnosis is (1) detection of hantavirus-specific immunoglobulin M or rising titers of hantavirus-specific immunoglobulin G, or (2) detection of hantavirus-specific ribonucleic acid sequence by polymerase chain reaction in clinical specimens, or (3) detection of hantavirus antigen by immunohistochemistry.

Communicability: Direct person-to-person transmission does not occur. When dealing with infected individuals, use standard contact precautions.

Normal Routes of Exposure: Inhalation; Ingestion; Abraded skin; Mucous membranes.

Infectious Dose: Unknown.

Secondary Hazards: Aerosols (blood, fluids, contaminated dust); Blood and body fluids (from mice); Fecal (from mice); Fomites (from mouse habitation).

Incubation: 2–48 days.

Signs and Symptoms: Initial symptoms are nonspecific and include fever, headache, chills, muscular pain, vomiting, diarrhea, and nausea followed by an abrupt onset of diffuse pulmonary edema, low oxygen levels in the blood (hypoxia), and low blood pressure. Most deaths occur within 24 h of hospital admission.

Suggested Alternatives for Differential Diagnosis: Acute respiratory distress syndrome, plague, congestive heart failure and pulmonary edema, HIV infection and AIDS, pneumonia, shock, phosgene, influenza, tularemia, phosphine toxicity, anthrax, silent myocardial infarction, and salicylate toxicity with pulmonary edema.

Mortality Rate (untreated): $\leq$36%.

C18-A059

Smallpox (Agent N1)

Poxviridae

ICD-10: B03

Although at onetime it was found worldwide, it was declared eradicated by the World Health Assembly in May 1980. There are only two known repositories of virus remaining, one in the United States and one in Russia. Humans are the only reservoir and vector. This is a biosafety level 4 agent.

It is on the HHS Select Agents list, the Australia Group Core list, and is listed as a Category A Potential Terrorism Agent by the CDC.

In People:

CDC Case Definition: An illness with acute onset of fever $\geq$101°F followed by a rash characterized by firm, deep seated vesicles or pustules in the same stage of development without other apparent cause. Clinically consistent cases are those presentations of smallpox that do not meet this classical clinical case definition: (1) hemorrhagic type, (2) flat type, and (3) *variola sine eruptione*. *Laboratory criteria* for diagnosis is (1) polymerase chain reaction (PCR) identification of variola DNA in a clinical specimen, or (2) isolation of smallpox (variola) virus from a clinical specimen (Level D laboratory only; confirmed by variola PCR).

Communicability: Direct person-to-person transmission is possible. It is communicable from the onset of the rash until the last scab falls off. When dealing with infected individuals, use airborne precautions.

Normal Routes of Exposure: Inhalation; Abraded skin; Mucous membranes.

Infectious Dose: Unknown.

Secondary Hazards: Aerosols (cough, sneeze); Contact; Blood and body fluids; Fomites.

Incubation: 7–17 days.

Signs and Symptoms: Initial symptoms include high fever, a vague feeling of bodily discomfort (malaise), head and body aches, and sometimes vomiting. A rash appears on the tongue and in the mouth that progress into open sores. The individual is now contagious. A macular rash appears on the face, spreads to the arms and legs, and then to the hands and feet. These progress to bumps that often have a depression in the center that looks like a bellybutton. The bumps become hard pustules and then scabs. Blisters and crusts are accompanied by severe itching. When the scabs fall off they leave a mark on the skin that will become a pitted scar. The individual is no longer contagious after all of the scabs have fallen off.

Suggested Alternatives for Differential Diagnosis: Acne, insect bites, drug eruptions, allergic dermatitis, acute leukemia, bullous pemphigoid, coxsackievirus, cytomegalovirus, ehrlichiosis, enteroviral infections, herpes simplex, impetigo, infectious mononucleosis, Kawasaki disease, chickenpox, measles, meningococcemia, parvovirus B19, rat-bite fever, scabies, scarlet fever, syphilis, Rocky Mountain spotted fever, rubella, molluscum contagiosum, cowpox, rickettsialpox, and monkeypox.

Mortality Rate (untreated): $\leq$25%.

C18-A060

St. Louis encephalitis

Flaviviridae
ICD-10: A83.3

It is transmitted by mosquitoes and normally found in the Caribbean, and North, Central, and South America. It is primarily a disease of humans but may also cause mild symptoms in horses. The natural reservoirs are birds and mosquitoes. Mosquitoes remain infective for life. This is a biosafety level 3 agent. It can survive in aerosol form for up to 6 h at room temperature.

It is on the Australia Group Core list.

In People:

CDC Case Definition: Arboviral infections may be asymptomatic or may result in illnesses of variable severity sometimes associated with central nervous system (CNS) involvement. When the CNS is affected, clinical syndromes ranging from febrile headache to aseptic meningitis to encephalitis may occur, and these are usually indistinguishable from similar syndromes caused by other viruses. Arboviral meningitis is characterized by fever, headache, stiff neck, and pleocytosis. Arboviral encephalitis is characterized by fever, headache, and altered mental status ranging from confusion to coma with or without additional signs of brain dysfunction (e.g., paresis or paralysis, cranial nerve palsies, sensory deficits, abnormal reflexes, generalized convulsions, and abnormal movements). *Laboratory criteria* for diagnosis is (1) fourfold or greater change in virus-specific serum antibody titer; or (2) isolation of virus from or demonstration of specific viral antigen or genomic sequences in tissue, blood, cerebrospinal fluid (CSF), or other body fluid; or (3) virus-specific immunoglobulin M (IgM) antibodies demonstrated in CSF by antibody-capture enzyme immunoassay (EIA); or (4) virus-specific IgM antibodies demonstrated in serum by antibody-capture EIA and confirmed by demonstration of virus-specific serum immunoglobulin G (IgG) antibodies in the same or a later specimen by another serologic assay (e.g., neutralization or hemagglutination inhibition).

Communicability: Direct person-to-person transmission does not occur. When dealing with infected individuals, use standard contact precautions.

Normal Routes of Exposure: Vectors (mosquitoes—*Culex* species).

Infectious Dose: Unknown.

Secondary Hazards: Aerosols (fluids); Body fluids (from animals); Fomites (from animal bedding).

Incubation: 7–21 days.

Signs and Symptoms: Range from mild fever and headache to high fever, headache, personality changes, confusion, stupor, tremors, coma, convulsions, and spastic paralysis. Recovery may be prolonged.

Suggested Alternatives for Differential Diagnosis: California encephalitis, eastern equine encephalitis, West Nile encephalitis, western equine encephalitis, herpes simplex encephalitis, meningitis, brain abscess, carcinomatous meningitis, CNS vasculitis, cerebrovascular disease.

Mortality Rate (untreated): ≤22%.

C18-A061

Swine vesicular disease

Picornaviridae

ICD-10: B08.8

It is normally found in Italy and some European countries. The natural reservoir is pigs. Virus may continue to be shed in the feces of pigs for up to 3 months after full recovery. Resistant to fermentation and smoking processes. May remain in hams for 6 months, dried sausages for more than a year, and in processed intestinal casings for more than 2 years.

It is on the USDA High Consequence list and the Australia Group Core list.

In People:

Significant disease does not occur in humans.

In Agricultural Animals:

Target species: Pigs.

Communicability: Direct transmission is possible.

Normal routes of exposure: Ingestion; Abraded skin; Mucous membranes.

Secondary Hazards: Blood and body fluids; Body tissue; Fecal matter; Vectors (mechanical—track out).

Incubation: 2–7 days.

Signs and Symptoms: Fever, vesicular lesions on the feet, mouth, lips, and/or snout. Pigs do not lose condition, and the lesions heal rapidly. Recovery is usually complete in less than 3 weeks.

Suggested Alternatives for Differential Diagnosis: Foot-and-mouth disease, vesicular exanthema of swine, vesicular stomatitis, foot rot, swine pox, chemical burns, and thermal burns.

Mortality Rate (untreated): 0%.

C18-A062

T3 coliphage

BW Simulant

A bacteriophage that infects the *E. coli* bacterium.

No significant harmful health effects to humans or animals are expected from exposure to this pathogen.

C18-A063

Tick-borne encephalitis complex

Flaviviridae
ICD-10: A84

It is normally found in many countries in Europe and Asia. The natural reservoir is ticks, which remain infected for life. The virus can be transmitted directly to the eggs and ticks are born infected. This is a biosafety level 2 agent.

It is on the HHS Select Agents list.

In People:

CDC Case Definition for arboviral encephalitis: Neuroinvasive disease requires the presence of fever and at least one of the following in the absence of a more likely clinical explanation: (1) Acutely altered mental status (e.g., disorientation, obtundation, stupor, or coma), or (2) other acute signs of central or peripheral neurologic dysfunction (e.g., paresis or paralysis, nerve palsies, sensory deficits, abnormal reflexes, generalized convulsions, or abnormal movements), or (3) increased white blood cell concentration in cerebrospinal fluid associated with illness clinically compatible with meningitis (e.g., headache or stiff neck). *Laboratory criteria* for diagnosis is (1) fourfold or greater change in virus-specific serum antibody titer; or (2) isolation of virus from or demonstration of specific viral antigen or genomic sequences in tissue, blood, CSF, or other body fluid; or (3) virus-specific immunoglobulin M (IgM) antibodies demonstrated in CSF by antibody-capture enzyme immunoassay (EIA); or (4) virus-specific IgM antibodies demonstrated in serum by antibody-capture EIA and confirmed by demonstration of virus-specific serum immunoglobulin G (IgG) antibodies in the same or a later specimen by another serologic assay (e.g., neutralization or hemagglutination inhibition). *Laboratory criteria* for a *probable case* is (1) stable (less than or equal to a twofold change) but elevated titer of virus-specific serum antibodies, or (2) virus-specific serum IgM antibodies detected by antibody-capture EIA but with no available results of a confirmatory test for virus-specific serum IgG antibodies in the same or a later specimen.

Communicability: Direct person-to-person transmission does not occur. When dealing with infected individuals, use standard contact precautions.

Normal Routes of Exposure: Ingestion; Abraded skin; Vectors (ticks—*Ixodes ricinus, Dermacentor marginatus*).

Infectious Dose: Unknown.

Secondary Hazards: Fomites (with vectors).

Incubation: 7–14 days.

Signs and Symptoms: Typically biphasic, the first phase consists of nonspecific flu-like symptoms that include fever, a vague feeling of bodily discomfort (malaise), anorexia, muscle pain (myalgia), headache, nausea, vomiting, low white blood cell count (leukopenia), low blood platelet count (thrombocytopenia) and last about a week. This is followed by a week long asymptomatic period. The third and final phase is an abrupt onset of meningitis or encephalitis, which affects the central nervous system phase producing fever, headache, stiff neck, tremors, dizziness, altered sensorium, and paralysis. Recovery is prolonged. May cause long lasting or permanent neuropsychiatric problems.

Suggested Alternatives for Differential Diagnosis: Meningitis, basilar artery blood clots (thrombosis), cardioembolic stroke, cavernous sinus syndromes, cerebral venous blood clots (thrombosis), confusional states and acute memory disorders, epileptic and epileptiform encephalopathies, febrile seizures, haemophilus meningitis, intracranial hemorrhage, leptomeningeal carcinomatosis, subdural pus (empyema), or bruise (hematoma).

Mortality Rate (untreated): ≤2%.

C18-A064

Variola minor

Poxviridae
ICD-10: B03

Mild form of smallpox caused by a less virulent form of the virus. The toxemia is less, lesions are more superficial, and healing time was more rapid. The natural reservoir is humans. This is a biosafety level 4 agent.

It is on the HHS Select Agents list and the Australia Group Core list (whitepox).

In People:

See smallpox (C18-A059).

Mortality Rate (untreated): ≤2%.

C18-A065
Venezuelan equine encephalitis (Agent NU)

Togaviridae
ICD-10: A92.2

It is a complex of viruses transmitted by mosquitoes and normally found in Central and South America. The natural reservoir is mosquitoes, horses, and birds. Horses may provide amplification of the virus. Mosquitoes remain infective for life. This is a biosafety level 3 agent.

It is on the HHS Select Agents list, the USDA High Consequence list, the Australia Group Core list, and is listed as a Category B Potential Terrorism Agent by the CDC.

In People:

CDC Case Definition: Arboviral infections may be asymptomatic or may result in illnesses of variable severity sometimes associated with central nervous system (CNS) involvement. When the CNS is affected, clinical syndromes ranging from febrile headache to aseptic meningitis to encephalitis may occur, and these are usually indistinguishable from similar syndromes caused by other viruses. Arboviral meningitis is characterized by fever, headache, stiff neck, and pleocytosis. Arboviral encephalitis is characterized by fever, headache, and altered mental status ranging from confusion to coma with or without additional signs of brain dysfunction (e.g., paresis or paralysis, cranial nerve palsies, sensory deficits, abnormal reflexes, generalized convulsions, and abnormal movements). *Laboratory criteria* for diagnosis is (1) fourfold or greater change in virus-specific serum antibody titer; or (2) isolation of virus from or demonstration of specific viral antigen or genomic sequences in tissue, blood, cerebrospinal fluid (CSF), or other body fluid; or (3) virus-specific immunoglobulin M (IgM) antibodies demonstrated in CSF by antibody-capture enzyme immunoassay (EIA); or (4) virus-specific IgM antibodies demonstrated in serum by antibody-capture EIA and confirmed by demonstration of virus-specific serum immunoglobulin G (IgG) antibodies in the same or a later specimen by another serologic assay (e.g., neutralization or hemagglutination inhibition).

Communicability: Direct person-to-person transmission does not occur. Humans can infect new mosquitoes during the fever phase. When dealing with infected individuals, use standard contact precautions and protect from vectors.

Normal Routes of Exposure: Vectors (mosquitoes—*Aedes* species, *Culex* species, *Psorophora* species, *Mansonia*, species, *Deinocerites species, Haemogogus* species, *Sabethes* species, *Anopheles* species).

Infectious Dose: 1 Plaque forming unit (Injection).

Secondary Hazards: Aerosols (blood, body fluids); Vector cycle.

Incubation: 1–6 days.

Signs and Symptoms: Initial symptoms are flu-like with abrupt onset of high fever, severe frontal headache, chills, muscle pain (myalgia), pain behind the eyes, nausea, and vomiting. Additional symptoms may include sensitivity to light (photophobia), diarrhea, and sore throat. Fever may be biphasic. CNS symptoms range from drowsiness to disorientation, convulsions, paralysis, coma, and death. May cause abortions or CNS malformations in the fetus.

Suggested Alternatives for Differential Diagnosis: Dengue fever, encephalitis, malaria, meningitis, yellow fever, St. Louis encephalitis, West Nile fever, Japanese encephalitis, western equine encephalitis, eastern equine encephalitis, arenaviruses, coxsackieviruses, cytomegalovirus, echoviruses, hepatitis, herpes simplex, infectious mononucleosis, influenza, leptospirosis, Listeria, lyme disease, malaria, meningitis, meningococcal infections, meningococcemia, Naegleria infection, Norwalk virus, Q fever, acute HIV infection, poliomyelitis, Colorado tick fever, measles.

Mortality Rate (untreated): $\leq$5%.

In Agricultural Animals:

Target species: Horses.

Communicability: Direct transmission does not occur.

Normal routes of exposure: Inhalation; Mucous membranes; Vectors (mosquitoes—*Culex* species).

Secondary Hazards: Aerosols (cough, sneeze); Contact; Vector cycle.

Incubation: 0.5–5 days.

Signs and Symptoms: A field diagnosis can rarely be made unless an epizootic of encephalitic disease is in progress and a prior etiologic diagnosis of Venezuelan equine encephalitis has been made. Symptoms include fever, weakness, depression, anorexia, colic, and diarrhea. Many die without showing neurologic signs.

Postmortem findings include changes in the central nervous system and possibly necrotic foci in internal organs such as pancreas, liver, and heart. There are no characteristic gross lesions.

Suggested Alternatives for Differential Diagnosis: Rabies, hepatoencephalopathy, leukoencephalomalacia, protozoal encephalomyelitis, equine herpes virus 1, verminous meningoencephalomyelitis, cranial trauma, botulism, meningitis, West Nile encephalitis, eastern equine encephalitis, and western equine encephalitis.

Mortality Rate (untreated): ≤75%.

C18-A066
Vesicular stomatitis fever

Rhabdoviridae
ICD-10: A93.8

It is normally found in North and Central America and northern South America. The natural reservoir is unknown. Black flies and sandflies remain infected for life. Infection is also transferred directly to their eggs. This is a biosafety level 2 agent. Capable of surviving for long periods of time at low temperatures.

It is on the USDA High Consequence list and the Australia Group Core list.

In People:

CDC Case Definition: None established.

Communicability: Direct person-to-person transmission does not occur. When dealing with infected individuals, use standard contact precautions.

Normal Routes of Exposure: Inhalation; Abraded skin; Mucous membranes; Vectors (mosquitoes—*Aedes* species; black flies—*Simuliidae*; sandflies—*Lutzomyia shannoni*).

Infectious Dose: Unknown.

Secondary Hazards: Aerosols (blood, body fluids from animals); Blood and body fluids; Body tissue; Body fluids and tissue (from animals).

Incubation: 24–48 h.

Signs and Symptoms: Produces flu-like symptoms including headache, fever, eye pain, a vague feeling of bodily discomfort (malaise), nausea, vomiting, diarrhea, sore throat as well as pain in the limbs and back. Blisters and/or lesions resembling herpes virus may appear in the mouth, throat, and occasionally on the hands. Illness may produce a prolonged mental depression.

Suggested Alternatives for Differential Diagnosis: No suggestions.

Mortality Rate (untreated): ≤1%.

In Agricultural Animals:

Target species: Horses, cattle, pigs.

Communicability: Direct transmission is possible.

Normal routes of exposure: Inhalation; Abraded skin; Mucous membranes; Vectors (black flies—*Simuliidae*; sandflies—*Lutzomyia shannoni*); Vectors (mechanical—rodents).

Secondary Hazards: Aerosols (body fluids); Body fluids; Body tissue; Fomites; Fomites (with vectors); Vector cycle; Vectors (mechanical—feeding and milking equipment).

Incubation: 1–21 days.

Signs and Symptoms: Easily confused with foot-and-mouth disease (C18-A017). Symptoms include fever, ulcers and erosions of the oral mucosa, sloughing of the epithelium of the tongue, and lesions at the mucocutaneous junctions of the lips, chewing movements and drooling (ptyalism), crusting lesions of the muzzle. Blisters, ulcers, and erosion of the coronary bands, and teats.

Suggested Alternatives for Differential Diagnosis: Foot-and-mouth disease, swine vesicular disease, vesicular exanthema of swine, rinderpest, infectious bovine rhinopneumonitis, bovine virus diarrhea, malignant catarrhal fever, bluetongue, bovine papular stomatitis, mycotic stomatitis, photosensitization, cowpox, pseudo-cowpox, pseudo-lumpy skin disease, bovine herpes mammillitis, Potomac Valley fever in horses, foot rot, chemical burns, and thermal burns.

Mortality Rate (untreated): "Very low."

C18-A067

West Nile fever

Flaviviridae
ICD-10: A92.3

It is normally found in the Middle East, Africa, and Asia, and is spreading through North America. The natural reservoirs are birds and mosquitoes. Other than sudden death, most birds exhibit little or no signs of disease. If present, typical symptoms include neurologic abnormalities and emaciation. Disease may also be present in household pets such as dogs and cats. This is a biosafety level 3 agent.

It is listed as a Category C Potential Terrorism Agent by the CDC.

In People:

CDC Case Definition: for arboviral encephalitis: Neuroinvasive disease requires the presence of fever and at least one of the following in the absence of a more likely clinical explanation: (1) Acutely altered mental status (e.g., disorientation, obtundation, stupor, or coma), or (2) other acute signs of central or peripheral neurologic dysfunction (e.g., paresis or paralysis, nerve palsies, sensory deficits, abnormal reflexes, generalized convulsions, or abnormal movements), or (3) increased white blood cell concentration in cerebrospinal fluid associated with illness clinically compatible with meningitis (e.g., headache or stiff neck). *Laboratory criteria* for diagnosis is (1) fourfold or greater change in virus-specific serum antibody titer; or (2) isolation of virus from or demonstration of specific viral antigen or genomic sequences in tissue, blood, CSF, or other body fluid; or (3) virus-specific immunoglobulin M (IgM) antibodies demonstrated in CSF by antibody-capture enzyme immunoassay (EIA); or (4) virus-specific IgM antibodies demonstrated in serum by antibody-capture EIA and confirmed by demonstration of virus-specific serum immunoglobulin G (IgG) antibodies in the same or a later specimen by another serologic assay (e.g., neutralization or hemagglutination inhibition). *Laboratory criteria* for a *probable case* is (1) stable (less than or equal to a twofold change) but elevated titer of virus-specific serum antibodies, or (2) virus-specific serum IgM antibodies detected by antibody-capture EIA but with no available results of a confirmatory test for virus-specific serum IgG antibodies in the same or a later specimen.

Communicability: Direct person-to-person transmission does not occur. When dealing with infected individuals, use standard contact precautions.

Normal Routes of Exposure: Abraded skin; Vectors (mosquitoes—*Culex* species, *Aedes* species, *Anopheles* species).

Infectious Dose: Unknown.

Secondary Hazards: Aerosols (blood, body fluids); Body fluids and tissue (from animals).

Incubation: 1–14 days.

Signs and Symptoms: Most infections are asymptomatic or produce a nonspecific flu-like illness. Symptoms include mild fever, headache, swollen lymph nodes, mental confusion, tremors, and flaccid paralysis. Liver and/or spleen may be enlarged. A maculopapular rash may be present on the trunk of the body. May progress to encephalitis and/or meningitis (meningoencephalitis) producing changes in mental status, seizures, and coma.

Suggested Alternatives for Differential Diagnosis: Hypertensive encephalopathy, brain abscess, brain tumor, meningitis, herpes simplex, Guillain–Barre syndrome, multiple sclerosis, stroke, nonsteroidal anti-inflammatory drugs, acute poliomyelitis, postpolio syndrome, catscratch disease, myasthenia gravis, hypoglycemia, leptospirosis, subarachnoid hemorrhage, Lyme diseases, Rocky Mountain spotted fever, toxoplasmosis, and tuberculosis.

Mortality Rate (untreated): ≤11%.

In Agricultural Animals:

Target species: Horses.

Communicability: Direct transmission does not occur.

Normal routes of exposure: Ingestion (in cats); Vectors (mosquitoes—*Culex species, Aedes species, Anopheles* species).

Secondary Hazards: Blood and body tissue (dead birds); Vector cycle.

Incubation: 10–14 days.

Signs and Symptoms: Symptoms include listlessness, stumbling, mild to severe incoordination (ataxia), and partial paralysis. Additionally, horses may exhibit weakness, muscle fasciculation, and cranial nerve deficits. Fever may or may not be present.

Suggested Alternatives for Differential Diagnosis: Eastern equine encephalitis, western equine encephalitis, Venezuelan equine encephalitis, Japanese encephalitis, equine protozoal myelitis, equine herpesvirus-1, Borna disease, and rabies. In birds Newcastle disease.

Mortality Rate (untreated): ≤40% (Horses).

C18-A068

Western equine encephalitis

Togaviridae

ICD-10: A83.1

It is normally found in the Americas. The natural reservoirs are birds and mosquitoes. Mosquitoes remain infective for life. Infection is also transferred directly to mosquito eggs. This is a biosafety level 2 agent.

It is on the Australia Group Core list and is listed as a Category B Potential Terrorism Agent by the CDC.

In People:

CDC Case Definition: Arboviral infections may be asymptomatic or may result in illnesses of variable severity sometimes associated with central nervous system (CNS) involvement. When the CNS is affected, clinical syndromes ranging from febrile headache to aseptic meningitis to encephalitis may occur, and these are usually indistinguishable from similar syndromes caused by other viruses. Arboviral meningitis is characterized by fever, headache, stiff neck, and pleocytosis. Arboviral encephalitis is characterized by fever, headache, and altered mental status ranging from confusion to coma with or without additional signs of brain dysfunction (e.g., paresis or paralysis, cranial nerve palsies, sensory deficits, abnormal reflexes, generalized convulsions, and abnormal movements). *Laboratory criteria* for diagnosis is (1) fourfold or greater change in virus-specific serum antibody titer; or (2) isolation of virus from or demonstration of specific viral antigen or genomic sequences in tissue, blood, cerebrospinal fluid (CSF), or other body fluid; or (3) virus-specific immunoglobulin M (IgM) antibodies demonstrated in CSF by antibody-capture enzyme immunoassay (EIA); or (4) virus-specific IgM antibodies demonstrated in serum by antibody-capture EIA and confirmed by demonstration of virus-specific serum immunoglobulin G (IgG) antibodies in the same or a later specimen by another serologic assay (e.g., neutralization or hemagglutination inhibition).

Communicability: Direct person-to-person transmission does not occur. Humans can infect new mosquitoes during the fever phase. When dealing with infected individuals, use standard contact precautions and protect from vectors.

Normal Routes of Exposure: Vectors (mosquitoes—*Culiseta tarsalis, Aedes melanimon).*

Infectious Dose: Unknown.

Secondary Hazards: Vector cycle.

Incubation: 5–15 days.

Signs and Symptoms: Difficult to diagnose because of the lack of specific symptoms. Once symptoms arise, the patient often deteriorates rapidly. Symptoms include headache, nausea or vomiting, confusion, seizures, semiconsciousness (somnolence), neck stiffness, a vague feeling of bodily discomfort (malaise) and weakness, cranial nerve palsies, sensitivity to light (photophobia), autonomic disturbances such as drooling (ptyalism), fever, chills, abdominal pain, diarrhea, sore throat, severe joint pain (arthralgias) and muscle pain (myalgia), and respiratory difficulty.

Suggested Alternatives for Differential Diagnosis: Bartonellosis, brucellosis, other causes of encephalitis, coxsackieviruses, cryptococcosis, cysticercosis, cytomegalovirus, histoplasmosis, legionellosis, leptospirosis, listeria, lyme disease, malaria, rabies, tuberculosis, mumps, stroke, metabolic encephalopathy, Reye syndrome, *Bartonella* infection, *Naegleria* infection, Ebstein-Barr virus, prion disease, toxic ingestions, and AIDS.

Mortality Rate (untreated): $\leq$4%.

In Agricultural Animals:

Target species: Horses, turkeys.

Communicability: Direct transmission does not occur.

Normal routes of exposure: Vectors (mosquitoes—*Culiseta tarsalis).*

Secondary Hazards: Vector cycle.

Incubation: 18–24 h.

Signs and Symptoms: Impaired vision, aimless wandering, head pressing, circling, inability to swallow, irregular uncoordinated (ataxic) gait, tremors, semiconsciousness (somnolence), paralysis, convulsions, and death. May also cause decreased egg production in turkeys.

Suggested Alternatives for Differential Diagnosis: *Horses*: Rabies, hepatoencephalopathy, leukoencephalomalacia, protozoal encephalomyelitis, equine herpes virus 1, verminous meningoencephalomyelitis, cranial trauma, botulism, and meningitis. *Turkeys*: Newcastle disease virus, avian encephalomyelitis virus, botulism, and listeriosis.

Mortality Rate (untreated): $\leq$30% (Horses); "high" (Turkeys).

C18-A069

Yellow fever (Agent UT)

Flaviviridae
ICD-10: A95

It is normally found in Africa and South America. The natural reservoirs are monkeys and mosquitoes. Mosquitoes remain infective for life. Infection is also transferred directly to mosquito eggs. This is a biosafety level 3 agent. It does not survive outside a host.

It is on the Australia Group Core list.

In People:

CDC Case Definition: A mosquito-borne viral illness characterized by acute onset and constitutional symptoms followed by a brief remission and a recurrence of fever, hepatitis, albuminuria, and symptoms and, in some instances, renal failure, shock, and generalized hemorrhages. Laboratory criteria for diagnosis is (1) fourfold or greater rise in yellow fever antibody titer in a patient who has no history of recent yellow fever vaccination and cross-reactions to other flaviviruses have been excluded or (2) demonstration of yellow fever virus, antigen, or genome in tissue, blood, or other body fluid.

Communicability: Direct person-to-person transmission does not occur. Humans can infect new mosquitoes during the fever phase. When dealing with infected individuals, use standard contact precautions and protect from vectors.

Normal Routes of Exposure: Vectors (mosquitoes—*Aedes* species, *Haemagogus* species, *Sabethes* species).

Infectious Dose: Unknown.

Secondary Hazards: Blood and body fluids; Vector cycle.

Incubation: 3–6 days.

Signs and Symptoms: Symptoms range from mild illness to hemorrhagic fever and death. Mild symptoms include mild fever and a vague feeling of bodily discomfort (malaise). Otherwise abrupt onset of general malaise, fever, chills, headache, lower back pain, nausea, dizziness, slow pulse (pulse-fever dissociation), conjunctival injection, facial flushing, decreased white blood cells (leukopenia). A period of remission (up to 24 h) is followed by a return fever, vomiting, abdominal pain, renal failure, and hemorrhage. Small hemorrhages (petechia), bruises (ecchymoses), nosebleed (epistaxis), and may ooze blood from gums. May progress to passage of black tar-like stools (melena), vomiting blood (hematemesis), bleeding from the uterus (metrorrhagia), cerebral edema, microscopic bleeding around the blood vessels (perivascular hemorrhage) and finally to delirium, stupor, coma, and death.

Suggested Alternatives for Differential Diagnosis: Acanthamoeba, louse-borne relapsing fever, dengue fever, Rift Valley fever, hemorrhagic fevers, leptospirosis, malaria, typhoid fever, typhus, liver failure, and hepatitis.

Mortality Rate (untreated): $\leq$70%.

References

Acha, Pedro N., and Boris Szyfres. *Zoonoses and Communicable Diseases Common to Man and Animals, Scientific and Technical Publication No. 580*. 3rd ed. Vol. 2, Chlamydioses, Rickettsioses, an Viroses. Washington, DC: Pan American Health Organization, 2003.

American Phytopathological Society. Plum Pox: A Devastating Threat to Peaches, Apricots, Plums, Nectarines, Almonds and Sweet and Tart Cherries. Undated Fact Sheet.

Borio, Luciana, Thomas Inglesby, C. J. Peters, Alan L. Schmaljohn, James M. Huges, Peter B. Jahrling, Thomas Ksiazek, Karl M. Johnson, Andrea Meyerhoff, Tara O'Toole, Michael S. Ascher, John Bartlett, Joel G. Breman, Edward. M. Eitzen, Jr., Margaret Hamburg, Jerry Hauer, D.A. Henderson, Richard T. Johnson, Gigi Kwik, Marci Layton, Scott Lillibridge, Gar J. Nabel, Michael T. Osterholm, Trish M. Perl, Philip Russell, and Kevin Tonat. "Hemorrhagic Fever Viruses as Biological Weapons: Medical and Public Health Management." *Journal of the American Medical Association* 287 (2002): 2391–405.

Brunt, Alan, Karen Crabtree, Michael Dallwitz, Adrian Gibbs, Leslie Watson, and Eric Zurcher, eds. *Plant Viruses Online: Descriptions and Lists from the VIDE Database*. http://md.brim.ac.cn/vide/refs.htm. 2006.

Burrows, W. Dickinson, and Sara E. Renner. "Biological Warfare Agents as Threats to Potable Water." *Environmental Health Perspectives* 107 (1999): 975–84.

California Department of Food and Agriculture. Animal Health and Food Safety Services. Animal Health Branch. *Biosecurity: Selection and Use of Surface Disinfectants*. Revision June 2002.

Canada Minister of National Health and Welfare. Laboratory Centre for Disease Control, Office of Biosafety. *Material Safety Data Sheet-Infectious Substances: Bluetongue Virus*. January 23, 2001.

———. *Material Safety Data Sheet-Infectious Substances: Cercopithecine Herpes Virus 1*. September 26, 2001.

———. *Material Safety Data Sheet-Infectious Substances: Chikungunya Virus*. April 23, 2001.

———. *Material Safety Data Sheet-Infectious Substances: Crimean-Congo Hemorrhagic Fever Virus*. January 23, 2001.

———. *Material Safety Data Sheet-Infectious Substances: Dengue Fever Virus (DEN 1, DEN 2, DEN 3, DEN 4)*. March 5, 2001.

———. *Material Safety Data Sheet-Infectious Substances: Eastern Equine Encephalitis Virus, Western Equine Encephalitis Virus*. March 5, 2001.

———. *Material Safety Data Sheet-Infectious Substances: Ebola Virus*. November 6, 2002.

———. *Material Safety Data Sheet-Infectious Substances: Hantavirus*. September 26, 2001.

———. *Material Safety Data Sheet-Infectious Substances: Influenza Virus*. September 26, 2001.

———. *Material Safety Data Sheet-Infectious Substances: Japanese Encephalitis Virus*. April 23, 2001.

———. *Material Safety Data Sheet-Infectious Substances: Junin Virus/Machupo Virus*. September 11, 1997.

———. *Material Safety Data Sheet-Infectious Substances: Kyasanur Forest Disease Virus*. March 2001.

———. *Material Safety Data Sheet-Infectious Substances: Lymphocytic Choriomeningitis Virus*. May 15, 2001.

———. *Material Safety Data Sheet-Infectious Substances: Marburg Virus*. September 11, 1997.

———. *Material Safety Data Sheet-Infectious Substances: Murray Valley Encephalitis*. April 23, 2001.

———. *Material Safety Data Sheet-Infectious Substances: Omsk Hemorrhagic Fever Virus*. May 23, 2001.

———. *Material Safety Data Sheet-Infectious Substances: O'Nyong-Nyong Virus*. March 2001.

———. *Material Safety Data Sheet-Infectious Substances: Powassan Encephalitis Virus*. May 14, 2001.

———. *Material Safety Data Sheet-Infectious Substances: Rabies Virus, Rabies-Related Viruses*. March 5, 2001.

———. *Material Safety Data Sheet-Infectious Substances: St. Louis Encephalitis*. April 23, 2001.

———. *Material Safety Data Sheet-Infectious Substances: Venezuelan Equine Encephalitis Virus*. September 25, 2001.

———. *Material Safety Data Sheet-Infectious Substances: Vesicular Stomatitis Virus*. September 27, 2001.

———. *Material Safety Data Sheet-Infectious Substances: West Nile Virus*. July 14, 2003.

———. *Material Safety Data Sheet-Infectious Substances: Yellow Fever Virus*. March 5, 2001.

Center for Infectious Disease Research & Policy, University of Minnesota. *Agricultural Biosecurity, Animal Diseases*. http://www.cidrap.umn.edu/cidrap/content/biosecurity/ag-biosec/anim-disease/index.html. 2006.

Centers for Disease Control and Prevention. "Biological and Chemical Terrorism: Strategic Plan for Preparedness and Response. Recommendations of the CDC Strategic Planning Workgroup." *Morbidity and Mortality Weekly Report 49* (RR-4) (2000): 1–14.

———. "Case Definitions for Infectious Conditions Under Public Health Surveillance." *Morbidity and Mortality Weekly Report 46* (RR-10) (1997): 12.

———. "Crimean-Congo Hemorrhagic Fever Fact Sheet." August 22, 2005.

———. "Fact Sheet: Basic Information About Monkeypox." June 12, 2003.

———. "Hemorrhagic Fever with Renal Syndrome Fact Sheet." Undated.

———. "Interim Case Definition for Animal Cases of Monkeypox." *Interim Document*, June 19, 2003.

———. "Monkeypox Infections In Animals: Updated Interim Guidance for Veterinarian." *Guidelines and Resources*, July 22, 2003.

———. "Tick-borne Encephalitis Fact Sheet." Undated.

Chin, James, ed. *Control of Communicable Diseases Manual*. 17th ed. Washington, DC: American Public Health Association, 2000.

College of Tropical Agriculture and Human Resources. *Banana Bunchy Top Virus*. University of Hawaii, December, 1997.

College of Veterinary Medicine. "Enterovirus Encephalomyelitis: Teschen Disease, Talfan Disease, Poliomyelitis Suum, Benign Enzootic Paresis." Iowa State University, August 5, 2005.

Committee on Foreign Animal Diseases of the United States Animal Health Association. *Foreign Animal Diseases*. Revised 1998. Richmond, VA: Pat Campbell & Associates and Carter Printing Company, 1998.

Compton, James A.F. *Military Chemical and Biological Agents: Chemical and Toxicological Properties*. Caldwell, NJ: The Telford Press, 1987.

Department of Agriculture. 9 CFR Part 121—"Possession, Use, and Transfer of Biological Agents and Toxins," 2005: 753–67.

Ehrlich, Richard, Sol Miller, and L.S. Idoine. "Effects of Environmental Factors on the Survival of Airborne T-3 Coliphage." *Applied Microbiology 12* (November 1964): 479–82.

European and Mediterranean Plant Protection Organization. *Diagnostic Protocols for Regulated Pests*. http://archives.eppo.org/EPPOStandards/diagnostics.htm. February 15, 2006.

Fediaevsky, Alexandre. *Manual for the Recognition of Exotic Diseases of Livestock: A Reference Guide for Animal Health Staff Food and Agriculture*. 2nd ed. Organization of the United Nations Secretariat of the Pacific Community. http://www.spc.int/rahs/manual/manuale.html. November 1, 2005.

Henderson, Donald A., Thomas V. Inglesby, John G. Bartlett, Michael S. Ascher, Edward Eitzen, Peter B. Jahrling, Jerome Hauer, Marcelle Layton, Joseph McDade, Michael T. Osterholm, Tara O'Toole, Gerald Parker, Trish M. Perl, Philip K. Russel, and Kevin Tonat. "Smallpox as a Biological Weapon: Medical and Public Health Management." *Journal of the American Medical Association* 281 (1999): 2127–137.

Henderson, Donald A., Thomas V. Inglesby, and Tara O'Toole, eds. *Bioterrorism: Guidelines for Medial and Public Health Management*. Chicago, IL: AMA Press, 2002.

Herenda, D. *Manual on Meat Inspection for Developing Countries*. Reprint 2000. Rome, Italy: Food and Agriculture Organization of the United Nations, 1994.

Kahn, Cynthia M., and Susan E. Aiello, eds. *The Merck Veterinary Manual*. 9th ed. Online Version. Whitehouse Station, NJ: Merck & Co., Inc., 2006.

Kortepeter, Mark, George Christopher, Ted Cieslak, Randall Culpepper, Robert Darling, Julie Pavlin, John Rowe, Kelly McKee, Jr., and Edward Eitzen, Jr., eds. *Medical Management of Biological Casualties Handbook*. 4th ed. Fort Detrick, MD: United States Army Medical Research Institute of Infectious Diseases, February 2001.

Lindler, Luther E., Frank J. Lebeda, and George W. Korch, eds. *Biological Weapons Defense: Infectious Diseases and Counterbioterrorism*. Totowa, NJ: Humana Press, Inc., 2005.

Pan American Health Organization. *Emergency Vector Control After Natural Disaster*. Scientific Publication No. 419. Washington, DC: Pan American Health Organization, 1982.

Public Health Service. 42 CFR Part 73—"Select Agents and Toxins," 2004: 443–58.
Richmond, Jonathan Y., and Robert W. McKinney, eds. *Biosafety in Microbiological and Biomedical Laboratories*. 4th ed. Washington, DC: US Government Printing Office, 1999.
Sidell, Fredrick R., Ernest T. Takafuji, and David R. Franz, eds. *Medical Aspects of Chemical and Biological Warfare, Textbook of Military Medicine Series, Part 1, Warfare, Weaponry, and the Casualty*. Washington, DC: Office of the Surgeon General, Department of the Army, 1997.
Sifton, David W., ed. *PDR Guide to Biological and Chemical Warfare Response*. Montvale, NJ: Thompson/Physicians Desk Reference, 2002.
Somani, Satu M., ed. *Chemical Warfare Agents*. New York: Academic Press, 1992.
United States Army Headquarters. *Potential Military Chemical/Biological Agents and Compounds, Field Manual No. 3-11.9*. Washington, DC: Government Printing Office, January 10, 2005.
———. *Technical Aspects of Biological Defense, Technical Manual No. 3-216*. Washington, DC: Government Printing Office, January 12, 1971.
United States Department of Agriculture. *Animal and Plant Health Inspection Service*. http://www.aphis.usda.gov/lpa/pubs/fsheet_faq_notice/fsfaqnot_animalhealth.html. 2004.
———. *Plant Protection and Quarantine*. http://www.aphis.usda.gov/ppq/. 2006.
World Health Organization. Fact Sheet #99 Rabies. Geneva: Health Communications and Public Relations, December 1995.
———. *Fact Sheet #100 Yellow Fever*. Geneva: Health Communications and Public Relations, December 2001.
———. *Fact Sheet #103 Ebola Hemorrhagic Fever*. Geneva: Health Communications and Public Relations, 1996.
———. *Fact Sheet #117 Dengue and Dengue Hemorrhagic Fever*. Geneva: Health Communications and Public Relations, May 1996.
———. *Health Aspects of Chemical and Biological Weapons: Report of a WHO Group of Consultants*. Geneva: World Health Organization, 1970.
———. *Public Health Response to Biological and Chemical Weapons: WHO Guidance*. Geneva: World Health Organization, 2004.
———. *WHO Slide Set on the Diagnosis of Smallpox*, October 24, 2001. (http://www.who.int/emc/diseases/smallpox/slideset/index.htm) February 3, 2006.
World Organization for Animal Health. *Manual of Diagnostic Tests and Vaccines for Terrestrial Animals 2004*. Updated July 22, 2005. http://www.oie.int/. January 2006.
———. "Technical Disease Card for African Horse Sickness." April 22, 2002.
———. "Technical Disease Card for Foot and Mouth Disease." April 22, 2002.
——— "Technical Disease Card for Highly Pathogenic Avian Influenza." April 22, 2002.
———. "Technical Disease Card for Classical Swine Fever (Hog Cholera)." April 22, 2002.
———. "Technical Disease Card for Rinderpest." April 22, 2002.
———. "Technical Disease Card for Newcastle Disease." April 22, 2002.
———. "Technical Disease Card for Swine Vesicular Disease." April 22, 2002.
———. "Technical Disease Card for Lumpy Skin Disease." April 22, 2002.
———. "Technical Disease Card for Peste des Petits Ruminants." April 22, 2002.
———. "Technical Disease Card for Rift Valley Fever." April 22, 2002.
———. "Technical Disease Card for Sheep Pox and Goat Pox." April 22, 2002.
———. "Technical Disease Card for Bluetongue." April 22, 2002.
———. "Technical Disease Card for African Swine Fever." April 22, 2002.
———. "Technical Disease Card for Vesicular Stomatitis." April 22, 2002.

19

Rickettsial Pathogens

19.1 General Information

Rickettsia are single-celled microorganisms that are less prevalent and produce fewer diseases than bacteria (Chapter 17). They are intermediate in size between bacteria and viruses (Chapter 18); approximately 0.3 μm in diameter and ranging from 0.3 to 0.5 μm in length. They are gram-negative, nonmotile, and nonsporing. They are easily killed by heat, dehydration, or common disinfecting agents.

Rickettsia are more difficult to produce in quantity than bacteria. Similar to viruses, they are strict obligate parasites and require living cells for growth. They cannot survive long outside a host. They also have a selective affinity for specific types of cells in the body. They are normally transmitted by an arthropod vector (i.e., ticks, lice, fleas, mites), which also serves as either the primary or intermediate host.

For biological warfare purposes, rickettsia have been primarily investigated as antipersonnel agents producing both lethal and incapacitating diseases. There are no known rickettsial diseases of plants. Pathogens deployed as antianimal biological weapons are generally used to produce lethal effects in an agriculturally significant species such as cows and sheep. There are no unclassified records of rickettsia being used as biological warfare simulants.

Rickettsia can be stored as freeze-dried powders. In this form, they are easy to disperse. However, because they are living organisms and can be killed during the dispersal process there are limitations to the methods that can be used. They can also be stored and dispersed via infected vectors (e.g., lice, ticks). In most cases, large-scale attacks will be clandestine and only detected through epidemiological analysis of resulting disease patterns. Localized or small-scale attacks may take the form of "anthrax" letters. Even in these cases, without the inclusion of a threat the attack may go unnoticed until the disease appears in exposed individuals (e.g., the initial 2001 anthrax attack at American Media Inc., which claimed the life of Robert Stevens).

It is possible that local insects can become both a reservoir and a vector for the pathogen. Under these circumstances, the pathogen can survive well after the initial release and can rapidly spread beyond the immediately affected area. In many cases, once a vector is infected, it is capable of transmitting the disease throughout its life span. Some pathogens are transmitted directly to the young of the vector so that the next generation is born infected. Response activities must also include efforts to contain and eliminate these vectors (e.g., application of pesticides).

Incubation times for diseases resulting from infection vary depending on the specific pathogen, but are generally on the order of days to weeks. Exposures to extremely high doses of some pathogens may reduce the incubation period to as short as several hours. The pathway of exposure (e.g., inhalation, bite) can also cause a significant change in the incubation time required as well as the clinical presentation of the disease. Diseases caused by rickettsia are not communicable and cannot be transferred directly from an infected individual to anyone else.

19.2 Response

19.2.1 Personal Protective Requirements

19.2.1.1 Responding to the Scene of a Release

A number of conditions must be considered when selecting protective equipment for individuals at the scene of a release. For instances such as "anthrax" letters when the mechanism of release is known and it does not involve an aerosol generating device, then responders can use Level C with high-efficiency particulate air (HEPA) filters.

If an aerosol generating device is employed (e.g., sprayer), or the dissemination method is unknown and the release is ongoing, then responders should wear a Level A protective ensemble. Once the device has stopped generating the aerosol or has been rendered inoperable, and the aerosol has settled, then responders can downgrade to Level B.

In all cases, there is a significant hazard posed by contact of contaminated material with skin that has been cut or lacerated, or through injection of pathogens by contact with debris. Appropriate protection to avoid any potential abrasion, laceration, or puncture of the skin is essential. Individuals with damaged or open skin should not be allowed to enter the contaminated area.

19.2.1.2 Working with Infected Individuals

Use infection control guidelines standard precautions. Avoid direct contact with wounds or wound drainage.

1. Wash hands after patient contact.
2. Wear gloves when handling potentially infectious items that are contaminated with blood, body fluids, or excreta.
3. Wear a mask and eye protection during procedures likely to generate splashes or sprays of infectious fluids.
4. Handle all contaminated and potentially contaminated equipment and linen in a manner that will prevent cross contamination.
5. Take care when handling sharps. When practical, use a mouthpiece or other ventilation device as an alternative to mouth-to-mouth resuscitation.

19.2.2 Decontamination

19.2.2.1 Food

All foodstuffs in the area of a release should be considered contaminated. Unopened items may be used after decontamination of the container. Opened or unpackaged items should be destroyed. Fruits and vegetables should be washed thoroughly with antimicrobial soap and water.

19.2.2.2 Casualties/Personnel

Infected individuals: Unless the individual is reporting directly from the scene of an attack (e.g., "anthrax" letter, aerosol release, etc.), then decontamination is not necessary.

Direct exposure: In the event that an individual is at the scene of a known or suspected attack (e.g., white-powder letter, aerosol release, etc.), have them wash their hands and face thoroughly with antimicrobial soap and water as soon as possible. If antimicrobial soap is not available, use any available soap or shampoo. They should also blow their nose to remove any agent particles that may have been captured by nasal mucous. Remove all clothing and seal in a plastic bag. To avoid further exposure of the head, neck, and face to the agent, cut off potentially contaminated clothing that must be pulled over the head. Shower using copious amounts of antimicrobial soap (if available) and water. Ensure that the hair has been washed and rinsed to remove potentially trapped agent. The CDC does not recommend that individuals use bleach or other disinfectants on their skin.

19.2.2.3 Animals

Unless the animals are at the scene of an attack then decontamination is not necessary.

Apply universal decontamination procedures using antimicrobial soap and water. If antimicrobial soap is not available, then use any available soap, shampoo, or detergent. In some cases, severe infection may require euthanasia of animals and/or herds. Consult local/state veterinary assistance office. If the pathogen has not been identified, then wear a fitted N95 protective mask, eye protection, disposable protective coverall, disposable boot covers, and disposable gloves when dealing with infected animals.

19.2.2.4 Property

Surface disinfectants: Compounds containing phenolics, chlorhexidine, quaternary ammonium salts (additional activity if bis-n-tributyltin oxide present), hypochlorites such as household bleach, alcohols such as 70–95% ethanol and isopropyl, potassium peroxymonosulfate, hydrogen peroxide, iodine/iodophores, and triclosan.

The military also identifies the following "nonstandard" decontaminants: Detrochlorite (thickened bleach mixture of diatomaceous earth, anionic wetting agent, calcium hypochlorite, and water), 3% aqueous peracetic acid solution, 1% aqueous hyamine solution, and a 10% aqueous sodium or potassium hydroxide solution.

Large area fumigants: Gases including formaldehyde, ethylene oxide, or chlorine dioxide. These materials are highly toxic to humans and animals, and fumigation operations must be adequately controlled to prevent unnecessary exposure. Additional methods include vaporized hydrogen peroxide and an ionized hydrogen peroxide aerosol.

Fomites: Some pathogens may be absorbed into clothing or bedding causing these items to become infectious and capable of transmitting the disease. Others may contain vectors (e.g., lice, ticks) that pose a transmission hazard. Deposit items in an appropriate biological waste container and send to a medical waste disposal facility.

Alternatively, cotton or wool articles can be boiled in water for 30 minutes, autoclaved at 253°F for 45 minutes, or immersed in a 2% household-bleach solution (i.e., 1 gallon of bleach in 2 gallons of water) for 30 minutes followed by rinsing.

19.3 Fatality Management

Unless the cadaver is coming directly from the scene of an attack (e.g., "anthrax" letter, aerosol release), process the body according to established procedures for handling potentially infectious remains.

Because of the nature of biological warfare agents, and rickettsia in particular, it is highly unlikely that a contaminated cadaver will be recovered from the scene of an attack unless it is from an individual who died from trauma or other complications while the attack was ongoing. If a fatality is grossly contaminated with a biological agent, wear disposable protective clothing with integral hood and booties, disposable gloves, and an N-95 respirator and eye protection or powered air-purifying respirator (PAPR) equipped with N-95 or HEPA filter.

Remove all clothing and personal effects. Items that will be retained for further processing should be double sealed in impermeable containers, ensuring that the inner container is decontaminated before placing it in the outer one. Otherwise, dispose contaminated articles at an appropriate medical waste disposal facility.

Thoroughly wash the remains with antimicrobial soap and water. Pay particular attention to areas where agent may get trapped, such as hair, scalp, pubic areas, fingernails, folds of skin, and wounds. If deemed appropriate, the cadaver can be treated with a surface disinfectant listed in Section 19.2.2.

If there is a potential that vectors may be involved, care must be taken to kill any vectors (e.g., lice, ticks) remaining either on the cadaver or residing in fomites. Remove all potentially infested clothing depositing it in a container that will trap and eliminate vectors. Dispose contaminated articles at an appropriate medical waste disposal facility.

Once the remains have been thoroughly decontaminated, process the body according to established procedures for handling potentially infectious bodies. Use appropriate burial procedures.

In 2004 (*Morbidity and Mortality Weekly Report 53*), the Centers for Disease Control and Prevention determined that the risks of occupational exposure to biological terrorism agents outweighed the advantages of embalming fatalities and recommended that the body should be buried without embalming.

C19-A

AGENTS

C19-A001

Rickettsia akari

Rickettsial pox
ICD-10: A79.1

Produces a mild, self-limiting disease. The natural reservoirs are house mice, rats, and mites. Mites remain infected for life. Infection is also transferred directly to mite eggs. Does not produce disease in animals. This is a biosafety level 3 agent.

In People:

CDC Case Definition: None established.

Communicability: Direct person-to-person transmission does not occur. When dealing with infected individuals, use standard contact precautions.

Normal Routes of Exposure: Vectors (mites—*Allodermanyssus sanguineus*).

Infectious Dose: Unknown.

Secondary Hazards: Aerosols (blood and body fluids); Blood and body tissue (from rodents).

Incubation: 10–21 days.

Signs and Symptoms: Sudden onset of high-grade fever, sore throat, chills, headaches, neck stiffness, muscle pain (myalgia), and sensitivity to light (photophobia). Lesions appear in the throat and mouth, and on the palms and soles of the feet. Lymph nodes near the site of the bite may be enlarged.

Suggested Alternatives for Differential Diagnosis: Boutonneuse fever, chickenpox, scrub typhus, hand-foot-and-mouth disease, meningitis.

Mortality Rate (untreated): 0%.

C19-A002

Rickettsia prowazekii (Agent YE)

Typhus

ICD-10: A75.0

The natural reservoirs are humans, and within the United States, flying squirrels. *R. prowazekii* is not transmitted by the bite of the louse, but rather by contamination of the bite, or other open wounds, by the infected feces of the louse. Lice begin releasing rickettsiae in their feces within 6 days after becoming infected. Lice die within 2 weeks after becoming infected. Rickettsiae may remain viable in dead lice for weeks. This is a biosafety level 3 agent.

It is on the HHS Select Agents list, the Australia Group Core list, and is listed as a Category B Potential Terrorism Agent by the CDC.

In People:

CDC Case Definition: None established.

Communicability: Direct person-to-person transmission does not occur. Humans can infect new lice during the fever phase. When dealing with infected individuals, use standard contact precautions and protect from vectors.

Normal Routes of Exposure: Vectors (lice—*Pediculus humanus corporis*).

Infectious Dose: $<$10 Organisms.

Secondary Hazards: Aerosols (contaminated material); Fomites (with vectors); Vector cycle.

Incubation: 7–14 days.

Signs and Symptoms: Sudden onset of headache, chills, prostration, fever, and general pains with a macular rash. May progress to delirium, stupor, and coma. Disease may reappear years after initial attack without involvement of lice (Brill–Zinsser disease).

Suggested Alternatives for Differential Diagnosis: Anthrax, brucellosis, dengue, ehrlichiosis, infectious mononucleosis, Kawasaki disease, leptospirosis, malaria, meningitis, meningococcemia, relapsing fever, Rocky Mountain spotted fever, syphilis, toxic shock syndrome, toxoplasmosis, tularemia, typhoid fever, rubella, measles.

Mortality Rate (untreated): $\leq$40%.

C19-A003

Rickettsia quintana

Trench Fever

ICD-10: A79.0

The natural reservoir is humans. This is a biosafety level 2 agent.

It is on the Australia Group Core list.

In People:

CDC Case Definition: None established.

Communicability: Direct person-to-person transmission does not occur. Humans can infect new lice for up to 1 year after the disease. When dealing with infected individuals, use standard contact precautions and protect from vectors.

Normal Routes of Exposure: Vectors (lice—*Pediculus humanus corporis*).

Infectious Dose: Unknown.

Secondary Hazards: Blood (transfusions); Fomites (with vectors); Vector cycle.

Incubation: 7–30 days.

Signs and Symptoms: Initial onset may be either sudden or slow. Symptoms include fever, headache with pain behind the eyes, conjunctivitis, a vague feeling of bodily discomfort (malaise), severe joint pain (arthralgias), enlarged spleen, and pain in the bones of the shins, neck, and back. Shins are especially painful and tender. May produce a transient macular rash. Symptoms may continue to reappear years after the primary infection.

Suggested Alternatives for Differential Diagnosis: Babesiosis, bacillary angiomatosis, cryptococcosis, Lyme disease, non-Hodgkin lymphoma, relapsing fever, Rocky Mountain spotted fever, tuberculosis, Ebstein–Barr virus, AIDS.

Mortality Rate (untreated): 0%.

C19-A004

Rickettsia rickettsii

Rocky Mountain Spotted Fever
ICD-10: A77.0

The natural reservoir is ticks. Ticks remain infected for life. Infection is also transferred directly to tick eggs. Rickettsiae not transmitted unless ticks are attached for 4–6 h. May also cause a lethal disease in dogs. This is a biosafety level 3 agent. Can survive for up to 1 year in tick tissue or blood; however, quickly inactivated by drying.

It is on the HHS Select Agents list and the Australia Group Core list.

In People:

CDC Case Definition: Laboratory criteria for diagnosis is (1) serological evidence of a significant change in serum antibody titer reactive with *Rickettsia rickettsii* antigens between paired serum specimens, as measured by a standardized assay conducted in a commercial, state, or reference laboratory; or (2) demonstration of *R. rickettsii* antigen in a clinical specimen by immunohistochemical methods; or (3) detection of *R. rickettsii* DNA in a clinical specimen by the polymerase chain reaction (PCR assay); or (4) isolation of *R. rickettsii* from a clinical specimen in cell culture. For confirmed cases, a significant change in titer must be determined by the testing laboratory; examples of commonly used measures of significant change include, but are not limited to, a fourfold or greater change in antibody titer as determined by indirect immunoflourescent antibody (IFA) assay or an equivalent change in optical density measured by enzyme-linked immunosorbent assay (EIA or ELISA).

Communicability: Direct person-to-person transmission does not occur. When dealing with infected individuals, use standard contact precautions.

Normal Routes of Exposure: Abraded skin; Mucous membranes; Vectors (ticks—*Dermacentor species*).

Infectious Dose: < 10 Organisms.

Secondary Hazards: Blood and body fluids (from ticks); Fecal (from ticks).

Incubation: 3–14 days.

Signs and Symptoms: Sudden onset of moderate to high fever, a vague feeling of bodily discomfort (malaise), deep muscular pain, severe headache, chills, and bloodshot eyes. Additional symptoms may include nausea, vomiting, and diarrhea. A macular rash typically appears on the third day. The rash may progress to small hemorrhages (petechia). In severe cases, the nervous system may be affected, producing agitation, insomnia, delirium, and/or coma. There may also be circulatory and pulmonary complications. If untreated, recovery can be prolonged.

Suggested Alternatives for Differential Diagnosis: Bronchitis, gastroenteritis, hepatitis, thrombocytopenic purpura, meningitis, mononucleosis, Kawasaki disease, measles, rubella, pneumonia, syphilis, ehrlichiosis, Lyme disease, Q fever, relapsing fever, tularemia, toxic shock syndrome, influenza, enterovirus infection, typhoid fever, bacterial sepsis, meningococcemia, disseminated gonococcal infection, drug hypersensitivity, immune complex vasculitis, staphylococcal sepsis, typhus, rickettsialpox.

Mortality Rate (untreated): ≤25%.

C19-A005

Rickettsia ruminantium

Heartwater
ICD-10:

It gets its name because it causes accumulation of fluid in the sac around the heart. The natural reservoirs are ruminants and ticks. Ticks remain infected for extended periods. Infected animals that recovered can be become chronic carriers for up to 8 months. It is extremely fragile and cannot persist outside a host for more than a few hours. It must be stored in dry ice or liquid nitrogen to preserve its infectivity.

It is on the USDA High Consequence list.

In People:
Does not occur in humans.

In Animals:

Target species: Cattle, sheep, goats.

Communicability: Direct transmission does not occur.

Normal routes of exposure: Vectors (ticks—*Amblyomma* species).

Secondary Hazards: Fomites (with vectors); Vector cycle.

Incubation: 10–28 days.

Signs and Symptoms: Fever, difficulty breathing (dyspnea), rapid breathing (tachypnea), lacrimation, lack of appetite, diarrhea, depression, and listlessness. Nervous signs may include sensitivity to stimulation (hyperesthesia), twitching eyelids, sticking out the tongue, clamping the jaw, high-stepping stiff gait, walking in circles, paddling with legs in recumbent animals, spasms, and convulsions. Once signs of the disease have developed, the prognosis is poor.

Postmortem lesions include excessive fluid in the sac surrounding the heart (hydropericardium); in the chest cavity (hydrothorax); pulmonary edema; hemorrhage in the abomasum and intestine; hemorrhagic gastroenteritis; enlarged liver, spleen, and lymph nodes; and edema and hemorrhage of the brain.

Suggested Alternatives for Differential Diagnosis: Anthrax, tetanus, rabies, meningitis, encephalitis, cerebral trypanosomiasis, piroplasmosis, theileriosis, listeriosis, parasitism; poisoning by strychnine, lead, organophosphates, arsenic, and various plants that affect the central nervous system.

Mortality Rate (untreated): Varies widely by species, overall ≤90%.

C19-A006

Rickettsia tsutsugamushi

Scrub typhus
ICD-10: A75.3

The natural reservoirs are mites and rodents. Mites remain infected for life. Infection is also transferred directly to mite eggs. This is a biosafety level 3 agent.

In People:

CDC Case Definition: None established.

Communicability: Direct person-to-person transmission does not occur. When dealing with infected individuals, use standard contact precautions.

Normal Routes of Exposure: Vectors (mites—*Leptotrombidium* species).

Infectious Dose: Unknown.

Secondary Hazards: Fomites (with vectors).

Incubation: 6–21 days.

Signs and Symptoms: Sudden onset of fever, headache, conjunctival congestion, generalized aches, swollen, painful lymph nodes, and inflammation of the lung tissue (pneumonitis). A dull red to dark purple maculopapular rash appears first on the trunk then spreads to the arms and legs. Can progress to pulmonary, encephalitic, and/or cardiac complications.

Suggested Alternatives for Differential Diagnosis: Brucellosis, dengue, ehrlichiosis, infectious mononucleosis, Kawasaki disease, leptospirosis, malaria, relapsing fever, Rocky Mountain spotted fever, syphilis, toxic shock syndrome, toxoplasmosis, tularemia, typhoid fever, rubella, measles.

Mortality Rate (untreated): ≤60%.

C19-A007

Rickettsia typhi

Endemic typhus
ICD-10: A75.2

The natural reservoirs are rats (*Rattus rattus, Rattus norvegicus, Rattus exulans) and fleas.* Fleas remain infected for life. *R. typhi* is not transmitted by the bite of the flea, but rather by contamination of the bite, or other open wounds, by the infected feces of the flea.

In People:

CDC Case Definition: None established.

Communicability: Direct person-to-person transmission does not occur. When dealing with infected individuals, use standard contact precautions.

Normal Routes of Exposure: Inhalation; Abraded skin; Mucous membranes; Vectors (fleas—*Xenopsylla cheopis, Ctenocephalides felis*).

Infectious Dose: Unknown.

Secondary Hazards: Aerosols (flea feces); Fecal (from fleas); Fomites (with vectors).

Incubation: 6–14 days.

Signs and Symptoms: Symptoms include fever, severe headache, pain in the muscles (myalgia), cough, nervousness, nausea, and vomiting. A macular rash appears first on the trunk then spreads to the arms and legs. The rash does not appear on the palms of the hands, soles of the feet, or the face. Recovery may be prolonged.

Suggested Alternatives for Differential Diagnosis: Brucellosis, dengue, ehrlichiosis, infectious mononucleosis, Kawasaki disease, leptospirosis, malaria, relapsing fever, Rocky Mountain spotted fever, syphilis, toxic shock syndrome, toxoplasmosis, tularemia, typhoid fever, rubella, measles.

Mortality Rate (untreated): $\leq$4%.

References

Acha, Pedro N., and Boris Szyfres. *Zoonoses and Communicable Diseases Common to Man and Animals, Scientific and Technical Publication No. 580*. 3rd ed. Vol. 2, Chlamydioses, Rickettsioses, and Viroses. Washington, DC: Pan American Health Organization, 2003.

Burrows, W. Dickinson, and Sara E. Renner. "Biological Warfare Agents as Threats to Potable Water." *Environmental Health Perspectives* 107 (1999): 975–84.

Canada Minister of National Health and Welfare. Laboratory Centre For Disease Control, Office of Biosafety. *Material Safety Data Sheet-Infectious Substances: Rickettsia akari*. March 5, 2001.

———. *Material Safety Data Sheet-Infectious Substances: Rickettsia prowazekii, Rickettsia canadensis (Formerly R. canada)*. March 5, 2001.

———. *Material Safety Data Sheet-Infectious Substances: Rickettsia rickettsii*. March 5, 2001.

Centers for Disease Control and Prevention. "Biological and Chemical Terrorism: Strategic Plan for Preparedness and Response. Recommendations of the CDC Strategic Planning Workgroup." *Morbidity and Mortality Weekly Report 49* (RR-4) (2000): 1–14.

———. "Case Definitions for Infectious Conditions Under Public Health Surveillance." *Morbidity and Mortality Weekly Report 46* (RR-10) (1997).

Chin, James, ed. *Control of Communicable Diseases Manual*. 17th ed. Washington, DC: American Public Health Association, 2000.

Committee on Foreign Animal Diseases of the United States Animal Health Association. *Foreign Animal Diseases*. Revised 1998. Richmond, VA: Pat Campbell & Associates and Carter Printing Company, 1998.

Compton, James A.F. *Military Chemical and Biological Agents: Chemical and Toxicological Properties*. Caldwell, NJ: The Telford Press, 1987.

Department of Agriculture. 9 CFR Part 121—"Possession, Use, and Transfer of Biological Agents and Toxins," 2005: 753–67.

Fediaevsky, Alexandre. *Manual for the Recognition of Exotic Diseases of Livestock: A Reference Guide for Animal Health Staff Food and Agriculture*. 2nd ed. Organization of the United Nations Secretariat of the Pacific Community. http://www.spc.int/rahs/manual/manuale.html. November 1, 2005.

Herenda, D. *Manual on Meat Inspection for Developing Countries*. Reprint 2000. Rome, Italy: Food and Agriculture Organization of the United Nations, 1994.

Kahn, Cynthia M., and Susan E. Aiello, eds. *The Merck Veterinary Manual*. 9th ed. Online Version. Whitehouse Station, NJ: Merck & Co., Inc., 2006.

Pan American Health Organization. *Emergency Vector Control After Natural Disaster*. Scientific Publication No. 419. Washington, DC: Pan American Health Organization, 1982.

Public Health Service. 42 CFR Part 73—"Select Agents and Toxins," 2004: 443–58.

Richmond, Jonathan Y., and Robert W. McKinney, eds. *Biosafety in Microbiological and Biomedical Laboratories*. 4th ed. Washington, DC: US Government Printing Office, 1999.

Sifton, David W., ed. *PDR Guide to Biological and Chemical Warfare Response*. Montvale, NJ: Thompson/Physicians Desk Reference, 2002.

United States Army Headquarters. *Potential Military Chemical/Biological Agents and Compounds, Field Manual No. 3-11.9*. Washington, DC: Government Printing Office, January 10, 2005.

———. *Technical Aspects of Biological Defense, Technical Manual No. 3-216*. Washington, DC: Government Printing Office, January 12, 1971.

United States Department of Agriculture. *Animal and Plant Health Inspection Service*. http://www.aphis.usda.gov/lpa/pubs/fsheet_faq_notice/fsfaqnot_animalhealth.html. 2004.

World Health Organization. *Fact Sheet #162 Epidemic Louse-Borne Typhus Fever*. Geneva: Health Communications and Public Relations, May 1997.

———. *Health Aspects of Chemical and Biological Weapons: Report of a WHO Group of Consultants*. Geneva: World Health Organization, 1970.

———. *Public Health Response to Biological and Chemical Weapons: WHO Guidance*. Geneva: World Health Organization, 2004.

World Organization for Animal Health. *Manual of Diagnostic Tests and Vaccines for Terrestrial Animals 2004*. Updated July 22, 2005. http://www.oie.int/. January 2006.

20

Fungal Pathogens

20.1 General Information

Fungi are unicellular or multicellular organisms that are more highly evolved than bacteria (Chapter 17). They are members of the plant kingdom and include molds, mildew, smuts, rusts, and yeasts. They range in size from 3 to 50 μm. With the exception of yeasts, they are usually rod shaped and arranged end-to-end in strands or filaments. Yeasts are usually oval.

Fungi reproduce by forming spores. Spores are part of the reproductive cycle and are not a protective mechanism as used by some bacteria when they are subjected to adverse conditions. Fungal spores can lie dormant, sometimes for decades, waiting for conditions that allow germination. Infections occur when a spore germinates on or in a host. Fungi are relatively easy to grow. Production, isolation, harvesting, and storage of spores are also relatively uncomplicated.

Although several notable antipersonnel fungal agents have been investigated as biological warfare agents, fungi have primarily been selected because of their ability to attack agriculturally significant crop species such as wheat, corn, or rice. Most antiplant agents are host specific, some are even specific to individual varieties of the host species. There is little potential for antiplant fungi to attack humans or animals.

A final group of fungi are those used as simulants to model the release of other, more hazardous agents. Pathogens employed as biological warfare simulants do not generally pose a significant risk to people, animals, or plants. However, individuals with respiratory illness or suppressed immune system may be at risk should they be exposed to an infectious dose of the agent.

Fungi can be stored as active cultures or isolated as spores. Spores are easy to disperse. However, because they are living organisms and can be killed during the dispersal process, there are limitations to the methods that can be used. In most cases, large-scale attacks will be clandestine and only detected through epidemiological analysis of resulting disease patterns. Localized or small-scale attacks may take the form of "anthrax" letters. Even in these cases, without the inclusion of a threat the attack may go unnoticed until the disease appears in exposed individuals (e.g., the initial 2001 anthrax attack at American Media Inc., which claimed the life of Robert Stevens).

Incubation times for diseases resulting from infection vary depending on the specific pathogen, but are generally on the order of days to weeks. Exposures to extremely high doses of some pathogens may reduce the incubation period to as short as several hours.

The pathway of exposure (e.g., inhalation, lacerations) can also cause a significant change in the incubation time required as well as the clinical presentation of the disease. Diseases caused by fungi are not communicable and cannot be transferred directly from an infected individual to anyone else.

20.2 Response

20.2.1 Personal Protective Requirements

20.2.1.1 Responding to the Scene of a Release

A number of conditions must be considered when selecting protective equipment for individuals at the scene of a release. For instances such as "anthrax" letters, when the mechanism of release is known and it does not involve an aerosol generating device, then responders can use Level C with high-efficiency particulate air (HEPA) filters.

If an aerosol generating device is employed (e.g., sprayer), or the dissemination method is unknown and the release is ongoing, then responders should wear a Level A protective ensemble. Once the device has stopped generating the aerosol or has been rendered inoperable, and the aerosol has settled, then responders can downgrade to Level B.

In all cases, there is a significant hazard posed by contact of contaminated material with skin that has been cut or lacerated, or through injection of pathogens by contact with debris. Appropriate protection to avoid any potential abrasion, laceration, or puncture of the skin is essential. Individuals with damaged or open skin should not be allowed to enter the contaminated area.

20.2.1.2 Working with Infected Individuals

Use infection control guidelines standard precautions. Avoid direct contact with wounds or wound drainage.

1) Wash hands after patient contact.
2) Wear gloves when handling potentially infectious items that are contaminated with blood, body fluids, or excreta.
3) Wear a mask and eye protection during procedures likely to generate splashes or sprays of infectious fluids.
4) Handle all contaminated and potentially contaminated equipment and linen in a manner that will prevent cross contamination.
5) Take care when handling sharps. When practical, use a mouthpiece or other ventilation device as an alternative to mouth-to-mouth resuscitation.

20.2.2 Decontamination

20.2.2.1 Food

All foodstuffs in the area of a release should be considered contaminated. Unopened items may be used after decontamination of the container. Opened or unpackaged items should be destroyed. Fruits and vegetables should be washed thoroughly with soap and water.

20.2.2.2 Casualties/Personnel

Infected individuals: Unless the individual is reporting directly from the scene of an attack (e.g., "anthrax" letter, aerosol release, etc.) then decontamination is not necessary.

Direct exposure: In the event that an individual is at the scene of a known or suspected attack (e.g., "anthrax" letter, aerosol release), have them wash their hands and face thoroughly with soap and water as soon as possible. They should also blow their nose to remove any agent particles that may have been captured by nasal mucous. Remove all clothing and seal in a plastic bag. To avoid further exposure of the head, neck, and face to the agent, cut off potentially contaminated clothing that must be pulled over the head. Shower using copious amounts of soap and water. Ensure that the hair has been washed and rinsed to remove potentially trapped agent. The CDC does not recommend that individuals use bleach or other disinfectants on their skin.

20.2.2.3 Animals

Unless the animals are at the scene of an attack then decontamination is not necessary.

Apply universal decontamination procedures using soap and water. In some cases, severe infection may require euthanasia of animals and/or herds. Consult local/state veterinary assistance office. If the pathogen has not been identified, then wear a fitted N95 protective mask, eye protection, disposable protective coverall, disposable boot covers, and disposable gloves when dealing with infected animals.

20.2.2.4 Plants

May require removal and destruction of infected plants. Incineration of impacted fields may be required. Consult local/state agricultural assistance office. Many fungi are easily spread by mechanical vectors (e.g., farm implements, track out, running water) and extreme care must be taken to avoid further contamination. Wear disposable protective coverall, disposable boot covers, and disposable gloves. Insects may also act as mechanical vectors for some plant fungi and, if appropriate, response activities must also include efforts to contain and eliminate these vectors.

20.2.2.5 Property

Surface disinfectants: Compounds containing phenolics, chlorhexidine, quaternary ammonium salts (additional activity if bis-*n*-tributyltin oxide present), alcohols such as 70–95% ethanol and isopropyl, potassium peroxymonosulfate, and iodine/iodophores.

The military also identifies the following "nonstandard" decontaminants: 3% aqueous peracetic acid solution, 1% aqueous hyamine solution, and a 10% aqueous sodium or potassium hydroxide solution.

Large area fumigants: Gases including formaldehyde, ethylene oxide, or chlorine dioxide. These materials are highly toxic to humans and animals and fumigation operations must be adequately controlled to prevent unnecessary exposure. Additional methods include vaporized hydrogen peroxide and an ionized hydrogen peroxide aerosol.

Fomites: Clothing or bedding may become contaminated with spores. Deposit items in an appropriate biological waste container and send to a medical waste disposal facility.

Alternatively, cotton or wool articles can be boiled in water for 30 minutes, autoclaved at 253°F for 45 minutes, or immersed in a 2% household-bleach solution (i.e., 1 gallon of bleach in 2 gallons of water) for 30 minutes followed by rinsing.

20.3 Fatality Management

Unless the cadaver is coming directly from the scene of an attack (e.g., "anthrax" letter, aerosol release), process the body according to established procedures for handling potentially infectious remains.

Because of the nature of biological warfare agents, it is highly unlikely that a contaminated cadaver will be recovered from the scene of an attack unless it is from an individual who died from trauma or other complications while the attack was ongoing. If a fatality is grossly contaminated with a biological agent, wear disposable protective clothing with integral hood and booties, disposable gloves, and an N-95 respirator and eye protection or powered air-purifying respirator (PAPR) equipped with N-95 or HEPA filter.

Remove all clothing and personal effects. Items that will be retained for further processing should be double sealed in impermeable containers, ensuring that the inner container is decontaminated before placing it in the outer one. Otherwise, dispose of contaminated articles at an appropriate medical waste disposal facility.

Thoroughly wash the remains with soap and water. Pay particular attention to areas where agent may get trapped, such as hair, scalp, pubic areas, fingernails, folds of skin, and wounds. If deemed appropriate, the cadaver can be treated with a surface disinfectant listed in Section 20.2.2.

Once the remains have been thoroughly decontaminated, process the body according to established procedures for handling potentially infectious bodies. Use appropriate burial procedures.

C20-A

AGENTS

C20-A001

Aspergillus fumigatus

BW Simulant

ICD-10: B44

Found worldwide in decaying vegetation or grains. Spores are commonly inhaled pollutants that do not cause any significant effects in healthy humans or animals. May cause disease (aspergillosis) if an individual's immune system is compromised. In animals, may cause inflammation of the nose and throat, difficulty breathing (dyspnea), pneumonia, and abortions. Spores survive in soil and decaying matter for a long time.

In People (immunocompromised):

CDC Case Definition: None established.

Communicability: Direct person-to-person transmission does not occur. When dealing with infected individuals, use standard contact precautions.

Normal Routes of Exposure: Inhalation.

Infectious Dose: Unknown.

Secondary Hazards: None.

Incubation: Days to weeks.

Signs and Symptoms: Symptoms in immunocompromised individuals may include fever, difficulty breathing (dyspnea), nonproductive cough, bloody sputum (hemoptysis), bloody nose (epistaxis), a vague feeling of bodily discomfort (malaise), pneumonia, weakness, chest pain, and anorexia. May progress to inflammation of the eyes (endophthalmitis), sensitivity to light (photophobia), and/or inflammation of the heart (endocarditis). May also cause abscesses in the heart, kidneys, liver, spleen, other soft tissue, or the bone. If the central nervous system becomes involved, can cause altered mental states and seizures.

Suggested Alternatives for Differential Diagnosis: Acute respiratory distress syndrome, allergic and environmental asthma, asthma, bronchiectasis, eosinophilia, pneumonia, granulocytopenia, abscesses of the lung or heart, pulmonary embolism, pulmonary eosinophilia, tuberculosis, sarcoidosis, wegener granulomatosis, other fungal infections, mucoid impaction.

Mortality Rate (untreated): ≤95% (Immunosuppressed individuals).

C20-A002

Coccidioides immitis (Agent OC)

Coccidioidomycosis
ICD-10: B38

It is a highly virulent soil-fungus that is thermally dimorphic, existing in both a mold and yeast form. It is normally found in soils in the southwest United States, northern Mexico, and certain areas in Central and South America. Does not cause clinical disease in cattle, sheep, and pigs. However, it may cause fatalities in dogs and sometimes in cats. This is a biosafety level 3 agent. Stable to drying.

It is on the HHS Select Agents list and the USDA High Consequence list.

In People:

CDC Case Definition: An illness characterized by one or more of the following: Influenza-like signs and symptoms (e.g., fever, chest pain, cough, myalgia, arthralgia, and headache); pneumonia or other pulmonary lesion, diagnosed by chest radiograph; erythema nodosum or erythema multiforme rash; involvement of bones, joints, or skin by dissemination; meningitis; and/or involvement of viscera and lymph nodes. Laboratory criteria for diagnosis is (1) cultural, histopathologic, or molecular evidence of presence of *C. immitis*; or (2) positive serologic test for coccidioidal antibodies in serum or cerebrospinal fluid by detection of coccidioidal immunoglobulin M (IgM) by immunodiffusion, enzyme immunoassay (EIA), latex agglutination, or tube precipitin, or by detection of rising titer of coccidioidal immunoglobulin G (IgG) by immunodiffusion, EIA, or complement fixation; or (3) coccidioidal skin-test conversion from negative to positive after onset of clinical signs and symptoms.

Communicability: Direct person-to-person transmission does not occur. When dealing with infected individuals, use standard contact precautions.

Normal Routes of Exposure: Inhalation; Abraded skin; Mucous membranes.

Infectious Dose: 1 spore.

Secondary Hazards: Aerosols (contaminated dust); Spores.

Incubation: 1–3 weeks.

Signs and Symptoms: Initial symptoms include cough, fever, night sweats, chills, chest pain, sputum production, and headache. May progress to breathing difficulty (dyspnea), bloody sputum (hemoptysis), and chest pain. May become disseminated causing destruction of the skin, subcutaneous tissues, bones, and joints. It may progress to meningitis if the central nervous system becomes involved.

Suggested Alternatives for Differential Diagnosis: Babesiosis, erythema multiforme, granuloma annulare, pyogenic granuloma, meningitis, lung neoplasms, pleural effusion, pneumonia, sarcoidosis, tuberculosis, histoplasmosis, blastomycosis, paracoccidioidomycosis, lung abscess, lymphoma.

Mortality Rate (untreated): ≤50% (Disseminated form).

C20-A003

Coccidioides posadasii

Coccidioidomycosis
ICD-10: B38

Until 2002, *C. posadasii* was believed to be a non-California variant of *C. immitis*. The two species can only be distinguished by genetic analysis and by the fact that *C. posadasii* grows more slowly in the presence of high salt concentrations. There is no apparent difference in pathogenicity between the two species. For more information on coccidioidomycosis, the disease caused by these two fungi, see *C. immitis* (C20-A002).

It is on the HHS Select Agents list.

C20-A004

Colletotrichum coffeanum* variant *virulans

Coffee berry Disease

The fungus lives in the bark of the coffee tree and produces spores that attack the immature or green coffee berries.

It is on the Australia Group Core list.

In Plants:

Target species: Coffee.

Normal routes of transmission: Windborne; Contact (introduction of infected seeds).

Secondary Hazards: Mechanical vectors (windblown rain); Spores.

Signs: Appears as small dark sunken spots that spread rapidly over the berry. Spots may have a pale pink crust on their surface. Berry may become mummified prior to full development. Otherwise, the berry ripens and the bean can become infected. Crop loss ranges from 20 to 80%.

C20-A005

Deuterophoma tracheiphila

Citrus wilt

Principal hosts are lemons although it has also been reported in citrons, bergamots, limes, sour oranges, and rough lemons.

It is on the Australia Group Awareness list.

In Plants:

Target species: Lemons, other citrus trees.

Normal routes of transmission: Windborne; Contact (through wounds).

Secondary Hazards: Mechanical vectors (windblown rain); Spores; Crop debris; Vectors (mechanical—birds, insects).

Signs: Raised black points within gray areas appear on withered twigs. Sprouts grow from the base of the affected branches. Ultimately the entire tree is affected and it dies. Interior wood of infected twigs has a characteristic salmon-pink or reddish orange color.

C20-A006

Erysiphe graminis

Powdery mildew of Cereals

Grows in cool, humid areas of the world. Overwinters on grasses and crop debris. It does not grow on synthetic media. Relatively cool and humid conditions are favorable for its growth.

In Plants:

Target species: Wheat, barley, oats, rye.
Normal routes of transmission: Windborne.
Secondary Hazards: Mechanical vectors (track out); Spores; Crop debris.
Signs: Grows on the surface of the foliage, especially the upper sides of the leaves. Can almost completely cover the plant. In later stages becomes gray-tan color with black bodies imbedded throughout. Because of a large genetic variability, it can often infect previously resistant plant varieties. Control through plowing or burning and crop rotation.

C20-A007

Helminthosporium oryzae (Agent E)

Brown spot of Rice

Grows in cool temperatures with high humidity. Leaves of the rice must remain wet for 8–24 h for infection to occur. It overwinters on crop debris.

It is on the Australia Group Core list.

In Plants:

Target species: Rice.

Normal routes of transmission: Windborne.

Secondary Hazards: Spores; Crop debris.

Signs: Leaf spots appear as small circular or oval spots on the first seedling leaves. As the plants grow, leaf spots can vary in size and shape. They can also vary in color ranging from light brown to reddish or dark brown. Spots may have a gray center and may also have a bright yellow halo surrounding the lesion. Severely infected leaves will produce lightweight or chalky kernels. The disease causes total blasting of kernels in cases of severe infection. When the invaded parts of the rice die, spores are formed and spread by the wind.

C20-A008

Histoplasma capsulatum

Histoplasmosis
ICD-10: B39.4

Normally found in soils around the world, and dung of bats and birds. Spores are less than 5 μm and are easily deposited in the lung. In addition to people, disease may be seen in dogs and cats. This is a biosafety level 3 agent. Spores are resistant to drying and may remain viable for long periods of time.

In People:

CDC Case Definition: None established.

Communicability: Direct person-to-person transmission does not occur. When dealing with infected individuals, use standard contact precautions.

Normal Routes of Exposure: Inhalation; Abraded skin; Mucous membranes.

Infectious Dose: Unknown.

Secondary Hazards: Aerosols (contaminated dust); Spores.

Incubation: 3–18 days.

Signs and Symptoms: Symptoms vary, depending on the number of spores inhaled, the host immune status, and any underlying lung disease. Generally, symptoms include a vague feeling of bodily discomfort (malaise), fever, chills, headache, diffuse muscle pain (myalgia), nonproductive cough, and chest pain. Additional symptoms include weight loss, fatigue, and night sweats. May progress to difficulty breathing (dyspnea), to low oxygen levels in the blood (hypoxemia), and to an adult respiratory distress syndrome (ARDS)-like illness. In disseminated form, may cause meningitis, seizures, focal neurologic deficits, and endocarditis. Histoplasma species may remain latent in healed granulomas and recur later without additional exposure.

Suggested Alternatives for Differential Diagnosis: Blastomycosis, coccidioidomycosis, aspergillosis, pneumonia, respiratory distress syndrome, mediastinal cysts, mycoplasma infections, Pancoast syndrome, sarcoidosis, tuberculosis, lung abscess, lung cancer, lymphoma.

Mortality Rate (untreated): "Low"—most cases resolve spontaneously.

C20-A009

Microcyclus ulei

South American leaf blight

Found naturally in the Caribbean and tropical Latin America.

It is on the Australia Group Core list.

In Plants:

Target species: Rubber plants.

Normal routes of transmission: Windborne; Rain borne.

Secondary Hazards: Mechanical vectors (rain); Spores.

Signs: Leaves are only susceptible to infection when they are less than 15 days old. Symptoms vary with the age of the leaves. In young leaves, causes discolored green masses that become gray. Spots grow together, consuming the leaf and causing it to die. In older leaves, the infection may cause holes in the leaves that are susceptible to further damage or infection.

C20-A010

Monilia rorei

Monilia pod rot

It is normally found in Central America and northwest South America. Only grows on the pods of the cocoa tree.

It is on the Australia Group Awareness list.

In Plants:

Target species: Cocoa tree.

Normal routes of transmission: Rain borne.

Secondary Hazards: Mechanical vectors (rain); Spores; Crop debris.

Signs: Conspicuous bumpy swellings on the pod surfaces progressing to a tan discoloration. Eventually the entire pod is covered by the fungus.

C20-A011

Peronosclerospora philippinensis

Philippine downy mildew

Has a high rate of infection when temperatures are above 77°F. Spore production requires high humidity, with at least a thin film of water on the infected leaf surface.

It is one of the USDA Select Agent Listed Plant Pathogens.

In Plants:

Target species: Corn, oats, sugarcane, sorghum.

Normal routes of transmission: Windborne; Rain borne.

Secondary Hazards: Spores; Crop debris.

Signs: Chlorotic stripes or overall yellowing on the leaves. Leaves may acquire a downy appearance. Spore formation is more abundant on the lower surface. As the plant ages, leaves may narrow, become abnormally erect, and appear somewhat dried out. Tassels may be malformed. Ears may be aborted. Early affected plants are stunted.

C20-A012

Phytophthora infestans (Agent LO)

Late blight of potato

Critically dependent on climate for growth. Greatest risk in high humidity when days are warm, nights are cool, and free moisture (e.g., dew) remains on the leaves for at least 4 h.

In Plants:

Target species: Potatoes; tomatoes.

Normal routes of transmission: Windborne.

Secondary Hazards: Mechanical vectors (planting); Spores; Crop debris.

Signs: Appears as large, water-soaked lesions in the leaves becoming a fine, whit mold-like growth. Fungus grows through the tissue causing disruption and rotting of the foliage and fruit.

C20-A013

Puccinia graminis (Agent IE)

Stem rust of cereals

Grows in moderate temperatures. Requires moisture on the plant surface, such as heavy dew, to initiate spore growth. Damages the foliage and stems that reduces photosynthesis and increases water loss.

It is on the Australia Group Core list.

In Plants:

Target species: Wheat, oats, barley, rye.

Normal routes of transmission: Windborne.

Secondary Hazards: None; Mechanical vectors (planting, harvesting, track out); Aerosols; Spores; Crop debris.

Signs: First appears as minute flecks, progressing into rough linear lesions with a brick-red color. Lesions eventually become black.

C20-A014

Puccinia striiformis

Stripe rust of wheat

Occurs mainly on wheat.

It is on the Australia Group Core list.

In Plants:

Target species: Wheat; barley; rye; triticale.

Normal routes of transmission: Windborne.

Secondary Hazards: Mechanical vectors (track out); Spores.

Signs: Yellow-orange pustules appear in long stripes on the leaves. Pustules rarely appear on the stems and heads. As the crop matures, the stripes turn from yellow to black.

C20-A015

Pyricularia grisea (Agent IE)

Rice blast

Primarily a disease of rice, but also infects other cereal grain plants. Most severe damage occurs in the seedling stage and when the head is developing. Greatest risk in high humidity when nights are cool, wind free, and free moisture (e.g., dew) remains on the leaves for 8–10 h.

It is on the Australia Group Core list.

In Plants:

Target species: Rice, wheat, rye, barley.

Normal routes of transmission: Windborne.

Secondary Hazards: Spores; Crop debris.

Signs: Lesions appear on the leaves of rice plants and vary in size. They are usually diamond shaped and have a gray or white center with a brown or reddish-brown border. Crop loss of 50–90% has been reported. Lesions also appear on the rice head but are brown or black in color. Rice grains do not develop properly. In severe neck infections, the stem will break and the head will drop off. The fungus can infect the roots and also invade the plant's vascular system blocking the transport of nutrients and water from the roots.

C20-A016

Sclerophthora rayssiae* var. *zeae

Downy mildew of corn

Initial infection is due to attack by spores in the soil. Warm soil temperatures are required for the initial infection to occur. Spores in air dried leaf tissue remain viable for up to 4 years.

It is one of the USDA Select Agent Listed Plant Pathogens.

In Plants:

Target species: Corn.

Normal routes of transmission: Windborne; Rain borne; Contact (spores in soil).

Secondary Hazards: Spores; Crop debris.

Signs: Lesions are well defined chlorotic or yellowish stripes. They are initially limited by the leaf veins. Lesions merge and become reddish or purple as the disease progresses. When disease occurs prior to flowering, seeds do not develop normally. Affected leaves remain intact and do not shred.

C20-A017

Synchytrium endobioticum

Potato wart disease

It is caused by a soil borne fungus that becomes active in the spring when soil temperatures rise above 46°F. It produces spores that can persist for up to 30 years.

It is one of the USDA Select Agent Listed Plant Pathogens.

In Plants:

Target species: Potatoes.

Normal routes of transmission: Contact (spores in soil).

Secondary Hazards: Mechanical vectors (planting, harvesting, track out); Spores; Crop debris; Fecal (animals that have fed on infected tubers).

Signs: Causes warty outgrowths on all tissues other than the roots. Most frequently affected are the stems, stolons, and tubers. Outgrowths can be tan, green, or brown. Currently there is no control for this disease. Once introduced into a field, the entire crop may be rendered unmarketable. Moreover, the fungus is so persistent that potatoes cannot be grown again safely for many years.

C20-A018

Tilletia indica

Wheat cover smut

It cannot establish in new locations where climatic conditions are not favorable. The most severe infections occur when there is cool and wet weather at the time the wheat is heading out.

It was examined by Iraq as a potential biological weapons in the late 1980s.

In Plants:

Target species: Wheat, rye, triticale.

Normal routes of transmission: Windborne; Rain borne.

Secondary Hazards: Mechanical vectors (harvesting); Spores.

Signs: It generally causes little direct loss in either quantity or quality of wheat seed or grain. The fungus invades the kernels and leaves behind waste products with a disagreeable fishy odor or taste that makes the kernels unpalatable for use in flour.

References

Acha, Pedro N., and Boris Szyfres. *Zoonoses and Communicable Diseases Common to Man and Animals.* Scientific and Technical Publication No. 580. 3rd ed. Vol. 1, Bacterioses and Mycoses. Washington, DC: Pan American Health Organization, 2003.

California Department of Food and Agriculture. Animal Health and Food Safety Services. Animal Health Branch. *Biosecurity: Selection and Use of Surface Disinfectants*. Revision June 2002.

Canada Minister of National Health and Welfare. Laboratory Centre for Disease Control, Office of Biosafety. *Material Safety Data Sheet-Infectious Substances: Aspergillus spp.* January 23, 2001.

———. *Material Safety Data Sheet-Infectious Substances: Coccidioides immitis.* January 23, 2001.

———. *Material Safety Data Sheet-Infectious Substances: Histoplasma capsulatum.* March 2001.

Centers for Disease Control and Prevention. "Biological and Chemical Terrorism: Strategic Plan for Preparedness and Response. Recommendations of the CDC Strategic Planning Workgroup." *Morbidity and Mortality Weekly Report* 49 (RR-4) (2000): 1–14.

———. "Case Definitions for Infectious Conditions Under Public Health Surveillance." *Morbidity and Mortality Weekly Report* 46 (RR-10) (1997).

Chin, James, ed. *Control of Communicable Diseases Manual.* 17th ed. Washington, DC: American Public Health Association, 2000.

Compton, James A.F. *Military Chemical and Biological Agents: Chemical and Toxicological Properties.* Caldwell, NJ: The Telford Press, 1987.

Department of Agriculture. 9 CFR Part 121—"Possession, Use, and Transfer of Biological Agents and Toxins," 2005: 753–67.

European and Mediterranean Plant Protection Organization. *Data Sheets on Quarantine Pests.* http://www.eppo.org/QUARANTINE/quarantine.htm. 2006.

———. *Diagnostic Protocols for Regulated Pests.* February 15, 2006. http://archives.eppo.org/EPPOStandards/diagnostics.htm. 2006.

Jepson, Susan B. "Fact Sheet on Late Blight of Potato and Tomato." Oregon State University Extension Service, 2005.

———. "Fact Sheet on Philippine Downy Mildew of Corn." Oregon State University Extension Service, 2005.

Kahn, Cynthia M., and Susan E. Aiello, eds. *The Merck Veterinary Manual.* 9th ed. Online Version. Whitehouse Station, NJ: Merck & Co., Inc., 2006.

Lenhart, Steven W., Millie P. Schafer, Mitchell Singal, and Rana A. Hajjeh. *Histoplasmosis: Protecting Workers at Risk.* Rev. ed. Centers for Disease Control and Prevention, December 2004.

Public Health Service. 42 CFR Part 73—"Select Agents and Toxins," 2004: 443–58.

Putnam, Melodie. "Fact Sheet on Brown Stripe Downy Mildew." Oregon State University Extension Service, 2005.

———. "Fact Sheet on Potato Wart." Oregon State University Extension Service, 2005.

Richmond, Jonathan Y., and Robert W. McKinney, eds. *Biosafety in Microbiological and Biomedical Laboratories.* 4th ed. Washington, DC: US Government Printing Office, 1999.

United States Army Headquarters. *Potential Military Chemical/Biological Agents and Compounds, Field Manual No. 3-11.9.* Washington, DC: Government Printing Office, January 10, 2005.

———. *Technical Aspects of Biological Defense, Technical Manual No. 3-216.* Washington, DC: Government Printing Office, January 12, 1971.

United States Department of Agriculture. Animal and Plant Health Inspection Service. *Plant Protection and Quarantine.* http://www.aphis.usda.gov/ppq/. 2006.

United States Department of Agriculture. Office of Pest Management National Plant Disease Recovery System. "Recovery Plan for Leaf Rust, Stem Rust, and Stripe Rust of Wheat Caused by *Puccinia triticina*, *Puccinia graminis*, and *Puccinia striiformis*, Respectively." August 28, 2006.

———. "Recovery Plan for Philippine Downy Mildew and Brown Stripe Downy Mildew of Corn Caused by *Peronosclerospora philippinensis* and *Sclerophthora rayssiae* var. *zeae*, Respectively." September 18, 2006.

University of Hawaii, Manoa. *EXTension ENTOmology & UH-CTAHR Integrated Pest Management Program.* http://www.extento.hawaii.edu/kbase/crop/Type/c_coffe.htm. 2006.

World Health Organization. *Health Aspects of Chemical and Biological Weapons: Report of a WHO Group of Consultants.* Geneva: World Health Organization, 1970.

———. *Public Health Response to Biological and Chemical Weapons: WHO Guidance.* Geneva: World Health Organization, 2004.

Section VI

Additional Information

21

Alphanumeric Indices

This chapter includes four indices: the Alphabetical index, the Chemical Abstract Service (CAS) numbers index, the International Classification of Diseases, 10th Revision (ICD-10) numbers index, and the Organization for the Prohibition of Chemical Weapons key (OPCW) numbers index. OPCW numbers are found in the "Handbook on Chemicals, version 2002," Appendix 2 in *Declaration Handbook 2002 for the Convention on the Prohibition of the Development, Production, Stockpiling, and Use of Chemical Weapons and on their Destruction*. OPCW numbers were developed to provide an easy method for tracking chemical warfare agents and precursors if CAS numbers were not available.

Each of these indices is in alphanumeric order and entries are cross-referenced to information about the specific agent through the handbook number. The first two digits of the handbook number following the "C" indicate the class of agent. The classes appear in order from Chapter 1 through Chapter 20 as the first 20 chapters of this handbook. They provide general information about each subgroup of agents. The letter following the hyphen (e.g., C01-**A**) indicates that the material is primarily used as an agent (A), component or precursor in the manufacture of that class of agents (C), or is a significant decomposition product or impurity of that class of agents (D). The three digits that follow the letter indicate the specific agent in the order that it appears in this handbook (e.g., C01-A**001**). These numbers proceed sequentially in each chapter. Materials that are not individually detailed in this handbook, typically due to the absence of published chemical or toxicological information, are only cross-referenced to the appropriate class. These materials do not have numbers after the letter following the hyphen (e.g., C01-A).

The list of synonyms included in the Alphabetical index is by no means exhaustive. Although it contains a large number of formal chemical and biological names, it primarily contains historical names, military code names, and common names for the agents, precursors, components, and degradation products included in this handbook. In some instances, names that have been popularized in the media but have no other historical connection to the agent have been included.

Some synonyms in the Alphabetical index are followed by bracketed notations. These notations provide additional clarifying information about the entry such as composition, modifications to the agents (e.g., thickened, dusty, binary), or a note for historical context. For example, "White Star" was a gas blend that was employed by the British in World War I consisting of 50% phosgene and 50% chlorine. The entry appears as:

White Star (British WWI Cylinder Gas) {Phosgene (50%) and Chlorine (50%) Mixture}

- A -

- B -

Blaukreuz (Blue Cross—German WWI Shell) {Diphenylchloroarsine (50–100%), and *N*-Ethylcarbazole (0–50%) Mixture with or without Phenyldichloroarsine (Solvent)}	C14-A
Blaukreuz 1 (Blue Cross 1—German WWI Shell) {Diphenylcyanoarsine with or without Phenyldichloroarsine (Solvent)}	C14-A
Blaukreuz 1 (Blue Cross 1—German WWI Shell)	C14-A002
Blauring 1 (Blue Ring 1—German WWI Shell)	C14-A003
Blauring 2 (Blue Ring 2—German WWI Shell) {Phenyldichloroarisne (50%), Diphenylchloroarsine (35%), Triphenylarsine (5%) and Arsenic Trichloride (5%) Mixture}	C14-A
Blauring 3 (Blue Ring 3—German WWI Shell)	C14-A002
Blausaeure	C07-A001
Blauwzuur	C07-A001
Blight Disease	C20-A015
Blight of Potato	C20-A012
Blister Gas No. 1	C03-A001
Blister Gas No. 2	C03-A001
Blister Gas No. 4	C03-A005
Blister Gas No. 5	C03-A011
Blood Gas	C07-A001
Blotchy Mottle Disease of Citrus	C17-A016
Blue Cross No. 1 (German WWI Shell) {Phenyldichloroarsine (60–50%) and Diphenylchloroarsine (40–50%) Mixture}	C04-A
Blue No. 1	C10-A003
Blue Star (British WWI Cylinder Gas) {Chlorine (80%) and Sulfur Chloride (20%) Mixture}	C10-A001
Blue-Green Algae Toxin	C16-A002
Blue-Green Algae Toxin	C16-A028
Bluetongue	C18-A006
Blue-X (Iraqi incapacitating agent. Thought to be BZ or a derivative.)	C12-A
Blumeria graminis	C20-A006
B. melitensis	C17-A004
Bn-Stoff	C13-A005
Bolivian Hemorrhagic Fever	C18-A031
Bomyl	C11-A002
Bonebreak Fever	C18-A012
Borane	C11-A095
Bordetella pertussis Toxin	C16-A030
Borer Sol	C11-A041
Boroethane	C11-A095
Boron Bromide	C11-A084
Boron Chloride	C11-A085
Boron Fluoride	C11-A086
Boron Hydride	C11-A095
Boron Tribromide	C11-A084
Boron Trichloride	C11-A085
Boron Trifluoride	C11-A086
Botox	C16-A005
Botulin Toxins	C16-A005
Botulinum Toxins	C16-A005
Botulism	C16-A005

- C -

- D -

- E -

- F -

- G -

- H -

- I -

- J -

- K -

- L -

- M -

- N -

- O -

- P -

- Q -

- R -

- S -

- T -

- U -

- V -

- W -

- X -

- Y -

- Z -

CAS #	Handbook	CAS #	Handbook	CAS #	Handbook
50-00-0	C11-A099	77-78-1	C03-A009	110-60-1	C15-A017
50-36-2	C12-A008	77-81-6	C01-A001	110-66-7	C15-A008
50-37-3	C12-A013	78-00-2	C11-A103	111-31-9	C15-A013
51-57-0	C12-A010	78-34-2	C11-A009	111-36-4	C11-A125
51-62-7	C12-A009	78-38-6	C01-C077	111-42-2	C03-D055
51-63-8	C12-A009	78-40-0	C01-D144	111-48-8	C03-C032
51-63-8	C12-A009	78-53-5	C01-A013	111-88-6	C11-A102
51-64-9	C12-A009	78-87-5	C11-A038	112-55-0	C15-A014
51-75-2	C03-A012	78-95-5	C13-A007	113-72-4	C12-A012
52-26-6	C12-A014	79-04-9	C11-A111	115-26-4	C11-A008
54-04-6	C12-A012	79-22-1	C11-A118	116-06-3	C11-A029
55-86-7	C03-A012	79-31-2	C15-A019	116-53-0	C15-A020
56-38-2	C11-A022	83-34-1	C15-A021	121-45-9	C01-C073
57-06-7	C11-A124	86-28-2	C13-A022	122-10-1	C11-A002
57-12-5	C07-A001	86-50-0	C11-A001	122-52-1	C01-C075
57-24-9	C11-A059	89-92-9	C13-A023	123-73-9	C11-A112
57-27-2	C12-A014	91-13-4	C13-A024	124-40-3	C01-C084
58-00-4	C14-A005	96-12-8	C11-A035	124-63-0	C11-A117
60-13-9	C12-A009	96-36-6	C01-C072	139-10-6	C12-A009
60-15-1	C12-A009	96-64-0	C01-A003	139-47-9	C12-A010
60-24-2	C15-A001	96-80-0	C01-C089	139-87-7	C03-C039
60-34-4	C11-A123	100-37-8	C01-C088	141-66-2	C11-A007
62-38-4	C11-A056	100-39-0	C13-A002	143-33-9	C07-C007
62-74-8	C11-A058	100-44-7	C13-A019	146-56-5	C12-A019
64-31-3	C12-A014	101-68-8	C11-A128	151-38-2	C11-A052
69-23-8	C12-A019	102-71-6	C03-C040	151-50-8	C07-C008
71-62-5	C16-A021	102-82-9	C01-C059	151-56-4	C11-A098
74-83-9	C11-A042	104-81-4	C13-A023	151-67-7	C12-A020
74-90-8	C07-A001	105-36-2	C13-A021	152-16-9	C11-A025
74-93-1	C11-A101	105-59-9	C03-C038	156-31-0	C12-A009
74-95-3	C11-A036	106-93-4	C11-A040	156-34-3	C12-A009
75-01-4	C11-A105	107-02-8	C13-A001	257-07-8	C13-A011
75-15-0	C11-A093	107-06-2	C11-A041	298-00-0	C11-A019
75-21-8	C11-A097	107-07-3	C03-C036	298-02-2	C11-A023
75-44-5	C10-A003	107-11-9	C11-A092	298-04-4	C11-A010
75-66-1	C15-A005	107-13-1	C11-A090	300-62-9	C12-A009
75-70-7	C10-A014	107-14-2	C11-A094	302-27-2	C16-A001
75-74-1	C11-A104	107-18-6	C11-A091	302-31-8	C12-A014
75-86-5	C11-A089	107-20-0	C11-A110	306-37-6	C11-A122
75-97-8	C01-C086	107-44-8	C01-A002	314-19-2	C14-A005
76-02-8	C11-A120	107-49-3	C11-A028	327-98-0	C11-A013
76-06-2	C10-A006	107-92-6	C15-A018	329-99-7	C01-A004
76-38-0	C12-A021	107-99-3	C01-C085	352-26-1	C03-A014
76-89-1	C12-C024	108-18-9	C01-D143	353-42-4	C11-A086
76-93-7	C12-C023	108-23-6	C11-A116	353-50-4	C11-A108
77-10-1	C12-A007	109-61-5	C11-A119	354-32-5	C11-A121
77-47-4	C11-A100	109-79-5	C15-A002	358-29-2	C01-A012
77-77-0	C03-D044	109-94-4	C11-A039	359-94-4	C01-C062

(Continued)

(Continued)

CAS #	Handbook	CAS #	Handbook	CAS #	Handbook
370-66-1	C03-A017	590-53-4	C01-A002	812-01-1	C10-A009
370-67-2	C03-A018	592-34-7	C11-A107	816-40-0	C13-A005
370-68-3	C03-A019	592-88-1	C15-A012	817-09-4	C03-A013
372-85-0	C11-A086	593-89-5	C04-A003	819-79-4	C01-D143
373-25-1	C03-A007	594-42-3	C10-A014	868-85-9	C01-C072
382-21-8	C10-A008	594-72-9	C11-A037	935-52-4	C06-A005
404-86-4	C13-A006	596-15-6	C12-A014	944-22-9	C11-A014
405-41-4	C12-A009	596-19-0	C12-A014	947-33-1	C11-A016
420-12-2	C03-D042	598-02-7	C01-D142	950-10-7	C11-A016
437-38-7	C12-A015	598-14-1	C04-A001	950-37-8	C11-A018
445-59-0	C03-A018	598-31-2	C13-A003	956-90-1	C12-A007
460-19-5	C07-A005	600-07-7	C15-A020	990-73-8	C12-A015
462-94-2	C15-A016	612-23-7	C13-A016	993-13-5	C01-C051
463-58-1	C11-A109	619-34-1	C03-A024	997-80-8	C01-D138
463-71-8	C13-A017	620-05-3	C13-A020	1005-93-2	C06-A002
464-07-3	C01-C087	620-13-3	C13-A023	1066-50-8	C01-C081
464-10-8	C10-A007	621-68-1	C03-A025	1077-43-6	C12-A009
470-90-6	C11-A005	622-44-6	C10-A013	1089-44-7	C12-A009
471-03-4	C03-D050	623-24-5	C13-A024	1115-15-7	C03-D045
502-39-6	C11-A053	623-48-3	C13-A012	1162-65-8	C16-A023
503-38-8	C10-A004	624-83-9	C11-A127	1165-39-5	C16-A023
505-29-3	C03-D054	624-92-0	C01-C054	1184-57-2	C11-A054
505-60-2	C03-A001	625-01-4	C10-A010	1189-87-3	C01-A005
506-59-2	C01-C084	627-51-0	C03-D046	1200-47-1	C12-A009
506-68-3	C07-A002	630-08-0	C09-A001	1223-69-4	C12-A009
506-77-4	C07-A003	630-81-9	C12-A014	1299-88-3	C11-A049
506-78-5	C07-A004	630-82-0	C12-A014	1302-45-0	C11-A044
513-44-0	C15-A004	637-39-8	C03-C040	1304-00-3	C11-A095
513-53-1	C15-A003	645-53-4	C12-A009	1305-99-3	C11-A044
532-27-4	C13-A008	650-20-4	C01-A011	1313-82-2	C03-C037
533-74-4	C11-A033	665-03-2	C01-A009	1314-80-3	C01-C071
537-46-2	C12-A010	673-97-2	C01-A010	1314-84-7	C11-A044
538-07-8	C03-A011	674-82-8	C11-A096	1327-53-3	C11-A047
538-75-0	C01-C049	676-83-5	C01-C057	1333-25-1	C11-A048
540-63-6	C03-D043	676-97-1	C01-C046	1333-83-1	C01-C066
540-73-8	C11-A122	676-98-2	C01-C058	1337-06-0	C11-A056
541-25-3	C04-A002	676-99-3	C01-C047	1341-49-7	C01-C067
541-31-1	C15-A010	677-42-9	C01-C060	1353-70-4	C16-A001
541-41-3	C11-A113	683-08-9	C01-C076	1385-95-1	C16-A023
542-88-1	C10-A011	693-13-0	C01-C050	1393-62-0	C16-A022
543-27-1	C11-A115	696-24-2	C04-A005	1402-68-2	C16-A023
544-40-1	C15-A007	696-28-6	C04-A004	1407-85-8	C12-A009
545-06-2	C11-A046	712-48-1	C14-A001	1445-75-6	C01-C048
555-77-1	C03-A013	716-16-5	C11-A032	1449-89-4	C06-A001
563-47-3	C11-A043	756-79-6	C01-C052	1462-73-3	C12-A009
578-94-9	C14-A003	762-04-9	C01-C074	1476-39-7	C15-A016
584-84-9	C11-A130	786-19-6	C11-A004	1498-40-4	C01-C080

CAS #	Handbook	CAS #	Handbook	CAS #	Handbook
1498-46-0	C01-C083	3689-24-5	C11-A026	7446-09-5	C11-A075
1538-69-8	C01-C078	3734-64-3	C03-A020	7446-11-9	C11-A076
1550-68-1	C11-A008	3734-96-1	C01-A013	7446-13-1	C11-A074
1563-66-2	C11-A031	3734-97-2	C01-A013	7446-18-6	C11-A060
1598-80-7	C03-A014	4170-30-3	C11-A112	7487-94-7	C11-A055
1609-86-5	C11-A129	4241-34-3	C01-A006	7488-97-3	C12-A013
1619-34-7	C12-C022	4282-87-5	C12-A009	7528-00-9	C12-A009
1679-09-0	C15-A011	4368-28-9	C16-A019	7550-45-0	C11-A082
1690-86-4	C12-A010	4478-53-9	C12-A001	7617-64-3	C03-A006
1794-86-1	C05-A001	4497-29-4	C10-A012	7637-07-2	C11-A086
1795-48-8	C11-A126	4584-46-7	C01-C085	7647-01-0	C11-A063
1832-53-7	C01-D138	4664-40-8	C16-A019	7647-19-0	C11-A087
1867-09-0	C03-C036	4664-41-9	C16-A019	7664-39-3	C11-A064
1867-10-3	C11-A094	4685-14-7	C11-A057	7664-41-7	C11-A061
1910-42-5	C11-A057	4708-04-7	C01-C082	7664-93-9	C11-A067
1931-26-6	C13-A014	4824-78-6	C11-A003	7681-49-4	C01-C064
2032-65-7	C11-A032	5002-47-1	C12-A019	7697-37-2	C11-A066
2084-19-7	C15-A009	5221-20-5	C12-A013	7704-34-9	C01-C053
2104-64-5	C11-A011	5244-34-8	C03-D041	7719-09-7	C03-C035
2130-41-8	C11-A096	5714-22-7	C10-A016	7719-12-2	C01-C068
2229-61-0	C16-A019	5798-79-8	C13-A004	7726-95-6	C10-A002
2270-40-8	C16-A038	5819-08-9	C03-D051	7778-44-1	C11-A048
2387-23-7	C01-D140	5967-37-3	C11-A084	7782-41-4	C11-A071
2404-73-1	C01-C079	5990-90-9	C03-A011	7782-50-5	C10-A001
2478-92-4	C01-A017	6009-81-0	C12-A014	7783-06-4	C07-A006
2511-10-6	C01-D139	6055-06-7	C12-A014	7783-07-5	C11-A077
2625-76-5	C03-A008	6089-42-5	C03-A014	7783-37-1	C01-C066
2641-09-0	C01-A017	6138-32-5	C03-A013	7783-61-1	C11-A079
2694-87-3	C07-A003	6143-52-8	C01-D143	7783-79-1	C11-A078
2694-88-4	C07-A003	6171-93-3	C01-A002	7783-80-4	C11-A081
2698-41-1	C13-A009	6171-94-4	C01-A002	7783-82-6	C11-A083
2699-79-8	C11-A045	6191-56-6	C14-A005	7784-34-1	C04-C006
2706-50-5	C12-A009	6211-15-0	C12-A014	7784-40-9	C11-A050
2746-81-8	C12-A019	6581-06-2	C12-A001	7784-42-1	C08-A001
2885-00-9	C15-A015	6700-54-5	C12-A009	7784-46-5	C11-A051
2937-50-0	C11-A106	6755-76-6	C03-A001	7786-34-7	C11-A020
2941-64-2	C11-A114	6795-23-9	C16-A023	7787-71-5	C11-A069
2981-31-9	C12-A007	6838-91-1	C01-D138	7789-23-3	C01-C063
3017-23-0	C07-A001	6842-10-0	C11-A096	7789-29-9	C01-C065
3019-04-3	C13-A013	6885-57-0	C16-A023	7789-30-2	C11-A068
3093-66-1	C12-A019	6899-11-2	C11-A059	7790-15-0	C01-C067
3148-09-2	C16-A039	6923-22-4	C11-A021	7790-91-2	C10-A015
3420-02-8	C15-A023	7086-02-4	C07-A001	7791-25-5	C11-A088
3563-36-8	C03-A002	7087-68-5	C01-D145	7803-51-2	C11-A044
3570-55-6	C15-A006	7220-81-7	C16-A023	7803-52-3	C11-A080
3590-07-6	C03-A011	7241-98-7	C16-A023	8000-97-3	C11-A006
3684-26-2	C12-C022	7327-58-4	C03-D047	8003-07-4	C11-A040

(Continued)

(Continued)

CAS #	Handbook	CAS #	Handbook	CAS #	Handbook
3687-31-8	C11-A050	7339-53-9	C11-A123	8005-38-7	C11-A099
8005-48-9	C11-A044	11050-21-8	C16-A008	14055-61-9	C01-C067
8006-07-3	C11-A099	11051-21-1	C16-A037	14287-82-2	C16-A038
8007-57-6	C11-A061	11061-96-4	C16-A024	14297-85-9	C12-A009
8007-58-7	C11-A066	11075-32-4	C11-A049	14297-86-0	C12-A009
8011-67-4	C04-C006	11077-03-5	C16-A015	14611-50-8	C12-A010
8013-13-6	C11-A099	11082-95-4	C01-C068	14713-46-3	C12-A009
8013-47-6	C11-A056	11100-45-1	C16-A033	14729-29-4	C16-A039
8014-90-2	C11-A065	11111-91-4	C11-A022	14875-95-7	C09-A003
8014-94-6	C11-A075	11114-31-1	C11-A018	14904-70-2	C07-A001
8014-95-7	C11-A067	11144-15-3	C07-A006	14945-01-8	C12-A009
8018-06-2	C11-A005	12002-03-8	C11-A049	14945-02-9	C12-A009
8023-77-6	C13-A006	12020-82-5	C01-C065	14945-03-0	C12-A009
8033-73-6	C03-D055	12057-74-8	C11-A044	14945-04-1	C12-A009
8050-82-6	C01-C053	12060-19-4	C11-A074	14945-09-6	C12-A009
8052-26-4	C11-A065	12066-62-5	C01-C071	14948-61-9	C07-A003
8053-16-5	C12-A014	12152-42-0	C13-A014	15000-32-5	C12-A009
8054-28-2	C11-A026	12255-53-7	C08-A000	15612-71-2	C11-A082
8056-95-9	C12-A021	12396-99-5	C11-A075	15636-95-0	C08-A001
8057-70-3	C11-A022	12422-22-9	C10-A015	15636-96-1	C08-A001
8058-73-9	C11-A006	12526-02-2	C10-A015	15700-19-3	C08-A001
8065-48-3	C11-A006	12542-90-4	C11-A053	15745-74-1	C07-C008
8067-67-2	C11-A005	12584-83-7	C16-A009	15798-64-8	C11-A112
9009-86-3	C16-A031	12585-40-9	C16-A033	15980-15-1	C03-D053
9012-63-9	C16-A025	12612-55-4	C09-A003	16096-32-5	C15-A022
9014-39-5	C16-A019	12626-86-7	C16-A019	16684-27-8	C07-A001
9035-99-8	C01-C053	12643-22-0	C11-A016	16752-77-5	C11-A034
9061-58-9	C16-A037	12673-82-4	C01-C053	16857-09-3	C01-C071
9067-26-9	C16-A031	12684-31-0	C01-C053	16890-46-3	C08-A001
10025-67-9	C03-C033	12687-39-7	C16-A007	16941-26-7	C01-C067
10025-87-3	C01-C070	12767-18-9	C11-A014	16970-81-3	C11-A095
10026-13-8	C01-C069	12767-24-7	C01-C053	17097-13-1	C01-C066
10031-59-1	C11-A060	12769-46-9	C16-A009	17107-61-8	C11-A067
10031-75-1	C11-A128	12778-32-4	C16-A007	17108-96-2	C12-A009
10034-85-2	C11-A065	12798-32-2	C11-A052	17279-36-6	C12-A009
10035-10-6	C11-A062	13004-56-3	C12-A001	17289-88-2	C16-A019
10102-43-9	C11-A072	13005-31-7	C09-A003	17522-62-2	C16-A019
10102-44-0	C11-A073	13025-64-4	C03-C039	17642-25-0	C01-C103
10103-62-5	C11-A048	13071-79-9	C11-A027	17642-26-1	C01-A036
10125-86-7	C03-A028	13127-85-0	C07-A001	17642-27-2	C01-A040
10265-92-6	C11-A017	13171-21-6	C11-A024	17642-28-3	C01-A041
10294-33-4	C11-A084	13214-11-4	C16-A023	17642-29-4	C01-A038
10294-34-5	C11-A085	13463-39-3	C09-A003	17642-30-7	C01-A042
10545-99-0	C03-C034	13463-40-6	C09-A002	17642-31-8	C01-A035
11003-08-0	C16-A023	13464-49-8	C08-A001	17642-32-9	C01-A039
11005-69-9	C16-A019	13464-50-1	C08-A001	17642-33-0	C01-C098
11017-04-2	C16-A016	13464-51-2	C08-A001	17642-34-1	C01-C114
11026-09-8	C16-A019	13637-63-3	C11-A070	17642-35-2	C01-C093

CAS #	Handbook	CAS #	Handbook	CAS #	Handbook
11032-79-4	C16-A007	13998-03-3	C07-C007	17650-48-5	C01-C096
11039-36-4	C16-A039	14012-92-1	C11-A060	17878-54-5	C16-A023
17878-56-7	C16-A023	22307-81-9	C01-C061	26979-72-6	C11-A083
17924-92-4	C16-A040	22618-91-3	C12-A009	26979-73-7	C11-A083
18005-40-8	C01-C061	22618-92-4	C12-A009	27261-02-5	C16-A023
18016-10-9	C01-A037	22618-94-6	C12-A009	27413-00-9	C07-A003
18046-08-7	C16-A039	22781-23-3	C11-A030	27413-01-0	C07-A003
18262-24-3	C01-C091	22916-10-5	C16-A037	27552-17-6	C16-A038
18262-25-4	C01-C094	22916-18-3	C16-A037	27640-92-2	C16-A037
18262-26-5	C01-C090	22956-47-4	C01-A003	28077-97-6	C11-A071
18262-27-6	C01-C104	22956-48-5	C01-A003	28175-96-4	C11-A008
18262-28-7	C01-C110	23233-25-2	C01-C102	28258-59-5	C13-A023
18262-29-8	C01-C111	23255-69-8	C16-A038	28380-38-3	C11-A047
18262-30-1	C01-C099	23255-72-3	C16-A038	28392-39-4	C16-A038
18262-31-2	C01-C107	23282-20-4	C16-A038	28454-61-7	C13-A014
18262-32-3	C01-C118	23509-16-2	C16-A004	28519-24-6	C11-A053
18262-33-4	C01-C101	23525-22-6	C14-A002	28841-71-6	C12-A009
18262-34-5	C01-C108	23724-81-4	C10-A002	29125-55-1	C12-A003
18421-60-8	C09-A001	24067-21-8	C12-A009	29630-03-3	C13-A014
18425-23-5	C01-C092	24147-91-9	C16-A023	29754-21-0	C11-A090
18623-82-0	C01-C065	24167-76-8	C11-A044	29801-27-2	C12-A009
18660-81-6	C16-A019	24271-46-3	C03-C040	29801-28-3	C12-A009
18695-28-8	C16-A040	24572-08-5	C12-A001	29924-74-1	C12-C022
18949-73-0	C12-A007	24582-93-2	C12-A009	30066-61-6	C16-A038
19034-08-3	C11-A097	24753-15-9	C01-A003	30321-74-5	C01-D143
19121-60-9	C01-C067	24753-16-0	C01-A003	30525-89-4	C11-A099
19149-77-0	C03-D052	24946-19-8	C01-C124	30567-80-7	C12-A009
19287-45-7	C11-A095	25059-10-3	C13-A014	31012-04-1	C11-A085
19287-88-8	C11-A095	25152-34-5	C16-A037	31223-79-7	C16-A023
20281-91-8	C01-C087	25311-71-1	C11-A015	32057-09-3	C11-A064
20309-96-0	C16-A023	25314-61-8	C13-A001	32204-36-7	C16-A038
20421-08-3	C16-A023	25482-84-2	C13-A014	32215-02-4	C16-A023
20421-10-7	C16-A023	25596-22-9	C07-A001	32315-10-9	C10-A005
20434-91-7	C10-A014	25596-52-5	C07-C007	34256-71-8	C01-D147
20677-21-8	C01-C051	25772-52-5	C03-A020	34461-56-8	C04-A002
20684-88-2	C11-A109	26102-97-6	C01-A043	34624-53-8	C11-A006
20751-44-4	C12-A009	26102-98-7	C01-A044	34902-96-0	C07-A003
20816-12-0	C11-A074	26102-99-8	C01-A045	35320-66-2	C16-A038
20820-80-8	C01-A017	26153-10-6	C16-A038	35489-04-4	C07-A002
20859-73-8	C11-A044	26198-63-0	C11-A038	35489-05-5	C07-A002
20978-13-6	C16-A023	26245-56-7	C01-D141	35495-69-3	C07-A002
21259-20-1	C16-A037	26400-47-5	C16-A037	35495-70-6	C07-A002
21462-53-3	C16-A039	26652-79-9	C03-D041	35523-89-8	C16-A016
21770-86-5	C01-A014	26758-53-2	C12-A003	35554-08-0	C16-A016
22128-62-7	C13-A010	26934-87-2	C16-A038	35554-08-6	C16-A016
22128-63-8	C13-A018	26979-63-5	C11-A083	36252-60-5	C09-A003
22135-69-9	C16-A038	26979-64-6	C11-A083	36519-25-2	C16-A038
22224-92-6	C11-A012	26979-66-8	C11-A083	36549-53-8	C03-C040

(Continued)

(Continued)

CAS #	Handbook	CAS #	Handbook	CAS #	Handbook
22224-93-7	C11-A034	26979-67-9	C11-A083	36549-54-9	C03-C040
36549-55-0	C03-C040	42126-46-5	C09-A003	54018-05-2	C16-A039
36585-70-3	C01-C061	42520-97-8	C03-A031	54060-15-0	C03-C038
36647-05-9	C03-A016	42542-10-9	C12-A011	54098-95-2	C13-A014
36647-06-0	C03-A016	42881-99-2	C12-A002	54182-73-9	C11-A001
36653-66-4	C16-A037	43086-52-8	C07-A002	54385-59-0	C16-A039
36659-79-7	C03-C040	44246-22-2	C11-A082	54390-94-2	C12-A005
36823-35-5	C09-A002	44584-76-1	C11-A129	54779-68-9	C03-A019
36913-09-4	C12-A009	45156-70-5	C01-D138	55070-94-5	C11-A084
37061-10-2	C01-D138	45156-72-7	C01-D138	55112-89-5	C03-A031
37061-11-3	C01-D138	45952-89-4	C12-A010	55146-10-6	C11-A033
37209-28-2	C16-A007	47106-99-0	C12-A014	55425-20-2	C07-A001
37220-42-1	C09-A002	47646-09-3	C12-A019	55425-21-3	C07-A001
37226-51-0	C11-A085	47861-26-7	C16-A001	55569-71-6	C11-A078
37244-63-6	C03-A012	50291-32-2	C12-A014	55803-44-6	C16-A016
37333-30-5	C11-A060	50361-06-3	C14-A004	56003-83-9	C11-A073
37338-78-6	C16-A004	50443-93-1	C11-A073	56449-52-6	C01-C053
37341-05-2	C11-A097	50642-23-4	C01-A002	56591-09-4	C01-C053
37359-35-6	C11-A019	50642-24-5	C01-A003	56645-30-8	C01-C053
37370-22-2	C11-A003	50722-37-7	C16-A038	57035-13-9	C01-C053
37830-21-0	C12-A004	50782-69-9	C01-A016	57168-28-2	C02-A012
37928-89-5	C01-A003	50888-64-7	C11-A124	57169-76-3	C02-A011
37928-91-9	C01-A014	50978-48-8	C11-A124	57533-99-0	C16-A016
37990-84-4	C01-C117	51005-19-7	C11-A063	57856-11-8	C01-C056
37990-97-9	C01-C116	51005-20-0	C11-A072	58149-55-6	C02-A027
38184-40-6	C01-A004	51005-21-1	C11-A072	58391-87-0	C11-A124
38521-66-3	C03-A025	51052-72-3	C06-A003	58619-61-7	C02-A013
38727-03-6	C12-A009	51128-68-8	C01-C061	58901-15-8	C03-C039
38738-57-7	C16-A038	51366-09-7	C01-A017	59111-23-8	C01-C065
38875-25-1	C12-A009	51481-10-8	C16-A038	59161-48-7	C12-A009
38875-35-3	C12-A009	51810-70-9	C11-A044	59161-49-8	C12-A009
39024-20-9	C12-A009	51848-47-6	C01-A016	59182-86-4	C11-A089
39287-69-9	C01-C064	51938-46-6	C16-A016	59217-74-2	C01-C063
39313-28-5	C16-A015	51943-82-9	C13-A014	59217-75-3	C01-C064
39342-49-9	C11-A044	51988-97-7	C03-A011	59392-53-9	C16-A027
39380-77-3	C01-C070	52019-39-3	C16-A017	59525-69-8	C12-A009
39445-43-7	C11-A065	52399-93-6	C11-A041	59569-98-1	C12-A009
39461-36-4	C03-C034	52697-21-9	C01-C067	60119-99-5	C16-A037
39465-37-7	C16-A020	52802-03-6	C03-A026	60124-79-0	C12-A007
39472-40-7	C03-A001	52803-29-9	C11-A113	60124-86-9	C12-A007
39920-04-2	C16-A019	52819-96-2	C16-A023	60195-52-0	C07-C008
39949-57-0	C12-A009	52907-24-1	C11-A015	60508-89-6	C16-A012
40334-69-8	C14-A004	53026-80-5	C11-A099	60537-65-7	C16-A012
40334-70-1	C04-C007	53034-67-6	C12-A002	60538-73-0	C16-A039
40405-61-6	C12-A009	53126-64-0	C16-A039	60558-95-4	C13-A024
40709-82-8	C03-D048	53800-40-1	C01-A016	60561-17-3	C12-A016
41372-20-7	C14-A005	53851-19-7	C11-A072	60577-40-4	C12-A009

CAS #	Handbook	CAS #	Handbook	CAS #	Handbook
41820-21-7	C12-A009	53950-09-7	C12-A001	60577-41-5	C12-A009
60748-39-2	C16-A012	65086-44-4	C01-C065	70957-06-1	C12-A001
60748-40-5	C16-A012	65087-26-5	C01-D143	70957-07-2	C12-A001
60748-41-6	C16-A012	65143-05-7	C01-A016	71171-49-8	C01-C089
60748-42-7	C16-A012	65167-63-7	C01-A016	71195-58-9	C12-A017
61134-73-4	C03-A023	65167-64-8	C01-A016	71293-89-5	C01-A020
61356-56-7	C16-A038	65332-44-7	C01-D137	71327-12-3	C09-A003
61370-44-3	C07-A001	65350-20-1	C12-A007	71751-04-7	C11-A044
61380-40-3	C12-A018	65350-21-2	C12-A007	72843-01-7	C12-A009
61380-41-4	C12-A018	65423-90-7	C08-A001	72870-45-2	C16-A027
61481-19-4	C06-A004	65580-81-6	C11-A011	72996-54-4	C12-A016
61533-57-1	C11-A030	65988-34-3	C16-A013	73062-48-3	C01-C054
61674-62-2	C11-A063	65988-88-7	C16-A029	73207-98-4	C01-D146
61711-63-5	C11-A015	66252-28-6	C11-A073	73603-71-1	C16-A019
61840-45-7	C11-A056	66429-59-2	C01-A003	73758-26-6	C12-A009
62086-97-9	C03-A009	66429-60-5	C01-A003	73790-51-9	C01-C061
62140-56-1	C11-A062	66432-32-4	C12-A009	74191-18-7	C11-A044
62306-10-9	C12-A009	66432-33-5	C12-A009	74192-15-7	C01-A004
62869-68-5	C12-A001	66432-34-6	C12-A009	74213-24-4	C05-A002
62869-69-6	C12-A001	66579-64-4	C11-A044	75321-25-4	C12-A006
63038-80-2	C16-A016	66767-39-3	C11-A014	75321-26-5	C12-A001
63038-81-3	C16-A016	66912-28-5	C01-D138	75321-27-6	C12-A001
63051-68-3	C01-C089	67112-29-2	C01-C064	75321-28-7	C12-A006
63419-72-7	C07-A002	67157-74-8	C12-A007	75491-76-8	C07-A001
63427-73-6	C12-A009	67181-74-2	C11-A044	75757-64-1	C16-A032
63427-74-7	C12-A009	68157-62-0	C03-A001	77103-99-2	C02-A017
63427-75-8	C12-A009	68190-07-8	C01-A003	77104-00-8	C02-A020
63490-86-8	C12-A004	68192-83-6	C07-A003	77104-01-9	C02-A014
63653-66-7	C11-A019	69020-37-7	C03-A001	77104-09-7	C02-A044
63705-05-5	C01-C053	69044-01-5	C13-A014	77104-13-3	C02-A044
63721-94-8	C12-A009	69153-76-0	C01-C085	77104-14-4	C02-A044
63815-55-4	C01-A008	69766-61-6	C07-A001	77104-59-7	C02-A048
63905-36-2	C03-A022	69777-01-1	C01-A017	77104-60-0	C02-A048
63905-44-2	C03-A030	69866-34-8	C16-A012	77104-61-1	C02-A048
63908-52-1	C11-A090	70135-17-0	C12-A001	77104-62-2	C02-A034
63915-56-0	C03-A011	70323-44-3	C16-A030	77104-63-3	C02-A037
63918-89-8	C03-A003	70470-07-4	C16-A002	77104-64-4	C02-A038
63933-47-1	C11-A101	70610-94-5	C07-A003	77104-65-5	C02-A038
64037-56-5	C03-A029	70610-95-6	C07-A003	77104-66-6	C02-A038
64037-57-6	C03-A028	70610-96-7	C07-A003	77104-68-8	C02-A033
64057-70-1	C12-A011	70610-98-9	C07-A002	77104-70-2	C02-A028
64285-06-9	C16-A002	70610-99-0	C07-A002	77104-71-3	C02-A028
64296-20-4	C16-A014	70735-29-4	C12-A009	77104-72-4	C02-A028
64296-25-9	C16-A012	70761-70-5	C12-A001	77111-71-8	C02-A040
64296-26-0	C16-A012	70876-63-0	C01-C065	77111-72-9	C02-A040
64314-16-5	C16-A002	70876-64-1	C01-C066	77111-73-0	C02-A040
64357-24-0	C12-A009	70879-28-6	C12-A017	77111-74-1	C02-A030

(Continued)

(Continued)

CAS #	Handbook	CAS #	Handbook	CAS #	Handbook
64684-45-3	C11-A056	70938-84-0	C01-A016	77111-75-2	C02-A030
77111-76-3	C02-A030	83857-41-4	C11-A021	99748-78-4	C15-A001
77169-87-0	C02-A029	85079-48-7	C16-A006	99932-75-9	C11-A097
77169-88-1	C02-A029	85087-27-0	C16-A006	99932-76-0	C11-A098
77223-00-8	C02-A029	85201-37-2	C12-A014	99991-06-7	C01-A024
77223-01-9	C02-A035	85473-32-1	C01-A007	99991-07-8	C01-A025
77223-02-0	C02-A039	85473-33-2	C01-A019	100454-47-5	C01-A026
77223-03-1	C02-A039	85514-42-7	C16-A002	100608-14-8	C03-A023
77223-04-2	C02-A039	87289-70-1	C03-A030	101043-37-2	C16-A028
77238-39-2	C16-A028	87670-17-5	C12-A009	101064-48-6	C16-A028
77504-99-5	C03-C039	87915-42-2	C16-A011	101078-31-3	C01-C056
77534-68-0	C01-C089	87993-82-6	C11-A060	101324-33-8	C11-A059
77734-91-9	C16-A015	88872-45-1	C01-A001	101756-17-6	C12-A012
77734-92-0	C16-A015	89125-89-3	C11-A075	101884-85-9	C01-A024
77874-59-0	C07-A002	89254-45-5	C01-A003	101884-86-0	C01-A025
78020-08-3	C07-A001	89254-46-6	C01-A003	102180-38-1	C01-A026
78246-05-6	C11-A073	89614-45-9	C16-A012	102388-57-8	C01-A010
78297-56-0	C02-A032	89674-98-6	C16-A012	102388-58-9	C01-A013
78682-76-5	C12-A009	89675-50-3	C12-A001	102388-60-3	C01-A019
78989-43-2	C11-A066	89930-68-7	C12-A013	102444-87-1	C01-A027
79351-10-3	C01-A013	89980-59-6	C03-A021	102444-88-2	C01-A027
79580-28-2	C16-A006	90452-29-2	C11-A072	102490-54-0	C01-A002
80150-98-7	C12-A009	90817-17-7	C12-A001	102490-55-1	C01-A001
80210-09-9	C11-A018	91134-95-1	C01-A022	102490-57-3	C01-A019
80226-62-6	C16-A012	91853-95-1	C12-A001	102490-59-5	C01-A019
80243-05-6	C11-A044	91933-11-8	C16-A036	102802-00-6	C01-C088
80248-94-8	C16-A012	92142-32-0	C16-A002	103170-78-1	C16-A003
80411-93-4	C12-A009	92168-03-1	C03-A021	104800-95-5	C03-A001
80523-80-4	C12-A009	92216-03-0	C16-A002	104800-97-7	C01-A002
80573-42-8	C12-A001	92519-34-1	C07-A001	104800-98-8	C01-A003
80655-66-9	C12-A001	92844-80-9	C16-A002	104801-07-2	C01-A002
80832-90-2	C12-A015	93117-56-7	C12-A009	104801-08-3	C01-A002
81032-32-8	C01-C053	93240-66-5	C01-A018	104801-09-4	C01-A003
81142-02-1	C03-D049	93384-43-1	C16-A005	104801-10-7	C01-A003
81142-25-8	C03-A001	93384-44-2	C16-A005	105567-86-0	C11-A035
81142-27-0	C03-A001	93384-46-4	C16-A005	105655-27-4	C03-C040
81254-41-3	C12-A009	93384-47-5	C16-A005	106955-89-9	C16-A006
81254-42-4	C12-A009	93957-08-5	C01-A001	107059-49-4	C01-A014
81254-44-6	C12-A009	93957-09-6	C01-A001	107231-15-2	C16-A005
81280-64-0	C01-D138	94274-22-3	C11-A116	107231-30-1	C11-A124
82063-46-5	C09-A001	96163-42-7	C07-A001	107663-00-3	C01-A013
82248-93-9	C16-A030	96332-84-2	C12-A009	107663-01-4	C01-A013
82326-63-4	C12-A003	96638-28-7	C16-A031	108202-65-9	C01-A003
82698-46-2	C12-A007	96740-33-9	C01-C067	108753-95-3	C01-A023
82810-43-3	C16-A012	97914-41-5	C12-A001	108776-13-2	C01-A014
82810-44-4	C16-A012	98112-41-5	C16-A006	108993-05-1	C01-A013
83008-56-4	C11-A075	98478-67-2	C03-A009	109100-20-1	C01-A023

CAS #	Handbook	CAS #	Handbook	CAS #	Handbook
83590-17-4	C16-A035	98543-25-0	C01-A021	109597-25-3	C01-A013
109644-82-8	C01-A027	112415-59-5	C11-A056	127848-41-3	C12-A001
109654-18-4	C01-A013	113402-23-6	C02-A024	127999-62-6	C11-A073
109704-87-2	C11-A086	113402-25-8	C02-A026	128399-00-8	C01-A013
109724-41-6	C01-A013	113402-82-7	C02-A022	128657-50-1	C16-A028
109973-92-4	C02-A031	113402-83-8	C02-A018	128981-16-8	C01-C121
109973-93-5	C02-A041	113584-74-0	C03-C037	128981-17-9	C01-C126
109973-94-6	C02-A047	113962-65-5	C11-A063	128981-18-0	C01-C122
109973-95-7	C02-A045	114700-90-2	C01-C119	128981-19-1	C01-C127
110180-29-5	C01-A013	114700-91-3	C01-C106	128981-20-4	C01-C125
110203-40-2	C02-A015	114700-92-4	C01-C115	128981-21-5	C01-C130
110232-11-6	C02-A049	114700-93-5	C01-C109	128981-22-6	C01-C129
110232-13-8	C02-A049	114700-94-6	C01-C100	128981-23-7	C01-C131
110255-20-4	C02-A049	114967-39-4	C01-C123	128981-24-8	C01-C134
110344-82-6	C02-A042	114985-24-9	C01-A013	128981-25-9	C01-C128
110344-83-7	C02-A036	115182-35-9	C11-A017	128981-26-0	C01-C133
110344-84-8	C02-A036	115725-44-5	C01-C053	128981-27-1	C01-C132
110344-85-9	C02-A036	115797-06-3	C16-A010	128981-28-2	C01-C135
110422-92-9	C01-A021	115825-61-1	C16-A038	129003-90-3	C01-C120
110501-56-9	C01-D136	115889-63-9	C16-A038	129497-92-3	C16-A019
110503-18-9	C15-A009	116047-10-0	C11-A057	129680-27-9	C01-C071
110616-89-2	C11-A022	116120-29-7	C01-C089	129868-34-4	C01-A002
110801-36-0	C02-A019	116163-63-4	C16-A038	130124-54-8	C01-A015
110801-37-1	C02-A023	116163-69-0	C16-A037	130810-27-4	C16-A013
110801-38-2	C02-A023	116163-75-8	C16-A037	130810-28-5	C16-A013
110801-39-3	C02-A016	117176-66-6	C11-A015	130810-29-6	C16-A013
110913-86-5	C02-A003	117208-80-7	C12-A007	130810-30-9	C16-A013
110913-90-1	C02-A007	117569-53-6	C02-A025	130810-31-0	C16-A013
110913-91-2	C02-A006	117585-55-4	C02-A021	130942-92-6	C01-C056
110913-92-3	C02-A005	117756-57-7	C07-A001	131566-30-8	C11-A044
110913-93-4	C02-A002	119039-59-7	C12-A009	132105-14-7	C01-A013
110913-94-5	C02-A010	119540-51-1	C11-A067	132412-52-3	C01-A020
110913-95-6	C02-A004	119990-11-3	C11-A073	132444-76-9	C16-A013
110913-96-7	C02-A008	120144-37-8	C01-C067	132620-99-6	C01-C071
110913-97-8	C02-A001	120857-66-1	C01-A005	132879-26-6	C15-A012
110913-98-9	C02-A009	120932-13-0	C01-C097	133098-03-0	C12-A020
110913-99-0	C02-A001	121951-54-0	C01-C105	133441-46-0	C11-A095
110914-01-7	C02-A043	121951-55-1	C01-C113	133708-36-8	C07-C008
110914-02-8	C02-A046	121951-56-2	C01-C112	134884-20-1	C11-A044
110914-03-9	C02-A043	122139-78-0	C16-A012	135154-84-6	C12-A009
110914-04-0	C02-A043	122452-72-6	C12-A007	135154-85-7	C12-A009
110953-03-2	C12-A007	122564-82-3	C16-A002	135268-27-8	C12-A009
111203-62-4	C01-C095	123210-68-4	C16-A010	135268-28-9	C12-A009
111422-20-9	C01-A004	126068-67-5	C03-C040	135446-53-6	C01-C067
111422-21-0	C01-A004	126298-90-6	C12-A001	136511-07-4	C11-A034
111755-37-4	C16-A028	126650-99-5	C07-A002	136765-27-0	C12-A009
112008-35-2	C07-A002	127026-25-9	C01-C067	136799-44-5	C11-A034

(Continued)

(Continued)

CAS #	Handbook	CAS #	Handbook	CAS #	Handbook
112068-71-0	C11-A099	127529-01-5	C11-A067	137038-30-3	C11-A087
137444-35-0	C12-A006	160902-51-2	C12-A001	243464-31-5	C11-A020
138847-51-5	C03-A001	161403-69-6	C01-C065	255841-99-7	C01-A003
139167-39-8	C07-A001	162342-48-5	C09-A001	255842-00-3	C01-A003
139341-09-6	C16-A008	164596-65-0	C07-A001	264195-70-2	C06-A005
139443-72-4	C03-A009	164790-60-7	C12-A016	288578-22-3	C01-A015
139881-57-5	C03-A020	165195-63-1	C07-A001	300666-46-0	C12-A009
139881-58-6	C03-A020	165892-23-9	C01-C063	300666-47-1	C12-A009
140187-99-1	C07-A001	167076-44-0	C11-A044	300666-48-2	C12-A009
140406-37-7	C01-C065	167416-30-0	C09-A001	326604-75-5	C11-A064
140623-70-7	C11-A067	168153-19-3	C01-C059	337913-53-8	C01-C071
141102-74-1	C01-A028	169565-13-3	C12-A009	341007-60-1	C01-C071
141102-75-2	C01-A030	171117-05-8	C07-A002	342401-47-2	C01-C071
141102-77-4	C01-A031	171117-06-9	C07-A002	344328-67-2	C10-A014
141102-78-5	C01-A032	171501-68-1	C12-A009	351011-43-3	C01-C052
141975-81-7	C07-A001	172672-28-5	C03-A005	355120-85-3	C11-A093
142185-85-1	C16-A008	172672-70-7	C03-A001	357264-41-6	C07-A001
143599-50-2	C12-A001	174593-75-0	C16-A024	360767-00-6	C12-A009
144705-55-5	C01-C065	176327-97-2	C03-A001	371151-90-5	C12-A009
145399-95-7	C12-A009	176971-71-4	C07-A001	373643-29-9	C07-A001
145427-90-3	C16-A038	177579-04-3	C16-A019	375375-20-5	C16-A018
145427-92-5	C16-A038	182321-38-6	C11-A044	378791-32-3	C03-A010
145427-93-6	C16-A037	184288-32-2	C11-A097	393581-39-0	C11-A044
147270-03-9	C07-A001	187108-44-7	C16-A018	398138-07-3	C13-A014
147527-08-0	C07-A001	187108-45-8	C16-A018	398138-08-4	C13-A014
148336-86-1	C07-A001	187939-22-6	C03-A001	398138-09-5	C13-A014
149007-17-0	C07-A001	192819-80-0	C09-A001	398138-10-8	C13-A014
149007-18-1	C07-A001	193090-30-1	C01-A007	398138-11-9	C13-A014
149007-19-2	C07-A001	193090-56-1	01-A007	398138-12-0	C13-A014
149097-10-9	C07-A001	195210-48-1	C03-A029	409314-70-1	C10-A014
150103-83-6	C01-C077	195210-49-2	C03-A029	409325-03-7	C01-C071
150296-72-3	C01-C089	195210-50-5	C03-A029	436859-78-8	C11-A097
151567-39-4	C07-A001	195210-51-6	C03-A029	440676-21-1	C12-A003
151567-40-7	C07-A001	201205-56-3	C05-A003	440676-22-2	C12-A003
152271-55-1	C01-A013	201295-24-1	C12-A016	440676-23-3	C12-A003
153226-48-3	C07-A002	201788-20-7	C01-A016	440676-24-4	C12-A003
153834-62-9	C16-A034	205437-60-1	C12-A009	464917-26-8	C03-C040
153929-54-5	C09-A001	205437-62-3	C12-A009	481717-87-7	C03-A001
155123-44-7	C11-A086	208990-07-2	C11-A061	496916-34-8	C16-A001
155399-52-3	C09-A001	214478-05-4	C11-A061	499126-35-1	C01-C089
155575-68-1	C01-A028	215812-86-5	C12-A016	503842-49-7	C11-A113
155590-03-7	C16-A019	218625-68-4	C11-A063	511262-76-3	C03-C038
155666-11-8	C16-A012	218625-70-8	C11-A066	524945-49-1	C16-A008
158847-17-7	C01-A033	220355-66-8	C16-A016	525584-40-1	C11-A059
158847-18-8	C01-A034	220718-02-5	C01-A014	540770-45-4	C09-A002
159431-44-4	C01-C067	220718-03-6	C01-A016	558443-52-0	C11-A061
159923-90-7	C03-A012	220718-05-8	C01-A021	560088-71-3	C11-A106

CAS #	Handbook	CAS #	Handbook	CAS #	Handbook
159939-87-4	C01-A015	223129-69-9	C03-A001	588806-89-7	C16-A018
603120-87-2	C07-A001	676570-37-9	C16-A018	781557-85-5	C02-A027
607368-40-1	C12-A013	679806-14-5	C13-A021	786681-36-5	C03-A016
620190-09-2	C16-A034	719993-31-4	C11-A071	786681-37-6	C03-A011
625084-37-9	C11-A040	724693-69-0	C09-A001	786681-38-7	C03-A028
629653-73-2	C16-A019	732189-13-8	C12-A009	854873-81-7	C03-A027
642494-56-2	C13-A014	740776-49-2	C09-A001	854873-82-8	C03-A027
657390-38-0	C08-A001	767612-49-7	C12-A009	858817-94-4	C03-A015
675600-78-9	C13-A015	777799-85-6	C03-A029	859060-27-8	C03-A020

ICD-10	Handbook	ICD-10	Handbook	ICD-10	Handbook
83.2	C18-A014	A23.1	C17-A004	A91	C18-A012
A00	C17-A025	A23.2	C17-A004	A92.0	C18-A010
A01.0	C17-A022	A24	C17-A020	A92.1	C18-A040
A03	C17-A024	A24.0	C17-A019	A92.2	C18-A065
A03.0	C17-A024	A24.1	C17-A020	A92.3	C18-A067
A03.1	C17-A024	A24.2	C17-A020	A92.4	C18-A051
A03.2	C17-A024	A24.3	C17-A020	A93.0	C18-A041
A03.3	C17-A024	A28.2	C17-A031	A93.8	C18-A066
A03.8	C17-A024	A35	C17-A009	A95	C18-A069
A04.0	C17-A011	A48.0	C17-A008	A95.0	C18-A069
A04.1	C17-A011	A48.1	C17-A014	A95.1	C18-A069
A04.2	C17-A011	A48.2	C17-A014	A96.0	C18-A025
A04.3	C17-A012	A70	C17-A005	A96.1	C18-A031
A04.3	C17-A011	A75.0	C19-A002	A96.2	C18-A027
A04.4	C17-A011	A75.2	C19-A007	A96.8	C18-A019
A05.0	C16-A033	A75.3	C19-A006	A96.8	C18-A055
A05.1	C17-A007	A77.0	C19-A004	A98.0	C18-A011
A05.2	C17-A008	A78	C17-A010	A98.1	C18-A039
A20	C17-A030	A79.0	C19-A003	A98.2	C18-A026
A20.0	C17-A030	A79.1	C19-A001	A98.3	C18-A033
A20.2	C17-A030	A82	C18-A050	A98.3	C18-A015
A20.7	C17-A030	A83.0	C18-A024	A98.4	C18-A033
A21	C17-A013	A83.1	C18-A068	A98.4	C18-A015
A21.0	C17-A013	A83.3	C18-A060	A98.5	C18-A013
A21.1	C17-A013	A83.4	C18-A036	A98.5	C18-A020
A21.2	C17-A013	A83.6	C18-A053	A98.5	C18-A049
A21.7	C17-A013	A83.8	C18-A036	A98.5	C18-A058
A21.8	C17-A013	A84	C18-A063	A98.5	C18-A056
A22	C17-A001	A84.0	C18-A054	B00.4	C18-A009
A22.0	C17-A001	A84.1	C18-A008	B03	C18-A059
A22.1	C17-A001	A84.8	C18-A028	B04	C18-A035
A22.2	C17-A001	A84.8	C18-A047	B08.8	C18-A017
A23	C17-A004	A87.2	C18-A030	B08.8	C18-A061
A23.0	C17-A004	A90	C18-A012	B30.8	C18-A037

(Continued)

(Continued)

ICD-10	Handbook	ICD-10	Handbook	ICD-10	Handbook
B33.8	C18-A021	B39.1	C20-A008	J10	C18-A042
B33.8	C18-A038	B39.3	C20-A008	J10.1	C18-A042
B38	C20-A002	B39.4	C20-A008	J11	C18-A042
B38.0	C20-A002	B39.5	C20-A008	J12.8	C18-A058
B38.1	C20-A002	B44	C20-A001	T61.0	C16-A008
B38.3	C20-A002	B44.0	C20-A001	T61.2	C16-A019
B38.7	C20-A002	B44.1	C20-A001	T61.2	C16-A016
B38.8	C20-A002	B44.7	C20-A001	T61.2	C16-A006
B39.0	C20-A008	B44.8	C20-A001		

OPCW #	Handbook	OPCW #	Handbook
(64-17-5)-E1A3-EMM(I)	C01-A021	(64-17-5)-M1A3-MII(I-)	C01-A016
(64-17-5)-E1A3-MMM(I)	C01-A021	(64-17-5)-M1A3-MME(I-)	C01-A017
(64-17-5)-E1A3-MMM(I-)	C01-A021	(64-17-5)-M1A3-MMM(I-)	C01-A017
(64-17-5)-E1A3-MMP(I)	C01-A021	(464-07-3)-M1A1	C01-A003
(64-17-5)-M1A3-EEE(I)	C01-A014	(464-07-3)-M1A1-C(-)	C01-A003
(64-17-5)-M1A3-EEE(I-)	C01-A014	(1516-08-1)-M1A1	C01-A010
(64-17-5)-M1A3-EEM(I)	C01-A014	(1516-08-1)-M1A3-I	C01-A016
(64-17-5)-M1A3-EMM(I)	C01-A017	(22739-76-0)-M1A1	C01-A002
(64-17-5)-M1A3-HII(Cl)	C01-A016	(66544-78-9)-M1A1	C01-A004
(64-17-5)-M1A3-IIM(I)	C01-A016	H-E1A3-MM	C01-D148
(64-17-5)-M1A3-MEE(I-)	C01-A014		

Additional References

3M Occupational Health and Environmental Safety Division. *The 3M 2005 Respirator Selection Guide.* St Paul, MN: 3M Inc., 2005.

American Conference of Governmental Industrial Hygienists. *2005 TLVs and BEIs Based on the Documentation of the Threshold Limit Values for Chemical Substances and Physical Agents & Biological Exposure Indices.* Cincinnati, OH: ACGIH Worldwide Signature Publications, 2005.

Bretherick, Leslie. *Bretherick's Handbook of Reactive Chemical Hazards.* 4th ed. Oxford, England: Butterworth-Heinemann, Ltd, 1990.

Büchen-Osmond, Cornelia. *ICTVdB: The Universal Virus Database of the International Committee on Taxonomy of Viruses.* http://www.ncbi.nlm.nih.gov/ICTVdb/index.htm. 2006.

Centers for Disease Control and Prevention. "Medical Examiners, Coroners, and Biologic Terrorism: A Guidebook for Surveillance and Case Management." *Morbidity and Mortality Weekly Report* 53 (No. RR-8) (2004).

———. http://www.cdc.gov/az.do, 2007.

Computer-Aided Management of Emergency Operations Version 1.1.2 (CAMEO 1.1.2). United States Environmental Protection Agency, and National Oceanic and Atmospheric Administration, October 2004.

Department of Health and Human Services. *National Toxicology Program, Technical Reports. August 9, 2005.* http://ntp.niehs.nih.gov/index.cfm?objectid=070E4598-9C8B-DF77-1C266EBE08732EB4. March 22, 2006.

Department of Justice and Department of Defense. *Medical Examiner/Coroner Guide For Mass Fatality Management of Chemically Contaminated Remains.* May 15, 2003.

DoctorFungus Corporation. www.doctorfungus.org. 2006.

DuPont Tychem Protective Apparel. *DuPont Personal Protection.* Wilmington, Delaware: E.I. du Pont de Nemours and Company, 2004.

eMedicine. *Clinical Knowledge Base.* http://www.emedicine.com. 2006.

Federal Emergency Management Agency, Chemical Stockpile Emergency Preparedness Program. *Will Duct Tape and Plastic Really Work? Issues Related to Expedient Shelter-in-Place. Report No. ORNL/TM-2001/154.* August 2001.

Ford, Marsha D., Kathleen A. Delaney, Louis J. Ling, and Timothy Erickson, eds. *Clinical Toxicology.* Philadelphia, PA: W.B. Saunders Company, 2001.

Haber, Ludwig Fritz. *The Poisonous Cloud: Chemical Warfare in the First World War.* Oxford: Clarendon Press, 1986.

Lawson, J. Randall, and Theodore L. Jarboe. *Aid for Decontamination of Fire and Rescue Service Protective Clothing and Equipment After Chemical, Biological, and Radiological Exposures, NIST Special Publication 981.* Washington, DC: Government Printing Office, 2002.

Lefebure, Victor. *The Riddle of the Rhine: Chemical Strategy in Peace and War.* New York: E.P. Dutton & Company, 1923.

Lewis, Richard J., Sr. *Hazardous Chemicals Desk Reference.* 5th ed. New York: John Wiley & Sons, 2002.

National Institute of Justice. *Guide for the Selection of Chemical and Biological Decontamination Equipment for Emergency First Responders, NIJ Guide 103-00.* Vol. 1. Washington, DC: Government Printing Office, October 2001.

National Institute of Standards and Technology. *NIST Chemistry WebBook: NIST Standard Reference Database Number 69.* June 2005 Release. http://webbook.nist.gov/chemistry/. 2005.

National Institutes of Health. *Integrated Risk Information System (IRIS).* http://toxnet.nlm.nih.gov/cgi-bin/sis/htmlgen?IRIS. 2004.

New Jersey Department of Health and Senior Services. *Right to Know Hazardous Substance Fact Sheets.* http://www.state.nj.us/health/eoh/rtkweb/rtkhsfs.htm. November 2005.

Organization for the Prohibition of Chemical Weapon. "Handbook on Chemicals, version 2002." In *Declaration Handbook 2002 for the Convention on the Prohibition of the Development, Production, Stockpiling and Use of Chemical Weapons and on their Destruction.* Appendix 2. http://www.opcw.org/handbook/html/app2.html. 2004.

Pohanis, Richard P., and Stanley A. Greene. *Wiley Guide to Chemical Incompatibilities.* 2nd ed. New York: John Wiley & Sons, 2003.

Robinson, Julian Perry. *The Rise of CB Weapons. The Problem of Chemical and Biological Warfare: A Study of The Historical, Technical, Military, Legal and Political Aspects of CBW, and Possible Disarmament Measures.* Vol. 1. Stockholm: Almqvist & Wiksell, 1971.

Robinson, Julian Perry. *CB Weapons Today. The Problem of Chemical and Biological Warfare: A Study of The Historical, Technical, Military, Legal and Political Aspects of CBW, and Possible Disarmament Measures.* Vol. 2. Stockholm: Almqvist & Wiksell, 1973.

Rossoff, Irving S. *Encyclopedia of Clinical Toxicology: A Comprehensive Guide and Reference to the Toxicology of Prescription and OTC Drugs, Chemicals, Herbals, Plants, Fungi, Marine Life, Reptiles and Insect Venoms, Food Ingredients, Clothing and Environmental Toxins.* London, England: The Parthenon Publishing Group, 2002.

The Sunshine Project. www.sunshine-project.org, 2007.

Transport Canada, United States Department of Transportation, Secretariat of Transport and Communications of Mexico. *2004 Emergency Response Guidebook.*

United States Army Headquarters. *Chemical Exposure Guidelines for Deployed Military Personnel, USACHPPM Technical Guide 230.* Washington, DC: Government Printing Office, January 2002.

———. *Chemical Exposure Guidelines for Deployed Military Personnel, A Companion Document to USACHPPM Technical Guide (TG) 230 Chemical Exposure Guidelines for Deployed Military Personnel, Reference Document 230, Version 1.3.* Washington, DC: Government Printing Office, Updated May 2003.

———. *Field Behavior of NBC Agents, Field Manual No. 3-6.* Washington, DC: Government Printing Office, November 1986.

———. *Health Service Support in a Nuclear, Biological, and Chemical Environment: Tactics, Techniques, and Procedures, Field Manual 4-02.7 (FM 8-10-7).* Washington, DC: Government Printing Office, October 2002.

———. *Military Chemistry and Chemical Agents, Technical Manual No. 3-215.* Washington, DC: Government Printing Office, August 1956.

———. *Military Chemistry and Chemical Agents, Technical Manual No. 3-215.* Washington, DC: Government Printing Office, December 1963.

———. *Military Chemistry and Chemical Compounds, Field Manual No. 3-9.* Washington, DC: Government Printing Office, October 1975.

———. *NATO Handbook on the Medical Aspects of NBC Defensive Operations AmedP-6(B), Field Manual No. 8-9.* Washington, DC: Government Printing Office, February 1996.

———. *NBC Decontamination, Field Manual No. 3-5.* Washington, DC: Government Printing Office, July 2000.

———. *NBC Handbook, Field Manual No. 3-7.* Washington, DC: Government Printing Office, September 1990.

———. *NBC Protection, Field Manual No. 3-4.* Washington, DC: Government Printing Office, May 1992.

———. *Potential Military Chemical/Biological Agents and Compounds, Field Manual No. 3-9.* Washington, DC: Government Printing Office, December 12, 1990.

———. *The Medical NBC Battlebook, USACHPPM Tech Guide 244.* Washington, DC: Government Printing Office, May 2000.

United States Army Intelligence Agency. *Dusty Agents: Implications for Chemical Warfare Protections (Declassified), Report No. AST-26602-055-88.* January 27, 1988.

United States Army Soldier and Biological Chemical Command. *Guidelines for Cold Weather Mass Decontamination During a Terrorist Chemical Agent Incident.* January 2002.

———. *Guidelines for Mass Fatality Management During Terrorist Incidents Involving Chemical Agents.* November 2001.

United States Defense Intelligence Agency. *Iraq–Kuwait: Chemical Warfare Dusty Agent Threat.* Serial Number: DSA 350-90. Filename: 73349033, October 10, 1990. http://desert-storm.com/Gulflink/950719dl.txt. 2003.

United States Environmental Protection Agency. *Acute Exposure Guidelines (AEGLs).* http://www.epa.gov/oppt/aegl/sitemap.htm. October 3, 2005.

United States Military Joint Chiefs of Staff. *Joint Tactics, Techniques, and Procedures for Mortuary Affairs in Joint Operations, Joint Publication No. 4-06.* Washington, DC: Government Printing Office, August 1996.

United States War Department. *Military Chemistry and Chemical Agents, Technical Manual No. 3-215.* Washington, DC: Government Printing Office, April 21, 1942.

University of California-Davis, Oregon State University, Michigan State University, Cornell University, and the University of Idaho. *EXTOXNET, Pesticide Information Profiles (PIPs).* http://ace.orst.edu/info/extoxnet/pips/ghindex.html. September 17, 2002.

Wireless Information System for Emergency Responders Version 2.2.94 (WISER 2.2). National Library of Medicine, April 23, 2006.

Index

D

E

F

G

H

I

M

N

O

Q

R

S

T

U

V

W

X

Y

Z